Communications in Computer and Information Science   965

Commenced Publication in 2007
Founding and Former Series Editors:
Phoebe Chen, Alfredo Cuzzocrea, Xiaoyong Du, Orhun Kara, Ting Liu,
Dominik Ślęzak, and Xiaokang Yang

More information about this series at http://www.springer.com/series/7899

Vladimir Voevodin · Sergey Sobolev (Eds.)

Supercomputing

4th Russian Supercomputing Days, RuSCDays 2018
Moscow, Russia, September 24–25, 2018
Revised Selected Papers

 Springer

Editors
Vladimir Voevodin ⓘD
Research Computing Center (RCC)
Moscow State University
Moscow, Russia

Sergey Sobolev ⓘD
RCC
Moscow State University
Moscow, Russia

ISSN 1865-0929   ISSN 1865-0937 (electronic)
Communications in Computer and Information Science
ISBN 978-3-030-05806-7   ISBN 978-3-030-05807-4 (eBook)
https://doi.org/10.1007/978-3-030-05807-4

Library of Congress Control Number: 2018963970

This Springer imprint is published by the registered company Springer Nature Switzerland AG
The registered company address is: Gewerbestrasse 11, 6330 Cham, Switzerland

Preface

The 4th Russian Supercomputing Days Conference (RuSCDays 2018) was held September 24–25, 2018, in Moscow, Russia. It was organized by the Supercomputing Consortium of Russian Universities and the Russian Academy of Sciences. The conference was supported by the Russian Foundation for Basic Research and our respected platinum educational partner (IBM), platinum sponsors (NVIDIA, T-Platforms, Intel), gold sponsors (Mellanox, Dell EMC, Hewlett Packard Enterprise), and silver sponsors (Lenovo, AMD, Xilinx, RSC). The conference was organized in partnership with the ISC High Performance conference series.

The conference was born in 2015 as a union of several supercomputing events in Russia and quickly became one of the most notable Russian supercomputing meetings. The conference caters to the interests of a wide range of representatives from science, industry, business, education, government, and students – anyone connected to the development or the use of supercomputing technologies. The conference topics cover all aspects of supercomputing technologies: software and hardware design, solving large tasks, application of supercomputing technologies in industry, exascale computing issues, supercomputing co-design technologies, supercomputing education, and others.

All papers submitted to the conference were reviewed by three referees in the first review round. The papers were evaluated according to their relevance to the conference topics, scientific contribution, presentation, approbation, and related works description. After notification of a conditional acceptance, the second review round was arranged aimed at the final polishing of papers and also at evaluating the authors' work after the referees' comments. After the conference, the 59 best papers were carefully selected to be included in this volume.

The proceedings editors would like to thank all the conference committee members, especially the Organizing and Program Committee members as well as the other reviewers for their contributions. We also thank Springer for producing these high-quality proceedings of RuSCDays 2018.

October 2018

Vladimir Voevodin
Sergey Sobolev

Organization

The 4th Russian Supercomputing Days Conference (RuSCDays 2018) was organized by the Supercomputing Consortium of Russian Universities and the Russian Academy of Sciences. The conference organization coordinator was Moscow State University Research Computing Center.

Steering Committee

V. A. Sadovnichiy (Chair)	Moscow State University, Russia
V. B. Betelin (Co-chair)	Russian Academy of Sciences, Moscow, Russia
A. V. Tikhonravov (Co-chair)	Moscow State University, Russia
J. Dongarra (Co-chair)	University of Tennessee, Knoxville, USA
A. I. Borovkov	Peter the Great Saint-Petersburg Polytechnic University, Russia
Vl. V. Voevodin	Moscow State University, Russia
V. P. Gergel	Lobachevsky State University of Nizhni Novgorod, Russia
G. S. Elizarov	NII Kvant, Moscow, Russia
V. V. Elagin	Hewlett Packard Enterprise, Moscow, Russia
A. K. Kim	MCST, Moscow, Russia
E. V. Kudryashova	Northern (Arctic) Federal University, Arkhangelsk, Russia
N. S. Mester	Intel, Moscow, Russia
E. I. Moiseev	Moscow State University, Russia
A. A. Moskovskiy	RSC Group, Moscow, Russia
V. Yu. Opanasenko	T-Platforms, Moscow, Russia
G. I. Savin	Joint Supercomputer Center, Russian Academy of Sciences, Moscow, Russia
A. S. Simonov	NICEVT, Moscow, Russia
V. A. Soyfer	Samara University, Russia
L. B. Sokolinskiy	South Ural State University, Chelyabinsk, Russia
I. A. Sokolov	Russian Academy of Sciences, Moscow, Russia
R. G. Strongin	Lobachevsky State University of Nizhni Novgorod, Russia
A. N. Tomilin	Institute for System Programming of the Russian Academy of Sciences, Moscow, Russia
A. R. Khokhlov	Moscow State University, Russia
B. N. Chetverushkin	Keldysh Institutes of Applied Mathematics, Russian Academy of Sciences, Moscow, Russia

| E. V. Chuprunov | Lobachevsky State University of Nizhni Novgorod, Russia |
| A. L. Shestakov | South Ural State University, Chelyabinsk, Russia |

Program Committee

Vl. V. Voevodin (Chair)	Moscow State University, Russia
R. M. Shagaliev (Co-chair)	Russian Federal Nuclear Center, Sarov, Russia
M. V. Yakobovskiy (Co-chair)	Keldysh Institutes of Applied Mathematics, Russian Academy of Sciences, Moscow, Russia
T. Sterling (Co-chair)	Indiana University, Bloomington, USA
S. I. Sobolev (Scientific Secretary)	Moscow State University, Russia
A. I. Avetisyan	Institute for System Programming of the Russian Academy of Sciences, Moscow, Russia
D. Bader	Georgia Institute of Technology, Atlanta, USA
P. Balaji	Argonne National Laboratory, USA
M. R. Biktimirov	Russian Academy of Sciences, Moscow, Russia
A. V. Bukhanovskiy	ITMO University, Saint Petersburg, Russia
J. Carretero	University Carlos III of Madrid, Spain
Yu. V. Vasilevskiy	Keldysh Institutes of Applied Mathematics, Russian Academy of Sciences, Moscow, Russia
V. E. Velikhov	National Research Center Kurchatov Institute, Moscow, Russia
V. Yu. Volkonskiy	MCST, Moscow, Russia
V. M. Volokhov	Institute of Problems of Chemical Physics of Russian Academy of Sciences, Chernogolovka, Russia
R. K. Gazizov	Ufa State Aviation Technical University, Russia
B. M. Glinskiy	Institute of Computational Mathematics and Mathematical Geophysics, Siberian Branch of Russian Academy of Sciences, Novosibirsk, Russia
V. M. Goloviznin	Moscow State University, Russia
V. A. Ilyin	National Research Center Kurchatov Institute, Moscow, Russia
V. P. Ilyin	Institute of Computational Mathematics and Mathematical Geophysics, Siberian Branch of Russian Academy of Sciences, Novosibirsk, Russia
S. I. Kabanikhin	Institute of Computational Mathematics and Mathematical Geophysics, Siberian Branch of Russian Academy of Sciences, Novosibirsk, Russia
I. A. Kalyaev	NII MVS, South Federal University, Taganrog, Russia
H. Kobayashi	Tohoku University, Japan
V. V. Korenkov	Joint Institute for Nuclear Research, Dubna, Russia
V. A. Kryukov	Keldysh Institutes of Applied Mathematics, Russian Academy of Sciences, Moscow, Russia

J. Kunkel	University of Hamburg, Germany
S. D. Kuznetsov	Institute for System Programming of the Russian Academy of Sciences, Moscow, Russia
J. Labarta	Barcelona Supercomputing Center, Spain
A. Lastovetsky	University College Dublin, Ireland
M. P. Lobachev	Krylov State Research Centre, Saint Petersburg, Russia
Y. Lu	National University of Defense Technology, Changsha, Hunan, China
T. Ludwig	German Climate Computing Center, Hamburg, Germany
V. N. Lykosov	Institute of Numerical Mathematics, Russian Academy of Sciences, Moscow, Russia
I. B. Meerov	Lobachevsky State University of Nizhni Novgorod, Russia
M. Michalewicz	University of Warsaw, Poland
L. Mirtaheri	Kharazmi University, Tehran, Iran
A. V. Nemukhin	Moscow State University, Russia
G. V. Osipov	Lobachevsky State University of Nizhni Novgorod, Russia
A. V. Semyanov	Lobachevsky State University of Nizhni Novgorod, Russia
Ya. D. Sergeev	Lobachevsky State University of Nizhni Novgorod, Russia
H. Sithole	Centre for High Performance Computing, Cape Town, South Africa
A. V. Smirnov	Moscow State University, Russia
R. G. Strongin	Lobachevsky State University of Nizhni Novgorod, Russia
H. Takizawa	Tohoku University, Japan
M. Taufer	University of Delaware, Newark, USA
V. E. Turlapov	Lobachevsky State University of Nizhni Novgorod, Russia
E. E. Tyrtyshnikov	Institute of Numerical Mathematics, Russian Academy of Sciences, Moscow, Russia
V. A. Fursov	Samara University, Russia
L. E. Khaymina	Northern (Arctic) Federal University, Arkhangelsk, Russia
T. Hoefler	Eidgenössische Technische Hochschule Zürich, Switzerland
B. M. Shabanov	Joint Supercomputer Center, Russian Academy of Sciences, Moscow, Russia
N. N. Shabrov	Peter the Great Saint-Petersburg Polytechnic University, Russia
L. N. Shchur	Higher School of Economics, Moscow, Russia
R. Wyrzykowski	Czestochowa University of Technology, Poland
M. Yokokawa	Kobe University, Japan

Industrial Committee

A. A. Aksenov (Co-chair)	Tesis, Moscow, Russia
V. E. Velikhov (Co-chair)	National Research Center Kurchatov Institute, Moscow, Russia
A. V. Murashov (Co-chair)	T-Platforms, Moscow, Russia
Yu. Ya. Boldyrev	Peter the Great Saint-Petersburg Polytechnic University, Russia
M. A. Bolshukhin	Afrikantov Experimental Design Bureau for Mechanical Engineering, Nizhny Novgorod, Russia
R. K. Gazizov	Ufa State Aviation Technical University, Russia
M. P. Lobachev	Krylov State Research Centre, Saint Petersburg, Russia
V. Ya. Modorskiy	Perm National Research Polytechnic University, Russia
A. P. Skibin	Gidropress, Podolsk, Russia
S. Stoyanov	T-Services, Moscow, Russia
N. N. Shabrov	Peter the Great Saint-Petersburg Polytechnic University, Russia
A. B. Shmelev	RSC Group, Moscow, Russia
S. V. Strizhak	Hewlett-Packard, Moscow, Russia

Educational Committee

V. P. Gergel (Co-chair)	Lobachevsky State University of Nizhni Novgorod, Russia
Vl. V. Voevodin (Co-chair)	Moscow State University, Russia
L. B. Sokolinskiy (Co-chair)	South Ural State University, Chelyabinsk, Russia
Yu. Ya. Boldyrev	Peter the Great Saint-Petersburg Polytechnic University, Russia
A. V. Bukhanovskiy	ITMO University, Saint Petersburg, Russia
R. K. Gazizov	Ufa State Aviation Technical University, Russia
S. A. Ivanov	Hewlett-Packard, Moscow, Russia
V. Ya. Modorskiy	Perm National Research Polytechnic University, Russia
S. G. Mosin	Kazan Federal University, Russia
I. O. Odintsov	RSC Group, Saint Petersburg, Russia
N. N. Popova	Moscow State University, Russia
O. A. Yufryakova	Northern (Arctic) Federal University, Arkhangelsk, Russia

Organizing Committee

Vl. V. Voevodin (Chair)	Moscow State University, Russia
V. P. Gergel (Co-chair)	Lobachevsky State University of Nizhni Novgorod, Russia

B. M. Shabanov (Co-chair)	Joint Supercomputer Center, Russian Academy of Sciences, Moscow, Russia
S. I. Sobolev (Scientific Secretary)	Moscow State University, Russia
A. A. Aksenov	Tesis, Moscow, Russia
A. P. Antonova	Moscow State University, Russia
A. S. Antonov	Moscow State University, Russia
K. A. Barkalov	Lobachevsky State University of Nizhni Novgorod, Russia
M. R. Biktimirov	Russian Academy of Sciences, Moscow, Russia
Vad. V. Voevodin	Moscow State University, Russia
T. A. Gamayunova	Moscow State University, Russia
O. A. Gorbachev	RSC Group, Moscow, Russia
V. A. Grishagin	Lobachevsky State University of Nizhni Novgorod, Russia
S. A. Zhumatiy	Moscow State University, Russia
V. V. Korenkov	Joint Institute for Nuclear Research, Dubna, Russia
I. B. Meerov	Lobachevsky State University of Nizhni Novgorod, Russia
D. A. Nikitenko	Moscow State University, Russia
I. M. Nikolskiy	Moscow State University, Russia
N. N. Popova	Moscow State University, Russia
N. M. Rudenko	Moscow State University, Russia
A. S. Semenov	NICEVT, Moscow, Russia
I. Yu. Sidorov	Moscow State University, Russia
L. B. Sokolinskiy	South Ural State University, Chelyabinsk, Russia
V. M. Stepanenko	Moscow State University, Russia
N. T. Tarumova	Moscow State University, Russia
A. V. Tikhonravov	Moscow State University, Russia
A. S. Frolov	NICEVT, Moscow, Russia
A. Yu. Chernyavskiy	Moscow State University, Russia
P. A. Shvets	Moscow State University, Russia
M. V. Yakobovskiy	Keldysh Institutes of Applied Mathematics, Russian Academy of Sciences, Moscow, Russia

Contents

Parallel Algorithms

A Parallel Algorithm for Studying the Ice Cover Impact onto Seismic
Waves Propagation in the Shallow Arctic Waters   3
  Galina Reshetova, Vladimir Cheverda, Vadim Lisitsa,
  and Valery Khaidykov

An Efficient Parallel Algorithm for Numerical Solution of Low Dimension
Dynamics Problems.   15
  Stepan Orlov, Alexey Kuzin, and Nikolay Shabrov

Analysis of Means of Simulation Modeling of Parallel Algorithms   29
  D. V. Weins, B. M. Glinskiy, and I. G. Chernykh

Block Lanczos-Montgomery Method over Large Prime Fields with GPU
Accelerated Dense Operations.   40
  Nikolai Zamarashkin and Dmitry Zheltkov

Comparison of Dimensionality Reduction Schemes for Parallel Global
Optimization Algorithms   50
  Konstantin Barkalov, Vladislav Sovrasov, and Ilya Lebedev

Efficiency Estimation for the Mathematical Physics Algorithms
for Distributed Memory Computers   63
  Igor Konshin

Extremely High-Order Optimized Multioperators-Based Schemes
and Their Applications to Flow Instabilities and Sound Radiation   76
  Andrei Tolstykh, Michael Lipavskii, Dmitrii Shirobokov,
  and Eugenii Chigerev

GPU-Based Parallel Computations in Multicriterial Optimization.   88
  Victor Gergel and Evgeny Kozinov

LRnLA Algorithm ConeFold with Non-local Vectorization
for LBM Implementation 101
  Anastasia Perepelkina and Vadim Levchenko

Numerical Method for Solving a Diffraction Problem of Electromagnetic
Wave on a System of Bodies and Screens 114
  Mikhail Medvedik, Marina Moskaleva, and Yury Smirnov

Parallel Algorithm for One-Way Wave Equation Based Migration
for Seismic Imaging   125
  *Alexander Pleshkevich, Dmitry Vishnevsky, Vadim Lisitsa,
  and Vadim Levchenko*

Parallel Simulation of Community-Wide Information Spreading
in Online Social Networks   136
  Sergey Kesarev, Oksana Severiukhina, and Klavdiya Bochenina

Technique for Teaching Parallel Programming via Solving a Computational
Electrodynamics Problems   149
  *Sergey Mosin, Nikolai Pleshchinskii, Ilya Pleshchinskii,
  and Dmitrii Tumakov*

The Algorithm for Transferring a Large Number of Radionuclide Particles
in a Parallel Model of Ocean Hydrodynamics   159
  Vladimir Bibin, Rashit Ibrayev, and Maxim Kaurkin

Supercomputer Simulation

Aerodynamic Models of Complicated Constructions Using Parallel
Smoothed Particle Hydrodynamics   173
  *Alexander Titov, Sergey Khrapov, Victor Radchenko,
  and Alexander Khoperskov*

Ballistic Resistance Modeling of Aramid Fabric with Surface Treatment   185
  *Natalia Yu. Dolganina, Anastasia V. Ignatova, Alexandra A. Shabley,
  and Sergei B. Sapozhnikov*

CardioModel – New Software for Cardiac Electrophysiology Simulation   195
  *Valentin Petrov, Sergey Lebedev, Anna Pirova, Evgeniy Vasilyev,
  Alexander Nikolskiy, Vadim Turlapov, Iosif Meyerov,
  and Grigory Osipov*

Examination of Clastic Oil and Gas Reservoir Rock Permeability Modeling
by Molecular Dynamics Simulation Using High-Performance Computing   208
  *Vladimir Berezovsky, Marsel Gubaydullin, Alexander Yur'ev,
  and Ivan Belozerov*

Hybrid Codes for Atomistic Simulations on the Desmos Supercomputer:
GPU-acceleration, Scalability and Parallel I/O   218
  Nikolay Kondratyuk, Grigory Smirnov, and Vladimir Stegailov

INMOST Parallel Platform for Mathematical Modeling and Applications   230
  Kirill Terekhov and Yuri Vassilevski

Maximus: A Hybrid Particle-in-Cell Code for Microscopic Modeling
of Collisionless Plasmas.   242
    Julia Kropotina, Andrei Bykov, Alexandre Krassilchtchikov,
    and Ksenia Levenfish

Microwave Radiometry of Atmospheric Precipitation: Radiative Transfer
Simulations with Parallel Supercomputers   254
    Yaroslaw Ilyushin and Boris Kutuza

Modeling Groundwater Flow in Unconfined Conditions of Variable
Density Solutions in Dual-Porosity Media Using the GeRa Code   266
    Ivan Kapyrin, Igor Konshin, Vasily Kramarenko, and Fedor Grigoriev

New QM/MM Implementation of the MOPAC2012 in the GROMACS   279
    Arthur O. Zalevsky, Roman V. Reshetnikov, and Andrey V. Golovin

Orlando Tools: Energy Research Application Development Through
Convergence of Grid and Cloud Computing.   289
    Alexander Feoktistov, Sergei Gorsky, Ivan Sidorov, Roman Kostromin,
    Alexei Edelev, and Lyudmila Massel

Parallel FDTD Solver with Static and Dynamic Load Balancing   301
    Gleb Balykov

Parallel Supercomputer Docking Program of the New Generation: Finding
Low Energy Minima Spectrum.   314
    Alexey Sulimov, Danil Kutov, and Vladimir Sulimov

Parallelization Strategy for Wavefield Simulation with an Elastic
Iterative Solver   331
    Mikhail Belonosov, Vladimir Cheverda, Victor Kostin,
    and Dmitry Neklyudov

Performance of Time and Frequency Domain Cluster Solvers Compared
to Geophysical Applications.   343
    Victor Kostin, Sergey Solovyev, Andrey Bakulin, and Maxim Dmitriev

Population Annealing and Large Scale Simulations
in Statistical Mechanics   354
    Lev Shchur, Lev Barash, Martin Weigel, and Wolfhard Janke

Simulation and Optimization of Aircraft Assembly Process Using
Supercomputer Technologies   367
    Tatiana Pogarskaia, Maria Churilova, Margarita Petukhova,
    and Evgeniy Petukhov

SL-AV Model: Numerical Weather Prediction at Extra-Massively
Parallel Supercomputer   379
  Mikhail Tolstykh, Gordey Goyman, Rostislav Fadeev,
  Vladimir Shashkin, and Sergei Lubov

Supercomputer Simulation Study of the Convergence of Iterative Methods
for Solving Inverse Problems of 3D Acoustic Tomography with the Data
on a Cylindrical Surface.   388
  Sergey Romanov

Supercomputer Technology for Ultrasound Tomographic Image
Reconstruction: Mathematical Methods and Experimental Results   401
  Alexander Goncharsky and Sergey Seryozhnikov

The Parallel Hydrodynamic Code for Astrophysical Flow with Stellar
Equations of State.   414
  Igor Kulikov, Igor Chernykh, Vitaly Vshivkov, Vladimir Prigarin,
  Vladimir Mironov, and Alexander Tutukov

Three-Dimensional Simulation of Stokes Flow Around a Rigid Structure
Using FMM/GPU Accelerated BEM   427
  Olga A. Abramova, Yulia A. Pityuk, Nail A. Gumerov,
  and Iskander Sh. Akhatov

Using of Hybrid Cluster Systems for Modeling of a Satellite
and Plasma Interaction by the Molecular Dynamics Method.   439
  Leonid Zinin, Alexander Sharamet, and Sergey Ishanov

High Performance Architectures, Tools and Technologies

Adaptive Scheduling for Adjusting Retrieval Process in BOINC-Based
Virtual Screening   453
  Natalia Nikitina and Evgeny Ivashko

Advanced Vectorization of PPML Method for Intel® Xeon® Scalable
Processors   465
  Igor Chernykh, Igor Kulikov, Boris Glinsky, Vitaly Vshivkov,
  Lyudmila Vshivkova, and Vladimir Prigarin

Analysis of Results of the Rating of Volunteer Distributed
Computing Projects.   472
  Vladimir N. Yakimets and Ilya I. Kurochkin

Application of the LLVM Compiler Infrastructure to the Program
Analysis in SAPFOR.   487
  Nikita Kataev

Batch of Tasks Completion Time Estimation in a Desktop Grid   500
  Evgeny Ivashko and Valentina Litovchenko

BOINC-Based Branch-and-Bound   511
  Andrei Ignatov and Mikhail Posypkin

Comprehensive Collection of Time-Consuming Problems for Intensive
Training on High Performance Computing   523
  Iosif Meyerov, Sergei Bastrakov, Alexander Sysoyev, and Victor Gergel

Dependable and Coordinated Resources Allocation Algorithms
for Distributed Computing   531
  Victor Toporkov and Dmitry Yemelyanov

Deploying Elbrus VLIW CPU Ecosystem for Materials Science
Calculations: Performance and Problems   543
  Vladimir Stegailov and Alexey Timofeev

Design Technology for Reconfigurable Computer Systems
with Immersion Cooling   554
  *Ilya Levin, Alexey Dordopulo, Alexander Fedorov,
  and Yuriy Doronchenko*

Designing a Parallel Programs on the Base of the Conception
of Q-Determinant   565
  Valentina Aleeva

Enumeration of Isotopy Classes of Diagonal Latin Squares of Small Order
Using Volunteer Computing   578
  *Eduard Vatutin, Alexey Belyshev, Stepan Kochemazov, Oleg Zaikin,
  and Natalia Nikitina*

Interactive 3D Representation as a Method of Investigating Information
Graph Features   587
  Alexander Antonov and Nikita Volkov

On Sharing Workload in Desktop Grids   599
  Ilya Chernov

On-the-Fly Calculation of Performance Metrics with Adaptive Time
Resolution for HPC Compute Jobs   609
  Konstantin Stefanov and Vadim Voevodin

Residue Logarithmic Coprocessor for Mass Arithmetic Computations   620
  Ilya Osinin

Supercomputer Efficiency: Complex Approach Inspired by Lomonosov-2
History Evaluation   631
    Sergei Leonenkov and Sergey Zhumatiy

Supercomputer Real-Time Experimental Data Processing: Technology
and Applications.   641
    *Vladislav A. Shchapov, Alexander M. Pavlinov,
    Elena N. Popova, Andrei N. Sukhanovskii, Stanislav L. Kalyulin,
    and Vladimir Ya. Modorskii*

The Conception, Requirements and Structure of the Integrated
Computational Environment.   653
    V. P. Il'in

The Elbrus-4C Based Node as Part of Heterogeneous Cluster for Oil
and Gas Processing Researches.   666
    *Ekaterina Tyutlyaeva, Igor Odintsov, Alexander Moskovsky,
    Sergey Konyukhov, Alexander Kalyakin, and Murad I. Neiman-zade*

The Multi-level Adaptive Approach for Efficient Execution of Multi-scale
Distributed Applications with Dynamic Workload   675
    *Denis Nasonov, Nikolay Butakov, Michael Melnik, Alexandr Visheratin,
    Alexey Linev, Pavel Shvets, Sergey Sobolev, and Ksenia Mukhina*

Using Resources of Supercomputing Centers with Everest Platform.   687
    Sergey Smirnov, Oleg Sukhoroslov, and Vladimir Voloshinov

Author Index   699

Parallel Algorithms

A Parallel Algorithm for Studying the Ice Cover Impact onto Seismic Waves Propagation in the Shallow Arctic Waters

Galina Reshetova[1,2]($\boxtimes$), Vladimir Cheverda[2], Vadim Lisitsa[2], and Valery Khaidykov[2]

[1] Institute of Computational Mathematics and Mathematical Geophysics SB RAS, Novosibirsk 630090, Russia
kgv@nmsf.sscc.ru
[2] Trofimuk Institute of Petroleum Geology and Geophysics SB RAS, Novosibirsk, Russia
{CheverdaVA,LisitsaVV,KhaidukovVG}@ipgg.sbras.ru

Abstract. The seismic study in the Arctic transition zones in the summer season is troublesome because of the presence of large areas covered by shallow waters like bays, lakes, rivers, their estuaries and so on. The winter season is more convenient and essentially facilitates logistic operations and implementation of seismic acquisition. However in the winter there is another complicating factor: intensive seismic noise generated by sources installed on the floating ice. To understand peculiarities of seismic waves and the origin of such an intensive noise, a representative series of numerical experiments has been performed. In order to simulate the interaction of seismic waves with irregular perturbations of underside of the ice cover, a finite-difference technique based on locally refined in time and in space grids is used. The need to use such grids is primarily due to the different scales of heterogeneities in a reference medium and the ice cover should be taken into account. We use the domain decomposition method to separate the elastic/viscoelastic model into subdomains with different scales. Computations for each subdomain are carried out in parallel. The data exchange between the two groups of CPU is done simultaneously by coupling a coarse and a fine grids. The results of the numerical experiments prove that the main impact to noise is multiple conversions of flexural waves to the body ones and vice versa and open the ways to reduce this noise.

Keywords: Seismic waves · Transition zones
Finite-difference schemes · Local grid refinement
Domain decomposition

1 Introduction and Motivation

One of the distinctive features of the Russian Far North (Fig. 1a) is the presence of large areas covered by shallow waters which bring a lot of troubles when implementing of seismic observations in the summer (Fig. 1b). At the same time, in the

© Springer Nature Switzerland AG 2019
V. Voevodin and S. Sobolev (Eds.): RuSCDays 2018, CCIS 965, pp. 3–14, 2019.
https://doi.org/10.1007/978-3-030-05807-4_1

winter these shoals are overlapped by the thick ice (up to a few meters), which significantly facilitates logistic operations and can be used to install seismic sources (Fig. 1c). It is worth mentioning that potential advantages of conducting seismic operations on the winter ice are motivating numerous investigations of the feasibility of shooting seismic from the ice cover [1]. In Fig. 2, one can see two field seismograms: for the onshore source position (Fig. 2a) and for a source placed onto the ice (Fig. 2b). Comparison of these seismograms reveals a high noise in seismic data for ice-placed sources, whereas their quality for land-based groups is quite acceptable and one can see the target reflections there even without preprocessing. Let us also pay attention to a sharp disappearance of the noise in Fig. 2b at a distance of 1000 m to the left from the source. In this place, the ice cover meets a solid ground (permafrost). It is well known that this noise is associated with flexural waves generated in the ice by seismic sources. These waves are one of the strongest known coherent noises. At the same time, such waves are much slower than the surface waves and seem to be easy canceled by f-k filtration. However, this type of filtration fails to suppress such a noise.

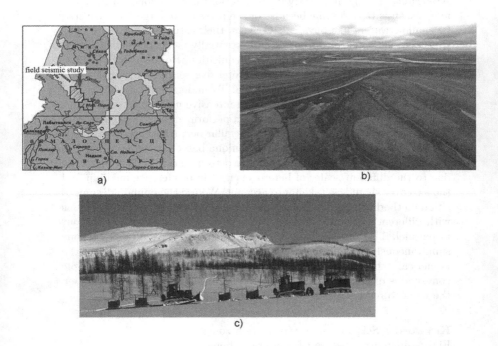

Fig. 1. (a) Geography of seismic field observations. (b) Top view of the area in summer. (c) The same area in winter.

There are some intuitive explanations of this fact, in particular, the bending of the ice produced by the impulse seismic source which generates multiples in the shallow water column as well as weakening of a signal by transferring the ice-water-mud interfaces. To understand the peculiarities of seismic waves and the origin of this high noise, a representative series of numerical experiments have been performed by means of a careful finite-difference simulation.

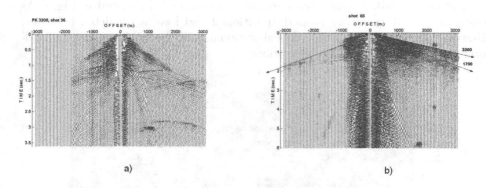

a)   b)

Fig. 2. Field seismograms. (a) Onshore source position. (b) Source on an iced shoals.

2 Mathematical Formulation

Our main suggestion is that the noise in seismograms is caused mainly by the flexural waves [5,8], and we are going to use the numerical simulation to justify this point. To do this, we use the Generalized Standard Linear Solid (GSLS) model [2] governing the seismic wave propagation in viscoelastic media: where ρ is the mass density; C_1 and C_2 are fourth order tensors, defining the model properties; u is the velocity vector; σ and ε are the stress and strain tensors, respectively; r^l are the memory variables. Note that the number of memory variable is L, which is typically two or three. The proper initial and boundary conditions are assumed.

3 Numerical Method

An explicit finite-difference standard staggered grid scheme (SSGS) approximating elastic/viscoelastic wave equations in the velocity-stress formulation with parameters modification is used to solve the problem. This numerical technique combines a high efficiency with a suitable accuracy [9,10].

We consider GSLS equations in order to take into account the seismic attenuation in the ice filling of the uppermost part of the model. For the ideal-elastic lower part of the model, the tensor C_2 is zero. Thus, memory variables do not appear in the equations in this area, and the system turns into that for an ideal-elastic wave equation.

4   Fitting a Geological Model to Data

To fit a geological model to real seismic data we have started with a simple 3D three layer ice/water/ground model, which possesses all the key components of the real life: solid ground with and without permafrost, ice, water and thawing area under water. The model was prepared together with experienced geologists and is presented in Fig. 3a (middle scale in depth) and Fig. 3b (the uppermost part of the model). A synthesized vertical component is presented in Fig. 4. As one can see, all waves are properly positioned and have a predicted behavior. However, the structure of the vertical component of the synthetic data considerably differs from the field observations.

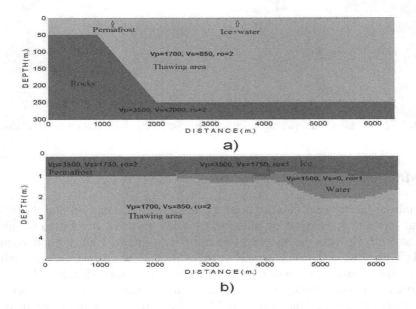

Fig. 3. The vertical slice of 3D realistic model of a shoal covered by the ice. (a) Middle scale depth (b) Shallow part.

The next modification of the model had the irregular underside of the ice imbedded into the previous model. The idea was found from the research of the U.S. Geological Survey. In the 1960-s they conducted a series of the winter research into the underside of the ice in the basin of the low St. Croix River, Wisconsin (see Fig. 5 presented in [3]). The measurements prove that typical scales of the water-ice interface are about 0.5 m × 0.5 m × 0.05 m. Hydrologically this area is similar to the Yamal peninsula and posseses approximately the same structure of the currents (compare with Fig. 2), but the latter is much farther to the north, hence, the winter season is much cooler and the ice is thicker. Therefore in the numerical experiments, perturbations of the underside of the ice was doubled and taken as a random distribution of 2 m × 2 m × 0.1 m prisms,

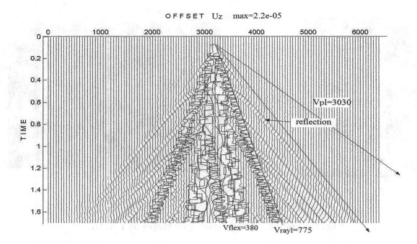

Fig. 4. The vertical displacement component of the on the free surface. One can clearly see all types of elastic waves.

which can be seen in Fig. 6. The vertical component of the synthetic seismogram for this model is presented in Fig. 7. Let us note that now it fits rather well to the real life field data, that is, there is a strong dispersive component which overlaps the target reflections. At the same time, for the receivers placed on the solid ground this noise disappears. Base on this fact the following conclusion can be made: the noise for seismic acquisitions onto the ice is associated with random perturbations of the ice/water interface.

5 Optimization of Computations

Let us indicate the upper part of the model as a subdomain Ω_1 (Fig. 6), where the full viscoelastic wave equation is used, while the remaining part of the medium Ω_2 is ideal-elastic.

The upper part of the geological model contains small-scale heterogeneities described by random perturbations of the ice-water interface and leads to representation with an extremely excessive detail. The straightforward implementation of finite-difference techniques brings about the necessity of using dramatically small grid steps to match the scale. From the computational point of view, this means a huge amount of memory required for the simulation and, therefore, extremely large computation time, even with the use of modern high-performance computers.

Our solution to this issue is to use two levels of optimization:

– To use the coupling of different elastic/viscoelastic equations, that is, different governing elastic/viscoelastic equations in subdomains Ω_1 and Ω_2. The local use of computationally expensive viscoelastic equations only in a narrow part of the subdomain Ω_1 of the model can essentially speed-up the simulation;

a)   b)

Fig. 5. The St. Croix river. (a) Map (b) Photo.

– To use multi-scale simulation, that is, different mesh sizes to describe different parts of the model: a coarse grid for the background model Ω_2 and a fine grid for describing the area Ω_1.

Let us take a closer look at each of the levels.

5.1 Coupling of Different Elastic/Viscoelastic Equations

Assume a subdomain Ω_1, where the full viscoelastic wave equation is used, while the ideal-elastic wave equation is valid over the rest of the space Ω_2. It is easy to prove that the conditions at the interface $\Gamma = \partial\,\Omega_1$ are the following:

$$[\sigma \cdot \boldsymbol{n}]|_{\Gamma=0}, \quad [u]|_{\Gamma=0}, \tag{1}$$

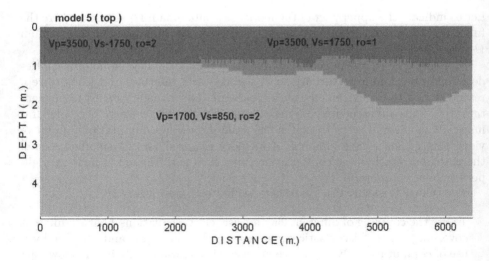

Fig. 6. The uppermost part of the 3D model slice with random perturbations of the ice-water interface (compare with Fig. 3).

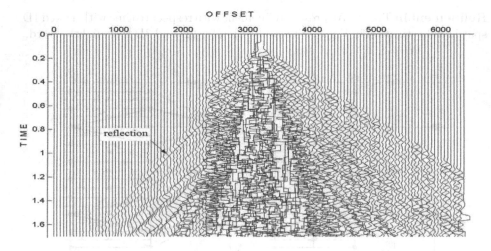

Fig. 7. Synthetic seismograms for the model presented in Fig. 6. Note strong partially correlated dispersive noise beneath the ice.

where n is the outward normal vector and $[f]$ denotes a jump of the function f at the interface Γ. These conditions are the same as those for the elastic wave equation at the interface. In our case, when the SSGS scheme is used, these conditions are automatically satisfied [7]. Thus, the coupling of different elastic/viscoelastic equations does not require any special treatment of conditions at the interface and can be implemented by allocating RAM for memory variables and solving equations for them only in the viscoelastic part of the model.

5.2 Multi-scale Simulation

A local mesh refinement is used only in the domain Ω_1 with a fine structure to perform the full waveform simulation of the long wave propagation through a model [6]. As was mentioned above, when an explicit finite differences are used, the size of a time step strongly depends on the spatial discretization, and so the time stepping should be local. As a result, the problem of the simulation of seismic wave propagation in models containing small-scale structures becomes a mathematical problem of a local time-space mesh refinement.

Let us consider how a coarse and a fine grids are coupled. The necessary features of the finite difference method, based on a local grid refinement, are the stability and an acceptable level of artificial reflections. The scattered waves have an amplitude of about 1% of the incident wave, thus the artifacts should be at most 0.1% of the incident wave. If we refine the grid simultaneously in time and in space, then the stability of the finite difference schemes can be provided via the coupling of a coarse and a fine grids based on the energy conservation, which bring about an unacceptable level (exceeding 1%) of artificial reflections [4]. We modify this approach so that the grid is refined in time and in space in turn on two different surfaces surrounding the target area Ω_1 with a microstructure.

Refinement in Time. We present refinement with respect to time with a fixed 1D spatial discretization in Fig. 8. Its modification for 2D and 3D is straightforward.

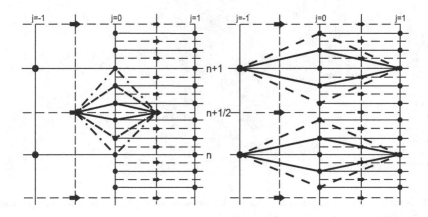

Fig. 8. Embedded stencils for the local time stem mesh.

Refinement in Space. In order to change the spatial grids, a FFT-based interpolation is used. Let us schematically explain the procedure. The mutual disposition of a coarse and a fine spatial grids one can see in Fig. 9, which corresponds to updating the stresses. As can be seen, to update the solution on a fine grid on the uppermost line it is necessary to know the wavefield at the points marked in black, these points not existing on a given coarse grid. Using the fact that all of the points are on the same line (plane in 3D), we seek for the values of missing nodes by the FFT-based interpolation. This procedure allows us to provide a required low level of artifacts (about 0.001 with respect to the incident wave) generated at the interface of these two grids.

6   Parallel Computations

The parallel implementation of the algorithm has been carried out using the static domain decomposition.

The use of the local space-time grid stepping makes it difficult to ensure a uniform work load for Processor Units (PU) in the domain decomposition.

The parallel computation is implemented using two groups of processors. The 3D heterogeneous background (a coarse grid Ω_2) is placed into one group, while the fine mesh describing Ω_1 is distributed in the other group. There is a need for interactions between processors within each group and between the groups as well. The data exchange within a group is done via faces of the adjacent subdomains by non-blocking iSend/iReceive MPI procedures. The interaction between the groups is designed for coupling a coarse and a fine grid.

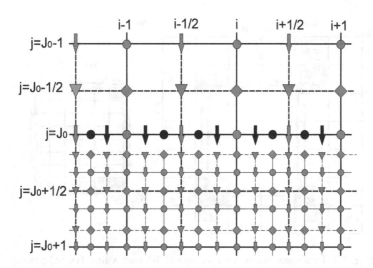

Fig. 9. Spatial steps refinement. Black marks correspond to points where the coarse-grid solution is interpolated.

From Coarse to Fine. The processors aligned to the coarse grid are grouped along each of the face in contact with the fine grid. At each of the faces a Master Processor (MP) gathers the computed current values of stresses/displacements, applies FFT and sends a part of the spectrum to the relevant MP on a fine grid (see Fig. 10). All the subsequent data processing, i.e. interpolation and inverse FFT, is performed by the relevant MP in the fine grid group. Subsequently, this MP sends the interpolated data to each processor in its subgroup.

Exchange of a part of FFT spectrum and interpolation performed by the MP of the second group essentially decreases the amount of sent/received data and, hence, idle time.

From Fine to Coarse. Processors from the second group compute the solution on the fine grid. For each face of the fine grid block, a MP is identified. This MP collects data from the relevant face, performs FFT, and sends a part of the spectrum to the corresponding MP of the first group (a coarse grid). Formally, FFT can be excluded and the data to be exchanged can be obtained as a projection of the fine grid solution onto the coarse grid, however the use of truncated spectra reduces the amount of data to be exchanged and ensures stability as it acts as a high-frequency filter.

Finally, the interpolated data are sent to the processors which need these data. The use of non-blocking procedures Isend/Irecv allow performing communications during updating the solution in the interior of the subdomains, thus a weak scaling of the algorithm is close to 93% for the use of up to 4000 cores.

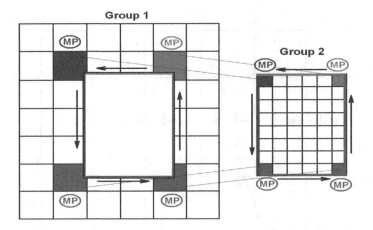

Fig. 10. Processor units for a coarse (left) and a fine (right) grids.

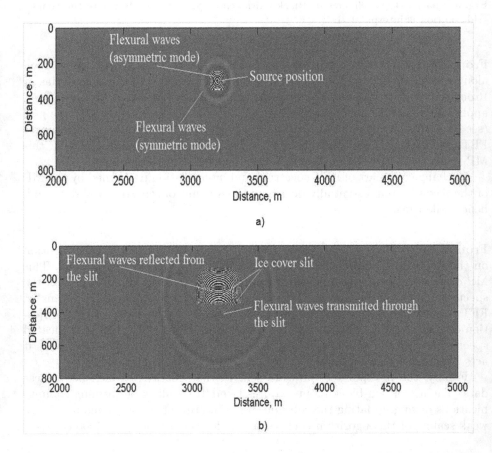

Fig. 11. Top view of the vertical component in the presence of a slit through the ice.

7   Possible Solutions to Reduce the Noise

One of the possible solutions to reduce the noise-to-signal ratio can be to slit the ice separating sources and receivers [5]. This slit stops asymmetric modes of the flexural waves and kills noise in the data beyond the slit. This effect is clearly seen in Fig. 11b as a remarkable damping of flexural waves. Another way to weaken flexural waves may be the construction of rib stiffness, some kind of artificial ice hummock. This will stiffen the ice and weaken flexural modes. The geometry of this rib can be chosen by numerical experiments. In contrast to the slit, such a hummock is cheaper and can be erected very quickly.

8   Conclusion

We have presented an original algorithm for the numerical simulation of seismic waves propagation in complex 3D media. The main idea of the approach proposed is the hybrid elastic/viscoelastic algorithm based on the local time-space mesh refinement. The use of independent domain decomposition for different regions with different system of equations and grid refinements results in a well-balanced algorithm and reduces need for of the computational resources up to several orders if compared with the fine grid simulations.

This approach was applied to simulate the interaction of seismic waves with irregular perturbations of the underside of the ice cover in order to understand the main peculiarities of seismic waves and the origin of a high seismic noise generated for acquisitions installed on the ice covering shallow waters.

A representative series of numerical experiments for realistic 3D models with allowance for the shape of the ice cover, batimetry, permafrost and thawing area have revealed the following roots of this noise:

- the vertical force generates intensive flexural waves which interact with heterogeneities of the ice-water interface and convert to fast symmetric modes;
- these modes propagate along the ice cover and interact again and again with the same heterogeneities and partially convert to slow flexural waves, and so on.

The amplitude of the primary flexural waves is extremely high; therefore the converted waves possess significant amplitudes and overlap the target reflections because they are multiply generated by random heterogeneities of the ice-water interface.

Acknowledgements. This work was supported by RSF (project No. 17-17-01128). The research was carried out using the equipment of the shared research facilities of HPC computing resources at Lomonosov Moscow State University, Joint Supercomputer Center of RAS and the Siberian Supercomputer Center.

References

1. Bailey, A.: Shooting seismic from floating ice. Pet. News **12**(5), 7–8 (2007)
2. Blanch, J., Robertson, A., Symes, W.: Modeling of a constant q: methodology and algorithm for an efficient and optimally inexpensive viscoelastic technique. Geophysics **60**, 176–184 (1995)
3. Carey, K.L.: Observed configuration and computed roughness of the underside of river ice, St. Croix River, Wisconsin. U.S.Geological Survey Research, Prof. Paper 550-B, pp. 192–198, Washington (1966)
4. Joly, P., Rodriguez, J.: An error analysis of conservative space-time mesh refinement methods for the one-dimensional wave equation. SIAM J. Numer. Anal. **43**(2), 825–859 (2006)
5. Henley, D.C.: Attenuating the ice flexural wave on arctic seismic data. Crewes Research Report 13, pp. 29–45 (2004)
6. Kostin, V., Lisitsa, V., Reshetova, G., Tcheverda, V.: Local time-space mesh refinement for simulation of elastic wave propagation in multi-scale media. J. Comput. Phys. **281**, 669–689 (2015)
7. Moczo, P., Kristek, J., Vavrycuk, V., Archuleta, R.J., Halada, L.: 3D heterogeneous staggered-grid finite-differece modeling of seismic motion with volume harmonic and arithmetic averagigng of elastic moduli and densities. Bull. Seismol. Soc. Am. **92**, 3042–3066 (2002)
8. Press, F., Ewing, M., Crary, A.P., Katz, S., Oliver, J.: Air-coupled flexural waves in floating ice, Part I and II. Geophysical Research Papers, vol. 6, 3–45 (1950)
9. Vishevsky, D., Lisitsa, V., Reshetova, G., Tcheverda, V.: Numerical study of the interface error of finite-difference simulation of seismic waves. Geophysics **79**(4), T213–T232 (2014)
10. Virieux, J., Calandra, H., Plessix, R.-E.: A review of the spectral, pseudo-spectral, finite-difference and finite-element modelling techniques for geophysical imaging. Geophys. Prospect. **59**, 794–813 (2011)

An Efficient Parallel Algorithm
for Numerical Solution of Low Dimension
Dynamics Problems

Stepan Orlov, Alexey Kuzin$^{(\boxtimes)}$, and Nikolay Shabrov

Computer Technologies in Engineering Department,
Peter the Great St. Petersburg Polytechnic University,
St. Petersburg, Russian Federation
majorsteve@mail.ru, kuzin_aleksei@mail.ru, shabrov@rwwws.ru

Abstract. Present work is focused on speeding up computer simulations of continuously variable transmission (CVT) dynamics. A simulation is constituted by an initial value problem for ordinary differential equations (ODEs) with highly nonlinear right hand side. Despite low dimension, simulations take considerable CPU time due to internal stiffness of the ODEs, which leads to a large number of integration steps when a conventional numerical method is used. One way to speed up simulations is to parallelize the evaluation of ODE right hand side using the OpenMP technology. The other way is to apply a numerical method more suitable for stiff systems. The paper presents current results obtained by employing a combination of both approaches. Difficulties on the way towards good scalability are pointed out.

Keywords: Continuously variable transmission · OpenMP
Numerical integration · Parallel algorithm

1 CVT Model Overview

The paper considers simulations of the dynamics of continuously variable transmission (CVT) consisting of two shafts (driving and driven ones), each bearing two pulleys, and the chain consisting of *rocker pins* and *plates* (Fig. 1). Pulleys have toroidal surfaces contacting with the convexo-convex end surfaces of pins. The torque is transmitted from the driving shaft to the driven one due to friction forces at pin-pulley contact points. One pulley at each shaft can move along the shaft axis, which allows to change CVT gear ratio. Rocker pins of the chain consist of two halves rolling over each other. A link of the chain is formed by 12–16 plates housing two pin halves.

Here we consider the mathematical model of CVT dynamics proposed in [1]. The model takes many details into account, which are the discrete chain structure; the extension and bending deformations of pins; the extension, bending, and torsion of plates; the elasticity of shafts, supports, and pulley-to-shaft attachments. The motion of the model is described by initial value problem for the

© Springer Nature Switzerland AG 2019
V. Voevodin and S. Sobolev (Eds.): RuSCDays 2018, CCIS 965, pp. 15–28, 2019.
https://doi.org/10.1007/978-3-030-05807-4_2

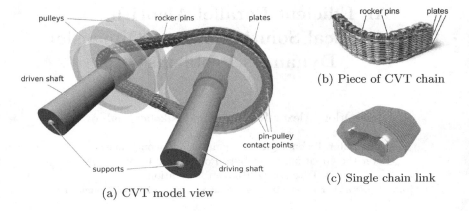

(a) CVT model view

(b) Piece of CVT chain

(c) Single chain link

Fig. 1. General view of CVT model; the chain; a single chain link

system of ordinary differential equations (ODEs), which is obtained by means of Lagrangian mechanics:

$$A\ddot{q} = F(q, \dot{q}, t), \qquad (1)$$

where q is the vector of generalized coordinates, A—positive-definite matrix of inertia, F—essentially non-linear function of q, $\dot{q}$ and t. The system can be transformed to normal form with obvious substitution: $u \equiv q$, $v \equiv \dot{q}$:

$$\dot{u} = v, \qquad A\dot{v} = F(u, v, t). \qquad (2)$$

Importantly, many contact interactions in the CVT model involve friction forces. The friction law used is close to the Coulomb's one, but is regularized: the friction coefficient f is constant at relative speeds greater than the saturation speed v_0, and depends on speed linearly at speeds less than v_0. Therefore, the friction law is piecewise linear. The non-smoothness of the friction law may affect the behavior of numerical methods investigated, that's why we also consider an alternative system in which the friction law is smoothed using a parabola connecting straight parts (Fig. 2).

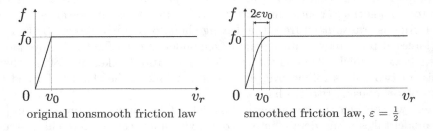

original nonsmooth friction law   smoothed friction law, $\varepsilon = \frac{1}{2}$

Fig. 2. Friction laws used in numerical experiments

Each step of numerical integration of the system (1) includes the evaluation of the right hand side F based on the current state vector $q, \dot{q}$, and the evaluation of accelerations by solving the system of linear algebraic equations $A\ddot{q} = F$. These steps are repeated 4 times per integration step when the RK4 integration method is used. Once the integration step is completed, the state vector at time instant $t + h$ is known. Then a *phase transition* procedure is run to detect pin—pulley contact state transition within the step. If transitions are found, the step is truncated, and the state is interpolated at the time of first transition.

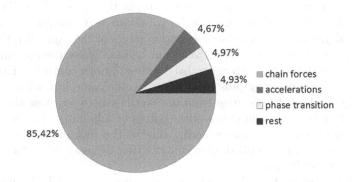

4,67%

4,97%

4,93%

▨ chain forces
■ accelerations
☐ phase transition
■ rest

85,42%

Fig. 3. CPU time consumption in CVT simulation. Sequential code

Parallelism can be introduced into the procedure in a natural way by evaluating the ODE right hand side in parallel. Presently the following parts of the integration step are parallelized: chain forces calculation, which takes almost all of the right hand side evaluation time, the reverse pass of Cholesky procedure of accelerations calculation, and the phase transition. Time costs of these separate operations in percents to overall simulation time obtained in sequential mode are shown in Fig. 3. The largest part of simulation time is spent on chain forces calculation, which is both the forces between pins and plates and the contact forces between pin tips and pulleys. Altogether with Cholesky reverse pass and phase transition it takes almost 95% of overall time. Therefore, the program contains about 5% of sequential code.

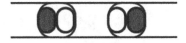

Fig. 4. Pin halves affected due to pin force application

Fig. 5. Pin halves affected due to link force application

An application of parallel technologies for the model of bearings, which is an example of periodic mechanical structure is presented in the work [2]. Also there are a lot of works dedicated to the parallel-in-time approach [3–5], which is to be perspective for the problems of small dimension. Therefore the approach can be considered as a field for future work for the problem considered while the present article shows the boundaries of speedup achieved using fine grained parallelism across the single integration step.

The chain forces that reside in the right hand side of (1) are generalized forces corresponding to the generalized coordinates of the pins. These forces emerge from interactions of two kinds. First of them are interactions between pin halves, and between pin halves and link plates; other forces, further referred to as *contact forces*, are result of contact between pin tips and pulleys. The forces of pins and links interactions also can be split into two groups: so called *pin forces* and *link forces*. Pin forces arise between halves of the same pin, these are, for example, contact interactions between pin halves. These interactions, when applied, give a contribution into generalized forces of both contacted halves and the pin halves of neighboring pins that belong to the same link with contacted ones (see Fig. 4). Link forces arise from interactions between pins that belong to the same link therefore these are outer halves of adjoining pins (see Fig. 5). These are, for example, the forces acting due to the link plates deformation. These interactions, when applied, give contribution to generalized forces of pin halves of the link only (Fig. 5).

Due to this specificity of interactions structure, the model of chain forces calculation chosen in the application seems to be natural. Let us consider chain consisting of N pins and N links. It can be split to M continuous parts by pin numbers n_i, $i \in [0; M]$, $(n_0 = n_M)$. These pins and links are distributed between M threads so each i-th thread takes $[n_i, n_{i+1})$ links and $[n_i + 1, n_{i+1})$ pins. Therefore M pins stay unassigned. Each thread calculates generalized forces in assigned pins independently; there is no synchronization needed. The calculation ends with a barrier, and then each i-th thread calculates pin forces in the unassigned pin n_{i+1}. Figure 6 demonstrates this procedure for the case of $M = 2$ and $N = 8$. The threads 0 and 1 take pins 1–3 and 5–7 respectively. Pin forces for unassigned pins 0 and 4 are evaluated after the barrier by threads 1 and 0 respectively. It is worth noticing that a portion of generalized forces is applied to pins 0 and 4 before the barrier. The threads add these generalized forces in parallel to the same pins but into the different halves.

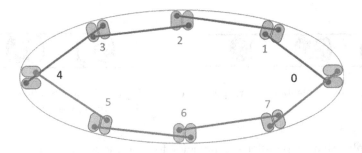

Fig. 6. An example of pins distribution between threads

Contact forces should be computed in a different way. These are link forces according to the schema of assignment, but they cannot be included into the procedure described because of calculation imbalance they produce. Therefore, pin halves in contact with pulleys are distributed uniformly between threads.

2 Results of Parallelization

Here the results of dynamics simulation of CVT with chain containing 84 pins are described. The simulations are performed on two nodes of "Polytechnic RSK Tornado" of Supercomputer Center Polytechnic of SPbPU, the parameters of the nodes are listed in Table 1.

Table 1. Parameters of computer used in simulations

Cores per socket	14
NUMA nodes	2
CPUs	Intel Xeon E5-2697 v3 2.60 GHz
Linux	CentOS Linux release 7.0.1406 (Core)
C++ compiler	Intel 2017.5.239

The CPUs affinity has been set up, so when $M \leq 14$ only one NUMA node is in use and when $M > 14$ the additional cores have been requested on the second node (M is the number of threads).

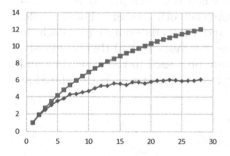
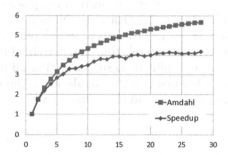

Fig. 7. Simulation speedup when acceleration calculation and phase transition are evaluated in parallel.

Fig. 8. Simulation speedup when acceleration calculation and phase transition are not evaluated in parallel.

The speedup of the whole simulation as a function of thread count is presented in Fig. 7. Also the chart contains ideal curve that corresponds to Amdahl's law with the fraction of serial work equal to 5%, which is the case presented in Fig. 3. The speedup is far from ideal mostly because of the bad scalability of the calculation of accelerations and phase transitions. One can see that analogous simulation with these parts evaluated sequentially has better correspondence to the ideal curve (Fig. 8), besides it loses in absolute speedup to the first simulation. The fraction of sequential code in this case, according to the Fig. 3, is about 14%, and Amdahl's curve lays significantly lower.

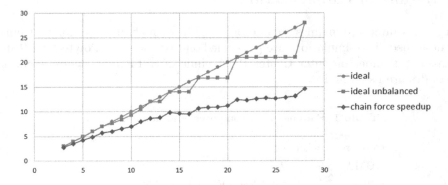

Fig. 9. Chain forces calculation speedup

The dependency of chain forces evaluation speedup on thread count is presented in Fig. 9. The chart contains *experimental* curve, ideal speedup (*ideal*), and ideal speedup taking into account thread non-ideal balancing (*ideal unbalanced*). The last curve is defined as follows. According to work distribution schema described above, each i-th thread takes contiguous part of the chain consisting of N_i pins. In the ideal case all values N_i are equal to each other for $i \in [0, M)$, but it obviously takes place only when the total pin count $N = 84$ is divisible by M. Otherwise N_i can differ from each other by one. For example, when $M = 11$, 7 threads process 8 pins each and 4 threads—only 7 pins each. In this case the overall time is determined by the "slowest" thread that, obviously, processes 8 pins (it is supposed that all pins are processed in the same time and this takes place in the simulation). The curve *ideal unbalanced* is calculated as a function of M in the following way:

$$f(M) = \frac{N}{N_{i,max}(M)}, \tag{3}$$

where $N_{i,max}(M) = \max_{i \in [0,M)} (N_i)$—the largest pin count per thread for the case of M threads. This is a piecewise line with horizontal segments. The segments consist of points where, despite the growth of thread count, $N_{i,max}$ remains the same. For example, when $M = 14$ all threads receive the same count of pins $N_i = 6$ and when $M = 15$, 9 threads receive 6 pins each and 6 threads—5 pins each, so there is no speedup in this transition. One can notice that the bends of experimental curve repeat those ones of *ideal unbalanced*. Therefore, the bends of experimental curve take place because of the irregularity of $N_{i,max}$ decrements as the thread count grows. The jumps of speedup take place when N is divisible by M: $M = 12, 14, 21, 28$. These are the cases of ideal balancing when all threads take the same count of pins. Analogous behavior takes place when $M = 17$. This is an "almost ideally" balanced case: 16 threads receive 5 pins each and one—4 pins each.

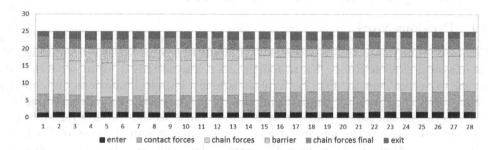

Fig. 10. Chain forces calculation time per thread

One can see that the experimental curve resides significantly lower than ideal ones, which can be explained by two factors. Firstly, there is an imbalance of loading between threads while contact forces are calculated; secondly, there is significant overhead time at parallel sections opening/closing. Both effects can be seen in Fig. 10. This diagram shows overall execution time of parallel code for each thread that evaluates chain forces for $M = 28$. Each bar represents overall time of one thread and is split into the stages. The stages *enter* and *exit* represent opening and closing of OpenMP section, *contact forces* is the stage of contact forces calculation, *chain forces* is the stage of pin/link forces calculation before the barrier, *barrier* is the stage that represents OpenMP barrier necessary before the rest M pin forces evaluation. And finally *chain forces final* is the stage of pin forces calculation of the rest M boundary pins.

One can notice that times of contact forces evaluation differ from one thread to another, while the times of pin/link forces evaluation are almost the same. This imbalance is noticeable, but one can see that more significant time is spent in sections *enter*, *exit* and *barrier*. This overhead cannot be explained only by load imbalance. Firstly, hotspot detection with Intel VTune Amplifier shows significant overhead time at OpenMP section opening/closing. Secondly, we have performed the same simulation with the application built with GCC compiler (GCC 5.4.0 and 7.2.0 have been used with the same results). This simulation has demonstrated much longer time (almost 1.5 times longer) of execution of sections *enter*, *exit*, and *barrier*, while durations of other sections have no significant changes. This fact also proves indirectly that the durations of sections *enter*, *exit* and *barrier* are determined in general by time spent in OpenMP code and not by unbalance of thread load.

Large overhead time is expectable with the schema of parallelism selected, when OpenMP sections are short and start and exit at each integration step.

Probably it is worth defining one common OpenMP section that would contain the whole ODE time integration procedure, but this is not straightforward and requires significant reorganization of integration scheme and can be a challenge for the future work.

3   Investigation of Numerical Methods

3.1   Previous Work

For a long time production versions of CVT simulation software have been using the classical explicit Runge–Kutta fourth order scheme (RK4) [6] to solve the initial value problem for Eq. (2). Explicit schemes impose limitations on step size h due to the stability requirement: the value $h\lambda$, where λ is an eigenvalue of ODE right hand side Jacobian matrix, must belong to the stability region of the method. That is the reason for long simulation times in the case of CVT dynamics equations. Previous work [7] presents the results of attempts to apply several other numerical integration methods, including explicit schemes (classical DOPRI45, DOPRI56, DOPRI78 methods), Gragg–Bulirsch–Stoer method (GBS), and Richardson's extrapolation applied to explicit Euler method; semi-implicit W-methods of Rosenbrock [8] type (SW24 [9] and Richardson's extrapolation applied to a W-method of first order); the trapezoidal rule implicit method (TRPZ). Among those methods, only the trapezoidal rule method allows to obtain sufficiently accurate numerical solution at steps much larger than the step chosen for RK4 (in the numerical example, step size for RK4 needs to be at most $5 \cdot 10^{-8}$ s, while the trapezoidal rule has allowed us to obtain numerical solution of the same accuracy at step $2 \cdot 10^{-6}$ s). Although the stability of implicit methods do not impose limitation on step size, such a limitation appears nevertheless: the nonlinear equation system that needs to be solved at each step has to be solved using a Newton-like iterative method; in our case, iterations only converge when the step size is not too large.

The trapezoidal rule has proved to be able to provide accurate numerical solution at time steps 40 times greater than steps we have to use with the RK4 scheme. However, the practical usefulness of the method is doubtful. In our numerical tests we found that the Newton's iterations converge at 5–10 steps only when the Jacobian matrix is fully recomputed at each iteration. Attempts to apply Broyden updates would break the sparsity of the Jacobian; attempts to limit Broyden's updates to current Jacobian sparsity stencil seem to work, but still recomputing the Jacobian may be needed more than once per time step, and the number of iterations increases up to tens. Jacobian updates proposed in [10] didn't work at all in our tests with CVT equations, although they worked well for other much simpler equation examples. Finally, not updating the Jacobian at all during the time step may increase the number of Newton's iterations up to hundreds, and the iterations may even fail to converge at all. Therefore, one practical approach to speedup simulations using the Newton's method is to compute the Jacobian matrix in parallel, probably employing the approach presented in [11] in order to reduce the number of ODE right hand side evaluations, and to use a parallel LU solver for linear systems at Newton's iterations. This approach hasn't been tried yet.

3.2   Eigenvalues of ODE Right Hand Side Jacobian Matrix

We continued our search for a numerical method that is more suitable for the numerical integration of CVT dynamics equations. To make the search more purposeful, we have investigated eigenvalues of ODE right hand side Jacobian matrix. Namely, the full eigenvalue problem was numerically solved at some typical state in a stationary regime of CVT operation. The eigenvalues found split into two groups: firstly, there are complex eigenvalues corresponding to damped oscillations; secondly, there are real negative eigenvalues corresponding to non-oscillatory dissipation processes. Absolute values of eigenvalue imaginary parts have maximum at approx. 10^6 s^{-1}, while absolute values of real parts— at approx. 10^8 s^{-1}. Figure 11 shows the computed eigenvalues of the Jacobian matrix in the linear scale (left) and the logarithmic scale (right). Notice that in the latter case, the scale is logarithmic everywhere except the vicinity of zero— the region limited by dashed lines; in that region, the scale is linear.

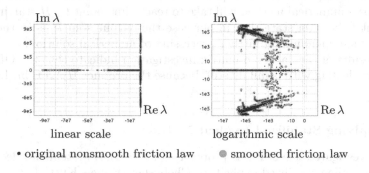

linear scale   logarithmic scale

• original nonsmooth friction law   ● smoothed friction law

Fig. 11. Eigenvalues of ODE right hand side Jacobian matrix

Further we proved that the largest negative eigenvalues are caused by friction forces acting between pin halves, and the operating points in the friction law are at the linear part ($v < v_0$). The idea of the proof is to change (e.g., increase 10 times) the saturation speed v_0 in the friction law, then recompute the Jacobian matrix and find its eigenvalues. The results of evaluation are shown in Fig. 12. In case (a), we increased the saturation speed v_0 10 times for the friction law between pin halves; in case (b), we also increased it 10 times for friction between pins and plates; in case (c), in addition, we increased it 10 times for friction between pins and pulleys.

Therefore, largest negative eigenvalues correspond to the friction between pin halves. This fact allows us to conclude that such eigenvalues are proportional to the tension force in the most loaded straight part of the chain, and, consequently, vary from one CVT operation regime to another.

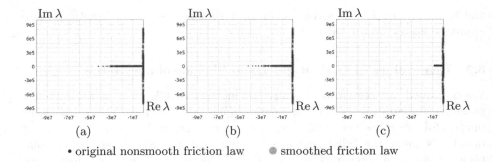

• original nonsmooth friction law   ● smoothed friction law

Fig. 12. Eigenvalues of ODE right hand side Jacobian matrix for modified friction law

The behavior and suitability of a numerical integration method strongly depend on the eigenvalues of the Jacobian matrix of the ODE right hand side. The product of a characteristic time span length H and the eigenvalue λ_* with maximum absolute value are of particular interest. The value H is the time between two neighboring states that we wish to have in the numerical solution, that is, the "output step". The product $H|\lambda_*|$ can be used to estimate how much steps a numerical method will take to reach time point $t + H$ starting from time point t. We can state that in our case the output step H is in the range 10^{-5}–10^{-4} s, depending on further processing of numerical solution.

The results of the Jacobian matrix investigation indicate that the ODE system we are dealing with is mildly stiff, because the product $H|\lambda_*|$ is in the range 10^3–10^4.

3.3 Applying Stabilized Explicit Method

Having investigated the stiffness properties of the ODE system, we decided to consider so called stabilized explicit, or Chebyshev–Runge–Kutta [12, ch. IV.2], numerical integration methods. Among those we picked the one called DUMKA3 [13]. The choice of that particular method was based on public availability of solver implementation code in C programming language.

A stabilized explicit scheme has stability region, defined by the condition $|R_s(h\lambda)| \leq 1$, extended into real negative direction of complex plane $h\lambda$. The function R_s is a polynomial of degree s, called the *stability polynomial*.

The DUMKA3 solver actually implements a family of s-stage Runge–Kutta schemes, each of which realizes stability polynomial of degree s varying from 3 up to 324. The solver implements automatic step size control, based on step local error estimation, and polynomial degree control, based on the estimation of ODE right hand side Jacobian matrix spectral radius. Preliminary tests have shown that solver performance with both control options enabled is far from optimal in our system. Besides, a numerical estimation of Jacobian spectral radius do not work well due to discontinuities of the ODE right hand side in the case of nonsmooth friction law, resulting sometimes in too large values. We disabled both control options. The motivation was to obtain best performance at fixed step for each fixed polynomial degree. Giving each polynomial an index k from 0 to 13 (DUMKA3 implements 14 polynomials), we denote corresponding schemes by suffixes $-Pk$. In particular, we tested degrees $s = 21$ ($k = 4$), $s = 27$ ($k = 5$), $s = 36$ ($k = 6$), $s = 48$ ($k = 7$), $s = 63$ ($k = 8$), $s = 81$ ($k = 9$). Polynomials $k \leq 4$ and $k \geq 9$ have shown poor performance: in the first case the stability region is too small in the real negative direction, and in the second case it is too small in the imaginary direction.

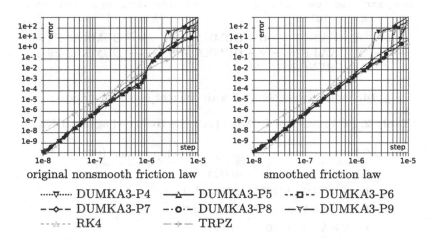

original nonsmooth friction law   smoothed friction law

······▽······ DUMKA3-P4   ──△── DUMKA3-P5   --□-- DUMKA3-P6
--◇-- DUMKA3-P7   --○-- DUMKA3-P8   ──▼── DUMKA3-P9
--✳-- RK4   --✳-- TRPZ

Fig. 13. Dependency of step local error norm on step size

Figure 13 shows the dependency of step local error norm on the step size. DUMKA3 results are compared against RK4 and the trapezoidal rule; the latter one is known [7] to give sufficiently accurate solution at $h = 2 \cdot 10^{-6}$ s. It can be shown that the slope for each curve at steps below 10^{-6} corresponds to the order of the scheme (3 for DUMKA3, 4 for RK4, and 2 for TRPZ). Notice also that at steps above 10^{-6} local error for all DUMKA schemes rises sharply at some point; additional error jumps can be see at step 10^{-6} for nonsmooth friction law. At large step sizes, local error for DUMKA schemes is approximately the same as for the trapezoidal rule, in the case of smoothed friction law; for nonsmooth friction law, the trapezoidal rule gives smaller error.

Figure 14 shows the sample curve (pin axial force when it enters the driving pulley set) computed numerically. It follows from the figure that schemes DUMKA3-P5 – DUMKA3-P8 give sufficiently accurate solution at step $h = 2 \cdot 10^{-6}$, and at step $h = 4 \cdot 10^{-6}$ only the scheme DUMKA3-P8 does so.

Comparing simulation times, we can conclude that the DUMKA3 solver can perform simulations several times faster than RK4, which is summarized in Table 2. Notice that the value n_{RHS} in the table is the total number of ODE right hand side evaluation in the test simulation (the same as used to obtain the sample curve).

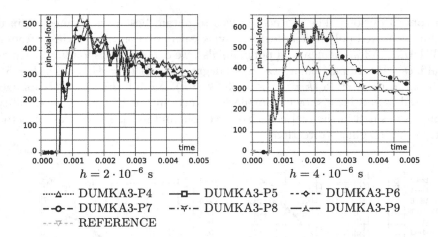

$$h = 2 \cdot 10^{-6} \text{ s} \qquad\qquad h = 4 \cdot 10^{-6} \text{ s}$$

······△······ DUMKA3-P4　　—□— DUMKA3-P5　　--◇-- DUMKA3-P6

--○-- DUMKA3-P7　　--▼-- DUMKA3-P8　　—▲— DUMKA3-P9

--▽-- REFERENCE

Fig. 14. Sample curve obtained with different polynomial degrees

Table 2. Performance of DUMKA3 schemes compared against RK4

Scheme	h, s	n_{RHS}	Speedup against RK4
RK4	$5 \cdot 10^{-8}$	400824	1
DUMKA-P5	$2 \cdot 10^{-6}$	75817	5.9
DUMKA-P7	$3 \cdot 10^{-6}$	95425	4.7
DUMKA-P8	$4 \cdot 10^{-6}$	104821	4.4

4　Conclusions and Future Work

The paper describes our activities in speeding up of simulation of nonlinear problem of dynamics modeling of CVT.

The procedure of ODEs integration over time can be parallelized in natural way when the right hand side is being calculated in parallel at each step. It can be done in this problem because of regular structure of the chain. Therefore the chain forces calculation can be distributed between threads by dividing the chain into continuous segments. This calculation procedure requires only one explicit barrier.

Another stages that can be calculated in parallel are accelerations evaluation and phase transition.

This approach leads to the problem of rather high overhead because of small parallel sections lengths. It can be overcome in the approach with one common parallel section that spans the whole integration procedure. We consider it as a perspective for future work.

The attempt to use stabilized explicit Runge–Kutta solver, DUMKA3, has proven to be successful in terms of performance. However, original polynomial order control implemented in the solver doesn't work: it tends to increase polynomial order, but in that case Jacobian eigenvalues with maximum imaginary parts fall outside the stability region, since it becomes a bit narrower in the imaginary direction. This motivates us to consider other stabilized explicit solvers, first of all RKC and SERK2 [14], and probably others, constructed according to the approach presented in [15], because there is a way to construct a method with stability region that best fits the spectrum of ODE right hand side Jacobian. Among all numerical integration methods tested so far for our problem, DUMKA3 has shown the best performance, for the case of sequential code. In the same time, estimations show that the trapezoidal rule method considered in [7] has the potential to gain the speedup of about 100 due to the parallelization of Jacobian calculation, which, however, does not reduce computational costs because the number of CPU cores required for that is about 600. The combination of the ODE right hand side parallelization and the DUMKA3 solver promises to achieve speedups of about 30.

References

1. Shabrov, N., Ispolov, Y., Orlov, S.: Simulations of continuously variable transmission dynamics. ZAMM **94**(11), 917–922 (2014). WILEY-VCH Verlag GmbH & Co. KGaA, Weinheim
2. Nordling, P., Fritzson, P.: Solving ordinary differential equations on parallel computers — applied to dynamic rolling bearings simulation. In: Dongarra, J., Waśniewski, J. (eds.) PARA 1994. LNCS, vol. 879, pp. 397–415. Springer, Heidelberg (1994). https://doi.org/10.1007/BFb0030169
3. Ruprecht, D., Krause, R.: Explicit parallel-in-time integration of a linear acoustic-advection system. Comput. Fluids **59**, 72–83 (2012)
4. Liu, J., Jiang, Y.-L.: A parareal waveform relaxation algorithm for semi-linear parabolic partial differential equations. J. Comput. Appl. Math. **236**(17), 4245–4263 (2012)
5. Kreienbuehl, A., Benedusi, B., Ruprecht, D., Krause, R.: Time parallel gravitational collapse simulation. Commun. Appl. Math. Comput. Sci. **12**(1), 109–128 (2015)
6. Hairer, E., Nørsett, S.P., Wanner, G.: Solving Ordinary Differential Equations I: Nonstiff Problems. Springer Series in Computational Mathematics, 2nd edn. Springer, New York (1993). https://doi.org/10.1007/978-3-540-78862-1
7. Orlov, S., Kuzin, A., Shabrov, N.: Two approaches to speeding up dynamics simulation for a low dimension mechanical system. Commun. Comput. Inf. Sci. **793**, 95–107 (2017)
8. Rosenbrock, H.H.: Some general implicit processes for the numerical solution of differential equations. Comput. J. **5**, 329–330 (1963)
9. Steihaug, T., Wolfbrandt, A.: An attempt to avoid exact Jacobian and nonlinear equations in the numerical solution of stiff differential equations. Math. Comp. **33**, 521–534 (1979)
10. Hart, W.E., Soesianto, F.: On the solution of highly structured nonlinear equations. J. Comput. Appl. Math. **40**(3), 285–296 (1992)
11. Ypma, T.J.: Efficient estimation of sparse Jacobian matrices by differences. J. Comput. Appl. Math. **18**(1), 17–28 (1987)
12. Hairer, E., Wanner, G.: Solving Ordinary Differential Equations II: Stiff and Differential-Algebraic Problems. Springer Series in Computational Mathematics. Springer, Heidelberg (1996). https://doi.org/10.1007/978-3-642-05221-7

13. Medovikov, A.A.: High order explicit methods for parabolic equations. BIT Numer. Math. **38**(2), 372–390 (1998)
14. Martín-Vaquero, J., Janssen, B.: Second-order stabilized explicit Runge-Kutta methods for stiff problems. Comput. Phys. Commun. **180**(10), 1802–1810 (2009)
15. Torrilhon, M., Jeltsch, R.: Essentially optimal explicit Runge-Kutta methods with application to hyperbolic-parabolic equations. Numerische Mathematik **106**(2), 303–334 (2007)

Analysis of Means of Simulation Modeling of Parallel Algorithms

D. V. Weins$^{(\boxtimes)}$ ⓘ, B. M. Glinskiy ⓘ, and I. G. Chernykh ⓘ

The Institute of Computational Mathematics
and Mathematical Geophysics SB RAS, Novosibirsk, Russia
vins@sscc.ru, gbm@opg.sscc.ru, chernykh@parbz.sscc.ru

Abstract. At the ICMMG, an integral approach to creating algorithms and software for exaflop computers is being developed. Within the framework of this approach, the study touches upon the scalability of parallel algorithms by using the method of simulation modeling with the help of an AGNES modeling system. Based on a JADE agent platform, AGNES has a number of essential shortcomings in the modeling of hundreds of thousands and millions of independent computing cores, which is why it is necessary to find an alternative tool for simulation modeling.

Various instruments of agent and actor modeling were studied in the application to modeling of millions of computing cores, such as QP/C++, CAF, SObjectizer, Erlang, and Akka. As a result, on the basis of ease of implementation, scalability, and fault tolerance, the Erlang functional programming language was chosen, which originally was developed to create telephony programs. Today Erlang is meant for developing distribution computing systems and includes means for generating parallel lightweight processes and their interaction through exchange of asynchronous messages in accordance with an actor model.

Testing the performance of this tool in the implementation of parallel algorithms on future exaflop supercomputers is carried out by investigating the scalability of the statistical simulation algorithm by the Monte Carlo methods on a million computing cores. The results obtained in this paper are compared with the results obtained earlier by using AGNES.

Keywords: Simulation modeling · Actor model · Scalability · Erlang

1 Introduction

According to the calculations given by D. Dongarra, the performance of supercomputers in exaflops will be reached by 2018–2020. Supercomputers will be able to serve 1 billion computing flows simultaneously. The number of cores will reach 100 million. The creation of exaflop supercomputers will require the development of parallel algorithms that can use tens and hundreds of millions of computing cores. The authors of the article develop an integral approach to developing algorithms and software for modern and future supercomputers. The approach is based on the technique for the development of algorithms and software for supercomputers of peta and exaflop levels, containing three related stages. The first stage is determined by co-design, which is understood as adapting the computational algorithm and mathematical method to a

V. Voevodin and S. Sobolev (Eds.): RuSCDays 2018, CCIS 965, pp. 29–39, 2019.
https://doi.org/10.1007/978-3-030-05807-4_3

supercomputer architecture at all stages of the problem solution. The second stage is the development of preventive algorithms and software for the most promising super-computers on the basis of simulation modeling for a given supercomputer architecture. The third stage is associated with estimating the energy efficiency of the algorithm for various implementations on this architecture or on various architectures [1].

This approach was tested on computationally complex problems of astrophysics, plasma physics, geophysics, and stochastic problems. The concept of co-design in the context of mathematical modeling of physical processes is understood as the construction of a physico-mathematical model of a phenomenon, numerical method, and parallel algorithm with its software implementation, which effectively uses the super-computer architecture [2, 3].

An important component of the integral approach is simulation modeling, which allows investigating the scalability of a parallel algorithm on a given supercomputer architecture, determining the optimal number of computational cores to implement computations, and revealing bottlenecks in its execution.

The problem of modeling of scalable algorithms is not new - many groups of researchers in the world are engaged in it. Among foreign studies, we note the ones carried out in the US (University of Urbana-Champagne, Illinois). They are mainly engaged in estimating the performance of algorithms implemented with the use of MPI [4]. One of the main projects of this team is the BigSim project. The project is aimed at creating an imitation environment that allows the development, testing, and adjustment through the modeling of future generations of computers, while allowing for computer developers to improve their design solutions with a special set of applications. Among domestic studies, we note the ones conducted at the Ivannikov Institute for System Programming of the Russian Academy of Sciences (Moscow). This team developed a parallel program model that can be effectively interpreted on an instrumental computer, enabling fairly accurate prediction of the time of real execution of a parallel program on a given parallel computing system. The model is designed for parallel programs with explicit messaging, written in Java language with the use of the MPI library, and is included in the ParJava environment.

It is worth noting that both of the projects considered do not take into account (at least explicitly) the issues of fault tolerance in the execution of large programs, while the use of tens of millions of computing cores at the same time is extremely urgent. The ParJava project, on the one hand, allows solving a wide range of problems for estimating the performance of parallel programs on promising computing systems, but, on the other hand, is tied to a specific programming language, which significantly reduces its capabilities.

The implementation of simulation modeling made it possible to investigate the scalability of algorithms for solving problems in astrophysics, plasma physics, geophysics, and problems using the Monte Carlo methods [1, 5]. The modeling was carried out on the basis of an AGNES multiagent system [6], with which it was possible to trace the behavior of algorithms up to several million computing cores. However, a further increase in the number of simulated computing cores proved to be difficult due to the limitations typical of the JADE platform, on the basis of which the AGNES system was built. Therefore, it became necessary to analyze other methods and tools, particularly based on an actor model. An actor model is a special technique for implementing agent modeling systems to reduce the overheads of agent communication [7].

2 Limitations of the AGNES Modeling System and Ways to Overcome Them

In the study of the possibility to scale different algorithms to a large number of computational cores, the task of simulating the execution and communication of hundreds of thousands and millions of computational threads arose. The perfect approach for this purpose is an agent-based approach to modeling systems containing autonomous and interacting intellectual agents [8]. As a tool, the AGNES modeling system [7] was used, based on the JADE multiagent modeling platform [9]. This system had several advantages and disadvantages.

Using the JADE platform in the separation of the agent functionality into a set of behaviors makes it easy to parallelize the execution of independent behaviors. By default, the agent manager of each agent within the platform has its own flow, and all behaviors of the agent are carried out within this flow. However, the behavioral infrastructure itself introduces additional computational costs, especially in a single-flow performance. Also, significant overheads are introduced by the messaging system. It is extremely flexible and allows for standard means to transfer complex types of data, but all this at the expense of performance.

The JADE platform also provides for the partitioning of agents within the platform into containers that can reside on different hosts. If it is possible to minimize the exchange of messages between containers, then an increase in the number of hosts enhances the performance of such a system. However, in modeling systems where "all agents are connected to each other", the system runs slower and slower.

In general, practice shows that the simultaneous launch of more than 1,000–2,000 simple interacting agents on a single average computing node causes a noticeable decrease in performance.

To overcome these shortcomings, various methods and techniques are used to run simulation models of algorithm behavior. The possibility of simulation of a large number of computational threads on a node requires that one agent simulates the behavior of a group of similar computational threads, each of which is an independent behavior of this agent and is executed in parallel. To minimize the exchange of messages between agents, the messages from different behaviors within the agent are formed into an array and forwarded to another agent as a single complex message. This allowed us to study the feasibility of scaling various algorithms into millions of computing cores.

Each algorithm under investigation has some special features. The special feature of algorithms for solving distributed statistical modeling problems using the Monte Carlo methods is the necessity for modeling an extremely large number of independent implementations. A feature of parallel grid methods in solving hyperbolic equations is the possibility of geometric decomposition of the computational domain and subsequent exchange of boundary values only between neighboring computing nodes. Below are the results of the study of the scalability of algorithms for solving the problems of astrophysics, plasma physics, and seismics and the problems using the Monte Carlo methods by means of the AGNES agent-based system (Fig. 1).

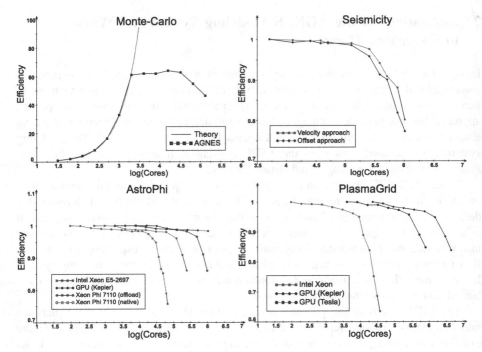

Fig. 1. Study of the scalability of different algorithms by means of the AGNES agent-based system

The upper section of the figure illustrates the results of the study of scalability of algorithms for solving the problem of direct statistical modeling by the Monte Carlo method and the seismic task in solving the problem in terms of displacement and stress velocities and in terms of displacements. Calculations were carried out on an NKS-30T cluster with graphic accelerators [10]. The bottom section of the figure shows the results of the study of scalability of the astrophysical code and the physics code of plasma [1]. In this case, the scaling of algorithms on clusters with MPP architecture and using GPU and Phi is under study. It can be seen from the figure that about 1 million computing cores can be effectively used to solve these problems, but the efficiency drops sharply with a greater number of cores. A similar situation is observed in the solution of stochastic problems, where the computations are independent and there are practically no exchanges [5]. For parallel grid methods, a decrease in efficiency with rising number of computing cores is associated with an avalanche-like increase in message exchanges between the cores. For distributed statistical modeling, this decrease is due to the fact that the assembler node does not manage to process a large number of messages with intermediate data. The question arises: is this behavior of algorithms is primarily due to the calculation scheme or is it a feature of the chosen simulation modeling system?

In this regard, the task is to find an alternative modeling technique that helps eliminating the shortcomings of the JADE platform, as well as tools for working with it.

3 Analysis of Means of Simulation Modeling of Parallel Algorithms

3.1 Actor Model

An actor model is a mathematical model of parallel computations that treats the concept of "actor" as a universal primitive of parallel numerical calculation: in response to received messages, the actor can make local decisions, create new actors, send messages, and determine how to respond to following messages. It is not assumed that there is a certain sequence of the above-described actions, and all of them can be performed in parallel. The model was created in 1973, and it was used as a basis for understanding the calculus of processes and as a theoretical basis for a number of practical implementations of parallel systems [11].

The actor model is characterized by the inherent parallelism of calculations within one actor and between actors, dynamic creation of actors, inclusion of actors' addresses in messages, and also interaction only through direct asynchronous messaging without any restrictions on the order of arrival of messages.

The actor model has some characteristic distinctive features:

1. Unlimited indeterminism. There is no global state in the actor model.
2. Messages in the actor model are not necessarily buffered. This is its difference from the rest of the approaches to a model of simultaneous computations. In addition, messages in the actor model are simply sent, and there is no requirement for a synchronous handshake with the recipient.
3. Creating actors and inclusion of addresses of participants into messages means that the actor model has potentially variable topology in their relationships with each other. According to the communication model, the message does not have any mandatory fields, they can all be empty. However, if the sender of the message wishes the recipient to have access to addresses that the sender does not have yet, the address should be sent in the message.
4. Unlike other approaches, based on the combination of sequential processes, the actor model was developed as a simultaneous model in its essence. As written in the theory of actor models, the sequence therein is a special case arising from simultaneous calculations.
5. The main innovation of the actor model is the introduction of the concept of behavior, defined as a mathematical function expressing the actor's actions when it processes messages, including determining a new behavior for processing the next message arrived. The behavior ensures the functioning of the mathematical model of parallelism and also frees the actor model from implementation details.

These and many other ideas introduced in the actor model are also used now in agent modeling systems. The actor model, in particular, is used in agent systems to minimize overheads in agent communication [7]. The key difference from agent modeling is that the system agent imposes additional restrictions on actors, usually requiring that they use commitments and goals. It is due to the advantages listed above that the use of the actor model to study the scalability of parallel algorithms seems very promising. Note that there are many papers on presentation and implementation of algorithms on the basis of the Actor model, but it has not yet been applied as a tool for investigating the scalability of parallel algorithms.

3.2   QP/C++, CAF and Sobjectizer

Let us consider tools and frameworks that allow implementing the concept of the actor model in C++. As C++ is a widely known native language, which makes it possible to switch easily from the lowest level close to hardware to a very high level, such as OOP and general programming. At the same time, this language is provided with a wide range of tools, books, and documentation.

Strictly speaking, there are not too many ready-made implementations of the actor model. Among popular and actively developing frameworks for C++, C++ Actor Framework (CAF) [12], QP/C++ [13], and SObjectizer (SO) [14] can be distinguished.

OpenSource project under the BSD license is C++ Actor Framework. Also known as CAF and libcppa, it is the most famous implementation of the actor model for C++. This is an easy-to-use framework with a specific syntax that maximally fully implements the principles of actor models. The scope of CAF is limited to the Linux platform. The CAF developers themselves describe it as the most productive framework. It also offers ready-made tools for creating distributed applications.

QP/C++ is a software product under a double license, designed to develop embedded software, including real-time systems and systems that can work directly on hardware implementation.

Actors in QP/C++ are called active objects and represent hierarchical finite-state machines. The code of actors can be typed in the form of ordinary C++ classes, and an actor can be drawn in a special tool for visual modeling and its code is generated automatically. Active objects in QP/C++ work on a context that QP allocates to them. Depending on the environment, active objects can work each on their own thread or they can share a common working context.

SObjectizer is an OpenSource project under the BSD license, has been developed since 2002, and is based on the ideas created and tested when building a small object-oriented SCADA system. SObjectizer is created specifically to simplify the development of multi-thread software in C++. Therefore, in SObjectizer, much attention is paid to compatibility. What SObjectizer does not currently provide is ready-made tools for constructing distributed applications.

Actors in SObjectizer are called agents. As in QP/C++, agents in SObjectizer are, as a rule, instances of individual C++ classes. Just like in QP/C++, agents are hierarchical finite-state machines. Just like in QP/C++, the working context for agents is provided by the framework. For this purpose, SObjectizer includes a dispatcher, which is a special entity that performs dispatching of agent events. SObjectizer strongly differs from the above-mentioned projects in that SObjectizer sends messages not directly to the recipient agents, but in mboxes (mailboxes). From an mbox, a message is delivered to those agents who are subscribed to it.

3.3 Erlang and Akka

When it comes to the actor model, one cannot help but mention Erlang, and, speaking of Erlang, one cannot help but talk about the actor model. Erlang is a functional programming language with strong dynamic typing, designed to create distributed computing systems [16]. It is developed and supported by Ericsson for writing programs for telephony. Erlang inherited its syntax and some concepts from the Prolog logical programming language. This language includes means of generating parallel lightweight processes and their interaction through the exchange of asynchronous messages in accordance with the actor model.

Erlang was purposefully designed for use in distributed, fault-tolerant, parallel real-time systems, for which, in addition to the language itself, there is a standard library of modules and library templates, known as an OTP framework. The program on Erlang is translated into bytecode, executed by virtual machines located on different nodes of a distributed computer network.

The popularity of Erlang began to grow due to the expansion of its application area (telecommunication systems) to highly loaded parallel distributed systems serving millions of WWW users, such as chats, content management systems, web servers, and distributed databases requiring scaling. A great number of products are developed on Erlang, and many companies use Erlang as a key tool.

Also speaking of the actor model, one cannot fail to mention the Akka framework for the Scala and Java languages [15]. Actors in Akka are instances of separate classes executed on JVM virtual machines located on different nodes of a distributed computer network. The actor in Akka consists of several interacting components: the actor's dispatcher and the actor itself. The dispatcher is responsible for placing messages in the queue leading to the mailbox of the actor and also orders this box to remove one or more messages from the queue (but only one at a time) and transfer them to the actor for processing. Last but not least: the actor is usually the only API that needs to be implemented, and it encapsulates the state and behavior. Akka does not allow direct access to the actor, so it guarantees that the only way to interact with the actor is through asynchronous messages.

Akka is widely used in the field of Web and online services (for example, Twitter and LinkedIn).

There are several key factors that explain the popularity of Erlang and Akka among modern developers:

- Simplicity of development. Using asynchronous messaging greatly simplifies the work when one has to deal with concurrent computing;
- Scaling. The actor model allows creating a huge number of actors, each of which is responsible for its particular task. The shared nothing principle and asynchronous messaging allow building distributed applications that are scaled horizontally as much as needed;
- Fault tolerance. A failure of one actor can be caught by other actors, who take appropriate actions to restore the situation. Also, if some lightweight process within Erlang VM performs division by zero, then Erlang VM simply closes one of these processes, and this will not affect the performance of other processes. However, if there is division by zero in one of the threads of a multithread application on C++, then the entire application crashes.

4 Experimental Study

To test the performance of the tools under consideration, a model of execution of the algorithm for solving the problem of distributed statistical modeling using the Monte Carlo methods is implemented on each of them. Unfortunately, due to problems with licensing, QP/C++ does not participate in this experiment. Such parameters as the amount of code for the description of actors, acceleration from parallelization, and the total simulation time are explored.

When simulating computations with the help of actors, threads-collectors and threads-calculators are simulated. Threads-calculators at each iteration of the calculation cycle determine the independent implementation and transmit the result of intermediate averaging to threads-collectors. The results obtained are averaged by thread-collectors. The calculation is carried out until the required level of the permissible relative statistical error is reached. Calculations on different threads-calculators are performed in an asynchronous mode. The calculation results are sent and received also in an asynchronous mode.

A comparison of the number of symbols needed to describe the actors/agents in the tools under consideration is shown in Fig. 2. As can be seen from the figure, the most concise description is obtained in Erlang and Akka. This can be explained by the fact that very many mechanisms for memory distribution and allocation, balancing, and messaging are already implemented within these tools.

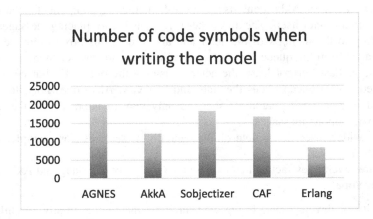

Fig. 2. Comparing the amount of programming code in the description of actors/agents in different tools

As is known, the theoretical acceleration for paralleling for statistical modeling methods is practically "ideal", i.e., the calculation time decreases in proportion to an increase in the computing nodes. This criterion is used to study the effectiveness of these computational algorithms.

We consider the resulting data on the scalability of the algorithm (Fig. 3) and data on the execution time of the model (Fig. 4) for various tools. For clarity, the charts show the theoretical acceleration for this algorithm and the results obtained earlier with the help of AGNES.

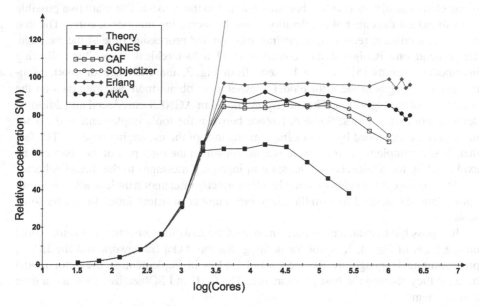

Fig. 3. Study of the scalability of the algorithm for solving the problem of direct statistical modeling using the Monte Carlo methods by means of various tools.

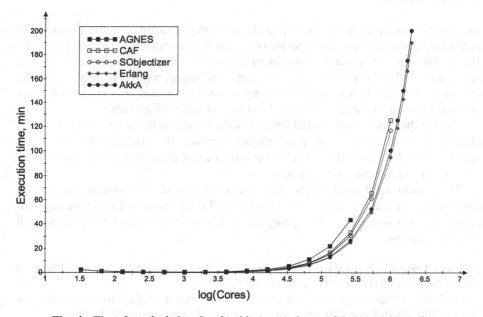

Fig. 4. Time for calculating the algorithm execution model on various tools.

Previously, similar behavior of the investigated algorithm (Fig. 3) was explained this way: as soon as the time for sending and processing messages from all calculators after one iteration of calculations begins to exceed the time for this iteration, the acceleration gain slows down and the scaling is no longer observed on a certain number of the effect calculators (the line becomes parallel to the X axis). There are two possible ways to extend the effect of acceleration from the increasing number of cores. The first one is via hardware: reduction of the transmission and processing time of the message. The second one is algorithmic: construction of a hierarchical model for collecting intermediate results [5]. As can be seen from Fig. 3, the exchange of short, asynchronous messages in the actor model allowed us to obtain more accurate data on the scalability of the algorithm. The messaging model in AGNES introduced an additional delay in communications. Some difference between the tools implementing the actor model can be explained by various implementations of the messaging process. The fact that, in some implementations, there is a slowdown on the right part of the chart can be explained by the avalanche-like increase in incoming messages to the thread-collector that the message queue either cannot handle correctly or cannot handle at all. It is worth noting that, the desired (one million or more) number of actors failed to run on some tools.

It is possible to evaluate the performance of the tools implementing the actor model on the basis of Fig. 4. It is not for nothing that the Akka framework and the Erlang programming language are the recognized leaders in implementing the actor model because they showed the best performance. The CAF and SObjectizer tools are a little behind them.

5   Conclusion

As experiments have shown, the use of the actor model to simulate the execution of parallel algorithms allows one to get rid of a number of significant disadvantages of the JADE platform. The process of modeling of millions of computing cores is significantly lesser affected due to the use of simple and asynchronous messages. And the idea of using multiple lightweight (memory-wise) actors makes it possible to achieve the simulation of execution of required millions of computing cores.

Among the tools, libraries, and languages implementing the actor model, a special place is occupied by the Erlang programming language. In it, the concept of the actor model is most fully implemented. This was confirmed in the course of the experimental study of the performance of these tools.

The experimental research has shown that, due to such parameters as ease of implementation, scaling, and fault tolerance, the Erlang functional programming language is most suitable for investigating the scalability of algorithms using simulation modeling methods.

Acknowledgments. This work was supported by the Russian Foundation for Basic Research (Grants No. 16-07-00434, 18-37-00279, and 18-07-00757).

The Siberian Supercomputer Center of the Siberian Branch of the Russian Academy of Sciences (SB RAS) is gratefully acknowledged for providing supercomputer facilities.

References

1. Glinskiy, B., Kulikov, I., Chernykh, I., Snytnikov, A., Sapetina, A., Weins, D.: The integrated approach to solving large-size physical problems on supercomputers. In: Voevodin, V., Sobolev, S. (eds.) RuSCDays 2017. CCIS, vol. 793, pp. 278–289. Springer, Cham (2017). https://doi.org/10.1007/978-3-319-71255-0_22
2. Glinskiy, B., Kulikov, I., Snytnikov, A., Romanenko, A., Chernykh, I., Vshivkov, V.: Co-design of parallel numerical methods for plasma physics and astrophysics. Supercomput. Front. Innovations 1(3), 88–98 (2014)
3. Glinsky, B., et al.: The co-design of astrophysical code for massively parallel supercomputers. In: Carretero, J., et al. (eds.) ICA3PP 2016. LNCS, vol. 10049, pp. 342–353. Springer, Cham (2016). https://doi.org/10.1007/978-3-319-49956-7_27
4. Hoefler, T., Schneider, T., Lumsdaine, A.: LogGOPSim - simulating large-scale applications in the LogGOPS Model
5. Glinsky, B., Rodionov, A., Marchenko, M., Podkorytov, D., Weins, D.: Scaling the distributed stochastic simulation to exaflop supercomputers. In: Proceedings of the 14th IEEE International Conference on High Performance Computing and Communications (HPCC-2012), pp. 1131–1136 (2012)
6. Podkorytov, D., Rodionov, A., Choo, H.: Agent-based simulation system AGNES for networks modeling: review and researching. In: Proceedings of the 6th International Conference on Ubiquitous Information Management and Communication (ACM ICUIMC 2012), p. 115. ACM (2012). https://doi.org/10.1145/2184751.2184883. ISBN 978-1-4503-1172-4
7. Mueong, J., Gul, A.: Agent framework services to reduce agent communication overhead in large-scale agent-based simulations. Simul. Model. Pract. Theory 4(6), 679–694 (2006)
8. Oren, T., Yilmaz, L.: On the synergy of simulation and agents: an innovation paradigm perspective. Int. J. Intell. Control Syst. 14(1), 4–19 (2009)
9. JADE Homepage. http://jade.tilab.com/. Accessed 10 Apr 2018
10. Glinskiy, B., Sapetina, A., Martynov, V., Weins, D., Chernykh, I.: The hybrid-cluster multilevel approach to solving the elastic wave propagation problem. In: Sokolinsky, L., Zymbler, M. (eds.) PCT 2017. CCIS, vol. 753, pp. 261–274. Springer, Cham (2017). https://doi.org/10.1007/978-3-319-67035-5_19
11. Карл Хьюитт, Питер Бишоп, Ричард Штайгер: Универсальный модульный формализм акторов для искусственного интеллекта. IJCAI, 1973
12. CAF - C++ Actor Framework. http://www.actor-framework.org/. Accessed 10 Apr 2018
13. QP/C++: About QP/C++. http://www.state-machine.com/qpcpp/. Accessed 10 Apr 2018
14. SObjectizer/Wiki/Home. https://sourceforge.net/p/sobjectizer/wiki/Home/. Accessed 10 Apr 2018
15. AkkA. https://akka.io/. Accessed 10 Apr 2018
16. Erlang Programming Language. http://www.erlang.org/. Accessed 10 Apr 2018
17. Cesarini, F., Thompson, S.: Erlang Programming. O'Reilly Media Inc., Sebastopol (2009). 498p.

Block Lanczos-Montgomery Method over Large Prime Fields with GPU Accelerated Dense Operations

Nikolai Zamarashkin[⊠] and Dmitry Zheltkov

INM RAS, Gubkina 8, Moscow, Russia
nikolai.zamarashkin@gmail.com,dmitry.zheltkov@gmail.com
http://www.inm.ras.ru

Abstract. Solution of huge linear systems over large prime fields is a problem that arises in such applications as discrete logarithm computation. Lanczos-Montgomery method is one of the methods to solve such problems. Main parallel resource of the method us the size of the block. But computational cost of dense matrix operations is increasing with block size growth. Thus, parallel scaling is close to linear only while complexity of such operations are relatively small. In this paper block Lanczos-Montgomery method with dense matrix operations accelerated on GPU is implemented. Scalability tests are performed (including tests with multiple GPU per node) and compared to CPU only version.

Keywords: Linear systems over prime fields · Parallel computations GPGPU

1 Introduction

The papers [1,2] describe the block Lanczos method for solving huge sparse linear systems over large prime finite fields that was developed in INM RAS. It was shown that the parallel efficiency of this method is limited by the unscalability of operations with the dense matrices and blocks. A qualitative explanation of this property is not so difficult.

Let K be a block size. Assume that the number of independent computational nodes is proportional to K. For the system size N the number of iterations does not exceed $\frac{N}{K}$. The complexity of operations with the dense $K \times K$ matrices and $N \times K$ blocks is proportional to NK^2. This kind of operations is performed on every iteration.

Thus, the following complexity estimate is valid:

$$\{\textbf{Time for dense operations}\} \sim \frac{N}{K}NK^2/K \sim N^2. \tag{1}$$

In spite of the fact that the calculations with dense matrices are usually ideally parallelized, the complexity estimate is *non-scalable by the number of nodes*.

© Springer Nature Switzerland AG 2019
V. Voevodin and S. Sobolev (Eds.): RuSCDays 2018, CCIS 965, pp. 40–49, 2019.
https://doi.org/10.1007/978-3-030-05807-4_4

In practice, things are as follows. While the block size K is small, the complexity of the dense block operations is hidden by the greater complexity of the huge sparse matrix by $N \times K$ block multiplications. Since the operation of sparse matrix by block multiplication has a significant parallel resource, for the small values of K the method has almost linear scalability [1,2]. However, the scalability is close to linear only up to block sizes about 10–20, which corresponds to the simultaneous use of up to 100–200 independent nodes [2]. The only way to extend the *linear scalability* area of the parallel block Lanczos method is *reducing the time for the operations with the matrices and blocks*.

In this paper we consider the use of GPUs to accelerate operations with dense matrices and blocks with elements from a large finite field. The principal possibility of significant acceleration of this kind of calculations using GPU was proved in [2]. The considered case of the large block size K varying from 100 to 1000 is important from the theoretical point of view. In practice, however, such block sizes are still very rare and purely exploratory in nature. At the same time, blocks of small sizes $K < 10$, are generally used. It will be shown that even in this case the use of GPUs allows to significantly speed up computations.

The paper describes the GPU implementations of two algorithms:

1. multiplication of "tall" $N \times K$ block by $K \times K$ matrix (Sect. 2);
2. multiplication of $K \times N$ by $N \times K$ blocks (Sect. 3).

From the mathematical point of view, we solve the problem of mapping an algorithm onto non-trivial parallel computing systems. And while choosing an algorithm the simplest one is preferable.

To reach high efficiency of GPU computations for our problem we must take in account properties of arithmetic operations in large prime fields, especially the multiplication of two field elements. Due to the design features of GPUs, their maximum performance is achieved if there is no conditional branches in the code. But the formal description of the multiplication and addition algorithms in the large prime fields implies the use of conditional branching operators, for example, in transfer bits [3]. The Sect. 3.3 describes some of the techniques that are used to exclude conditional branches. The corresponding part of software implementation is written in Cuda PTX.

For more efficient use of GPUs computations are performed asynchronously, where possible. Also, support of multiple GPUs per node is implemented.

The numerical experiments (see Sect. 4) were performed with and without CUDA. To distinguish between the implementations the following notations are used:

- the software without CUDA is denoted by **P0**.
- the software using single GPU per node is denoted by **P1**.
- the software using 2 GPUs per node is denoted by **P2**.

2   Multiplication of Dense $N \times K$ Block by $K \times K$ Matrix for Small K

2.1   General Algorithm Structure

Let $A \in \mathbb{F}^{N \times K}$ be a block and $B \in \mathbb{F}^{K \times K}$ be a square matrix with elements in a large prime field $\mathbb{F}$. We assume that the elements of $\mathbb{F}$ are specified using W computer words (in the experiments $W = 8$, and the size of the machine word is 64 bits), and that A, B, and the resulting block $C = AB$ are stored in the global memory of the GPU.

The purpose of this section is to describe the effective implementation of the *naive algorithm* of multiplying A by B on GPUs.

As a rule, effective implementation on GPU implies a large number of independent *executable blocks*. This is our immediate goal.

We represent the block A as a union of blocks $A_i \in \mathbb{F}^{64 \times K}$ (see Fig. 1). Each $64 \times K$ matrix block is associated with an independent executable 64-thread block on GPU. Elements of A_i are loaded into the shared memory by column vectors of 64 elements; and from B just one element is loaded. The independent threads multiply 64-column of A by the number from B, and the result is collected in a column of C.

Now we give a detailed description of the $N \times K$ block by $K \times K$ matrix multiplication algorithm.

Algorithm 1. $N \times K$ **block by** $K \times K$ **matrix multiplication. "Naive" approach.**

1. *Loop over the block size;*
2. *Load a column of A and an element of B into the shared memory;*
3. *each of the 64 threads loads on the registers of its Stream processor (SP) the following two $\mathbb{F}$ numbers: the number stored in the shared memory, and the column element corresponding to the thread number;*
4. *each thread executes the multiplication of its pair, and adds it to the current value of the result;*
5. *The Montgomery reduction is performed once at the end of all calculations. The necessary constants are loaded from the constant memory.*

2.2   Some Details of Block-by-matrix Multiplication

In order to achieve the optimal performance, one should pay attention to (a) the organization of data loading on the GPU registers; (b) the organization of downloading from the registers into memory. The optimal loading from the global memory is possible only if the loaded data is stored in 32-byte blocks. And the efficiency of loading becomes higher if the neighboring threads of the same warp use different memory banks. In other words, the numbers of machine words used by warp should give the maximum possible number of different residues when dividing by 32.

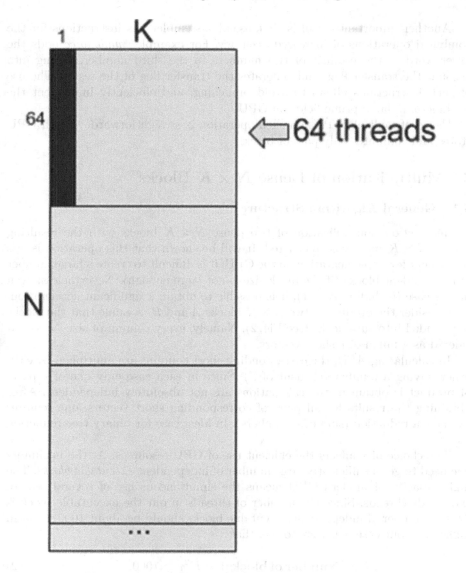

Fig. 1. Representation of A for the executable blocks.

A similar observation is true for loading data from shared memory to the registers. If the threads of one warp load either the same element, or elements from different memory banks, the best results are obtained.

To make loads from global memory faster data matrices is stored columnwise. Indeed, in this case each of 64 SP could load at once one of the consecutive 64 machine words. Thus, all threads in warp use different memory banks. After performing such operation W times all needed 64 large numbers are stored in shared memory.

Another important detail is that pseudo-assembler has instructions for the combined operations of increased accuracy. For example, *madc.lo.cc* adds the lower word of the product of two numbers to the third number, taking into account the transfer flag, and generates the transfer flag of the result. The use of such instructions allows to avoid branching, and efficiently implement the arithmetic in large prime fields on GPU.

Usage of multiple GPUs on such operation is straightforward — each GPU stores and computes only part of block.

3   Multiplication of Dense $N \times K$ Blocks

3.1   General Algorithm Structure

In this section, multiplication of two dense $N \times K$ blocks with the resulting square $K \times K$ matrix is considered. It will be shown that this operation is less convenient for implementation on the GPU: it is difficult to create a large number of independent blocks (if the block size is not large enough). Nevertheless, even in this case (including $K = 1$), it is possible to obtain a significant acceleration.

Consider the product of two $N \times K$ blocks A and B. Assume that the blocks are divided into sub-blocks (see Fig. 2). Namely, every column of size N is considered as a union of r short vectors.

In calculating $A^T B$, the corresponding short columns are multiplied by each other, giving a number (element of $\mathbb{F}$). Since in each case only one of r parts of product is obtained, the calculations are not absolutely independent. After obtaining the results for all pairs of corresponding short vectors, one requires to make a reduction (sum of r numbers). In ideal case for binary tree reduction $r = 2^k$.

The choice of r affects the efficient use of GPU resources. At the beginning one need to get a sufficiently large number of independent *executable blocks*. The high capacity of modern GPU means the simultaneous use of several tens of thousands threads. Since the number of threads in our the executable block is 64, the number of independent executable blocks should be about 1000 (or even higher). In our case it is easy to see, that

$$\{\textbf{Number of blocks}\} = K^2 r \geq 1000. \tag{2}$$

Therefore, for $K < 30$ one should not split blocks into parts. Note that for convenience, we choose the block size that is multiple of the number of threads (64), that is:

$$N \approx 64 l K^2 r. \tag{3}$$

with integer l that defines the number of multiplications of long words computed by each threads.

Note that the algorithm consists of three successive parts:

1. Each thread calculates the sum of l products of long numbers, and the incomplete Montgomery reduction (the result is the number of length $W + 1$). The number obtained by each thread is written to its place in the global memory.

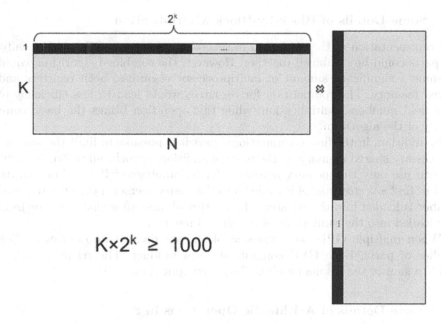

Fig. 2. Division of independent blocks for calculation $A^T B$.

2. Parallel binary tree reduction. Each computational block sums all the numbers necessary for the result. The resulting number (with in length $W + 1$) is written to the global memory.
3. Finding the remainder of dividing each element by prime number of length W.

The second part of the algorithm is standard; its implementation is known. The third one is trivial. In addition, it is assumed that $K \ll N$, and therefore the second and third parts have relatively low complexity. Thus, only the first part is of interest.

Algorithm 2. Multiplication of $N \times K$ blocks.

1. *In loop for l;*
2. *Loading the corresponding bits of the 64×1 vectors of the blocks A and B from the device memory into shared memory of the stream multiprocessor (SM);*
3. *Each of the 64 threads loads two numbers into the registers of its stream processor (SP);*
4. *Each thread executes the multiplication of its pair, and adds it to the current value of the result;*
5. *The incomplete Montgomery reduction is performed once at the end of all calculations; the necessary constants are loaded from the constant memory.*

3.2   Some Details of Block-by-Block Multiplication

Our representation of the algorithm in several parts is not arbitrary. Formally, the parts could be combined together. However, the combined algorithm would consume a significant amount of multiprocessor resources, both registers and shared memory. The competition for resources would lead to less efficiency of large field numbers multiplication, while this operation brings the basic complexity of the algorithm.

In addition, in the first computational part it is possible to limit the amount of necessary shared memory by the price of additional synchronization. Namely, one can use only the memory necessary for 64 numbers of W machine words. That is, first a vector from A is loaded into the shared memory; then the required number is loaded into the registers of each thread; and after that a vector from B is loaded into the same place of the shared memory.

When multiple GPUs are available blocks are divided into corresponding number of parts. Each GPU compute its own product. The result (which is equal to sum of the results on all GPUs) is computed on CPU.

3.3   Some Details of Arithmetic Operations in $\mathbb{F}$

Commonly the descriptions of Montgomery reduction, and the reduction modulo a prime number use conditional branches. As for GPUs, the conditions depending on the processed data lead to execution of all paths of the branch by the warp. This significantly reduces the performance of the device. In addition, processing of the different paths of the branch requires additional registers. As a result, fewer threads can created on the Stream multiprocessor, so data and instruction loading is worse hidden. This leads to an additional decrease in performance.

Let us explain on an example how to get rid of branching in Montgomery reduction. A typical case of branching here is the following:

if $a > b$, then subtract b from a, else do not change a.

Now let's perform the subtraction $a - b$ with the transfer flag generating. First we sum two zeros with the transfer flag, and set the result c (so now c contains the flag value). Then we add $c * b$ to $a - b$, getting the correct value of $a - b$. Note that in multiplication c by b, one may calculate just the lower word of the products.

4   Numerical Experiments

Before describing the numerical experiments, let's focus on the properties of the block Lanczos method implementation, created in the INM RAS. Namely, the implementations have two main parallel resources.

First one is the efficient parallel procedure for sparse matrix by vector multiplication [6]. The matrix is split into blocks, which are processed in parallel. However, this parallelism is limited. For example, the time for data exchanges increases as the node number grows.

Table 1. Cluster **L**. Time spent to calculate one vector of Krylov subspace. The optimal block size K is in brackets.

Nodes\Prog.	P0, M1, ms	P1, M1, ms	P0, M2, s	P1, M2, s
1	56.9 (1)	41.1 (1)	3.81 (1)	3.10 (1)
2	29.2 (1)	21.5 (2)	2.07 (1)	1.61 (2)
4	21.8 (1)	11.2 (4)	1.14 (1)	0.815 (4)
8	13.8 (2)	6.9 (8)	0.709 (1)	0.417 (8)
16	8.0 (4)	4.3 (16)	0.401 (4)	0.231 (16)
32	–	–	0.216 (8)	0.128 (32)

Table 2. Cluster **L**. Acceleration compared to one node

Nodes\Prog.	P0, M1	P1, M1	P0, M2	P1, M2
1	1	1	1	1
2	1.95	1.91	1.84	1.93
4	2.61	3.67	3.34	3.80
8	4.12	5.96	5.37	7.43
16	7.11	9.56	9.5	13.42
32	–	–	17.64	24.22

Table 3. Cluster **L2**. Time spent to calculate one vector of Krylov subspace. The block size K is in brackets.

Nodes\Prog.	P0, M1, ms	P1, M1, ms	P0, M2, s	P1, M2, s
1	38.3 (1)	29.4 (1)	1.66 (1)	1.41 (1)
2	26.1 (1)	19.2 (2)	0.87 (1)	0.725 (2)
4	15.8 (1)	9.4 (4)	0.500 (1)	0.381 (4)
8	9.9 (1)	5.8 (8)	0.314 (2)	0.202 (8)
16	7.3 (1)	4.0 (16)	0.193 (4)	0.119 (16)
32	6.5 (2)	2.8 (16)	0.116 (8)	0.0698 (32)

The second parallelism resource is connected with the block size K (note that the known implementations [4, 5] do not have it). The independent computations are executed for each individual vector in the block. In [1] an implementation proposed such that increasing K leads to the decrease of data exchanging time proportional to K. Thus, the block size is not only an independent parallel resource, but also expands the parallelism of sparse matrix by vector multiplication. Increasing K is restricted by the growth of algorithmic complexity for operations with dense matrices and blocks at each iteration. It means that acceleration of block operations is *a priority task for the most efficient parallel implementation.*

Table 4. Cluster **L2**. Acceleration compared to one node

Nodes\Prog.	P0, M1	P1, M1	P0, M2	P1, M2
1	1	1	1	1
2	1.47	1.53	1.91	1.95
4	2.42	3.13	3.32	3.70
8	3.87	5.07	5.29	6.98
16	5.25	7.35	8.6	11.85
32	5.89	10.5	14.31	20.2

Table 5. Cluster **T**. Time spent to calculate one vector of Krylov subspace. The block size K is in brackets.

Nodes\Prog.	P0, M1, ms	P1, M1, ms	P2, M1, ms	P0, M2, s	P1, M2, s	P2, M2, s
1	87.9 (1)	52.2 (1)	49.8 (1)	5.91 (1)	4.7 (1)	4.65 (1)
2	50.5 (1)	29.9 (2)	28.2 (2)	3.19 (1)	2.48 (2)	2.43 (2)
4	31.1 (1)	16.8 (4)	15.3 (4)	1.79 (2)	1.28 (4)	1.25 (4)

Table 6. Cluster **T**. Acceleration compared to one node.

Nodes\Prog.	P0, M1	P1, M1	P2, M1	P0, M2	P1, M2	P2, M2
2	1.74	1.75	1.77	1.85	1.90	1.91
4	2.82	3.11	3.25	3.30	3.67	3.72

We used the following two matrices for the numerical experiments:

1. Matrix 1 (**M1**): (a) size 64446×65541; (b) number of nonzeros 1588524; (c) average number of nonzeros in a row $\rho = 24.65$; (d) 5 dense blocks.
2. Matrix 2 (**M2**): (a) size 2097152×2085659; (b) number of nonzeros 182117529; (c) average number of nonzeros in a row $\rho = 86.84$; (d) 5 dense blocks.

Numerical experiments were performed on the following computer systems:

1. GPU cluster of INM RAS (**T**): each node is equipped with 4-core processor Intel Core i7-960 3,2 GHz and 2 graphical adapters Nvidia Tesla C2070. The nodes are interconnected with Infiniband 10 Gbit/s.
2. Supercomputer "Lomonosov" (**L**): each node is equipped with two 8-cores processors Intel Xeon X5570 2,93 GHz. Some nodes in addition are equipped with graphical adapters Nvidia Tesla X2070. The nodes are interconnected with Infiniband 40 Gbit/s.
3. Supercomputer "Lomonosov-2" (**L2**): each node is equipped 14-cores processors Intel Xeon E5-2697v3 2,6 GHz and with graphical adapter Nvidia Tesla K40M. The nodes are interconnected with Infiniband 56 Gbit/s.

The results are shown in the Tables 1, 2, 3, 4, 5, and 6.

Table 1 gives the time required to compute one vector of A-orthogonal basis of the Krylov space on cluster **L**. The first column of table indicates the number of nodes. As we see, implementation **P1** with CUDA allows block size increasing easier than implementation **P0**. Thus, the parallel resource associated with the procedure for multiplying extra-large sparse matrix by vector is preserved and can be exploited later. Thus, the almost linear scalability of **P1** persists wider. This conclusion is vividly confirmed in Table 2. Indeed, the results for **P1** show better acceleration relative to the calculations on one node. That is, **P1** has better parallel properties.

Similar results are obtained for the cluster **L2**. Finally, we note that the difference in the results of **P0** and **P1** becomes more and more evident with the growth of K.

On cluster **T** results shows effect of multiple GPU usage. For cluster with more nodes available impact of multiple GPU usage is expected to be much more significant and to allow close to linear parallel scaling for larger K.

Acknowledgments. The work was supported by the RAS presidium program №1 "Fundamental Mathematics and its applications".

References

1. Zamarashkin, N., Zheltkov, D.: Block Lanczos-Montgomery method with reduced data exchanges. In: Voevodin, V., Sobolev, S. (eds.) RuSCDays 2016. CCIS, vol. 687, pp. 15–26. Springer, Cham (2016). https://doi.org/10.1007/978-3-319-55669-7_2
2. Zamarashkin, N., Zheltkov, D.: GPU acceleration of dense matrix and block operations for Lanczos Method for systems over large prime finite field. In: Voevodin, V., Sobolev, S. (eds.) RuSCDays 2017. CCIS, vol. 793, pp. 14–26. Springer, Cham (2017). https://doi.org/10.1007/978-3-319-71255-0_2
3. Zheltkov, D.A.: Effectivnie basovye operacii lineinoi algebry dlya reshenia bolshyh razrezennyh sistem nad konechnymi polyami. RuSCDays (2016, in Russian)
4. Dorofeev, A.: Reshenie sistem lineinyh uravnenii pri vichislenii logarifmov v konechnom prostom pole. Matematicheskie voprosy kriptographii **3**(1), 5–51 (2012). (in Russian)
5. Popovyan, I., Nesterenko, Yu., Grechnikov, E.: Vychislitelno sloznye zadachi teorii chisel. MSU Publishing (2012, in Russian)
6. Zamarashkin, N.: Algoritmy dlya razrezennyh sistem lineinyh uravnenii v GF(2). MSU Publishing (2013, in Russian)
7. Nath, R., Tomov, S., Dongarra, J.: An improved MAGMA GEMM for Fermi graphics processing units. Int. J. High Perform. Comput. Appl. **24**(4), 511–515 (2010)
8. Cuda C Programming guide. http://docs.nvidia.com/cuda/cuda-c-programming-guide

Comparison of Dimensionality Reduction Schemes for Parallel Global Optimization Algorithms

Konstantin Barkalov, Vladislav Sovrasov$^{(\boxtimes)}$, and Ilya Lebedev

Lobachevsky State University of Nizhni Novgorod, Nizhni Novgorod, Russia
{konstantin.barkalov,ilya.lebedev}@itmm.unn.ru, sovrasov.vlad@gmail.com

Abstract. This work considers a parallel algorithms for solving multi-extremal optimization problems. Algorithms are developed within the framework of the information-statistical approach and implemented in a parallel solver Globalizer. The optimization problem is solved by reducing the multidimensional problem to a set of joint one-dimensional problems that are solved in parallel. Five types of Peano-type space-filling curves are employed to reduce dimension. The results of computational experiments carried out on several hundred test problems are discussed.

Keywords: Global optimization · Dimension reduction
Parallel algorithms · Multidimensional multiextremal optimization
Global search algorithms · Parallel computations

1 Introduction

In the present paper, the parallel algorithms for solving the multiextremal optimization problems are considered. In the multiextremal problems, the opportunity of reliable estimate of the global optimum is based principally on the availability of some information on the function known *a priori* allowing relating the probable values of the optimized function to the known values at the points of performed trials. Very often, such an information on the problem being solved is represented in the form of suggestion that the objective function $\varphi(y)$ satisfies Lipschitz condition with the constant L not known a priori (see, for example, [14–16]). At that, the objective function could be defined by a program code i.e. could represent a "black-box"-function. Such problems are presented in the applications widely (problems of optimal design of objects and technological processes in various fields of technology, problems of model fitting according to observed data in scientific research, etc.).

Many methods destined to solving the problems of the class specified above reduce the solving of a multidimensional problem to solving the one-dimensional subproblems implicitly (see, for example, the methods of diagonal partitions [12] or simplicial partitions [13]). In the present work, we will use the approach developed in Lobachevsky State University of Nizhni Novgorod based on the

© Springer Nature Switzerland AG 2019
V. Voevodin and S. Sobolev (Eds.): RuSCDays 2018, CCIS 965, pp. 50–62, 2019.
https://doi.org/10.1007/978-3-030-05807-4_5

idea of the dimensionality reduction with the use of Peano space-filling curves $y(x)$ mapping the interval $[0, 1]$ of the real axis onto an n-dimensional cube continuously and unambiguously.

Several methods of constructing the evolvents approximating the theoretical Peano curve have been proposed in [2,3,6,8]. These methods were implemented in the Globalizer software system [11]. The goal of the present study was comparing the properties of the evolvents and the selecting the most suitable ones for the use in the parallel global optimization algorithms.

2   Statement of Multidimensional Global Optimization Problem

In this paper, the core class of optimization problems, which can be solved using Globalizer, is formulated. This class involves the multidimensional global optimization problems without constraints, which can be defined in the following way:

$$\varphi(y^*) = \min\{\varphi(y) : y \in D\}, \tag{1}$$
$$D = \{y \in \mathbb{R}^N : a_i \leq y_i \leq b_i, 1 \leq i \leq N\}$$

with the given boundary vectors a and b. It is supposed, that the objective function $\varphi(y)$ satisfies the Lipschitz condition

$$|\varphi(y_1) - \varphi(y_2)| \leq L\|y_1 - y_2\|, y_1, y_2 \in D, \tag{2}$$

where $L > 0$ is the Lipschitz constant, and $\|\cdot\|$ denotes the norm in $\mathbb{R}^N$ space.

Usually, the objective function $\varphi(y)$ is defined as a computational procedure, according to which the value $\varphi(y)$ can be calculated for any vector $y \in D$ (let us further call such a calculation a *trial*). It is supposed that this procedure is time-consuming.

3   Methods of Dimension Reduction

3.1   Single Evolvent

Within the framework of the information-statistical global optimization theory, the Peano space-filling curves (or evolvents) $y(x)$ mapping the interval $[0, 1]$ onto an N-dimensional hypercube D unambiguously are used for the dimensionality reduction [1,2,4,5].

As a result of the reduction, the initial multidimensional global optimization problem (1) is reduced to the following one-dimensional problem:

$$\varphi(y(x^*)) = \min\{\varphi(y(x)) : x \in [0, 1]\}. \tag{3}$$

It is important to note that this dimensionality reduction scheme transforms the Lipschitzian function from (1) to the corresponding one-dimensional function $\varphi(y(x))$, which satisfies the uniform Hölder condition, i.e.

$$|\varphi(y(x_1)) - \varphi(y(x_2))| \leq H|x_1 - x_2|^{\frac{1}{N}}, x_1, x_2 \in [0, 1], \tag{4}$$

where the constant H is defined by the relation $H = 2L\sqrt{N+3}$, L is the Lipschitz constant from (2), and N is the dimensionality of the optimization problem (1).

The algorithms for the numerical construction of the Peano curve approximations are given in [5].

The computational scheme obtained as a result of the dimensionality reduction consists of the following:

- The optimization algorithm performs the minimization of the reduced one-dimensional function $\varphi(y(x))$ from (3),
- After determining the next trial point x, a multidimensional image y is calculated by using the mapping $y(x)$,
- The value of the initial multidimensional function $\varphi(y)$ is calculated at the point $y \in D$,
- The calculated value $z = \varphi(y)$ is used further as the value of the reduced one-dimensional function $\varphi(y(x))$ at the point x.

3.2   Shifted Evolvents

One of the possible ways to overcome the negative effects of using a numerical approximation of evolvent (it destroys the information about the neighbor points in $\mathbb{R}^N$ space, see [3]) consists in using the multiple mappings

$$Y_L(x) = \left\{y^0(x),\ y^1(x), ...,\ y^L(x)\right\} \tag{5}$$

instead of single Peano curve $y(x)$ (see [3,5,7]).

Such set of evolvents can be produced by shifting the source evolvent $y^0(x)$ by $2^{-l}, 0 \le l \le L$ on each coordinate. Each evolvent has it's own corresponding hypercube $D_l = \left\{y \in R^N : -2^{-1} \le y_i + 2^{-l} \le 3 \cdot 2^{-1},\ 1 \le i \le N\right\},\ 0 \le l \le L$.

In Fig. 1a the image of the interval $[0,1]$ obtained by the curve $y^0(x)$, $x \in [0,1]$, is shown as the dashed line. Since the hypercube D from (1) is included in the common part of the family of hypercubes D_l, having introduced an additional constraint function

$$g_0(y) = \max\left\{|y_i| - 2^{-1} : 1 \le i \le N\right\}, \tag{6}$$

one can present the initial hypercube D in the form

$$D = \left\{y^l(x) :\ x \in [0,1],\ g_0(y^l(x)) \le 0\right\},\ 0 \le l \le L,$$

i.e., $g_0(y) \le 0$ if $y \in D$ and $g_0(y) > 0$ otherwise. Consequently, any point $y \in D$ has its own preimage $x^l \in [0,1]$ for each mapping $y^l(x)$, $0 \le l \le L$.

Thus, each evolvent $y^l(x)$, $0 \le l \le L$, generates its own problem of the type (1) featured by its own extended (in comparison with D) search domain D_l and the additional constraint with the left hand part from (6)

$$\min\left\{\varphi(y^l(x)) :\ x \in [0,1],\ g_j(y^l(x)) \le 0,\ 0 \le j \le m\right\},\ 0 \le l \le L. \tag{7}$$

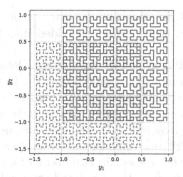
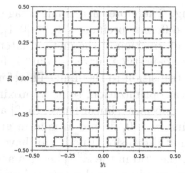

(a) Two shifted evolvents on the hyper-cubes D_0 and D_1

(b) Two rotated evolvents on the same plane

Fig. 1. Multiple evolvents built with low density

3.3   Rotated Evolvents

The application of the scheme for building the multiple evolvents (hereinafter called the shifted evolvents or S-evolvents) described in Subsect. 3.2 allows to preserve the information on the nearness of the points in the multidimensional space and, therefore, to provide more precise (as compared to a single evolvent) estimate of Lipschitz constant in the search process. However, this approach has serious restrictions, which narrow the applicability of the parallel algorithms, designed on the base of the S-evolvents (see the end of the Sect. 5.1).

To overcome complexity of the S-evolvent and to preserve the information on the nearness of the points in the N-dimensional space, one more scheme of building of the multiple mappings was proposed. The building of a set of Peano curves not by the shift along the main diagonal of the hypercube but by rotation of the evolvents around the coordinate origin is a distinctive feature of the proposed scheme [8]. In Fig. 1b two evolvents being the approximations to Peano curves for the case $N = 2$ are presented as an illustration. Taking into account the initial mapping, one can conclude that current implementation of the method allows to build up to $N(N - 1) + 1$ evolvents for mapping the N-dimensional domain onto the corresponding one-dimensional intervals. Moreover, the additional constraint $g_0(y) \leq 0$ with $g_0(y)$ from (6), which arises in shifted evolvents, is absent. This method for building a set of mappings can be "scaled" easily to obtain more evolvents (up to 2^N) if necessary.

3.4   Non-univalent Evolvent

As it has been already mentioned above (Sect. 3.2), the loss of information on the proximity of the points in the multidimensional space could be compensated in part by the use of multiple mappings $Y_L(x) = \{y^1(x), ..., y^L(x)\}$. However, the Peano-type curve preserves a part of this information itself: it is not an injective mapping. Therefore, if a single image $y(x) \in \mathbb{R}^N$ is available, one can obtain several different preimages $t_j \in [0,1], t_j \neq x$, which could be added into the search information of the method later.

The Peano-type curve used in (3) for the dimensionality reduction is defined via the transition to the limit. Therefore, it cannot be computed directly. In the numerical optimization, some approximation of this curve is used, and it is an injective piecewise-linear curve. In [2] a non-univalent mapping of a uniform grid in the interval $[0,1]$ onto a uniform grid in a hypercube D has been proposed. Each multidimensional node can have up to 2^N one-dimensional preimages. In Fig. 2b, the grid in the $\mathbb{R}^2$ space is marked by the crosses, for two nodes of which the corresponding one-dimensional preimages from $[0,1]$ are pointed (marked by the squares and circles). Each node mentioned above has 3 preimages.

A potentially large number of preimages (up to 2^N) and the inability to use the parallel scheme for the multiple mappings form Sect. 4.2 are the disadvantages of the non-univalent evolvent.

3.5 Smooth Evolvent

The methods of constructing the evolvents considered in the previous paragraphs produce the curve $y(x)$, which in not a smooth one (see Fig. 1a). The absence of smoothness may affect the properties of the reduced one-dimensional function $\varphi(y(x))$ adversely since a smooth curve reflect the information on the growth/decay of the initial function better. On the basis of initial algorithm of constructing the non-smooth evolvent, a generalized algorithm allowing constructing a smooth space-filling curve has been proposed [6]. As an illustration, a smooth evolvent for the two-dimensional case is presented in Fig. 2a. An increased computational complexity (several times as compared to the piecewise-linear curves) is a disadvantage of the smooth evolvent. This caused by computing of the nonlinear smooth functions.

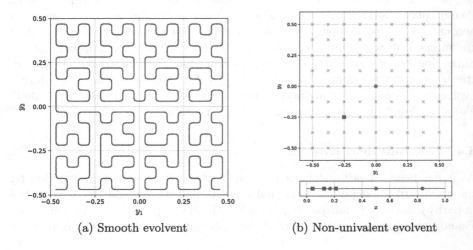

(a) Smooth evolvent   (b) Non-univalent evolvent

Fig. 2. Different evolvents built with low density

4   Parallel Computations for Solving Global Optimization Problems

4.1   Core Multidimensional Algorithm of Global Search (MAGS)

The optimization methods applied in Globalizer to solve the reduced problem (3) are based on the MAGS method, which can be presented as follows — see [2,5].

The initial iteration of the algorithm is performed at an arbitrary point $x^1 \in (0,1)$. Then, let us suppose that k, $k \geq 1$, optimization iterations have been completed already. The selection of the trial point x^{k+1} for the next iteration is performed according to the following rules.

Rule 1. Renumber the points of the preceding trials by the lower indices in order of increasing value of coordinates $0 = x_0 < x_1 < ... < x_{k+1} = 1$.

Rule 2. Compute the characteristics $R(i)$ for each interval (x_{i-1}, x_i), $1 \leq i \leq k+1$.

Rule 3. Determine the interval with the maximum characteristic $R(t) = \max_{1 \leq i \leq k+1} R(i)$.

Rule 4. Execute a new trial at the point x^{k+1} located within the interval with the maximum characteristic from the previous step $x^{k+1} = d(x_t)$.

The stopping condition, which terminated the trials, is defined by the inequality $\rho_t < \varepsilon$ for the interval with the maximum characteristic from Rule 3 and $\varepsilon > 0$ is the predefined accuracy of the optimization problem solution. If the stopping condition is not satisfied, the index k is incremented by 1, and the new global optimization iteration is executed.

The convergence conditions and exact formulas for decision rules $R(i)$ and $d(x)$ of the described algorithm are given, for example, in [5].

The numerical experiments, the results of which are presented in [17,18] demonstrate that the method, at least, is not worse than the well-known global optimization algorithms DIRECT [14] and DIRECT*l* [15], and even overcome these ones with respect to some parameters.

4.2   Parallel Algorithm Exploiting a Set of Evolvents

Using the multiple mapping allows solving initial problem (1) by parallel solving the problems

$$\min\{\varphi(y^s(x)) : x \in [0,1]\}, 1 \leqslant s \leqslant S$$

on a set of intervals $[0,1]$ by the index method. Each one-dimensional problem is solved on a separate processor. The trial results at the point x^k obtained for the problem being solved by particular processor are interpreted as the results of the trials in the rest problems (in the corresponding points $x^{k_1}, ..., x^{k_S}$). In this approach, a trial at the point $x^k \in [0,1]$ executed in the framework of the s-th problem, consists in the following sequence of operations.

1. Determine the image $y^k = y^s(x^k)$ for the evolvent $y^s(x)$.
2. Inform the rest of processors about the start of the trial execution at the point y^k (the blocking of the point y^k).
3. Determine the preimages $x^{k_s} \in [0,1], 1 \leqslant s \leqslant S$, of the point y^k and interpret the trial executed at the point $y^k \in D$ as the execution of the trials in the S points $x^{k_1}, \ldots, x^{k_s}$
4. Inform the rest of processors about the trial results at the point y^k.

The decision rules for the proposed parallel algorithm, in general, are the same as the rules of the sequential algorithm (except the method of the trial execution). Each processor has its own copy of the software realizing the computations of the problem functions and the decision rule of the index algorithm. For the organization of the interactions among the processors, the queues are created on each processor, where the processors store the information on the executed iterations in the form of the tuples: the processor number s, the trial point x^{k_s}.

The proposed parallelization scheme was implemented with the use of MPI technology. Main features of implementation consist in the following. A separate MPI-process is created for each of S one-dimensional problems being solved, usually, one process per one processor employed. Each process can use p threads, usually one thread per an accessible core.

At every iteration of the method, the process with the index $s, 0 \leqslant s < S$ performs p trials in parallel at the points $x^{(s+iS)}, 0 \leqslant i < p$. At that, each process stores all S_p points, and an attribute indicating whether this point is blocked by another process or not is stored for each point. Let us remind that the point is blocked if the process starts the execution of a trial at this point.

At every iteration of the algorithm, operating within the s-th process, determines the coordinates of p "its own" trial points. Then, the interchange of the coordinates of images of the trial points $y^{(s+iS)}, 0 \leqslant i < p, 0 \leqslant s < S$ is performed (from each process to each one). After that, the preimages $x^{(q+iS)}, 0 \leqslant q < S, q \neq s$ of the points received by the s-th process from the neighbor ones are determined with the use of the evolvent $y^s(x)$. The points blocked within the s-th process will correspond to the preimages obtained. Then, each process performs the trials at the non-blocked points, the computations are performed in parallel using OpenMP. The results of the executed trials (the index of the point, the computed values of the problem functions, and the attribute of unblocking of this point) are transferred to all rest processes. All the points are added to the search information database, and the transition to the next iteration is performed.

5   Results of Numerical Experiments

The computational experiments have been carried out on the Lobachevsky supercomputer at State University of Nizhni Novgorod. A computational node included 2 Intel Sandy Bridge E5-2660 2.2 GHz processors, 64 GB RAM. The CPUs had 8 cores (i.e. total 16 cores were available per a node). All considered algorithms and evolvents were implemented using C++ within the Globalizer software system [11]. In order to enable the parallelism, OpenMP was used on a single node, and MPI was used for the parallelization on several nodes.

The comparison of the global optimization algorithms was performed by the evaluation of the quality of solving a set of problems from some test class. In the present paper, the test class generated by GKLS (Gaviano, Kvasov, Lera, Sergeyev) generator [9] was considered. The generator creates objective functions by distorting a convex quadratic function by polynomials in order to introduce local minima. Thus, GKLS allows constructing the complex multiextremal problems of various dimensions. In the present work, the series of 100 problems from the classes of the dimensions of 2, 3, 4, and 5 were considered. Each class had two degrees of complexity — *Simple* and *Hard*. These the classes have different radius of the attraction region of the global minimizer (*Hard* has smaller region) and distance from the global minimizer to the vertex of the quadratic function (*Simple* has smaller distance). The parameters of the generator for the considered classes were given in Ref. [9].

In order to evaluate the efficiency of an algorithm on a given set of 100 problems, we will use the operating characteristics [10], which are defined as a curve, showing the dependency of number of solved problems vs the number of iterations.

5.1   Comparison of the Sequential Evolvents

In order to understand whether any type of evolvents listed above has an essential advantage as compared to other ones, the operating characteristics of the index method with different types of evolvents have been obtained for the classes GKLS 2d Simple and GKLS 3d Simple. The global minimum was considered to be found if the algorithm generates a trial point y^k in the δ-vicinity of the global minimizer, i.e. $\left\|y^k - y^*\right\|_\infty \leq \delta$. The size of the vicinity was selected as $\delta = 0.01 \left\|b - a\right\|_\infty$. In case of GKLS $\delta = 0.01$.

In all experiments, the evolvent construction density parameter $m = 12$. The minimum value of the reliability parameter r was found for each type of evolvents by scanning over a uniform grid with the step 0.1.

On the GKLS 2d Simple class at the minimum r, the non-univalent evolvent and the smooth one provide a faster convergence (Fig. 3b). The same was observed at $r = 5.0$ as well (Fig. 3a). In the latter case, the shifted evolvent and the rotating one begin to lag behind the rest since the value $r = 5.0$ is too big for them.

On the GKLS 2d Simple class at the minimum r, the non-univalent evolvents and multiple ones have a considerable advantage over the single evolvent (Fig. 4b). The value $r = 4.5$ is too big for the rotated evolvents and for the shifted one (Fig. 4a).

Overhead Costs When Using the Shifted Evolvents. In all experiments presented above, the number of computations of the objective function from the GKLS class was taken into account when plotting the operating characteristics. However,

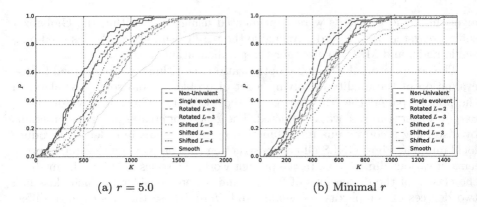

(a) $r = 5.0$ (b) Minimal r

Fig. 3. Operating characteristics on GKLS 2d Simple class

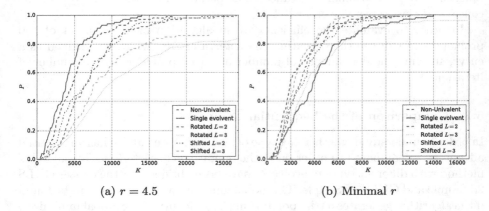

(a) $r = 4.5$ (b) Minimal r

Fig. 4. Operating characteristics on GKLS 3d Simple class

in the case of the shifted evolvent, the index method solves the problem with the constraint g_0 from (6). At the points where g_0 is violated, the value of the objective function is not computed. Nevertheless, these points are stored in the search information producing the additional computational costs. In Table 1, the averaged numbers of calls to g_0 and to the objective function are presented. At $L = 3$, the constraint g_0 was computed almost 20 times more than the objective function φ i.e. 95% of the whole search information account for the auxiliary points. Such overhead costs are acceptable when solving the problems of small dimension with the computation costly objective functions. However, when increasing dimensionality and total number of trials other types of evolvents are preferred.

Table 1. Averaged number of computations of g_0 and of φ when solving the problems from GKLS 3d Simple class using the shifted evolvent

L	$calc(g_0)$	$calc(\varphi)$	$\frac{calc(g_0)}{calc(\varphi)}$ ratio
2	96247.9	6840.14	14.07
3	153131.0	7702.82	19.88

5.2 Parallel Rotated Evolvents

In order to evaluate the efficiency of the parallel algorithm from Sect. 4.2, the numerical experiments on the GKLS 4d (Hard, Simple) classes and on the GKLS 5d (Hard, Simple) ones were conducted. The value of r in all experiments was equal to 5.0, the size of the δ-vicinity of the known solution was increased up to 0.3. When solving the series of problems, up to 8 cluster nodes and up to 32 computational threads on each node were employed.

In Table 2, an averaged number of iterations when solving 100 problems from each considered class is presented. The number of iterations is reduced considerably with increasing the number of nodes and the number of threads on each node (except the GKLS 4d Simple class at the transition from 1 node to 4 ones in the single thread mode).

If one assumes the costs of parallelization to be negligible as compared to the costs of computing the objective functions in the optimization problems, the speedup in time due to the use of the parallel method would be equal to the speedup with respect to the number of iterations. However, actually this suggestion is not always true. In all numerical experiment the time of computing the objective function was approximately 10^{-3} s. In Table 3, the speedups in iterations and in time (in the pendent brackets) are presented. In the first row of the table corresponding to the sequential mode, the averaged time of solving a single problem is presented in the pendent brackets. One can see from the table that for the GKLS 4d classes it is more efficient to utilize a single node in the multithread mode whereas for solving more complex five-dimensional problems, the use of several nodes is better, each node is operating in the parallel mode.

Table 2. Averaged numbers of iterations executed by the parallel algorithm for solving the test optimization problems

		p	$N = 4$		$N = 5$	
			Simple	Hard	Simple	Hard
I	1 cluster node	1	12167	25635	20979	187353
		32	328	1268	898	12208
II	4 cluster nodes	1	25312	11103	1472	17009
		32	64	913	47	345
III	8 cluster nodes	1	810	4351	868	5697
		32	34	112	35	868

Table 3. Speedup of parallel computations executed by the parallel algorithm

		p	$N = 4$		$N = 5$	
			Simple	Hard	Simple	Hard
I	1 cluster node	1	12167 (10.58 s)	25635 (22.26 s)	20979 (22.78 s)	187353 (205.83 s)
		32	37.1 (18.03)	20.2 (8.55)	23.3 (8.77)	15.4 (9.68)
II	4 cluster nodes	1	0.5 (0.33)	2.3 (0.86)	14.3 (6.61)	11.0 (6.06)
		32	190.1 (9.59)	28.1 (1.08)	446.4 (19.79)	543.0 (43.60)
III	8 cluster nodes	1	15.0 (6.05)	5.9 (2.36)	24.2 (17.56)	32.9 (24.87)
		32	357.9 (2.36)	228.9 (2.64)	582.8 (20.96)	793.0 (33.89)

6   Conclusions

In the present work, 5 different Peano curve-type mappings applied to the dimensionality reduction in the global optimization problems were considered. From the preliminary comparison conducted in Sect. 5.1, one can make the following conclusions:

- the smooth evolvent and the non-univalent one demonstrate the best result in the problems of small dimensionality and can be applied successfully in solving the problems with the computational costly objective functions. The properties of these evolvents don't allow developing the optimization algorithms scalable onto several cluster nodes based on these ones.
- the shifted evolvents introduce large overhead costs on the operation of the method due to the requirement to adding an auxiliary functional constraint into the problem (1). The experiments have demonstrated that up to 95% of the search information account for the points, in which the auxiliary constraint is computed only. The shifted evolvents can be used as the base for the parallel algorithm from Sect. 4.2. However, the costs of processing the auxiliary points would result likely in a small speedup from the parallelization. However, if the objective function is computation-costly enough, the use of these evolvents could make sense.
- the rotated evolvents have provided an acceptable speed of convergence in the problems of small dimensionality in the sequential mode. The use of these ones don't result in the introduction of the auxiliary constraints that allows constructing an efficient parallel algorithm based on these evolvents.

In Sect. 5.2 the results of the numerical experiments are presented, which have demonstrated the algorithm from Sect. 4.2 based on the rotated evolvents allowed obtaining the speedup up to 43 times when solving the problem series employing several nodes of the computer cluster. It is worth noting that the objective functions in the considered problems are not computation-costly (the averaged computation time was 10^{-3} s). In the case of more complex problems, the speedup in time could approach the speedup with respect to the number of iterations.

Acknowledgements. The study was supported by the Russian Science Foundation, project No 16-11-10150.

References

1. Sergeyev, Y.D., Strongin, R.G., Lera, D.: Introduction to Global Optimization Exploiting Space-Filling Curves. SpringerBriefs in Optimization. Springer, New York (2013). https://doi.org/10.1007/978-1-4614-8042-6
2. Strongin, R.G.: Numerical Methods in Multi-Extremal Problems (Information-Statistical Algorithms). Nauka, Moscow (1978. in Russian)
3. Strongin, R.G.: Algorithms for multi-extremal mathematical programming problems employing a set of joint space-filling curves. J. Glob. Optim. **2**, 357–378 (1992)
4. Strongin, R.G., Gergel, V.P., Grishagin, V.A., Barkalov, K.A.: Parallel Computations for Global Optimization Problems. Moscow State University, Moscow (2013). (in Russian)
5. Strongin, R.G., Sergeyev, Y.D.: Global Optimization with Non-convex Constraints: Sequential and Parallel Algorithms. Kluwer Academic Publishers, Dordrecht (2000). (2nd edn. 2013, 3rd edn. 2014)
6. Goryachih, A.: A class of smooth modification of space-filling curves for global optimization problems. In: Kalyagin, V., Nikolaev, A., Pardalos, P., Prokopyev, O. (eds.) NET 2016. PROMS, vol. 197, pp. 57–65. Springer, Cham (2017). https://doi.org/10.1007/978-3-319-56829-4_5
7. Strongin, R.G.: Parallel multi-extremal optimization using a set of evolvents. Comput. Math. Math. Phys. **31**(8), 37–46 (1991)
8. Strongin, R.G., Gergel, V.P., Barkalov, K.A.: Parallel methods for global optimization problem solving. J. Instrum. Eng. **52**, 25–33 (2009). (in Russian)
9. Gaviano, M., Kvasov, D.E., Lera, D., Sergeyev, Y.D.: Software for generation of classes of test functions with known local and global minima for global optimization. ACM Trans. Math. Softw. **29**(4), 469–480 (2003)
10. Grishagin, V.A.: Operating characteristics of some global search algorithms. Probl. Stat. Optim. **7**, 198–206 (1978). (in Russian)
11. Gergel, V.P., Barkalov, K.A., Sysoyev, A.V.: Globalizer: a novel supercomputer software system for solving time-consuming global optimization problems. Numer. Algebra Control Optim. **8**(1), 47–62 (2018)
12. Sergeyev, Y.D., Kvasov, D.E.: A deterministic global optimization using smooth diagonal auxiliary functions. Commun. Nonlinear Sci. Numer. Simul. **21**(1–3), 99–111 (2015)
13. Paulavičius, R., Žilinskas, J.: Simplicial Lipschitz optimization without the Lipschitz constant. J. Glob. Optim. **59**(1), 23–40 (2014)
14. Jones, D.R.: The direct global optimization algorithm. In: Floudas, C.A., Pardalos, P.M. (eds.) The Encyclopedia of Optimization, 2nd edn, pp. 725–735. Springer, Heidelberg (2009). https://doi.org/10.1007/978-0-387-74759-0

15. Gablonsky, J.M., Kelley, C.T.: A locally-biased form of the DIRECT algorithm. J. Glob. Optim. **21**(1), 27–37 (2001)
16. Evtushenko, Y., Posypkin, M.: A deterministic approach to global box-constrained optimization. Optim. Lett. **7**(4), 819–829 (2013)
17. Gergel, V., Lebedev, I.: Heterogeneous parallel computations for solving global optimization problems. Proc. Comput. Sci. **66**, 53–62 (2015)
18. Gergel, V., Sidorov, S.: A two-level parallel global search algorithm for solution of computationally intensive multiextremal optimization problems. In: Malyshkin, V. (ed.) PaCT 2015. LNCS, vol. 9251, pp. 505–515. Springer, Cham (2015). https://doi.org/10.1007/978-3-319-21909-7_49

Efficiency Estimation for the Mathematical Physics Algorithms for Distributed Memory Computers

Igor Konshin[1,2(✉)]

[1] Dorodnicyn Computing Centre of FRC CSC RAS, Moscow 119333, Russia
igor.konshin@gmail.com
[2] Marchuk Institute of Numerical Mathematics of the Russian Academy of Sciences, Moscow 119333, Russia

Abstract. The paper presents several models of parallel program runs on computer platforms with distributed memory. The prediction of the parallel algorithm efficiency is based on algorithm arithmetic and communication complexities. For some mathematical physics algorithms for explicit schemes of the solution of the heat transfer equation the speedup estimations were obtained, as well as numerical experiments were performed to compare the actual and theoretically predicted speedups.

Keywords: Mathematical physics · Parallel computing
Parallel efficiency estimation · Speedup

1 Introduction

There are many works and Internet resources, which describe the properties of computational algorithms, parallel computing models, and also give recommendations on writing the most effective applications [1–3]. In this paper, an attempt is made to develop a constructive model of parallel computations, on the basis of which it is possible to predict the parallel efficiency of an algorithm implementation on a particular computing system.

When working on the shared memory computers, it is possible to construct such a model based on the Amdahl law (see [4,5]), while on distributed memory computers, it is necessary to take into account the parallel properties of both the implemented algorithm and computer system. In the present paper, we will perform a detailed analysis of the message transfer rate depending on the length of the message. If the message initialization time is ignored in the model, then it is possible to obtain more compact formulas for parallel efficiency estimates, while taking it into account, they become somewhat more complex, but also constructive and meaningful.

In this paper, we will focus our attention on the estimation of parallel efficiency for mathematical physics problems. In the scientific literature there are descriptions of a huge number of results on the achieved parallel efficiency for

© Springer Nature Switzerland AG 2019
V. Voevodin and S. Sobolev (Eds.): RuSCDays 2018, CCIS 965, pp. 63–75, 2019.
https://doi.org/10.1007/978-3-030-05807-4_6

mathematical physics problems, but there are no theoretical estimates of what efficiency could be achieved in practice, and thus a comparison of theoretical and actual ones.

This paper is organized as follows. Sections 2 and 3 give estimates of the algorithms parallel efficiency and their application to mathematical physics problems. Section 4 briefly describes the configuration of the computational cluster used. In Sects. 5 and 6, the numerical experiments on combination of asynchronous interprocessor data exchanges and calculations, as well as the dependance of data transmission rate on the message length are studied in detail. In Sect. 7, the parallel efficiency estimates are refined on the basis of the results obtained, as well as their application to the mathematical physics algorithms is considered. Section 8 describes the results of numerical experiments, while the conclusion sums up the main results of the paper.

2   Estimates of the Algorithms Parallel Efficiency

To obtain an estimate of the parallel algorithm run time, the key point is to estimate the transmission rate of the message. Let us exploit for this purpose the widely used formula

$$T_c = \tau_0 + \tau_c L_c, \tag{1}$$

where τ_0 is the message initialization time, τ_c is the message transfer rate (i.e., the message transmission time for the unit message length), and T_c is the message transmission time for the message length L_c. A detailed study of the values of τ_0 and τ_c will be carried out later in Sect. 6, and now it will suffice for us to assume that the length of the messages in the algorithms under investigation is large enough and therefore, for simplicity, we can assume that $\tau_0 = 0$. This allows us to substantially simplify formula (1):

$$T_c = \tau_c L_c. \tag{2}$$

Let us estimate the speedup that can be achieved by using p processors in the implementation of some parallel algorithm. Let $T(p)$ be the time of solving the problem on p processors, then the speedup obtained using this algorithm will be expressed by the formula:

$$S = T(1)/T(p).$$

Following [4,5], for further estimates we denote by L_a the total number of arithmetic operations of the algorithm, and by τ_a the execution time of one characteristic arithmetic operation. Similarly, let L_c be the total length of all messages, and τ_c is the already introduced time for transmitting a message of unit length. Then, the execution time of all arithmetic operations can be expressed by the formula $T_a = \tau_a L_a$, and the transmission time of all messages by $T_c = \tau_c L_c$.

Additionally, you can introduce a value

$$\tau = \tau_c/\tau_a, \tag{3}$$

which expresses the characteristic of the 'parallelism' property of the used computer, in other words meaning how many arithmetic operations can be performed during the transfer of one number to another processor.

Similarly, we introduce the value

$$L = L_c/L_a, \tag{4}$$

which expresses the characteristic of the 'parallelism' property of the algorithm under investigation, i.e. is the reciprocal of the number of arithmetic operations performed during the algorithm execution process to transfer of one number to another processor.

Now, in estimating the speedup, we can write:

$$S = S(p) = T(1)/T(p) = T_a/(T_a/p + T_c/p) = pT_a/(T_a + T_c)$$
$$= p/(1 + T_c/T_a) = p/(1 + (\tau_c L_c)/(\tau_a L_a)) = p/(1 + \tau L). \tag{5}$$

In this connection, the estimate of the algorithm's parallel performance is written as follows:

$$E = S/p = 1/(1 + \tau L). \tag{6}$$

Remark 1. We note that a twofold decrease in the efficiency of the algorithm occurs when $L = 1/\tau$.

In papers [4,5], a detailed discussion of the applicability of the obtained estimates (5), (6) can be found, but we shall confine ourselves to explaining at first glance the strange fact that the formula for evaluating the efficiency of (6) does not include the number of processors used. In fact, the value of L depends on the total transmission length L_c, which, in turn, depends on the number of processors p.

3   Estimates for Mathematical Physics Algorithms

Now, following [6], we can proceed to the estimate of the parallel efficiency of mathematical physics algorithms. For simplicity, we restrict our consideration to explicit schemes used for nonstationary problems described by some finite-difference equations.

We consider the problem in a d-dimensional cube ($d = 1, 2, 3$) with side in n cells, the total number of d-dimensional cubic cells is equal to $N = n^d$. Let V denote the number of unknown functions per computational cell (for example, $V = 5$ for three velocities u, v, w, as well as pressure and temperature). To calculate the values at a new time step for each cell, it is required to know the values in the nearest neighboring cells, which means using the $(2d + 1)$-points d-dimensional discretization stencil (more complex discretizations, especially for equations with cross derivatives, can contain in the stencil up to 3^d points). We denote by C the average number of arithmetic operations per cell when calculating the solution at a new time step. Now, almost everything is ready for estimating the speedup when solving the described problem on p processors, but the resulting speedup will depend essentially on the way the cells are assigned to the processors.

Figure 1 shows three different types of data distribution by processors, $D = 1, 2, 3$, using r layers in one, two, and three directions, respectively. The total number of processors in these cases is equal to $p = r^D$.

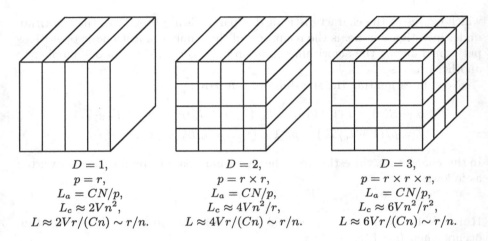

$$D = 1,$$
$$p = r,$$
$$L_a = CN/p,$$
$$L_c \approx 2Vn^2,$$
$$L \approx 2Vr/(Cn) \sim r/n.$$

$$D = 2,$$
$$p = r \times r,$$
$$L_a = CN/p,$$
$$L_c \approx 4Vn^2/r,$$
$$L \approx 4Vr/(Cn) \sim r/n.$$

$$D = 3,$$
$$p = r \times r \times r,$$
$$L_a = CN/p,$$
$$L_c \approx 6Vn^2/r^2,$$
$$L \approx 6Vr/(Cn) \sim r/n.$$

Fig. 1. Distribution of data by processors for the three-dimensional problem of mathematical physics for r layers in each of one, two, and three directions.

Assuming that $D \leqslant d$, we can begin to estimate the computational and communication costs.

The arithmetic cost per processor for all the cases under consideration is $L_a = CN/p = Cn^d/r^D$. Communication costs will depend on the number of boundary cells which need to be sent (received) on each of the processors, and will be $L_c = (2 - 2/r)DVn^{d-1}/r^{D-1}$. Note that $L_c = 0$ for $r = 1$ as required.

In this way, the main characteristic of the parallelism for the algorithm (4) is calculated as follows:

$$L = L_c/L_a = (2 - 2/r)DVn^{d-1}r^D/(Cn^2r^{D-1}) = (2 - 2/r)DVC^{-1} \cdot r/n \sim r/n. \tag{7}$$

This value is inversely proportional to the number of cells of the given processor in the direction of partitioning by processors.

Substituting the resulting expression for L into (6), we obtain

$$E = 1/(1 + (2 - 2/r)DVC^{-1}\tau \cdot r/n), \qquad S = pE \tag{8}$$

or to clarify the dependence on the original problem dimension d:

$$E(d, D) = 1/(1 + (2 - 2p^{-1/D})DVC^{-1}\tau \cdot p^{1/D}N^{-1/d}), \qquad S = pE(d, D). \tag{9}$$

To illustrate the specificity of (9), we give a short Table 1, which shows theoretical estimates of the parallel efficiency depending on the number of processors p and the type of domain decomposition by processors D for the three-dimensional problem with the following set of parameters:

$$d = 3, \quad N = n^3, \quad n = 1000, \quad V = 5, \quad C = 30, \quad D = 1, 2, 3, \quad \tau = 10. \quad (10)$$

The obtained efficiency values show the importance of using the right type of decomposition D, especially when utilizing a large number of processors.

4 INM RAS Cluster

Numerical experiments were performed on the cluster [7] of the Marchuk Institute of Numerical Mathematics of the Russian Academy of Sciences. Configuration of computational nodes from the 'x6core' segment used for calculations:

- Compute Node Asus RS704D-E6;
- 12 cores (two 6-core Intel Xeon processor X5650@2.67 GHz);
- RAM: 24 GB;
- Operating system: SUSE Linux Enterprise Server 11 SP1 (x86_64);
- Network: Mellanox Infiniband QDR 4x.

To build the code, we used the Intel C compiler version 4.0.1, with support for MPI version 5.0.3 [8].

5 Asynchronous Data Exchanges

The availability in the MPI standard the possibility of carrying out the asynchronous communications allows, after initializing the exchanges, without waiting for them to complete, immediately proceed with the calculations for which the data required is already available on the processor. The most of parallel computation guides [2] are permeated by the spirit of carrying out asynchronous data exchanges. However, the effect of overlapping calculations and data exchanges directly depends on specifics of architecture and implementation of MPI. Let we try to deal with this issue in more detail.

A special program was developed that implements test 1 from [9]. When carrying out the intensive calculations for the vector of numbers, there were sending the portions of processed values to another processor, which was used in its calculations at the next iteration. Asynchronous data exchanges were performed using the MPI_Isend and MPI_Irecv functions, as well as the MPI_Waitall function was used for completing all the data transmissions. The length of the vector was chosen to be $M = 2^{25}$, and the number of iterations was taken equal to 10. The number of portions to which the vector was partitioned before it was sent to another processor was chosen to be 1, 8, and 64. The calculations were performed on the INM RAS cluster [7] (see Sect. 4) in the 'x6core' segment.

Table 1. Theoretical estimate of the parallel efficiency (9) using the parameters set (10).

p	$D = 1$	$D = 2$	$D = 3$
1	1.00	1.00	1.00
10	0.97	0.98	0.98
64	0.82	0.95	0.96
729	0.29	0.85	0.92

Table 2. Time of test 1 on asynchronous data exchanges in the 'x6core' segment.

N_{drops}	$p = 1$	$p = 2$	$p = 13$
1	15.42	17.18	18.52
8	15.43	15.58	15.78
64	15.43	15.63	15.77

The results of numerical experiments on asynchronous data exchanges are given in Table 2. Deviations at time measurements at various iterations were insignificant and were equal to from 1 to 5 percent. Let us consider the obtained results in more detail. Calculations and data exchanges were carried out at the first and last of the used processes. The parameter p means the number of cores used, thus, in the second column $p = 1$ the results of calculations without data exchanges are shown. The time measurements in this column are almost the same, we will use them as a reference point for our further observations. In the first line for $N_{\mathrm{drops}} = 1$, we have a situation where all the data is sent in one portion, so the exchanges are synchronous and no overlap occurs with the calculations. The $p = 2$ column reflects the situation where two exchanging processes are physically located on one compute node and, despite the use of the MPI library, exchanges are actually conducted within the shared memory. One can see that due to synchronous exchanges ($N_{\mathrm{drops}} = 1$), the total computation time increased by about 10%. The $p = 13$ column contains data when two exchanging processes are physically located on different compute nodes. Indeed, one computing node of the 'x6core' segment consists of two 6-core processors, and since calculations and exchanges are performed in the first and last core, the communications were carried out through the interprocessor network. Due to synchronous exchanges ($N_{\mathrm{drops}} = 1$), the total computational time in this case increased by about 20%. With the increase in the number of portions N_{drops}, the total computational time decreases, tending to the respective time on 1 processor.

The obtained results allow to conclude that on the particular computer the effect of asynchronous exchanges is observed, but it is not significant and determinative.

It should also be noted that the theoretical consideration of the contribution of asynchronous exchanges to the overall solution time is extremely difficult, since, assuming that the delay time due to exchanges is negligible, the speedup of all algorithms should be considered linear (which is very far from reality even with an ideally uniform CPU usage). Our goal is to develop simple and constructive efficiency estimates for parallel applications for mathematical physics problems.

6   A Detailed Analysis of the Data Transmission Rate

To refine the estimates (5), (6) and (9), it is necessary to perform a detailed analysis of the message transfer rate, taking into account the initialization time of the message τ_0 from formula (1). Let us describe test 2 from [9], which investigates this dependence.

The described numerical experiment consisted in transmitting a message with the total length of $L_c = M = 2^m$ words of the type 'double', and the message was divided into $n_{c,i} = 2^i$ portions of length $L_{c,i} = L_c/n_{c,i} = 2^{m-i}$, $i = 0, ..., m$. For more statistical confidence, the test was repeated several times.

Thus $m + 1$ values of $T_{c,i}$ was obtained for the total message transmission time.

Let us derive the theoretical estimates of $T_{c,i}$, taking into account the time of initialization of the messages τ_0 in (1):

$$T_c(L_c) = T_c n_c = (\tau_0 + \tau_c L_c)n_c, \quad n_c = M/L_c. \tag{11}$$

For numerical experiments, the parameters $m = 25$ and $M = 2^m = 33554432$ were chosen, and the values $T_c(L_{c,i})$ for $L_{c,i} = 2^i$, $i = 0, ..., m$, were obtained for the 'x6core' segment. As the measured initialization time of the message we can take the value

$$\tau_0 = \max_{i=0,...,m} T_c(L_{c,i})/M = T_c(1)/M = 10.0/M \approx 3.0 \cdot 10^{-7}, \tag{12}$$

and as the average transmission rate of one number, we can use the value

$$\tau_c = \min_{i=0,...,m} T_c(L_{c,i})/M = 0.10/M \approx 3.0 \cdot 10^{-9}. \tag{13}$$

The values of the constants 10.0 and 0.10 were obtained from the described numerical experiment conducted in the 'x6core' segment, for simplicity, the values of the constants were rounded. You can see that for the values obtained, the ratio τ_0/τ_c is 100. In other words, we can conclude that the time $T_c(0)$ of transmitting an empty message will be only 2 times less than the time $T_c(100)$ of transferring 100 numbers of the type 'double'.

Figure 2 presents the results of comparing the experimental data with the theoretical ones, calculated from (11). The measurement data are statistically reliable: the average statistical deviation for most measurements is less than 1–5% for 10 repetitions of the test described. In Fig. 2a it can be seen that the theoretical curve practically coincides with the experimental data, which indicates the rationality of the assumptions made regarding the transmission rate, in spite of the significant rounding-off of the constants (12) and (13).

It is interesting to note that in Fig. 2b some retardation of transmission rate is observed when using a too long messages. This is probably related to the size of the internal buffers of the MPI library, which is selected when it is installed. In addition, for small messages with $L_c < 14$, the data transmission rate on one computing node is in most cases slightly more than for two nodes, and for

$L_c \geqslant 14$, on the contrary. On average, the difference is about 10%. Probably, this effect is related to the specifics of MPI implementation and configuration of the nodes from the 'x6core' segment.

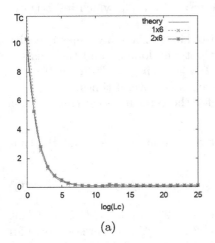

(a)

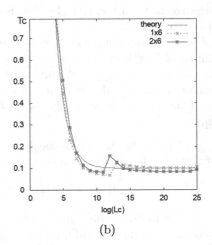

(b)

Fig. 2. The total time for sending a message of 2^{25} double words by portions of length L_c. A theoretical estimate from (11) and actual calculations in the 'x6core' segment.

7   Estimation Refinement and Its Usage for the Mathematical Physics Algorithms

Taking into account (1), we write down the time of data transmission:

$$T_c = n_c \tau_0 / q + \tau L_c, \qquad (14)$$

where L_c, as before, denotes the total length of all communications per processor, q is the number of overlapping subdomains (the necessary exchanges are performed once per q time steps), and n_c is the total number of communications per each processor. Note that in most algorithms n_c usually does not modified with increasing number of processors p.

We write down the time spent on arithmetic operations as follows:

$$T_a = (1 + Q)\tau_a L_a, \qquad (15)$$

where L_a, as before, denotes the total number of arithmetic operations performed per one processor, the new parameter Q expresses the proportion of the increase in the number of arithmetic operations, if the duplication of some arithmetic operations was performed. This duplication may be due to attempt of reduce the number or length of communications. If, as in the previously considered algorithms, there is no duplication, then $Q = 0$ and the formula (15) goes into the previously used one $T_a = \tau_a L_a$, while if the computations are duplicated, then Q takes some small positive value, depending on the algorithm features.

Using (14) and (15) in estimating the algorithm speedup, we obtain

$$
\begin{aligned}
S = S(p, \tau_0, Q) &= T(1)/T(p) = T_a(1)/(T_a(p) + T_c(p)) \\
&= \tau_a L_a/((1 + Q)\tau_a L_a/p + n_c \tau_0/q + \tau_c L_c/p) \\
&= p/(1 + \tau L + Q + pn_c \tau_0/(q\tau_a L_a)).
\end{aligned} \tag{16}
$$

Here, as before in formulas (3) and (4), the values τ and L mean the parallelism characteristics of the computer and the algorithm, respectively, while their product in the denominator characterizes the decrease in the efficiency due to data exchanges. The value Q, if it is different from 0, expresses the loss of speedup due to duplication of calculations; and the ratio of the delays in initializing of n_c transmissions $(n_c \tau_0/q)$ to the execution time of arithmetic operations on each processor $(\tau_a L_a/p)$ contributes to the speedup loss due to message initialization. The value τ_0, involved in the estimate (16), can be calculated rather accurately, as it was done in (12).

Let us turn again to the mathematical physics algorithms to estimate the remaining unknowns in (16). The number n_c of data transmissions per one time step will be equal or proportional to the number of subdomains connected to the specific processor. In the simplest case, in the notation of Sect. 3, depending on the method of decomposition of the domain D, we can take $n_c = 2D$.

Suppose now that instead of one layer of neighboring cells, we want to exchange q layers of cells at once, in order to perform exchanges in q times less often. This can lead to some reduction in the calculation time due to a decrease in the transmission initialization time, although it leads to some duplication of calculations. The total length of each transmission will increase by q times to qL_c, although the total length of transmissions for q time steps will remain practically unchanged. Note that duplicated computations will be performed at $(q - 1)$ time steps exactly in the cells that participated in the transmissions, in this case the number of additional arithmetic operations being $L_c q(q-1)/2$. Thus, the estimate of the last unknown quantity in (16) is obtained:

$$
Q = \tfrac{1}{2}q(q-1)L_c/L_a = \tfrac{1}{2}q(q-1)L. \tag{17}
$$

Note that in the traditional 'high-communication no-extra-computation' scheme (HCNC) without duplicating the calculations (i.e., for $q = 1$), we get $Q = 0$. The case $q > 1$ corresponds to the 'low-communication high-extra-computation' scheme (LCHC) with $Q > 0$.

Remark 2. For further analysis of formula (17), suppose that the user is aimed to avoid double duplication of computations (i.e. $Q = 1$) and the associated double decrease in the computations efficiency, then it is sufficient to select the value of q not more than $\sqrt{2/L}$. Turn to formula (6), one can see that the same efficiency drop can be observed for the case $\tau L = 1$. Therefore, the previous restriction can also be rewritten as $q < \sqrt{2\tau}$. The computers encountered by the author are usually characterized by the parameter τ from the range from 10 to 30 (sometimes up to 100), therefore it is recommended to take $q < 5$.

Returning to the speedup estimate and substituting the values derived into (16), we obtain the final relation

$$S = p/\left(1 + (\tau_{ca} + \tfrac{1}{2}q(q-1))(2 - 2p^{-1/D})DVC^{-1}p^{1/D}N^{-1/d} + 2DC^{-1}q^{-1}pN^{-1}\tau_{0a}\right), \quad (18)$$

where $\tau_{ca} = \tau = \tau_c/\tau_a$ and $\tau_{0a} = \tau_0/\tau_a$ are denoted.

The impact on the speedup of various quantities has already been discussed, we note only that the computer-dependent quantity $\tau = \tau_c/\tau_a$ contributes to the second term of the denominator, and the ratio τ_0/τ_a is involved in the third term. We also note that the second term increases with increasing of q, while the third one decreases.

Remark 3. Rewriting (18) in the form

$$S = p/\left(1 + (\tau_{ca} + \tfrac{1}{2}q(q-1))C_1 + C_2\right),$$
$$C_1 = (2 - 2p^{-1/D})DVC^{-1}p^{1/D}N^{-1/d}, \quad C_2 = 2DC^{-1}q^{-1}pN^{-1}\tau_{0a}) \quad (19)$$

we can take the derivative of S over q to estimate the optimal overlap size. In this way for $S'(q) = 0$ we obtain the cubic equation $2C_1q^3 - C_1q^2 - 2C_2 = 0$. For $q > 1$ there is only one root $q_* \approx (C_1/C_2)^{1/3}$, or

$$q_*^3 \approx C_1/C_2 = 2V^{-1}p^{1-1/D}N^{1/d-1}\tau_{0a}. \quad (20)$$

Substituting to (20) some reasonable values $D = d = 3$, $V = 5$, $p = 10^3$, $N = 10^6$, and $\tau_{0a} = 10^4$ we obtain $q_* \approx 3$.

At the end of this section, it should be noted that the estimates (9) and (18) can easily be generalized to the case of a region in the form of a rectangle or a parallelepiped. In the case of complex shape regions and/or arbitrary data distribution over processors, all the unknown values $\tau = \tau_c/\tau_a$ and $L = L_c/L_a$ can be easily calculated, for example, directly in the designed application just before the computations.

8   Results of Numerical Experiments

In Sect. 4 the specification of the computational cluster exploited for all numerical experiments made in this paper. We now turn to the description of the model problem to be solved.

As a model problem, the solution of the heat transfer equation in the d-dimensional cubic region $d \leqslant 3$ with the same number of cells in each direction was chosen. The cells in the computational domain were distributed over processors in D dimensions, $D \leqslant d$. The overlap of subdomains in q layers was considered, which allowed only one stage of data exchanges for q time steps. We

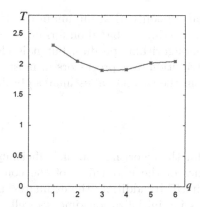

Fig. 3. The solution time depending on the subdomain overlap size $q = 1, ..., 6$.

focus on the standard finite-difference discretization of the heat transfer equation with an explicit scheme in time.

Figure 3 shows the results of computations for the model problem of dimension $100p \times 100 \times 100$, $d = 3$, $D = 1$, $p = 64$, using the subdomain overlap of size $q = 1, ..., 6$. The number of time steps was equal to 120. It can be noted that the minimal solution time is obtained for the case $q = 3$, which is in full agreement with the theoretical estimates made in Sect. 7.

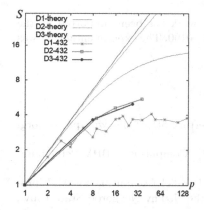

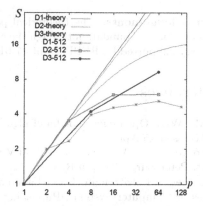

Fig. 4. Speedup for the $432 \times 432 \times 432$ problem for $D = 1, 2, 3$.

Fig. 5. Speedup for the $512 \times 512 \times 512$ problem for $D = 1, 2, 3$.

Figures 4 and 5 contain the results for two problems of dimension $432 \times 432 \times 432$ and $512 \times 512 \times 512$, respectively. We analyzed all possible one-, two- and three-dimensional ($D = 1, 2, 3$) distributions over processors, for all variants of processor ratio, in which all subdomains have the same size. We considered a conventional version with a minimum overlap of size $q = 1$, with 120 time steps. Theoretical estimates based on formula (18) give a slightly more optimistic forecast of the algorithm speedup, but it is clear that the use of a larger dimension for distribution over processors in most cases provides greater speedup, as it follows from the theoretical estimates obtained in Sects. 3 and 7.

9   Conclusion

Constructive estimates for the algorithm parallel efficiency are obtained, which include the characteristics of the parallelism of the computer and the parallelism of the algorithm itself. In addition, the estimates are obtained that take into account the transmission initialization time, as well as the ability to group messages. For explicit discretization schemes applied to mathematical physics problems, estimates are obtained that, in addition to the geometric parameters of the problem and the type of cell distribution over processors, also include the number of unknown functions per cell, as well as the number of arithmetic operations per cell at one time step. The dependence of the message transfer rate on message length is analyzed in detail, and also the expediency of performing calculations at the same time as the asynchronous communications. For the sample heat transfer problem, a direct comparison of the experimental results with the theoretical estimates on the parallel efficiency was performed. Confirmed the conclusion about the profitability of using the higher dimension type of data distribution over processors.

Acknowledgements. The theoretical part of this work has been supported by the Russian Science Foundation through the grant 14-11-00190. The experimental part was partially supported by RFBR grant 17-01-00886.

References

1. AlgoWiki: Open encyclopedia of algorithm properties. http://algowiki-project.org. Accessed 15 Apr 2018
2. Voevodin, V.V., Voevodin, Vl.V.: Parallel Computing. BHV-Petersburg, St. Petersburg (2002, in Russian)
3. Gergel, V.P., Strongin, R.G.: Fundamentals of Parallel Computing for Multiprocessor Computer Systems. Publishing House of the Nizhny Novgorod State Univ., Nizhny Novgorod (2003, in Russian)
4. Konshin, I.N.: Parallel computational models to estimate an actual speedup of analyzed algorithm. In: Proceedings of the International Conference on Russian Supercomputing Days, 26–27 September 2016, Moscow, Russia, pp. 269–280. Moscow State University, Moscow (2016, in Russian) http://2016.russianscdays.org/files/pdf16/269.pdf
5. Konshin, I.: Parallel computational models to estimate an actual speedup of analyzed algorithm. In: Voevodin, V., Sobolev, S. (eds.) RuSCDays 2016. CCIS, vol. 687, pp. 304–317. Springer, Cham (2016). https://doi.org/10.1007/978-3-319-55669-7_24

6. Konshin, I.N.: Parallelism in computational mathematics. International Summer Supercomputer Academy. Track: Parallel Algorithms of Algebra and Analysis and Experiments of Supercomputer Modeling. MSU, Moscow (2012, in Russian). http://academy2012.hpc-russia.ru/files/lectures/algebra/0704_1_ik.pdf. Accessed 15 Apr 2018
7. INM RAS cluster (2018, in Russian). http://cluster2.inm.ras.ru. Accessed 15 Apr 2018
8. MPI: The Message Passing Interface standard. http://www.mcs.anl.gov/research/projects/mpi/. Accessed 15 Apr 2018
9. Bajdin, G.V.: On some stereotypes of parallel programming. Vopr. Atomn. Nauki Tekhn., Ser. Mat. Model Fiz. Prots., no. 1, pp. 67–75 (2008, in Russian)

Extremely High-Order Optimized Multioperators-Based Schemes and Their Applications to Flow Instabilities and Sound Radiation

Andrei Tolstykh$^{(\boxtimes)}$, Michael Lipavskii, Dmitrii Shirobokov, and Eugenii Chigerev

Dorodnicyn Computing Center, Federal Research Center "Computer Science and Control" of Russian Academy of Sciences, Moscow, Russian Federation
tol@ccas.ru

Abstract. Multioperators-base schemes up to 32nd-order for fluid dynamics calculations are described. Their parallel implementation is outlined. The results of applications of their versions to instability and sound radiation problems are presented. The extension to strongly discontinuous solutions is briefly outlined.

Keywords: Multioperators-based schemes
Euler and Navier-stokes equations · Parallel methodology
Instability and sound radiation · Jets · Discontinuous solutions

1 Introduction

The multioperators technique proposed in [1] is a way to construct arbitrary high-order numerical analysis formulas and, in particular, arbitrary high-order approximations to fluid dynamics equations. High orders are obtained via increasing numbers of *basis operators* with fixed stencils rather than by enlarging stencils or polynomial orders (that is, by increasing numbers of *basis functions*). The basis operators are generated by one-parameter families of compact approximations $L_h(c)$ to a target linear operator L and the resulting multioperators look like

$$L_M(c_1, c_2, \ldots, c_M) = \sum_{i=1}^{M} \gamma_i L_h(c_i) \qquad (1)$$

where $c_1, c_2, \ldots, c_M$ are the input values of parameter c. The c_i values uniquely define the γ_i coefficients making approximation orders proportional to either M or $2M$. Considering them as free parameters, one can control the multioperators properties.

As follows from Eq. (1), calculations of the multioperators actions on known grid functions involve performing similar arithmetic operations for each parameter c_i. Thus multioperators-based numerical analysis formulae can be calculated

V. Voevodin and S. Sobolev (Eds.): RuSCDays 2018, CCIS 965, pp. 76–87, 2019.
https://doi.org/10.1007/978-3-030-05807-4_7

in a parallel manner. That property can be used also for constructing parallel algorithms for Computational Fluid Dynamics (CFD) applications.

Several types of multioperators were investigated and used in CFD algorithms, the main tasks being 2D and 3D Navier-Stokes problems for compressible gas flows . Some theoretical topics can be fond in particular in [2]. Various types of high-order multioperators-based numerical analysis formulae are described in [3]. In [5], a family of extremely high order multioperators based on two-point compact approximations to derivatives is presented. The resulting conservative linear and non-linear schemes can be used for smooth and discontinuous solutions. The complete theory of multioperators can be found in [4].

Below a brief overview of the latest both theoretical and numerical results is presented.

2   Schemes Outlines

Using the uniform mesh $\omega_h : (x_j = jh, j = 0, \pm 1, \pm 2, \ldots)$, $h = const$, the multioperators family under consideration can be created in the following way. First, the two-point operators depending on parameter c are introduced in the form

$$R_l(c) = I + c\Delta_-, \quad R_r(c) = I - c\Delta_+ \tag{2}$$

where Δ_- and Δ_+ are the two-point left and right differences. Then the upwind-downwind pair for approximating the first derivatives with the truncation orders $O(h)$ are defined by

$$L_l(c) = \frac{1}{h}R_l(c)^{-1}\Delta_-, \quad L_r(c) = \frac{1}{h}R_r(c)^{-1}\Delta_+. \tag{3}$$

Assuming the Hilbert space of bounded grid functions with the inner product defined by the summation over grid points, they have the same skew-symmetric component and the self-adjoint components with opposite signs. It follows from the Eq. (3) that very simple two-diagonal inversions of the operators from (2) are needed to calculate the actions of operators L_l and L_r on known grid functions.

Now the skew-symmetric second order operator $L_h(c)$ in Eq. (1) is defined as $L_h(c) = (L_l(c) + L_r(c))/2$. Fixing M values $c_1, c_2, \ldots c_M$ and solving the linear system for γ_i coefficients, one obtains the skew-symmetric multioperator $L_M(c_1, c_2, \ldots, c_M)$ providing the approximation order $O(h^{2M})$ for derivatives of sufficiently smooth functions. Additionally, we construct the self-adjoint multioperator defined by

$$D_{M_1}(\bar{c}_1, \bar{c}_2, \ldots, \bar{c}_{M_1}) = \sum_{i=1}^{M_1} \bar{\gamma}_i L_1(\bar{c}_i), \quad \sum_{i=1}^{M_1} \bar{\gamma}_i = 1 \tag{4}$$

where $L_1 = L_l(c) - L_r(c))$ and M_1 is possibly differs from M; in the following it is however assumed that $M_1 = M$. For fixed values $\bar{c}_i, i = 1, 2, \ldots M_1$, the $\bar{\gamma}_i$ coefficients can be obtained to give $D_{M_1}[u]_j = O(h^{2M_1-1})$ where $[u]_h$ is a sufficiently smooth function projected into mesh ω_h. To simplify the multioperators investigations, we always suppose that the parameters values are linearly distributed inside the intervals $[c_{min}, c_{max}]$ and $[\bar{c}_{min}, \bar{c}_{max}]$. In this way, the multioperators become two-parameter dependent.

In the case of model equation

$$\frac{\partial u}{\partial t} + \frac{\partial f(u)}{\partial x} = 0, \tag{5}$$

the semi-discretized scheme in the index-free form reads

$$\frac{\partial u}{\partial t} + L_M(c_{min}, c_{max})f(u) + C\,D_M(\bar{c}_{min}, \bar{c}_{max})u = 0, \quad C \geq 0. \tag{6}$$

Setting $f(u) = au$, $a = const$ and using the Fourier transform, the c_{min}, c_{max} values can be used to control the spectral properties. The $\bar{c}_{min}, \bar{c}_{max}$ values were considered as admissible if $D_M(\bar{c}_{min}, \bar{c}_{max}) \geq 0$. Then scheme (6) is stable in the L_2 norm as a scheme with a non-negative operator. The term with D_M is the high-order dissipation mechanism which can be used to damp possible spurious oscillations.

Scheme Eq. (6) can be readily extended to multidimensional cases by constructing the multioperators for each spatial coordinate independently. Consider, for example, K-dimensional conservation laws in the case of vector valued functions $\mathbf{u}$, $\mathbf{f}_k(\mathbf{u})$, $k = 2, 3$

$$\frac{\partial \mathbf{u}}{\partial t} + \sum_{k=1}^{K} \frac{\partial \mathbf{f}_k(\mathbf{u})}{\partial x_k} = 0. \tag{7}$$

Using uniform meshes ω_k and the above defined operators for each spatial coordinate x_k, the kth multioperator looks as

$$L_M^{(k)}(c_{min}, c_{max}) = \sum_{i=1}^{M} \gamma_i L_0^{(k)}(c_i), \quad D_M^{(k)}(\bar{c}_{min}^{(k)}, \bar{c}_{max}^{(k)}) = \sum_{i=1}^{M_k} \bar{\gamma}_i^{(k)} L_1^{(k)}(\bar{c}_i) \tag{8}$$

where the basis operators $L_0^{(k)}$ and $L_1^{(k)}$ are the L_0 and L_1 ones corresponding to the ω_k mesh. It is suggested in Eq. (8) that dissipative multioperators $D_{M_k}(\bar{c}_{min}^{(k)}, \bar{c}_{max}^{(k)})$ can be different for different coordinates x_k while the main skew-symmetric multioperators are defined uniquely for each coordinate.

The semi-discretized scheme for Eq. (7) now reads

$$\frac{\partial \mathbf{u}}{\partial t} + \sum_{k=1}^{K} L_M^{(k)} \mathbf{f}_k(\mathbf{u}) + \sum_{k=1}^{K} C_k\,D_M^{(k)} \mathbf{u} = 0, \quad C_k \geq 0. \tag{9}$$

Considering the Hilbert space of vector-valued functions, one can prove that the scheme under some assumptions concerning the Jacobian matrices is stable in the frozen coefficients case. It can be readily cast in the flux form.

The Euler equations may be viewed as a particular case of the governing equations Eq. (7). In the case of the Navier-Stokes equations, various approximations to the viscous terms can be added to Eq. (9). For example, they can be compact or multioperators ones.

2.1   High Resolution During Long-Time Integration

Returning to the scalar 1D case, the dependence of the multioperators on two parameters allows one to control their spectral properties. The well accepted way to characterize the properties of approximations to convection terms is to consider the exact numerical solutions of the semi-discretized advection equation obtained from (5) by setting $f(u) = au$, $a = const$. In our case, the exact solution of Eq. (6) with $u(0, x) = exp(ikx)$ reads

$$u_j = \exp\left(-Cdt/h\right)\exp(ik(hj - a_*t)), \quad a_* = a\hat{L}_M(\alpha, c_{min}, c_{max})/\alpha,$$
$$d = h\hat{D}_M(\alpha, \bar{c}_{min}, \bar{c}_{max}).$$

The difference between the numerical phase velocity a_* and the exact one a can be viewed as the phase errors while function $d(\alpha) > 0$ is responsible for the harmonics damping. It can be viewed as a measure of amplitude errors. Clearly, even small phase errors can produce large solution errors for large time values t. Using the free parameters c_{min}, c_{max}, the optimizing procedure minimizing the phase errors for as large intervals of the dimensionless wave numbers $\alpha = kh$ as possible can be readily carried out. This was done for the 16th,20th,32nd and 36th - order schemes.

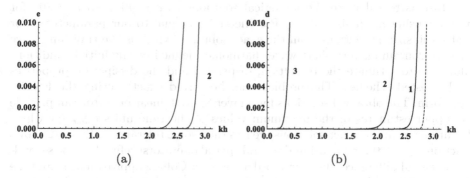

Fig. 1. a: Phase errors vs. wave numbers: curves 1,2 correspond to 16th- and 32nd-order multioperators, b: Dissipation exponent vs. wave numbers; curves 1,2,3 correspond to various choice of $\bar{c}_{min}, \bar{c}_{max}$ of the 15th-order multioperator.

Figure 1a shows the phase errors $e(\alpha) = |a_*/a - 1|$ for the 16th- and the 32th -order multioperators. They correspond to the near-optimal values of c_{min} and c_{max} obtained by calculating the functions for selected points in the two-dimensional space of the parameters. As follows from the Figure, the range of dimensionless wave numbers $[0, \alpha_*]$ for which the phase errors are small is noticeably greater for the 32th-order multioperator than that for the 16th-order one. Moreover, the "small" values of the errors shown by both curves in the Figure can differ by orders of magnitude indicating the advantages of the 32th-order multioperator. As an illustration, Table 1 presents the phase errors e for the selected values of the dimensionless wave numbers.

Table 1. Phase errors e for the selected values of α.

α	1.5	2.0	2.2	2.5
16th order	1.70e−6	1.8e−5	1.7e−5	1.2e−3
32th order	4.9e−11	5.6e−8	5.3e−7	7.7e−6

The amplitude errors introduced by multioperator D_M are characterized by the dissipation exponent $d(\bar{c}_{min}, \bar{c}_{max})$. Their dependance on wave numbers are shown in Fig. 1b for several parameter pairs. Curves1,2,3 correspond to the 15th-order multioperator. They look as cut-off filters of high wave numbers harmonics with various cut-off values. Choosing $\bar{c}_{min}, \bar{c}_{max}$, one can control the dissipation property of schemes. The dashed curve in Fig. 4 is obtained for near maximum cut-of value of the 31st-order multioperator.

To estimate the ability of the schemes to preserve high resolution during long-time integration, consider the benchmark problem [6] for the advection Eq. (5) with $f(u) = u$ and the initial condition

$$u(0, x) = [2 + \cos(\beta x)][\exp(-\ln 2\,(x/10)^2)].$$

The task is to calculate the numerical solutions at $t = 400$ and $t = 800$ for $\beta = 1.7$ using mesh size $h = 1$. Parameter β is equal to our parameter α for that mesh size. Deviations from the exact solution (which is the travelling wave package containing very short waves harmonics defined by the initial condition) allow one to estimate the resolution, dispersion and the dissipation properties of the tested schemes. The problem can be solved exactly using the Fourier transform. The obtained solutions for a given spatial linear operator can provide the upper estimates of the maximum values of the time units t_{max} for which deviations from the exact solutions are less than some tolerance ε. Figure 2 shows functions $t_{max}(\beta)$ for $\varepsilon = 0.1$ and several spatial operators(defined by the second-order central difference, the fourth-order compact Collatz approximation and the 16th and 36th-order multioperators). As seen, large phase errors for high wave numbers can prevent reasonable description of harmonics advection during large time intervals in the case of non-optimized relatively low-order schemes. Clearly, filtering and time stepping devices can decrease t_{max} values.

The calculations were carried out for the higher than required wave number ($\beta = 2.1$ instead of $\beta = 1.7$) using the 32nd-order multioperator. Both numerical and exact solutions are presented in Fig. 3 at $t = 15000$ (markers and solid lines correspond to the numerical and the exact solutions) showing very good dispersion-preserving property of the optimized multioperator.

3   Parallel implementation

The important multioperators property is the possibility to calculate their actions on known grid functions by parallel calculations of the actions of their

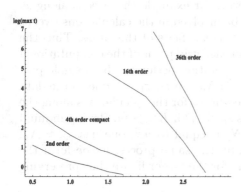

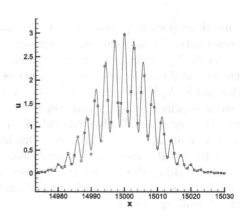

Fig. 2. Maximum time units for which solution errors are less than 10%.

Fig. 3. Numerical solution at $t = 15000$ obtained with 32th-order multi-operator.

basis operators. It can be used for example in the framework of MPI when performing calculations with multicore PC. In the case of massively parallel system for solving 3D CFD problems, it can be combined with domain decomposition approaches exploiting considerable amount of left and right sweeps. Their number for a line along which the left and right sweeps are carried out can be estimated as $K = 5M$ where M is the number of parameters defining the multi-operator. For example, one has $K = 20$ for the 16th-order scheme. Each sweep consists of calculating a current value with known previous one differing only in the parameters values and the functions which derivatives are approximated. Considering for example, the general form of the left sweep for a grid value v_i with $i = 1, 2, \ldots, N$, the process looks as

$$v_i = a(c_j)v_{i-1} + b(c_j)f_i, \ j = 1, 2, \ldots, M \tag{10}$$

where f_i is a known grid function. Thus it is possible to use m processors by partitioning the interval $i \in [0, N]$ into m equal parts with transferring the value calculated by the k-th processor to the $(k + 1)$-th one, $k = 1, 2, \ldots m - 1$. Using the idea, the calculations for one-line left sweeps with m processors can be schematically outlined in the form of Table 1 where "sweep k" means the calculations according Eq. 10.

As seen, some processors are idle at some stages. However the duration of $(K + m)$ stages is $(K + m)\tau$ where τ is time needed for each part of the decomposed space interval. Neglecting the data transfer expenses and comparing with the time $Km\tau$ in the case of a single processor, one obtains the speed up s equal to $Km/(K + m)$. It increase approximately linear if $m << K$ giving, for example, $s = 8$ for $K = 40$, $m = 10$. The above "one-line" idea applied to the 3D Euler or Navier-Stokes equations looks as follows. Supposing for the sake of argument that there are m^3 processors and a $N_x \times N_y \times N_z$ mesh, the computational domain is partitioned into $m \times m \times m$ cubes with m cubes in each spatial directions (say, x, y and z ones). Considering for example the sweeps along x-coordinate, all processors are supposed to be involved in the calculations sweeps along $N_y N_z / m^2$ lines intersecting their $x = const$ faces of the cubes. Thus the time needed to calculate the x-derivatives in each grid point of the computational domain is $N_y N_z (K + m)\tau / m^2$. The same operation performed by a single processor requires $N_y N_z m\tau$ time (in reality, it is $N_y N_z m\tau_1$, $\tau_1 < \tau$ due to the data transferring loses). It gives the "ideal" speed-up s_x for the calculations along the x-coordinate $s_x = K m^3 / (K + m)$. One has $s_x \approx m^3$ if $K >> m$. Having in mind that the total number of the processors N_p is equal to m^3, one has $s_x \approx N_p$. Upon finishing the job for the x-coordinate, the same processors perform the calculations for the y-coordinate and finally for the z-coordinate thus preserving the "ideal" speed-up.

Table 2. Organizing parallel calculations for K sweeps

Processors	1	2	3	4	...	m
stage 1	sweep 1					
stage 2	sweep 2	sweep 1				
stage 3	sweep 3	sweep 2	sweep1			
...	...	...	...	...	...	
stage K	sweep K	sweep K-1	sweep K-2	...	...	sweep K-m+1
stage K +1		sweep K	sweep K-1	...	...	sweep K-m
...	...	...	...	...	...	
stage K+m						sweep K

In Table 2, the calculation times for our 3D Euler calculations of jets instability with several number of the processors of the Lomonosov supercomputer of the Moscow State University are presented, the MPI programming being used. The mesh was $360 \times 100 \times 100$. The processors were distributed for three spatial coordinates as $m \times m \times m$ with m ranging from 2 to 10.

Table 3. Execution times per time step and acceleration

Number of processors	8	27	64	125	216	1000
Distributions	$2 \times 2 \times 2$	$3 \times 3 \times 3$	$4 \times 4 \times 4$	$5 \times 5 \times 5$	$6 \times 6 \times 6$	$10 \times 10 \times 10$
Time per step, sec	113	27.45	12.64	6.34	3.99	1.70
Acceleration	1	4.12	8.94	17.8	40.3	66.5

In the above Table 3, the acceleration is defined as the time decrease when comparing with the case of $m = 2$.

4 Numerical Examples

4.1 Applications to Jets Instabilities

The constructed schemes fit neatly into computational aeroacoustic requirements. They allow to describe properly pressure pulsations which amplitudes are about $10^{-6} - 10^{-7}$ of mean pressure levels when solving the Euler or the Navier-Stokes equations, no linearization or introducing base flows being needed.

Calculations with 10th-order multioperators schemes were carried out for cold and hot axisymmetric jets using both Cartesian and cylindrical coordinates analytically transformed to condense grid points near shear layers. The boundary conditions at nozzles lips were posed either as the results of the nozzles flow calculations or using analytical expressions (as in [7]) providing various initial shear layers thicknesses. Instabilities resulting in the break down of the steady state of the jets with vortex rings formation and sound radiations were investigated using both axisymmetric and 3D forms of the governing equations. It was found that the axisymmetric formulation is incomplete due to the excitation of azimuthal modes appearing in the 3D calculations. The snapshots of the vorticity fields in both cases are shown in Figs. 4 and 5 with more regular structure of the vorticity in the axisymmetric case.

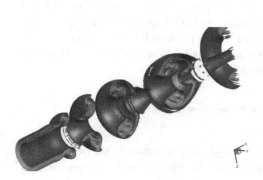 

Fig. 4. Snapshot of the vorticity field. Axisymmetric formulation

Fig. 5. Snapshot of the vorticity field. 3D formulation

Figures 6 and 7 shows the examples of the acoustic pressure spectra calculated for the microphones placed at points with polar coordinates $r = 20R, \theta$ where R is the initial jet radius, the origin being places at the center of the initial cross section of the jets. In the Figures, the curves differ in that they correspond to the boundary conditions at the nozzle lips specified in analytical forms (labelled as "synthetic", red lines) and obtained via flow calculations in the nozzle (green lines). The results of the calculations with the Cartesian grid are also shown in the Figures (blue lines).

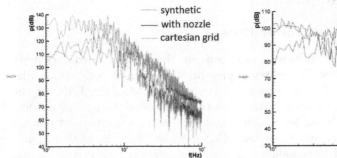

Fig. 6. Spectra of acoustics pressures for 3D calculations; $r = 20R$, $\theta = 10^0$ (Color figure online)

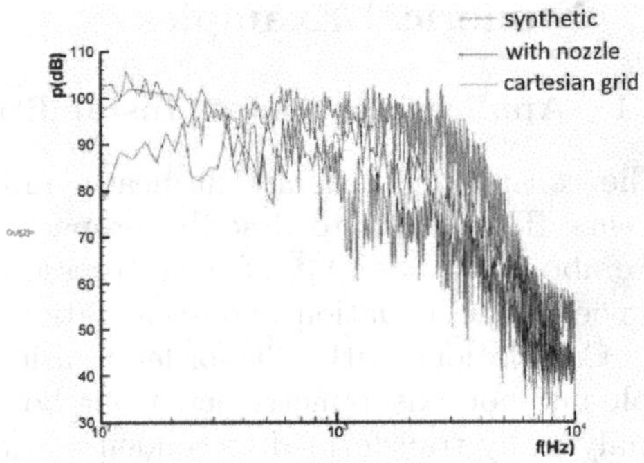

Fig. 7. Spectra of acoustics pressures for 3D calculations; $r = 20R$, $\theta = 40^0$ (Color figure online)

The thicknesses of the shear layers at the jets boundaries were different in all presented cases causing different sound pressures levels seen in the figures. However the general form of the spectra looks very similar.

Calculations with the multioperators schemes were carried out also for the 3D underexpanded jets in the case of narrow rectangular nozzles. They were aimed at the direct numerical simulation of the screech effect previously considered in the case of 2D nozzles [9]. The obtained acoustics fields generated by the unsteady behavior of the shock cells were found to correlate well with the 2D results. The calculated spectra with the main peaks close to the experimental ones looks very similar to those presented in [9].

4.2   Compact and Multioperators Schemes with Immersed Boundary Method (IBM)

High approximation orders of multioperators-based schemes is entirely due to exact solution smoothness allowing to get high-order terms in the corresponding Taylor expansion series. Thus smooth meshes are needed to provide peak performances of the schemes. Having in mind possible complexities when constructing smooth meshes for complex geometries, the IBM offers very good opportunity for using compact and multioperators-based schemes with the Cartesian coordinates when solving the Navier-Stokes equations. Skipping the extensive relevant literature, its earlier formulations were presented for example in [8].

In the present study, the direct forcing version of the IBM was applied to the compressible Navier-Stokes equations. The idea behind this is to get at least the second-order accurate solutions inside boundary layers at solid (in general moving) boundaries and highly accurate solutions away from them using non-adaptive to solid boundaries smooth meshes (for example, the Cartesian ones).

The 16th-order multioperators were used to approximate the inviscid terms of the compressible Navier-Stokes equations written for the Cartesian coordinates.

Having in mind low-order representation of the boundary forcing terms, the viscous terms were discretized via the second-order centered differences.

In the present study, calculations for Mach number $M = 0.2$ were carried out using non-local high-order approximations (5th-order compact and 16h-order multioperators-based) with the sweeps along the Cartesian coordinate lines, the presence of the solid body being modelled by the forcing terms only. Bilinear or Radial Basis Functions (RBF) interpolants were used to define the values of the dependent variables at the near-boundary points outside or inside the cylinder. Several meshes were used to verify the mesh-convergence and good agreements with existing experimental and numerical data for low Reynolds data.

In the near-resolved case $Re \leq .400$, the calculation clearly showed the Von Karman vortex streets behind t5he cylinder. Figure 8 presents the vorticity field for $Re = 400$ while Fig. 9 shows the acoustic spectrum for that case.

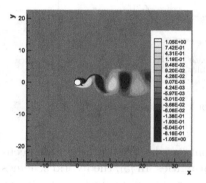

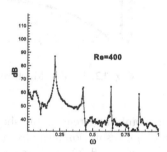

Fig. 8. $Re = 400$: snapshot of the vorticity field.

Fig. 9. $Re = 400$: acoustic spectrum.

Figure 10 shows good agreement between calculated Strouhal numbers and various numerical and experimental data. The acoustic spectrum in the case of underresolved boundary layer ($Re = 10^8$) is displayed in Fig. 11.

4.3 Using the 32nd- and 16th-order Multioperators in the Case of Strong Discontinuities

The conservative property of the multioperators-based schemes with dissipation mechanisms allows one to use them in the case of relatively small Mach numbers supersonic flows. In those cases, high-order dissipation mechanisms can be sufficient to suppress spurious oscillations. An example is the results of numerical simulations of underexpanded jets at $M \leq 1.5$ [9] with clear pictures of the screech waves. However monotonization devices are needed in the high Mach number cases. Following the well known ways, non-linear schemes can be constructed. The main aim in the multioperators context is to obtain numerical

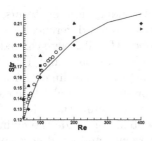

Fig. 10. Strouhal number vs. Reynolds number.

Fig. 11. $Re = 10^8$: acoustic spectrum.

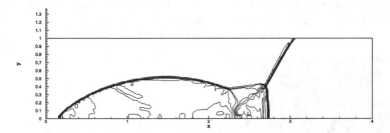

Fig. 12. Double Mach reflection problem; density contours.

solutions combining reasonable shocks and contacts descriptions and high accuracy and high resolution away from the discontinuities. In [5], the hybrid schemes with the 16th- and 32th-order multioperators are tested against 1D problems (discontinuous solutions of the Burgers equation, extremal Riemann problems). Extending testing calculations, the double Mach reflection problem [10] was considered. Figure 12 shows the calculated density field resulting from a shock front hitting a ramp which is inclined by 30°. The calculation were carried out using the 16th-order hybrid scheme from [5] and the setup described in [10] but with coarser mesh (the mesh sizes are $h_x = h_y = 1/60$).

References

1. Tolstykh, A.I.: Multioperator high-order compact upwind methods for CFD parallel calculations. In: Emerson, D.R., et al. (Eds.) Parallel Computational Fluid Dynamics, pp. 383–390. Elsevier, Amsterdam (1998)
2. Tolstykh, A.I.: Development of arbitrary-order multioperators-based schemes for parallel calculations. 1. Higher-than-fifth order approximations to convection terms. J. Comput. Phys. **225**, 2333–2353 (2007)
3. Tolstykh, A.I.: On the use of multioperators in the construction of high-order grid approximations. Comput. Math. Math. Phys. **56**(6), 932–946 (2016)
4. Tolstykh, A.I.: High Accuracy Compact and Multioperators Approximations for Partial Differential Equations. Nauka, Moscow (2015)

5. Tolstykh, A.I.: On 16th and 32th order multioperators-based schemes for smooth and discontinuous fluid dynamics solutions. Commun. Comput. Phys. **22**, 572–598 (1977)
6. Tam, C.K.W.: Problem 1-aliasing. In: Fourth Computational Aeroacoustics (CAA) Workshop on Benchmark Problems, NASA/CP-2004-2159 (2004)
7. Lesshafft, L., Huerre, P., Sagaut, P.: Frequency selection in globally unstable round jets. Phys. Fluids **19**, 054108-1–054108-10 (2007)
8. Lai, M.-C., Peskin, C.S.: An immersed boundary method with formal second-order accuracy and reduced numerical viscosity. J. Comput. Phys. **160**, 705–719 (2000)
9. Tolstykh, A.I., Shirobokov, D.A.: Fast calculations of screech using highly accurate multioperators-based schemes. J. Appl. Acoustics. **74**, 102–109 (2013)
10. Woodward, P., Colella, P.: The numerical simulation of two-dimensional fluid flow with strong shocks. J. Comput. Phys. **54**, 115–173 (1984)

GPU-Based Parallel Computations in Multicriterial Optimization

Victor Gergel$^{(\boxtimes)}$ and Evgeny Kozinov

Lobachevsky State University of Nizhni Novgorod, Nizhni Novgorod, Russia
gergel@unn.ru, evgeny.kozinov@itmm.unn.ru

Abstract. In the present paper, an efficient approach for solving the time-consuming multicriterial optimization problems, in which the optimality criteria could be the multiextremal ones and computing the criteria values could require a large amount of computations is proposed. The proposed approach is based on the reduction of the multicriterial problems to the scalar optimization ones with the use of the minimax convolution of the partial criteria, on the dimensionality reduction with the use of the Peano space-filling curves, and on the application of the efficient information-statistical global optimization methods. An additional application of the block multistep scheme provides the opportunity of the large-scale parallel computations with the use of the graphics processing units (GPUs) with thousands of computational cores. The results of the numerical experiments have demonstrated such an approach to allow improving the computational efficiency of solving the multicriterial optimization problems considerably – hundreds and thousands.

Keywords: Decision making · Multicriterial optimization
Global optimization · High performance computations
Dimensionality reduction · Criteria convolution
Global search algorithm · Computational costs

1 Introduction

The multicriterial optimization (MCO) problems are classified as the most general statements of the decision-making problems – the statement of MCO problems covers many classes of optimization problems, including unconstrained optimization, nonlinear programming, global optimization, etc. The opportunity to set several criteria is very useful in the formulating of the complex decision-making problems and is used in the applications widely. Such practical importance has caused a high activity of research in the field of the MCO problems. As a result of the performed investigations, a large number of the efficient methods of solving the MCO problems have been proposed and a great number of the applied problems have been solved – see, for example, the monographs [1–4] and the reviews of the scientific and practical results in the field [5,6].

The present paper is devoted to the solving of the MCO problems, which are used for formulating the decision-making problems in the design of complex

© Springer Nature Switzerland AG 2019
V. Voevodin and S. Sobolev (Eds.): RuSCDays 2018, CCIS 965, pp. 88–100, 2019.
https://doi.org/10.1007/978-3-030-05807-4_8

technical objects and systems. In such applications, the partial criteria could take a complex multiextremal form, and computing the values of the criteria could require a large amount of computations. In such conditions, finding even a single efficient decision requires a significant amount of computations whereas finding several decisions (or the complete set of these ones) becomes a problem of high computation costs.

Among the directions used for solving the MCO problems widely, the scalarization approach utilizing some methods of the partial criteria convolution to a single scalar criterion is applied – see, for example, [2,4,7]. Among such approaches, there are the methods of finding the decisions, which are the closest to the ideal one or to the compromised ones, or to the prototypes existing actually, etc. Among such algorithms, there exists the method of successive concessions, in which some tolerances to possible values of criteria are introduced. The scalarization of a vector criterion allows reducing the solving of a MCO problem to solving a series of the multiextremal optimization problems and, therefore, utilizing all existing highly efficient global search algorithms for the multicriterial optimization.

One of the promising approaches to solving the time-consuming global optimization problems consists in utilizing the graphics processing units (GPUs). At present, a GPU is a high-performance flexible programmable massive parallel processor, which can provide solving many complex computational problems [15]. However, the use of the GPU computational potential in the field of global optimization is quite limited. As a rule, GPUs are used for the parallelization of the algorithms, which are based on the random search concept in any way (see [16–18]). A review of this direction is given in [19].

Further structure of the paper is as follows. In Sect. 2, the multicriterial optimization problem statement is given and the basics of the developed approach are considered, namely the reduction of the multicriterial problems to the scalar optimization ones using the minimax convolution of the partial criteria and the dimensionality reduction using the Peano space-filling curves. In Sect. 3, the parallel global search algorithm for solving the reduced scalar optimization problems is described and the block multistep scheme of the dimensionality reduction, which provides the opportunity to use the GPUs with thousands of computational cores is presented. Section 4 includes the results of numerical experiments confirming the proposed approach to be a promising one. In Conclusion, the obtained results are discussed and main possible directions of further investigations are outlined.

2   Problem Statement

The multicriterial optimization (MCO) can be defined as follows:

$$f(y) = (f_1(y), f_2(y), \ldots, f_s(y)) \to min, y \in D, \tag{1}$$

where $y = (y_1, y_2, \ldots, y_N)$ is the vector of the varied parameters, N is the dimensionality of the multicriterial optimization problem being solved, $f(y)$ is the vector efficiency criterion, and D is the search domain representing an N-dimensional hyperparallelepiped

$$D = y \in R^N : a_i \leq y_i \leq b_i, 1 \leq i \leq N \qquad (2)$$

at given boundary vectors a and b.

Without loss of generality, the values of partial criteria in the problem (1) are supposed to be non-negative, and the decreasing of these ones corresponds to increasing efficiency of the decisions $y \in D$.

In the present work the problem (1) will be considered in application to the most complex decision-making problems, in which the partial criteria $f_i(y)$, $1 \leq i \leq s$ could be multiextremal, and obtaining the criteria values at the points of the search domain $y \in D$ could require a considerable amount of computations. Let us suppose also the partial criteria $f_i(y)$ to satisfy the Lipschitz condition

$$|f_i(y') - f_i(y'')| \leq L_i \|y' - y''\|, y', y'' \in D, 1 \leq i \leq s. \qquad (3)$$

where L_i is the Lipschitz constant for the functions $f_i(y)$, $1 \leq i \leq s$ and $\| * \|$ denotes the Euclidean norm in R^N.

The general approach to solving the MCO problem applied in the present work consists in the reduction of solving the MCO problems to the solving of a series of one-dimensional optimization problems:

$$\min \phi(x) = F(\lambda, y(x)), x \in [0,1], \qquad (4)$$

where

$$F(\lambda, y(x)) = \max\left((\lambda_i f_i(y(x)), 1 \leq i \leq s\right) \qquad (5)$$

is the minimax convolution of the partial criteria of the MCO problem with the use of the vector of the convolution coefficients

$$\lambda = (\lambda_1, \lambda_2, \ldots, \lambda_s) \in \Lambda \subset R^s : \sum_{i=1}^{s} \lambda_i = 1, \lambda_i \geq 0, 1 \leq i \leq s \qquad (6)$$

and $y(x)$ is a continuous and unambiguous mapping of the interval $[0,1]$ onto the N-dimensional search domain D – see, for example, [8–10].

3   GPU-Based Parallel Computations for Solving the Multicriterial Optimization Problems

The convolution of the partial criteria applied within the framework of the developed approach and the dimensionality reduction allow reducing the solving of the MCO problem (1) to solving a series of the reduced multiextremal problems (4). And, therefore, the problem of the development of the methods for solving the MCO problems is resolved by the opportunity of a wide use of the global search algorithms. The state of the art in the field of global optimization is presented comprehensively enough, for example, in [8,20,21].

In the present work, the global search algorithm developed within the framework of the information-statistical theory of the multiextremal optimization is proposed to use for solving the reduced problems (4). This theory served as a basis for the development of a large number of algorithms, which have been substantiated mathematically, have demonstrated high efficiency, and allowed solving many complex optimization problems in various fields of application [8,9,22–25].

The approach proposed for the organization of the parallel computations when solving the time-consuming multiextremal optimization problems is based on the simultaneous computing of the partial criteria values in the MCO problem (1) at several different points of the search domain D. Such an approach provides the parallelization of the most time-consuming part of the global search and is a general one – it can be applied for many global search methods in various global optimization problems.

3.1 Parallel Algorithm of Global Search for Finding the Efficient Decisions in the Multicriterial Optimization Problems

Within the framework of this approach, the multidimensional generalized parallel algorithm of global search (PAGS) for finding the efficient decisions of the multicriterial optimization problems constitutes the basis for the developed optimization methods. The general computational scheme of the algorithm can be described as follows [8–10].

Let p is the number of employed parallel computers (processors or cores) of a computational system with shared memory. The initial two iterations of the algorithm are performed at the boundaries of the interval $x^0 = 0$, $x^1 = 1$. Besides these boundary points, the algorithm should perform additional iterations at the points x^i, $1 < i \le p$, which can be defined a priori or computed by any auxiliary computational procedure. Then, let k, $k > p$ global search iterations have been completed, at each of which the computing of the value of the minimized function $\phi(x)$ from (4) (hereafter called a trial) has been performed. The choice of the points for trials performed within the next iteration in parallel is determined by the following rules.

Rule 1. Renumber the trials points of the completed search iterations by the lower indices in the order of increasing coordinate values

$$0 = x_0 < x_1 < \cdots < x_i < \cdots < x_k = 1. \tag{7}$$

Rule 2. Compute current estimate of the Hölder constant of the reduced function $\phi(x)$:

$$m = \begin{cases} rM, & M > 0 \\ 1, & M = 0 \end{cases}, M = \max_{1 \le i \le k} \frac{|z_i - z_{i-1}|}{\varrho_i} \tag{8}$$

where $z_i = \phi(x_i)$, $\varrho_i = \sqrt[N]{x_i - x_{i-1}}$, $1 \le i \le k$. The constant r, $r > 1$ is the *reliability parameter* of the algorithm.

Rule 3. Compute *the characteristic* $R(i)$ for each interval (x_{i-1}, x_i), $1 \le i \le k$ according to the expression

$$R(i) = \varrho_i + \frac{(z_i - z_{i-1})^2}{m^2 \varrho_i} - 2\frac{(z_i + z_{i-1})}{m}, 1 \le i \le k, \qquad (9)$$

Rule 4. Arrange the characteristics of the intervals (x_{i-1}, x_i), $1 \le i \le k$ obtained according to (9) in the decreasing order

$$R(t_1) \ge R(t_2) \ge \cdots \ge R(t_{k-1}) \ge R(t_k) \qquad (10)$$

and select p intervals with the indices t_j, $1 \le j \le p$ having the maximum values of the characteristics.

Rule 5. Perform new trials at the points x^{k+j}, $1 \le j \le p$ placed in the intervals with the maximum characteristics from (10) according to the expressions

$$x^{k+j} = \frac{x_{t_j} + x_{t_j-1}}{2} - sign(z_{t_j} - z_{t_j-1})\frac{[\frac{|z_{t_j} - z_{t_j-1}|}{m}]^N}{2r}, 1 \le j \le p. \qquad (11)$$

Stopping condition for the algorithm, according to which the execution of the algorithm is terminated, consists in checking the lengths of the intervals, in which the scheduled trials are performed, with respect to the required *accuracy* of the problem solution i.e.

$$\varrho_t \le \varepsilon, 1 \le t_j \le p. \qquad (12)$$

Various modifications of this algorithm and the corresponding theory of convergence are presented in [8,9].

3.2   Multilevel Decomposition of the Parallel Computations

Further development of the methods of the parallel computations in the multicriterial global optimization problems and, therefore, the expansion of the possible quantity of the employed processors/cores can be ensured by the use of one more dimensionality reduction method in the decomposition scheme of the MCO problems (1) – the multistep scheme of decomposition of the optimization problems [8,9,25,26]. According to this scheme, the solution of a multidimensional optimization problem can be obtained by solving a series of   nested   one-dimensional problems:

$$\min\{\phi(y) : y \in D\} = \min_{a_1 \le y_1 \le b_1} \min_{a_2 \le y_2 \le b_2} \cdots \min_{a_N \le y_N \le b_N} \phi(y). \qquad (13)$$

The original multistep reduction scheme (13) can be generalized for the use in combination with the dimensionality reduction scheme based on the Peano curves [27]. According to the generalized block multistep scheme, the vector of variables $y \in D$ of the global optimization problem (1) is considered as a set of the block variables

$$y = (y_1, y_2, \ldots, y_N) = (u_1, u_2, \ldots, u_M), \tag{14}$$

where the ith block variable u_i is a vector with the dimensionality N_i of the elements of the vector y taken sequentially i.e.

$$u_i = (y_{n_i+1}, y_{n_i+2}, \ldots, y_{n_i+N_i}),$$
$$n_0 = N_0 = 0, n_i = n_{i-1} + N_{i-1}, 1 \le i \le M \tag{15}$$

and $N_1 + N_2 + \cdots + N_M = N$.

Using the new variables, main equation of the multistep reduction scheme (13) can be rewritten in the form

$$\min\{\phi(y) : y \in D\} = \min_{u_1 \in D_1} \min_{u_2 \in D_2} \ldots \min_{u_M \in D_M} \varphi(y). \tag{16}$$

where the subdomains D_i, $1 \le i \le M$ are the projections of the initial search domain D onto the subspace corresponding the variables u_i, $1 \le i \le M$. As a result, in the generalized block multistep reduction scheme, the nested subproblems

$$\phi_i(u_1 \ldots u_i) = \min_{u_{i+1} \in D_{i+1}} \phi_{i+1}(u_1 \ldots u_i, u_{i+1}), 1 \le i \le M - 1 \tag{17}$$

are the multidimensional ones, and the dimensionality reduction method based on the Peano curves can be applied to solve these ones.

To provide the parallel computations in the block multistep reduction scheme, at each decomposition level one can generate several optimization problems simultaneously for the parallel solving of these ones [26,27] (see Fig. 1). The resulting set of problems to be solved in parallel can be controlled by means of the predefined parallelization vector

$$\pi = (\pi_1, \pi_2, \ldots, \pi_M), \tag{18}$$

where π_i, $1 \le i < M$ is the number of subproblems being solved in parallel at the $(i+1)^{th}$ level of decomposition arising as a result of performing the parallel iterations at the ith level. For the Mth level, the quantity π_M means the number of parallel trials in the course of minimization of the function

$$\phi_M(u_1, \ldots, u_M) = \phi(y_1, \ldots, y_N) \tag{19}$$

with respect to the variable u_M at fixed values $u_1, \ldots, u_{M-1}$, i.e. the number of values of the objective function $\phi(y)$ computed in parallel. Then, the total number of the employed processors/cores will be

$$\prod = 1 + \sum_{i=1}^{M-1} \prod_{j=1}^{i} \pi_j \tag{20}$$

The resulting multilevel scheme of parallel computations allows ensuring the efficient employment of all processors/nodes available in the high-performance systems with a large number of the computational nodes (including the ones with the distributed memory). The generation of a large number of optimization problems solved in parallel initiates a promising direction on a wide employment of GPUs with a large number of computational cores. This direction has been tested in solving the time-consuming global optimization problems [28–30]. In the present work, the possibility of utilizing the GPUs for solving the multicriterial optimization problems has been evaluated.

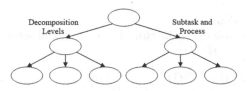

Fig. 1. General scheme of generating the parallel problems using the block multistep dimensionality reduction scheme

The PAGS algorithm combined with the block multistep scheme of dimensionality reduction will be called hereafter Generalized Parallel Algorithm of Global Search (GPAGS).

4   Results of Numerical Experiments

The numerical experiments have been carried out on the Lobachevsky supercomputer at State University of Nizhni Novgorod (operating system – CentOS 6.4, management system – SLURM). Each supercomputer node included 2 Intel Sandy Bridge E5-2660 2.2 GHz, 64 Gb RAM processors. The central processor units were the 8-core one (i. e. total 16 CPU cores per a node were available). At each node, three NVIDIA Kepler K20X GPUs were installed. To provide parallel computations MPI and CUDA technologies are applied.

The evaluation of the efficiency of the developed approach for solving the MCO problems without using the computational accelerators has already been performed earlier [11–14]. Let us consider, for example, the results of experiments from [13]. For comparison, a bicriterial test problem proposed in [31] was used:

$$f_1(y) = (y_1 - 1)y_2^2 + 1, f_2(y) = y_2, 0 \le y_1, y_2 \le 1. \tag{21}$$

As a solution of this MCO problem, the construction of a numerical approximation of the Pareto domain (PDA) was considered. For evaluating the quality of approximation, the completeness and the uniformity of the coverage of the Pareto domain were compared with the use of the following two indicators [13,31]:

- The *hypervolume index* (HV) defined as the volume of the subdomain of the values of the vector criterion $f(y)$ dominated by the points of the Pareto domain approximation. This indicator characterizes the completeness of the Pareto domain approximation (a higher value corresponds to more complete coverage of the Pareto domain).
- The *distribution uniformity* index (DU) of the points from the Pareto domain approximation. This indicator characterizes the uniformity of coverage of the Pareto domain (a lower value corresponds to more uniform coverage of the Pareto domain).

Within the described experiment, five multicriterial optimization algorithms were compared: the Monte-Carlo (MC) method, the genetic algorithm SEMO from the PISA library [5,32], the Non-uniform coverage (NUC) method [5], the bi-objective Lipschitz optimization (BLO) method proposed in [32], and the serial variant of the GPAGS algorithm proposed in the present paper.

Total for the GPAGS algorithm 50 subproblems were solved at various values of the convolution coefficients λ distributed in Λ uniformly. The results of experiments from [13] are presented in Table 1.

Table 1. Results of numerical experiments from [13] for the test problem (21)

Method	Iterations	PDA points	HV	DU
MC	500	67	0.300	1.277
SEMO	500	104	0.312	1.116
NUC	515	29	0.306	0.210
BLO	498	68	0.308	0.175
GPAGS	370	100	0.316	0.101

The results of performed experiments have demonstrated the GPAGS algorithm to have a considerable advantage as compared to considered multicriterial optimization methods even in solving relatively simple MCO problems.

In the present paper, the numerical experiments were conducted in order to evaluate the efficiency of the developed approach in solving the MCO problems with the use of the GPUs. In the conducted series of experiments, the solving of the bicriterial six-dimensional MCO problems (i.e. $N = 6$, $s = 2$) has been performed. As the test problem criteria, the multiextremal functions obtained with the use of the GKLS generator [33] were used.

In the course of experiments, 50 multicriterial problems of this class have been solved. In each problem, the search of the Pareto-optimal decisions was performed for 10 convolution coefficients λ from (4) distributed in Λ uniformly (i.e. 500 global optimization subproblems have been solved). In solving the problems, two levels of the block dimensionality reduction scheme were used. At the first level of the reduction scheme, the optimization with respect to the first two variables was performed (i.e. $u_1 = (y_1, y_2)$, $N_1 = 2$). The optimization with respect to the rest variables was performed at the second decomposition level (i.e. $u_2 = (y_3, y_4, y_5, y_6)$, $N_2 = 4$). Trials at the second decomposition level are executed on GPU. Computations are implemented in accordance with the master/slave scheme which is not required any data communication between GPUs. Trial points are sent by CPU just before every iteration and are stored in the GPU global memory.

As the parameters the accuracy of the method $\varepsilon = 0.025$ and the reliability of the method $r = 6.5$ for the first level of the block reduction scheme and $r = 4.5$ for the second level of decomposition were used. The results of the numerical experiments are presented in Table 2.

Table 2. Comparison of the times of solving of the six-dimensional bicriterial MCO problems

Nodes	P	Th	P*Th	Core type	Time, s	Speedup
1	1	16	16	CPU	7 186.4	1.0
16	16	16	256	CPU	957.3	7.5
16	32	4 032	129 024	GPU	529.9	13.6
16	64	2 016	129 024	GPU	291.8	24.6
16	128	1 008	129 024	GPU	272.4	26.4
32	128	2 016	258 048	GPU	214.9	33.4
32	256	1 008	258 048	GPU	253.2	28.4

In Table 2, the column "Nodes" shows the number of the supercomputer nodes employed, the column P shows the number of the parallel subproblems generated at the first level of the block reduction scheme (the parameter π_1 from (18)), the column Th shows the number of generated points of trials performed in parallel at the second decomposition level (the parameter π_2 from (18)).

The results of experiments presented in the first row show the averaged time of solving of the MCO problems with the use a single computational node of the supercomputer ($\pi_1 = 1$, $\pi_2 = 16$). In the second row of the table, the averaged time of solving of the MCO problems with the use of sixteen computational nodes of the supercomputer ($\pi_1 = 16$, $\pi_2 = 16$) is given. As follows from the presented results, the resulting speedup of the computations was 7.5 times. In the rows 3–7, the averaged times of solving the MCO problems with the use of 16 and 32 supercomputer nodes are given. At each node, 3 GPUs were employed. The number of parallel subproblems generated at the first level of the block reduction scheme (the parameter π_1 from (18)) ranged from 32 to 256 whereas the number of generated points of trials performed in parallel at the second decomposition level (the parameter π_2 from (18)) ranged from 1008 to 4032. Total number of employed GPU cores was 129024 with using 16 nodes, and 258048 with using 32 nodes. The maximal speedup of computations achieved was 33.4 times.

It is worth noting that the speedup of computations in the time depends on the time of computing the values of the criteria of the MCO problem being solved. This time is relatively small in the test optimization problems, but it can be essential when solving the applied problems in various scientific and technical applications. As a result, along with the evaluation of the achieved speedup of computations in the time, it is reasonable to evaluate the speedup of computations with respect to the reduction of the number of iterations performed in the course of computations. The results of experiments performed to evaluate such speedup are presented in Table 3.

Table 3. Comparison of the number of iterations when solving the six-dimensional bicriterial MCO problems

Nodes	P	Th	P*Th	Core type	Iterations	Speedup
1	1	16	16	CPU	12 279 179.8	1.0
16	16	16	256	CPU	808 858.8	15.2
16	32	4 032	129 024	GPU	3 086.5	3 978.4
16	64	2 016	129 024	GPU	2 426.9	5 059.6
16	128	1 008	129 024	GPU	2 910.2	4 219.4
32	128	2 016	258 048	GPU	1 581.5	7 764.4
32	256	1 008	258 048	GPU	2307.5	5 321.4

The results of experiments presented above demonstrate the speedup of computations with respect to the reduction of the number of performed global search iterations to be considerable. Thus, when employing 16 computer nodes, the speedup can be more than 5000 times whereas when employing 32 computer nodes – more than 7700 times.

5   Conclusion

In the present paper, an efficient approach for solving the complex multicriterial optimization problems, in which the criteria of optimality can be the multiextremal ones, and the computing of the criteria values can require a large amount of computations has been proposed. The proposed approach is based on the reduction of the multicriterial problems to the scalar optimization ones by means of the minimax convolution of the partial criteria, the dimensionality reduction with the use of Peano space-filling curves, and the application of the efficient information-statistical methods of global optimization.

The opportunity of the large-scale parallel computations is provided by application of the block multistep dimensionality reduction scheme and by the use of the GPUs with many thousands computational cores. The results of numerical experiments have shown the developed approach to allow reducing the computational costs of solving the multicriterial optimization problems considerably – by hundreds and thousands times.

The results of the performed experiments have demonstrated the developed approach to be a promising one and to require further investigations. First of all, it is necessary to continue carrying out the numerical experiments on solving the multicriterial optimization problems at larger number of the partial criteria of efficiency and for larger dimensionality of the optimization problems being solved.

Acknowledgements. This research was supported by the Russian Science Foundation, project No 16-11-10150 "Novel efficient methods and software tools for time-consuming decision-making problems using supercomputers of superior performance".

References

1. Marler, R.T., Arora, J.S.: Multi-Objective Optimization: Concepts and Methods for Engineering. VDM Verlag, Saarbrücken (2009)
2. Ehrgott, M.: Multicriteria Optimization. Springer, Heidelber (2005). https://doi.org/10.1007/3-540-27659-9. (2nd ed., 2010)
3. Collette, Y., Siarry, P.: Multiobjective Optimization: Principles and Case Studies (Decision Engineering). Springer, Heidelberg (2011)
4. Pardalos, P.M., Žilinskas, A., Žilinskas, J.: Non-Convex Multi-Objective Optimization. Springer, Cham (2017). https://doi.org/10.1007/978-3-319-61007-8
5. Hillermeier, C., Jahn, J.: Multiobjective optimization: survey of methods and industrial applications. Surv. Math. Ind. **11**, 1–42 (2005)
6. Cho, J.-H., Wang, Y., Chen, I.-R., Chan, K.S., Swami, A.: A survey on modeling and optimizing multi-objective systems. IEEE Commun. Surv. Tutor. **19**(3), 1867–1901 (2017)
7. Eichfelder, G.: Scalarizations for adaptively solving multi-objective optimization problems. Comput. Optim. Appl. **44**, 249–273 (2009)
8. Strongin, R., Sergeyev, Y.: Global Optimization with Non-convex Constraints. Sequential and Parallel Algorithms. Kluwer Academic Publishers, Dordrecht (2000). (2nd ed. 2013, 3rd ed. 2014)
9. Strongin, R., Gergel, V., Grishagin, V., Barkalov, K.: Parallel computations for global optimization problems, Moscow State University Press (2013). (in Russian)
10. Sergeyev, Y.D., Strongin, R.G., Lera, D.: Introduction to Global Optimization Exploiting Space-filling Curves. Springer, New York (2013). https://doi.org/10.1007/978-1-4614-8042-6
11. Gergel, V.P., Kozinov, E.A.: Accelerating parallel multicriterial optimization methods based on intensive using of search information. Procedia Comput. Sci. **108**, 1463–1472 (2017)
12. Gergel, V., Kozinov, E.: Parallel computing for time-consuming multicriterial optimization problems. In: Malyshkin, V. (ed.) PaCT 2017. LNCS, vol. 10421, pp. 446–458. Springer, Cham (2017). https://doi.org/10.1007/978-3-319-62932-2_43
13. Gergel, V., Kozinov, E.: Efficient methods of multicriterial optimization based on the intensive use of search information. In: Kalyagin, V., Nikolaev, A., Pardalos, P., Prokopyev, O. (eds.) NET 2016. PROMS, vol. 197, pp. 27–45. Springer, Cham (2017). https://doi.org/10.1007/978-3-319-56829-4_3

14. Gergel, V., Kozinov, E.: An approach for parallel solving the multicriterial optimization problems with non-convex constraints. In: Voevodin. V., Sobolev, S. (eds.) RuSCDays 2017. CCIS, vol. 793, pp. 121–135. Springer, Cham (2017). https://doi.org/10.1007/978-3-319-71255-0_10

15. Cai, Y., See, S. (eds.): GPU Computing and Applications. Springer, Singapore (2015). https://doi.org/10.1007/978-981-287-134-3

16. Ferreiro, A.M., Garcia, J.A., Lopez-Salas, J.G., Vazquez, C.: An efficient implementation of parallel simulated annealing algorithm in GPUs. J. Glob. Optim. **57**(3), 863–890 (2013)

17. Zhu, W.: Massively parallel differential evolution–pattern search optimization with graphics hardware acceleration: an investigation on bound constrained optimization problems. J. Glob. Optim. **50**(3), 417–437 (2011)

18. Garcia-Martinez, J.M., Garzon, E.M., Ortigosa, P.M.: A GPU implementation of a hybrid evolutionary algorithm: GPuEGO. J. Supercomput (2014). https://doi.org/10.1007/s11227-014-1136-7

19. Langdon, W.B.: Graphics processing units and genetic programming: an overview. Soft. Comput. **15**(8), 1657–1669 (2011)

20. Locatelli, M., Schoen, F.: Global Optimization: Theory, Algorithms, and Applications. SIAM (2013)

21. Floudas, C.A., Pardalos, M.P.: Recent Advances in Global Optimization. Princeton University Press, Princeton (2016)

22. Gergel, V.P., Strongin, R.G.: Parallel computing for globally optimal decision making. In: Malyshkin, V.E. (ed.) PaCT 2003. LNCS, vol. 2763. pp. 76–88. Springer, Heidelberg (2003). https://doi.org/10.1007/978-3-540-45145-7_7

23. Gergel, V.P., Kuzmin, M.I., Solovyov, N.A., Grishagin, V.A.: Recognition of surface defects of cold-rolling sheets based on method of localities. Int. Rev. Autom. Control. **8**(1), 51–55 (2015)

24. Modorskii, V.Y., Gaynutdinova, D.F., Gergel, V.P., Barkalov, K.A.: Optimization in design of scientific products for purposes of cavitation problems. In: AIP Conference Proceedings, vol. 1738, p. 400013 (2016). https://doi.org/10.1063/1.4952201

25. Grishagin, V., Israfilov, R., Sergeyev, Y.: Convergence conditions and numerical comparison of global optimization methods based on dimensionality reduction schemes. Appl. Math. Comput. **318**, 270–280 (2018). https://doi.org/10.1016/j.amc.2017.06.036

26. Sergeyev, Y., Grishagin, V.: Parallel asynchronous global search and the nested optimization scheme. J. Comput. Anal. Appl. **3**(2), 123–145 (2001)

27. Barkalov, K.A., Gergel, V.P.: Multilevel scheme of dimensionality reduction for parallel global search algorithms. In: Proceedings of the 1st International Conference on Engineering and Applied Sciences Optimization, pp. 2111–2124 (2014)

28. Gergel, V., Lebedev, I.: Heterogeneous parallel computations for solving global optimization problems. Procedia Comput. Sci. **66**, 53–62 (2015)

29. Gergel, V., Sidorov, S.: A two-level parallel global search algorithm for solution of computationally intensive multiextremal optimization problems. In: Malyshkin, V. (ed.) PaCT 2015. LNCS, vol. 9251, pp. 505–515. Springer, Cham (2015). https://doi.org/10.1007/978-3-319-21909-7_49

30. Gergel, V.: An unified approach to use of coprocessors of various types for solving global optimization problems. In: 2nd International Conference on Mathematics and Computers in Sciences and in Industry, MCSI, vol. 7423935, pp. 13–18 (2016)

31. Evtushenko, Y.G., Posypkin, M.A.: A deterministic algorithm for global multiobjective optimization. Optim. Methods Softw. **29**(5), 1005–1019 (2014)

32. Žilinskas, A., Žilinskas, J.: Adaptation of a one-step worst-case optimal univariate algorithm of bi-objective Lipschitz optimization to multidimensional problems. Commun. Nonlinear Sci. Numer. Simul. **21**, 89–98 (2015)
33. Gaviano, M., Kvasov, D.E., Lera, D., Sergeyev, Y.D.: Software for generation of classes of test functions with known local and global minima for global optimization. ACM Trans. Math. Softw. **29**(4), 469–480 (2003)

LRnLA Algorithm ConeFold with Non-local Vectorization for LBM Implementation

Anastasia Perepelkina$^{(\boxtimes)}$ and Vadim Levchenko

Keldysh Institute of Applied Mathematics RAS, Moscow, Russia
mogmi@narod.ru, lev@keldysh.ru

Abstract. We have achieved a $\sim$0.3 GLUps performance on a 4 core CPU for the D3Q19 Lattice Boltzmann method by taking an advanced time-space decomposition approach. The LRnLA algorithm ConeFold was used with a new non-local mirrored vectorization. The roofline model was used for the performance estimation and parameter choice. There are many expansion possibilities, so the developed kernel may become a foundation for more complex LBM variations.

Keywords: Lattice Boltzmann method · LRnLA algorithms
Parallel computation

1 Introduction

One of the reasons for the popularity of the Lattice Boltzmann Method (LBM) [17] for Computational Fluid Dynamics (CFD) is the ease of its efficient computer implementation. However, some issues exist. LBM implementations remain memory-bound, and the vectorization is complicated since misaligned writes or reads require significant overhead. To try to reach the maximum performance efficiency authors vary the data storage method, data layout, propagation (streaming) algorithms. The efficient data synchronization for massively parallel implementations is also in a high demand since CFD problems are radically multiscale.

GPU makes the aforementioned issues more prominent. However, it provides better performance results than CPU in many cases. At the same time, CPU codes remain relevant. CPU codes are more flexible for multiphysics frameworks, as we see in the famous packages such as waLBerla [3] and OpenLB [5]. For large supercomputer simulations CPU has more memory so that the potentially bigger problems may be solved. And since the GPU computers are essentially heterogeneous, efforts are made to offload some work to the CPU kernels [15].

The aim of this work is to break the performance records of CPU implementation with the use of LRnLA algorithms [7]. LRnLA algorithms may be seen as an advanced temporal blocking method. Temporal blocking was previously applied to LBM [4,12]. Indeed, it seems to grant better parallelization

© Springer Nature Switzerland AG 2019
V. Voevodin and S. Sobolev (Eds.): RuSCDays 2018, CCIS 965, pp. 101–113, 2019.
https://doi.org/10.1007/978-3-030-05807-4_9

efficiency [19]. For GPU, it was used for host-device and intra-device communications [16]. While the apparent similarity in using the space-time parallelism exists, LRnLA method is different. Its base lies in analyzing the dependency and influence conoids in the dependency graph, and the optimization is conducted with the account for memory and parallelism hierarchy of the computer. In fact, we disagree with [12] that the blocking method that is presented there is most efficient, and with the idea that 3D blocking is undesired.

In this paper, we show how the LRnLA algorithm ConeFold is built and implemented for LBM on CPU, how its performance may be estimated with the use of the roofline model. This implementation actually gives the performance per node record that surpasses every CPU result we found in the published work.

2   Lattice Boltzmann Method

In LBM, the simulation domain is split into $Nx \times Ny \times Nz$ cubic cells. In each cell, the probability distribution function is known for a set of discrete velocities $\mathbf{c}_{ijk}$. The specific method is denoted by a word like D3Q19, where the first number is the dimensionality of the model and the second number is the number of velocities. Discrete velocities are chosen as vectors that point from the center of the cell to the centers of its neighbors, and a zero velocity. In D3Q27, there is a set of vectors that point to each cell in a $3 \times 3 \times 3$ cube. In D3Q19, the longest vectors of D3Q27 are pruned.

For each velocity the update rule for its Distribution Function (DF) is split into two sub-steps: streaming $f_{ijk}(\mathbf{r}_{ijk}, t + \Delta t) \leftarrow f_{ijk}(\mathbf{r}_{000}, t)$, and collision $f_{ijk} \leftarrow f_{ijk} - (f_{ijk} - f_{ijk}^{eq})/\tau$; $i,\, j,\, k = -1, 0, 1$ for D3Q27. Streaming copies the f_{ijk} from cell with coordinates $\mathbf{r}_{000}$ to the cell with the relative position $\mathbf{r}_{ijk} = \mathbf{r}_{000} + \mathbf{c}_{ijk}\Delta t$, $\mathbf{c}_{ijk} = (i, j, k)$. The collision operates with the DF in the same spatial coordinates. The expression for the equilibrium DF $f_{ijk}^{eq}(\rho, \mathbf{u})$ $(\rho = \sum f_{ijk}$, $\mathbf{u} = \sum \mathbf{c}_{ijk} f_{ijk})$ is taken as the most commonly used second-order polynomial in $\mathbf{u}$ [17] to make the performance comparison easier, but any expression that operates on the data inside one LBM cell may be used in the current implementation.

3   ConeFold Algorithm

3.1   Algorithm as a Decomposition of a Dependency Graph

The core idea in the LRnLA method revolves around the existence of the influence and dependency region in space-time for each cell [8]. The dependencies in an LBM stencil fill a cube, so the data in one cell is influenced by a 4D pyramid in space-time. The whole simulation region may be decomposed in such regions.

That is why we illustrate algorithms as shapes in a dependency graph space with a subdivision rule [9]. A dependency graph consists of nodes (operations) and directed links (data dependencies between operations). A shape covers some nodes, so an algorithm described by this shape should perform all the operations of these nodes. It may be subdivided by planes with one restriction: if there are dependencies directed from one side of the plane to the other, there should not be any that are directed backward. It ensures that if one part is influenced by the other, there is no backward dependency. This way the algorithm is decomposed into several sub-steps that should be processed in a sequence, that is determined by the direction of the dependencies. If there are no dependencies between the parts, the algorithms may be processed asynchronously. The decomposition continues until the shapes cover only one node. Thus, this definition of the algorithm is recursive, and leads to the existence of non-local asynchronous elements.

For example, the algorithm of parallel implementation of LBM for N_t time steps on 2 nodes with 4 cores and SIMD support.

- Decompose $N_t \times Nx \times Ny \times Nz$ domain (i.e. the whole dependency graph) into N_t flat (in time) layers with size $Nx \times Ny \times Nz$. They have data dependencies pointing upwards, so they need to be processed in sequence. We call algorithms that start with this kind of decomposition 'stepwise' or 'orthodox'.
- Decompose the layers into two rectangles by a vertical (in time) plane $x = const$. They have no data dependencies between each other, so they may be processed in parallel by the two processors.
- Decompose the rectangles into two smaller rectangles by vertical planes $y = const$. They have no data dependencies between each other, so they may be processed in parallel by the 4 cores.
- Decompose the rectangles into separate cells. These are traversed these in any loop sequence, which may be vectorized by the compiler.

This kind of implementation is far from optimal. The loop traversal over large blocks of data in the lowest level of decomposition cause the memory-bound limitation: at each time step all DF data should be loaded and stored in memory. At the multi-core stage cache conflicts may arise, their resolution may lead to overhead. The necessity of using two lattice copies for the propagation scheme arise here. For multi-node parallelism, the data exchange should be performed at each time step, and this becomes the bottleneck for the parallel scaling. The point where this algorithm went wrong is the first subdivision into flat time steps. There are plentiful possibilities of the dependency graph subdivision that lead to better data access locality and do not make the result incorrect.

In LRnLA, the choice of the subdivision planes comes from natural requirements. First, we have planes with $t = const$, the synchronization instants, where data may be visualized or analyzed in any other way. Second, we choose hyperplanes $Ct = x + const$, $Ct = y + const$ and $Ct = z + const$ where C is the discrete information propagation speed. Here, $C = \Delta x / \Delta t$.

The dependency graph for D1Q3 LBM with a 'swap' propagation scheme shown in Fig. 1(a). The subdivision planes in it surround an elementary ConeFold algorithm shape.

$Ct = -x + const$, $Ct = -y + const$, $Ct = -z + const$ hyperplanes may also be considered. This leads to a subdivision into pyramids, diamonds and other shapes to complement them in 4D space-time [7]. Otherwise, all simulation domain may be tiled by just one shape. It has many advantages, and the easier coding is just one of them.

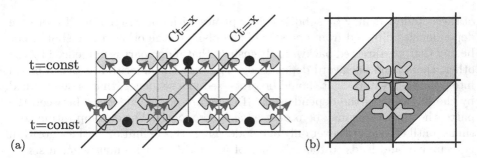

Fig. 1. (a) Dependency graph for LBM with 'swap' propagation scheme. Subdivision planes surround an elementary ConeFold (LRnLA cell). Operations are denoted by square markers. Red: non-local swap. Blue: local swap and collision.(b) The DFs that participate in a non-local swap in an LRnLA cell in D3Q27 case. The figure looks similar in x-y, y-z and z-x axis, the colored area is the CF projection (Color figure online)

3.2 Implementation of LBM with ConeFold

The ConeFold (CF) algorithm is implemented with recursive templates in C++.

1D Case. The level of recursive subdivision of a CF is parametrized by an integer which is called rank. One 1D CF with rank $r = R$ (denoted by CF$\langle$d $= 1, R\rangle$) is a function call of 4 CF$\langle 1, R - 1\rangle$ in the order which satisfies data dependencies (Fig. 2(a)). On rank $r = 0$ the CF is an elementary update according to the scheme stencil (Fig. 1(a)). It is a portion of the dependency graph that is defined as an LRnLA cell for CF.

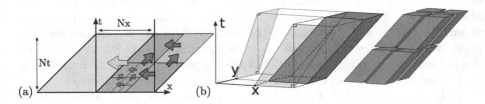

Fig. 2. (a) Two CF$\langle 1, \mathtt{MaxRank}\rangle$ cover the domain and are recursively decomposed into smaller CFs. Arrows show data dependencies. (b) ConeFold for d = 2.

If the top or bottom base is outside the computation region, the CF is specified as a right or left boundary CF. If $r > 0$ these call one CF inside the domain and two boundary CFs with $r - 1$ and a same type (left or right). If $r = 0$ the boundary condition of the scheme is applied. The periodic boundaries are not possible with this algorithm without additional techniques.

The computation is started with a right boundary $\mathtt{CF}\langle 1, \mathtt{MaxRank}\rangle$. Assuming the domain has $N_x = 2^{\mathtt{MaxRank}}$ cells, after it is finished, the cells evolved to the number of steps from 1 to $N_t = 2^{\mathtt{MaxRank}}$, in a linear progression from left to right. Only the rightmost cell has been updated N_t times. The left boundary $\mathtt{CF}\langle 1, \mathtt{MaxRank}\rangle$ is required to update all other cells to the same time step.

Recursive d-binary subdivision of the CF algorithm makes it natural to use a Morton Z-curve [10] for data storage. Thus, the data storage cells are organized in a recursive structure, and the indices to the neighboring cells are computed accordingly. Inside one CF, the pointer to the data in its bottom base, and array offset to the data in the projection of its top base are known. Thus, one $\mathtt{CF}\langle 1, 0\rangle$ has access to 2 data structure cells.

We need to find a LRnLA cell so that it uses the available data and homogeneously tiles the whole dependency graph of the domain evolution for $2^{\mathtt{MaxRank}}$ steps, with a possible exception for boundary conditions.

For example, in D1Q3 the following variation of the propagation scheme would suffice (Fig. 1(a)). Each data storage cell contains the data for f_{-1}, f_0, f_1, there is access to the two adjacent cells: c_0 and c_1. The update rule is:

1. swap f_1 of cell c_0 and f_{-1} of the cell c_1 to the right of it (non-local swap);
2. swap f_1 and f_{-1} of c_1 (local swap).
3. collision in one cell c_1.

Note that before this $\mathtt{CF}\langle 1, 0\rangle$ is performed, the f_1 value of c_1 would contain f_{-1} from the cell c_2 to the right of it. This is because the $\mathtt{CF}\langle 1, 0\rangle$ had been already performed for cells c_1 and c_2, and its step 1 had swapped f_1 of c_1 and f_{-1} of c_2. This is why c_1 contains the required post-stream data. The local swap may be merged with the collision.

2D, 3D Case. The d-dimensional algorithm is constructed by a direct product of 1D algorithms (Fig. 2(b)). There are the following changes.

– $\mathtt{CF}\langle d, R\rangle$ is subdivided into 2^{d+1} CFs with R-1.
– There are 3^d types of boundary CFs. It is necessary to specify CFs with $r = 0$ and with $r > 0$ for all cases of boundary: faces, edges, corners. Obviously, a code generator is used for this task.
– $(3^d - 1)$ boundary CFs of the maximum rank are required to progress from one synchronization instant to the next one.
– The $\mathtt{CF}\langle d, \mathtt{MaxRank}\rangle$ base projection covers $2^{d \cdot \mathtt{MaxRank}}$ cells.
– The top base is shifted from the bottom one by $2^{\mathtt{MaxRank}}$ cells in d directions.

Note that the dimensionality of the CF is independent of the dimensionality of the model. If $d = 2$ for D3Q27 scheme, one LRnLA cell would be redefined to include the update for Nz LBM cells. In one data storage cell there are 27 N_z sized arrays for f_{ijk}.

Other Stencils. The other possible reason to redefine an LRnLA cell is the existence of more extended dependencies since one $CF\langle d, 0\rangle$ has access only to 2^d cells. For example, the LRnLA cell for D1Q5 would update 2 full sets of distribution functions:

- swap f_1 with f_{-1} and f_2 with f_{-2} in c_0 and in c_1 (local swap);
- perform the collision in c_0, c_1;
- swap f_2 from c_0 with f_{-2} of c_1, f_2 from c_1 with f_{-2} of c_2;
- swap f_{-2} of c_1 with f_2 in c_1 and in c_2;
- swap f_1 from c_1 with f_{-1} of c_2, f_1 from c_2 with f_{-1} of c_3;
- swap f_2 from c_1 with f_{-2} of c_2, f_2 from c_2 with f_{-2} of c_3;

Here, local swap for all cells is required in initialization and in output. Some steps may be merged in optimization.

The velocities that are swapped in a 3D case of D3Q27 $CF\langle 3, 0\rangle$ are depicted in Fig. 1(b). D3Q19, D3Q15, D3Q7 are devised from this pattern by pruning.

For clarity, we repeat the three terms for cells that are used here:

- LBM cell is the lattice node with the corresponding DFs;
- LRnLA cell is the portion of the dependency graph and consists of a number of operations for an elementary update;
- data storage cell is a set of variables that are organized as an element of a data structure that is used.

Non-local Vectorization. There are several possibilities of vectorization in CF.

1. 2D CF may be used for 3D computation, as in [13]. In each cell, $27N_z$ DFs would be stored in a SoA manner, and in one $CF\langle 2, 0\rangle$ they would be processed in a loop, that may be vectorized by a compiler or manually. The fact that one axis is detached from the LRnLA decomposition becomes the main issue here, since it brings back the problems of stepwise algorithms to this dimension.
2. The cell of the data structure may be redefined to contain $2 \times 2 \times 2$ LBM cell data. The 8 values for each DF in it would be collected in one SIMD vector. This vectorization is local, but requires many reshuffle operations, especially in the collision step. It may be useful to keep this method in mind for possible GPU implementation, but for CPU the performance is limited by the overhead.

In the first case the automatic vectorization is local as well, and the misalignment issues need to be resolved. The non-local vectorization may be implemented in a way that requires less overhead. For vectors of length 4, the cell data for cells $\{i, i + Nz/4, i + 2Nz/4, i + 3Nz/4\}$ with $i = 0, 1, ..Nz/4 - 1$ are combined into vectors manually. This way, the domain is folded 4 times, the reshuffle is only required on the folds (Fig. 3).

Here a third method is proposed, which results in a non-local vectorization for 3D CF, and, moreover, fixes the issue of the impossibility of periodic boundaries.

For the explanation of the basic idea let us assume SIMD vector length 2 in D1Q3. Take a domain with $2 \cdot 2^{\text{MaxRank}}$ cells. Pack the data from cell $(2^{\text{MaxRank}} - 1 - i)$ with the data from cell $(2^{\text{MaxRank}} + i)$ for $i = 0, 1, .., (2^{\text{MaxRank}} - 1)$ into one LRnLA cell. Combine all pairs of values into SIMD vector. Then, construct a CF for the 2^{MR} cells of the resulting data structure. The code for LBM inside the domain stays the same but operates with SIMD vectors instead of scalars.

As a result, for the second value of the pairs, the computation is the same as for the non-vectorized case. For the first value of the pairs, the computation is mirrored. It has no impact on the streaming step. In the collision step, we need to change the sign of the directed macro-values, such as $\mathbf{u}$. The constant SIMD vector $\{-1, 1\}$ is used as a coefficient in the scheme.

Thus, f_1 in the left half of the domain represents propagation to the left. f_1 in the right half of the domain represents propagation to the right.

The right boundary $\text{CF}\langle 3, 0 \rangle$ links the cell at $(2^{\text{MaxRank}} - 1)$ with the cell at 2^{MaxRank}. The swap is performed between two values in SIMD vector for f_1. The left boundary $\text{CF}\langle 3, 0 \rangle$ links the cell 0 with cell $(2 \cdot 2^{\text{MaxRank}} - 1)$ and the swap is performed between values in f_{-1}.

Thus, this is a non-local mirrored vectorization (Fig. 3).

To enable the use of SIMD vectors with size 4 or 8 the domain is either mirrored in another axis, or has more reflected copies along the same axis. Here we present simulation with SIMD length 8 for single precision. With the use of AVX512 the same may be done for double precision.

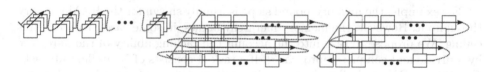

Fig. 3. Local, non-local, and non-local mirrored vectorization. Arrow shows the coordinate axis direction, blue segment shows SIMD vector pack. Vector length is 4. (Color figure online)

Parallel Algorithm. The TLP parallelization with CF is implemented with the TorreFold LRnLA algorithm [14]. We define an integer parameter nLArank (non-Locally Asynchronous rank). The TorreFold shape is similar to the shape of the CF with $r = \text{MaxRank}$, but the decomposition rule is different (Fig. 4). It is decomposed into $2^{(d+1)\text{nLArank}}$ CFs. Some of these are independent and may be processed by different threads. The asynchronous CFs are shown in a 1D case in Fig. 4. In 3D there is also asynchrony in an x–y–z–t diagonal cross-section.

The specific implementation may differ, the following is used in the current code. The CFs which stand on top of each other are collected in a tower (Cone-Torre). All ConeTorres are distributed between threads in the Z-curve order. The dependencies between them are ensured by semaphores. Each tower has d

semaphores. Initially, all semaphores are locked. The thread is assigned to some ConeTorre, and it processes CFs with $r = \text{MR} - \text{nLArank}$ in it one by one. Before starting a CF, the thread waits for one semaphore in each of the towers that are influencing it. After a thread finishes one CF it unlocks all its semaphores.

Some ConeTorres start outside of the domain, but the CFs that are outside are just skipped.

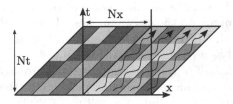

Fig. 4. TorreFold algorithm in 1D. Same color CFs may be processed in parallel. Threads (curled lines) process CFs in one ConeTorre. (Color figure online)

3.3   Roofline Model

The roofline model [18] helps to analyze bottlenecks in the implementation on a given computer. It does not account for all possible code capabilities and hardware limitations, but still, it is valuable for its simplicity.

For example, the memory-bound slope is usually shown for the RAM memory throughput. However, the memory hierarchy allows, in some cases, to break this ceiling. We propose to take into the account the caching ability of the hardware by a divide-and-conquer approach [8]. If a task A consists of N similar sub-tasks $B1$, $B2$, ..., BN, and all task A data is small enough to fit some level of cache, then the memory bound limit of tasks B is determined by the throughput of a higher level of cache. However, if we assume that each sub-task is carried out one-by-one, the load on the memory throughput is not less than the sum of the data required by each B task individually.

This conforms to the recursive definition of the algorithm as a decomposition of a task into subtasks. For a given algorithm we need to estimate the arithmetic intensity and the amount of data, that needs to be loaded into the cache.

The $\text{CF}\langle d, r\rangle$ consists of $2^{(d+1)r}$ LRnLA cells, so it has $O(r, d) = o2^{(d+1)r}$ operations, where o is the number of operations in an LRnLA cell. During its execution, the total amount of data loaded is $L(r, d) = N^d + N((N + 1)^d - N^d)$ data storage cells, where $N = 2^r$. The amount of data stored is $S(r, d) = (N - 1)^d + N \cdot N_T(N^d - (N - 1)^d)$.

The arithmetic intensity is $O(r, d)/(L(r, d) + S(r, d))$. The cached data size is $L(r, d)$.

In the presented code the recursive subdivision is as follows:

- a 4D cube that covers $2^{3\text{MaxRank}}$ LBM cells for 2^{MaxRank} updates is decomposed into $2^{3(\text{nLArank}+1)}$ ConeTorres;

– ConeTorres are decomposed into $2^{\texttt{nLArank}}$ CFs with rank $r^* = \texttt{MaxRank} - \texttt{nLArank}$;
– CFs are recursively decomposed into the same shapes of smaller rank until the LRnLA cell is reached.

In this estimation, the boundary effects are not considered.

The Roofline for our implementation on Intel Core i5-6400 is shown in Fig. 5. The arithmetic intensity increases with r. However, the algorithm performance is limited by the roofline of all its smaller parts. Several arrows from the higher levels of decomposition to the lower levels are plotted one by one from right to left. If an arrow reaches memory bound slope, arrows to the left of it are not allowed to be higher than this arrow. The color of the arrow shows the color of the roofline that defines its memory throughput limit.

$\texttt{nLArank}$ is set equal to 4. This choice is determined by the roofline: the advantages of the $(d+1)$-binary recursive subdivision are evident only for ranks smaller than $\texttt{MaxRank} - 4 = 4$.

With this model, we can estimate the disadvantage of the lower dimensionality of LRnLA decomposition. If a $\texttt{CF}\langle 2, 0 \rangle$ contains N_z cell updates, it, and all the CFs of higher rank, require more memory, and may not be localized in higher levels of cache. For large N_z the arrows for $R = 0, 1, 2$ would all be limited by the RAM roofline. This goes in contrast with the guidelines from [12], and also suggests that the non-local vectorization without mirroring the domain decreases the efficiency.

The local vectorization with $d = 3$ would make the code compute-bound according to this roofline model. Still, we do not see an implementation solution for it without significant overhead.

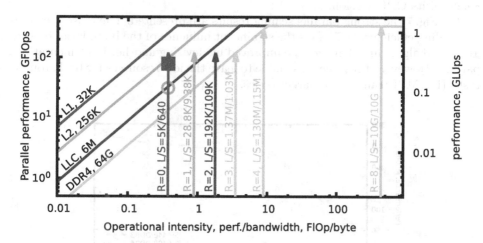

Fig. 5. The roofline for Intel Xeon i5-6400. The red marker shows the highest performance achieved in our implementation, the green marker shows the maximum achieved one thread performance. (Color figure online)

3.4   Performance Results

The described algorithm was implemented in code with the use of C++ (gcc compiler version 6.3) with parallelisation with POSIX threads. Code generation tools are made with Python3.6. Data visualization for verification of results uses aiwlib library [6]. The performance scaling of D3Q19 LBM implementation was verified on the Intel Core i5-4440 (2ch 32GB DDR3 RAM) and the Intel Core i5-6400 CPU (2ch 64GB DDR4 RAM).

On Fig. 6 the cube shaped domain is scaled up to the maximum size that still fits RAM memory. Starting from ∼2 MB data size the performance of about 25% from the peak is reached. Despite the fact that low-budget CPU is used, the achieved performance of >0.25 GLUps (billions of Lattice Update per second) may even be comparable to some GPU implementations. Note that the performance rises with the increase of the data size. With stepwise approaches without any kind of temporal blocking the performance drops each time the data size exceeds some level of cache. For comparison, in [15] CPU kernels for 12 core CPU Xeon E5-2690v3 reached ∼0.22 GLUps performance, while GPU kernels reach 3 GLUps. Walberla framework reaches ∼0.08 GLUps on 8 core Intel Xeon E5-2680 [3].

We have performed performance tests on one node of the K60 cluster [1] (Intel Xeon E5-2690 v4, 8ch 256GB DDR4 RAM), and achieved the performance up to 1.2 GLUps. Although the strong scaling efficiency is unsatisfactory in this case, we see how the result may be improved. The LRnLA algorithm for NUMA architecture is developed [20], but not implemented in the current code. Nevertheless, the result of our CPU code for one node of K60 cluster is better than the CPU version in [15] for one node of the Piz Daint supercomputer, and ∼42% of its GPU version.

On Fig. 7 the strong parallel scaling results are presented. The reason for the low scaling efficiency (∼72%) is the significant influence of the boundaries on the TorreFold algorithm. This may be improved by devising another implementation pattern. However, it may be reasonable to leave the idle resources for background tasks (i.e. MPI transfers) in larger codes.

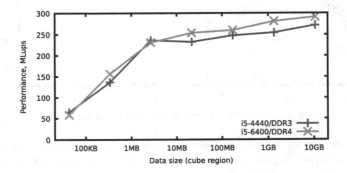

Fig. 6. Performance dependency on the data size.

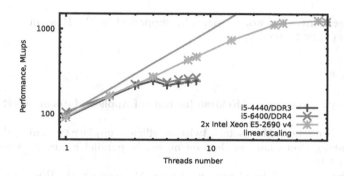

Fig. 7. Performance dependency on the number of POSIX threads.

4   Conclusion

We have used the LRnLA algorithm ConeFold to make a high-performance LBM simulation code. This approach optimizes the space-time traversal to take advantage of the memory hierarchy and all available levels of parallelism. The algorithm has been augmented by the non-local vectorization method, which not only increases the performance but also fixes the inability of performing simulation in periodic domains.

The LBM method proved to be simple for implementation, and the propagation method that conforms with the ConeFold decomposition has been found. It is interesting that the most memory-efficient propagation schemes of the step-wise codes [2,11] share the similarity with the method that arose naturally from the scheme dependencies and data access locality requirements.

The roofline model of the target CPU was built and the estimation of the code limits was aided by the LRnLA theory of algorithm construction. From this analysis, we see the necessity of using 4D localization in the algorithm and choose the parameters for parallelization.

We have measured the performance of the code on some low-cost CPU, and the results that were expected from the roofline construction were achieved. Namely, we have achieved >0.100 GLUps performance per core, ~0.3 GLUps per CPU, which exceeds the results that were found in the published work.

This result proves the advantages of the LRnLA approach. Namely, the algorithmic optimization is more important than the low-level considerations. The algorithms are built independent of the numerical method and hardware but are successfully adapted by taking account of dependency propagation speed and operations count from the numerical method side and the hierarchy of memory and parallelism from the hardware side.

As a further study, we aim to apply the method to more complex variations of the LBM method, like free surface LBM or double distribution function methods. Further, the optimizations of memory managing for sparse geometries and non-uniform grids are apparent.

Acknowledgement. The work is partially supported by the Russian Science Foundation (project #18-71-10004).

References

1. Computational resources of Keldysh Institute of Applied Mathematics RAS. www. kiam.ru
2. Geier, M., Schönherr, M.: Esoteric twist: an efficient in-place streaming algorithmus for the lattice Boltzmann method on massively parallel hardware. Computation **5**(2), 19 (2017)
3. Godenschwager, C., Schornbaum, F., Bauer, M., Köstler, H., Rüde, U.: A framework for hybrid parallel flow simulations with a trillion cells in complex geometries. In: Proceedings of the International Conference on High Performance Computing, Networking, Storage and Analysis, p. 35. ACM (2013)
4. Habich, J., Zeiser, T., Hager, G., Wellein, G.: Enabling temporal blocking for a lattice Boltzmann flow solver through multicore-aware wavefront parallelization. In: 21st International Conference on Parallel Computational Fluid Dynamics, pp. 178–182 (2009)
5. Heuveline, V., Latt, J.: The OpenLB project: an open source and object oriented implementation of lattice Boltzmann methods. Int. J. Mod. Phys. C **18**(04), 627–634 (2007)
6. Ivanov, A., Khilkov, S.: Aiwlib library as the instrument for creating numerical modeling applications. Sci. Vis. **10**(1), 110–127 (2018)
7. Levchenko, V.D.: Asynchronous parallel algorithms as a way to archive effectiveness of computations (in Russian). J. Inf. Tech. Comp. Syst. (1), 68 (2005)
8. Levchenko, V.D., Perepelkina, A.Y.: Locally recursive non-locally asynchronous algorithms for stencil computation. Lobachevskii J. Math. **39**(4), 552–561 (2018)
9. Levchenko, V.D., Perepelkina, A.Y., Zakirov, A.V.: DiamondTorre algorithm for high-performance wave modeling. Computation **4**(3), 29 (2016)
10. Morton, G.M.: A computer oriented geodetic data base and a new technique in file sequencing (1966)
11. Neumann, P., Bungartz, H.J., Mehl, M., Neckel, T., Weinzierl, T.: A coupled approach for fluid dynamic problems using the PDE framework peano. Commun. Comput. Phys. **12**(1), 65–84 (2012)
12. Nguyen, A., Satish, N., Chhugani, J., Kim, C., Dubey, P.: 3.5-D blocking optimization for stencil computations on modern CPUs and GPUs. In: High Performance Computing, Networking, Storage and Analysis (SC), pp. 1–13. IEEE (2010)
13. Perepelkina, A.Y., Levchenko, V.D., Goryachev, I.A.: Implementation of the kinetic plasma code with locally recursive non-locally asynchronous algorithms. J. Phys. Conf. Ser. **510**, 012042 (2014)
14. Perepelkina, A.: 3D3V kinetic code for simulation of magnetized plasma (in Russian). Ph.D. thesis, Keldysh Institute of Applied Mathematics RAS, Moscow (2015)
15. Riesinger, C., Bakhtiari, A., Schreiber, M., Neumann, P., Bungartz, H.J.: A holistic scalable implementation approach of the lattice Boltzmann method for CPU/GPU heterogeneous clusters. Computation **5**(4), 48 (2017)
16. Shimokawabe, T., Endo, T., Onodera, N., Aoki, T.: A stencil framework to realize large-scale computations beyond device memory capacity on GPU supercomputers. In: Cluster Computing (CLUSTER), pp. 525–529. IEEE (2017)
17. Succi, S.: The Lattice Boltzmann Equation: For Fluid Dynamics and Beyond. Oxford University Press, Oxford (2001)

18. Williams, S., Waterman, A., Patterson, D.: Roofline: an insightful visual performance model for multicore architectures. Commun. ACM **52**(4), 65–76 (2009)
19. Wittmann, M.: Hardware-effiziente, hochparallele Implementierungen von Lattice-Boltzmann-Verfahren für komplexe Geometrien (in German). Ph.D. thesis, Friedrich-Alexander-Universität Erlangen-Nürnberg (2016)
20. Zakirov, A.V., Levchenko, V.D.: The code for effective 3D modeling of electormagnetic wavesevolution in actual electrodynamics problems. Keldysh Institute Preprints (28) (2009)

Numerical Method for Solving a Diffraction Problem of Electromagnetic Wave on a System of Bodies and Screens

Mikhail Medvedik[ID], Marina Moskaleva[✉][ID], and Yury Smirnov[ID]

Penza State University, Penza, Russia
_medv@mail.ru, m.a.moskaleva1@gmail.com, mmm@pnzgu.ru

Abstract. The three-dimensional vector problem of electromagnetic wave diffraction by systems of intersecting dielectric bodies and infinitely thin perfectly conducting screens of irregular shapes is considered. The original boundary value problem for Maxwell's equations is reduced to a system of integro-differential equations. Methods of surface and volume integral equations are used. The system of linear algebraic equations is obtained using the Galerkin method with compactly supported basis functions. The subhierarchical method is applied to solve the diffraction problem by scatterers of irregular shapes. Several results are presented. Also we used a parallel algorithm.

Keywords: Boundary value problem · Inverse problem of diffraction
Permittivity tensor · Tensor Green's function
Integro-differential equation

1 Introduction

The important area in modern electrodynamics is three-dimensional problems of electromagnetic wave diffraction on systems of dielectric bodies and infinitely thin perfectly conducting screens of various shapes. In such diffraction problems, it is necessary to find solutions of Maxwell's equations that satisfy certain boundary or transmission conditions and radiation conditions at infinity.

Works of Samokhin [1] and Costabel [2,3] are devoted to problems of diffraction of electromagnetic waves by dielectric bodies. In these works, as well as in the work of Colton and Kress [4], a theory of solvability of vector diffraction problems is developed: existence and uniqueness of the solutions are proved and numerical methods are presented.

A theory of solvability of three-dimensional electrodynamics problems on nonclosed surfaces is developed by Ilyinsky and Smirnov [5]. Existence and uniqueness of the solution (in suitable spaces) are proved.

In papers [6–8] the diffraction problem of electromagnetic wave by screens of various forms is considered. The statement of the boundary-value problem of the diffraction of an electromagnetic wave by a system of nonintersecting bodies and screens is presented in [9,10].

© Springer Nature Switzerland AG 2019
V. Voevodin and S. Sobolev (Eds.): RuSCDays 2018, CCIS 965, pp. 114–124, 2019.
https://doi.org/10.1007/978-3-030-05807-4_10

It should be noted that the theory of solvability of electromagnetic diffraction problems on a system of dielectric bodies and infinitely thin perfectly conducting screens is far from completion. However, recent advantages in this theory have been presented in [9,10] for the case of nonintersecting bodies and screens.

In this paper we present numerical results of the solution of the diffraction problem on a system of intersecting bodies and screens, which is new in comparison with [5–11].

Let us consider various numerical methods for solving electromagnetic waves diffraction problems on bodies of various configurations, such as finite element methods (FEM), finite-difference methods (FDM) and methods of surface and volume integral equations. FEM and FDM also are used in various application package for solving electrodynamic problems. However, these methods have some disadvantages. For example, its application is possible only if the region in which the problem is solved is made finite. This limitation of the region leads to incorrect results. In order to avoid this, it is necessary to artificially increase the size of the region in which the problem is solved. Such algorithm leads to the appearance of sparse matrices of sufficiently large order $(10^5 - 10^6)$. The boundary value problem also is not elliptic in the general case. Therefore, the use of traditional method of proof of convergence of projection methods is excluded. Alternatively, the method of surface and volume integro-differential equations, free from described above disadvantages, can be applied. Then the equation is solved in the region of inhomogeneity inside the body and on the screen. Thus, after the discretization of the problem, we obtain a finite-dimensional system of equations with a dense matrix of order $(10^3 - 10^4)$, that is substantially smaller than in the case FEM or FDM.

In this paper we apply methods of surface and volume integro-differential equations for the numerical solution of the diffraction problem on a system of intersecting bodies and screens. Also we use the subhierarchical approach [6,7] to solve the problem on a system of intersecting bodies and screens of complex shapes. The problem is reduced to a system of integro-differential equations. The system of linear algebraic equations is obtained using the Galerkin method. Solutions of the diffraction problem of electromagnetic waves on a system consisting of a screen that is intersected with an inhomogeneous body are presented.

2   Statement of the Problem

Let Θ be a system of infinitely thin perfectly conducting screens Ω_k $k = 1 \ldots K$ and dielectric bodies Q_j $j = 1 \ldots J$. The system Θ is placed in $\mathbb{R}^3$. An example of Θ, where $K = 1$, $J = 1$, is shown in Fig. 1.

The screens Ω_k are connected orientable nonclosed disjoint bounded surfaces of class C^∞ in $\mathbb{R}^3$ [5]. The boundary $\partial \Omega_k := \overline{\Omega}_k \backslash \Omega_k$ of screen Ω_k is a piecewise smooth curve that consist of a finite number of simple arcs without self-intersections. We use the notation $\Omega = \cup_k \Omega_k$, and $\partial \Omega = \cup_k \partial \Omega_k$.

We assume that Q_j is a bounded domain having the boundary $\partial Q_j := \overline{Q}_j \backslash Q_j$, $j = 1 \ldots J$. It is assumed the ∂Q_j is a piecewise smooth closed orientable surface

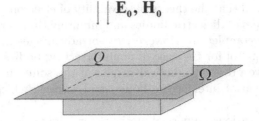

Fig. 1. The body Q and the screen Ω.

consisting of a finite number of surfaces of class C_1. The bodies Q_j can be inhomogeneous and anisotropic. Thus the inhomogeneity is described by the tensor

$$\hat{\varepsilon} = \begin{cases} \varepsilon_e \hat{I}, & x \in \mathbb{R}^3 \backslash (\bar{Q} \bigcup \bar{\Omega}), \\ \hat{\varepsilon}_j(x), & x \in \bar{Q}_j, \end{cases}$$

$\hat{\varepsilon}_j = \hat{\varepsilon}_j^T$, $\operatorname{Im} \hat{\varepsilon}_j \geq 0$.

The screens Ω_k and bodies Q_j are intersecting.

Let $\mathbf{E_0}$, $\mathbf{H_0}$ be an electromagnetic wave with a harmonic time dependence of the form $e^{-i\omega t}$. We consider diffraction of the wave $\mathbf{E_0}$, $\mathbf{H_0}$ on the system Θ. The source of the incident field $\mathbf{E_0}$, $\mathbf{H_0}$ can be a current $j_{0,E}$ localized in such a way that $supp(j_{0,E}) \cap (\bar{Q} \bigcup \bar{\Omega}) = \varnothing$.

We look for the (complete) electromagnetic field $(\mathbf{E}, \mathbf{H})$ satisfying: the Maxwell equations

$$\begin{cases} \operatorname{rot} \mathbf{H} = -i\omega \hat{\varepsilon} \mathbf{E} + j_{0,E}, \\ \operatorname{rot} \mathbf{E} = i\omega \mu_e \mathbf{H}, \end{cases} \tag{1}$$

in $x \in \mathbb{R}^3 \backslash (\bar{Q} \bigcup \bar{\Omega})$, continuity conditions for the tangential components on the boundary

$$[\mathbf{E}_\tau]|_{\partial Q} = [\mathbf{H}_\tau]|_{\partial Q} = 0, \tag{2}$$

the boundary conditions in the internal points of Ω

$$\mathbf{E}_\tau|_{\partial \Omega} = 0, \tag{3}$$

the energy finiteness condition in any bounded domain

$$\mathbf{E}, \mathbf{H} \in L_{2,loc}(\mathbb{R}^3) \tag{4}$$

and the Sommerfeld radiation condition at infinity

$$\frac{\partial (\mathbf{E}_s, \mathbf{H}_s)}{\partial r} - i k_e (\mathbf{E}_s, \mathbf{H}_s) = o\left(\frac{1}{r}\right), \ r \to \infty \tag{5}$$

$\mathbf{E}_s = \mathbf{E} - \mathbf{E_0}$ and $\mathbf{H}_s = \mathbf{H} - \mathbf{H_0}$ is the scattered field, $r = | x |$, $x \in \mathbb{R}^3$.

The following uniqueness theorem is proved in [11].

Theorem 1. *The diffraction problem* (1)–(5) *has not more than one quasi-classical solution.*

3 System of Integro-Differential Equations

The problem (1)–(5) is reduced to the system of integro-differential equations [11]:

$$\hat{\xi}\mathbf{J} - \left(k_e{}^2 + \text{grad div}\right)\int_Q G(x,y)\mathbf{J}(y)dy-$$

$$-\frac{1}{i\omega\varepsilon_e}\left(k_e{}^2 + \text{grad div}\right)\int_\Omega G(x,y)\mathbf{u}(y)ds_y = \mathbf{E}_{0,Q}(x), \qquad x \in Q$$

$$\left(-\left(k_e{}^2 + \text{grad div}\right)\int_Q G(x,y)\mathbf{J}(y)dy-\right.$$

$$\left.-\frac{1}{i\omega\varepsilon_e}\left(k_e{}^2 + \text{grad div}_\tau\right)\int_\Omega G(x,y)\mathbf{u}(y)ds_y\right)_\tau = \mathbf{E}_{0,\tau}(x), \qquad x \in \Omega,$$

$$(6)$$

where $\hat{\xi} = \left(\frac{\hat{\varepsilon}(x)}{\varepsilon_e}{}^{-1} - \hat{I}\right)$, $\mathbf{J} = \left(\frac{\hat{\varepsilon}(x)}{\varepsilon_e} - \hat{I}\right)\mathbf{E}$ is the unknown polarization current vector in Q, $k_e = \omega\sqrt{\varepsilon_e, \mu_e}$ is the wave number of a free space, $\text{Im}\,\varepsilon_e \geq 0$, $\text{Im}\,\mu_e \geq 0$, $\text{Im}\,k_e \geq 0$, $G(x,y) = \frac{1}{4\pi}\frac{e^{ik_e|x-y|}}{|x-y|}$ is the known Green function, $\mathbf{u}$ is an unknown surface current density on Ω, $\mathbf{E}_{0,\tau}$ is the tangential component of the incident field on the screen Ω.

Rewrite (6) in the operator form as follows:

$$\hat{L}(\mathbf{V}) = \mathbf{f}; \qquad\qquad (7)$$

where

$$\hat{L} = \hat{L}_1 + \hat{L}_2 = \begin{pmatrix} A & 0 \\ 0 & S \end{pmatrix} + \begin{pmatrix} 0 & K_1 \\ K_2 & 0 \end{pmatrix},$$

the operators A, S, K_1 and K_2 are defined as follows:

$$A\mathbf{J} := \hat{\xi}\mathbf{J}(x) - \left(k_e{}^2 + \text{grad div}\right)\int_Q G(x,y)\mathbf{J}(y)dy,$$

$$S\mathbf{u} := \left(-\frac{1}{i\omega\varepsilon_e}\left(k_e{}^2 + \text{grad div}\right)\int_\Omega G(x,y)\mathbf{u}(y)ds_y\right)_\tau,$$

$$K_1\mathbf{u} := -\frac{1}{i\omega\varepsilon_e}\left(k_e{}^2 + \text{grad div}\right)\int_\Omega G(x,y)\mathbf{u}(y)ds_y,$$

$$K_2\mathbf{J} := \left(-\left(k_e{}^2 + \text{grad div}\right)\int_Q G(x,y)\mathbf{J}(y)dy\right)_\tau,$$

$\mathbf{V} = (\mathbf{J}, \mathbf{u})$, the right-hand side is the vector $\mathbf{f} = (\mathbf{E}_{0,Q}, \mathbf{E}_{0,\tau})$, where $\mathbf{E}_{0,Q}$ is the restriction of the incident field.

Using non-coinciding integration points, we avoid singularity in the Green function.

4   Numerical Method

We suggest a numerical algorithm for solving Eqs. (6). The Eq. (6) is discretized. Let the system Θ consist of a plane rectangular screen Ω, and a rectangular parallelepiped Q. We construct on the system Θ a generalized computational grid [6]. The computational grid is regular. Generalized computational grids allow us to introduce basis functions of different types [6].

We divide the screen Ω and the body Q into elementary cells that are called finite elements. For the screen Ω, the finite elements are rectangles

$$P_{k_1 k_2} = \{x = (x_1, x_2),\ k_l h_l < x_l < (k_l + 1)h_l,\ l = 1, 2\},$$

where $k = (k_1, k_2)$, $k_l = 0, \ldots, n - 1$.

For the body Q the finite elements are rectangular parallelepipeds

$$\Pi^1_{j_1 j_2 j_3} = \{x : x_{1,j_1-1} < x_1 < x_{1,j_1+1},\ x_{2,j_2} < x_2 < x_{2,j_2+1},\ x_{3,j_3} < x_3 < x_{3,j_3+1}\},$$

$$x_{1,j_1} = a_1 + \frac{a_2 - a_1}{n} j_1,\ x_{2,j_2} = b_1 + \frac{b_2 - b_1}{n} j_2,\ x_{3,j_3} = c_1 + \frac{c_2 - c_1}{n} j_3,$$

where $j_1 = 0, \ldots, n - 2$; $j_2, j_3 = 0, \ldots, n - 1$. The number of supports on the screen Ω is $N_1 = 2n(n - 1)$, the number of supports on the body Q is $N_2 = 3n^2(n - 1)$. Thus, the number of supports for the system Θ is $N = N_1 + N_2$.

Let us consider the screen Ω. As the basis functions, we will use the "rooftop" functions, which are defined on a pair of adjacent rectangles of the grid with a shared edge. For each edge k there exists a support consisting of two rectangles with a shared edge k. The basis function $\varphi(x_1, x_2, x_3)$ corresponding to the edge k is determined as follows

$$\varphi_k(x_1, x_2, x_3) = \begin{cases} (x_1 - x_{1,k-1}, x_2 - x_{2,k-1}, x_3 - x_{3,k-1})\frac{l_k}{S_k^+} \text{ in } P_k^+, \\ (x_{1,k+1} - x_1, x_{2,k+1} - x_2, x_{3,k+1} - x_3)\frac{l_k}{S_k^-} \text{ in } P_k^-, \end{cases}$$

where l_k is the length of the edge k, S_k^+ and S_k^- are areas of P_k^+ and P_k^-, respectively.

Let us consider the body Q. Supports of basis functions are pairs of adjacent elementary parallelepipeds belonging to the body and having a shared face. The parallelepipeds of a support are located along one of the coordinate axes.

We define basis functions on the body. Let $h^1 := |x_{1,j_1} - x_{1,j_1-1}|$. The function $\psi^1_{j_1,j_2,j_3}$ is defined as follows

$$\psi^1_{j_1,j_2,j_3}(x_1, x_2, x_3) = \begin{cases} 1 - \frac{1}{h^1}|x_1 - x_{1,j_1}| & \text{in } \Pi^1_{j_1,j_2,j_3} \\ 0, & \text{otherwise.} \end{cases}$$

The functions $\psi^2_{j_1,j_2,j_3}$ and $\psi^3_{j_1,j_2,j_3}$ are defined in the same way. The function $\psi^i_{j_1,j_2,j_3}$ is piecewise-linear in the Ox_i direction and piecewise-constant in two other directions.

Applying the Galerkin method [12] to the system (7), we obtain the equation

$$\mathbf{LV} = \mathbf{f}; \tag{8}$$

where $\mathbf{L}$ is the generalized matrix of system of linear algebraic equations (SLAE), $\mathbf{V} = (\mathbf{J}, \mathbf{u})$ is the column of unknown coefficients for basis functions, $\mathbf{f}$ is a known.
    The matrix of the SLAE has a block form

$$\mathbf{L} = \begin{pmatrix} \mathbf{L}^{11} & \mathbf{L}^{12} \\ \mathbf{L}^{21} & \mathbf{L}^{22} \end{pmatrix}.$$

    The matrix block $\mathbf{L}^{11}$ corresponds to the solution of the diffraction problem only on the body. The matrix block $\mathbf{L}^{22}$ corresponds to the solution of the diffraction problem only on the screen.
    The matrix elements, which correspond to the solution of the problem on the body Q, are obtained by calculating threefold and sixfold integrals over the body

$$\mathbf{L}^{11}_{pq} = \int\limits_Q \hat{\xi}(x)\psi_q(x)\psi_p(x)dx - \int\limits_Q (k_e^2 + \mathrm{grad}_x\ \mathrm{div}_x) \int\limits_Q G(x,y)\psi_q(y)dy\psi_p(x)dx.$$

    The matrix elements, which correspond to the solution of the problem on the screen Ω, are obtained by calculating fourfold integrals over the screen

$$\mathbf{L}^{22}_{pq} = \frac{1}{\omega^2\varepsilon_e^2} \int\limits_\Omega \left((k_e^2 + \mathrm{grad}_x\ \mathrm{div}_x) \int\limits_\Omega G(x,y)\varphi_q(y)ds_y \right) \varphi_p(x)ds_x.$$

    The blocks $\mathbf{L}^{12}$ and $\mathbf{L}^{21}$ correspond to the interaction of fields on the body and screen; the matrix elements in these blocks are obtained by calculating fivefold integrals over the body and the screen

$$\mathbf{L}^{12}_{pq} = -\frac{1}{i\omega\varepsilon_e} \int\limits_Q (k_e^2 + \mathrm{grad}_x\ \mathrm{div}_{\tau,x}) \int\limits_\Omega G(x,y)\varphi_q(y)ds_y\psi_p(x)dx,$$

$$\mathbf{L}^{21}_{pq} = -\frac{1}{i\omega\varepsilon_e} \int\limits_\Omega \left((k_e^2 + \mathrm{grad}_x\ \mathrm{div}_x) \int\limits_Q G(x,y)\psi_q(y)dx \right) \varphi_p(x)ds_y.$$

    The generalized matrix is symmetrical. The block $\mathbf{L}^{21}$ coincides with the transposed block $\mathbf{L}^{12}$: $\mathbf{L}^{21} = \left(\mathbf{L}^{12}\right)^T$.

5   Numerical Results

An example of graphical results of calculations for a system of an inhomogeneous parallelepiped body and a rectangular screen (Fig. 2) is presented. The wave number k_e is 1. The size of the computational grid for each axis are 24 steps for the screen and 13 steps for the body.
    The screen is located in the plane Ox_1x_2, ($x_3 = 0$). The center of the body coincides with the center of the coordinate system. Thus

$$\Omega = \left\{ x \in \mathbb{R}^3 : x_1, x_2 \in \left(-\frac{\lambda}{2}, \frac{\lambda}{2} \right); \ x_3 = 0 \right\},$$

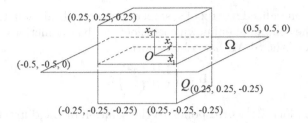

Fig. 2. The system Θ.

$$Q = \left\{ x \in \mathbb{R}^3 : x_k \in \left(-\frac{\lambda}{4}, \frac{\lambda}{4} \right) \right\}, \quad k = 1, 2, 3.$$

The incident field is a plane wave; its direction vector of which is collinear to the axis Ox_1. The body Q consists of two parts with different permittivities. The integration accuracy is 8 knots per support, λ is wavelength.

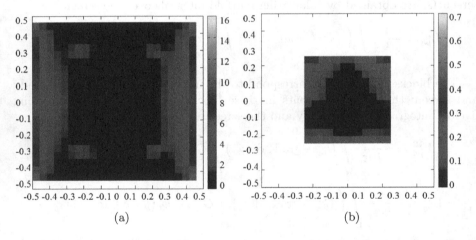

Fig. 3. Results of solving the diffraction problem on the system Θ on the different computation grids.

Figure 3(a) illustrates the distribution of surface current modules on the screen. Figure 3(b) illustrates the distribution of the field inside the body Q on the middle layer of the computational grid perpendicular to the axis Ox_1.

Figure 3(a) shows that the tangential component of the surface currents increases at the boundary of the intersection of the body and screen. Figure 3(b) shows the "jump" of the field inside the body at the interface of the two parts with different permittivities. Also Fig. 3b shows that the field inside the body increases at the boundary of the intersection of the body and screen.

We will use the subhierarchical method [6] to solve the diffraction problem one a system of inhomogeneous bodies and screens of complex shapes. Following the proposed method, at the first step the diffraction problem is solved most accurately on the canonical system of bodies and screens. The generalized computational grid is used. In the next step, using the obtained matrix SLAE, we select a new system of bodies and screens, or only bodies, or only screens that is contained in the canonical system. Further, without repeating calculations in the SLAE matrix, we find the solution of the system of integro-differential equations corresponding to the new problem.

Now let us consider the systems Θ_1 consisting of the screen Ω, and the system Θ_2 consisting of the body Q. Thus $\Theta = \Theta_1 \cup \Theta_2$.

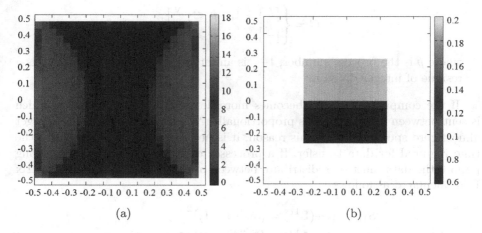

(a)   (b)

Fig. 4. Results of solving the diffraction problem on Θ_1 and Θ_2 on the different computation grids.

Figure 4(a) illustrates the distribution of surface current modules on the screen, Fig. 4(b) illustrates the distribution of the field inside the body Q on the middle layer of the computed grid perpendicular to the axis Ox_1.

The amount of memory allocated to the generalized matrix is about 2 GB. The results presented in the articles [6–8] for a screen and in the article [13] for a body are in good agreement the results presented in figure 4a and b. We observe a qualitative coincidence of the results on the graphs presented in Figs. 3(a)–(b) and 4(a)–(b), with increasing grid steps on the system. This corresponds to the internal convergence of the solutions.

6 Parallel Algorithm

Due to large computations we use a parallel algorithm. About 80% of time is used to the calculations of the matrix entries, which are calculated independently. The matrices have about 10^4 entries that are threefold and sixfold volume integrals, fourfold surface integrals, fourfold volume and surface integrals. The main computational complexity is to calculate the matrix block $\mathbf{L}^{11}$.

Parallel algorithm:

1. We convert the matrix into a 1-dimensional array $\{A_I\}$ of size N^2;
2. It is necessary to uniformly distribute coefficients C of the array $\{A_I\}$ which are computed at each process [14]:

$$
C = \begin{cases} \left[\frac{N}{p}\right] + 1, & p < \left\{\frac{N}{p}\right\}, \\ \left[\frac{N}{p}\right], & p \geqslant \left\{\frac{N}{p}\right\}, \end{cases}
$$

where p is the process' number, $\left[\frac{N}{p}\right]$ is an integer part of the ratio, $\left\{\frac{N}{p}\right\}$ is a residue of integer division.

If the computational grid becomes more dense, then the data level, which is sent between cores, increases proportionally. Neglecting this fact, one can get almost zero speed gain. For this reason, it is necessary to pay attention to the time required for data transfer. If a processor does not respond during a long period, the data must be redistribute between other cores. Size S of the matrix $\mathbf{L}$ is calculated as follows

$$
\begin{aligned}
S_{11} &:= \text{size}(\mathbf{L}^{11}) = \left(3n^2(n-1)\right)^2, \\
S_{12} &:= \text{size}(\mathbf{L}^{12}) = \left(3n^2(n-1)\right)\left(2n(n-1)\right), \\
S_{21} &:= \text{size}(\mathbf{L}^{21}) = \left(3n^2(n-1)\right)\left(2n(n-1)\right), \\
S_{22} &:= \text{size}(\mathbf{L}^{22}) = \left(2n(n-1)\right)^2,
\end{aligned}
$$

$$
S := S_{11} + S_{12} + S_{21} + S_{22}.
$$

It is seen that memory capacity needed for solution to such a problem increases drastically. For example, with $n = 10$ the matrix requires 124 Mb memory, with $n = 20$ the matrix requires 8.3 Gb memory, with $n = 30$ the matrix requires 95 Gb memory.

Below we show results of speed gain for different number of cores:

1. with $p = 8$ the computation speed increases approximately 8 times;
2. with $p = 16$ the computational speed increases approximately 15 times;
3. with $p = 32$ the computational speed increases approximately 26 times;
4. with $p = 64$ the computational speed increases approximately 40 times.

The parallel algorithm is quite simple in use and allows reducing calculating time up to 40 times. We used a 64-core processor to solve the problem.

7 Conclusion

The method of solving the diffraction problem on a system of intersecting dielectric bodies and screens of complex shape is suggested. To solve the problem, the computational algorithm is parallelized, the supercomputer complex of the Penza State University was used for calculations. The methods proposed in this paper can be used in practice to study the behavior of the field reflected from intersecting (dielectric) bodies and (infinitely thin perfectly conducting) screens. Numerical results are presented and visualized. In the neighborhood of the screen edge and the body surface the behavior of the reflected field is in qualitatively agreement with the theory [5, 13].

Acknowledgments. This study is supported by the Ministry of Education and Science of the Russian Federation [project 1.894.2017/4.6] and by the Russian Foundation for Basic Research [project 18-31-00108]. The research is carried out using the equipment of the shared research facilities of HPC computing resources at Lomonosov Moscow State University.

References

1. Samokhin, A.B.: Integral Equations and Iteration Methods in Electromagnetic Scattering. VSP, Utrecht (2001)
2. Costabel, M.: Boundary integral operators on curved polygons. Ann. Mat. Pura Appl. **133**, 305–326 (1983)
3. Costabel, M., Darrigrand, E., Kon'e, E.: Volume and surface integral equations for electromagnetic scattering by a dielectric body. J. Comput. Appl. Math. **234**, 1817–1825 (2010)
4. Colton, D., Kress, R.: Integral Equation Methods in Scattering Theory. Wiley, New York (1983)
5. Ilyinsky, A.S., Smirnov, Y.G.: Electromagnetic Wave Diffraction by Conducting Screens. VSP, Utrecht (1998)
6. Medvedik, M.Y., Moskaleva, M.A.: Analysis of the problem of electromagnetic wave diffraction on non-planar screens of various shapes by the subhierarchic method. J. Commun. Technol. Electron. **60**(6), 543–551 (2015)
7. Medvedik, M.Y.: Solution of integral equations by the subhierarchic method for generalized computational grids. Math. Models Comput. Simul. **7**(6), 570–580 (2015)
8. Medvedik, M.Y.: Numerical solution of the problem of diffraction of electromagnetic waves by nonplanar screens of complex geometric shapes by means of a subhierarchic method. J. Commun. Technol. Electron. **58**(10), 1019–1023 (2013)
9. Smirnov, Y.G., Tsupak, A.A.: Integrodifferential equations of the vector problem of electromagnetic wave diffraction by a system of nonintersecting screens and inhomogeneous bodies. Adv. Math. Phys. **2015** (2015). https://doi.org/10.1155/2015/945965
10. Valovik, D.V., Smirnov, Y.G., Tsupak, A.A.: On the volume singular integro-differential equation approach for the electromagnetic diffraction problem. Appl. Anal. **2015**, 173–189 (2015). https://doi.org/10.1080/00036811.2015.1115839

11. Smirnov, Y.G., Tsupak, A.A.: Diffraction of acoustic and electromagnetic waves by screens and inhomogeneous solids: mathematical theory. M. RU-SCIENCE (2016)
12. Kress, R.: Linear Integral Equations. Applied Mathematical sciences, vol. 82, 2nd edn. Springer, Hedielberg (1989). https://doi.org/10.1007/978-3-642-97146-4
13. Kobayashi, K., Shestopalov, Y., Smirnov, Y.: Investigation of electromagnetic diffraction by a dielectric body in a waveguide using the method of volume singular integral equation. SIAM J. Appl. Math. **70**(3), 969–983 (2009)
14. Medvedik, M.Y., Smirnov, Y.G., Sobolev, S.I.: A parallel algorithm for computing surface currents in a screen electromagnetic diffraction problem. Numer. Methods Program. **6**, 99–108 (2005)

Parallel Algorithm for One-Way Wave Equation Based Migration for Seismic Imaging

Alexander Pleshkevich[1], Dmitry Vishnevsky[2], Vadim Lisitsa[3(✉)],
and Vadim Levchenko[4]

[1] Central Geophysics Expedition, Moscow, Russia
psdm3d@yandex.ru
[2] Institute of Petroleum Geology and Geophysics SB RAS, Novosibirsk, Russia
vishnevskydm@ipgg.sbras.ru
[3] Institute of Petroleum Geology and Geophysics SB RAS, Novosibirsk State
University, Novosibirsk, Russia
lisitsavv@ipgg.sbras.ru
[4] Institute of Applied Mathematics RAS, Moscow, Russia
vadimlevchenko@mail.ru

Abstract. Seismic imaging is the final stage of the seismic processing
allowing to reconstruct the internal subsurface structure. This procedure
is one of the most time consuming and it requires huge computational
resources to get high-quality amplitude-preserving images. In this paper,
we present a parallel algorithm of seismic imaging, based on the solution
of the one-way wave equation. The algorithm includes parallelization of
the data flow, due to the multiple sources/receivers pairs processing.
Wavefield extrapolation is performed by pseudo-spectral methods and
applied by qFFT - each dataset is processed by a single MPI process.
Common-offset vector images are constructed using all the solutions from
all datasets thus by all-to-all MPI communications.

Keywords: One-way wave equation · Pseudo-spectral methods
qFFT · CUDA · Nested OMP · MPI

1 Introduction

Modern development of the supercomputers with hybrid architecture, especially
use of GPUs, leads to the revision of the numerical methods and algorithms
which were computationally inefficient and hard to implement using conventional cluster architecture. Seismic imaging is the procedure to reconstruct the
interior structure of the subsurface which is based on the correlation of solutions
of direct and adjoint wave propagation problems. Traditionally three types of
approaches are considered. The first one is based on the asymptotic solution
of scalar or elastic wave equation; i.e. based on the ray theory [1]. The advantage of this technique is the numerical efficiency of the solution construction,

© Springer Nature Switzerland AG 2019
V. Voevodin and S. Sobolev (Eds.): RuSCDays 2018, CCIS 965, pp. 125–135, 2019.
https://doi.org/10.1007/978-3-030-05807-4_11

possibility to store and combine a high number of ray paths and travel times, thus selective images for different source-receiver offsets and azimuth as well as for different inclination angles of the reflecting surface can be constructed. In addition, ray tracing in models with local velocity decrease or in anisotropic media where triplication of the wavefield may appear is troublesome. The second group of methods is based on the full waveform simulation; i.e. on the solution of the wave equation for all positions of the sources and receivers [2,14,15]. This group is antipode to the first one, it is extremely computationally intense but provides complete information about the wavefield. The third group includes methods based on the solution of the one-way wave equation (OWE) to generate the solution of direct and adjoint problems [3–5,7,12,13,17]. This approach is dealing with the pseudo-differential operator possessing solutions propagating in the positive direction with respect to depth. OWE based techniques are more computationally efficient than full-waveform simulations but free from the drawbacks of the ray-based methods. However, use of conventional CPU architecture made these type of approaches unattractive, because they still required rather intense computations, but nowadays with intense part can be done using GPUs, moreover, amount of available memory (either on device and host) makes it possible to solve OWEs for a large number of right-hand sides forming images for different source-receiver offsets and azimuths. In this paper, we present an algorithm implementing the third type of methods. We use the pseudo-spectral method, based on the intense use of FFT procedures, to solve OWE, thus we achieve high performance by implementing CUDA technology and qFFT.

2   Mathematical Background

2.1   One-Way Wave Equation

One-way wave equation is a non-local pseudo-differential operator, governing down-going wave propagation:

$$
\begin{aligned}
&\frac{\partial u}{\partial z} + i\sqrt{\frac{\omega^2}{v^2(x,y,z)} + \frac{\partial^2}{\partial x^2} + \frac{\partial^2}{\partial y^2}}\,u = 0, \\
&u(\omega, x, y, 0) = u_0(\omega, x, y), \\
&u(\omega, x, y, z) \to 0|_{x\to\pm\infty}, \\
&u(\omega, x, y, z) \to 0|_{y\to\pm\infty}.
\end{aligned}
\tag{1}
$$

We consider this equation defined in a domain $D = \{(x, y, z)| \ x \in R, \ y \in R, \ z \in [0, Z_1]\}$. In this notations x and y denote the coordinates in horizontal direction, z is vertical coordinate, u is a wavefield, $\omega \in [\Omega_1, \Omega_2]$ is the temporal frequency, $0 < V_1 \le v(x, y, z) \le V_2 < \infty$ is the wave propagation velocity. Throughout the paper we will assume that the velocity is smoothly varying, because dealing with discontinuous coefficients for wave propagation requires additional study and special treatment of the coefficients [6,8–10,16].

Our goal is to solve problem (1) inside the domain $D_0 = \{(x, y, z)| \ x \in [X_0, X_1], \ y \in [Y_0, Y_1], \ z \in [0, Z_1]\} \subseteq D$ so that the solution to be accurate for propagation direction deviating from vertical by the angle $\alpha \le 50°$.

Solution of the one-way wave equation can be computed layer-by-layer with respect to depth. In particular, we suggest using the pseudo-spectral method with the sixth order of approximation with respect to the direction of wave propagation (inclination from the vertical direction) [11,12]. Assume the wavefield at the level $z = z_0$ is known then the solution at the level $z = z_0 + \Delta z$ is computed by the following rule:

$$\tilde{u}(\omega, x, y, z_0 + \Delta z) = \alpha_0 u(\omega, x, y, z_0)$$
$$+ \alpha_j(\omega, x, y, z_0) u_j(\omega, x, y, z_0 + \Delta z) + \alpha_{j+1}(\omega, x, y, z_0) u_{j+1}(\omega, x, y, z_0 + \Delta z),$$
$$(2)$$

where velocity in the particular point (x, y, z) belongs to an interval between two reference velocities; i.e. $v(x, y, z) = v_0 \in (v_j, v_{j+1})$. Functions u_j and u_{j+1} are the solutions of the one-way wave equation with constant velocities v_j and v_{j+1}; computed by the formulae:

$$u_j(\omega, x, y, z_0 + dz) = F_{x,y}^{-1} \left[e^{ik_z^j dz} F_{x,y} \left[u(\omega, x, y, z_0) \right] \right], \tag{3}$$

with

$$k_z^j = sign(\omega) \sqrt{\frac{\omega^2}{v_j^2} - k_x^2 - k_y^2}. \tag{4}$$

Coefficients α_j and α_{j+1} are

$$\alpha_j = \frac{v_0 \left[v_{j+1}^2 \left(1 - \frac{i\omega dz}{v_{j+1}}\right) - v_0^2 \left(1 - \frac{i\omega dz}{v_0}\right) \right]}{v_j \left[v_{j+1}^2 \left(1 - \frac{i\omega dz}{v_{j+1}}\right) - v_j^2 \left(1 - \frac{i\omega dz}{v_j}\right) \right]} e^{i\frac{\omega dz}{v_0} - i\frac{\omega dz}{v_j}}, \tag{5}$$

$$\alpha_{j+1} = \frac{v_0 \left[v_0^2 \left(1 - \frac{i\omega dz}{v_0}\right) - v_j^2 \left(1 - \frac{i\omega dz}{v_j}\right) \right]}{v_{j+1} \left[v_{j+1}^2 \left(1 - \frac{i\omega dz}{v_{j+1}}\right) - v_j^2 \left(1 - \frac{i\omega dz}{v_j}\right) \right]} e^{i\frac{\omega dz}{v_0} - i\frac{\omega dz}{v_{j+1}}}, \tag{6}$$

$$\alpha_0 = e^{i\frac{\omega dz}{v_0}} - \alpha_j e^{i\frac{\omega dz}{v_j}} - \alpha_{j+1} e^{i\frac{\omega dz}{v_{j+1}}}. \tag{7}$$

Thus to compute solution at a depth level $z = z_0 + \Delta z$ with variable velocity one needs to update the a set reference solutions and then construct linear combinations in each point at the plane $(x, y, z_0 + \Delta z)$.

As discussed in [12] wavefield depth extrapolation by pseudospectral method and its modifications, such as phase-shift plus interpolation approach, can be applied using very coarse discretization in vertical direction. In particular, in our algorithm we use $\Delta z = 50\,\text{m}$. Whereas the images are constructed on a mesh with the step as small as $5\,\text{m}$. To interpolate the solution we use the following formulae:

$$u(x, y, z_0 + \gamma \Delta z, \omega)$$
$$= (1 - \gamma) u(x, y, z_0, \omega) e^{\frac{i\omega}{v(x,y,z_0)} \gamma \Delta z} + \gamma u(x, y, z_0 + \Delta z, \omega) e^{\frac{i\omega}{v(x,y,z_0)}(1-\gamma)\Delta z}$$

here $\gamma \in [0, 1]$ and $v(x, y, z)$ is assumed to be independent of z within the slab $[z_0, z_0 + \Delta z]$.

2.2 Imaging Condition

To construct the seismic image at a point (x, y, z) one needs to precompute the Green's functions for source and receiver positions $G(x, y, z, x_s, y_s, \omega)$ and $G(x, y, z, x_r, y_r, \omega)$ respectively, and then convolve them with the wavefield, recorded in the receiver $\phi(x_s, y_s, x_r, y_r, \omega)$; i.e.

$$
\begin{aligned}
&I(x, y, z, x_s, y_s, x_r, y_r) \\
&= \int_{\omega_0}^{\omega_1} \bar{G}(x, y, z, x_s, y_s, \omega) \phi(x_s, y_s, x_r, y_r, \omega) \bar{G}(x, y, z, x_r, y_r, \omega) d\omega,
\end{aligned}
\tag{8}
$$

where ω_0 and ω_1 define the frequency range, and $\bar{G}$ denotes the complex conjugate of the Green's functions.

Seismic data acquisition typically includes dozens to hundreds thousands of sources and receivers, thus different images can be computed; i.e.

$$
\begin{aligned}
&I^k(x, y, z) \\
&= \int_{(x_s, y_s) \in \Omega_s^k} \int_{(x_r, y_r) \in \Omega_r^k} I(x, y, z, x_s, y_s, x_r, y_r) dx_s dy_s dx_r dy_r,
\end{aligned}
\tag{9}
$$

where Ω_r^k and Ω_s^k where can be defined according to the particular needs of seismic processing. Moreover, we use superscript k to denote that usually more than one image is constructed, depending on the processing procedures.

One of the most widely-used seismic images are the common-offset vector images, where the distance between source and receiver positions is fixed whereas parametrization of the image is implemented via mid-points between the sources and receivers. So, that fixed superscript k means that for all points $(x, y, 0)$ one considers pairs of sources and receivers, for which this point is the midpoint and the distance and direction between the pair is fixed, One computes images for all such pairs and then integrates over all points $(x, y, 0)$.

3 Description of the Algorithm

3.1 Dataflow

Before proceeding to the description of the computational aspects of the algorithm let us consider the data flow, and estimate the typical amount of data. The input data are seismograms acquired in the field. These data are usually viewed as a 5D cube; i.e. for each source position (2D space), for each receiver position (2D) a single trace is recorded (1D). Typical acquisition system includes dozens to hundreds of thousands of sources and receivers. Let us denote the number of sources and receivers by N_s nad N_r respectively, usually $N_s \approx N_r = 10^4 - 10^5$. The length of the trace record N_t is about 10^3. Thus the volume of the input data varies from 1 TB to 100 TB.

Output data are the set of common-offset vector images. For modern acquisitions, more than 100 images are constructed. Estimate the size of the single image. Typically the domain is about 15 to 20 Km in horizontal directions and

5 Km in the vertical one. Using the grids steps equal to 25 m in horizontal and 5 m in vertical direction one finishes up with the size of the domain of $0.5 \cdot 10^9$ points, or 2 Gb per a single image.

One of the principal aspects of the seismic imaging is that each source/receiver is used to compute all the output images. Thus even if data can be split into several datasets, each dataset will be used to compute all images, and the best hypothetical algorithm would be able to process whole data and provide the complete set of images. However, this is not affordable by modern computers, and we suggest algorithm which operates independently with several datasets, intensively use MPI communications, and provide several images corresponding to these datasets. A sketch of the data flow is provided in Fig. 1.

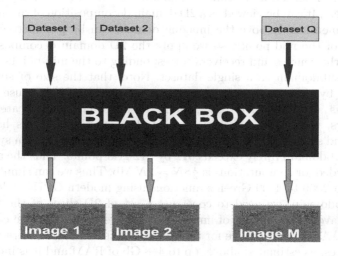

Fig. 1. A sketch of the data flow in the seismic imaging algorithm

3.2 Hypothetical Sequential Version

To simplify the description the parallel version of the algorithm for seismic imaging, let us consider a simple sequential version first. Assume a whole input seismic data, Green's functions for all sources and receivers positions, and all constructed seismic images can be stored in RAM. In this case, the algorithm will have the following structure. The outer loop is the integration with respect to the time frequencies. For each time-frequency we use a loop with respect to depth; i.e. we compute the Green's functions for all sources and receivers positions, combine them and multiply by the corresponding signal. Thus at fixed depth level we construct all possible images $I^k(x, y, z_l, \omega_m)$. Next we interpolate images within the slab $[z_{l-1}, z_l]$. Repeating this procedure for all frequencies and all depth layers we obtain all common-offset vector images at the same time.

However, as it follows from the estimates, presented in the previous section, the sequential version of the algorithm cannot be implemented using modern computers. Because, it requires storage of all fixed-depth 2D cross-section of the Green's functions Let us remind that the number of the sources and receivers is about 10^5, the size of single 2D cross-section of the Green's function is about 10^6 points, which leads to the estimate of 8 Mb per Green's function (single precision complex numbers). Thus storing all Green's functions would require 800 Gb, which is unacceptable. To overcome this and design parallel algorithm we suggest consider relatively small datasets, which fits the devise memory and process each dataset in parallel.

3.3   Datasets Construction

Construction of the datasets can be explained on the physical aspects of the solved problem. It can be viewed as a 2D domain decomposition of the acquisition system. As mentioned above the imaging domain can be parametrized by the coordinates of the mid-points, so we apply the 2D domain decomposition and consider all the sources and receivers corresponding to the mid-points belonging to a single subdomain as a single dataset. Note, that the size of subdomains is restricted by the amount of the memory per single GPU because the main computations are done on GPUs. Assume a single have N allocated sources and receivers, thus N 2D cross-sections of the Green's functions have to be computed and stored. Due to the typical geometry of the acquisition system, the size of the subdomain rarely exceeds 512 by 512 grid points. Thus the amount of memory, needed for computations is $\frac{1}{4}8N = 2N$ Mb. Thus we can simultaneously operate from 2000 to 4000 Green's functions using modern GPUs. Additionally, for each subdomain, we need to construct a set of 2D slices of the images at each depth layer. The number of images hardly exceed 100, thus we can neglect this amount. A single 3D image for the whole domain should be stored in RAM, which requires, as estimated above, up to 4–8 Gb of RAM and it is independent on the datasets.

Division of the data leads to the significant loss of the algorithm scaling if compared with the hypothetical sequential one. This happens because some of the sources/receivers belong to several datasets, thus we need to compute the Green's functions for these sources/receivers several times, whereas in the hypothetical algorithm we would do it only ones. In our, numerical experiments, we achieve the number of the Green's functions recomputations as low as 2.3 in average. Examples of the datasets for synthetic SEG Salt marine data and for onshore field data are provided in Fig. 2.

3.4   Parallel Algorithm MPI+OMP+CUDA

The parallel algorithm is similar to the sequential one, described above, but each MPI process operates with a single dataset. The block-scheme of the algorithm is provided in Fig. 3. So, the input data for a single MPI process is a single dataset. We perform the integration with respect to time-frequency in the outer loop. Then we use the loop with respect to the depth layers. We use GPU to compute the set of 2D cross-sections of the Green's functions (let us assume we

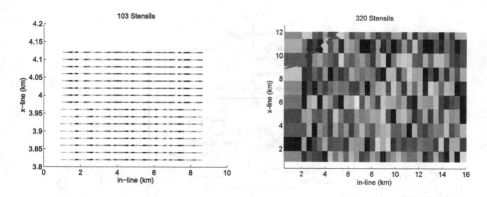

Fig. 2. Examples of the datasets (coordinates of the mid-points) for SEG Salt model (left), and real onshore seismic data (right).

proceed from layer z_j to z_{j+1}). Computations are based on the intense use of 2D qFFT. The set of images at the layer z_{j+1} are also computed by GPU. After that 2D slices of images are passed to RAM. While GPU constructs the solutions and images at the layer z_{j+1}, CPU sorts the 2D cross-sections of the images, use MPI send/recv to exchange the 2D cross-sections of images with other MPI processes, so that single-offset vector images for all datasets are accumulated by a single MPI process. After that, we use CPU to interpolate the accumulated image within the slab $[z_{j-1}, z_j]$ and sum it up to the final image. So, the results of one step with respect to the depth for a single MPI process are: set of 2D cross-sections of the Green's functions (all offsets, single dataset) at layer $z = z_{j+1}$; full set of 2D cross-sections of the images (all offsets, single dataset) at layer $z = z_{j+1}$; a single common offset image for all datasets on a fine grid within a slab $[z_{j-1}, z_j]$.

4 Numerical Experiments

4.1 Weak Scaling

Structure of the algorithm makes it difficult to estimate the strong scaling in its classical meaning; i.e. fixing a size of the problem and increase the number of MPI processes. However, we can estimate weak scaling, because increasing the number of MPI processes we process a larger number of datasets. We consider the SEG Salt model, for which we compute 17 images, having 2600 datasets. We perform simulations using 17, 34, 51, and 68 MPI processes. This means that a single node deals with one dataset at a time, but only 17 of them compute images. Measured times are provided in Table 1. The loss of the efficiency is caused by the MPI exchanges, which are implemented with enforced synchronization. Moreover, this algorithm allows using non-blocking procedures Isen/Irecv which can be performed during computations, however, in the presented version of the algorithm, it is not implemented.

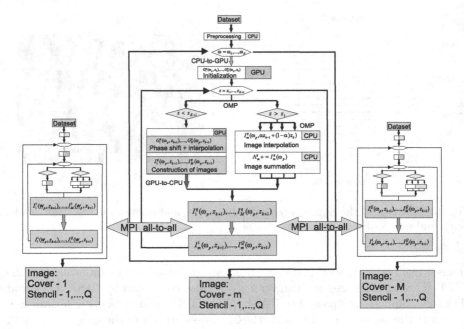

Fig. 3. A block-scheme of the parallel algorithm

Table 1. Computation time for weak scaling estimation.

	17 proc	34 proc	51 proc	68 proc
Time (hours)	1.91	2.22	2.64	3.02
W scaling (%)	-	86	72	63

4.2   Images Construction for SEG Salt Model

We use the algorithm to compute the image for the SEG Salt model with pre-computed seismic data. The size of the model is 7980 m InLine (x-direction) and 7920 m CrLine (y-direction). The depth of imaging is 4000 m. We use the grid with the steps 20 by 20 m in horizontal directions. Computations are done on a grid with step 50 m in the vertical direction, whereas the image is constructed on a grid with the step of 5 m in the vertical direction. We compute 17 common-offset vector images using 2600 datasets. We perform simulations using 51 nodes. Total computation time, was slightly less than the estimate, coming from Table 1. So, the total computation time is 5032 core-hours. We provide the 2D cross-sections of the model and the image in Fig. 4.

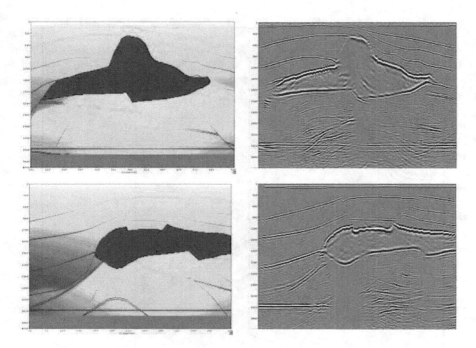

Fig. 4. Cross-sections of the SEG Salt model (left) and corresponding image (right), along a cross-line (upper row) and in-line (lower raw).

4.3   Real-Data Example

The second numerical experiment is done to construct the seismic image using real onshore seismic data. The size of the model is 15525 m InLine (x-direction) and 11250 m CrLine (y direction), depth is 5000 m. We use the grid with the steps 25 by 25 by 50 m for computations and 25 by 25 by 5 m for image construction. We compute 40 common-offset vector images, using 320 datasets. The geometry of the datasets is provided in Fig. 2. We use 40 nodes for simulations; i.e. equal to the number of images. One simulation takes almost 40 h, thus total wall-clock time is about 320 h, which leads to the 11520 node-hours to construct the set of 40 images. A slalom-line section of the obtained 3D seismic image is presented in Fig. 5.

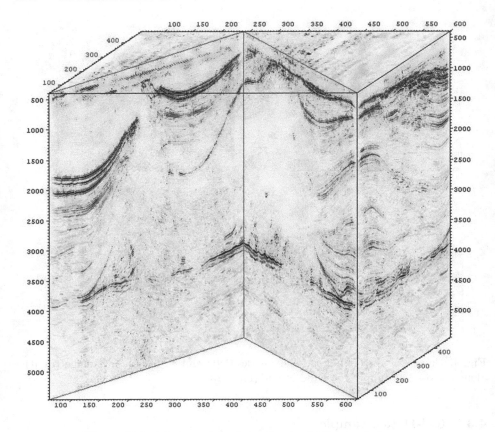

Fig. 5. Slalom-line section of the obtained 3D seismic image.

5   Conclusion

We presented an original algorithm of seismic migration, based on the solution of one-way wave equation. The algorithm combines MPI, OMP, and CUDA technologies. Dataflow is parallelized via MPI, so that each node deals with a single dataset. For computations of the Green's functions and the images for the dataset are performed by GPU. After that, the images are passed between the nodes using MPI. Additional, computations and I/O are implemented via OMP technology. As the result, we are able to compute the common-offset vector images by means of solving one-way wave equation, which is more accurate and informative than images constructed by conventional ray-based techniques.

Acknowledgements. This research was initiated and sponsored by Central Geophysical Expedition JSC of Rosgeo. V. Lisitsa and D. Vishnevsky are also thankful to Russian Foundation for Basic Research for partial financial support of this work, grans no. 18-05-00031, 18-01-00579, 16-05-00800. The research is carried out using the equipment of the shared research facilities of HPC computing resources at Lomonosov Moscow State University and cluster NKS-30T+GPU of the Siberian supercomputer center.

References

1. Claerbout, J.F.: Imaging the Earth's Interior. Blackwell Scientific Publication, Oxford (1985)
2. Etgen, J., Gray, S.H., Zhang, Y.: An overview of depth imaging in exploration geophysics. Geophysics **74**(6), WCA5-WCA17 (2009)
3. Fu, L.-Y.: Wavefield interpolation in the Fourier wavefield extrapolation. Geophysics **69**(1), 257–264 (2004)
4. Gazdag, J.: Wave equation migration with the phase-shift method. Geophysics **43**(7), 1342–1351 (1978)
5. Gazdag, J., Sguazzero, P.: Migration of seismic data by phase shift plus interpolation. Geophysics **49**(2), 124–131 (1984)
6. Lisitsa, V., Podgornova, O., Tcheverda, V.: On the interface error analysis for finite difference wave simulation. Comput. Geosci. **14**(4), 769–778 (2010)
7. Liu, H.W., Liu, H., Tong, X.-L., Liu, Q.: A Fourier integral algorithm and its GPU/CPU collaborative implementation for one-way wave equation migration. Comput. Geosci. **45**, 139–148 (2012)
8. Moczo, P., Kristek, J., Vavrycuk, V., Archuleta, R.J., Halada, L.: 3D heterogeneous staggered-grid finite-differece modeling of seismic motion with volume harmonic and arithmetic averaging of elastic moduli and densities. Bull. Seismol. Soci. Am. **92**(8), 3042–3066 (2002)
9. Moczo, P., Kristek, J., Galis, M.: The finite-difference modelling of earthquake motion: waves and ruptures. Cambridge University Press (2014)
10. Muir, F., Dellinger, J., Etgen, J., Nichols, D.: Modeling elastic fields across irregular boundaries. Geophysics **57**(9), 1189–1193 (1992)
11. Pleshkevich, A., Vishnevsky, D., Lisitsa, V.: Explicit additive pseudospectral schemes of wavefield continuation with high-order approximation. SEG Techn. Program Expanded Abs. **36**(1), 5546–5550 (2017)
12. Pleshkevich, A.L., Vishnevsky, D.M., Licitsa, V.V.: Development of pseudospectral amplitude-preserving 3D depth migration. Russian Geophys. (S), 94–101 (2017)
13. Stoffa, P.L., Fokkema, J.T., de Luna, F.R.M., Kessinger, W.P.: Split-step Fourier migration. Geophysics **55**(4), 410–421 (1990)
14. Virieux, J., Calandra, H., Plessix, R.-E.: A review of the spectral, pseudo-spectral, finite-difference and finite-element modelling techniques for geophysical imaging. Geophys. Prospect. **59**(5), 794–813 (2011)
15. Virieux, J., et al.: Seismic wave modeling for seismic imaging. Lead. Edge **28**(5), 538–544 (2009)
16. Vishnevsky, D., Lisitsa, V., Tcheverda, V., Reshetova, G.: Numerical study of the interface errors of finite-difference simulations of seismic waves. Geophysics **79**(4), T219–T232 (2014)
17. Zhang, J.-H., Wang, S.-Q., Yao, Z.-X.: Accelerating 3d Fourier migration with graphics processing units. Geophysics **74**(6), WCA129-WCA139 (2009)

Parallel Simulation of Community-Wide Information Spreading in Online Social Networks

Sergey Kesarev$^{(\boxtimes)}$, Oksana Severiukhina, and Klavdiya Bochenina

ITMO University, Saint Petersburg, Russia
kesarevs@gmail.com, oseveryukhina@gmail.com, k.bochenina@gmail.com

Abstract. Models of information spread in online social networks (OSNs) are in high demand these days. Most of them consider peer-to-peer interaction on a predefined topology of friend network. However, in particular types of OSNs the largest information cascades are observed during the community-user interaction when communities play the role of superspreaders for their audience. In the paper, we consider the problem of the parallel simulation of community-wide information spreading in large-scale (up to dozens of millions of nodes) networks. The efficiency of parallel algorithm is studied for synthetic and real-world social networks from VK.com using the Lomonosov supercomputer (Moscow State University, Russian Federation).

Keywords: Parallel simulation · Model of information spread
Online social networks

1 Introduction

Wide spread of online social networks (OSNs) gave rise to a variety of models aimed at simulating and forecasting processes of information assimilation and transmission between users. While first models of this type were purely abstract (e.g. linear threshold models of opinion dynamics operating on syntetic graphs like Erdos-Renyi or Barabasi-Albert), current availability of large amounts of OSN data allows to create more realistic, data-driven models of information cascades. These models avoid the assumption of the homogeneity of nodes that was intrinsic to earlier ones; instead, these models are aimed to reproduce aggregated dynamics of information spread from reactions of diverse individuals acting in frames of a given network topology. The main difference then is that friendship graph and parameters of users' behavior are extracted directly from the observed data.

Data-driven approach not only refines the basic models but also states new problems arising from the nature of data observed. In this paper, we consider the example of such a problem related to the existence of 'super-hubs' (communities) in OSN which numbers of subscribers are orders of magnitude larger than for

© Springer Nature Switzerland AG 2019
V. Voevodin and S. Sobolev (Eds.): RuSCDays 2018, CCIS 965, pp. 136–148, 2019.
https://doi.org/10.1007/978-3-030-05807-4_12

individuals (ordinary users). For example, a median number of subscribers for an individual in Russian OSN vk.com is equal to 180 while communities can have size of several hundreds thousands and even millions subscribers. Thus, simulating community-wise information spread requires consideration of large-scale networks. Together with the rapidity of informational processes in OSNs this leads to the necessity of applying parallel algorithms to obtain results of a forecast in time.

In this study, we present the parallel algorithm of the community-wise information spread supporting the flow of heterogeneous news (with account of daily rhythms) through the network of community subscribers, friends of subscribers etc. Presented algorithm (Sect. 3), on the one hand, is general in a sense that it supports embedding of arbitrary models of activities and reactions of OSN entities, but, on the other hand, is tuned to the specific properties of OSN. Section 4 presents the results of experiments on scalability of the algorithm for single- and multi-community cases, for real-world and synthetic test cases.

2 Related Work

The modeling of the process of information dissemination in social networks has a wide range of applications: the study of factors that affect the process of information dissemination, the prediction of reaction, the maximization of influence, rumor controlling, the evaluation of public opinion, etc.

In the task of modeling processes on large networks, researchers and developers are faced with a large number of problems associated with the heterogeneity of vertex types, the amount of data transferred, and the distribution of vertices by processes. Load balancing algorithms on graphs can be divided into several groups depending on the problem under consideration: (i) type of process (spreading information, epidemiological processes), (ii) network topology (real world network or artificial like small-world, scale-free), (iii) possibility of generalization (algorithms for specific tasks or multi-purpose systems for parallel graph processing).

Systems of the first class take into account features of the simulated process. The most common problem is the modeling of epidemiological processes. Examples of systems of this class are EpiFast [1], EpiSimdemics [2] and Indemics [3]. EpiFast has a master-slave computation model. Experiments show that it shows scalability to 224 processes for approximately 16 million of entities. However, major drawback which has an impact on scalability is the arbitrary assignment of vertices to processes containing approximately the same number of edges. Next example is EpiSimdemics, which uses semantics of disease evolution and disease propagation in large networks (up to 100 million) and considers geospatial constraints. The above mentioned approaches use the MPI standard for parallelisation. Another solution that takes into account features of the process under consideration is Indemics. It has an interface for work and functionality for stop the simulation at any point, find out the state of the simulated system and add additional interventions.

Modeling and simulation of social networks focuses more on other processes like reproducing complex social phenomena, such as spreading of information and its impact on others, maximization influence, identification of opinion leaders and what-if tasks. Hou [4] proposed framework named SUPE-Net for parallel discrete event simulation. It has utilities for social network generation, algorithms (PageRank and the SIR model in study), which show the scalability and effectiveness of this framework. To distribute the vertices of the network by processes, the authors use the algorithm CommPar [5], which has a community-based multilevel graph partition strategy. However, for this algorithm it is required to know the number of available processes. Nevertheless, it efficiently reduces the overhead of communications between processors. Another example is using parallel algorithms for community detection for networks, for example, work [6] can process billions of edges in short time (50M edges per second for the fastest algorithm).

Thus, solutions of this class most often provide scalability of parallel algorithms with additional constraints on the interaction of vertices, which significantly reduces the communication complexity of modeling.

In the second approach, modeling techniques are considered on networks with a special topology. For example, authors [7] consider a stochastic Kronecker graph, which allows to reproduce real-world networks and keep their important topological properties. Experiments were carried out for a sparse graph with a size up to one billion nodes.

Last group of systems are general systems for parallel processing of graphs. They use different computing models to implement parallelization. Pregel [8] system uses Bulk Synchronous Parallel model. Apache Giraph [9] is based on previous mentioned system Pregel, however, it has several improvements like master computation, sharded aggregators, edge-oriented input. Another approach is using asynchronous model in a shared-memory setting, for example, GraphLab [10].

In work [11], Aydın Buluc reviewed various algorithms for partitioning graphs into parts: global algorithms for small graphs or as local search in multilevel algorithms, iterative improvement heuristics which consistently improve the solution, multilevel graph partitioning and evolutionary methods. Besides this, there is solution for streaming graph partitioning [12], which able to compute an approximately balanced partitioning of graph's vertex. This solution uses degree-based criteria and reaches the results for single pass over the input data.

If there is an uneven distribution of vertices on different processes, there is an uneven computing load between the nodes. However, if different processes have a large number of common edges, then much time is required to transfer data between them. Thus, in order to reduce the running time of the program, it is necessary to distribute the vertices between processes in such a way that the most related components are processed in one process.

The existing algorithms do not take into account the presence of superspreaders that arise during modeling processes on networks. So for the task of spreading information from, superspreaders are communities, which have edges with a large number of subscribers, and individual users with a large number of subscribers.

3   Method

3.1   Model Description

In order to reproduce information spreading processes in cyberspace more naturally it is necessary to consider a lot of details such as types of entities, their possible actions and interactions between them. The model used for our experiments is described in detail in our previous paper [13]. Below there is a brief description of it.

The model consists of three main entities: informational messages (IMs), communities and users. Networks for information spreading include communities and users as vertices and relations between them (subscriptions or friendship relations) as edges. IM represents the post, which can be transferred between vertices.

The behavior of presented entities is defined by three internal models: model of IM's generation, model of activity and model of reaction. First model determines the time for the creation of new messages. Each message has following characteristics: topic, publication time, virality coefficient. Model of activity defines the status of each agent: active or inactive. The last internal model is responsible for the result of the user's interaction with messages: inaction, approval (like), the generation of a text message (comment), participation in the dissemination of information (share). Each message has counters that reflect the number of users' responses to the message at the current time.

The presented approach allows to use various independent internal models to adjust the necessary process for modeling. In addition, each model can be specified in the appropriate input file. For the generative model input can contain the distribution of the probability for generating messages depending on the day of the week and time of day. In the presented model, each user can have his own parameters for responding to messages, thus providing the heterogeneity of agents.

Thus, the presented model allows you to simulate various processes and configure model parameters, for e.g. using processed data from social networks.

3.2   Parallelization

The efficiency of the parallel simulation of the information spreading on a social network depends a lot on the uniformity of computational and communication load for different workers. In our previous work, a similar problem was already examined for the task of epidemiological process modeling (SIRS) on stochastic Kronecker graphs [7]. That study has shown that Master-Slave parallelizing approach shows the best parallel efficiency. A reincarnation of this approach is used in the current study to handle parallelization challenges of a social network graph.

Detailed algorithms for Slave and Master processes are presented in Algorithm 1 and Algorithm 2 respectively. The simulation is discrete and operates according to 3 internal models that were mentioned in the previous section: generative model G_m, activity model A_m and reaction model R_m. The algorithm is implemented in C++ using MPI standard for message exchange.

Masters and Slaves are two types of computational nodes, differing by their functionality. Slave nodes host subsets of a social graph and perform iterative updates of the system state according to internal models. The Master node is responsible for fast and non-redundant data forwarding between subnetworks. The responsibility to generate news and store user reaction statistics for them also rests with Master node. Primary reasons for choosing Master node instead of Slave nodes here are to circumvent the need to develop the synchronous news generating engine on each Slave node and to reduce synchronization time between Slave processes.

The goal of the simulation is to represent the process of the information spread step-by-step, not omitting any of transition states. This implies the necessity to maintain the relationship between each user and each piece of news for each moment of time. On the other side, memory usage in the process of the simulation should be as small as possible. Having this limitation in mind, we propose the news-oriented storing system. That is, for each piece of news on each Slave node there are defined three boolean masks that describe the state of publication regarding users on this node:

- potential viewers – stores 1 in positions corresponding to users who haven't yet seen this publication, but have it included in their news feed; for example, for some post that is published a moment ago, all subscribers of the community of this post are potential viewers of it (if they were not reading their news feed at the very moment of publication);
- spreaders – users who decided to share this piece of news with all their subscribers;
- viewers – users who have already seen this publication.

On each iteration, Slave process receives the list of generated news from the Master node (line 2 in Algorithm 1). Then each publication is processed in the chronological order, from newest to oldest posts (line 3).

In the first phase, the list of potential viewers of this publication is examined (line 4): we attempt to show the publication to the user and gather his feedback about it. If the user is not active according to the activity model A_m, the algorithm moves on to the next potential viewer, and this one stays in the list to be checked on the next iteration (lines 6). If the user is active, the flow continues. Algorithm checks according to reaction model R_m whether the user likes the publication or wants to comment it (lines 8–10), and if he does, the algorithm increases corresponding counters for the publication. The next check is whether the user wants to share the publication (lines 12), and if he does, he is added to the bit mask of spreaders of this publication. When the reaction is gathered, the user is deleted from the bit mask of potential viewers (line 13) and is added to the list of those who saw this publication (line 14).

The second phase of processing each publication is processing of the list of spreaders (which was constructed while examining potential viewers). Each subscriber of each spreader from the list is offered this publication, i.e., is

Input : Activity model A_m, reaction model R_m, news feed in the
chronological order N, a number of iterations T, M — master node
for worker w, **all_masters** — the list of all Master processes

```
1   for t from 1 to T :
2       syncNews(M);
3       foreach n ∈ N :
```

4	**foreach** $p \in potentialViewers(n)$:	15	**foreach** $s \in spreaders(n)$:
5	**if not** $A_m.isActive(p,t)$:	16	**for** $e \in s.edges$:
6	continue	17	**if** dest$(e) \neq w$:
7	**if** $R_m.isLike(n,p,t)$:	18	send_pools[dest(e)] $\leftarrow (n,e)$
8	$addLike(n,p,t)$	19	**else:**
9	**if** $R_m.isComment(n,p,t)$:	20	**if not** $isViewed(n,e)$:
10	$addComment(n,p,t)$	21	$addPotentialViewer(n,e)$
11	**if** $R_m.isRepost(n,p,t)$:	22	$deleteAllSpreaders(n)$;
12	$addRepost(n,p,t)$	23	$sendInfoMessageToMaster(M)$;
13	$deletePotentialViewer(n,p)$;	24	$sendPoolsToMaster(M)$;
14	$addViewer(n,p)$;	25	**for** $each\ (n,e)\ in$
			$recv\_pools($**all_masters**$)$:
		26	**if not** $isViewed(n,e)$:
		27	$addPotentialViewer(n,e)$

Algorithm 1. Parallel simulation scheme for the information spreading process (for a Slave worker w)

added to the list of potential viewers of this publication (lines 15–21). If a particular subscriber is not hosted on the current computational node, information about him and this publication is added to the message that will be sent to the Master node (lines 18). If a subscriber is on the current node, then, if he has not already seen the publication, he is added to the list of potential viewers of this publication (line 21). By this moment the publication is set to be offered to all subscribers, and the list of spreaders can be deleted (line 22).

When the algorithm processed in the described way all publications, it is time to share results with other parts of the simulation system. After processing spreaders of each news we obtained a set of pairs (n, e) of a user e from some other node who became a potential viewer of a post n (see line 21). First of all, the Slave node sends an informational message to Master node (line 23). This message shows how many changes $((n, e)$ pairs) should be sent to the each other Slave node in the system. Then a message containing all pairs grouped by hosting computational nodes is sent (line 24). When the Master node has processed this message along with others (processing details are thoroughly described below), it sends back a message with the list of changes for the current Slave node. This message contains pairs (n, e) obtained in the process of information spreading on other Slave nodes. Each pair is processed as usual: if e has not yet seen n, then e is added to the list of potential viewers of n. Only after processing all these messages a node can move on to the next iteration.

Input : $i*$ – an index of a master process, G_m – generative model, A_m – activity model, N – news feed in the chronological order, T – number of iterations, `all_leafs` – the list of all Slave nodes, `own_leafs` – the list of Slaves that obey directly to this Master node

1 **for** i *from 1 to* T :
2   **if** $i* = 0$:
3      $generteNews(t)$;
4      $syncNewsWithLeafs($`all_leafs`$)$;
5   **foreach** $l \in$ `own_leafs` :
6      $infos \leftarrow recvInitInfo(l)$;
7    $consolidatedPool \leftarrow allocateMemory(infos)$;
8   **foreach** $l \in$ `own_leafs` :
9      $nonBlockingRecv(l, consolidatedPool, infos)$;
10   $waitForAllRecvs()$;
11   $types \leftarrow createMpiDatatypes(infos)$;
12   $dispatchPool($`all_leafs`$, types, consolidatedPool)$;

Algorithm 2. Parallel simulation scheme for the information spreading process (for a Master worker w)

The primary task of a Master process on each iteration is to synchronize states of its Slaves. Generally speaking, there may be several Master processes, and each Master node cares only for a subset of Slaves (which is called `own_leafs` in the Algorithm 2). However, if there are several Master nodes, we still have to choose the only one which will be responsible for news generation. For this selected Master node each iteration starts with generating news according to generative model G_m and syncing them with all Slave nodes (lines 2–4). The next step is receiving informational messages from Slave nodes (line 5). Master node allocates memory for the consolidated pool according to the received informational message (line 7). When the memory is allocated, Master node receives detailed information from its Slaves to this memory (lines 8–9).

The only thing left is to send to each Slave node all changes that are related to this node. For each message that will be sent to the Slave node, we are creating an MPI datatype according to informational messages received before (line 11). This approach allows avoiding additional memory allocation. Then these messages are sent to Slave nodes and are processed by them (line 12).

4   Experiments

4.1   VK Dataset Description

The first dataset is the example crawled from the popular Russian social network VK.com (hereafter it will be called VK dataset). The crawled graph consists of one big charity community (294,345 members), all its members and all their subscribers. The graph contains 33,768,036 vertices (i.e. accounts in social network), and only 227,473 of them have at least one outgoing edge (i.e. at least one friend). Note that this is not because other 33 millions accounts do not have friends, but because the fact of having friends other than community members is not recorded. That means there are at least 66,872 community members that do not have subscribers at all and that there are 5,262,115 nodes who are friends to two or more community members at the same time. A more thorough description is given in our recent work [13].

4.2   Artificial Dataset Description

Characteristics of the VK dataset allow to conclude that an n-ary tree might be a good rough approximation of VK graph. Let us think about the community as a tree root which has a big pre-specified number of child nodes: these will be community members. Each of these community member nodes has a fixed pre-specified number of child nodes, which will represent their friends. These friend sets are not intersected (i.e. there are no nodes that are subscribed to more than one community member). A forest of such trees was used as an artificial dataset to test some hypotheses about algorithm efficiency with respect to the number of communities. The graph scheme is presented in Fig. 1.

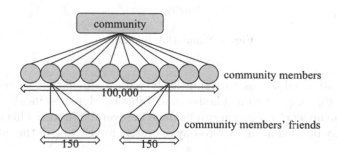

Fig. 1. Artificial dataset scheme

4.3   Results on the Natural Data

All experiments were performed using Lomonosov supercomputer (Moscow State University, Russian Federation) which has 4,096 cluster nodes with Intel Xeon X5570 2.93 GHz processor and 12 GB of RAM per node. Each cluster node can hold up to 8 processes.

The algorithm from the previous section was applied at first to the VK dataset and run for 3,000 iterations (500 h of model time) in several paral-lelization settings: a sequential version and parallel version on 3, 8, 16, 32 or 64 processes.

Simulation dynamics are presented in Fig. 2. Computational load does not significantly increase in time despite the growing number of publications in the system, even though publication processing is the main source of computational load in the system.

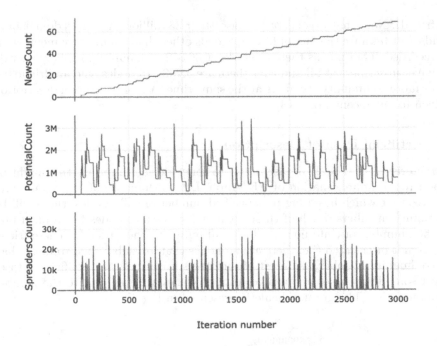

Fig. 2. Simulation dynamics

Parallelization allows us to significantly reduce simulation time (which is shown in the Fig. 3, exact timing is shown in Table 1). More specifically, modeling of 500 h of the network evolution can be done in several minutes. This means the algorithm may be used in the day-to-day operative forecasting of the information flow.

However, even with sufficient computation-to-communication ratio (around $2/3$ of iteration time is spent on computations, mostly processing potential viewers, and $1/3$ on communication, see Fig. 4) the parallel efficiency of the algorithm is low. This is because the computational load is not increasing during simulation.

Table 1. Simulation time by the number of processes on VK dataset

Number of processes	Simulation time, seconds
1	19820.20
3	13187.40
8	10180.00
16	6959.38
32	3513.66
64	1793.57

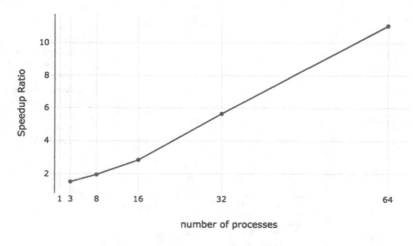

Fig. 3. Speedup ratio on the VK dataset for the different number of processes

Computational load increases when information cascades become deeper and more intensive. Intensity could be increased either by increasing probabilities of the news generating model or by increasing the number of communities in the dataset. Increasing probabilities, however, makes the model less plausible. Therefore for deeper understanding of the algorithm properties, it was run on an artificial network with several communities. This network was described in details in the Sect. 4.2.

4.4   Results on Artificial Data

The goal of the experiment with artificial data was to understand the relation between parallel efficiency and the number of communities in the network.

Parallel efficiency for different numbers of processes for 3,000 iterations is presented in the Fig. 5. Exact timings are available in the Table 2. It is clearly seen there that the parallel efficiency of the algorithm increases with the number of communities in the graph. Almost exponentially decreasing simulation time when the number of processes increases from 16 to 32, 64 and 128 for any number of communities can mark the insufficient computational load in the simulation.

Table 2. Simulation time by the number of processes on artificial dataset, seconds

Processes:	1	3	8	16	32	64	128
Communities							
5	28420.10	16017.30	8001.07	5391.60	2893.30	1570.42	978.86
4	20393.20	11523.10	7263.05	4387.40	2246.57	1212.05	759.22
3	14509.60	8239.09	5966.98	3485.21	1683.66	896.08	634.81

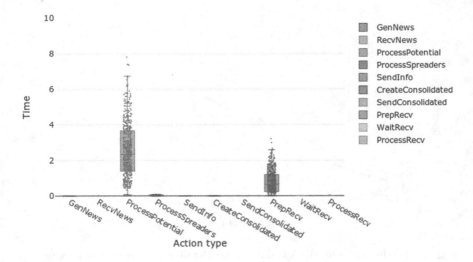

Fig. 4. Time spent on different stages of simulation (8 processes)

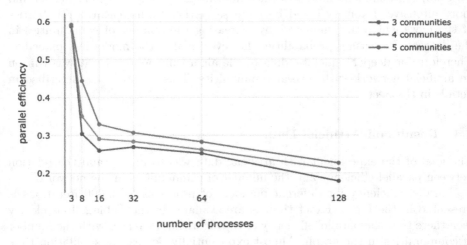

Fig. 5. Parallel efficiency on artificial data

That is why future research directions are towards modeling a multi-community landscape with mixing audience, creating specialized load balancing algorithms for such communities and involving Master processes in the active computations.

5 Conclusion

We presented an algorithm for parallel simulation of community-oriented information spread which allows easy modification of behavioral patterns of users and communities via probabilistic models. The algorithm was applied to the subset of the VK.com, the largest Russian social network. The subset contained all members of a big charity community and all their friends. Experimental results show the applicability of this algorithm for day-to-day modeling of planned information cascades. However, experiments also revealed insufficient computational load provided by this dataset. For the detailed study of algorithm properties it was applied to the artificial dataset of similar structure that included several communities. The increasing number of communities increases the number of news and therefore rises up computational load, allowing the proposed method to finally prove its applicability.

Acknowledgements. This research was supported by The Russian Scientific Foundation, Agreement #14-21-00137-П (02.05.2017). The research was carried out using the equipment of the shared research facilities of HPC computing resources at Lomonosov Moscow State University.

References

1. Bisset, K.R., Chen, J., Feng, X., Kumar, V.S.A., Marathe, M.V: EpiFast: a fast algorithm for large scale realistic epidemic simulations on distributed memory systems. In: Proceedings of the 23rd International Conference on Supercomputing, pp. 430–439 (2009)
2. Barrett, C., Bisset, K.R., Eubank Stephen, G., Feng, X., Marathe, M.: EpiSimdemics: an efficient and scalable framework for simulating the spread of infectious disease on large social networks. In: International Conference for High Performance Computing, Networking, Storage Analysis (2008). https://doi.org/10.1145/1413370.1413408
3. Bisset, K.R., Chen, J., Deodhar, S., Feng, X., Ma, Y., Marathe, M.V.: Indemics: an interactive high-performance computing framework for data intensive epidemic modeling. ACM Trans. Model. Comput. Simul. **24**, 1–32 (2014). https://doi.org/10.1145/2501602
4. Hou, B., Yao, Y., Wang, B., Liao, D.: Modeling and simulation of large-scale social networks using parallel discrete event simulation. Simulation **89**, 1173–1183 (2013). https://doi.org/10.1177/0037549713495752
5. Hou, B., Yao, Y.: COMMPAR: A community-based model partitioning approach for large-scale networked social dynamics simulation. In: Proceedings - IEEE International Symposium on Distributed Simulation and Real-Time Applications, DSRT. pp. 7–13. IEEE (2010)
6. Staudt, C.L., Meyerhenke, H.: Engineering parallel algorithms for community detection in massive networks. IEEE Trans. Parallel Distrib. Syst. **27**, 171–184 (2016). https://doi.org/10.1109/TPDS.2015.2390633
7. Bochenina, K., Kesarev, S., Boukhanovsky, A.: Scalable parallel simulation of dynamical processes on large stochastic Kronecker graphs. Futur. Gener. Comput. Syst. **78**, 502–515 (2017). https://doi.org/10.1016/j.future.2017.07.021

8. Malewicz, G., et al.: Pregel: a system for large-scale graph processing. In: Proceedings of the 28th ACM Symposium on Principles of Distributed Computing - PODC 2009, p. 6 (2009)
9. Apache Giraph. http://giraph.apache.org/
10. Low, Y., Gonzalez, J., Kyrola, A., Bickson, D., Guestrin, C., Hellerstein, J.M.: Distributed GraphLab: A Framework for Machine Learning in the Cloud, pp. 716-727 (2012). https://doi.org/10.14778/2212351.2212354
11. Buluc, A., Meyerhenke, H., Safro, I., Sanders, P., Schulz, C.: Recent Advances in Graph Partitioning. CoRR. abs/1311.3, (2013). https://doi.org/10.1007/978-3-319-49487-6_4
12. Tsourakakis, C.E.: Streaming Graph Partitioning in the Planted Partition Model (2014). https://doi.org/10.1145/2817946.2817950
13. Severiukhina, O., Bochenina, K., Kesarev, S., Boukhanovsky, A.: Parallel data-driven modeling of information spread in social networks. In: Shi, Y., Fu, H., Tian, Y., Krzhizhanovskaya, V.V., Lees, M.H., Dongarra, J., Sloot, P.M.A. (eds.) ICCS 2018. LNCS, vol. 10860, pp. 247–259. Springer, Cham (2018). https://doi.org/10.1007/978-3-319-93698-7_19

Technique for Teaching Parallel Programming via Solving a Computational Electrodynamics Problems

Sergey Mosin, Nikolai Pleshchinskii, Ilya Pleshchinskii, and Dmitrii Tumakov(✉)

Kazan Federal University, Kazan, Russia
dtumakov@kpfu.ru

Abstract. Three-dimensional problems of computational electrody-namics for the regions of complex shape can be solved within the rea-sonable time only using multiprocessor computer systems. The paper discusses the process of converting sequential algorithms into more effi-cient programs using some special techniques, including object-oriented programming concepts. The special classes for data storage are recom-mended to use at the first stage of programming. Many objects in the program can be destroyed when optimizing the code. Special attention is paid to the testing of computer programs. As an example, the problem of the electromagnetic waves diffraction by screens in three-dimensional waveguide structures and its particular cases are considered. The tech-nique of constructing a parallel code for solving the diffraction problem is used in teaching parallel programming.

Keywords: Parallel programming teaching · Effective program
Computational electrodynamics

1 Introduction

Parallel programming is becoming more widespread with increasing requirements for resources needed to solve problems and with development of computer tech-nology. Currently, the three most widely used parallel programming technologies are OpenMP [1–3], MPI [3–5] and CUDA [6–8]. Each of them is oriented to a certain type of multiprocessor computer systems. All these three technologies are important for training professional programmers at bachelor and master levels. The programmer should be able to chose the most suitable technology and take into account a particular computer system for solving the considered problem.

There is a large amount of literature devoted to algorithms for the paralleliza-tion of classical problems, for example [9–12]. However, the issues of "correct" training in parallel programming technology remain relevant to this day. Stu-dents, who study parallel programming for the first time, deal with difficulty in

V. Voevodin and S. Sobolev (Eds.): RuSCDays 2018, CCIS 965, pp. 149–158, 2019.
https://doi.org/10.1007/978-3-030-05807-4_13

understanding some specific features due to essential difference between implementations of basic methods using parallel and logical/structured programming. Different approaches to teaching are considered, starting from the students of the first years [13] and ending with the Ph.D. students [14]. It is also obvious that the training should take into account the peculiarities in the preparation of students for certain specializations [15].

In this paper, we propose a methodology for teaching bachelor and master students the basics of parallel programming via the problems of electrodynamics. Computational electrodynamics contains a wide range of problems, for the numerical solution of which the methods of parallel programming are successfully used. The key problems of waveguide electrodynamics are problems for electromagnetic waves diffraction by thin conducting screens located in waveguide structures. In the three-dimensional case, the numerical solution of such problems is associated with high computational cost. We will show on a specific example (and its special cases) that (1) algorithms for numerical solution for the problems of the electromagnetic wave diffraction by screens allow effective parallelization and (2) development and debugging of computer programs for solving such problems contribute to the mastering the modern parallel programming technologies.

2   Infinite Systems of Linear Algebraic Equations in Diffraction Problems

The problem of wave diffraction by the screen in the general formulation is as follows. Suppose that an infinitely thin perfectly conducting screen is placed in the cross section of a cylindrical waveguide with metal walls. The electromagnetic wave falls on the screen. It is necessary to find the electromagnetic field that arises from its diffraction.

Any electromagnetic field in a cylindrical waveguide with conducting walls can be represented as an overlay of a superposition set of eigenwaves of TE- and TM-polarization [16]. Therefore, the problem of determining the diffracted electromagnetic field is reduced to determining the coefficients of the expansion of this field in the eigenvectors of the waveguide.

The work [17] shows that the diffraction problem under consideration is equivalent to two independent infinite systems of linear algebraic equations (ISLAE) with respect to the unknown coefficients A_k^-, b_k^- of the field expansion:

$$-a_k^- \lambda_k + \sum_{m=0}^{+\infty} a_m^- \gamma_m \lambda_m \sum_{n=0}^{+\infty} \frac{1}{\gamma_n} I_{m,n}^\varphi J_{n,k}^\varphi = \sum_{m=0}^{+\infty} a_m^0 \lambda_m I_{m,k}^\varphi, \quad k = 0, 1, \dots \quad (1)$$

$$-b_k^- \chi_k + \sum_{m=0}^{+\infty} b_m^- \delta_m \chi_m \sum_{n=0}^{+\infty} \frac{1}{\delta_n} I_{m,n}^\psi J_{n,k}^\psi = \sum_{m=0}^{+\infty} b_m^0 \chi_m I_{m,k}^\psi, \quad k = 0, 1, \dots \quad (2)$$

Here

$$I^{\varphi}_{m,n} = \iint\limits_{\mathcal{M}} \varphi_m(\xi,\eta)\,\varphi_n(\xi,\eta)\,d\xi\,d\eta, \quad J^{\varphi}_{m,n} = \iint\limits_{\mathcal{N}} \varphi_m(\xi,\eta)\,\varphi_n(\xi,\eta)\,d\xi\,d\eta,$$

$$I^{\psi}_{m,n} = \iint\limits_{\mathcal{M}} \psi_m(\xi,\eta)\,\psi_n(\xi,\eta)\,d\xi\,d\eta, \quad J^{\psi}_{m,n} = \iint\limits_{\mathcal{N}} \psi_m(\xi,\eta)\,\psi_n(\xi,\eta)\,d\xi\,d\eta,$$

(3)

where $\psi_m(x,y)$ and χ_m, $m = 0,1,\ldots$ are eigenfunctions and eigenvalues of the eigenvalue problem:

$$\Delta\psi_m + \chi_m\,\psi_m = 0, \quad \psi_m\big|_C = 0, \tag{4}$$

$\varphi_m(x,y)$ and λ_m, $m = 0,1,\ldots$ are eigenfunctions and eigenvalues of the eigenvalue problem:

$$\Delta\varphi_m + \lambda_m\,\varphi_m = 0, \quad \frac{\partial\varphi}{\partial\nu}\Big|_C = 0, \tag{5}$$

$\gamma_m = \sqrt{k^2 - \lambda_m}$, $\delta_m = \sqrt{k^2 - \chi_m}$, C is the boundary of the waveguide cross section where the cross section consists of two parts: $\mathcal{M}$ corresponds to the screen and $\mathcal{N}$ is the remaining part of the section.

Thus, the first obvious step when parallelizing the algorithm is the definition of TE- and TM-fields using independent computing nodes.

3 Parallel Calculations and Verification of Results of the Calculations

The diffraction problem is reduced to two independent ISLAE. An approximate solution of each of them can be found by the truncation method: it is necessary to retain a finite number of unknowns (assuming that the remaining ones are equal to zero) and the same number of equations. Thus, the solution of the original problem reduces to the solution of two systems of linear algebraic equations (SLAE).

Note that the truncation order can not be established a priori. Even if estimates of the approximate solution error were obtained (this could be done only in a few simple special cases), then they are very crude. Therefore, it is assumed that a series of calculations will be performed, during which it becomes possible to establish reasonable values for the truncation parameters.

The computational algorithm consists of three main steps during the numerical solving of the SLAE. Firstly, the integrals I and J are calculated. Then the values of the coefficients of the SLAE matrix and the vector of the right-hand sides are calculated. Finally, a solution to the SLAE is obtained. It is easy to see that each stage of the algorithm allows parallelization.

It should be expected that the preparation of coefficients matrices and vectors of the right-hand sides of the SLAE will require significantly more CPU time than solving the final SLAE. Therefore, the analysis of existing and the development of new optimal parallel methods for solving the SLAE are not of primary importance.

The reliability of the computing results should be placed to the foreground. Therefore, the writing and debugging of the program are reasonable to start with the simple special cases of the general problem. Obtained result should be tested at each step of the computational experiment. Series of tests are developed for this purpose.

The following tests can be recommended for the computational problems of the class under consideration.

1. With increasing the order of approximation (that is, with an increase in the number of unknowns and the number of equations). The convergence of the sequence of approximate solutions to the exact solution must be observed. Since the exact value is unknown, it is to be expected that the sought values will change insignificantly as the dimension of the SLAE increases, beginning with some certain dimension.
2. The law of conservation of energy must be fulfilled: how much energy a wave brought from an external source to an inhomogeneity, the same amount in total must go from heterogeneity to infinity. The energy balance should improve when the dimension of the SLAE increases. The empirical experience shows that this happens when the accuracy of calculating the coefficients matrix of SLAE is increased.
3. In limiting cases, from physical considerations, it is possible to determine what solution of the problem should be like. For example, if the conducting screen in the section of the waveguide structure has a small area, then the field scattered from the screen should become insignificant. Conversely, if the screen occupies almost the entire cross-section, then the original wave should almost completely be reflected from the screen.
4. Finally, if we are dealing with problems of the electromagnetic waves diffraction by conducting bodies, then it is most important to check whether the boundary condition on the metal is satisfied (the tangential component of the electric vector must be zero).

The first three checks are based only on the necessary conditions for the reliability of the computing results. The fourth test is more important, because if we are seeking a solution of the diffraction problem in the form of expansions in the waveguide waves, then this guarantees that a number of requirements are automatically satisfied. These requirements include the Maxwell equations, the conditions on the waveguide walls, and the conditions for the coupling of the fields in the waveguide section outside the screen. In fact, the coefficients of these expansions must be found in such a way that the conditions on the conducting screen are satisfied. It is clear, that these conditions will never be fulfilled exactly due to approximate calculations.

4   Fast Programming or Effective Code?

There are two parameters which are considered at a program implementation of the computational algorithms. The first one is operationability of the proposed algorithm and the second one is effectiveness of the program providing minimal time cost on a computation.

The two-dimensional diffraction problem of a electromagnetic eigenwave by a transverse partition in a plane waveguide with metal walls is considered as a simple special case of the general diffraction problem. The problem is reduced to SLAE (after truncating the ISLAE) for the TE-polarization field.

$$- a_k + \sum_{n=1}^{N} a_n \gamma_n \sum_{m=1}^{M} \frac{1}{\gamma_m} I_{n,m} J_{m,k} = I_{l,k}, \quad k = 1..N, \tag{6}$$

a_n, $n = 1..N$ are the unknown coefficients. The propagation constants (complex)

$$\gamma_n = \sqrt{k^2 - \left(\frac{\pi n}{h}\right)^2} \tag{7}$$

are defined according to the following condition

$$\mathrm{Re}\,\gamma_n > 0, \quad or \quad \mathrm{Im}\,\gamma_n > 0. \tag{8}$$

$k = 2\pi/\lambda$ is the wavenumber, λ is the wavelength, h is the thickness of the waveguide. All these quantities must have the same order and $\lambda < 2h$ in order to have, at least, γ_1 being real. The integrals in (6) have the following form:

$$I_{m,n} = \int_{\mathcal{M}} s_n(t)\, s_m(t)\, dt, \quad J_{m,k} = \int_{\mathcal{N}} s_m(t)\, s_k(t)\, dt = \delta_{m,k} - I_{m,k}, \tag{9}$$

where

$$s_n(t) = \sqrt{\frac{2}{h}} \sin \frac{\pi n}{h} t. \tag{10}$$

Here $\mathcal{M}$ is the part of the segment $[0, h]$ occupied by the screen, and $\mathcal{N}$ is the remaining part of $[0, h]$. In the simplest case, $\mathcal{M} = [\alpha, \beta] \subset [0, h]$; in the more general case, $\mathcal{M}$ includes several segments. The numbers M and N are parameters of the method; l is the number of an eigenwave incident on the partition.

Taking this case study as a starting point, it is convenient to begin studying the features of using the OpenMP parallel programming technology.

The following procedure is recommended in the work [18]. C++ programming language is used for the algorithms description. All the initial data are declared as global constants. Functions which return the values γ_n, I_n, J_n are considered at the first stage of programming. The standard C++ libraries for working with complex numbers may be used since almost all values in the program are the complex numbers. Alternatively the own proprietary software may be prepared as a special library. For example, we need to describe the class "comp" for declaring complex-valued variables. We describe the necessary operations on the complex numbers as functions of the class members.

When using standard libraries, we should pay attention, for example, to how the branches of multi-valued functions are selected. For example, in our problem, either $\operatorname{Re} \gamma_n > 0$, or $\operatorname{Im} \gamma_n > 0$. Is this possible if we extract the root of a complex number using the standard function?

The classes "covect" and "comatr" are prepared in order to store the complex values of the elements of vectors and matrices. The indices of elements are changed in the same way as in the original formulas in order to reduce the probability of error in accessing the elements of such arrays. It also makes sense to overload the operator "()" to access these elements.

If the first version of the program confirms the operability of the algorithm, then the multiple calls of the functions γ_n, I_n, J_n will be replaced by the calls to array elements with the same names. Of course, these arrays must be filled before calculating the coefficients of the SLAE matrix. At the same time, the computing time will be significantly reduced, but more RAM space will be required.

Now one can proceed to convert the serial code to parallel code. OpenMP technology allows to do this most simply. The first skills of parallel programming are easy to acquire if we use the directive "#pragma omp parallel for", understand the difference between common and private data and learn the reduction (when computing internal sums) in the example under consideration.

It should not be expected the significant decreasing of the computation time when the loops are parallelized in such a simple problem. Rather, it is even the other way around. In fact, the cost of creating the additional threads, allocating local memory and some other actions overlap the gain from acceleration when parallelizing loops with a number of iterations less than 2000.

The CUDA technology is also well mastered via this simple example. Since the solving SLAE requires significantly less time than preparing the coefficient matrix and the right-hand side vector, it makes sense to transfer only these preliminary calculations to the device. One or more cores should calculate the arrays γ_n, I_n, J_n and calculate the elements of the coefficients matrix and the vector of the right-hand sides. Then, the matrix and the vector are sent to the host and the SLAE is solved.

The time for transferring data from the device's memory to the host's memory and back is the critical parameter for CUDA technology. The advisability of having the dimensions of arrays as multiples of the dimension for the threads (warp) has to be assessed. We also recommend teaching how to use the constant and shared memory of the device.

5   Benefit and Harm of Object Programming

As already mentioned, at the first stages of programming, it makes sense to define special classes for storing data with encapsulated operations on them. This approach significantly reduces the time spent on the debugging programs, and simplifies the perception of their text. In addition, the probability of error is reduced, since the programmer does not have to rewrite (copy) several fragments of the code.

At the same time, the use of object programming ideas often increases the computing time and, therefore, reduces the effectiveness of programs. This is most noticeable when using CUDA technology. There is an opinion that object programming is not compatible with the technology of computing on a GPU. But this is not so; convincing examples are given in the book [6]. We only need to have two assignment-compatible class descriptions, one for the host, and one for the device.

If the problem of the electromagnetic wave diffraction in a rectangular waveguide is considered, then it is customary to enumerate the eigenfunctions and eigenvalues using two indices:

$$\psi_{mn} = \sin \frac{\pi m x}{a} \sin \frac{\pi n y}{b}, \quad m, n = 1, 2, \dots,$$

$$\varphi_{mn} = \cos \frac{\pi m x}{a} \cos \frac{\pi n y}{b}, \quad m, n = (0), 1, \dots, \tag{11}$$

$$\chi_{mn} = \lambda_{mn} = \sqrt{k^2 - (\pi m/a)^2 - (\pi n/b)^2}. \tag{12}$$

Independent ISLAE in the diffraction problem have the same form as the system (1), (2), but all sums become double. In this case, it is necessary to provide two classes of constructors for the corresponding classes and to arrange an access to the elements of matrices and vectors in two ways: by double number and by number in the linear list. Here is one of the examples:

```
class covect {
    comp* p;
protected:
    int imi, ima, ile;
    int jmi, jma, kmi, kma, jle;
public:
    covect(int imii, int imaa);
    covect(int jmii, int jmaa, int kmii, int kmaa);
    ~covect() delete [] p;
    comp& operator ()(int i) return p[i-imi];
    comp& operator ()(int j, int k) return p[j-jmi+(k-kmi)*jle];
};
```

The main drawback of the object-oriented approach is related to the need to take care of address alignment when accessing global memory at optimization CUDA programs. Otherwise, the time required to transfer data becomes too large. In this sense, the structure of the arrays is better than the array of structures. In our case, this should be understood as follows: a pair of vectors of real and imaginary parts of complex numbers will be processed faster than a vector of complex numbers.

Experience shows that object-oriented programming is very effective for testing the efficiency of a numerical method (especially if the programmer already has well-established libraries with a description of "live" data). But when we need to optimize the code, we have to "disassemble" the classes into separate parts. As a rule, after this we have to search for errors in the program for a long time.

Similar examples appear in the numerical solving of problems of diffraction by screens in layered media [20], by spherical screens [19] and by doubly periodic gratings [18].

6   Summary

Let us single out the main advices for parallel programming of electrodynamics problems.

1. Write a serial code for solving the problem.
2. Use object-oriented programming to improve understanding and facilitate debugging when writing the code.
3. Carry out experiments. For example, the influence of the truncation parameters on the solution of the problem has to be monitored at solving the ISLAE. Check the satisfaction of physical laws for the solution obtained.
4. Redefine the library functions for complex data types, if necessary. For example, it can be the extraction of the square root.
5. Provide an access to both of the two indices and one index (the number in the linear list) for problems with two-index arrays.
6. Abandon the object-oriented data, especially when using CUDA for parallelizing.
7. Use the array structure instead of an array of structures, especially in CUDA programs.

7   Conclusions

The advices for getting started with OpenMP technology and some recommendations for teaching CUDA programming are given in the paper. The teaching technology developed in parallel programming is used at the Department of Applied Mathematics of the Institute of Computational Mathematics and Information Technologies of Kazan Federal University when conducting classes with first-year undergraduates in the specialization "Applied Mathematics and Informatics". The total number of students is ~30 people. The content of the lectures is in line with the textbooks [3,8]. In the laboratory classes, various problems are offered for unassisted solution. Parallel programming technology can be also used for R&D.

We recommend to use Visual Studio tools or open-source g++ compilers (MinGW package) for the Linux platform for mastering the parallel programming on the base of C++ language. Note also that you can improve the training program, supplementing it with the study of specialized patterns [21].

Acknowledgements. The work is performed according to the Russian Government Program of Competitive Growth of Kazan Federal University.

References

1. Antonov, A.S.: Parallel Programming using OpenMP technology: textbook. Izd-vo MGU, Moscow (2009). [in Russian]
2. Levin, M.P.: Parallel programming with OpenMP: textbook. Laboratoriya znanij, Moscow, BINOM (2012). [in Russian]
3. Pleshchinskii, N.B., Pleshchinskii, I.N.: Multiprocessor computing systems. Parallel programming technologies: textbook. Izd-vo Kazan. un-ta, Kazan (2018). [in Russian]
4. Korneev, V.D.: Parallel programming in MPI. IKI, Moscow-Izhevsk (2003). [in Russian]
5. Grishagin, V.A., Svistunov, A.N.: Parallel programming based on MPI. Textbook. Izd-vo NNGU im. N.I, Lobachevskogo, Nizhny Novgorod (2005). [in Russian]
6. Sanders, J., Kandrot, E.: CUDA by Example an Introduction to General-purpose GPU Programming. Addison-Wesley (2010)
7. Boreskov, A.V., Kharlamov, A.A.: Parallel computing on the GPU. Architecture and software model of CUDA: textbook. Izd-vo MGU, Moscow (2012) [in Russian]
8. Tumakov, D.N., Chickrin, D.E., Egorchev, A.A., Golousov, S.V.: CUDA programming technology: textbook. Izd-vo Kazan. un-ta, Kazan (2017). [in Russian]
9. Herlihy, M., Shavit, N.: The Art of Multiprocessor Programming. Elsevier, San Francisco (2008)
10. Trobec, R., Vajteršic, M., Zinterhof, P.: Parallel Computing. Numerics, Applications, and Trends. Springer, London (2009). https://doi.org/10.1007/978-1-84882-409-6
11. Voevodin, V., Voevodin, : Vl.: Parallel computing. BKhV-Peterburg, SPb (2002). [in Russian]
12. Gergel, V.: High-performance computing for multi-processor multi-core systems, Izd-vo MGU, Moscow (2010). [in Russian]
13. Grossman, M., Aziz, M., Chi, H., Tibrewal, A., Imam, S., Sarkar, V.: Pedagogy and tools for teaching parallel computing at the sophomore undergraduate level. J. Parallel Distrib. Comput. **105**, 18–30 (2017). https://doi.org/10.1016/j.jpdc.2016.12.026
14. Antonov, A., Popova, N., Voevodin, Vl.: Computational science and HPC education for graduate students: paving the way to exascale. J. Parallel Distrib. Comput. (2018). https://doi.org/10.1016/j.jpdc.2018.02.023 [In Print]
15. Shemetova, A.: Techniques for parallel programming teaching. J. Appl. Informatics **11**(6), 43–48 (2016). [In Russian]
16. Samarskii, A.A., Tichonov, A.N.: The representation of the field in a waveguide in the form of the sum of TE and TM modes. Zhurn. Teoretich. Fiziki **18**(7), 971–985 (1948). [in Russian]
17. Pleshchinskii, N.B.: On boundary value problems for Maxwell set of equations in cylindrical domain. SOP Trans. Appl. Math. **1**(2), 117–125 (2014). https://doi.org/10.15764/AM.2014.02011
18. Pleshchinskii, I., Pleshchinskii, N.: Software implementation of numerical algorithms of solving the electromagnetic wave diffraction problems by periodical gratings. J. Fundam. Appl. Sci. **9**(1S), 1602–1614 (2017). https://doi.org/10.4314/jfas.v9i1s.809
19. Pleshchinskii, N.B., Tumakov, D.N.: A new approach to investigation of Maxwell equations in spherical coordinates. Lobachevskii J. Math. **36**(1), 15–27 (2015). https://doi.org/10.1134/S1995080215010114

20. Pleshchinskaya, I.E., Pleshchinskii, N.B.: On parallel algorithms for solving problems of scattering of electromagnetic waves by conducting thin screens in layered media. Vestnik Kazansk. gos. tekhnol. un-ta **16**(17), 38–41 (2013). [in Russian]
21. Capel, M.I., Tomeu, A.J., Salguero, A.G.: Teaching concurrent and parallel programming by patterns: an interactive ICT approach. J. Parallel Distrib. Comput. **105**, 42–52 (2017). https://doi.org/10.1016/j.jpdc.2017.01.010

The Algorithm for Transferring a Large Number of Radionuclide Particles in a Parallel Model of Ocean Hydrodynamics

Vladimir Bibin[4,5(✉)], Rashit Ibrayev[1,2,3], and Maxim Kaurkin[1,2,3]

[1] Institute of Numerical Mathematics RAS, Moscow, Russia
ibrayev@mail.ru, kaurkinmn@gmail.com
[2] P.P. Shirshov Institute of Oceanology RAS, Moscow, Russia
[3] Hydrometeorological Centre of Russia, Moscow, Russia
[4] Bauman Moscow State Technical University, Moscow, Russia
devbibinva@gmail.com
[5] Nuclear Safety Institute RAS, Moscow, Russia

Abstract. The aim of the research is concerned with the description of algorithm of transferring a large, up to 10^6, number of radionuclide particles in a general circulation model of the ocean, INMIO. Much attention is paid to the functioning of the algorithm in conditions of the original model parallelism. The order of the information storage necessary in the course of model calculations is given in this paper. The important aspects of saving calculated results to external media are revealed. The algorithm of radionuclide particles decay is described. The results of the experiment obtained by calculation of the original model based on the configuration of the Laptev Sea are presented.

Keywords: Lagrangian model · Parallel computing · Particles transfer
Radioactive decay

1 Introduction

In solutions of gas- and hydrodynamics problems, the two following approaches are traditionally used – the Lagrangian and the Eulerian methods. These methods are effective for various classes of problems and both of them possess a wide scope of application [1, 2].

The Lagrangian method got an impetus to development after its application in conjunction with dynamical system models which allowed obtaining fundamentally new results. With this approach, in order to solve nonlinear differential equations in general case, the phase space is introduced that coincides with the physical space of the particles being transported. From the utilization of the Lagrangian method, there emerge a number of additional opportunities for a detailed study of dynamic structures, for example, oceanic eddies.

Our interest in the problem of transport modeling with utilization of the Lagrangian method arises from the studies of radionuclide transfer in marine environment from potential sources located on the sea bottom. A number of studies are dedicated to the

V. Voevodin and S. Sobolev (Eds.): RuSCDays 2018, CCIS 965, pp. 159–170, 2019.
https://doi.org/10.1007/978-3-030-05807-4_14

solution of this problem [3, 4]. The utilization of the Lagrangian method in solving the problem of radionuclide transfer possesses a number of evident advantages over the utilization of the Eulerian method, particularly in the description of the contamination consisting of various radionuclides, each having a different half-life, with the possibility of further formation of new radionuclides, with the possibility of the radionuclide transfer to contiguous environments (ice, atmosphere), with individual buoyancy characteristics, etc. The model of the Lagrangian transport (LT) can be divided into two classes: online models that synchronize with the hydrodynamic models and offline models in which the particles are transported with pre-calculated fields of flow. The following models should be mentioned as offline: TRACMASS [5, 6], Ariane [7], CMS [8]. In the study [9], an attempt of development of online Lagrangian transport model and the ocean circulation model has been made, however, this model is a prototype and currently works as an offline model. Also it should be noted that code of the model is not parallel. Latest review on the problem of Lagrangian analysis of ocean currents [10] mentions that the studies on particle trajectories calculation, in which instantaneous velocities and diffusion tensor are used, have not come to the authors' notice.

The aim of this research is to create an algorithm for solving the LT problem by three-dimensional ocean currents, which has the following features. First, the transport model should work in synchronism with the parallel model of ocean dynamics. Secondly, the number of particles being transported is sufficiently large, up to 1 million. Thirdly, the particles can have individual properties, for example, lifetime, buoyancy, etc.

The INMIO model [11] is used as the oceanic component of the online model. An essential feature of the INMIO software implementation is that it operates under the control of the computational modelling platform CMF2.0 [12], which is the development of the high-level driver idea. Originally CMF2.0 was developed to create online models of the Earth system and was limited by the coupler function that provides interprocessor exchanges in a completely parallel mode, multilevel interpolation of data and asynchronous work with the file system [12]. Later it turned out that the utilization of a computational modelling platform allows to effectively solve the problems of data assimilation [13], nesting [14], etc. While creating an online LT model and the INMIO ocean dynamics model, the tasks of information input/output, as well as interprocessor exchanges, were solved with the use of CMF2.0.

The plan of the article looks as follows. In Sect. 2, the algorithm for interprocessor exchange for Lagrangian particles and synchronization with the INMIO model are covered. Issues of information storage and output to external storage media are discussed in Sects. 3 and 4. An example of the program work is given in Sect. 5. Section 6 describes the algorithm for radioactive decay. In Conclusion part we give conclusions and prospects for further development.

2   Interprocessor Exchanges in the Oceanic Component

As already mentioned, the LT model is being developed to work in conjunction with the INMIO parallel model of ocean hydrodynamics. Therefore, the numerical scheme, the implementation of parallel computations, the input-output algorithms in the LT model must be coordinated with the INMIO model.

Offline LT models use Runge-Kutta schemes (4th and higher orders), since the discreteness of the data about flows is likely to be quite large, and to ensure the accuracy of the particle's trajectory reproduction, a high order of approximation is required. In the case when the online LT and flow dynamics model is used, the requirement for a high order of the time scheme becomes not rigorous, and one can use the Euler method (1st order scheme) without significant loss of accuracy. Nevertheless, the results of numerical experiments on the LT in a circular flow field show that, regardless of the scheme (Euler, Runge-Kutta), the time integration step $\Delta t^{\text{lagrange}}$ must be noticeably less than Δt^{ocean} in order to achieve an acceptable accuracy of reproduction of the circular particle's trajectory (not published, see also [10]). Therefore, we assume that

$$\Delta t^{\text{lagrange}} = \Delta t^{\text{ocean}}/p \tag{1}$$

Where p is integer and according to the results of test calculations $p \sim (10)$.

In the INMIO model of hydrodynamics, two-dimensional decomposition of the simulated space is accepted (Fig. 1a), which provides parallel calculations on multi-processor computers with distributed memory. The INMIO model, at each step in time, exchanges data of cells on the lateral faces of the calculated subregions, including the corner cells. Parallel system efficiency, including input-output algorithms and interpolation between different grids, was investigated in [15]. The LT model fill arrays for exchange with data about particles that left the processor's subregion and transfers these arrays to neighboring processors.

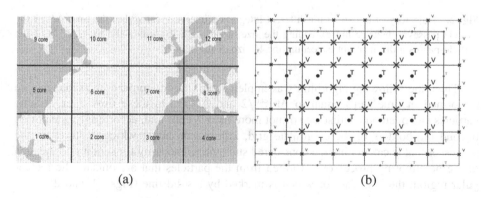

9 core	10 core	11 core	12 core
5 core	6 core	7 core	8 core
1 core	2 core	3 core	4 core

(a)   (b)

Fig. 1. Two-dimensional domain decomposition and grid in the INMIO model (a), core subregion in the INMIO model (b). The large marks and font denote the nodes that belong to the considered subregion, the small ones are the nodes belonging to the neighboring subregions (b). The information from the neighboring sub-areas is transmitted to the considered sub-area during interprocessor exchanges.

Let us consider the particle's trajectory on the grid of the ocean dynamics model, Fig. 2. Let the particle at the initial moment be in an oceanic cell $V_{i,j,k}$. In p steps in time, i.e. in Δt^{ocean}, the particle will move for a distance no greater than the size of the cell, because in the ocean model, the Courant-Friedrichs-Lewy stability condition is fulfilled. If the particle is located close to the border-cell V, then it can cross the dashed line (Fig. 2a). Note that beyond the dashed line, within the considered subregion, the velocities in the four nodes surrounding the particle are not determined. Accordingly, at the last few steps of the particle transfer, velocity at the particle location point cannot be found using the regular interpolation algorithm for the inner cells of the subregion. However, in $p/2$ steps in time, i.e. in $\Delta t^{ocean}/2$, the maximum distance that the particles can overcome is the half the cell $V_{i,j,k}$ and, correspondingly, the velocity at the particle location point can be found using the regular algorithm of velocity interpolation for the inner cells of the subdomain in the four nodes surrounding the particle (Fig. 2b).

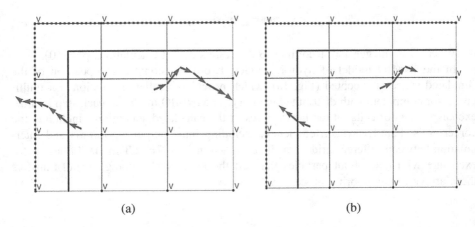

(a)   (b)

Fig. 2. Trajectory of a particle for different integration intervals: (a) within the time Δt^{ocean} the particle passes a distance not exceeding the size of one cell; (b) within the time $\Delta t^{ocean}/2$ the particle passes a distance not exceeding the ½ size of the cell.

Thus, the following algorithm was implemented: the interprocessor exchanges are performed in time steps equal to $\Delta t^{ocean}/2$ and Δt^{ocean}. In these time intervals, the particles can overcome a distance of not more than ½ cells and, accordingly, cannot go beyond the borders of the coordinate grid, information about which is stored in the processor that is calculating the considered subregion. The array for sending to the area of the neighboring processor is formed from the particles that are outside the rectangular region, the boundary of which is marked by a solid line (Figs. 1b and 2).

3   Order of Information Storage

Since the developed model of Lagrange transfer should work together with the model of ocean dynamics, it is necessary to solve the problem dealt with the memory required for storing information about particles. There are two possible options: it is either to reserve memory for global arrays of Lagrangian particles on each calculated core of computational models of the ocean or to create local arrays about Lagrangian particles for each subarea. Obviously, this or that choice is determined by the number of particles. Practice of computations with the INMIO model on computers of parallel architecture with distributed memory shows that the optimum loading of computing cores is achieved when the domain is decomposed into local areas of 2000–4000 nodes in the horizontal plane. In this case, there is enough free memory on the compute node to accommodate the global arrays for 10^6 particles.

So, there is a vector $f\left(N^{\text{attribute}}, N^{\text{lagrange}}\right)$ with the properties of particles where $N^{\text{attribute}}$ is the property of a particle, N^{lagrange} is the total number of particles, each particle $n^{\text{particle}} \in \left[1, N^{\text{lagrange}}\right]$.

The vector of properties associated with the particle allows one to store all the necessary information in the course of calculations. It includes the following characteristics:

- 3 coordinates characterizing the current position of the particle in space;
- 3 coordinates of the cells in which the particle is currently located;
- particle state flag which in the case of finding a particle on a particular core takes the value 1 or 0.

On each of the cores three two-dimensional arrays are also allocated for each component of the coordinate for storing the particles' trajectory: $g\left(N^{\text{lagrange}}, N^{\text{saved\_elements}}\right)$ where N^{lagrange} is the total number of particles, each particle $n^{\text{particle}} \in \left[1, N^{\text{lagrange}}\right]$ and $N^{\text{saved\_elements}}$ is the number of trajectory points to be stored.

After writing to the file considered arrays are reset and reused, which provides significant saving of RAM during the calculations.

4 Checkpoints and Output of Trajectories to External Storage Media

One of the main issues is the problem of output of the information received during calculations to external storage media. As it has already been mentioned, in the ocean model, the work with the file system is provided by means of the CMF2.0 coupler [12], the main tasks of which are to synchronize the interaction between various components of the model and to solve auxiliary problems, one of which is the output of the model to external storage media. In case when the I/O task associated with LT is performed by one of the oceanic cores, there is a substantial imbalance of the computational loads in the model in general: while the ocean master-core is in the process of writing to a file, other cores cannot start their work on the next oceanic step. Thus, the delegation of the input-output task to the core of the coupler is a necessity. Delegating the I/O task to the coupler provides an explicit division of responsibilities between the cores within the model: the oceanic cores directly calculate the physics of the processes of hydrodynamics, the cores of the coupler are in the role of the manager of calculations in the model and solve the auxiliary problems in obtaining the results.

The software code written in the coupler solves the problem of receiving information from the cores of other components and its output to a netCDF format file. However, this system has a number of limitations that make it advisable to write an additional module that solves the problem of writing data about the particle trajectories to a file. One of the restrictions is the presence of arrays that are registered in the coupler system and have dimensions corresponding to the size of the coordinate grid and cannot provide a unified reception of data on particles because their number for each experiment is variable.

To solve this problem, a software module Particle I/O was developed as a part of the CMF2.0 platform, the functionality of which allows one to save the trajectory of particles to a file with a user-defined frequency. By means of Particle I/O the information is collected from the cores that execute the ocean model code. The work of this module and the LT model are synchronized, which allows to exclude the possibility of deadlock and undefined behavior during the model calculations. Each ocean core sends data about the particles that are in the subdomain of the grid of this computational node at the time of saving to a file. Thus, an array containing information about the trajectory of all active particles is formed for subsequent recording to an external storage media. The Particle I/O module is an auxiliary tool in the work of CMF2.0 and can be deactivated for conducting experiments that do not require solving the LT problem.

5   Example of the Work

The computations were carried out on MVS-10P at Joint Supercomputer Center of the Russian Academy of Sciences (JSCC RAS). The current peak performance of this system, which includes 208 computing nodes based on RSC Tornado architecture with direct liquid cooling, Intel server boards, Intel Xeon E5-2690 processors and Intel Xeon Phi coprocessors, is 523.8 tflops.

To test the work of the LT module, an experiment was conducted using the Laptev Sea flow model. The INMIO model belongs to the class 3DPEM [11]. It solves the equations of three-dimensional thermohydrodynamics of the ocean by the finite volume method in Boussinesq approximations and hydrostatics on a grid of type B in vertical z-coordinates. The dimensions of the calculated area are $160 \times 80 \times 49$, the time step is 2 min. Here it is more important for us to make sure that the LT model works correctly rather than to go into details of the problem statement.

To set the horizontal components x, y of the initial coordinates, a random number generator is used which produces a set of numbers X where $\forall r \in X, r \in [0, 1]$. Thus, the formulas, by which the horizontal components of the initial coordinates of an arbitrary particle are calculated, look as follows:

$$x = x_c + (r_1 - 0.5) * r_2 * R \qquad (2)$$

$$y = y_c + (r_3 - 0.5) * r_4 * R \qquad (3)$$

Where $r_1, r_2, r_3, r_4 \in X$ and x_c, y_c, R are constants that set region of grid where particles are located.

The vertical component z of the initial coordinates is initialized so that the particle is on the surface of the sea. At these initial coordinates, the calculation is made for a period of two model months, with the integration step Δt^{ocean} equal to 2 min and

$\Delta t^{\text{lagrange}}$ equal to 1 min, i.e. every minute the next coordinates of the particles are calculated. The current coordinates of the particles are stored to the file at a frequency of 1 h of the model time (Fig. 3).

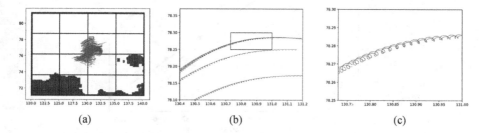

Fig. 3. The trajectory of particles in the Laptev Sea current model. The considered region is zoomed; With rectangles in Figures (a) and (b) the areas are highlighted that are shown in Figures (b) and (c) respectively.

Figure 4 shows the trajectory of a particle crossing the subregion boundary. The trajectory does not undergo discontinuities and continues the expected dynamics of transferring from one subregion to another, which indicates the correctness of the interprocessor exchange of particles.

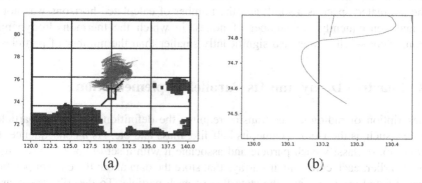

Fig. 4. The trajectory of particles in the Laptev Sea current model. Interprocessor exchange. Figure (b) that is highlighted in Figure (a) with rectangle shows the trajectory of a particle that twice crossed the boundary between neighboring subdomains.

To test the model for successful parallelization of calculations, a series of experiments was performed for 10^6 particles uniformly distributed over the surface of the simulated space. The LT model program code was run on a different number of cores (Fig. 5). As a result, the expected, close to linear, dependence of the execution time of one model step on the number of cores on which the experiment was run was obtained.

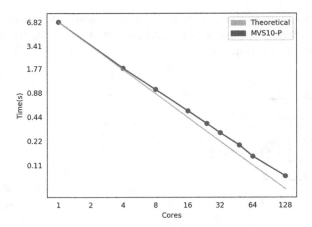

Fig. 5. Execution time of one LT model step in seconds depending on the cores number on the MVS10-P supercomputer. The X axis - number of cores used by the LT model, and on the Y - time, spending on the calculations performed by this model.

Thus, in this experiment, the time for calculating the Lagrangian transfer on the model day takes 60 s on 128 cores, if a single processor core were used, then its simulation would take about 4900 s, which is not acceptable.

The time for modelling general dynamics of the ocean is 22% of the total time, for 10^6 radionuclides transfer - 78% of the total time. These numbers are due to the fact that the simulated space is a shelf, and the number of calculated horizons was not so great, and consequently the number of nodes, in which the thermohydrodynamics equations were solving, was also significantly smaller than the number of particles.

6   Radioactive Decay and Its Parallel Implementation

The description of radionuclides transfer requires the definition of specific particles' properties, such as the isotope name, its half-life, mass and etc. As a consequence, the LT model must classify each particle and associate it with a set of specific properties. As noted earlier, each core contains arrays that store the data about the current position, the accumulated trajectory and the state flag of each particle. To describe the characteristics of radionuclides, the following arrays are additionally introduced: ParticleType(N_p) where N_p is the dimension of the array, equal to the number of particles in the original ParticleTypeDescription(N_c), where N_c is the dimension of the array, equal to the number of particle types in the original model.

The array ParticleType contains the indices of the array ParticleTypeDescription (Fig. 6), which is essentially an example of relational database model [16], where the value in the array ParticleType is the key to the information stored in the ParticleTypeDescription. Thus, each particle is associated with a set of defined specific properties and access to the set of these characteristics is guaranteed in the process of program execution.

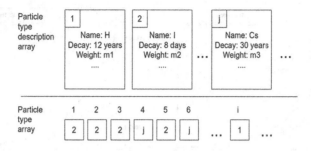

Fig. 6. Classification of particles.

The description of radioactive decay is an important part of the radionuclide transfer analysis, since each radionuclide particle can cease to exist at any time. The probability of this event is determined by the half-life. The algorithm describing radioactive decay and its implementation in the form of a program code must satisfy the following conditions:

- radioactive decay is described by fundamental laws;
- possible conversion of a single particle in a number of others;
- correct operation of the algorithm in parallel computing.

To solve this problem, the following algorithm was developed:

1. The cycle starts by all particles that are present in the model.
2. If the n-th particle is processed by the current processor, then go to the steps (3–7), otherwise, the particle is absent in this processor subdomain and does not need to be processed.
3. The particle's lifetime is increased by a model time step.
4. The moment of checking for the n-th particle's decay occurs when the following condition is fulfilled: $L(n) \bmod C(n) < \mathrm{dt}$ where $L(n)$ is the current lifetime of the particle, $C(n)$ is the period of checking for decay, dt is the time step in the original model. The check period for decay is specified by the user.
5. If the moment for checking has come, i.e. the condition from block 4 has been fulfilled, a random number x from 0 to 1 is generated which has a uniform distribution over this interval.
6. If $x < 1 - 2^{(-C(n)/T(n))}$ where $T(n)$ is the half-life of the n-th particle, then it is considered that the particle has decayed and ceased to exist.
7. If the particle has decayed, certain procedures are performed to ensure the elimination of the n-th particle from the computational process.
8. End of the cycle for all particles in the model.

The described algorithm was tested in a numerical experiment on modelling the transport of Lagrangian particles for two months in the dynamics model of the Laptev Sea. Figure 7 shows the results of the work of the model, illustrating the process of radioactive decay in time.

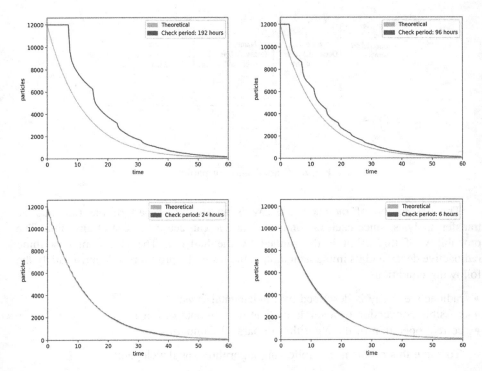

Fig. 7. Carrying out the experiment for different checking periods for decay for element I with a half-life of 8 days. Yellow curve shows the theoretical decay, blue - the result obtained in the course of the algorithm. (Color figure online)

The considered algorithm ensures correctness of the decay description regardless of the particle distribution over the processor subdomains of the coordinate grid. Thus, it is guaranteed that if the experiment is carried out on N processors, at the same time X_i is the number of particles with a half-life period T in the region of the i processor at the time step $t = 0$, then at the time step $t = T$, in the model will remain $\frac{\sum_{i=1}^{N} X_i}{2} + \epsilon$ particles where ϵ is some error that sets a deflection from the theoretical value which is caused by the use of a random number generator.

Theoretical decay (Fig. 7) was described by the formula: $N = N_0 e^{(-\lambda t)}$, where N_0 is the number of atoms in the initial time $t = 0$.

The described algorithm gives the expected results and repeats the form of a theoretical curve (Fig. 7). When the check period for decay decreases, the curve describing the dependence of the particles number on time tends to theoretical values, which also indicates the correctness of the algorithm.

7 Conclusion

This paper presents an algorithm for transferring a large, up to 10^6, number of particles of radionuclides. The algorithm makes it possible to obtain the trajectory of particles together with online LT model INMIO. The developed algorithm for interprocessor exchange which ensures the transfer of data on particles that left the subregion of one processor and moved to the subdomain of another, guarantees the possibility of moving particles across the entire grid of the ocean model.

The online model of LT and of ocean hydrodynamics minimizes the number of calls to external memory in comparison with similar offline models. Delegating the task of I/O information to the LT model cores with the CMF2.0 coupler [12] eliminates the possibility of imbalance related to the operations of I/O during the calculations. The efficiency of the described algorithm is confirmed experimentally.

The work was carried out at the Nuclear Safety Institute of The Russian Academy of Sciences with the financial support of the project RNF No. 17-19-01674.

References

1. Zhang, Z., Chen, Q.: Comparison of the Eulerian and Lagrangian methods for predicting particle transport in enclosed spaces. Atmos. Environ. **41**(25), 5236–5248 (2007)
2. Durst, F., Milojevic, D.: Eulerian and Lagrangian predictions of particulate two-phase flows: a numerical study. Appl. Math. Model. **8**, 101–115 (1984)
3. Bilashenko, V.P., et al.: Prediction and evaluation of the radioecological consequences of a hypothetical accident on the sunken nuclear submarine B-159 in the Barents Sea Antipov. At. Energ. **119**(2), 132–141 (2015)
4. Heldal, H.E., Vikebo, F., Johansen, G.O.: Dispersal of the radionuclide Caesium-137 from point sources in the barents and norwegian seas and its potential contamination of the arctic marine food chain: coupling numerical ocean models with geographical fish distribution. Environ. Pollut. **180**, 190–198 (2013)
5. Döös, K., Kjellsson, J., Jönsson, B.: TRACMASS—a lagrangian trajectory model. In: Soomere, T., Quak, E. (eds.) Preventive Methods for Coastal Protection, pp. 225–249. Springer, Heidelberg (2013). https://doi.org/10.1007/978-3-319-00440-2_7
6. Döös, K., Jönsson, B., Kjellsson, J.: Evaluation of oceanic and atmospheric trajectory schemes in the TRACMASS trajectory model v6.0. Geosci. Model Dev. **10**, 1733–1749 (2017)
7. Blanke, B., Raynaud, S.: Kinematics of the pacific equatorial undercurrent: an eulerian and lagrangian approach from GCM results. J. Phys. Oceanogr. **27**, 1038–1053 (1997)
8. Paris, C.B., Helgers, J., van Sebille, E., Srinivasan, A.: Connectivity Modeling System: A probabilistic modeling tool for the multiscale tracking of biotic and abiotic variability in the ocean. Environ. Modell. Softw. **42**, 47–54 (2013)
9. Lange, M., Sebille, E.: Parcels v0.9: prototyping a Lagrangian ocean analysis framework for the petascale age. Geosci. Model Dev. **10**, 4175–4186 (2017)
10. van Sebille, E., Griffies, S.M., Abernathey, R., et al.: Lagrangian ocean analysis: Fundamentals and practices. Ocean Model. **121**, 49–75 (2018). (total 35 authors)
11. Ibrayev, R.A., Khabeev, R.N., Ushakov, K.V.: Eddy-resolving 1/10° model of the world ocean. Izvestiya Atmos. Oceanic Phys. **48**(1), 37–46 (2012)

12. Kalmykov, V.V., Ibrayev, R.A.: A framework for the ocean-ice-atmosphere-land coupled modeling on massively-parallel architectures. Numer. Methods Program. **14**(2), 88–95 (2013). (in Russian)
13. Kaurkin, M.N., Ibrayev, R.A., Belyaev, K.P.: Data assimilation ARGO data into the ocean dynamics model with high spatial resolution using Ensemble Optimal Interpolation (EnOI). Oceanology **56**(6), 774–781 (2016)
14. Koromyslov, A., Ibrayev, R., Kaurkin, M.: The technology of nesting a regional ocean model into a global one using a computational platform for massively parallel computers CMF. In: Voevodin, V., Sobolev, S. (eds.) Supercomputing. RuSCDays 2017. Communications in Computer and Information Science, vol. 793, pp. 241-250. Springer, Cham (2017). https://doi.org/10.1007/978-3-319-71255-0_19
15. Kalmykov, V.V., Ibrayev, R.A., Kaurkin, M.N., Ushakov, K.V.: Compact modeling framework v3.0 for high-resolution global ocean-ice-atmosphere models. Geosci. Model Dev. Discuss. https://doi.org/10.5194/gmd-2017-294
16. Date, C.J.: Introduction to Database Systems (2003)

Supercomputer Simulation

Aerodynamic Models of Complicated Constructions Using Parallel Smoothed Particle Hydrodynamics

Alexander Titov, Sergey Khrapov, Victor Radchenko,
and Alexander Khoperskov[✉]

Volgograd State University, Volgograd, Russia
{alexandr.titov,khrapov,viktor.radchenko,khoperskov}@volsu.ru

Abstract. In current paper we consider new industrial tasks requiring of air dynamics calculations inside and outside of huge and geometrically complicated building constructions. An example of such constructions are sport facilities of a semi-open type for which is necessary to evaluate comfort conditions depending on external factors both at the stage of design and during the further operation of the building. Among the distinguishing features of such multiscale task are the considerable size of building with a scale of hundreds of meters and complicated geometry of external and internal details with characteristic sizes of an order of a meter. Such tasks require using of supercomputer technologies and creating of a 3D-model adapted for computer modeling. We have developed specialized software for numerical aerodynamic simulations of such buildings utilizing the smoothed particle method for Nvidia Tesla GPUs based on CUDA technology. The SPH method allows conducting through hydrodynamic calculations in presence of large number of complex internal surfaces. These surfaces can be designed by 3D-model of a building. We have paid particular attention to the parallel computing efficiency accounting for boundary conditions on geometrically complex solid surfaces and on free boundaries. The discussion of test simulations of the football stadium is following.

Keywords: Computational fluid dynamics · Nvidia Tesla
CUDA · Smooth particle hydrodynamics · Multiscale modeling

1 Introduction

The supercomputer methods qualitatively enhance our capabilities in the design of complex and non-standard constructions. The determination of aerodynamic fields in the immediate vicinity of the object and study of possible environment impacts on it or its parts are the main aims of many studies. One of the most powerful tools in studying of the aerodynamic properties of projected constructions are the wind tunnel experiments [1]. However, the huge sizes, complexity of constructions and lack of hydrodynamic similarity in some cases limit the possibilities of such approach [2].

© Springer Nature Switzerland AG 2019
V. Voevodin and S. Sobolev (Eds.): RuSCDays 2018, CCIS 965, pp. 173–184, 2019.
https://doi.org/10.1007/978-3-030-05807-4_15

Here we list several tasks for which numerical gas-dynamic simulations for building constructions seem necessary. For the effective and safe missiles launching it is very important to determine the shock-wave loads on the launcher and other surrounding objects [3,4]. Technical improvements of design of the missile body, launch container, ballistic missile silo launcher requires the analysis of gas-dynamic and shock-wave processes at missile launch [3,5]. The problem complicates in the case of above-water or submerged missile launches [6]. Examples of outstanding applied scientific achievements are technologies of computational aerodynamics of moving objects in the field of missile construction [7], aircraft building [8–10], design of cars and other land transport [11–13].

The interaction with air flow should be accounted for in practice of design and building of large constructions. The calculation of wind load for high-rise buildings or the entire microdistrict with complex landscape features is an independent problem [14]. An important trend in the development of such models connects with calculations of wind loads on new architectural solutions with complex free geometry [15]. In all tasks listed above the main aim is the definition of the strength characteristics of the constructions.

In current paper, we notice another class of industrial problems requiring of air motion calculation outside and inside of complex and very large constructions depending on meteorological conditions. Sport stadiums of a semi-open type can be a good example due to the need of comfortable conditions creation depending on external factors. And we have to solve such problem at the design stage. The task is similar to the determination of the air motion inside of large industrial buildings, where a variety of technological equipment is the source of flows [16, 17]. The differences are caused with scales of the computational domain and determinant influence of external meteorological factors in our case. The massive transition of numerical algorithms and parallel software to GPUs is a modern trend [18]. Thus, that is our key aim as well.

2   Aerodynamic Numerical Model

2.1   Mathematical Model

We consider the problem of calculating the velocity field inside and outside a complex large structure, depending on the wind regime, air temperature. The gas compressibility calculation for the complete system of gas dynamics equations, gravity and vertical inhomogeneity of the density ϱ, pressure p, entropy $s \propto \ln(p/\varrho^\gamma)$ allows us to model convective instability when necessary conditions arise. We use the equations of hydrodynamics in the following form:

$$\frac{\partial \varrho}{\partial t} + \nabla(\varrho \mathbf{U}) = 0 , \quad \frac{\partial \mathbf{U}}{\partial t} + (\mathbf{U}\nabla)\mathbf{U} = -\frac{\nabla p}{\varrho} + \mathbf{g} , \quad \frac{\partial \varepsilon}{\partial t} + \mathbf{U}\nabla\varepsilon + \frac{p}{\varrho}\nabla\mathbf{U} = 0 , \quad (1)$$

where ϱ is the gas mass density, $\mathbf{U} = \{u_x; u_y; u_z\}$ is the velocity vector, $\nabla = \{\partial/\partial x; \partial/\partial y; \partial/\partial z\}$, $|\mathbf{g}| = 9.8 \, \text{m·s}^{-2}$, p is the pressure, ε is the internal energy.

Fig. 1. The common view of the 3D stadium model is shown from the outside (top) and inside (bottom)

The thermodynamic parameters are related by the equation of state of an ideal gas

$$p = \frac{\varrho}{\mu}\mathcal{R}T \quad \text{or} \quad \varepsilon = \frac{p}{(\gamma - 1)\,\varrho}, \tag{2}$$

where $\mathcal{R}$ is the ideal gas constant, $\gamma = 1.4$ is the adiabatic index, $\mu = 29\,\text{g/mol}$ is air the molar mass, and T is the temperature. The standard gas impenetrable boundary conditions should been set on solid surfaces.

2.2   3D-Model of Construction

The construction complexity is caused by the large number of different passways, the complex geometry of internal system of storey and stairways, the tribunes and the adjoining terrain (Fig. 1). We set the boundary conditions on the gas-dynamic functions at all solid surfaces, thus the quality of the aerodynamic model is substantially defined by the accuracy of reproduction of construction structural features.

We propose using of the standard technologies of 3D modeling of solids, for which powerful software tools exist. The key idea is a 3D model created at the stage of building design using the drawings in the formats .dwg, .dxf, .pdf, etc. The spatial accuracy of such model lies within the range 0.01–0.1 m. The 3D model includes all design features of the construction, internal partitions, passways to the interior rooms, tribunes and field (See Fig. 1).

The further consideration and all the test calculations have been carried out for the football stadium model of the Volgograd-Arena type. The original accurate high-resolution 3D-model requires adaptation for aerodynamic modeling. Our task is concordance of the design parameters with the SPH model resolution taking into the account the finite particle radius h. In our 3D-model we have removed the smallest girders, columns and structural details (see Fig. 1b) for hydrodynamic simulations.

Using standard 3D modeling methods in Blender the spatial volume for all building elements such as roof, storeys, internal stairways and tribunes have been assigned. All components were combined into a single 3D model by Boolean operations. Finally, a particular attention has been paid to the modeling of through passways to the field and tribunes, on which air flows significantly affect. The working format of the 3D model is *.stl.

2.3   Boundary Conditions

Using of fixed particles along a solid surface is one of the methods of solid boundaries modeling in SPH approach [19,20]. Particles are placed on a boundary or inside a solid body depending on a distance between confining surfaces and radius of particle, h. These Fixed Particles have all the properties of gas particles (density, energy, pressure), but their velocity is zero and they always remain at their initial positions. Recalculation of density and energy at each time step increases the calculation accuracy in the vicinity of solid surfaces. Such fixed particles take part in the through calculation ensuring the parallelization efficiency.

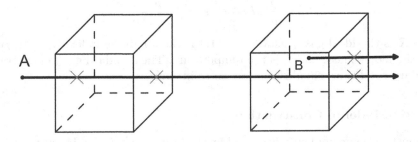

Fig. 2. The distribution of Fixed Particles by the raytracing method

Our specially written Python script for Blender provides the Fixed Particles configuration formation. We use an algorithm based on ray tracing determining whether a point belongs to a complex object (Fig. 2). In 3D model the parity of number of planes intersections is verified by an arbitrary ray for a particle at the point A or B (see Fig. 2). The function returns "False" in case of an even number of intersections and "True" for an odd number of intersections.

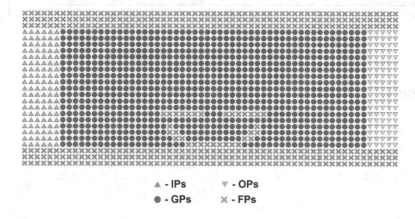

Fig. 3. The SPH particles types and the configuration of the computational domain

To model the wind flow in a Cartesian coordinates along the x-coordinate, we set free boundary conditions on the left and right side of the computational domain (see Fig. 3). It is convenient to define different wind directions $\mathbf{V}_a$ by a corresponding rotation of the 3D-model in a fixed coordinate system. We introduce three additional types of particles (IPs, OPs and DPs) to realize the free boundary conditions ensuring the inflow/outflow of air into the computational domain. As a result, our model contains five types of particles: Gaseous Particles (GPs), Fixed Particles (FPs), Outflow Particles (OPs), Inflow Particles (IPs), Dead Particles (DPs).

The inflow region at free boundary (left part of Fig. 3) consists of several layers of uniformly distributed IPs particles. For such particles all the gas dynamic parameters are fixed and the condition $\mathbf{U}_i = \mathbf{V}_a$ fulfills for their velocity until the moment then the particle crosses the inflow region. After the particle coming into the outflow zone it moves with constant velocity and fixed parameters (at the right side of Fig. 3). When the particle leaves the outflow region through the right border and ceases to affect the GPs it receives the "Dead Particle" status (Fig. 4).

2.4   The Numerical Model for the GPU

The SPH approach asserts wide opportunities for computational code parallelization for a variety of applications using GPUs. Using specially developed

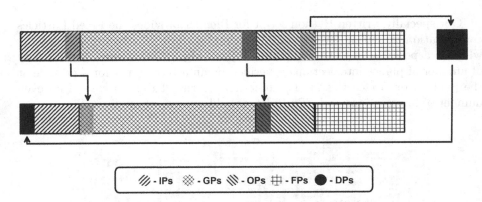

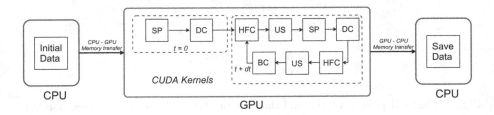

Fig. 4. The redistribution of particles of different types for free boundaries

Fig. 5. Flow diagram for the parallel code

algorithms for solving the resource-intensive problem of N-bodies [21, 22] the wonderful results were obtained for astrophysical flows [23, 24]. The SPH method evinces considerable capabilities for modeling the dynamics of liquid free surface in the Earth's gravitational field [25].

The program written for problem solve from Ref. [22] without self-gravitation which accounts for new conditions of our task has been our base of program code. The boundary conditions are calculated in the CUDA BC (Boundary Conditions) kernel, while presence of different particle types is accounted in the CUDA SP (Sorting Particles), DC (Density Calculating), HFC (Hydrodynamics Forces Calculating) and US (Update Systems) kernels (Fig. 5). BC CUDA block consists of two CUDA-kernels:

(1) kernelInsert_Dead $<<<N/BlockSize, BlockSize>>>$ indexes particles that have left the calculation domain (DPs).
(2) kernelAdd_Inflow $<<<(Nin + BlockSize - 1)/BlockSize, BlockSize>>>$ creates a new layer of Inflow Particles from the Dead Particles.

The corresponding snippet of code is shown below.

The code for CUDA-core: kernelInsertDead and kernelAdd_inflow

```
__global__ void kernelInsert_Dead(int*
            SPH_newinflow_ind, int* type, int *count)
{
        int i = threadIdx.x + blockIdx.x * blockDim.x;
        if (type[i] == type_dead) {
                int ndx = atomicAdd(count, 1);
                SPH_newinflow_ind[ndx] = i;
        }
}
__global__ void kernelAdd_Inflow(int* SPH_newinflow_ind,
                Real4* Pos, Real4* Vel, Real4* Post,
                Real4* Velt, Real2* Mass_hp, int* type)
{       int i = threadIdx.x + blockIdx.x * blockDim.x;
        if (i < dd.N_inflow) {
                int ndx = SPH_newinflow_ind[i];
                int kz = i / dd.Ny_inflow;
                int ky = i - kz * dd.Ny_inflow;
                Real x = dd.x_inflow;
                Real y = dd.y_inflow + ky * dd.step;
                Real z = dd.z_inflow + kz * dd.step;
                Pos[ndx].x = x;  Pos[ndx].y = y;  Pos[ndx].z = z;
        Update_Particles_Characteristics(Pos, Vel, Post, Velt,
                                        Mass_hp, type);
        }
}
```

The efficiency rise of an algorithm of the nearest neighbor search is an important direction in the parallel SPH-method development [26]. For the nearest neighbor search the hierarchical grids and cascading particle sorting algorithm have been applied using the partial sums parallel calculation in SP CUDA kernels. The sorting algorithm have been specified in Ref. [22]. Figure 6 shows the corresponding common scheme of the nearest neighbor search for the SPH method.

The analysis reveals that a relative contribution of the SP and BC CUDA kernels to the overall SPH algorithm runtime is less than 0.3%. The latter indicates the efficiency of the parallel implementation of these CUDA kernels. We analyzed the scalability of computations on models with different number of particles N and for different Nvidia Tesla GPUs (Table 1). The memory of the GPU K20 is not enough for calculation with $N = 2^{25}$. Our SPH algorithm uses only global memory, so increasing the runtime of Cuda kernels by 1.7 times can be explained by more efficient access to global memory on the K80 compared to the K40. The memory bandwidth is 480 Gbit/s on K80 and 288 Gbit/s on K40.

We tested the code on standard problems, for example, the decay of an arbitrary pressure discontinuity with the formation of a shockwave, a rarefaction

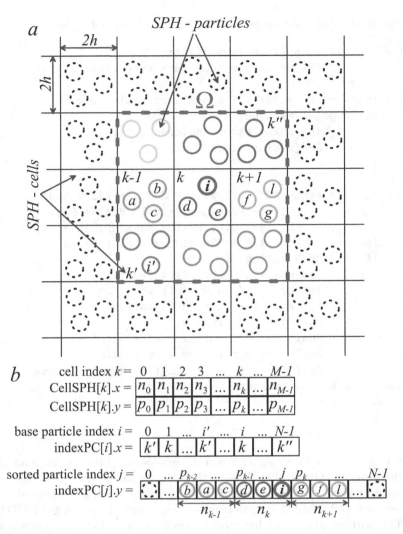

Fig. 6. The scheme of the nearest-neighbor search. (a) The distribution of SPH particles in grid cells. Ω is the subdomain determining the i-th particle interaction with neighboring particles. (b) The arrays for sorting particles (M is the number of cells, n_k is the number of particles at the k-th cell, $p_k = \sum_{l=0}^{k} n_l$, and N is total number of particles.

Table 1. The execution time of CUDA kernels on GPUs $[t_{SPH}] = $ sec, $[Memory] = $ Gb.

	K20 (1GPU)		K40 (1GPU)		$\frac{1}{2} \times$ K80 (1GPU)	
$N \times 1024$	t_{SPH}	$Memory$	t_{SPH}	$Memory$	t_{SPH}	$Memory$
4096	6.02	0.618	4.95	0.618	2.47	0.618
32768	–	–	69.2	6.243	41.6	6.243

wave and a contact discontinuity, the gas flow from the vessel into the vacuum through a small hole, Couette flow, and others.

2.5   Simulation Results and Discussion

We have considered the structure of the wind flow on the Volgograd-Arena stadium 3D-model (see Sect. 2.2). The velocity and direction of wind, $\mathbf{V}_a$, the SPH particle radius, h, the number of particles, N have been test parameters to determine the efficiency and quality of our numerical model. Figure 7 represents the flow structure in the xOy plane for different heights z in the model with the particles number $N = 2^{25}$.

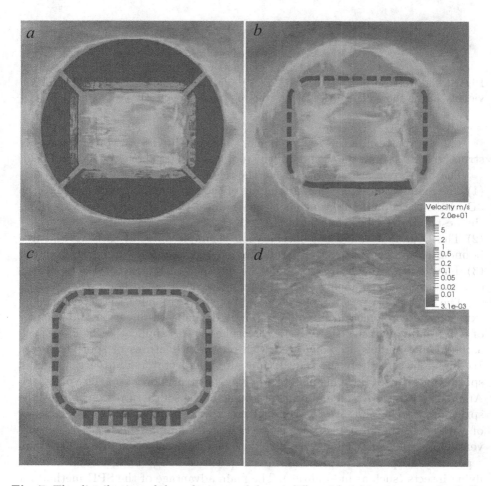

Fig. 7. The distribution of the velocity modulus at different heights above the football field level: (a) the ground level layer, $z = 2$ m, (b) at the second-floor height, $z = 12$ m, (c) under the roof $z = 21$ m, (d) above the roof $z = 40$ m.

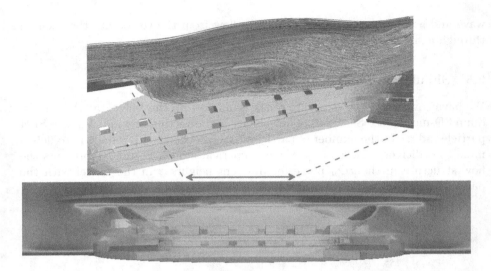

Fig. 8. The vertical flow pattern ($|\mathbf{U}(x,z)|$). The insert shows the flow structure in vicinity of the roof. The vortices below the roof plane are clearly visible.

The flow pattern in the vertical plane is presented in Fig. 8, where the vortex structures are seen in the vicinity of the roof inner edges above the football field.

The SPH approach advantages are the following.

(1) The manufacturability of constructing of the aerodynamic models due to the 3D-model of the building, which is always built at the building design stage and the best way describes the real object.

(2) The boundary conditions specification is simple on solid surfaces and free boundaries and provides better parallelization in comparison with grid methods.

(3) The SPH approach disclose incredible prospects for structural components elastic movements modeling under the influence of the wind flow.

The air mobility inside of open and semi-open sport facilities is determined by external meteorological factors and strongly depends on direction and velocity of wind and convective state of an atmosphere. The application fields of such aerodynamic models are connected with operation features of sport facilities. We highlight separately the security measures of spectators and participants of sport events caused by flying insects problem, which is relevant for the Volgograd-Arena, the Rostov-Arena and other stadiums in the southern regions during the spring-summer period. We can conduct an expert examination of the efficiency of various methods of technical protection, for example the usage of special ventilation and aspiration devices to reduce insects penetration into the stadium separately for passively spreading insects (such as imago midges) and actively flying insects (such as mosquitoes). The main advantage of the SPH method for our problem is a simple procedure for specifying solids using the 3D model of a structure instead of generating 3D meshes with a complex topology.

Acknowledgments. We used the results of numerical simulations carried out on the supercomputers of the Research Computing Center of M.V. Lomonosov Moscow State University. AK and SK are grateful to the Ministry of Education and Science of the Russian Federation (government task No. 2.852.2017/4.6). VR is thankful to the RFBR (grants 16-07-01037).

References

1. Egorychev, O.O., Orekhov, G.V., Kovalchuk, O.A., Doroshenko, S.A.: Studying the design of wind tunnel for aerodynamic and aeroacoustic tests of building structures. Sci. Herald Voronezh State Univ. Archit. Civil Eng. **2012**(4), 7–12 (2012). Construction and Architecture
2. Sun, X., Liu, H., Su, N., Wu, Y.: Investigation on wind tunnel tests of the Kilometer skyscraper. Eng. Struct. **148**, 340–356 (2017)
3. Peshkov, R.A., Sidel'nikov, R.V.: Analysis of shock-wave loads on a missile, launcher and container during launches. Bull. South Ural State Univ. Ser. Mech. Eng. Ind. **15**, 81–91 (2015)
4. Kravchuk, M.O., Kudryavtsev, V.V., Kudryavtsev, O.N., Safronov, A.V., Shipilov, S.N., Shuvalova, T.V.: Research on gas dynamics of launch in order to ensure the development of launching gear for the Angara A5 Rocket at Vostochny Cosmodrome. Cosmonautics Rocket Eng. **7**(92), 63–71 (2016)
5. Fu, D., Hao, H.: Investigations for missile launching in an improved concentric canister launcher. J. Spacecraft Rockets **52**, 1510–1515 (2015)
6. Yang, J., Feng, J., Li, Y., Liu, A., Hu, J., Ma, Z.: Water-exit process modeling and added-mass calculation of the submarine-launched missile. Pol. Marit. Res. **24**, 152–164 (2017)
7. Dongyang, C., Abbas, L.K., Rui, X.R., Guoping, W.: Aerodynamic and static aeroelastic computations of a slender rocket with all-movable canard surface. Proc. Inst. Mech. Eng. Part G J. Aerosp. Eng. **232**, 1103–1119 (2017)
8. Liang, Y., Ying, Z., Shuo, Y., Xinglin, Z., Jun, D.: Numerical simulation of aerodynamic interaction for a tilt rotor aircraft in helicopter mode. Chin. J. Aeronaut. **29**(4), 843–854 (2016)
9. Lv, H., Zhang, X., Kuang, J.: Numerical simulation of aerodynamic characteristics of multielement wing with variable flap. J. Phys. Conf. Ser. **916**, 012005 (2017)
10. Guo, N.: Numerical prediction of the aerodynamic noise from the ducted tail rotor. Eng. Lett. **26**, 187–192 (2018)
11. Janosko, I., Polonec, T., Kuchar, P., Machal, P., Zach, M.: Computer simulation of car aerodynamic properties. Acta Universitatis Agriculturae et Silviculturae Mendelianae Brunensis **65**, 1505–1514 (2017)
12. Janson, T., Piechna, J.: Numerical analysis of aerodynamic characteristics of a of high-speed car with movable bodywork elements. Arch. Mech. Eng. **62**, 451–476 (2015)
13. Saad, S., Hamid, M.F.: Numerical study of aerodynamic drag force on student formula car. ARPN J. Eng. Appl. Sci. **11**, 11902–11906 (2016)
14. Zhang, Z., Sien, M., To, A., Allsop, A.: Across-wind load on rectangular tall buildings. Struct. Eng. **95**, 36–41 (2017)
15. Jendzelovsky, N., Antal, R., Konecna, L.: Determination of the wind pressure distribution on the facade of the triangularly shaped high-rise building structure. MATEC Web Conf. **107**, 00081 (2017)

16. Butenko, M., Shafran, Y., Khoperskov, S., Kholodkov, V., Khoperskov, A.: The optimization problem of the ventilation system for metallurgical plant. Appl. Mech. Mater. **379**, 167–172 (2013)
17. Averkova, O.A., Logachev, K.I., Gritskevich, M.S., Logachev, A.K.: Ventilation of aerosol in a thin-walled suction funnel with incoming flow. Part 1. Development of mathematical model and computational algorithm. Refract. Ind. Ceram. **58**, 242–246 (2017)
18. Kopysov, S., Kuzmin, I., Nedozhogin, N., Novikov, N., Sagdeeva, Y.: Scalable hybrid implementation of the Schur complement method for multi-GPU systems. J. Supercomput. **69**, 81–88 (2014)
19. Monaco, A.D., Manenti, S., Gallati, M., Sibilla, S., Agate, G., Guandalini, R.: SPH modeling of solid boundaries through a semi-analytic approach. Eng. Appl. Comput. Fluid Mech. **5**, 1–15 (2011)
20. Valizadeh, A., Monaghan, J.J.: A study of solid wall models for weakly compressible SPH. J. Comput. Phys. **300**, 5–19 (2015)
21. Bedorf, J., Gaburov, E., Zwart, S.P.: A sparse octree gravitational N-body code that runs entirely on the GPU processor. J. Comput. Phys. **231**, 2825–2839 (2012)
22. Khrapov, S., Khoperskov, A.: Smoothed-particle hydrodynamics models: implementation features on GPUs. Commun. Comput. Inf. Sci. **793**, 266–277 (2017)
23. Buruchenko, S.K., Schafer, C.M., Maindl, T.I.: Smooth particle hydrodynamics GPU-acceleration tool for asteroid fragmentation simulation. Proc. Eng. **204**, 59–66 (2017)
24. Khrapov, S.S., Khoperskov, S.A., Khoperskov, A.V.: New features of parallel implementation of N-body problems on GPU. Bull. South Ural State Univ. Ser. Math. Model. Program. Comput. Softw. **11**, 124–136 (2018)
25. Afanas'ev, K.E., Makarchuk, R.S.: Calculation of hydrodynamic loads at solid boundaries of the computation domain by the ISPH method in problems with free boundaries. Russ. J. Numer. Anal. Math. Model. **26**, 447–464 (2011)
26. Winkler, D., Rezav, M., Rauch, W.: Neighbour lists for smoothed particle hydrodynamics on GPUs. Comput. Phys. Commun. **225**, 140–148 (2017)

Ballistic Resistance Modeling of Aramid Fabric with Surface Treatment

Natalia Yu. Dolganina[(✉)] ⓘ, Anastasia V. Ignatova,
Alexandra A. Shabley, and Sergei B. Sapozhnikov ⓘ

South Ural State University, 76 Lenin prospekt, Chelyabinsk 454080, Russia
{dolganinani, ignatovaav, shableiaa, ssb}@susu.ru

Abstract. The minimization of mass and reducing the value of deflection of the back surface of an armored panel, which will lower the level of trauma to the human body, are crucial tasks in the current development of body armors. A significant part of the bullet energy is dissipated due to the friction of pulling-out yarns from ballistic fabrics in the body armor. We present a method for controlling the process of dry friction between yarns – surface treatment with various compositions (PVA suspension, rosin). This procedure causes only a slight increase weighting of the fabric. We investigated an impact loading of aramid fabrics of plain weave P110 with different types of surface treatment and without it (the samples were located on the backing material – technical plasticine). The indenter speed was in the range of 100–130 m/s. We also developed a model of an impact loading of considered samples in explicit FE code LS-DYNA. The surface treatment of the fabric in the model was taken into account by only one parameter – the coefficient of dry friction. We considered several methods of the task parallelizing. Numerical experiments were conducted to study the problem scalability. We found that the surface treatment reduces deflection of fabric up to 37% with an increase in weight up to 5.1%. The numerical values of the depths of the dents in the technical plasticine are in good agreement with the experimental data.

Keywords: Supercomputer modelling · FEA model
Aramid fabric · Impact · Surface treatment · Frictional coefficient
Technical plasticine · LS-DYNA

1 Introduction

Minimization of armored panels mass while maintaining a given level of protection is the main task in their designing. In outer layers of armored panels, which contact with the high-speed bullet, the dynamic phase predominates, and in back layers – the friction-based and low speed phase. In a fabric armored panel the most part of the kinetic energy of a bullet is dissipated due to the yarns pull-out from the fabric, frictional interaction, and the rest of the impact energy is spent on straining and failure the yarns [1–3]. Thus, to reduce the energy transferred to the protected object, i.e. to reduce the deflection of the rear side of the armored panel, the armored panel should dissipate the possible maximum of the bullet kinetic energy. The deflection of the rear

© Springer Nature Switzerland AG 2019
V. Voevodin and S. Sobolev (Eds.): RuSCDays 2018, CCIS 965, pp. 185–194, 2019.
https://doi.org/10.1007/978-3-030-05807-4_16

side of armored panels resulting from local impacts can be lowered by the modernization of fabric panels: the combination of fabrics with different types of weave (fabric with minimal curvature of yarns in outer layers, and with maximum curvature – in rear layers) [4]; the use of fabric layers made of polyamide fibers, which reduce the speed of sound and extend the work dynamic phase of the back layers of panel, between the fabrics of aramid fiber [4]; through-the-thickness stitching of package [5–9]; polymer layer covering [10, 11]; the use of non-Newtonian fluids [8, 9, 11–15]; abrasive particles insertion [10, 16]. However, these methods have drawbacks: the fabrics combination complicates logistics and increases the product cost; the stitching reduces the flexibility and comfort of wearing; abnormally viscous liquids increase the fabrics surface density by many times; polymer covering significantly increases the fabric mass and prevent the yarns pulling out (local impact results in yarns failure); abrasive particles work only for the case of a puncture. At the same time, fabric surface treatment allows us to increase the frictional interaction of yarns and to reduce the deflection of fabric after a local impact with the minimal increase of weight. To investigate the mechanism of deformation and failure of fabrics under local impact, both numerical approach and experimental study were used [1, 11, 17]. Unfortunately, experiments cannot reveal the influence of individual factors on the impact interaction of the bullet with the armored panel. By means of the finite element method fabrics were modeled by a continuous medium [18], individual yarns modeled by beams [3, 19], shells [14, 20] and solids [1, 21] finite elements. Models of a continuous medium do not allow us to investigate the yarn pull-out from a fabric. In models with beam finite elements, it is impossible to take into account the contact interaction between the yarns. Fabric models with solid anisotropic finite element yarns require large computational resources. In our opinion, models with shell finite elements [22, 23] have great prospects. This approach allow us to take into account multiple contact interactions, the anisotropy of yarns and require an order of magnitude less computational resources in comparison with models with solid anisotropic finite elements.

The work is structured as follows. The problem is formulated in Sect. 2. Section 3 describes the ballistic tests. Section 4 contains ballistic impact simulations. In Sect. 5 the key results are provided.

2 Problem Statement

We considered aramid fabrics of plain weave P110 with a surface density of 110 g/m^2. Methods of fabric surface treatment were chosen based on the following conditions: the processing should allow us to control the frictional interaction between the yarns – increase or decrease of the coefficient of friction; the weighting of the fabric should be negligible; the connections between the yarns should be not too strong to exclude yarn failure in the process of pulling out.

We considered the following surface treatments:

1. Fabric with no treatment.
2. Water emulsion PVA, solid content 38%, fabric weighting increase on 5.1%.

3. PVA-T – water emulsion PVA, solid content 38%, fabric weighting increase on
   5.1%. Further temperature treatment of the fabric at +98 °C.
4. Pine rosin B10, fabric weighting increase on 3.1%.

We performed experiments and supercomputer modeling to determine the deflection of the back side of two layers of ballistic fabric with and without surface treatment after local impact. Samples were located on the backing material. The edges of the fabrics were not fixed (Fig. 1). We chose two layers of fabric to ensure that the depth of a dent on the technical plasticine surface could be measured clearly. In calculations and experiments, the fabric size constituted 100 × 100 mm, the size of backing material equaled to 100 × 100 × 30 mm, the indenter was a ball with the diameter of 4.5 mm, the mass of 0.5 g, and the velocity of 100–130 m/s. Several ways of problem parallelizing were considered, and we obtained speedup graphs.

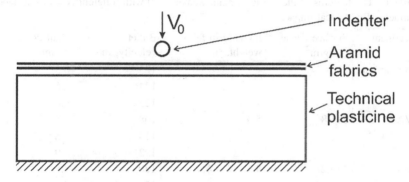

Fig. 1. Schematic representation of the problem

3   Ballistic Tests

Two layers of the fabric were placed on the surface of the backing material without fixing. Shots were produced by steel balls at right angle using pneumatic pistol IZH53M. The speed of the ball was fixed with the S04 chronograph with an accuracy of 1 m/s. We measured the depth of the dent in the technical plasticine, which was left by the fabrics after impact. The impact velocities were not sufficient to failure yarns in the fabric, so the ball kinetic energy was dissipated by yarn pull-out and the plasticine deformation.

The treatment allowed us to reduce the deflection of the ballistic fabric back surface: treatment with rosin (with addition of 3.1% of weight) reduced the deflection by up to 32%, water suspension PVA-T and PVA (with addition of 5.1% weight) – by 35% and 37.4%, respectively. The fabric and plasticine deformations under local impact are shown in Fig. 2. The features of the fabric after different types of surface treatment are presented in Table 1.

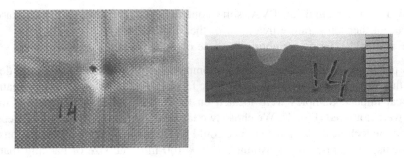

Fig. 2. Deformation of two layers of the fabric of plain weave P110 with PVA treatment and technical plasticine after the local impact

Table 1. The features of the fabric of plain weave P110 with different types of surface treatment after a local impact

Treatment	Surface density ρ, kg/m²	Increase in fabric weight, %	Bullet velocity, m/s	Dent depth w, mm
No treatment	110	–	123	6.5
			124	6.6
			125	6.8
PVA	115.8	5.3	108	3.5
			113	3.8
			122	3.9
PVA-T	115.2	4.7	120	3.8
			122	4.1
			128	4.8
Pine rosin B10	114.1	3.7	110	3.7
			111	3.8
			114	4.5

4 Ballistic Impact Simulations

4.1 Model Description

We developed a model of deformation and failure of ballistic fabrics, which consist of individual yarns with and without surface treatment, using the LS-DYNA software package. The geometry of the yarn was simplified and was represented by a piecewise linear set of shell elements with a constant width and thickness (Fig. 3). This representation provides us with the minimum of geometric parameters and numerical efficiency – minimal time for computer calculations. In the model the yarns have relative freedom of movement with the possibility of pulling-out with dry friction. The thickness of the yarns (shells) is 50 μm, width – 410 μm.

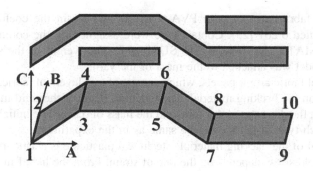

Fig. 3. The geometry of the yarn (plain weave)

The yarn material was orthotropic (*MAT_ENHANCED_ COMPOSITE_-DAMAGE) [24]. The material characteristics are shown in Table 2. The modulus of elasticity along the yarn (E_A) was determined experimentally [25], the remaining modules of elasticity (E_B, E_C), two shear modules (G_{AB}, G_{BC}) and three Poisson's ratios were chosen according to the literature recommendations [14, 26–28].

Table 2. Yarns material characteristics

Parameter	Symbol	Value
Modules of elasticity, Pa	E_A	$1.4 \cdot 10^{11}$
	E_B, E_C	$1.4 \cdot 10^{10}$
Density, kg/m³	ρ	1 440
Poisson's ratios	μ_{AB}, μ_{BC}, μ_{AC}	0.001
Shear modules, Pa	G_{AB}, G_{BC}	$1.4 \cdot 10^{10}$
	G_{CA}	$4 \cdot 10^7$

Aramid yarns in P110 fabric consist of many fibers with a diameter of 10–15 μ and a small twist and have a weak bending resistance. Therefore, to consider the bending in the model, three integration points over the thickness were used. The resistance to bending is affected by the transversal shear modulus G_{CA}. We found this parameter from preliminary calculations by comparing the calculated and experimental dependences of the load on displacement when pulling-out a yarn from a fabric without surface treatment [25].

For a fabric without treatment the static coefficient of friction was determined experimentally, its value equals to 0.174 [25]. Surface treatment of fabrics in models was incorporated by a corresponding change in the value of the static dry friction coefficient. Therefore, we made calculations for pulling-out the yarn from the fabric with different static friction coefficient so that the calculated force-displacement dependencies coincided with the experimental for all types of surface treatment. We found that for fabrics treated with PVA, PVA-T, and Rosin the coefficient of dry friction constituted 0.261 [25]. Contact type were assigned by the command *CONTACT_AUTOMATIC_SURFACE_TO_SURFACE. The weight of the surface treatment in the model was attached to the mass of the yarns.

In the model fabric armor panels, which consist of two layers of fabric, were placed without fixing on the backing material. The indenter, made of the rigid material, had a spherical shape, the diameter of 4.5 mm, and the mass of 0.5 g. The initial speed of the indenter for each type of fabric was the same as in the experiment.

The material of the backing material – technical plasticine is elastic–plastic model, in which the yield stress depends on the rate of strain. From the list of materials in the software package LS-DYNA for plasticine we chose *MAT_STRAIN_RATE_DEPENDENT_PLASTICITY, which allowed us to take into account the dependence of the yield stress on the strain rate in a tabular form [24]. The parameters for the material model were determined experimentally [29]. The elastic modulus E of plasticine is 20 MPa, the Poisson's ratio μ is 0.45, and the density ρ is 1400 kg/m^3. The mesh of the finite elements of the developed model is shown in Fig. 4.

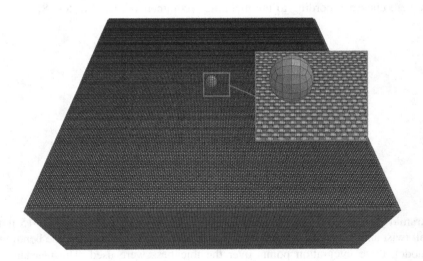

Fig. 4. Finite element mesh

4.2   Results of the Simulation

Calculations were performed on the supercomputer "Tornado SUSU" [30]. We considered three different methods of the model parallelization for fabrics without treatment. Firstly, we used the automatic parallelization of the model (Fig. 5a). In the second case, the model was divided into bands lying along one side of the model and passing through the entire thickness (Fig. 5b). Finally, we divided the model in the cylindrical coordinate system (Fig. 5c).

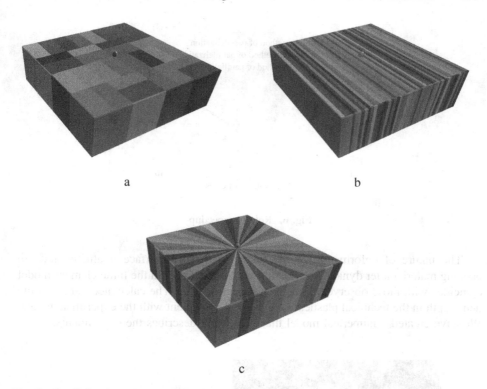

Fig. 5. Parallelization of the models on 48 cores (a – automatic, b – the bands passing through the entire thickness of the model, c – parallelization in the cylindrical coordinate system)

The speedup graphs are shown in Fig. 6. This task was calculated maximum to 48 cores. The number of cores was limited by the existing license for the software package LS-DYNA. The time for solving the problem on one core is 269,793 s. We found that, with the increase in the number of cores, the first parallelization method – automatic parallelization of the model gives a lowest speedup. In this case, fabrics and technical plasticine which contacting between themselves are distributed on different supercomputer cores. The time for calculating the task is increased by increasing the amount of data to transfer messages between these cores. The second method of parallelization gives the best speedup. Model was divided into bands lying along one side of the model and passing through the entire thickness. Contact surfaces are not distributed to different supercomputer cores. The intensity of data transfer between the cores in this case is lower than, in the first method of parallelization. The third method of parallelization gives an average speedup. In this case, we divided the model in the cylindrical coordinate system, parts of the model are divided through the entire thickness. The intensity of data transfer is higher here than in the second method of parallelization. This is due to the fact that the center of the model where there is intensive interaction between the bullet, fabrics and technical plasticine is divided into all the cores of the supercomputer.

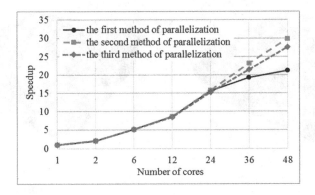

Fig. 6. Relative speedup

The nature of deformation of the fabrics with PVA surface treatment and the backing material after dynamic interaction with the indenter in the finite element model coincides with those observed in the experiment (Fig. 7). The calculated values of the dent depth in the technical plasticine are in good agreement with the experimental data. We have created a numerical model that adequately describes the experiments.

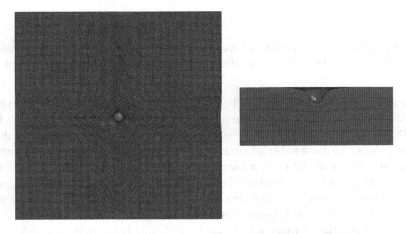

Fig. 7. The dent in the fabrics with PVA surface treatment and in the backing material after dynamic interaction with the indenter

5 Conclusion

We experimentally determined the depth of a dent in the backing material in case of a local impact on ballistic fabrics P110 with and without surface treatments. We showed that the treatment with pine rosin reduced the deflection up to 32% with the increase weighting of the fabric on 3.1%, the water suspension PVA-T and PVA – by 35% and by 37.4%, respectively, with the increase weighting on 5.1%.

We developed numerically effective models, which allowed us to calculate the dent depth in considered cases. The surface treatment of the fabric in the model was taken into account by changing one parameter – the coefficient of dry friction. The calculated and experimental values of the dent depths in technical plasticine agree with each other. Three methods of parallelization of the model are considered and speedup graphs are obtained.

Acknowledgements. The reported study was funded by RFBR according to the research project № 17-08-01024 A.

References

1. Zhu, D., Soranakom, C., Mobasher, B., Rajan, S.D.: Experimental study and modeling of single yarn pull-out behavior of Kevlar® 49 fabric. Compos. Part A **42**, 868–879 (2011). https://doi.org/10.1016/j.compositesa.2011.03.017
2. Nilakantan, G., Merrill, R.L., Keefe, M., Gillespie Jr., J.W., Wetzel, E.D.: Experimental investigation of the role of frictional yarn pull-out and windowing on the probabilistic impact response of Kevlar fabrics. Compos. Part B **68**, 215–229 (2015). https://doi.org/10.1016/j.compositesb.2014.08.033
3. Das, S., Jagan, S., Shaw, A., Pal, A.: Determination of inter-yarn friction and its effect on ballistic response of para-aramid woven fabric under low velocity impact. Compos. Struct. **120**, 129–140 (2015). https://doi.org/10.1016/j.compstruct.2014.09.063
4. Dolganina, N.Yu.: Deformirovanie i razrushenie sloistykh tkanevykh plastin pri lokal'nom udare. Ph.D. thesis [Deformation and fracture layered tissue plates under local impact. Ph.D. thesis], Chelyabinsk, 128 p. (2010)
5. Bhatnagar, A.: Lightweight Ballistic Composites, p. 429. Woodhead Publishing Limited, Cambridge (2006)
6. Ahmad, M.R., Ahmad, W.Y.W., Salleh, J., Samsuri, A.: Effect of fabric stitching on ballistic impact resistance of natural rubber coated fabric systems. Mater. Des. **29**, 1353–1358 (2008). https://doi.org/10.1016/j.matdes.2007.06.007
7. Karahan, M., Kus, A., Eren, R.: An investigation into ballistic performance and energy absorption capabilities of woven aramid fabrics. Int. J. Impact Eng. **35**, 499–510 (2008). https://doi.org/10.1016/j.ijimpeng.2007.04.003
8. Kang, T.J., Lee, S.H.: Effect of stitching on the mechanical and impact properties of woven laminate composite. J. Compos. Mater. **28**(16), 1574–1587 (1994). https://doi.org/10.1177/002199839402801604
9. Park, J.L., Yoon II, B., Paik, J.G., Kang, T.J.: Ballistic performance of p-aramid fabrics impregnated with shear thickening fluid; Part I – effect of laminating sequence. Text. Res. J. **82**(6), 527–541 (2012). https://doi.org/10.1177/0040517511420753
10. Ahmad, M.R., Ahmad, W.Y.W., Samsuri, A., Salleh, J., Abidin, M.H.: Blunt trauma performance of fabric systems utilizing natural rubber coated high strength fabrics. In: Proceeding of the International Conference on Advancement of Materials and Nanotechnology, ICAMN 2007, Langkawi, 29 May–1 June 2007, vol. 1217, pp. 328–334 (2010)
11. Gawandi, A., Thostenson, E.T., Gilllespie Jr., J.W.: Tow pullout behavior of polymer-coated Kevlar fabric. J. Mater. Sci. **46**, 77–89 (2011). https://doi.org/10.1007/s10853-010-4819-3
12. Majumdar, A., Butola, B.S., Srivastava, A.: Development of soft composite materials with improved impact resistance using Kevlar fabric and nano-silica based shear thickening fluid. Mater. Des. **54**, 295–300 (2014). https://doi.org/10.1016/j.matdes.2013.07.086

13. Lee, B.-W., Kim, I.-J., Kim, Ch.-G.: The influence of the particle size of silica on the ballistic performance of fabrics impregnated with silica colloidal suspension. J. Compos. Mater. **43**(23), 2679–2698 (2009). https://doi.org/10.1177/0021998309345292
14. Lee, B.-W., Kim, C.-G.: Computational analysis of shear thickening fluid impregnated fabrics subjected to ballistic impacts. Adv. Compos. Mater. **21**(2), 177–192 (2012). https://doi.org/10.1080/09243046.2012.690298
15. Hassan, T.A., Rangari, V.K., Jeelani, S.: Synthesis, processing and characterization of shear thickening fluid (STF) impregnated fabric composites. Mater. Sci. Eng. A **527**, 2892–2899 (2010). https://doi.org/10.1016/j.msea.2010.01.018
16. Mayo Jr., J.B., Wetzel, E.D., Hosur, M.V., Jeelani, S.: Stab and puncture characterization of thermoplastic-impregnated aramid fabrics. Int. J. Impact Eng. **36**, 1095–1105 (2009). https://doi.org/10.1016/j.ijimpeng.2009.03.006
17. Bazhenov, S.L., Goncharuk, G.P.: A Study of Yarn Friction in Aramid Fabrics. Polym. Sci. Ser. A **54**(10), 803–808 (2012). https://doi.org/10.1134/S0965545X12090015
18. Lim, C.T., Shim, V.P.W., Ng, Y.H.: Finite-element modeling of the ballistic impact of fabric armor. Int. J. Impact Eng. **28**, 13–31 (2003). https://doi.org/10.1016/S0734-743X(02)00031-3
19. Tan, V.B.C., Ching, T.W.: Computational simulation of fabric armor subjected to ballistic impacts. Int. J. Impact Eng. **32**(11), 1737–1751 (2006). https://doi.org/10.1016/j.ijimpeng.2005.05.006
20. Barauskasa, R., Abraitiene, A.: Computational analysis of impact of a bullet against the multilayer fabrics in LS-DYNA. Int. J. Impact Eng. **34**, 1286–1305 (2007). https://doi.org/10.1016/j.ijimpeng.2006.06.002
21. Ha-Minh, C., Imad, A., Kanit, T., Boussu, F.: Numerical analysis of a ballistic impact on textile fabric. Int. J. Mech. Sci. **69**, 32–39 (2013). https://doi.org/10.1016/j.ijmecsci.2013.01.014
22. Sapozhnikov, S.B., Forental, M.V., Dolganina, N.Yu.: Improved methodology for ballistic limit and blunt trauma estimation for use with hybrid metal/textile body armor. In: Proceeding of Conference "Finite Element Modelling of Textiles and Textile Composites", vol. 1. CD-ROM, St-Petersburg (2007)
23. Gatouillat, S., Bareggi, A., Vidal-Sallé, E., Boisse, P.: Meso modelling for composite preform shaping – simulation of the loss of cohesion of the woven fibre network. Compos. Part A **54**, 135–144 (2013). https://doi.org/10.1016/j.compositesa.2013.07.010
24. LS-DYNA R7.0 Keyword user's manual. http://www.lstc.com. Accessed 11 Apr 2018
25. Ignatova, A.V., Dolganina, N.Yu., Sapozhnikov, S.B., Shabley, A.A.: Aramid fabric surface treatment and its impact on the mechanics of yarn's frictional interaction. PNRPU Mech. Bull. **4**, 121–137 (2017). https://doi.org/10.15593/perm.mech/2017.4.09
26. Nilakantan, G., Nutt, S.: Effects of clamping design on the ballistic impact response of soft body armor. Compos. Struct. **108**, 137–150 (2014). https://doi.org/10.1016/j.compstruct.2013.09.017
27. Nilakantan, G., Wetzel, E.D., Bogetti, T.A., Gillespie, J.W.: Finite element analysis of projectile size and shape effects on the probabilistic penetration response of high strength fabrics. Compos. Struct. **94**(5), 1846–1854 (2012). https://doi.org/10.1016/j.compstruct.2011.12.028
28. Nilakantan, G., Keefe, M., Wetzel, E.D., Bogetti, T.A., Gillespie, J.W.: Effect of statistical yarn tensile strength on the probabilistic impact response of woven fabrics. Compos. Sci. Technol. **72**(2), 320–329 (2012). https://doi.org/10.1016/j.compscitech.2011.11.021
29. Sapozhnikov, S.B., Ignatova, A.V.: Mechanical properties of technical plasticine under static and dynamic loadings. PNRPU Mech. Bull. **2**, 201–219 (2014)
30. Kostenetskiy, P.S., Safonov, A.Y.: SUSU supercomputer resources. In: Proceedings of the 10th Annual International Scientific Conference on Parallel Computing Technologies (PCT 2016), Arkhangelsk, vol. 1576, pp. 561–573 (2016)

CardioModel – New Software for Cardiac Electrophysiology Simulation

Valentin Petrov[1], Sergey Lebedev[1], Anna Pirova[1],
Evgeniy Vasilyev[1], Alexander Nikolskiy[1,2], Vadim Turlapov[1],
Iosif Meyerov[1], and Grigory Osipov[1(✉)]

[1] Lobachevsky State University of Nizhni Novgorod, Nizhny Novgorod, Russia
valentin.s.petrov@gmail.com,
sergey.a.lebedev@gmail.com, anna.pirova@itmm.unn.ru,
eugene.unn@gmail.com, yahtingman@rambler.ru,
vadim.turlapov@gmail.com, meerov@vmk.unn.ru,
grosipov@gmail.com
[2] Nizhni Novgorod State Medical Academy, Nizhny Novgorod, Russia

Abstract. The rise of supercomputing technologies during the last decade has enabled significant progress towards the invention of a personal biologically relevant computer model of a human heart. In this paper we present a new code for numerical simulation of cardiac electrophysiology on supercomputers. Having constructed a personal segmented tetrahedral grid of the human heart based on a tomogram, we solve the bidomain equations of cardiac electrophysiology using the finite element method thus achieving the goal of modeling of the action potential propagation in heart. Flexible object-oriented architecture of the software allows us to expand its capabilities by using relevant cell models, preconditioners and numerical methods for solving SLAEs. The results of numerical modeling of heart under normal conditions as well as a number of simulated pathologies are in a good agreement with theoretical expectations. The software achieves at least 75% scaling efficiency on the 120 ranks on the Lobachevsky supercomputer.

Keywords: Heart simulation · Cardiac electrophysiology · Bidomain model
Finite element method · Numerical analysis · Parallel computing

1 Introduction

The unique capabilities of the mathematical modeling method combined with the growth of the performance of supercomputers open new horizons in biomedical research. One of the directions of such studies is the invention of a personal biologically relevant 3D computer model of a human heart. There are a lot of practical applications of such model, such as the extraction of new knowledge about the processes in the heart, study of the effects of drugs, generation of synthetic data for the development and tuning of novel diagnostics methods, and the construction of realistic training simulators for educational purposes. Studies in the field of numerical modeling of processes in the human heart using supercomputers are being actively pursued in

© Springer Nature Switzerland AG 2019
V. Voevodin and S. Sobolev (Eds.): RuSCDays 2018, CCIS 965, pp. 195–207, 2019.
https://doi.org/10.1007/978-3-030-05807-4_17

many research groups [1–12]. Current research areas include the development of methods for reconstructing a personal 3D model of heart and vessels according to computer tomography, automated segmentation of CT images, the creation of a mathematical and computer electrophysiological and electro-mechanical model of the whole human heart and its basic elements. The development of methods for the effective use of supercomputers for solving problems of numerical modeling of processes in the heart is one of the state-of-the-art topics of the current research.

In this paper, we present CardioModel – new software for numerical modeling of the human heart electrophysiology using supercomputers. The main motivation for development of this software is the abundance of open scientific problems the two main topics: (i) personalization and increasing the relevance of models of heart, (ii) increasing the efficiency of computational resources usage. Our software is based on the numerical solution of monodomain or bidomain equations [14, 15] that describe electrical activity in heart. We discretize the equations in time and space. Time discretization is performed using an explicit scheme. Space discretization is done by representing the patient's heart tissue domain with a three-dimensional tetrahedral grid. The structure of the grid is given by a finite-element Atlas model of the human heart [28]. The Atlas model is anatomically segmented by marking the affiliation of the grid nodes to specific anatomical segments (left, right, ventricles, atria, etc.) [29]. The conduction system of heart is represented as a graph and is connected with the grid. In order to achieve personalization of the Atlas model we merge the vertices of the tetrahedral mesh with an array of the patient's heart CT or MRI data. To increase the reliability of simulation results we employ biologically relevant cell models available from the CellML repository taking into account the segmentation of the grid to specific regions. Our software is implemented in C++ and has a flexible object-oriented architecture. That allows us easily extending the code with cell models and numerical schemes to take into account additional physical effects. We employ the MPI technology to parallelize the code on distributed memory. The ParMetis library [16] is used to distribute the grid over nodes of a supercomputer. For the SLAE solution we use iterative methods from the PETSc library [17] or the implementations of direct methods developed at Intel [18] and at UNN [19]. The project is based on the experience gained earlier in the development of the Virtual Heart software [20].

The rest of the paper is organized as follows. Section 2 provides an overview of similar works. Section 3 describes a mathematical model. The numerical scheme is formulated in Sect. 4. A parallel algorithm is given in Sect. 5. Section 6 provides an implementation overview. In Sect. 7 we present selected results of numerical simulation and discuss performance data.

2   Related Work

In 1993 the Physiome Project (http://physiomeproject.org) was presented aimed at "a quantitative description of the physiological dynamics and functional behavior of the intact organism" [1]. The CellML model repository [2] is one of the most important components of the project. It contains computer implementations of mathematical models of human and animal cells for use in numerical simulations.

The cardiac computational modeling software Alya Red CCM [3] is created in the Barcelona Supercomputing Center. The model underlying the code includes electrophysiological and electromechanical activity of the heart, as well as blood flow through its cavity. Electrophysiological and mechanical components are simulated on the same grid and the modeling of the blood flow is associated with the mechanics of the heart. The rabbit ventricular mesh developed in the University of Oxford [4] is used in simulations. The tool is part of the Alya software – the multiphysics simulation code [5] being developed by BSC.

Essential results of electrophysiology and electromechanics modeling of the heart are obtained in the Computational Cardiology Lab in the Johns Hopkins University [6]. The laboratory implements full heart simulation. Much attention is paid to the applications of the model: the reproduction of normal and pathological cardiac activity. Studies are underway to personalize the heart model [7].

Research centers of IBM in conjunction with the Lawrence Livermore National Laboratory carry out the Cardioid Cardiac Modeling project. The software developed by the team simulates the heart at several levels: electrophysiology of the heart tissue, electromechanics of the whole heart, and also modeling of the tissue fibers [8].

In the REO laboratory of INRIA a group of researchers is working on the modeling of electrophysiology of the heart using the model based on the description of surfaces [9]. The proposed approach allows reproducing typical heart pathologies. The group also deals with the inverse problem of electrophysiology – determining the parameters of the model from a given cardiogram, numerical modeling of cardiograms and creating simplified models of electrophysiology.

Remarkable results in numerical simulation of the heart are achieved by a team of researchers at the University of Oxford. The team develops the Chaste software aimed at solving computationally intensive problems in biology and physiology on supercomputers [13]. The part of the software, Cardiac Chaste, is capable of doing simulations of electro-mechanical processes in the heart. The software is publicly available.

Research in this direction is done in the institutes of the Russian Academy of Sciences (RAS) and in the universities. In the Institute of Numerical Mathematics of the RAS and MIPT, studies are underway to model the human cardiovascular system [10, 11]. Research on the human heart left ventricle modeling is performed in the Institute of Mathematics and Mechanics of the Ural branch of the RAS. The team proposed a new model of the left ventricle of heart which allows personalization according to MRI data. The software for simulation was also developed [12].

In general, there is a considerable interest in developing a biologically relevant personalized computer model of the human heart. One of the steps in this direction requires the development of parallel algorithms and their implementations capable of carrying out extremely computationally intensive calculations on supercomputers.

3 Model

The bidomain model is one of the most common continuous mathematical models describing the electrical activity of the heart. The model was proposed by Schmidt in 1969 [14] and was first formalized by Tang in 1978 [15]. The model assumes that the heart tissue consists of two continuous domains separated by a cell membrane: the inner space of the cells of the cardiac tissue (intracellular domain) and the space between the cells (extracellular domain). Interaction between domains is described by the flux through the cell membrane. The description of each domain includes the relationship between the electric field obtained from the potential, the current density, and the conductivity tensor.

Next, we give a formal description of the bidomain model. Let some domain Ω be filled with cardiac tissue. The tissue is considered as a set of two continuous domains – intracellular and extracellular, interacting through the cell membrane. Let u be a set of variables describing the tissue cell, φ_m and φ_e – intra- and extracellular potentials, $V = \varphi_m - \varphi_e$ – the transmembrane voltage, σ_m and σ_e – intra- and extracellular conductivity tensors, I_m^{st} and I_e^{st} – stimulus currents inside and outside the cell per unit area, χ – surface area to volume ratio, C_m – membrane capacity per unit area, I_{ion} – ion current per unit area. The functions I_{ion} and $\frac{\partial u}{\partial t}$ are defined by the model of the cell. The bidomain model is then given by the following set of equations:

$$\chi\left(C_m \frac{\partial V}{\partial t} + I_{ion}(u, V)\right) - \nabla \cdot (\sigma_m \nabla (V + \varphi_e)) = I_m^{st}$$
$$\nabla \cdot ((\sigma_m + \sigma_e)\nabla \varphi_e + \sigma_m \nabla V) = -I_m^{st} - I_e^{st}$$
$$\frac{\partial u}{\partial t} = f(u, V)$$
$$n \cdot (\sigma_m \nabla (V + \varphi_e)) = 0; n \cdot (\sigma_e \nabla \varphi_e) = 0$$

$$(1)$$

4   Method

We employ the following numerical scheme to get the solution of (1) in time. The bidomain equations are discretized in time and space. Time discretization is performed using an explicit scheme. Space discretization is done by dividing the heart tissue domain with a three-dimensional tetrahedral mesh; the finite element method is applied. Numerical solution of the equations is performed using the operator splitting method [24], which allows solving linear and nonlinear parts of the system (1) independently. This approach is commonly used to simulate electrophysiological activity [25–27]. According to the operator splitting method, the first equation of system (1) can be represented as follows: $\frac{\partial V}{\partial t} = (\Gamma_1 + \Gamma_2)V$, where Γ_1 and Γ_2 are nonlinear and linear differential operators, respectively. In the bidomain model $\Gamma_1 = \frac{-I_{ion}}{C_m}$, $\Gamma_2 = \frac{\nabla \cdot (\sigma_m \nabla (V + \varphi_e)) + I_m^{st}}{\chi C_m}$. Then the solution of system (1) is calculated as follows.

1. Integrate the ODE system (2) over time with a step size $\frac{\Delta t}{2}$ using computed values in time t as initial conditions:

$$\begin{cases} \frac{\partial V}{\partial t} = \Gamma_1 V \\ \frac{\partial u}{\partial t} = f(u, V) \end{cases}$$

$$(2)$$

2. Integrate the PDE $\frac{\partial V}{\partial t} = \Gamma_2 V$ over time with a step size Δt using the values computed during the step 1 as initial conditions.
3. Integrate the ODE system (2) over time with a step size $\frac{\Delta t}{2}$ using the values computed during the step 2 as initial conditions.

We employ the fourth order Runge–Kutta method to solve the ODE system, the finite element method to solve PDE system, and the Euler method for time discretization. The SLAE systems arising during the application of the finite element method are solved by means of relevant iterative methods. Additional information on the numerical integration of the bidomain equations can be obtained in the reviews [21–23].

The described approach allows us to find the solution of the bidomain equations. However, in order to perform fine tuning of the model behavior one has to adjust multiple parameter values in a large number of equations. Therefore, we developed a special scheme to simplify tuning. In this regard we represent a conduction system (CS) of the heart as a graph connecting a sinoatrial node, an atrioventricular node, left and right legs of the His bundle and Purkinje fiber. Vertices of the graph correspond to the regions of the heart, and the edges show the presence of conduction pathways up to the terminal nodes. Additional graph vertices can be added to improve the quality of the model. Each vertex of the graph is associated with the biological model of the cell from the CellML repository, the coupling coefficients with other nodes of the conduction system, and also the region of influence on other cells of the heart.

Each cell in the conduction system is an oscillator with its own natural frequency. Tuning those frequencies along with the coupling strengths (gap junction conductance) allows achieving the globally synchronous regime in the conduction system with the desired delays of the pulse propagation on different segments of the graph.

The dynamics of the conduction system is simulated in parallel with the main solver pipeline. The cells of the conduction system in heart are electrically coupled with myocardium through gap junctions. This gives a way to connect the dynamics of the conduction system with the dynamics of the excitable cells of cardiac tissue obtained in the main solver pipeline. Particularly, each time step of simulation the membrane potentials of the conduction system cells are coupled with the membrane potentials of the tissue cells with the diffusive coupling terms. Each node of the conduction system graph has a parameter R that describes the area of tissue this node corresponds to. Each tissue cell that is located inside the sphere of radius R around the conduction system node will have additional term in the equation for the rate of membrane potential: $D_{t-cs}(V_{cs} - V_t)$ where V_{cs} and V_t are membrane potentials of the conduction system cell and tissue cell respectively. Including the conduction system in the heart model allows us to fit the model behavior with theoretical expectations and to simulate typical heart pathologies. The corresponding simulation results are given in Sect. 7 below.

5 Parallel Algorithm

The parallel algorithm propagates the voltage V and the potential φ_e over time. The main numerical scheme is as follows:

1. *Data initialization step.* We read parameters from the configuration file, initialize MPI, allocate memory, read and initialize the mesh, the voltage $V(0)$, the potential $\varphi_e(0)$, the cells state u, the currents I_e^{st} and I_m^{st}.
2. *Mesh partitioning step.* The mesh is partitioned by means of the parallel ParMetis software to reduce MPI communications during the main computational loop.
3. *Matrix assembly step.* We generate FEM matrices in parallel: the mass matrix and the stiffness matrix.
4. *Pre-compute step.* Each MPI process solves of the ODE system (2) with the time step $\frac{\Delta t}{2}$ by the fourth order Runge-Kutta method for the heart cells located in the mesh vertexes belonging to the process. The initial conditions are set to default values (rest states for the tissue models) or are loaded from file.
5. *Main computational loop.* During this step we propagate the voltage, potential and cells state over time by means of the Operator splitting method with a time step Δt. Every iteration of this loop is fully parallel and is as follows:
   a. The software calculates diffusion $\nabla \cdot (\sigma_m \nabla V)$ for the voltage $V(t)$ by the finite element method. In this regard we need finding a solution of a sparse SLAE by iterative or direct method. Both approaches make it possible to find the solution of the system in parallel on the distributed memory. In most experiments we use iterative methods with appropriate preconditions, if necessary.
   b. We compute the total current $I_{sum} = I_m^{st} + I_e^{st} + \nabla \cdot (\sigma_m \nabla V)$ in parallel.
   c. We solve the Poisson equation $\nabla \cdot ((\sigma_m + \sigma_e)\nabla\varphi_e) = -I_{sum}$ with respect to $\varphi_e(t+dt)$ by the finite element method. We solve a sparse SLAE in parallel.
   d. The software computes diffusion $(\sigma_m \nabla \varphi_e)$, where φ_e has been found before. We employ the finite element method and solve a sparse SLAE in parallel.
   e. For each cell (mesh vertex) that has a coupling with the conduction system node compute additional current term: $I_{cs} = D_{t-cs}(V_{cs} - V_t)$.
   f. Each MPI process solves ODE $\frac{\partial V}{\partial t} = \frac{\nabla \cdot (\sigma_m \nabla (V+\varphi_e)) + I_m^{st} + I_{cs}}{\chi C_m}$ with respect to the voltage V by the Euler method with the time step Δt. The results of computation at step A are used as initial conditions.
   g. Each MPI process solves ODEs describing the dynamics of conduction system with RungeKutta 4th order approximation for time step Δt.
   h. Each MPI process solves the ODE system (2) with the time step Δt by the fourth order Runge-Kutta method for the heart cells located in the mesh cells belonging to the process. The initial conditions are the results of calculating V at step f.
6. *Post-compute step.* Each MPI process solves the ODE system (2) with the time step $\frac{\Delta t}{2}$ by the fourth order Runge-Kutta method for the heart cells located in the mesh cells belonging to the process. The initial conditions are the results from the last step of the main computation loop.
7. *Finalization step.* The results of computations are stored, the memory is released.

6   Implementation Overview

There are many approaches for the heart activity simulation. Therefore, when designing the CardioModel software, we considered the possibility of expanding the software with new models and numerical schemes as one of the key requirements. In this regard, we make use of the object-oriented methodology to design our application. The following main subsystems and interfaces of their communication have been identified: Modeling of electrical activity of heart, Mathematical models, Configuration, Numerical schemes, Post-processing, and Graphical user interface. Thereby we can easily choose different cell models, try new methods of solving SLAEs and specific preconditioners, and add new modules for processing simulations results.

The software is implemented in the C++ programming language. Parallelization for distributed memory systems is performed using MPI technology. The ParMetis parallel library is used to distribute the mesh along the compute processes in an optimal manner. We employ distributed SLAE solvers and preconditioners from the PETSc library. The results of simulations are stored using the VTK library to ensure compatibility with the most common professional 3D visualization tools, for example, ParaView. We also implement storing the current calculations state and restarting from the previously saved checkpoint. The CMake software is used to provide the build infrastructure.

7   Numerical Results

7.1   Application

The personal three-dimensional tetrahedral finite element model of the human heart tissue is constructed by combining the vertices of a reference finite element Atlas model of a human heart with the patient's heart tissue on a tomogram (Fig. 1).

The Atlas model is anatomically segmented by marking the affiliation of the mesh nodes to specific anatomical segments (left, right ventricles, atria, etc.). This markup is then transferred to a tomogram. The fitting is performed by the Coherent Point Drift resulting in a FEM mesh of heart which is a prerequisite for the simulation of electrical activity. The conduction system of the heart is represented as a graph and is connected with the mesh. The pre-segmented Atlas model is constructed from a three-dimensional surface model of the heart as follows. Firstly, we employ the NetGen software to generate a tetrahedral grid of the heart bounded with its outer surface. Next, by means of smoothing we remove cavities, limited by the inner surface of the chambers (atria and ventricles). Finally, we make the segmentation of the finite-element mesh of the heart. At this step, the inner surfaces are expanded until they come into contact with elements of another compartment or with outer surfaces. Our software performs fast validation of the vertices of the finite element mesh.

Below we present some simulation results. First, we consider three regimes of heart: the regular mode (Fig. 2a), atrial fibrillation due to a single spiral wave (Fig. 2b), and atrial fibrillation with spiral wave brake up and the formation of spiral chaos (Fig. 2c). The snapshots are taken at specific times, corresponding to the initial, intermediate and final part of the simulation. The color scale corresponds to a change in voltage V from −100 mV to 25 mV. In the bottom right inset one can see the wave front pattern computed with a threshold transform.

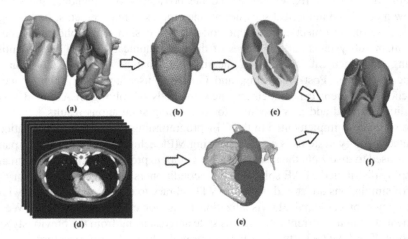

Fig. 1. The data preparation cycle. (a) The Atlas model of the heart surface. (b) The unified tetrahedral mesh of the heart. (c) The segmented mesh. (d) 3D image data. (e) The heart segmented by machine learning methods. (f) The personalized model of the human heart.

In the case of normal heart activation (Fig. 2a) one can see how the action potential originates from the sino-atrial node and propagates over atria. After that the excitation is delivered down to the ventricular apex through the Hiss bundle and eventually goes up covering left and right ventricles with a help of Purkinje fibers. The second set of results (Fig. 2b) demonstrates the simulation of atrial fibrillation which is believed to be stipulated by the spiral wave of action potential that persists in the atrial media and generates high frequency pulses. The rate of excitation may reach 500 bpm due to spiral wave curvature. At the same time AVN serves as a filter only allowing limited number of excitations to go down into the ventricles. Then (Fig. 2c) we demonstrate how the single spiral wave may become unstable and break up into a dynamic regime called spiral chaos which is also considered to be a background for atrial fibrillation. One can observe how the increase in the wave width leads to the spiral instability and to the consequent wave front break up. Multiple spiral rotors occur in the atria. They are irregular and persistent and lead to incoherent high frequency excitation.

Lastly (Fig. 3), we demonstrate the simulated electrical activity of heart under infarction conditions. The area of infarction was introduced in the left ventricle of the heart that otherwise showed a normal activation pattern. The Figure has three insets: top left illustrates the 3D image of the electrical action potential with the colormap depicting the membrane voltage (mV). The snapshot is taken at the moment of time when the ventricles finished depolarization. The area of infarction (and hence a conduction block) is clearly observed. Top right inset shows the snapshot of the dynamics of the conduction system at the moment the terminal pieces of Purkinje fiber are excited. The time series in the inset below shows the electrical activity of significant

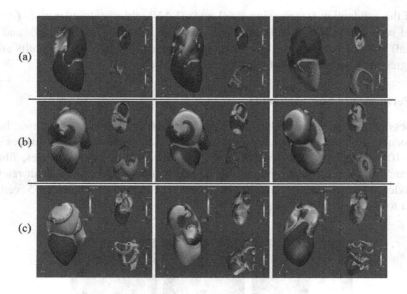

Fig. 2. Simulation results. (a) V in the regular mode. (b) Atrial fibrillation, a single spiral wave. (c) Atrial fibrillation, spiral chaos.

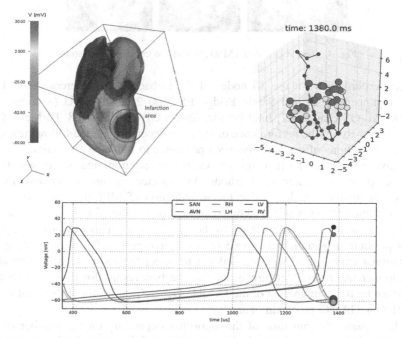

Fig. 3. Simulation results. Electrical activity of the heart under infarction conditions.

cells of the conduction system: sino-atrial node (SAN), atrio-ventricular node (AVN), terminal cells of the Hiss bundle located at the apex of the heart (LH, RH), and last (terminal) cells of the Purkinje fiber of both ventricles (LV, RV). The results are in good agreement with theoretical expectations.

7.2   Performance and Scalability

We assess performance and scalability of the CardioModel software as follows. In the experiments we use the finite element mesh (Fig. 4) with 998 151 vertices and 5 140 810 tetrahedrons. The left and right atrium, left and right ventricles, fibrous tissue, and ventricular septum are considered as separate compartments with relevant cell models. The conduction system is represented as the graph with 55 vertices mapped to the mesh and 54 edges.

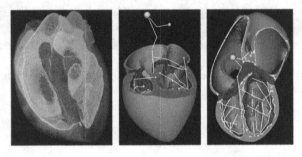

Fig. 4. The segmented 1M mesh and the conduction system.

In the experiments we use 15 nodes of the Lobachevsky supercomputer. Every node is equipped by 2 Intel Sandy Bridge E5-2660 2.2 GHz CPU (8 cores each), 64 GB RAM, OS Centos 7.2, Intel Parallel Studio 2017, PETSc 3.8, ParMetis 4.0.3.

First, we analyzed the performance of the code depending on the following factors: possible distributions of the mesh between processes to reduce MPI communications; SLAE solver selection; SLAE solver preconditioner selection; and proper selection of the starting point for the iterative methods. As expected, the use of ParMetis dramatically speeds up the calculation due to the reduction of MPI communications. The solution obtained at the previous iteration of the computational loop was chosen as a starting point for the iterative methods of the SLAE solver. This obvious strategy has made it possible to significantly speed up the convergence of methods. Large-scale computational experiments were carried out to select the most suitable method and preconditioner. As a result, it turned out that the best results for the given accuracy were shown by the combination of the Flexible Generalized Minimal Residual method (FGMRES) + GAMG preconditioner.

Table 1 shows the run time of the algorithm depending on the number of MPI processes used. The first column shows the number of processes, in the second – the run time of the ODE solver. The third and fourth columns give the time for solving the SLAE in calculating the diffusion and solving the Poisson equation, respectively. The proportion of time that has come to other stages is negligible and is shown in the sixth column. Lastly, the total calculation time is presented.

The results of the experiments show that the stage of the ODE solution scales well enough up to 120 processes. Contrary, the SLAE solution requires a significant amount of MPI communications, which leads to a decrease in the strong scaling efficiency of these stages. Overall strong scaling efficiency calculated from 8 to 64 processes is 96.2%. The further increase in the number of processes to 120 still speeds up the computations, the efficiency is 74.6% of the theoretical optimum.

Table 1. Performance results when propagating the bidomain equations. The number of time steps is set to 100. The PETSc tolerance parameter is equal to 1e–10. Time is given in seconds.

# Processes	ODE	Diffusion solver	Poisson equation solver	Other	Total
8	24.28	24.08	76.67	4.39	129.41
16	11.82	9.87	34.22	1.49	57.40
32	6.00	5.24	17.78	1.19	30.22
64	3.09	3.04	10.03	0.65	16.82
96	2.17	3.06	9.14	0.59	14.96
120	1.76	2.32	7.07	0.42	11.57

8 Conclusion and Future Work

We presented the description of the CardioModel software, intended for numerical modeling of human heart electrophysiology on supercomputers. The software uses a tetrahedral finite-element mesh of the heart as input data. The mesh is constructed according to the results of the patient's computer tomogram and is automatically segmented. Using this mesh we integrate bidomain equations describing the propagation of an electrical signal in cardiac tissues. The paper presents the results of numerical modeling of several regimes of the heart, including the development of pathologies. Software shows acceptable (75%) scalability when using 120 MPI ranks on the mesh of ∼1 million vertices and ∼5 million tetrahedrons. Further developments include improving personalization of models and increasing the scalability of the software. We plan to develop custom load balancing schemes when using different cell models.

Acknowledgements. The study was supported by the Ministry of Education of Russian Federation (Contract # 02.G25.31.0157, date 01.12.2015).

References

1. Crampin, E.J., et al.: Computational physiology and the physiome project. Exp. Physiol. **89**(1), 1–26 (2004)
2. Lloyd, C.M., et al.: The CellML model repository. Bioinformatics **24**(18), 2122–2123 (2008)
3. Vázquez, M., et al.: Alya red CCM: HPC-based cardiac computational modelling. In: Klapp, J., Ruíz Chavarría, G., Medina Ovando, A., López Villa, A., Sigalotti, L. (eds.) Selected Topics of Computational and Experimental Fluid Mechanics. ESE, pp. 189–207. Springer, Cham (2015). https://doi.org/10.1007/978-3-319-11487-3_11

4. Bishop, M.J., et al.: Development of an anatomically detailed MRI-derived rabbit ventricular model and assessment of its impact on simulations of electrophysiological function. Am. J. Physiol. Heart Circ. Physiol. **298**(2), H699–H718 (2010)
5. Vázquez, M., et al.: Alya: multiphysics engineering simulation toward exascale. J. Comput. Sci. **14**, 15–27 (2016)
6. Trayanova, N.A.: Whole-heart modeling. Circ. Res. **108**(1), 113–128 (2011)
7. Arevalo, H.J., et al.: Arrhythmia risk stratification of patients after myocardial infarction using personalized heart models. Nat. Commun. **7**, 11437 (2016)
8. Richards, D.F., et al.: Towards real-time simulation of cardiac electrophysiology in a human heart at high resolution. Comput. Methods Biomech. Biomed. Eng. **16**(7), 802–805 (2013)
9. Chapelle, D., Collin, A., Gerbeau, J.-F.: A surface-based electrophysiology model relying on asymptotic analysis and motivated by cardiac atria modeling. Math. Models Methods Appl. Sci. **23**(14), 2749–2776 (2013)
10. Vassilevski, Y., Danilov, A., Ivanov, Y., Simakov, S., Gamilov, T.: Personalized anatomical meshing of the human body with applications. In: Quarteroni, A. (ed.) Modeling the Heart and the Circulatory System. MS&A, vol. 14, pp. 221–236. Springer, Cham (2015). https://doi.org/10.1007/978-3-319-05230-4_9
11. Danilov, A.A., et al.: Parallel software platform INMOST: a framework for numerical modeling. Supercomput. Frontiers Innov. **2**(4), 55–66 (2015)
12. Pravdin, S., et al.: Human heart simulation software for parallel computing systems. Procedia Comput. Sci. **66**, 402–411 (2015)
13. Mirams, G.R., et al.: Chaste: an open source C++ library for computational physiology and biology. PLoS Comput. Biol. **9**(3), e1002970 (2013)
14. Schmitt, O.H.: Biological information processing using the concept of interpenetrating domains. In: Leibovic, K.N. (ed.) Information Processing in The Nervous System, pp. 325–331. Springer, Heidelberg (1969). https://doi.org/10.1007/978-3-662-25549-0_18
15. Tung, L.: A bi-domain model for describing ischemic myocardial dc potentials. Massachusetts Institute of Technology (1978)
16. Karypis, G., Kumar, V.: Parallel multilevel k-way partitioning scheme for irregular graphs. SIAM Rev. **41**(2), 278–300 (1999)
17. Balay, S.: PETSc Users Manual, ANL-95/11–Revision 3.8. Argonne National Lab (2017)
18. Intel MKL. Sparse solver routines. https://software.intel.com/en-us/mkl-developer-reference-fortran-sparse-solver-routines. Accessed 1 Mar 2018
19. Lebedev, S., Akhmedzhanov, D., Kozinov, E., Meyerov, I., Pirova, A., Sysoyev, A.: Dynamic parallelization strategies for multifrontal sparse cholesky factorization. In: Malyshkin, V. (ed.) PaCT 2015. LNCS, vol. 9251, pp. 68–79. Springer, Cham (2015). https://doi.org/10.1007/978-3-319-21909-7_7
20. Bastrakov, S., et al.: High performance computing in biomedical applications. Procedia Comput. Sci. **18**, 10–19 (2013)
21. Linge, S., et al.: Numerical solution of the bidomain equations. Philos. Trans. R. Soc. A Math. Phys. Eng. Sci. **367**(1895), 1931–1950 (2009)
22. Clayton, R.H., et al.: Models of cardiac tissue electrophysiology: progress, challenges and open questions. Progr. Biophys. Mol. Biol. **104**(1), 22–48 (2011)
23. Pathmanathan, P., et al.: A numerical guide to the solution of the bidomain equations of cardiac electrophysiology. Progr. Biophys. Mol. Biol. **102**(2), 136–155 (2010)
24. Strang, G.: On the construction and comparison of difference schemes. SIAM J. Numer. Anal. **5**(3), 506–517 (1968)
25. Bernabeu, M.O., et al.: Chaste: a case study of parallelisation of an open source finite-element solver with applications to computational cardiac electrophysiology simulation. Int. J. HPC Appl. **28**(1), 13–32 (2014)

26. Sundnes, J., Lines, G.T., Tveito, A.: An operator splitting method for solving the bidomain equations coupled to a volume conductor model for the torso. Math. Biosci. **194**(2), 233–248 (2005)
27. Santos, R.W., et al.: Parallel multigrid preconditioner for the cardiac bidomain model. IEEE Trans. Biomed. Eng. **51**(11), 1960–1968 (2004)
28. Vasiliev, E.: Generation of an atlas-based finite element model of the heart for cardiac simulation. Int. Sci. J. Math. Model. **4**, 207–209 (2017)
29. Lachinov, D., Belokamenskaya, A., Turlapov, V.: Precise automatic cephalometric landmark detection algorithm for CT images. In: Proceedings of Graphicon 2017, pp. 275–278 (2017)

Examination of Clastic Oil and Gas Reservoir Rock Permeability Modeling by Molecular Dynamics Simulation Using High-Performance Computing

Vladimir Berezovsky$^{(\boxtimes)}$ (iD), Marsel Gubaydullin, Alexander Yur'ev,
and Ivan Belozerov

M.V. Lomonosov Northern (Arctic) Federal University, Arkhangelsk, Russia
v.berezovsky@narfu.ru

Abstract. "Digital rock" is a multi-purpose tool for solving a variety of tasks in the field of geological exploration and production of hydrocarbons at various stages, designed to improve the accuracy of geological study of subsurface resources, the efficiency of reproduction and usage of mineral resources, as well as applying of the results obtained in industry. This paper presents the results of numerical calculations and their comparison with the full-scale natural examination. Laboratory studies have been supplemented with petrographic descriptions to deepen an insight into behaviors of the studied rock material. There is a general tendency to overestimate the permeability, which may be associated with the application of a rather crude resistive model for assessing permeability and owing to the porous cement has not been considered.

Keywords: Digital rock model · High-Performance computing
Clastic oil and gas reservoir's rock · Molecular dynamics

1 Introduction

Petroleum well cores examinations are among the most costly stages of development of oil and gas. Its partial replacement by the numerical experiments can provide significant economic benefits [1]. 3D-reconstruction of the rock microstructure problem arises within a framework of mathematical modeling of porous media macroscopic properties [2–4]. And the relation between the geometry of the microstructure and macroscopic physical properties, recently, arises interest in research teams. Permeability of porous materials is one of the macroscopic parameters of practical interest, and its measurement is important for predicting flows on macrolevel [5]. Significant advances in obtaining maps of interstitial space have been made recently. The usage of high-performance computing technology has accelerated the development and the use of tools "digital rock" in addition to the physical laboratory measurements.

The method of molecular dynamics is one of the promising approaches for building the mathematical model of the macroscopic properties of porous media and for the 3D-reconstruction of the rock microstructure. The results of numerical calculations and their comparison with the full-scale examination are provided in the paper. A general

© Springer Nature Switzerland AG 2019
V. Voevodin and S. Sobolev (Eds.): RuSCDays 2018, CCIS 965, pp. 208–217, 2019.
https://doi.org/10.1007/978-3-030-05807-4_18

tendency to overestimate the permeability persists. It may be associated with the application of a rather crude resistive model for assessing permeability and neglecting taking into account the porous cement.

2 Method

Simulation of the rock cores pore space is carried out in several stages. Primitives of the shapes of grains, granulometric composition, and rock texture are modeled during the first stage. The coefficients of filling the pore space with clay cement are given. The parameters of the physical model, the calibration algorithm, and the presentation of the results are indicated at the last stage. As a result, a model of porous medium is designed. The obtained model is used for modeling of filtration processes in porous formation environment.

A developed digital rock model consists of several embedded models:

- rock microstructure model
- numerical permeability models and filtration-capacitance rock properties
- numerical models of asphaltene, paraffin and resin precipitation.

The first one is the construction of the 3D geometry of the rock microstructure, the second one is the simulation of micro-flow physics, and the third one is chemistry of processes occurring under reservoir conditions. These components are integrated into a single digital rock model, designed to provide a means of predicting behavior of fluids under different conditions. This report examines the approaches to constructing the geometry of the microstructure of a rock.

A rock microstructure model obtained by the stochastic [6–8] or simulated packing algorithm of microparticles with compaction is used as a basis for the subsequent analysis of pore space. Having made the transition from the representation of particle packing to the representation of the pore network model, the permeability of single channels can be calculated using molecular dynamics. Thus, for the network model of the pore collectors, a system of linear equations is compiled with respect to the pressure in each pore. The evaluation of a sample pressure drop makes it possible to calculate the absolute permeability according to Darcy's law [9].

In the report we have suggested a simplified model of the geometric structure of rock core space, i.e. a box densely filled with some balls of different diameters. The distribution of the diameter of the balls can be obtained by the laboratory studies of the rock core (Fig. 1). Packing of balls is made as a result of simulation of molecular dynamics. The box has fixed walls and bottom. Simulation of molecular dynamics has been done with atoms having the radius of the van der Waals interaction corresponded to the distribution, determined as a result of laboratory rock core studies (Fig. 1).

The Lennard-Jones (LJ) potential (1) is chosen as an interatomic interaction potential because of its simplicity and because it has never led to physically unacceptable results. It can be applied to model not only atoms and molecules with spherical symmetry but also other non-polar substances:

$$U(r) = 4\epsilon\left[\left(\frac{\sigma}{r}\right)^{12} - \left(\frac{\sigma}{r}\right)^{6}\right] \tag{1}$$

Where ϵ is the depth of the potential well, σ is the finite distance at which the inter-particle potential is zero, r is the distance between the particles.

The parameters σ are used according to the radius of a corresponding ball, and the cross-section interaction parameters have been defined using the Lorentz-Berthelot mixing rule (2):

$$\epsilon_{ij} = \left(\epsilon_{ii}\epsilon_{jj}\right)^{\frac{1}{2}}, \sigma_{ij} = \frac{1}{2}\left(\sigma_{ii} + \sigma_{jj}\right) \tag{2}$$

Gravitational force has also been applied. To save computing time the LJ potential has been truncated at a cut-off distance of and to avoid a jump discontinuity at cut-off distance, the LJ potential has been shifted. During the simulation the temperature linearly decreases passing the melting point down to the low values. For the temperature maintenance the Berendsen thermostat have been used.

Passing the coordinate space of the box with a specific test atom having specified van der Waals radius is sufficient to determine the pore space. The result is a system of micro-channels, which can be used within the electrodynamic analogy, presenting trends in the form of a network-related electrical resistances. Having solved the problem of finding the impedance, one can found permeability.

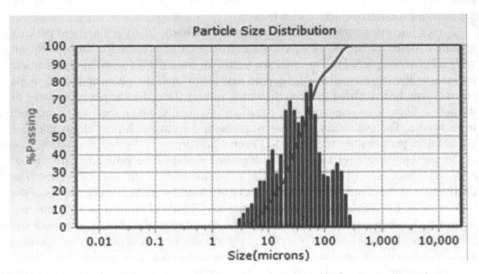

Fig. 1. The bar graph shows the percentage distribution of the particle size of the core sample. A graph of the cumulative volume is shown by the green line. The value of cumulative data is the percentage of particles whose size is less than the specified. The graph of the cumulative volume with the axis "% Passing" (% of the recorded data) and the axis "Size (microns)" is displayed on a semilogarithmic scale. Table 1 shows the results of an experimental study of the core sample and a numerical experiments with the fitted model. (Color figure online)

3 Computing Details

The NArFU Computing cluster [11] with a peak performance of 17.6 TFLOPS, has a hybrid architecture consisting of twenty 10-core dual-processor nodes with an Intel Xeon processor, eight of which have Intel Xeon Phi coprocessors. The nodes are connected by the high-performance interconnect Infiniband 56. The computing cluster has the following characteristics:

- 20 computing nodes.
- Each node has two 10-core Intel Xeon E5-2680v2 (2,8 GHz) processors and 64 GB of RAM.
- At eight nodes, the Intel Xeon Phi 5110P (8 GB, 1.053 GHz, 60 core) math coprocessors are additionally installed.
- Internal computer network for calculations: Infiniband 56 Gb/s.
- Network file system FEFS (Fujitsu Exabyte File System) with a capacity of more than 50 TB and a throughput of 1.67 GB/s (13.36 Gb/s).
- The cluster performance on the CPU in the LINPACK test is 8.02 Tflops; on the CPU+Xeon Phi 7.68 Tflops, the cumulative 15.7 Tflops.

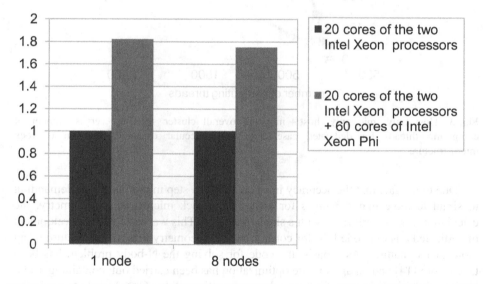

Fig. 2. Acceleration with the Intel® Xeon Phi™ coprocessor in offload LAMMPS numerical simulation

The modeling of the geometric structure of the rock core space have been carried outout using molecular dynamics methods, using the application software package LAMMPS [12]. LAMMPS parallel simulator of atomic, meso or continuous scale atomic models. It can be executed both on one processor system, and using the technique of message passing (MPI) with the help of spatial partitioning of the calculation area. Since 2015, the support for the using of Intel® Xeon Phi™ coprocessors has appeared as a part of the code being redistributed to date. A number of LAMMPS routines are optimized for operation in the offload mode of the Xeon Phi™ operation.

The comparison of computational performance with and without applying a coprocessor is illustrated in Fig. 2.

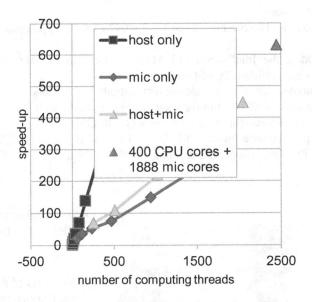

Fig. 3. Host only, mic only, host + mic and overall cluster speed-up versus number of computing threads for dedicated case code for molecular dynamics creation of rock microstructure

Due to the fact that the accuracy in calculating the step motion is not so demanding to simulate molecular dynamics for creating the rock microstructure geometry, the calculation can be performed with a single precision. This was used in the development of dedicated custom code [13] for constructing the geometry of a microstructure using molecular dynamics. As a basis, the code for solving the N-body problem has been taken from [14] and an appropriate optimization has been carried out, containing scalar tuning, vectorization with SoA and memory optimization. OpenMP threads on the coprocessor have been started started in offload mode. Data between nodes have been synchronized with MPI processes running on CPU nodes. Such an approach has yielded approximately 12 times the performance in comparison with LAMMPS. Scaling properties of code are shown in Fig. 3.

4 Core Samples Investigation

The experimental studies of the terrigenous reservoir, represented by a single core sample of standard size 30 mm in diameter and 30 mm in length, have been carried out. The coefficient of open porosity have been determined on the sample by the method of liquid saturation in accordance with (GOST 26450.1-85. The method of determining the coefficient of open porosity by liquid saturation) and the coefficient of absolute permeability in accordance with (GOST 26450.2-85. Gas permeability for stationary and non-stationary filtration). Residual water saturation - by capillarimetry method in accordance with GOST 39-204-86 Oil Method for determination of residual water saturation pit oil and gas.). After that, the sample has been saturated with kerosene and the core permeability has been determined by fluid in the formation conditions at the PEC-5 (7) installation.

Table 1. Results of natural and numerical investigations

	Connected porosity, %	Permeability for kerosene, 10^{-3} um^2
Natural rock core samples	20.11	55.66
Stochastic packing	25.82	68.45
Molecular dynamics simulation packing	22.27	60.92

After carrying out the filtration experiments, the input end of the sample have been photographed, and subsequently have been cut by 1 mm along the length of the sample by grinding and again photographed. The operation have been carried out to the full "attrition" of the sample. In the course of these experiments an array of image layers has been obtained, which can be used in the validation of a digital model of the rock microstructure. It is important to determine the granulometric composition of sandy-silty rocks to study out their properties as oil and gas collectors [10]. Determination of the granulometric composition of the sample has shown that the average particle diameter by the type of distribution of their number is 5.82 um, the average particle diameter by the form of their distribution is 54.52 um. The obtained percentage distribution of the particle size of the core sample is shown on Fig. 1. Table 1 shows a comparison of the results obtained by the methods described above.

Petrographic description of 10 thin sections (Fig. 4) of the terrigenous sandstone rocks of Berea Sandstone field has been performed to obtain lithological and petrographic information. The investigated samples are medium-fine-grained and finely-mediumgrained light-gray sandstones which are the reservoirs with mainly average usable storage capacity pore-type. Their porosity according to petrophysical analyzes varies in the range of 17–20%. In terms of material composition: the content of quartz in rocks varies within the limits of 70–85%, feldspars 2–5%, fragments of rocks of different genesis of 10–20%, mica 1–2%. Cement mixed: locally developed cementation of contact type with conformal contacts of individuals; regenerative quartz does not exceed 1%; pore-film cement clay 2–5%; Authigenic carbonate plays the role of porous

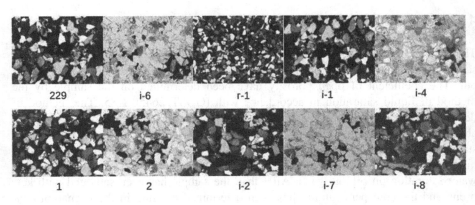

Fig. 4. Photos of the sections of core samples. 100x magnification. Polarized light for 229, r-1, i-1, 1, i-2, i-8. and passing light for i-6, i-4, 2, i-7.

corrosive cement (2–5%). The key factor determining the porosity and permeability of terrigenous reservoirs is the dimension of the grains of rock-forming minerals and their sorting. A direct relationship between the grain size and the porosity of the rocks along the sections has been shown in natural and simulated examinations. The larger the grain size, the greater the porosity. It should also be emphasized that in these rocks quartz grains and rock fragments have an irregular shape and semi-entangled particles, thus they are laid less densely, which leads to an increase in porosity.

A dependence of permeability on the temperature of the rocks (Table 2, Fig. 5) has been shown in natural and simulated examinations. The graphs obtained during examinations of the permeability of ten samples are presented. Two graphs of permeability calculated for two simulated microstructures and comparison are also presented: the first one is for the microstructure obtained by stochastic packing, the second one is for the microstructure from molecular dynamics simulations. The permeability coefficient of clastic rocks is affected by: the granulometric composition of the rocks, sorting, the shape of the grains and packaging. A comparatively close direct relationship between

Table 2. Open porosity and absolute permeability of investigated core samples.

Sample	Length, cm	Diameter, cm	Absolute permeability, 10^{-3} um^2	Open porosity, %
r-1	3,068	3,003	101,01	20,11
i6	2,682	2,997	125,64	20,42
i1	3,110	2,999	130,25	17,99
i2	2,722	2,996	96,43	18,89
i4	2,743	3,002	85,82	17,48
i7	2,854	3,003	91,64	18,55
229	2,974	2,995	121,13	19,63
1	3,580	2,995	101,032	18,07
i8	2,967	2,997	97,22	18,52
2	3,630	3,003	83,64	18,79

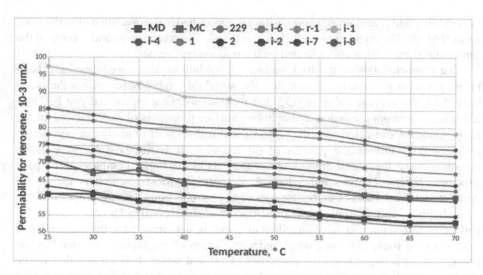

Fig. 5. Natural and numerical core sample investigation of permeability dependence from the temperature. Circles indicates data from natural experiments. Square indicates data from simulations. MD for molecular dynamics and MC for stochastic packing.

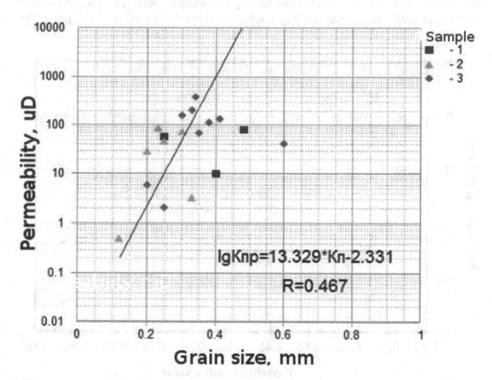

Fig. 6. Dependence of permeability on average grain diameter for core samples from different deposits

permeability and granulometric composition is established in the investigated terrigenous rocks (Fig. 6). The correlation coefficient R is 0.467. The composition and content of the cementitious material is also reflected on the reservoir properties.

The characteristics of the rock surface, depending on the wettability have been investigated. In general, for all samples, the wettability index varies within the range 0.03–0.94, the average value of 0.55, the rocks of the samples studied can be characterized as hydrophilic and as hydrophobic. Surface properties of rocks are one of the most important characteristics, they affect the distribution of fluids in the pore space and largely determine the process of oil recovery. It is not possible to evaluate unambiguously the wettability of terrigenous rocks having a complex pore space structure, a heterogeneous mineralogical composition. Most researchers are of the opinion that the surface of carbonate rocks has heterogeneous wetting properties, that is, hydrophilic and hydrophobic regions coexist simultaneously. The research of this collection has supported it. Wettability of rocks is determined by the material that makes up the rock, but can change under the action of the liquids contained in the pores, that is, it depends on the properties of oil and water. It is known that in the formation oil there are surface-active and polar substances that can be adsorbed by the rock. Therefore, the integral wettability of rocks is a complex function that depends on the properties of the rock and liquids that saturate the pore space (Fig. 7).

According to the graph, there is no clear correlation between the core permeability for oil and the wettability index. Both hydrophilic samples and hydrophobic samples are present and have a permeability coefficient from 10^{-3} μm^2 to 10^{-1} μm^2. The

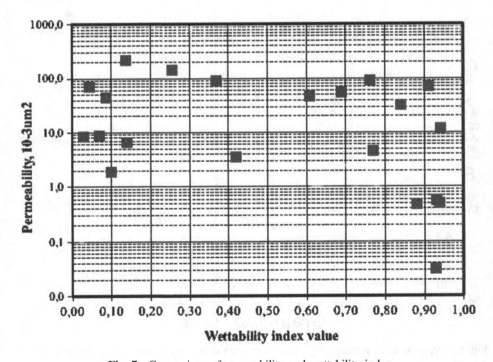

Fig. 7. Comparison of permeability and wettability index

question of how the wettability of the collector surface changes when it is saturated with oil has not been clarified at the present time. In this connection, the hydrophobic ability of oil in contact with the surface of reservoir rocks is an object of interest.

5   Conclusions

An approach to the solution of the problem of mathematical modeling of macroscopic properties of porous media is proposed in which the molecular dynamics method is applied for 3D reconstruction of the rock microstructure. The results of numerical calculations and their comparison with the full-scale experiments are presented. There is a general tendency to overestimate the permeability, which may be associated with the usage of a fairly coarse resistive model to assess permeability and the fact that porous cement is not taken into account.

Acknowledgements. The research was carried out with the financial support of the Russian Foundation for Basic Research (RFBR) within the framework of the scientific project No. 16- 29-15116 All models has been simulated used HPC environment at NArFU(HPC NArFU).

References

1. Renard, P., Genty, A., Stauffer, F.: Laboratory determination of the full permeability tensor. J. Geophys. Res. **106**, 443–452 (2001)
2. Andra, H., et al.: Digital rock physics benchmarks-part II: computing effective properties. Comput. Geosci. **50**, 33–43 (2013)
3. Andrew, M., Bijeljic, B., Blunt, M.J.: Pore-scale imaging of geological carbon dioxide storage under in situ conditions. Geophys. Res. Lett. **40**, 3915–3918 (2013)
4. Blunt, M.J., et al.: Pore-scale imaging and modelling. Adv. Water Resour. **51**, 197–216 (2013)
5. Bear, J.: Dynamics of Fluids in Porous Media. American Elsevier Publishing Company, New York (1972)
6. Dong, H., Blunt, M.J.: Pore-network extraction from micro-computerized-tomography images. Phys. Rev. E **80**, 036307 (2009)
7. Garcia, X., Akanji, L.T., Blunt, M.J., Matthai, S.K., Latham, J.P.: Numerical study of the effects of particle shape and polydispersity on permeability. Phys. Rev. E **80**, 021304 (2009)
8. Mizgulin, V.V., Kosulnikov, V.V., Kadushnikov, R.M.: An optimization approach to simulation of microstructures. Comput. Res. Model. **5**(4), 597–606 (2013)
9. Whitaker, S.: Flow in porous media I: a theoretical deviation of Darcy's law. Transp. Porous Media **1**, 3–25 (1986)
10. Hanin A.A.: Reservoir rocks of oil and gas and their study. Nedra (1969)
11. HPC NArFU. http://narfu.ru/imikt/science/cluster. Accessed 13 Nov 2016
12. LAMMPS Molecular Dynamics Simulator. http://lammps.sandia.gov. Accessed 13 Nov 2016
13. Berezovsky, V.: GEOSCI. https://github.com/valber-8/geosci. Accessed 13 Nov 2017
14. Vladimirov, A., Asai, R., Karpusenko, V.: Parallel Programming and Optimization with Intel Xeon Phi Coprocessors. Colfax International (2015). ISBN 978-0-9885234-0-1

Hybrid Codes for Atomistic Simulations on the Desmos Supercomputer: GPU-acceleration, Scalability and Parallel I/O

Nikolay Kondratyuk[1,2,3], Grigory Smirnov[1,2,3], and Vladimir Stegailov[1,2,3(✉)]

[1] Joint Institute for High Temperatures of RAS, Moscow, Russia
[2] Moscow Institute of Physics and Technology, Dolgoprudny, Russia
stegailov.vv@mipt.ru
[3] National Research University Higher School of Economics, Moscow, Russia

Abstract. In this paper, we compare different GPU accelerators and algorithms for classical molecular dynamics using LAMMPS and GRO-MACS codes. BigDFT is considered as an example of the modern *ab initio* code that implements the density functional theory algorithms in the wavelet basis and uses effectively GPU acceleration. Efficiency of distributed storage managed by the BeeGFS parallel file system is analysed with respect to saving of large molecular-dynamics trajectories. Results have been obtained using the Desmos supercomputer in JIHT RAS.

Keywords: Molecular dynamics · Density functional theory
GPU acceleration · Strong scaling · Parallel I/O

1 Introduction

Hybrid codes that use GPU acceleration provide an effective way to cost-effective and energy-effective calculations. GPU-based Summit and Sierra supercomputers are going to be, perhaps, the fastest supercomputers in 2018, each with about 150 PFlops of peak performance. Their development is a clear illustration of the success achieved by hybrid computing methods that have been growing intensively since 2007 when Nvidia introduced the CUDA framework for GPU programming. The complexity of hybrid parallelism and the diversity of GPU hardware motivate the detailed studies of the hardware-software matching and efficiency, especially for widely used popular software packages.

Desmos (see Fig. 1) is a supercomputer targeted to MD calculations that has been installed in JIHT RAS in December 2016. Desmos is the first application of the Angara interconnect for a GPU-based MPP system [1]. In 2018 one half of the nodes have been upgraded to AMD FirePro S9150 accelerators.

Modern MPP systems can unite up to 10^5 nodes for solving one computational problem. For this purpose, MPI is the most widely used programming

© Springer Nature Switzerland AG 2019
V. Voevodin and S. Sobolev (Eds.): RuSCDays 2018, CCIS 965, pp. 218–229, 2019.
https://doi.org/10.1007/978-3-030-05807-4_19

model. The architecture of the individual nodes can differ significantly and is usually selected (co-designed) for the main type of MPP system deployment. The most important component of MPP systems is the interconnect that properties stand behind the scalability of any MPI-based parallel algorithm. In this work, we describe performance results related to the Desmos supercomputer based on 1CPU+1GPU nodes connected by the Angara interconnect.

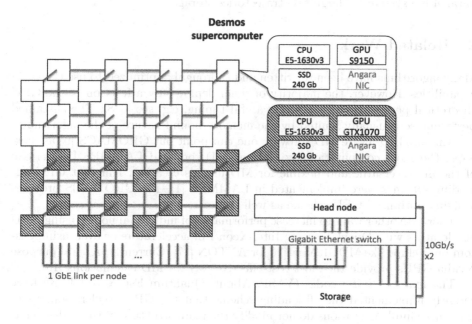

Fig. 1. The scheme of the Desmos supercomputer. 32 computational single-socket nodes based on Intel Xeon E5-1650v3 CPUs are connected by the Angara interconnect with torus topology (16 nodes are equipped with Nvidia GTX1070 GPUs and 16 nodes are equipped with AMD FirePro S9150 GPUs).

The Angara interconnect is a Russian-designed communication network with torus topology. The interconnect ASIC was developed by JSC NICEVT and manufactured by TSMC with the 65 nm process. The Angara architecture uses some principles of IBM Blue Gene L/P and Cray Seastar2/Seastar2+ torus interconnects. The torus interconnect developed by EXTOLL is a similar project [2]. The Angara chip supports deadlock-free adaptive routing based on bubble flow control [3], direction ordered routing [4,5] and initial and final hops for fault tolerance [4].

The results of the benchmarks confirmed the high efficiency of commodity GPU hardware for MD simulations [1]. The scaling tests for the electronic structure calculations also showed the high efficiency of the MPI-exchanges over the Angara network.

In the first part of the paper, we consider several GPU accelerators installed in the chassis of the Desmos supercomputer. We compare performance of different classical MD algorithms in LAMMPS and GROMACS. Additionally, we make the similar benchmarks on the IBM Minsky server and compare the results. In the second part of the paper, the performance of the BigDFT code is considered in the CPU-only variant and in the GPU-accelerated version using AMD FirePro S9150 accelerators installed in Desmos. In the third part of the paper, we consider the efficiency of distributed storage of Desmos nodes managed by the BeeGFS parallel file system for large MD trajectories storage.

2   Related Work

Many algorithms have been rewritten and thoroughly optimized to use the GPU capabilities. However, the majority of them deploy only a fraction of the GPU theoretical performance even after careful tuning, e.g. see [6–8]. The sustained performance is usually limited by the memory-bound nature of the algorithms.

Among GPU-aware MD software one can point out GROMACS [9] as, perhaps, the most computationally efficient MD tool and LAMMPS [10] as one of the most versatile and flexible for MD models creation. Different GPU off-loading schemes were implemented in LAMMPS [11–14]. GROMACS provides a highly optimized GPU-scheme as well [15].

There are other ways to increase performance of individual nodes: using GPU accelerators with OpenCL, using Intel Xeon Phi accelerators or even using custom built chips like MDGRAPE [16] or ANTON [17]. Currently, general purpose Nvidia GPUs provide the most cost-effective way for MD calculations [18].

The main *ab initio* codes (VASP, Abinit, Quantum Espresso, CP2K) have recently implemented the off-loading scheme that use GPU-acceleration. However these implementations do not modify the main structure of the codes. That is why only limited speed-up of calculations can be achieved (limited in comparison with huge GPU peak performance). All the codes require GPU computing in double precision (here we can mention the success of TeraChem package [19] that illustrates the amazing perspectives of single precision GPU usage for quantum chemistry). BigDFT code is a rare example of the code that (1) has been developed with GPU acceleration from the very beginning and (2) uses mainly OpenCL for its GPU parts that widens the spectrum of the GPU hardware one could consider.

The ongoing increase of data that are generated by HPC calculations leads to the requirement for parallel file system for rapid I/O operations. However, benchmarking parallel file system is a complicated (and usually expensive!) task that is why accurate results of particular case studies are quite rare (see e.g. [20]).

3   Classical Molecular Dynamics with LAMMPS and GROMACS

The performance of the different node specifications is tested on the molecular dynamics benchmarks that represent the most common models of the matter.

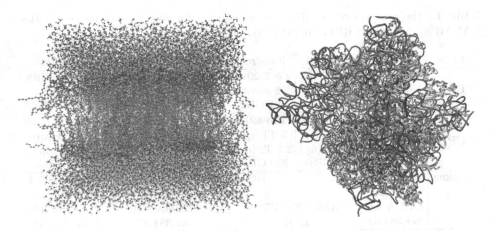

Fig. 2. The structure of the benchmark models MEM (left) and RIB (right, water molecules not shown).

The accuracy of the single precision of GPU is enough for such simulations in contrast to the quantum calculations that are described below. Therefore, the results presented in this section show the performance in the terms of single precision.

We use two wide-spread molecular dynamics programs: LAMMPS compiled with GPU package [21–23] and GROMACS which is highly optimized for the simulations of biological systems. In LAMMPS, the hydrocarbon liquid $C_{30}H_{62}$ [24] is used as a benchmark system. In GROMACS, the bio-membrane (MEM) and ribosome (RIB) systems [18] are considered (see Fig. 2). In all cases, the performance is measured in nanoseconds of the simulated physical time per day.

Table 1 represents the tests performed on one Desmos node (Intel E5-1650v3 + Nvidia GTX 1070) as well as on the separate test nodes: the FirePro1 node (Intel E5-1650v3 + AMD FirePro S9150), the FirePro2 node (Intel E5-2620v4 + AMD FirePro S9150), the Tesla1 node (Intel E5-1650v3 + Nvidia Tesla V100), the Tesla2 node (Intel E5-2620v4 + Nvidia Tesla V100) and the Jupiter node at the supercomputer center of FEB RAS (2x IBM POWER8 + 2x Nvidia Tesla P100). The number of atoms in the model of hydrocarbon liquid in LAMMPS is about 700 k. In the GROMACS benchmarks, MEM consists of 100 k atoms and RIB is about 2 M atoms.

For MEM test with LAMMPS, the Desmos node shows 2x better result than the FirePro1 node which has nearly the same peak floating point performance in single precision. The reason is that GPU package in LAMMPS is better optimized for CUDA than for OpenCL. The Tesla1 node exceeds the Desmos node due to the higher GPU performance.

LAMMPS can be also built with the GPU capabilities implemented with the KOKKOS library (this is an alternative for the GPU package). The KOKKOS library requires double precision. The advantage of the KOKKOS package is that the most of calculations are performed on GPU, hence it is not significant with

Table 1. The performance of different node types for the 3 benchmarks: $C_{30}H_{62}$ (LAMMPS), MEM and RIB (GROMACS).

Node	Desmos	FirePro1	FirePro2	Tesla1	Tesla2	Jupiter
CPU	E5-1650v3 6 cores 3.5 GHz	E5-1650v3 6 cores 3.5 GHz	E5-2620v4 8 cores 2.1 GHz	E5-1650v3 6 cores 3.5 GHz	E5-2620v4 8 cores 2.1 GHz	2xPOWER8 2x10 cores 2.86 GHz
GPU	GTX 1070 5.8 TF(sp) 256 GB/s	FPro S9150 5 TF(sp) 2.5 TF(dp) 320 GB/s	FPro S9150 5 TF(sp) 2.5 TF(dp) 320 GB/s	Tesla V100 14 TF(sp) 7 TF(dp) 900 GB/s	Tesla V100 14 TF(sp) 7 TF(dp) 900 GB/s	2xTesla P100 10 TF(sp) 5 TF(dp) 732 GB/s
Options	SLES11 gcc 4.8 CUDA 8.0 ver.384.69	Ubuntu 16.04 gcc 5.3 AMDGPU-PRO 17.40		CUDA 9.1 ver.384.98		CentOS 7.3 xlc 13.1.5 CUDA 8.0 ver.384.59
LAMMPS (ns/day)						
$C_{30}H_{62}$	0.53	0.26		0.70		1.04 (0.52)
GROMACS (ns/day)						
MEM	35.5	27	23.8	50	44.4	44
RIB	2.1	2.4	2.5	3.5	3.2	4.7

respect to performance which kind of CPU is used as a host device. We varied the number of CPU cores used in the benchmark on the Jupiter node, the ideal choice is 2 cores (from different sockets) per 2 GPUs (Fig. 3). The performance is two times less than in the GPU package case (see the value 0.52 ns/day in brackets in Table 1). The reason is that Tesla P100 has two times lower floating point peak performance in double precision than in single precision.

The MEM benchmark results on the Desmos and FirePro1, FirePro2 nodes show that GROMACS loses less in performance when it runs using OpenCL than LAMMPS. Despite lower peak floating point performance, the Tesla1 and Tesla2 nodes overpower one Jupiter node in the case of a smaller system (MEM). It is caused by the higher memory bandwidth for Tesla V100 than for Tesla P100.

The FirePro nodes win the Desmos node in the RIB benchmark which is big enough to use FirePro S9150 at the limit of its performance. The peak performance of Jupiter node (maximum among the systems considered) is visible on such a big model as well since it results in the fastest calculation speed.

4   Density Functional Theory with BigDFT

Kohn-Sham density functional theory (DFT) is a well-established method to calculate electronic structure of atomistic systems by reducing many-body Schrödinger equation to a one-body problem. This approach is relatively simple and accurate, so it became common in chemistry, physics and biology. However, algorithm is computationally demanding, modern supercomputer is required even for hundreds of atoms. Computational power of the CPUs stimulated the development of many

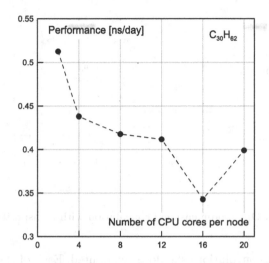

Fig. 3. The performance of the Jupiter node for $C_{30}H_{62}$ model using the KOKKOS package in LAMMPS (the runs are performed with two Tesla P100 GPUs). The best performance is achieved using two CPU cores for two GPUs.

free and commercial DFT codes in the past decades. Unfortunately, most of them do not have GPU-accelerated version or GPU support is limited (e.g. only single GPU) due to difficulties with many-cores hybrid parallelization.

One of the key properties of a DFT code is the form of basis set functions to expand the Kohn-Sham orbitals. The most popular choices are plane waves and Gaussian functions. Plane waves codes (e.g. ABINIT, CPMD, Quantum ESPRESSO, VASP) are mainly used for periodic and homogeneous systems, but they are less efficient for isolated systems, as high-energy cutoff and many plane waves are required for accurate expanding of localized functions. So Gaussian functions are preferred for electronic structure calculations of isolated molecules. Examples of such codes are GAMESS, GAUSSIAN, ORCA, TeraChem. Siesta is another package developed for periodic systems. Mixing of plane waves and localized orbitals in one basis is possible. Such a mixed Gaussian and plane wave method is implemented in CP2K and allows to perform efficient calculations of complex systems.

BigDFT [25, 26] is a unique DFT code as it uses Daubechies wavelets basis for Kohn-Sham orbitals. Such method is well suited for GPU acceleration, not only single, but multiple cards. Both CUDA and OpenCL versions has been developed, the code is distributed under GNU GPL licence. Daubechies wavelets have all desired properties for effective parallelization, they form a systematic orthogonal and smooth basis, localized both in real and Fourier spaces. Some calculations in this basis can be done analytically, the other operations are based on convolutions with short-range filters, which can be effectively calculated on GPU.

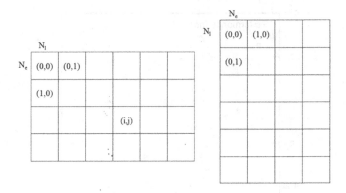

Fig. 4. One unidimensional convolution with transposition.

N independent convolutions should be computed. Each of the lines is split in chunks of size N_e, each multiprocessor of the GPU computes a group of N_l different chunks. After, $N_e N_l$ elements are copied in the corresponding part of the transposed output array. Three-dimensional convolutions can be expressed as a succession of three unidimensional convolution/transposition. Figure 4 shows the data distribution on the grid of blocks during the transposition. The GPU versions of the convolutions are about ten times faster than the CPU ones, though theoretical peak performance can not be achieved due to memory transfers. Some tricks are also used to reduce CPU-GPU and GPU-GPU communications.

In order to benchmark GPU-acceleration, we consider two models. One model represents a ball-like molecule of B_{80} (Fig. 5). Another example represents a unit cell of f.c.c. Ag crystal with 4 atoms and a k-mesh of $15 \times 15 \times 15$ (Fig. 6).

Benchmark results show that GPU-acceleration can be quite substantial. Both models have quite good strong scaling for up to 16 nodes of Desmos equipped with FirePro accelerators.

One remarkable difference between the models is that for B_{80} the best speed corresponds to 1 MPI process per CPU. But in the case of the Ag crystal using multiple MPI processes with one GPU is beneficial. The subset of points on Fig. 6 shows the decrease of the time-to-solution with increasing number of cores.

BigDFT code combines different parallel programming techniques: OpenMP, MPI and OpenCL/CUDA. At the same time, DFT algorithm has several strategies of parallel data mapping (parallelism for orbitals, for k-points, for basis functions). These two aspects make difficult the choice of the most efficient software-hardware combination of a particular problem. The benchmarks presented in this work is an attempt to pave the way in solving this complicated problem.

5   Parallel File System Benchmarks

Many HPC codes generate huge amounts of data. For example, for classical molecular dynamics (MD) the limits of the system size are trillions of atoms [27].

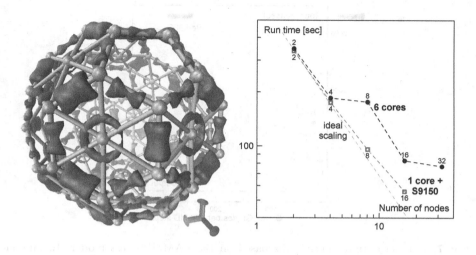

Fig. 5. The visual representation of the B_{80} molecule (red surfaces correspond to the electron density 0.15 e/A^3) and the strong scaling of the calculation of the electronic structure of this molecule for Desmos nodes w/ and w/o deploying the GPU (a grey dashed line shows ideal scaling, numbers of nodes are shown near symbols). (Color figure online)

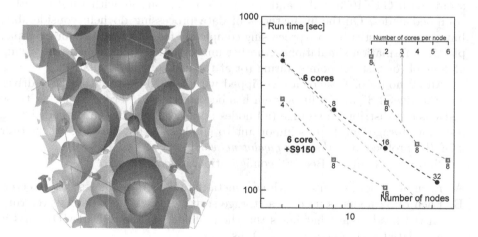

Fig. 6. The visual representation of the f.c.c. Ag crystal (red surfaces correspond to the electron density 0.009 e/A^3, several replicated unit cells are shown for demonstration of periodicity) and the strong scaling of the calculation of the electronic structure of this system for Desmos nodes w/ and w/o deploying the GPU (a grey dashed line shows ideal scaling, numbers of nodes are shown near symbols, the additional set of data shows the dependence on the number of cores per node that is the number of MPI processes using the same GPU simultaneously). (Color figure online)

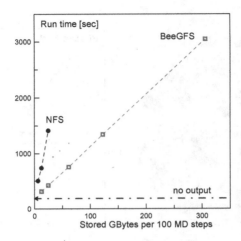

Fig. 7. Parallel output benchmarks based on the LAMMPS test model (16 million atoms, 2500 timesteps). The black circles show the case when the NFS-mounted /home folder from the head node is used for storage. The red squares represent the case when BeeGFS file system is mounted on all Desmos nodes and on the head node. This BeeGFS file system unites all 32 SSD disks in the Desmos nodes (with xfs file system). (Color figure online)

Desmos allows GPU-accelerated modelling of MD system with up to 100 millions atoms with GTX1070 nodes and up to 200 million atoms with FirePro S9150 equipped nodes. On-the-fly methods of data processing do help considerably, but can not substitute post-processing completely for such tasks as, for example, plastic deformation simulations. Another unavoidable requirement is the saving of control (or restart) points during (or at the end) of the calculation.

All 32 nodes of Desmos are equipped with fast XFS formatted SSD drives and the BeeGFS parallel file system has been installed in order to use all these disks as one distributed storage (all nodes boot over ethernet so these disks are used for storage only). It is important to emphasize that BeeGFS works over TCP/IP protocol and *the slow gigabit network*.

Two variants of the BeeGFS configuration are considered.

A. Each of 32 nodes is used both as a meta-data server and as a storage server.
B. Each of 32 nodes is used as a storage server only. A meta-data server runs on the head node (that hosts the /home folder on the SSD RAID1) and is associated with a dedicated SSD disk.

The standard Lennard-Jones benchmark is used with the LAMMPS molecular dynamics package (the benchmark is based on the replicated example "melt"). LAMMPS has different variants for output large amounts of data. Here, we present the results obtained with MPI-IO.

Figure 7 presents the benchmarks for different frequencies of writing MD trajectory data. The NFS storage (/home folder of the head node) becomes prohibitively slow for higher writing frequencies. The distributed storage governed by BeeGFS demonstrates saturation and does not scale well but, at least, gives the possibility to store necessary data in a reasonable time. No difference between A and B configurations is found for these tests.

However, there is a considerable difference in speed of copying file from /mnt/beegfs to /home on the head node. For benchmarking, we use

```
dd iflag=direct,fullblock bs=10G \
                if=/mnt/beegfs/dump.melt.mpiio of=/dev/null
```

that results in 200 MB/s reading speed for the A configuration and 315 MB/s reading speed for the B configuration. It shows that the configuration of the meta-data servers plays an important role and can significantly increase the reading speed from the distributed storage.

6   Conclusions

The representative set of CPU-GPU combinations have been benchmarked for classical MD GPU-algorithms implemented in LAMMPS and GROMACS. The results show that performance of the Desmos nodes is on par with other configurations (newer and/or more expensive). An important observation is that the GPU-oriented MD algorithm based on the KOKKOS library does not require a high performance CPU for getting maximum efficiency from the GPU.

The GPU-acceleration of the BigDFT code provides about 2x speed-up for the benchmarks considered. The benchmarks of an isolated DFT model and a periodic DFT model scale on Desmos well.

Distributed storage united by BeeGFS can improve performance in comparison with a local NFS storage for large MD data files even over slow gigabit ethernet network. The proper choice for the configuration of the meta-data servers can give 1.5x increase in the reading speed from the distributed storage.

Acknowledgments. The work is supported by the Russian Science Foundation (grant No. 14-50-00124). HSE and MIPT helped in accessing some hardware used in this study. We acknowledge Shared Resource Center of CC FEB RAS for the access to the Jupiter supercomputer (http://lits.ccfebras.ru). We are grateful to Nvidia for providing us with one Tesla V100 card for the benchmarks.

References

1. Stegailov, V., et al.: Early performance evaluation of the hybrid cluster with torus interconnect aimed at molecular-dynamics simulations. In: Wyrzykowski, R., Dongarra, J., Deelman, E., Karczewski, K. (eds.) PPAM 2017. LNCS, vol. 10777, pp. 327–336. Springer, Cham (2018). https://doi.org/10.1007/978-3-319-78024-5_29
2. Neuwirth, S., Frey, D., Nuessle, M., Bruening, U.: Scalable communication architecture for network-attached accelerators. In: 2015 IEEE 21st International Symposium on High Performance Computer Architecture (HPCA), pp. 627–638, February 2015

3. Puente, V., Beivide, R., Gregorio, J.A., Prellezo, J.M., Duato, J., Izu, C.: Adaptive bubble router: a design to improve performance in torus networks. In: Proceedings of the 1999 International Conference on Parallel Processing, pp. 58–67 (1999)

4. Scott, S.L., Thorson, G.M.: The Cray T3E network: adaptive routing in a high performance 3D torus. In: HOT Interconnects IV, Stanford University, 15–16 August 1996

5. Adiga, N.R., et al.: Blue Gene/L torus interconnection network. IBM J. Res. Dev. **49**(2), 265–276 (2005)

6. Smirnov, G.S., Stegailov, V.V.: Efficiency of classical molecular dynamics algorithms on supercomputers. Math. Models Comput. Simul. **8**(6), 734–743 (2016)

7. Stegailov, V.V., Orekhov, N.D., Smirnov, G.S.: HPC hardware efficiency for quantum and classical molecular dynamics. In: Malyshkin, V. (ed.) PaCT 2015. LNCS, vol. 9251, pp. 469–473. Springer, Cham (2015). https://doi.org/10.1007/978-3-319-21909-7_45

8. Rojek, K., Wyrzykowski, R., Kuczynski, L.: Systematic adaptation of stencil-based 3D MPDATA to GPU architectures. Concurr. Comput. Pract. Exp. **29**, e3970 (2017)

9. Berendsen, H.J.C., van der Spoel, D., van Drunen, R.: GROMACS: a message-passing parallel molecular dynamics implementation. Comput. Phys. Commun. **91**(1–3), 43–56 (1995)

10. Plimpton, S.: Fast parallel algorithms for short-range molecular dynamics. J. Comput. Phys. **117**(1), 1–19 (1995)

11. Trott, C.R., Winterfeld, L., Crozier, P.S.: General-purpose molecular dynamics simulations on GPU-based clusters. ArXiv e-prints, September 2010

12. Brown, W.M., Wang, P., Plimpton, S.J., Tharrington, A.N.: Implementing molecular dynamics on hybrid high performance computers - short range forces. Comput. Phys. Commun. **182**(4), 898–911 (2011)

13. Brown, W.M., Kohlmeyer, A., Plimpton, S.J., Tharrington, A.N.: Implementing molecular dynamics on hybrid high performance computers - particle-particle particle-mesh. Comput. Phys. Commun. **183**(3), 449–459 (2012)

14. Edwards, H.C., Trott, C.R., Sunderland, D.: Kokkos: enabling manycore performance portability through polymorphic memory access patterns. J. Parallel Distrib. Comput. **74**(12), 3202–3216 (2014). Domain-Specific Languages and High-Level Frameworks for High-Performance Computing

15. Abraham, M.J., et al.: GROMACS: high performance molecular simulations through multi-level parallelism from laptops to supercomputers. SoftwareX **12**, 19–25 (2015)

16. Ohmura, I., Morimoto, G., Ohno, Y., Hasegawa, A., Taiji, M.: MDGRAPE-4: a special-purpose computer system for molecular dynamics simulations. Phil. Trans. R. Soc. A **372**, 20130387 (2014)

17. Piana, S., Klepeis, J.L., Shaw, D.E.: Assessing the accuracy of physical models used in protein-folding simulations: quantitative evidence from long molecular dynamics simulations. Curr. Opin. Struct. Biol. **24**, 98–105 (2014)

18. Kutzner, C., Pall, S., Fechner, M., Esztermann, A., de Groot, B.L., Grubmuller, H.: Best bang for your buck: GPU nodes for gromacs biomolecular simulations. J. Comput. Chem. **36**(26), 1990–2008 (2015)

19. Luehr, N., Ufimtsev, I.S., Martínez, T.J.: Dynamic precision for electron repulsion integral evaluation on graphical processing units (GPUs). J. Chem. Theor. Comput. **7**(4), 949–954 (2011). PMID: 26606344

20. Nicholas, M., Feltus, F.A., Ligon III, W.B.: Maximizing the performance of scientific data transfer by optimizing the interface between parallel file systems and advanced research networks. Fut. Gener. Comput. Syst. **79**(Part 1), 190–198 (2018)
21. Plimpton, S.J., Tharrington, A.N., Brown, W.M., Wang, P.: Implementing molecular dynamics on hybrid high performance computers - short range forces. Comput. Phys. Commun. **182**, 898–911 (2011)
22. Plimpton, S.J., Tharrington, A.N., Brown, W.M., Kohlmeyer, A.: Implementing molecular dynamics on hybrid high performance computers - particle-particle particle-mesh. Comput. Phys. Commun. **183**, 449–459 (2012)
23. Masako, Y., Brown, W.M.: Implementing molecular dynamics on hybrid high performance computers - three-body potentials. Comput. Phys. Commun. **184**, 2785–2793 (2013)
24. Kondratyuk, N.D., Norman, G.E., Stegailov, V.V.: Self-consistent molecular dynamics calculation of diffusion in higher n-alkanes. J. Chem. Phys. **145**(20), 204504 (2016)
25. Genovese, L., et al.: Daubechies wavelets as a basis set for density functional pseudopotential calculations. J. Chem. Phys. **129**(1), 014109 (2008)
26. Genovese, L., Ospici, M., Deutsch, T., Méhaut, J.-F., Neelov, A., Goedecker, S.: Density functional theory calculation on many-cores hybrid central processing unit-graphic processing unit architectures. J. Chem. Phys. **131**(3), 034103 (2009)
27. Eckhardt, W., et al.: 591 TFLOPS multi-trillion particles simulation on superMUC. In: Kunkel, J.M., Ludwig, T., Meuer, H.W. (eds.) ISC 2013. LNCS, vol. 7905, pp. 1–12. Springer, Heidelberg (2013). https://doi.org/10.1007/978-3-642-38750-0_1

INMOST Parallel Platform
for Mathematical Modeling
and Applications

Kirill Terekhov[1](✉) and Yuri Vassilevski[1,2,3]

[1] Marchuk Institute of Numerical Mathematics of the Russian Academy of Sciences,
Moscow 119333, Russia
kirill.terehov@gmail.com
[2] Moscow Institute of Physics and Technology, Dolgoprudny 141701, Russia
[3] Sechenov University, Moscow 199991, Russia

Abstract. In the present work we present INMOST, the programming
platform for mathematical modelling and its application to a couple of
practical problems. INMOST consists of a number of tools: mesh and
mesh data manipulation, automatic differentiation, linear solvers, support for multiphysics modelling. The application of INMOST to black-oil
reservoir simulation and blood coagulation problem is considered.

Keywords: Open-source library · Linear solvers
Automatic differentiation · Reservoir simulation · Blood coagulation

1 Introduction

Nowadays the necessity in parallel modelling of complex phenomena with multiple stiff physical processes is very acute. This puts a significant burden on the
programmer who has not only to cope with various computational methods but
also to manage the unstructured grid, data exchanges with MPI, assembly of
large distributed linear systems, solve the resulting linear and nonlinear systems
and finally postprocess the result. The INMOST [1] is an open-source library
that alleviates the most of the complexity from the programmer and provides a
unified set of tools to address each of the aforementioned issues. We have used
the INMOST platform to implement the fully implicit black-oil reservoir model
and fully coupled and implicit blood coagulation problem. The black-oil reservoir model involves simultaneous solution of three Darcy laws that describe the
mixture of water, oil and gas. The blood coagulation model couples the Navier-
Stokes equations with a Darcy term and nine additional advection-diffusion-
reaction equations that participate in reactive cascade during coagulation of the
blood. The numerical results for both models are presented.

Among notable alternatives for multiphysics modelling, commercial are
COMSOL [2], ANSYS Fluent [3], Star-CD [4] and open-source are Dumux [5],
OPM [6], Elmer [7], OOFEM [8], OpenFOAM [9], SU2 [10], CoolFluid [11] and

© Springer Nature Switzerland AG 2019
V. Voevodin and S. Sobolev (Eds.): RuSCDays 2018, CCIS 965, pp. 230–241, 2019.
https://doi.org/10.1007/978-3-030-05807-4_20

many others. A comparison of some of these packages is available in [12]. Perspectives for multiphysics software are discussed in [13]. Should be noted, that all of these packages provide modelling environment with integrated set of computational methods. The methods are implemented with certain assumptions and has certain limitations on physics, time stepping and couplings. An attempt to apply OpenFOAM package to blood coagulation model didn't allow to simulate the problem in reasonable time due to impossibility to construct fully-coupled approach from provided methods.

At present INMOST does not contain integrated computational methods but provides a programming platform to implement them. Among programming platforms the notable alternatives are Dune [17], Trilinos [15] and PETSc [14]. Dune is famous for distributed mesh management. Trilinos and PETSc are famous for a collection of built-in parallel linear solvers and seamless integration of third-party linear solvers; both also provide excellent nonlinear solvers, but rely on third party libraries for mesh management. Trilinos provides Sacado package for automatic differentiation. The framework for coupling of multiple physics modules is under active development in Dune (within Dumux project [5]) and in Trilinos (wihtin Amanzi project [41]) Functionality of all of these packages could be successfully used in large extent to build a simulator. The first version of the black-oil simulator reported in this work was based on linear and nonlinear solvers from PETSc and our own mesh management tool.

There is a number of C/C++ libraries that solve a particular task. For mesh management there are MSTK, MOAB, libMesh, FMDB and many other libraries. For linear system assembly and solution there are Trilinos, PETSc, SuperLU, MUMPS, Hypre and many others. For automatic differentiation there are Sacado (Trilinos), ADOL-C, FAD, ADEL, Adept and so on. For nonlinear solvers there are Trilinos, PETSc, SUNDIALS, Ipopt, Snopt and so on. For linear algebra integration there is Eigen library. The advantage of using separate tools could be in a greater level of maturity of popular libraries, the disadvantage is the absence of tight integration that is inevitably needed for construction of multiphysics framework.

2   Mesh and Data

The algorithms and functionality that form the basis of mesh manipulation module in INMOST were previously reported in [19,20,22]. Parallel mesh refinement instruments were reported in [21]. We develop further the mesh modification functionality to support general mesh modification and synchronization in parallel and to preserve layers of ghost cells minimizing computational work.

The primary functionality required in this work are:

- load a mesh and associated data;
- compute partitioning of the mesh;
- redistribute the mesh;
- build multiple layers of ghost cells;

- access mesh elements, status of elements, mesh data;
- save the results in parallel VTK format.

Among novel functionalities of the mesh module we mention two novelties. First, the support of the Eclipse simulator mesh format *grdecl* is used in the INMOST black-oil reservoir model. Second, the ability to store on a mesh data containing partial derivatives is used in the implicit computation of a Jacobian matrix by the module of Automatic differentiation.

3   Automatic Differentiation

The algorithms used in the automatic differentiation module were previously reported in [21]. The basis for the module are:

- a C++ class *'variable'* to represent a value of function with its first order partial derivatives;
- expression templates that form a templated tree of classes corresponding to operations on variables;
- a C++ class *'variable_expression'* that can store expression template and evaluate it on demand;
- a dense linked-list structure *'Sparse::RowMerger'* is used for fast addition of sparse vectors, corresponding to the partial derivatives;

A novel functionality of INMOST is a templated class *'Matrix'* from the dense linear algebra module that allows for operations on dense matrices of values with specified type. Many *'blas'* and *'lapack'* operations are implemented in this class: sub-matrix access, addition, subtraction, multiplication, system solution, matrix inversion, singular value decomposition, pseudo-solve and pseudo-inverse and so on. The whole functionality is supported for matrices consisting first partial derivatives values. A set of simple rules prescribes the outcome when two matrices with different types of elements are involved in operations, *e.g.* multiplication of a matrix of doubles with a matrix of *'variable*'s results in a matrix of *'variable*'s). This functionality is heavily used in the blood coagulation model, see Sect. 7.

To connect the mesh data and the automatic differentiation, we introduce an *'Automatizator'* class. An *'AbstractEntry'* sub-class allows to group the mesh data into blocks of unknowns and register them with the object of the *'Automatizator'* class. An index data is created and associated to each entry of a block of unknowns and enumeration for unknowns is performed on every mesh element containing the data related to the unknowns. Each block is enumerated consequently on each element of the mesh. This leads to the block-structured organization of the Jacobian matrix. Based on this, one can split the Jacobian matrix into blocks of physical processes and apply multigrid linear solvers [23–26] to blocks. In the future the information of the block structure of the matrix will be transfered to linear solvers.

The main functionality of the '*AbstractEntry*' class covers:

– *Value* provides on mesh element a matrix of values of unknowns;
– *Unknown* provides on mesh element a matrix of partial derivatives values;
– *Index* provides on mesh element a matrix of indices of unknowns, these indices are positions of unknowns in the Jacobian matrix;
– *MatrixSize* determines the size of the returned matrix.

Each value, unknown and index can be accessed individually if it is undesirable to get the whole matrix, the assembly of the matrix is being avoided in this case.

Various scenarios of unknowns organization is covered by sub-classes with extended functionality:

– '*SingleEntry*' is unknown as a single entry of datum on mesh elements;
– '*VectorEntry*' is unknown as multiple entries of datum on mesh elements (possibly with variable length);
– '*BlockEntry*' is unknown as fixed size data on mesh elements;
– '*StatusBlockEntry*' allows to change the status of unknowns in the block;
– '*MultiEntry*' is a union of entries of different types, *e.g.*, an object of '*SingleEntry*' can be used together with an object of '*VectorEntry*'.

Each of these sub-classes is inherited from the '*AbstractEntry*' class and retains the original functionality.

To facilitate the assembly of the residual vector and the Jacobian matrix, we introduced a '*Residual*' class. This class contains a sparse matrix and a residual vector (classes '*Sparse::Matrix*' and '*Sparse::Vector*', respectively). During the assembly stage a sparse matrix is stored in INMOST as a set of sparse vectors corresponding to rows of the matrix (class '*Sparse::Row*'). Each sparse vector is expandable and allows fast modifications. It is the same vector that is used to store partial derivatives values in the automatic differentiation module.

When an object of the class '*Residual*' is accessed by its index (as an array via square brackets $a[i]$), an object of the class '*variable_reference*' with references to corresponding entry in the residual vector and the row in the sparse matrix is returned. The object of this class can enter expression templates from automatic differentiation module and thus allows for all the same operations as an object of the '*variable*' class. Assignment to an object of this class results in modification of the sparse matrix and the residual vector stored in the object of the '*Residual*' class.

On top of that, an object of the class '*Residual*' can be accessed by a matrix of indices (*i.e.* $a[\mathbb{I}]$ where $\mathbb{I}$ is the matrix), then it returns a matrix with the type '*variable_reference*'. On assignment the underlaying sparse matrix and residual vector are modified in the block-structured fashion. In the future we shall introduce block-structured sparse matrices and use them in the '*Residual*' class to take advantage of the fast block-structured assembly.

All of this allows for seamless integration between the automatic differentiation module and the sparse linear algebra module. The next user's step is to solve the arising distributed linear system.

4 Linear Solver

INMOST linear solver module provides integration of multiple open-source solver libraries: Trilinos [15], PETSc [14], SuperLU [16]. It also contains a number of integrated solvers. Certain comparison of linear solvers was provided in [21]. One of the most widely used solvers is described below.

We solve the problem using preconditioned iterative method BiCGStab(L) [27]. This method makes L BiCG steps and fits a polynomial function to accelerate convergence of preconditioned residual to zero value.

The Additive Schwarz method with a prescribed overlap is used for the parallelization of the preconditioner.

For each local matrix extended by the overlap, we use the Crout-ILU method [28] with the dual threshold τ, τ_2 [29]. During elimination we estimate the condition number of the inverse factors $|L^{-1}|$ and $|U^{-1}|$ and adjust both thresholds accordingly to [31].

Before the elimination step, the matrix is preprocessed by reordering and rescaling in three steps. First, we maximize the product on the diagonal using the Dijkstra algorithm [33, 34]. Second, we use the reverse Cuthill-Mckee algorithm to reduce the fill-in. At last, we use a rescaling to balance the second norms of all rows and all columns to unity [30].

In the future we plan to add local 2×2 pivoting [35], block-structured elimination and OpenMP parallelization [32].

5 Multiphysics

At present, INMOST contains trial implementation of the multi-physics module. The idea of the module is to provide basic functionality that allows one to split the problem into physical processes. Each physical process is responsible for a subset of unknowns and assembly of equations corresponding to these unknowns. Such a physical process is represented by an '*AbstractSubModel*' class. The programmer has to inherit from this class and implement the following functions:

- '*PrepareEntries*' introduces unknowns of a process to the model;
- '*FillResidual*' computes the residual for the process;
- '*UpdateMultiplier*' performs backtracking of an update to meet constraints;
- '*UpdateSolution*' updates unknowns during nonlinear iterations;
- '*UpdateTimeStep*' proceeds to the next time step;
- '*AdjustTimeStep*' computes an optimal time step for the process;
- '*RestoreTimeStep*' returns back in case of nonlinear solver failure.

Coupling between two physical processes introduces coupling terms into equations involving unknowns of both processes. This requires access to unknowns and equations of both processes.

Everything is managed by an object from a '*Model*' class. It incorporates an object of the '*Automatizator*' class and named arrays of meshes, entries of unknowns, sub-models and couplings. The coupling is represented by '*Abstract-Coupling*' that has functions '*PrepareEntries*' and '*FillResidual*'.

This module uses heavily all the previously introduced modules and a nonlinear solver module which should guide the convergence of the multiphysics model to the solution. The latter module is not implemented in public repository yet.

6 Black-Oil Model

For the black-oil model we assume that the mixture of water, oil and gas fills all the voids, *i.e.* $S_o + S_w + S_g = 1$, the mixture in the heterogeneous porous media is guided by the Darcy law with capillary pressure (Fig. 1). There are three unknowns in the model: the water pressure p, the oil saturation S_o and either the gas saturation S_g or the bubble point pressure p_b depending on whether the gas phase is present or is fully dissolved in the oil phase, respectively. The oil and gas pressures are connected to water pressure p through the capillary pressures Pc_o and Pc_g. The system of equations takes the form:

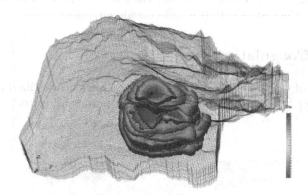

Fig. 1. An example of water injection into reservoir with real structure featuring layered heterogeneous media.

$$\frac{\partial \rho_w \theta S_w}{\partial t} - \nabla \cdot \lambda_w \mathbb{K} \left(\nabla p - \rho_w g \nabla z \right) = q_w,$$

$$\frac{\partial \rho_o \theta S_o}{\partial t} - \nabla \cdot \lambda_o \mathbb{K} \left(\nabla p - \nabla Pc_o - \rho_o g \nabla z \right) = q_o,$$

$$\frac{\partial \rho_{go} \theta S_o + \rho_g \theta S_g}{\partial t} - \nabla \cdot \lambda_g \mathbb{K} \left(\nabla p + \nabla Pc_g - \rho_g g \nabla z \right)$$

$$- \nabla \cdot \lambda_{go} \mathbb{K} \left(\nabla p + \nabla Pc_g - \rho_{go} g \nabla z \right) = q_g. \quad (1)$$

The equations are nonlinear due to the dependence of all the terms on unknowns of the problem. Here $\theta(p)$ is the porosity of the media, $\rho_\alpha(p)$ is the density of phase α, $\rho_{go}(p_b) = \rho_g(p_b) Rs(p_b)$ is the density of gas dissolved in oil, $Rs(p_b)$ describes the solubility of gas in the oil at a given bubble point pressure p_b, $\lambda_\alpha(p, S_\alpha) = \frac{\rho_\alpha(p)kr_\alpha(S_\alpha)}{\mu_\alpha(p)}$ is the mobility of phase α that accounts change in density, relative permeability and viscosity, $\lambda_{go}(p_b) = \frac{\rho_{go}(p_b)kr_g(S_g)}{\mu_g(p_b)}$ is the mobility for gas dissolved in oil phase, $Pc_o(S_o)$ and $Pc_g(S_g)$ are capillary pressures for oil and gas phase, respectively, q_α is the source or sink for phase α (conventionally the Peacman's well formula is used).

We use a nonlinear two-point flux approximation method [36,37] to calculate Darcy velocity $\mathbb{K}(\nabla \dots)$ terms. Once the velocity is obtained, a single point upstream discretization is used for the mobility. The whole system is approximated in the fully implicit manner by the backward Euler time-stepping method.

At each time step, the nonlinear system is solved by the Newton method. If on nonlinear iteration l the gas fully dissolves into oil $\widetilde{S}_g^l < 0$ or the gas is emitted $\widetilde{p}_b^l > p$, we solve a local nonlinear problem on the bubble point pressure p_b^l: $\rho_{go}\left(p_b^l\right) S_o^l = \rho_{go}\left(p^l\right) S_o^l + \rho_g(p^l)\widetilde{S}_g^l$ or on amount of the emitted gas S_g^l: $\rho_{go}\left(\widetilde{p}_b^l\right) S_o^l = \rho_{go}\left(p^l\right) S_o^l + \rho_g(p^l)S_g^l$, respectively. This step allows us to obtain physically reasonable quantities for the bubble point pressure and the released gas on the state switch. We also consider the same model without the gas phase, i.e. $S_g = 0$ and the last equation is not considered.

7   Blood Coagulation

The dynamics of the blood plasma in the present work is described by the incompressible Navier-Stokes equations with a Darcy permeability term (Fig. 2):

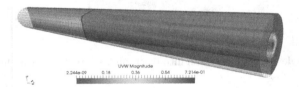

Fig. 2. An example of clot formation in the vessel.

$$\begin{cases} \dfrac{\partial \rho \mathbf{u}}{\partial t} + \operatorname{div}\left(\rho \mathbf{u}\mathbf{u}^T - \mu \nabla \mathbf{u} + p\mathbb{I}\right) = -\dfrac{1}{K_f}\mathbf{u}, \\ \operatorname{div}\mathbf{u} = 0. \end{cases} \tag{2}$$

The density ρ and viscosity μ are assumed to be constant. The equations (2) are coupled to the following concentrations of reactive components in the flow: prothrombin PT, thrombin T, anti-thrombin A, blood clotting factor FX Ba, fibrin F, fibrinogen F_g, fibrin polymer F_p, free platelets ϕ_f in the flow, platelets

ϕ_c trapped in the clot. All the components except for the fibrin polymer and platelets satisfy the advection-diffusion equation: $\frac{\partial C}{\partial t} + \nabla \cdot (C\mathbf{u} - D\nabla C) = R_C$ for a component C. The platelets satisfy the nonlinear advection-diffusion equation $\frac{\partial C}{\partial t} + \nabla \cdot k(\phi_c)\,(C\mathbf{u} - D_c\nabla C) = R_C$ with $k(\phi_c) = \tanh{(\pi(1 - \phi_c/\phi_{max}))}$, $C = \phi_f, \phi_c$. Polymerised fibrin does not move with the flow. The set of kinetic reactions between components of the flow is taken from [40]. The parameter K_f (2) describes the formation of porous media due to fibrin polymer and platelets similar to [40].

All the unknowns (velocity vector $\mathbf{u}$, pressure p and all the additional components) are collocated at the centres of the cells. For the advection and diffusion terms we use the second order finite-volume approximation, the inertia term is being approximated nonlinearly. Collocation of velocities and pressures at cell centers requires special stabilization for discretization of the flux related to the gradient of pressure and divergence of velocity. The finite volume discretization results in a system composed of 13 unknowns and equations per computational cell. The whole system is solved simultaneously in the implicit manner by the BDF2 time-stepping scheme.

8 Numerical Results

For the black oil model we use the secondary recovery problem on the Norne field [39] with one injection well and two production wells. The problem runs for 100 modelling days with at maximum 1 day step. The partition of the mesh is demonstrated in Fig. 3 (left). The parameters are taken from SPE9 test. For the blood coagulation model we simulate 15 seconds of the clot formation observed in the experiment [38]. The partition of the mesh is demonstrated in Fig. 3 (right). The parameters are taken from [40].

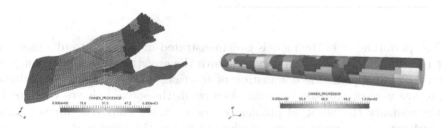

Fig. 3. Partitioning of the computational mesh among processors, the black-oil model on the Norne field (left), the blood coagulation model in a cylindric vessel corresponding to experiment [38].

In both problems the number of overlap layers in the mesh was set to 2 for assembly of fluxes and in additive Schwartz method was set to 1, the solver dropping parameters are $\tau = 10^{-2}$ and $\tau_2 = 10^{-3}$.

Table 1. Performance of the models on different numbers of processors.

Two-phase model						
Processors	Assembly		Solution		Total		Equations
1	1028	-	1544	-	2581	-	89830
8	172	6x	383	4x	558	5x	11229
16	89	12x	255	6x	345	8x	5615
32	53	20x	135	12x	189	14x	2808
64	30	35x	71	22x	101	26x	1404
128	19	55x	65	24x	84	31x	702

Three-phase model						
Processors	Assembly		Solution		Total		Equations
1	2783	-	5368	-	8171	-	134745
8	449	6x	3295	2x	3749	2x	16844
16	252	11x	1656	3x	1911	4x	8422
32	152	19x	472	11x	626	13x	4211
64	88	32x	325	17x	415	20x	2106
128	59	47x	154	35x	213	38x	1053

Blood clotting model						
Processors	Assembly		Solution		Total		Equations
1	2781	-	2156	-	4938	-	318500
8	461	6x	214	10x	680	7x	39812
16	253	11x	117	18x	373	13x	19906
32	148	19x	80	27x	234	21x	9954
64	90	31x	46	47x	141	35x	4977
128	46	60x	29	74x	77	64x	2489

The performance of the models is demonstrated in Table 1. Results indicate that the clotting model scales very well with the growth of the number of processors. This happens due domination of reactions, the problem is hyperbolic due to low viscosity of blood plasma. As a result the solver behaves very good.

On contrary the black oil problem contains strong elliptic component and the performance of linear solver deteriorates faster. For good performance this problem require a specific solver [25] that can extract elliptic part of the problem and apply multi-grid on it. In this problem the assembly of the matrix does not ideally scale, since we have to perform certain operations on overlap which may become significant with large number of processors. Still it remains reasonable to increase number of processors to reduce total computational time.

The calculations were performed on cluster of INM RAS [18], we run the job on nodes with different types of processors.

9   Conclusion

We have presented briefly the open-source platform INMOST for parallel mathematical modelling. The platform allows to greatly facilitate programming of complex coupled models. Parts of the INMOST platform are still under active development. In the future we plan to advance the multiphysics tools for the solution of nonlinear systems of equations and abstractions for seamless modular integration of independent physical models.

Acknowledgement. This work was supported by the RFBR grants 17-01-00886, 18-31-20048.

References

1. INMOST - a toolkit for distributed mathematical modeling. http://www.inmost. org. Accessed 15 Apr 2018
2. COMSOL Multiphysics Reference Manual, version 5.3, COMSOL Inc. www. comsol.com. Accessed 30 May 2018
3. ANSYS   FLUENT.   https://www.ansys.com/products/fluids/ansys-fluent. Accessed 30 May 2018
4. STAR-CD. https://mdx.plm.automation.siemens.com/star-cd. Accessed 30 May 2018
5. Flemisch, B., et al.: DuMux: DUNE for multi-phase, component, scale, physics,... flow and transport in porous media. Adv. Water Resour. **34**(9), 1102–1112 (2011)
6. The Open Porous Media (OPM) initiative encourages open innovation and reproducible research for modeling and simulation of porous media processes. https:// opm-project.org. Accessed 30 May 2018
7. Elmer - Finite Element Solver for Multiphysical Problems. https://www.csc.fi/ web/elmer. Accessed 30 May 2018
8. Patzák, B.: OOFEM-an object-oriented simulation tool for advanced modeling of materials and structures. Acta Polytech. **52**(6), 59–66 (2012)
9. OpenFOAM is the free, open source CFD software. www.openfoam.com. Accessed 30 May 2018
10. SU2 is an open-source collection of software tools written in C++ and Python for the analysis of partial differential equations (PDEs) and PDE-constrained optimization problems on unstructured meshes with state-of-the-art numerical methods. https://su2code.github.io. Accessed 30 May 2018
11. COOLFluiD is a component based scientific computing environment that handles high-performance computing problems with focus on complex computational fluid dynamics (CFD) involving multiphysics phenomena. https://github.com/ andrealani/COOLFluiD/wiki. Accessed 30 May 2018
12. Babur, Ö., Smilauer, V., Verhoeff, T., van den Brand, M.: A survey of open source multiphysics frameworks in engineering. Procedia Comput. Sci. **51**, 1088–1097 (2015)
13. Keyes, D.E., et al.: Multiphysics simulations: challenges and opportunities. Int. J. High Perform. Comput. Appl. **27**(1), 4–83 (2013)
14. PETSc is a suite of data structures and routines for the scalable (parallel) solution of scientific applications modeled by partial differential equations. http://www. mcs.anl.gov/petsc. Accessed 15 Apr 2018

15. Trilinos - platform for the solution of large-scale, complex multi-physics engineering and scientific problems. http://trilinos.org/. Accessed 15 Apr 2018
16. SuperLU is a general purpose library for the direct solution of large, sparse, nonsymmetric systems of linear equations. http://crd-legacy.lbl.gov/xiaoye/SuperLU/. Accessed 15 Apr 2018
17. Distributed and Unified Numerics Environment. https://dune-project.org/. Accessed 30 May 2018
18. INM RAS cluster. http://cluster2.inm.ras.ru. Accessed 15 Apr 2018
19. Terekhov, K.M.: Application of unstructured octree grid to the solution of filtration and hydrodynamics problems. Ph.D. thesis, INM RAS (2013). (in Russian)
20. INMOST - programming platform and graphical environment for development of parallel numerical models on general grids. Vassilevski, Yu.V., Konshin, I.N., Kopytov, G.V., Terekhov, K.M.: Moscow University Press, 144 p. (2013). (in Russian)
21. Bagaev, D.V., Burachkovskii, A.I., Danilov, A.A., Konshin, I.N., Terekhov, K.M.: Development of INMOST programming platform: dynamic grids, linear solvers and automatic differentiation. Russian Supercomputing Days 2016 (2016)
22. Danilov, A.A., Terekhov, K.M., Konshin, I.N., Vassilevski, Y.V.: Parallel software platform INMOST: a framework for numerical modeling. Supercomput. Front. Innov. 2(4), 55–66 (2015)
23. Hu, J.J., Prokopenko, A., Siefert, C.M., Tuminaro, R.S., Wiesner, T.A.: MueLu multigrid framework. http://trilinos.org/packages/muelu. Accessed 15 Apr 2018
24. SAMG - Efficiently solving large Linear Systems of Equations. https://www.scai.fraunhofer.de/en/business-research-areas/fast-solvers/products/samg.html
25. Lacroix, S., Vassilevski, Y.V., Wheeler, M.F.: Decoupling preconditioners in the implicit parallel accurate reservoir simulator (IPARS). Numer. Linear Algebr. Appl. 8(8), 537–549 (2001)
26. Castelletto, N., White, J.A., Tchelepi, H.A.: Accuracy and convergence properties of the fixed-stress iterative solution of two-way coupled poromechanics. Int. J. Numer. Anal. Methods Geomech. 39(14), 1593–1618 (2015)
27. Sleijpen, G.L.G., Fokkema, D.R.: BiCGstab (l) for linear equations involving unsymmetric matrices with complex spectrum. Electron. Trans. Numer. Anal. 1(11), 11–32 (1993)
28. Li, N., Saad, Y., Chow, E.: Crout versions of ILU for general sparse matrices. SIAM J. Sci. Comput. 25(2), 716–728 (2003)
29. Kaporin, I.E.: High quality preconditioning of a general symmetric positive definite matrix based on its UTU+ UTR+ RTU-decomposition. Numer. Linear Algebr. Appl. 5(6), 483–509 (1998)
30. Kaporin, I.E.: Scaling, reordering, and diagonal pivoting in ILU preconditionings. Russ. J. Numer. Anal. Math. Model. 22(4), 341–375 (2007)
31. Bollhöfer, M.: A robust ILU with pivoting based on monitoring the growth of the inverse factors. Linear Algebr. Appl. 338(1–3), 201–218 (2001)
32. Aliaga, J.I., Bollhöfer, M., Marti, A.F., Quintana-Orti, E.S.: Exploiting thread-level parallelism in the iterative solution of sparse linear systems. Parallel Comput. 37(3), 183–202 (2011)
33. Olschowka, M., Neumaier, A.: A new pivoting strategy for Gaussian elimination. Linear Algebr. Appl. 240, 131–151 (1996)
34. Duff, I.S., Koster, J.: The design and use of algorithms for permuting large entries to the diagonal of sparse matrices. SIAM J. Matrix Anal. Appl. 20(4), 889–901 (1999)
35. Bunch, J.R., Kaufman, L.: A computational method for the indefinite quadratic programming problem. Linear Algebr. Appl. 34, 341–370 (1980)

36. Nikitin, K., Terekhov, K., Vassilevski, Yu.: A monotone nonlinear finite volume method for diffusion equations and multiphase flows. Comput. Geosci. **18**(3–4), 311–324 (2014)

37. Terekhov, K.M., Mallison, B.T., Tchelepi, H.A.: Cell-centered nonlinear finite-volume methods for the heterogeneous anisotropic diffusion problem. J. Comput. Phys. **330**, 245–267 (2017)

38. Shen, F., Kastrup, C. J., Liu, Y., Ismagilov, R.F.: Threshold response of initiation of blood coagulation by tissue factor in patterned microfluidic capillaries is controlled by shear rate. Arter. Thromb. Vasc. Biol. **28**(11), 2035–2041 (2008)

39. Norne: the full Norne benchmark case, a real field black-oil model for an oil field in the Norwegian Sea. https://opm-project.org/?page_id=559. Accessed 15 Apr 2018

40. Bouchnita, A.: Mathematical modelling of blood coagulation and thrombus formation under flow in normal and pathological conditions. Ph.D. thesis, Université Lyon 1 - Claude Bernard; Ecole Mohammadia d'Ingénieurs - Université Mohammed V de Rabat - Maroc (2017)

41. Coon, E.T., Moulton, J.D., Painter, S.L.: Managing complexity in simulations of land surface and near-surface processes. Environ. Model. Softw. **78**, 134–149 (2016)

Maximus: A Hybrid Particle-in-Cell Code for Microscopic Modeling of Collisionless Plasmas

Julia Kropotina[1,2]([✉]), Andrei Bykov[1,2], Alexandre Krassilchtchikov[1], and Ksenia Levenfish[1]

[1] Ioffe Institute, St. Petersburg, Russia
`juliett.k@gmail.com`, {`byk,kra,ksen`}`@astro.ioffe.ru`
[2] Peter the Great St. Petersburg Polytechnic University, St. Petersburg, Russia

Abstract. A second-order accurate divergence-conserving hybrid particle-in-cell code *Maximus* has been developed for microscopic modeling of collisionless plasmas. The main specifics of the code include a constrained transport algorithm for exact conservation of magnetic field divergence, a Boris-type particle pusher, a weighted particle momentum deposit on the cells of the 3d spatial grid, an ability to model multispecies plasmas, and an adaptive time step. The code is efficiently parallelized for running on supercomputers by means of the message passing interface (MPI) technology; an analysis of parallelization efficiency and overall resource intensity is presented. A *Maximus* simulation of the shocked flow in the Solar wind is shown to agree well with the observations of the Ion Release Module (IRM) aboard the Active Magnetospheric Particle Tracer Explorers interplanetary mission.

Keywords: Hybrid particle-in-cell modeling
High-performance computing · Shocked collisionless plasmas
The Solar wind

1 Introduction

Collisionless plasmas are ubiquitous in space on a vast range of scales from the Earth magnetosphere to galaxy clusters – the largest bound objects in the universe. The growing amount and enhancing quality of multiwavelength observational data require adequate models of the observed objects and structures to be developed and confronted with the data. The required modeling is a very complicated task, as it has to account with sufficient accuracy and detalization for physical processes occurring on a very broad range of spatial and temporal scales, such as self-consistent acceleration of charged particles along with the

Electronic supplementary material The online version of this chapter (https://doi.org/10.1007/978-3-030-05807-4_21) contains supplementary material, which is available to authorized users.

V. Voevodin and S. Sobolev (Eds.): RuSCDays 2018, CCIS 965, pp. 242–253, 2019.
https://doi.org/10.1007/978-3-030-05807-4_21

evolution of the underlying bulk plasma flows and multiscale electromagnetic fields.

As substantial energy of the bulk flows can be converted into accelerated particles, whose dynamical role and back-reaction on the structure of the flows is significant, the modeling can not be performed within the frame of magnetic hydrodynamics – it has to be done on a microscopic kinetic level, and even further, within a particle-in-cell approach, where dynamics of individual particles (ions and electrons) is modelled on a spatial grid with piecewise-constant values of the electromagnetic field.

Hybrid models are a special class of particle-in-cell models of plasma flows, where the ions are treated as individual particles, while the electrons are considered as a collisionless massless fluid [1–3]. Such an approach allows one to resolve nonlinear collisionless effects on ion scales, at the same time saving substantial computational resources due to gyro-averaging of fast small-scale motions of the electrons. Hybrid codes are employed to study various astrophysical processes, such as evolution of ion-ion instabilities or shocked collisionless plasma flows [4–10], with the advantage of much greater time and spatial scales compared to those of the full particle-in-cell approach.

Here we present a detailed description of the numerical scheme and parallelization technique of *Maximus*[1], a second-order accurate divergence-conserving hybrid particle-in-cell code intended for modeling of nonlinear evolution of collisionless plasma flows with shocks. Previous versions of the code are briefly described in [11,12]. The main features of *Maximus* are: a divergence-conserving constrained transport approach based on finite-volume discretization, a linearized Riemann solver with parabolic reconstruction, an ability to treat multi-species flows, a Boris-type particle pusher. In the new version of the code, the numerical scheme and parallelization technique are advanced, allowing for better performance, stability, and energy conservation. The recent improvements include:

(i) an account for the non-zero electron pressure;
(ii) an advanced particle pusher with time-centered electromagnetic fields (instead of previously used "current advance method" [1] combined with a simple particle mover based on Taylor expansion);
(iii) a triangular-shaped cloud (TSC) particle model used instead of the nearest-grid point (NGP) model for charge deposit and force interpolation [13,14];
(iii) 2d parallelization in physical space with adaptive sizes of areas, assigned to particular MPI processes;
(iv) adaptive time steps.

The structure of the paper is as follows. Equations governing dynamics of the ions and the electromagnetic field are discussed in Sect. 2. The overall numerical approach is outlined in Sect. 3 and briefly compared with the approaches of codes *dHybrid* [15] and *Pegasus* [16]. The technique and efficiency of code

---

[1] The code is named after out late colleague, Maxim Gustov (1978–2014), who developed its first version in 2005.

parallelization is illustrated in Sect. 4, where some estimates of computational resources needed to reach the scales required for typical astrophysical applications are given. It is shown in Sect. 5 how *Maximus* can be used for modeling of particle acceleration in the Solar wind. The simulated particle spectra agree with in-situ measurements of the AMPTE/IRM interplanetary mission [17].

2 Dymanics of the Ions and Evolution of the Electromagnetic Field

A standard set of equations used for hybrid modeling of ion dynamics and evolution of the electromagnetic field (see, e.g., [1]) can be formulated as follows:

$$\frac{d\boldsymbol{r}_k}{dt} = \boldsymbol{V}_k \tag{1}$$

$$\frac{d\boldsymbol{V}_k}{dt} = \frac{Z_k}{m_k}\left(\boldsymbol{E} + \boldsymbol{V}_k \times \boldsymbol{B}\right) \tag{2}$$

$$\frac{\partial \boldsymbol{B}}{\partial t} = -\bigtriangledown \times \boldsymbol{E} \tag{3}$$

$$\boldsymbol{E} = \frac{1}{\rho_c}\left(\bigtriangledown \times \boldsymbol{B}\right) \times \boldsymbol{B} - \frac{1}{\rho_c}\left(\boldsymbol{j}_{\text{ions}} \times \boldsymbol{B}\right) - \bigtriangledown P_e/\rho_c \tag{4}$$

$$\boldsymbol{j}_{\text{ions}} = \sum_{\text{cell}} S(\boldsymbol{r}_k)Z_kV_k, \quad \rho_c = \sum_{\text{cell}} S(\boldsymbol{r}_k)Z_k \tag{5}$$

Here $\boldsymbol{r}_k$ and $\boldsymbol{V}_k$, Z_k and m_k are positions, velocities, charges, and masses of individual ions, $\boldsymbol{E}$, $\boldsymbol{B}$ denote the electric and magnetic field vectors, ρ_c and j_c are the ion charge and the density of the electric current integrated over a cell of the numerical grid. The latter are derived from positions of the ions via a weighting function $S(\boldsymbol{r}_k)$. The electron pressure P_e is derived from an isothermal gyrotropic Maxwellian (this can be justified by expanding the electric Vlasov-Landau kinetic equation by the square root of the electron-to-ion mass ratio [18]). Similar treatment of the electrons is employed in *Pegasus*. In the early versions of *dHybrid* the electron pressure is neglected, later versions include it in the adiabatic limit [7]. Actually, both the adiabatic and isothermal approaches are somewhat artificial and seem to have little impact on the ion scales. Some recent developments consider finite-mass electrons [19], which is more physical and will be implemented in *Maximus* mainly for stability reasons.

The generalized Ohm's law (Eq. 4) can be represented as:

$$E_j = -\frac{1}{\rho_c}\frac{\partial P_{ij}}{\partial x_i} - \frac{1}{\rho_c}\left(\boldsymbol{j}_{\text{ions}} \times \boldsymbol{B}\right)_j, \tag{6}$$

where $P_{ij} = \left(B^2/2 + P_e\right)\delta_{ij} - B_iB_j$ is the pressure tensor.

These equations are further normalized by a number of natural scale units: the proton inertial length $l_i = c\sqrt{m_p/4\pi n_0 e^2}$, the strength of the unperturbed large-scale magnetic field B_0, the inverse proton gyrofrequency $\Omega = eB_0/m_pc$, and the Alfven speed $V_A = B_0(4\pi\rho_0)^{-1/2}$ (n_0 and ρ_0 are the unperturbed plasma number and mass density respectively[2]).

3 The Numerical Approach

To model evolution of the ions from some initial state in a predefined spatial area, Eqs. (1–5) are numerically solved (advanced in time) on a three-dimensional Cartesian grid via the standard loop of procedures, which consists of the following steps:

1. Moment Collector (calculate charges and currents from Eq. 5)
2. Field Solver (solve Eqs. 3 and 4 to follow the evolution of the electromagnetic field)
3. Particle Mover (solve Eqs. 1 and 2 to move individual ions through the grid cells)

These steps are described in more detail below.

3.1 Moment Collector

At the Moment Collector step the velocities and positions of all particles are used to calculate charge and current densities in each of the grid cells.

According to the standard technique of particle-in-cell modeling, a "macro-particle" approximation is used, with the macro-particles occupying certain areas in the discrete phase space, i.e., representing an ensemble of real ("physical") particles, which move together. This approximation is inevitable due to a huge difference between the number of particles in real physical systems and the maximum number of simulated particles allowed by the available computational resources. The main reason to use the finite-sized macro-particles with charge and mass distributed over multiple cells is suppression of the strong numerical noise caused by the limited number of particles per grid cell (*ppc*). The shapes of the macro-particles are defined by the weighting function $S(r_k)$ (see Eq. 5). The impact of a weighting function on characteristic collision and heating time was studied by [14]. Compared to the case of point-like particles, with a second-order accurate triangular-shaped-cloud (TSC) weighting function we achieve an order of magnitude improvement of energy and momentum conservation, which could not be achieved by any reasonable increase of *ppc*. On the other hand, finite-sized particles can suppress the instabilities with wavelengths less than the particle size. However, the TSC length is only twice that of a cell and 75% of its weight is contained within one cell length, so the minimal wavelengths are still close to the grid resolution. Similar weighting is employed in the hybrid codes *dHybrid* and *Pegasus*. The TSC function is used to deposit charge and current densities (hereafter "cell moments") onto the mesh. The cell moments are further used to update the values of the electromagnetic field at the Field Solver step of the algorithm.

---

[2] *Maximus* allows treatment of multiple sorts of ions. However, the normalization units are always defined for the "reference" pure hydrogen plasma of the same initial mass density ρ_0 as the particular modeled composition.

3.2   Field Solver

In order to keep the divergence of magnetic field vector equal to zero, a Godunov-type constrained transport scheme [20] has been implemented. It employs the staggered grid with edge-centered edge-aligned electric fields and face-centered face-normal magnetic fields. Additional time splitting is used to simultaneously solve the interconnected Eqs. 3 and 4. Advancing physical time (with cell moments kept constant) requires the values of electric and magnetic field to be leap-frogged several times in the following way:

1. Spatial derivatives of the face-centered values of the magnetic field are found. In order to keep the total variation diminishing (TVD) condition, the monotonized central-difference limiter is used at this step (see, e.g., [21]).
2. A piecewise-parabolic reconstruction of magnetic field inside each cell is made, employing the values of the normal component of B and their spatial derivatives. The reconstruction coefficients are calculated according to [20].
3. A linearized Riemann solver [22] is used to obtain E on the cell edges from Eq. 4.
4. The face-centered magnetic field is updated from Eq. 3.

In the middle of this step the time- and cell-centered values of E and B are obtained, to be further used at the Particle Mover step. For this the face-centered values of the pressure tensor P_{ij} are calculated with a linearized Riemann solver near the cell faces. The electric field in the cell center is then found from Eq. 6 via a numerical differentiation of P_{ij}. At the same time, the magnetic field is found via cell-averaging of the current parabolic reconstruction. Compared to *Maximus*, *dHybrid* seems to lack a constrained transport algorithm, while in *Pegasus* the cell-centered values of the magnetic field are simple face averages, and the evaluation of edge electric fields is not specified [16].

3.3   Particle Mover

A Boris-type particle pusher [2,13] is used to propagate the ions. The electromagnetic field at each particle position is calculated by interpolation of time- and cell-centered field values with the same weighting function as during the Moment Collector step. This ensures the second order accuracy in time and space together with the absence of self-forces. To increase the accuracy, the following predictor-corrector algorithm is used:

1. The "initial" Lorentz force is calculated from current particle positions and velocities using field interpolation.
2. The "predicted" particle positions and velocities at half time-step are obtained from Eqs. 1 and 2.
3. The "predicted" Lorentz force is found from "predicted" particle positions and velocities.
4. The particles are moved from the "predicted" to final positions using the "predicted" Lorentz force.

3.4 Implementation

The code is written in C++. The following structure types are used: PARTICLE, SORT, CELL, FACE, EDGE, VERTEX. As particles are extremely memory-consuming due to their large number, each PARTICLE structure contains only coordinates and velocity components, but neither charge, nor mass. Particles of the same sort are organized in lists inside each cell of the grid, so that they are promptly accessible. Their charge, mass, and abundance are stored only once in a separate SORT structure. Cell-centered field values and cell moments are stored in CELLs, while FACEs contain values of B normal components and P_{ij} components, EDGEs contain the values of the electric field. VERTEXes are used to store some temporary intermediate values, which are used at the Field Solver step.

The numerical scheme of *Maximus* is illustrated in Fig. 1. All the variables are divided into 5 color-coded categories, corresponding to the structures mentioned above. CELLs moments and fields are separated for clarity.

Each time step starts at t^n with known particle positions and normal components of field vectors. First cell moments for the same moment of time are calculated at the Moment Collector step (substep 1). Then edges of E and faces of B are leap-frogged to half-timestep via the constrained transport algorithm (substep 2.1). Next, the same algorithm is used to find the half-timestep pressure tensor at cell faces (substep 2.2) and the cell-centered fields are stored (substep 2.3). Then the edges of E and the faces of B are leap-frogged until the end of the time step (substep 2.4). Finally, the particles are moved to their new positions (substep 3), and the next step t^{n+1} starts. Due to a similar field solver, the scheme of *Pegasus* is much the same, though *Maximus* takes the advantage of just one Particle Mover and Moment Collector per step, at the same time keeping the time-centering scheme.

The size of a time step can be automatically adjusted to comply with the numerical stability criteria [15] and the Courant condition, for the particles not to cross more than one cell during one time step. Such an adaptive time step seems to be surprisingly absent in other publicly known hybrid codes, though it can lead to a substantial speed-up of the computations.

The program sequence is realized by means of a sorted list, which allows one to flexibly change the set of procedures (for example, switch off the Moment Collector and Field Solver to study test particle movement in external fields).

4 Parallelization Technique and Efficiency

Usual tasks for hybrid simulations are highly resource-intensive, as the modelled systems must be resolved on scales of ion inertial lengths and time periods less than their inverse gyrofrequencies. At the same time, sizes and lifetimes of typical astrophysical objects, like supernova remnants, are many orders of magnitude larger. Hence hybrid simulations of such objects as a whole are not realistic. Even though only a tiny part of the real object is modelled, its size and modeling time still must be substantial to study long-wave, slowly growing instabilities, transport of energetic particles, and their back-reaction on the background plasma.

Another crucial parameter is the ppc number. The numerical noise of the electromagnetic field values, which is due to the limited number of macro-particles, typically scales as $\delta B/B \sim ppc^{-1/2}$. Hence ppc typically has to exceed 100 to guarantee the accuracy of simulation at least at the 10% level. Simulation of some finer effects may demand the ppc to exceed several thousand ([23–25]). The effects of limited ppc on simulations of astrophysical shocks were investigated in [12]. It should be noted that sometimes hybrid simulations of space plasmas are performed at $ppc = 8$ or even $ppc = 4$ [7, 15]. However, such simulations obviously require an artificial suppression of the numerical noise via low-pass filtering. Great care should be taken in such cases, as conservation of energy and spectra of the electromagnetic field are likely violated.

Due to their high resource-intensity, hybrid codes can be efficiently run only on multi-core computers and clusters. The Message Passing Interface (MPI) parallelization technology was chosen for *Maximus*: different MPI processes operate on different spatial parts of the simulation box. Various tests have shown that the inclusion of the third spatial dimension is usually less important for the modelled physical systems than maintaining substantial simulation box sizes in 2d. So *Maximus* is usually run within 2d boxes, though all the three components of velocity and field vectors are considered. Hence, the division of the modelled physical area between the MPI processes is done in two dimensions.

The directions of such division are automatically chosen along the two maximum simulation box sizes for each particular run. The number of processes per each direction is adjusted so that each area is made square-like. Notably, when modeling a shock, the area of r times higher density (i.e., higher ppc) appears downstream (r is the shock compression, which equals 4 for a typical strong shock in a monoatomic gas). Hence, the MPI regions in the shock downstream are made 4 times smaller to balance the load. When modeled shock is launched at $t = 0$ and then moves across the simulation box, the downstream region grows with time. In this case the processes domains are apriori made 4 times smaller in the region, which will be in downstream at the end of simulation.

The numerical scheme assumes that each cell influences not only its 26 closest 3d neighbours (for implementation of TSC particles), but also the neighbours of neighbours (to interpolate forces at the predicted positions of particles during the Particle Mover step, see Sect. 3). So the spatial region of each process is surrounded by two layers of ghost cells, used for processes communication (see Fig. 2). One layer of ghost faces is also introduced for the Field Solver step. It should be noted that in the case of small spatial regions number of ghost cells exceeds number of cells inside the domain, and memory and time for communication become relatively significant. However, this is usually not the case in large astrophysical setups.

In order to access the parallelization efficiency, an illustrative setup of *Maximus* was run on several sets of processes (from 1 to $28 \times 8 = 224$) on the

"Tornado" cluster of SPbPU, which has 612 28-core nodes, each with 64 Gb RAM and 2 × Intel Xeon CPU E5-2697 v3 (2.60 GHz). The setup was chosen large enough to eliminate relative communication expenses. Also the box size was taken to be divisible by the number of cores per node, so that the processes treated equal space regions. In total, 28000 spatial cells were modelled with single sort $ppc = 100$ and periodic boundary conditions applied for particles and fields. The Moment Collector, Field Solver, and Particle Mover procedures were executed with fixed time step $dt = 0.005\,\Omega^{-1}$ until $t = 1000\,\Omega^{-1}$ (i.e., 200000 time steps were made). Two simulation box shapes were chosen: 1d ($28000 \times 1 \times 1$ cells) and 2d ($280 \times 1 \times 100$ cells).

Figure 2 shows the decrease of total run time with the increase of the processes number. One can see that the run time scales well with the number of processes, even if more than one node is employed. This indicates the relatively small contribution of internode communications to the total resource consumption, which is expected for large box sizes. Similar estimates for *dHybrid* [15] (with box size of 96 cells) showed substantial communication expenses due to small domain sizes. The difference between the 1d and 2d configurations is not substantial, though 1d appears slightly quicker. Overall, the characteristic run time per particle per step is almost constant and can be estimated as $t_{ps} \sim 500\,\text{ns}$.

Consider a typical problem of high energy astrophysics – simulation of particle injection and acceleration at collisionless shocks formed in energetic space objects, which harbour powerful outflows (supernova remnants, pulsar wind nebulae, active galactic nuclei, etc.). The number of spatial cells required to appropriately describe such an object, can be of order 10^8. If at least two sorts of ions (say, hydrogen and helium) are taken into account, about 100 *ppc* are need, i.e., ten billion particles are required. Being distributed among 100 nodes (2800 cores), it averages to about 350000 particles per MPI process. Hence, one step would take about 0.2 s, which translates to about twenty-four hours for 500000 time steps. In fact, even such number of steps would not be enough to model a complex nonlinear configuration, because much smaller time steps may be required in that case. One may conclude, that most of the relevant astrophysical problems to be investigated with *Maximus* demand substantial computational power for relatively long periods of time (from weeks to months).

5 Simulation of a Collisionless Shock in the Solar Wind

The obvious way to verify numerical models is to confront their results to experimental data. In order to test the predictions of *Maximus*, we simulated distributions of accelerated ions in the Solar wind and compared them with the in-situ measurements made with the AMPTE/IRM interplanetary mission [17]. The simulation was run with the following parameters, taken from the direct observational data: the upstream Mach number (in the downstream frame) $M_A = 3.1$, the magnetic field inclination angle $\theta = 10°$, the ratio of thermal to magnetic pressure $\beta = 0.04$. The standard Solar wind composition was considered: 90.4%

H (+1) (by mass), 8.1% He (+2), and about 1.5% of heavy ions, represented by O(+6). A shock was initialized in a rectangular box sized $10000 \times 400l_i$ via the "rigid piston" method (reflection of a superalfvenic flow from a conducting wall). The simulated spectra of ions in the shock downstream are shown in Fig. 3 together with the experimental data. A good agreement of the model and the observations can be seen; the small differences may be due to variations of shock parameters. The acceleration Fermi process is illustrated in the supplementary video, where individual H (+1) (white circles) and He (+2) (green circles) ions positions are placed on the protons $x - E$ phase spaces, evolving with time. Both sorts of ions are reflected during the first shock encounter and subsequently gain energy by scattering upstream and downstream of the shock.

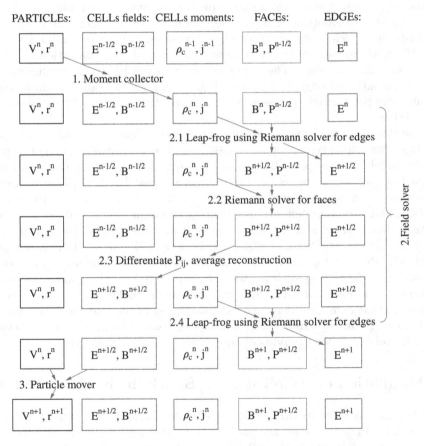

Fig. 1. A schematic illustration of the algorithm employed to update the variables from $t = t^n$ to $t = t^{n+1}$ (see Sect. 3 for a detailed description). The variables are shown as 5 color-coded categories. The red arrows indicate the substeps performed. (Color figure online)

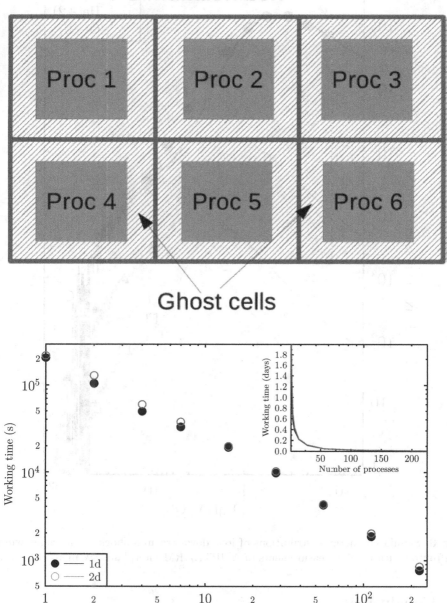

Fig. 2. Upper panel: *Maximus* parallelization scheme. The area allocated for each of the cores in the physical space is shown in blue, while ghost cells used for interconnection are hatched. **Lower panel:** Scalability of *Maximus*. Run time of the same tasks in 1d and 2d for different numbers of employed processor cores. The top-right inside panel shows the same time, measured in days, with linear scale.

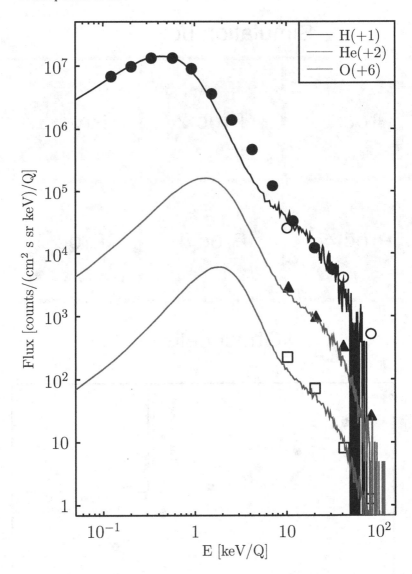

Fig. 3. Simulated energy distributions of ions downstream a shock in the Solar wind confronted with in-situ measurements of AMPTE/IRM interplanetary mission [17].

6   Conclusions

The 3d second-order accurate in time and space divergence-conserving hybrid code *Maximus* is presented. The code has been verified to keep energy and momentum conservation in modelled systems. It can be efficiently run on multi-core computers and clusters due to implementation of the MPI parallelization technology. For an illustrative case of the Solar wind, a good agreement of the modeling with the observational data of AMPTE/IRM has been demonstrated. *Maximus* can be effectively used to model nonlinear energetic processes in astrophysical plasmas on relevant spatial and time scales once substantial numerical resources are provided (hundreds of computational cores for time periods from weeks to months).

Acknowledgments. Some of the presented modeling was performed at the "Tornado" subsystem of the supercomputing center of St. Petersburg Polytechnic University, whose support is readily acknowledged.

References

1. Matthews, A.P.: J. Comput. Phys. **112**, 102–116 (1994)
2. Lipatov, A.S.: The Hybrid Multiscale Simulation Technology, p. 403. Springer, Heidelberg (2002). https://doi.org/10.1007/978-3-662-05012-5
3. Winske, D., Omidi, N.: Hybrid codes. Methods and applications. Computer Space Plasma Physics: Simulation Techniques and Software, pp. 103–160 (1993)
4. Quest, K.B.: J. Geophys. Res. **93**, 9649 (1988)
5. Giacalone, J., Burgess, D., Schwartz, S.J., Ellison, D.C., Bennett, L.: J. Geophys. Res. **102**, 19789 (1997)
6. Burgess, D., Möbius, E., Scholer, M.: Space Sci. Rev. **173**, 5 (2012)
7. Caprioli, D., Spitkovsky, A.: Astrophys. J. **783**(2), 91 (2014)
8. Caprioli, D., Spitkovsky, A.: Astrophys. J. **794**(2), 46 (2014)
9. Caprioli, D., Spitkovsky, A.: Astrophys. J. **794**(2), 47 (2014)
10. Marcowith, A., et al.: Rep. Progress Phys. **79**(4), 046901 (2016)
11. Kropotina, Yu.A, Bykov, A.M., Gustov, M.Yu., Krassilchtchikov, A.M., Levenfish, K.P.: JTePh **60**(2), 231–239 (2016)
12. Kropotina, Yu.A., Bykov, A.M., Krasil'shchikov, A.M., Levenfish, K.P.: JTePh, **61**(4), 517–524 (2016)
13. Birdsall, C.K., Langdon, A.B.: Plasma Physics via Computer Simulation, p. 373. IOP Publishing, Bristol (1991)
14. Hockney, R.W.: Measurements of collision and heating times in a two-dimensional thermal computer plasma. J. Comput. Phys. **8**(1), 19–44 (1971)
15. Gargaté, L., et al.: Comput. Phys. Commun. **176**, 419 (2007)
16. Kunz, M.W., Stone, J.M., Bai, X.-N.: J. Comput. Phys. **259**, 154 (2014)
17. Ellison, D.C., Moebius, E., Paschmann, G.: Astrophys. J. **352**, 376 (1990). Thermal Computer Plasma. Journal of computational physics 8, pp. 19–44 (1971)
18. Rosin, M.S., et al.: Mon. Not. R. Astron. Soc. **413**, 7 (2011)
19. Amano, T., Higashimori, K., Shirakawa, K.: J. Comput. Phys. **275**, 197 (2014)
20. Balsara, D.S.: Astrophys. J. **151**, 149–184 (2004)
21. LeVeque, R.J., Mikhalas, D., Dorfi, E.A., Muller, E.: Computational Methods for Astrophysical Fluid Flow, p. 481. Springer, Heidelberg (1997). https://doi.org/10.1007/3-540-31632-9
22. Balsara, D.S.: Astrophys. J. **116**, 119–131 (1998)
23. Florinsky, V., Heerikhuisen, J., Niemec, J., Ernst, A.: Astrophys. J. **826**, 197 (2016)
24. Niemec, J., Florinsky, V., Heerikhuisen, J., Nishikava, K.-I.: Astrophys. J. **826**, 198 (2016)
25. Kropotina, J.A., Bykov, A.M., Krasilshchikov, A.M., Levenfish, K.P.: JPCS **1038**, 012014 (2018)

Microwave Radiometry of Atmospheric Precipitation: Radiative Transfer Simulations with Parallel Supercomputers

Yaroslaw Ilyushin[1,2](✉) and Boris Kutuza[2]

[1] Physical Faculty, Moscow State University, GSP-2, 119992 Moscow, Russia
ilyushin@phys.msu.ru
[2] Kotel'nikov Institute of Radio Engineering and Electronics,
Russian Academy of Sciences, Moscow 125009, Russia
kutuza@cplire.ru

Abstract. In the present paper, the problems of formation and observation of spatial and angular distribution of thermal radiation of raining atmosphere in the millimeter wave band are addressed. Radiative transfer of microwave thermal radiation in three-dimensional dichroic medium is simulated numerically using high performance parallel computer systems. Governing role of three dimensional cellular inhomogeneity of the precipitating atmosphere in the formation of thermal radiation field is shown.

Keywords: Microwave radiometry · Precipitation · Radiative transfer

1 Introduction

Investigation of atmospheric precipitation with space borne instrumental observations (see the Fig. 1) is one of most important problems in remote sensing. The possibility of retrieving rain intensity from microwave observations was demonstrated in 1968 with the results of Cosmos-243 space-borne experiment [1]. Rain precipitation zones over the sea were identified and rain intensity was estimated from radio brightness temperatures at wavelengths 0.8, 1.35 and 3.2 cm.

Further development of techniques of space-borne precipitation observations was related to the DMSP satellite with the microwave radiometer SSM/I [2] on board, operating in the wavelength band 0.35–1.6 cm. Precipitation zones have been determined as low radio brightness areas at 0.35 cm wavelength because of high single scattering albedo of big rain droplets. Also, microwave radiometry of rain precipitation has been performed from the TRMM satellite. At present, the GPM (Global Precipitation Mission) project is under development.

Due to this and other directions of research, the interaction of microwave radiation with precipitation and clouds of various types is now extensively studied [3,4]. Many of them consist of non-spherical particles, having prevalent orientation like falling raindrops. In contrast to acroscopically isotropic media, in

© Springer Nature Switzerland AG 2019
V. Voevodin and S. Sobolev (Eds.): RuSCDays 2018, CCIS 965, pp. 254–265, 2019.
https://doi.org/10.1007/978-3-030-05807-4_22

these media dichroism plays a key role in the generation and propagation of the radiation. These effects together with spatial inhomogeneity of the medium and underlying surface require consideration of radiation fields in three-dimensional inhomogeneous dichroic media.

The flat layered medium model, which has been widely used for simulations during long time, is now well studied and developed [5–8]. However, it does not completely describe radiative fields in spatially inhomogeneous atmospheric precipitation. Understanding of impact of the three-dimensional effects on the microwave radiative transfer (RT) is critically important for correct interpretation of the passive radiometric measurements.

Three-dimensional RT codes were first applied to the microwave atmospheric studies [9]. Since that, there have been several published studies considering microwave RT in space-borne remote sensing, and how it is effected by inhomogeneity [3] and dichroism [6].

However, to the authors' knowledge, only a very few numerical studies of vectorial RT in three dimensional anisotropic scattering media are published [4]. This paper is motivated by the need for theoretical assessments of intensity and polarization of thermal radiation of rain precipitation, observed by a space-borne microwave radiometer. The focus of the present study is on rain instead of cloud ice mostly studied by [4]. This work also differs from [4] in that it was largely an algorithm comparison study focused on ground based viewing geometry of rain, rather than space-borne one. In addition, several different wavelengths are used in simulations to investigate capabilities of multi-spectral radiometry for retrievals of rain intensity.

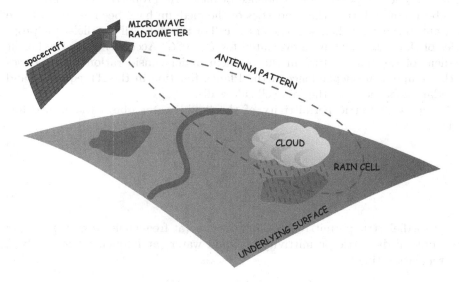

Fig. 1. Schematic view of the space-borne radiometry experiment.

2   Physical Model of Radiative Characteristics of the Medium

In the present study, two models of raining atmosphere (uniform flat layer and three dimensional cubic rain cell) are investigated and compared.

In the rain cell model, the contribution of vertical side walls in radiative balance between the cell and surrounding medium is essential. Following to [4], we chose a cubic rain cell model ($3 \times 3 \times 3$ km), uniformly filled with falling raindrops. These dimensions are close to the typical size of realistic rain cells [10] and characteristic scales of spatial inhomogeneity in stochastic RT models in rain [11].

Physical (thermodynamic) temperature in the atmosphere is non-uniform and decreases with height $T_2 = (300 - 7z) K$, where z is a height in km.

The underlying surface (Fig. 2) is approximated with a horizontal flat nearly isotropically radiating gray surface with partial lambertian reflection. Thermal radiative emission of the heated underlying surface is slightly vertically polarized [12]. In the millimeter wave band, this model provides a reasonable approximation for most terrestrial soils and vegetation covers. Radiative absorption by the liquid phase water clouds [13] and atmospheric gases is taken into account separately.

Underlying surface in the flat layer model is either black (lambertian reflection coefficient $R = 0$) or gray ($R = 0.25$). In the cubic rain cell model, the underlying surface within the square bottom of the cell is also black or gray as well as in the uniform slab model. Outside the bottom surface, it is totally black ($R = 0$). The temperature dependence of dielectric properties of liquid water have been ignored, thus the properties of the medium have been assumed to be constant in the whole height range of the cell or the slab. All the dielectric properties of the water have been evaluated for $T = 0°C$. Accounting for the height gradient of the temperature in the problem under consideration thus reduces to the temperature dependence of the Planck function in the RT equation and boundary conditions on the rain cell side walls.

Complex dielectric permittivity of the liquid water obeys the Debye formula [14]

$$\varepsilon = \varepsilon_0 + \frac{\varepsilon_s - \varepsilon_0}{1 + i\Delta\lambda/\lambda}, \tag{1}$$

where

$$\Delta\lambda = 2\pi\tau_p \frac{\varepsilon_s + 2}{\varepsilon_0 + 2}, \tag{2}$$

ε_s – staticdielectric permittivity of liquid water (at frequencies $\nu \ll 1/\tau_p$), $\varepsilon_0 \approx 4.9$ – optical dielectric permittivity of liquid water (at frequencies $\nu \gg 1/\tau_p$), and relaxation time

$$\tau_p = \exp\left\{9.8\left(\frac{273}{T + 273} - 0.955\right)\right\} \cdot 10^{-12} \text{ s.} \tag{3}$$

The static dielectric permittivity if the liquid water is described by an approximate formula [12]

$$\varepsilon_s(T) = 88.045 - 0.4147\,T + 6.295 \cdot 10^{-4}\,T^2 + 1.075 \cdot 10^{-5}\,T^3. \tag{4}$$

Geometrical model of non-spherical falling raindrops shape can be found, e.g. in [15]. With a reasonable degree of approximation falling raindrops can be regarded to be oblate spheroids with vertically oriented rotational axis of symmetry. The ratio of axes of the spheroid approximately is [5]

$$\frac{b}{a} = 1 - 0.091\bar{a}, \tag{5}$$

where $\bar{a}$ – is the radius of the spherical raindrop of the equivalent volume in mm.

Raindrop sizes are distributed statistically according to Marshall-Palmer distribution [16]

$$n(\bar{a}) = N_0 \exp(-q\bar{a}), \tag{6}$$

where $N_0 = 16000\,\mathrm{m}^{-3}\mathrm{mm}^{-1}$, $q = 8.2R^{-0.21}$, R – rain intensity, mm/h.

Extinction and absorption matrices of the spheroidal particles of fixed orientation can be evaluated by publicly available T-matrix codes [17].

Thus, radiative properties of the rain precipitation medium (volume extinction and absorbtion matrices) were evaluated within the physical model formulated above. Obtained radiative properties are close to those evaluated in [4, 18]. It can be seen that the volume extinction coefficient decreases with the increase of the wavelength. Simultaneously, single scattering albedo decreases, which is in qualitative agreement with the known results for the small absorbing particles (in clouds). In addition, the longer the wavelength, the more pronounced is dichroism (difference of the absorption and extinction between the horizontal and vertical polarization).

3   Radiative Transfer in the Anisotropic Scattering Medium

Spatial and angular distribution of the intensity and polarization of thermal radiation in the rain precipitation medium is governed by the vectorial radiative transfer equation

$$(\boldsymbol{\Omega} \cdot \nabla)\mathbf{I}(\boldsymbol{r}, \boldsymbol{\Omega}) = -\hat{\sigma}_\varepsilon \mathbf{I}(\boldsymbol{r}, \boldsymbol{\Omega}) + \bar{\sigma}_a B_\lambda(T_2(z)) + \frac{1}{4\pi} \int_{4\pi} \hat{x}(\boldsymbol{\Omega}, \boldsymbol{\Omega}\,)\mathbf{I}(\boldsymbol{r}, \boldsymbol{\Omega}')d\boldsymbol{\Omega}', \tag{7}$$

where $\boldsymbol{\Omega} = (\mu_x, \mu_y, \mu_z)$ – unit vector of arbitrary direction, $\mathbf{I}(z, \boldsymbol{\Omega}) = \{I, Q, U, V\}$ – Stokes parameters vector of polarized radiation, $\hat{\sigma}_\varepsilon = \hat{\sigma}_\varepsilon(\boldsymbol{\Omega})$ – extinction matrix of polarized radiation in the medium, $\bar{\sigma}_a$ – vector of true absorption in the medium, $\hat{x}(\boldsymbol{\Omega}, \boldsymbol{\Omega}')$ – scattering matrix, $B_\lambda(T)$ – Planck black body radiation function. Formulae for evaluation of the true absorption vector components $\bar{\sigma}_a$ can be found, e.g., in [16].

In the millimeter wave band at temperatures about 300 K, typical for lower terrestrial atmosphere, Planckian black body radiation at the given wavelength is roughly proportional to the black body temperature (Rayleigh-Jeans law). This allows the Stokes parameters to be expressed immediately through the equivalent black body temperature (radio brightness temperature).

Boundary problem for the vectorial radiative transfer equation in three dimensional cubic rain cell consists of the Eq. (7) and boundary conditions for the incoming radiation at bottom and top sides of the cubic cell, respectively, and boundary conditions at the side walls of the cube. Boundary condition on the top surface of the rain cell is just zero boundary condition for the incoming radiation. Boundary conditions on side walls and bottom of the rain cell account for the thermal radiation coming from the heated underlying surface and lossy gray surrounding atmosphere. For their exact formulation the reader is referred to [19].

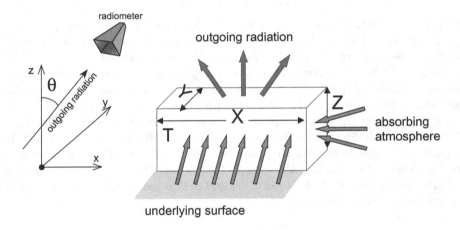

Fig. 2. Schematic view of the rain cell model geometry.

4   Radiative Transfer Equation in Three Dimensional Rain Cell

The boundary problem for the vectorial radiative transfer equation in a three dimensional cubic rain cell consists of the Eq. (7) and boundary conditions for the incoming radiation on the bottom and top sides of the cubic cell, and boundary conditions at the side walls of the cube. Boundary conditions on the top and bottom sides of the cube are

$$\mathbf{I}(0, \boldsymbol{\Omega} \cdot \mathbf{z} > 0) = \{2T_1 + 2T_D, Q_1, 0, 0\}, \tag{8}$$

$$\mathbf{I}(z_0, \boldsymbol{\Omega} \cdot \mathbf{z} < 0) = \{0, 0, 0, 0\}, \tag{9}$$

where $\mathbf{z}$ – unit vector of positive direction of z axis, T_1 and Q_1– radio brightness temperature and polarization of the underlying surface emission, T_D – radio brightness temperature of diffuse reflection from the underlying surface, defined by an equation

$$T_D \int_0^1 \mu \, d\mu = -R \int_{-1}^0 T(\mu)\mu \, d\mu \tag{10}$$

at $z = 0$.

Boundary conditions on the side walls determine intensity and polarization of the radiative fluxes, incident on the wall of the cube from the underlying surface and heated absorbing atmosphere, surrounding the cell from side directions, as is schematically shown in the Fig. 3. If the physical temperature of the atmosphere is stratified vertically, fluxes going up consist of attenuated thermal radiation coming from the periphery of the underlying surface and thermal radiation of the atmosphere itself,

$$I(Z) = I_1(0) \exp\left(-\int_0^Z \frac{\kappa dz}{\mu_z} \right) + \int_0^Z 2T_2(z)\kappa \exp\left(-\int_z^Z \frac{\kappa dz'}{\mu_z} \right) dz, \tag{11}$$

$$Q(Z) = Q_1(0) \exp\left(-\int_0^Z \frac{\kappa dz}{\mu_z} \right), \tag{12}$$

where κ is the volume extinction coefficient in the surrounding atmosphere. Corresponding fluxes going down consist of the atmospheric thermal radiative emissions only

$$I(Z) = \int_{z_0}^Z 2T_2(z)\kappa \exp\left(-\int_z^Z \frac{\kappa dz'}{\mu_z} \right) dz, \tag{13}$$

$$Q(Z) = 0. \tag{14}$$

Horizontal fluxes, incident on the side walls of the cube, are

$$I(Z) = 2T_2(Z), \tag{15}$$

$$Q(Z) = 0. \tag{16}$$

For linear dependence of the atmospheric temperature T_2 from the height

$$T_2(z) = T_2(0) + gradT_2 \cdot z, \tag{17}$$

the formulae given above yield explicit expressions, convenient for practical usage in computer simulations.

The vectorial radiative transfer equation was solved by the DO method. Corresponding discretized three dimensional equation is

$$\mu_{xi}\frac{\partial}{\partial x}\mathbf{I}_i + \mu_{yi}\frac{\partial}{\partial y}\mathbf{I}_i + \mu_{zi}\frac{\partial}{\partial z}\mathbf{I}_i = -\hat{\sigma}_\varepsilon \mathbf{I}_i + 2\pi \sum_{l,j} x(\mathbf{\Omega}_i, \mathbf{\Omega}_j)a_j\mathbf{I}_j + SF_i(z), \tag{18}$$

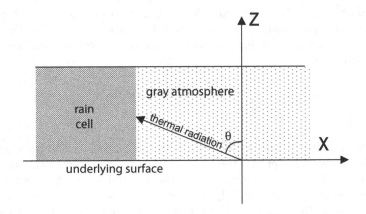

Fig. 3. To the formulation of the side wall boundary condition for the rain cell model.

where $\Omega_i = \{\mu_{xi}, \mu_{yi}, \mu_{zi}\}$ is a unit vector of direction of i − th node of the quadrature formula.

In practice, observed radiometric data are in fact a convolution of the observed radiation field with the receiving antenna pattern of the instrument. [20] concluded that inhomogeneity effects could be largely explained by the beam-filling effect, which is essentially a straightforward consequence of averaging over radiances that have a non-linear response to hydrometeors. In [21] this was also demonstrated for polarized RT by comparison with the independent pixel approximation. However, the footprint size of the space-borne microwave radiometer [22–25] is notably larger than typical size of the rain cell [10]. For this reason, averaged contribution of rain cell in the registered radio brightness temperature can be well approximated by an integral over the whole projection area of the top and two side surfaces of the cube, as is shown in the Fig. 2:

$$
\begin{aligned}
\bar{\mathbf{I}}(\Omega) &= \frac{\underset{(1)}{\int} \mathbf{I}(x,y,Z,\Omega)\mu_z\,dxdy + \underset{(2)}{\int} \mathbf{I}(X,y,z,\Omega)\mu_x\,dzdy + \underset{(3)}{\int} \mathbf{I}(x,Y,z,\Omega)\mu_y\,dxdz}{\underset{(1)}{\int} \mu_z\,dxdy + \underset{(2)}{\int} \mu_x\,dzdy + \underset{(3)}{\int} \mu_y\,dxdz} \\[2mm]
&= \frac{\underset{(1)}{\int} \mathbf{I}(x,y,Z,\Omega)\mu_z\,dxdy + \underset{(2)}{\int} \mathbf{I}(X,y,z,\Omega)\mu_x\,dzdy + \underset{(3)}{\int} \mathbf{I}(x,Y,z,\Omega)\mu_y\,dxdz}{XY\mu_z + XZ\mu_y + YZ\mu_x},
\end{aligned}
\tag{19}
$$

where the indices (1), (2) and (3) denote integration over the top and two side surfaces of the rectangular cell, respectively, X, Y, Z are horizontal dimensions of the cell. For the cubic cell, which is considered here, $X = Y = X = 3\,\mathrm{km}$.

Assuming typical values of total water vapor and liquid water content $2\,\mathrm{g/cm}^2$ and $0.5\,\mathrm{kg/m}^2$, respectively, and accounting for the absorption by the molecular oxygen [19], additional absorption coefficients 0.33, 0.066 and $0.013\,km^{-1}$ for wave lengths 3, 8 and 22.5 mm have been assumed for simulations. These values are relatively small and do not play significant role in millimeter wave band thermal radiation field of rain, except of the weakest ones (less than several mm/h).

5 Simulation Results

The set of simultaneous differential equations (18) together with the proper discretized boundary conditions for all the sides of the cubic cell has been solved by the finite difference (FD) scheme (upwind differences [26]) by iterative convergence to the stationary regime. Scattering integral has been approximated by the Gaussian spherical quadrature formula of 29th order G29 [27] with 302 discrete ordinates.

Numerical algorithm was implemented and run on parallel computers "Tshchebyshev" and "Lomonosov" of the Scientific and Research Computing Center of the Moscow State University [28] with C++ programming language. Cyclically repeated calculations of the iteration scheme were parallelized with OpenMP parallel computing standard. Each scenario, dependent of rain intensity and the wave length, has been simulated on a separate individual node of the parallel cluster. That means, 8 or 12 processing cores were allocated for each scenario. On the nodes with processors supporting hyper-threading technology (HTT), it corresponds to 16 or 24 threads, respectively. Size of memory and number of computing cores of each individual node of both parallel clusters allowed for fully polarized radiative transfer simulations on the three-dimensional rectangular grid as large as $64 \times 64 \times 64$ nodes with $4 \times 302 = 1208$ unknown parameters at each node. Since solution in each node of the grid is updated independently at every iteration of the FD scheme, computer code parallelization requires a minimal effort, with the speedup nearly proportional to the number of processing cores. Running time limit (72 h) allowed to run the process until it converged to the accuracy of the Stokes parameters $10^{-4} K$ uniformly over the whole three-dimensional domain.

There has been performed large series of numerical simulations of microwave RT in slab layer and rectangular rain cell with different parameters (rain intensity, additional atmospheric absorption and underlying surface diffuse reflection coefficient) [29]. Angular and spatial distributions of the intensity and polarization of the outgoing thermal radiation at different wavelength has been simulated for a number of values of rain intensity, reflectivity of underlying surface and some other parameters, which required significant computing time on parallel cluster with many nodes. Results of complete series of calculations are available from the official site of the project [30]. Typical simulated three-dimensional views of rain cell are shown in the Figs. 4 and 5.

Distributions of the polarization parameters Q and U of the cell with absolutely black underlying surface (the Fig. 4) do not differ significantly from the ones of partially reflective underlying surface (the Fig. 5). On the other hand, the radio brightness temperature (the intensity, Stokes parameter I, the Fig. 4) is notably impacted by the Lambertian reflectivity of the surface beneath the cell. Roughly speaking, the cold cell bottom is partially seen through the cell's side walls.

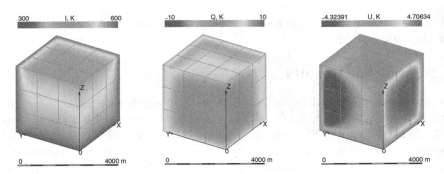

Fig. 4. (Color online) 3D rain cell view projection. Wave length $\lambda = 3$ mm , rain intensity 20 mm/h, Lambertian reflectance $R = 0$.

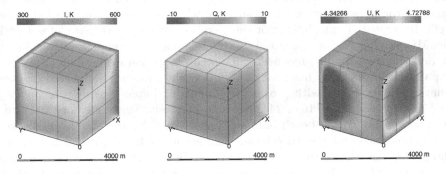

Fig. 5. (Color online) 3D rain cell view projection. Wave length $\lambda = 3$ mm , rain intensity 20 mm/h, Lambertian reflectance $R = 0.25$

From the simulation results it turns out that geometrical model of isolated cubic rain cell, which is rough simplification of true inhomogeneous distribution of realistic rain precipitation in the atmosphere, showed inhomogeneous distribution of thermal radiation intensity and polarization. It worth also noting appearance of the third Stokes parameter U, which is observed experimentally [31], in the simulated radiative field of the rain cell. This parameter U, which characterizes the tilt angle of the radiation polarization plane, does not appear in the layered slab medium model with the vertically oriented axes of the rotationally symmetric particles (raindrops). The polarization plane of the thermal radiation of layered slab medium can be tilted, if the rotational axes of the raindrops are systematically tilted themselves [31]. The mechanisms of such raindrops' axes inclination (wind etc.) are, however, still questionable, and are out of the scope of this paper. The rain cell model is therefore able to qualitatively explain the polarization plane tilt without the tilted raindrops hypothesis being involved.

Thus, RT simulations with the cubic cell model discovered limitations of flat layered slab model, which still largely remains the basic and most used computational model in atmospheric radiation studies.

6   Conclusions and Final Remarks

In the study presented here, extensive numerical simulations of thermal radiation field in precipitating atmosphere in millimeter wave band have been performed. Commonly used uniform slab model of rain atmosphere is validated against three-dimensional rain cell model. Simulation results for intensity and polarization of the thermal radiation have been presented and interpreted.

Obtained results can appear to be helpful for interpretation of space-borne microwave radiometry observational data. Calculations confirm linear polarization of rain cell thermal radiation, which is about 2–3 K in average over observable surface area of the cell.

Spatial resolution of existing space-borne millimeter wave radiometers is now about 15–20 km, which well exceeds typical size of the rain cell. Thus, several rain cells with different rain intensity are typically covered by the radiometer's field of view. Because of large contribution of underlying surface to the intensity and polarization, it is not possible to separate it from the contribution of precipitation (rain). To do that, one needs to improve significantly spatial resolution of the radiometer, to make its field of view comparable to or smaller than typical rain cell dimension. This can be achieved with millimeter wave synthetic aperture interferometer, or with large size antenna systems with capability of electrical or mechanical scanning.

Acknowledgements. The research is carried out using the equipment of the shared research facilities of HPC computing resources at Lomonosov Moscow State University. Support from the Russian Fundamental Research Fund with grants 13-02-12065 ofi-m and 15-02-05476 is also kindly acknowledged.

References

1. Basharinov, A.E., Gurvich, A.S., Egorov, S.T.: Radio Emission of the Earth as a Planet. Nauka, Moscow (1974)
2. Spencer, R., Goodman, H., Hood, R.: Precipitation retrieval over land and ocean with the SSM/I: Identification and characteristics of the scattering signal. J. Ocean. Technol. **6**, 254–273 (1989)
3. Roberti, L., Haferman, J., Kummerow, C.: Microwave radiative transfer through horizontally inhomogeneous precipitating clouds. J. Geophys. Res. **99**(D8), 16707–16718 (1994)
4. Battaglia, A., Davis, C., Emde, C., Simmer, C.: Microwave radiative transfer intercomparison study for 3-D dichroic media. J. Quant. Spectrosc. Radiat. Transf. **105**(1), 55–67 (2007)
5. Evtushenko, A.V., Zagorin, G., Kutuza, B.G., Sobachkin, A., Hornbostel, A., Schroth, A.: Determination of the Stokes vector of the microwave radiation emitted and scattered by the atmosphere with precipitation. Izv.-Atmos. Ocean. Phys. **38**(4), 470–476 (2002)
6. Emde, C., Buehler, S.A., Davis, C., Eriksson, P., Sreerekha, T.R., Teichmann, C.: A polarized discrete ordinate scattering model for simulations of limb and nadir longwave measurements in 1-D/3-D spherical atmospheres. J. Geophys. Res. Atmos. **109**(D24), D24207 (2004)

7. Ilyushin, Y., Seu, R., Phillips, R.: Subsurface radar sounding of the Martian polar cap: radiative transfer approach. Planet. Space Sci. **53**(14–15), 1427–1436 (2005)
8. Ilyushin, Y.A.: Radiative transfer in layered media: Application to the radar sounding of Martian polar ices. II. Planet. Space Sci. **55**(1–2), 100–112 (2007)
9. Weinman, J.A., Davies, R.: Thermal microwave radiances from horizontally finite clouds of hydrometeors. J. Geophys. Res. Ocean. **83**(C6), 3099–3107 (1978)
10. Begum, S., Otung, I.E.: Rain cell size distribution inferred from rain gauge and radar data in the UK. Radio Sci. **44**(2) (2009). RS2015
11. Tsintikidis, D., Haferman, J.L., Anagnostou, E.N., Krajewski, W.F., Smith, T.F.: A neural network approach to estimating rainfall from spaceborne microwave data. IEEE Trans. Geosci. Remote. Sens. **35**(5), 1079–1093 (1997)
12. Ulaby, F.T., Moore, R.K., Fung, A.K.: Microwave Remote Sensing: Active and Passive, vol. 1. Addison-Wesley, Reading (1981)
13. Kutuza, B.G., Smirnov, M.T.: The influence of clouds on the radio-thermal radiation of the 'atmosphere-ocean surface' system. Issledovanie Zemli iz Kosmosa **1**(3), 76–83 (1980)
14. Basharinov, A.E., Kutuza, B.G.: Determination of temperature dependence of the relaxation time of water molecules in clouds and possibilities for assessing the effective temperature of drop clouds by uhf radiometric measurements. Izv. Vyssh.Uchebn. Zaved., Radiofiz. **17**(1), 52–57 (1974)
15. Czekala, H., Havemann, S., Schmidt, K., Rother, T., Simmer, C.: Comparison of microwave radiative transfer calculations obtained with three different approximations of hydrometeor shape. J. Quant. Spectrosc. Radiat. Transf. **63**(2–6), 545–558 (1999)
16. Czekala, H., Simmer, C.: Microwave radiative transfer with nonspherical precipitating hydrometeors. J. Quant. Spectrosc. Radiat. Transf. **60**(3), 365–374 (1998)
17. Moroz, A.: Improvement of Mishchenko's T-matrix code for absorbing particles. Appl. Opt. **44**(17), 3604–3609 (2005)
18. Hornbostel, A.: Investigation of Tropospheric Influences on Earth-satellite Paths by Beacon, Radiometer and Radar Measurements/Doctoral thesis (1995)
19. Ilyushin, Y.A., Kutuza, B.G.: Influence of a spatial structure of precipitates on polarization characteristics of the outgoing microwave radiation of the atmosphere. Izv.-Atmos. Ocean. Phys. **52**(1), 74–81 (2016)
20. Kummerow, C.: Beamfilling errors in passive microwave rainfall retrievals. J. Appl. Meteorol. **37**(4), 356–370 (1998)
21. Davis, C., Evans, K., Buehler, S., Wu, D., Pumphrey, H.: 3-D polarised simulations of space-borne passive mm/sub-mm midlatitude cirrus observations: a case study. Atmos. Chem. Phys. **7**(15), 4149–4158 (2007)
22. Kutuza, B.G., Hornbostel, A., Schroth, A.: Spatial inhomogeneities of rain brightness temperature and averaging effect for satellite microwave radiometer observations, vol. 3, pp. 1789–1791 (1994)
23. Kutuza, B.G., Zagorin, G.K., Hornbostel, A., Schroth, A.: Physical modeling of passive polarimetric microwave observations of the atmosphere with respect to the third Stokes parameter. Radio Sci. **33**(3), 677–695 (1998)
24. Kutuza, B.G., Zagorin, G.K.: Two-dimensional synthetic aperture millimeter-wave radiometric interferometric for measuring full-component Stokes vector of emission from hydrometeors. Radio Sci. **38**(3), 8055 (2003)
25. Volosyuk, V.K., Gulyaev, Y.V., Kravchenko, V.F., Kutuza, B.G., Pavlikov, V.V., Pustovoit, V.I.: Modern methods for optimal spatio-temporal signal processing in active, passive, and combined active-passive radio-engineering systems. J. Commun. Technol. Electron. **59**(2), 97–118 (2014)

26. Richtmyer, R.D., Morton, K.W.: Difference Methods for Init.al-Value Problems. Interscience Publishers, New York (1967)

27. Lebedev, V.: Quadrature formulas for a sphere of the 25–29th order of accuracy. Sib. Mat. Zh. **18**(1), 132–142 (1977)

28. Sadovnichy, V.A., Tikhonravov, A., Voevodin, V., Opanasenko, V.: "lomonosov": supercomputing at moscow state university. In: In Contemporary High Performance Computing: From Petascale toward Exascale, pp. 283–307. Chapman & Hall/CRC Computational Science, Boca Raton, USA, CRC Press (2013)

29. Ilyushin, Y.A., Kutuza, B.G.: New possibilities of the use of synthetic aperture millimeter-wave radiometric interferometer for precipitation remote sensing from space. Proceedings -: International Kharkov Symposium on Physics and Engineering of Microwaves. Millimeter and Submillimeter Waves, MSMW (2013), pp. 300–302 (2013)

30. http://vrte.ru/16X2014KutuzaVDO3D/testSKIF1/htmGlobal/index.html

31. Evtushenko, A., Zagorin, G., Kutuza, B., Sobachkin, A., Hornbostel, A., Schroth, A.: Determination of the Stokes vector of the microwave radiation emitted and scattered by the atmosphere with precipitation. Izv.-Atmos. Ocean. Phys. **38**(4), 470–476 (2002)

Modeling Groundwater Flow in Unconfined Conditions of Variable Density Solutions in Dual-Porosity Media Using the GeRa Code

Ivan Kapyrin[1,2]([✉]), Igor Konshin[2,3], Vasily Kramarenko[2],
and Fedor Grigoriev[1,2]

[1] Nuclear Safety Institute of the Russian Academy of Sciences (IBRAE),
Moscow 115191, Russia
ivan.kapyrin@gmail.com, grig-fedor@yandex.ru
[2] Marchuk Institute of Numerical Mathematics of the Russian Academy of Sciences
(INM RAS), Moscow 119333, Russia
igor.konshin@gmail.com, kramarenko.vasiliy@gmail.com
[3] Dorodnicyn Computing Centre (FRC CSC RAS), Moscow 119333, Russia

Abstract. Flow of variable density solution in unconfined conditions and transport in dual-porosity media mathematical model is introduced. We show the application of the model to a real object which is a polygon of deep well liquid radioactive waste injection. Several assumptions are justified to simplify the model and it is discretized. The model is aimed at assessment of the role of density changes on the contaminant propagation dynamics within the polygon. The method of parallelization is described and the results of numerical experiments are presented herein.

Keywords: Parallel computing · Density-driven flow
Dual-porosity media · Unconfined flow · Liquid waste deep injection

1 Introduction

Simulation of groundwater (GW) flow in unconfined regime with allowance for density convection and the effect of double porosity is quite an exotic task, poorly studied in the scientific literature. The necessity to develop this model was stipulated by the practical task of assessing the effect of the density driven convection effect on the dynamics of the spreading of contaminants on the polygon of liquid radioactive waste (LRW) deep disposal "Severny" [1,2]. The "Severny" polygon is located in the Krasnoyarsk Territory. In the deep (more than 400 m) I-st horizon, since 1967, injection of solutions with a high mineralization (up to 200 g/l) has been carried out. In the geoflow-geomigration model of the "Severny" polygon, GEOPOLIS, developed by the Nuclear Safety Institute of the RAS, the process of density driven convection was not taken into account, and in the process of improving this model, the question arose about the need to simulate it.

V. Voevodin and S. Sobolev (Eds.): RuSCDays 2018, CCIS 965, pp. 266–278, 2019.
https://doi.org/10.1007/978-3-030-05807-4_23

The transition from the conventional flow and transport model to the model, taking into account the density driven convection, is associated with a significant increase in the complexity of the model and slowing down the calculations, as the processes of flow and transport in porous media become strongly coupled thus no longer allowing to divide the solution of the problem into two (flow and transfer) at each time step. Note that the rate of computations is quite critical for the polygon model, since when calibrating the model parameters calculations have to be performed repeatedly. In addition, to ensure the necessary accuracy of calculations and the stability of the calculation schemes (using the explicit low-dissipation scheme of discretization of the advection operator), the time step for this model should not exceed 10 days for a total simulation period of 50 years for epigenetic calculations. Thus, it is necessary to determine how essential the process of density driven convection is for the rate of migration of radionuclides in the underground aquifers of the polygon. Based on the final results of this study, the process of density driven convection must either be included in the model, or be reasonably ignored in the simulation, which will save the computational time.

In the present paper, a mathematical model is derived, simplifying assumptions are justified, and parallel calculations are performed to demonstrate the applicability and efficiency of the model. The model is implemented in the GeRa (**Ge**omigration of **Ra**dionuclides) computational code developed by IBRAE and INM RAS for solving the problems of radioactive waste disposal safety assessment [3,4] on unstructured 3D grids [5]. Parallelization is done using MPI, which is a common trend for similar contemporary hydrogeological codes, such as PFLOTRAN [6], Amanzi [7] and TOUGH2-MP [8].

2   Mathematical Model

Let us formulate the complete system of equations of the GW flow model under conditions of variable saturation and transport taking into account double porosity and density driven convection:

$$
\begin{cases}
\dfrac{\partial(S_{\mathrm m}\varphi_{\mathrm m}\rho_{\mathrm m})}{\partial t} + \nabla\cdot(\rho_{\mathrm m}\boldsymbol{u}) = Q\rho_s - \rho_{\mathrm{exch}}\Gamma_w, \\[4pt]
\dfrac{\partial\theta_{\mathrm{im}}}{\partial t} = \Gamma_w, \\[4pt]
\dfrac{\partial\big((\theta_{\mathrm m}+\rho_{b_{\mathrm m}}k_{d_{i,\mathrm m}})C_{i,\mathrm m}\big)}{\partial t} + \lambda\left(\theta_{\mathrm m} + \rho_{b_{\mathrm m}}k_{d_{i,\mathrm m}}\right)C_{i,\mathrm m} + \nabla\left(\boldsymbol{u}C_{i,\mathrm m}\right) \\[4pt]
\quad - \nabla D_{i,\mathrm m}\nabla C_{i,\mathrm m} + \zeta\left(C_{i,\mathrm m} - C_{i,\mathrm{im}}\right) = QC_{i,s} - C_{\mathrm{exch}}\Gamma_w, \\[4pt]
\dfrac{\partial\big((\theta_{\mathrm{im}}+\rho_{b_{\mathrm{im}}}k_{d_{i,\mathrm{im}}})C_{i,\mathrm{im}}\big)}{\partial t} + \lambda\left(\theta_{\mathrm{im}} + \rho_{b_{\mathrm{im}}}k_{d_{i,\mathrm{im}}}\right)C_{i,\mathrm{im}} \\[4pt]
\quad = \zeta\left(C_{i,\mathrm m} - C_{i,\mathrm{im}}\right) + C_{\mathrm{exch}}\Gamma_w, \qquad i = 1,\dots,N_{\mathrm{ccmp}}.
\end{cases}
\tag{1}
$$

In system (1), the third and fourth equations are written out for each of the components carried in the solution, the index i denotes the component number, N_{comp} is the total number of components. The subscripts "m" and "im" mean that the variable or coefficient belongs to the mobile and immobile zones, respectively. Here the following notations are used (in square brackets the dimension of the quantity is indicated: M, L, and T stand for mass, length, and time, respectively, while [-] means dimensionless value): S is saturation, [-]; θ is the moisture content, [-] associated with the saturation by the formula:

$$S = \frac{\theta}{\varphi}; \tag{2}$$

where φ is the porosity of the medium, [-], composed of the porosity of the flow zone; φ_m and φ_{im} are the mobile and immobile zone porosities; ρ is the density of the solution, [ML^{-3}]; u is the Darcy velocity vector, [LT^{-1}]; ρ_s is the density of the solution in sources or sinks, [ML^{-3}]. Q is the volumetric flow rate of the fluid sources or sinks (usually wells), [T^{-1}]; Γ_w is the intensity of moisture exchange between the mobile and immobile zones, [T^{-1}]; ρ_{exch} and C_{exch} is the density of the liquid and the concentration of the impurity in the solution flowing from the mobile to the immobile zone:

$$\rho_{exch} = \begin{cases} \rho_m, \Gamma_w \geq 0, \\ \rho_{im}, \Gamma_w < 0; \end{cases} \qquad C_{exch} = \begin{cases} C_m, \Gamma_w \geq 0, \\ C_{im}, \Gamma_w < 0; \end{cases}$$

ρ_b is the rock density, [ML^{-3}]; k_d is the distribution coefficient, [M^{-1}L^3]; C is the contaminant concentration, [ML^{-3}]; D is the diffusion-dispersion tensor, [L^{2}T^{-1}]; ζ is the mass exchange coefficient between the mobile and immobile zones, [T^{-1}].

Note that in this case the first order rate dual-porosity model [9] is used. The system of equations (1) must be supplemented by a number of constitutive relations. First, it is the generalized Darcy's law for the dependence of Darcy velocity on the equivalent freshwater head h, [L] and density:

$$u = -K \left(\nabla h_m + \frac{\rho - \rho_0}{\rho_0} \nabla z \right), \tag{3}$$

where K is the hydraulic conductivity, [LT^{-1}]; ρ_0 is the reference density of the liquid in the absence of contaminants, [ML^{-3}]; z is the vertical coordinate, [L].

Note that here the equivalent freshwater head (later on simply called head) is used, expressed in meters of the water column of the reference liquid (a solution with a reference density ρ_0, without impurities). In fact, if p is pressure, then

$$h = \frac{p}{\rho_0 g} + z,$$

where g is the gravity constant, [LT^{-2}].

Secondly, the density of the liquid depends on the concentration. Let's put it linear:

$$\rho = \rho_0 + \sum_{i=1}^{N_{comp}} \varkappa_{vol,i} C_i. \tag{4}$$

Here $\varkappa_{vol,i}$ is the volumetric expansion coefficient, [-]. The linear dependence of this type is quite justified not only for small impurity concentrations, but in some cases also for high concentrations. Specifically for the "Severny" polygon, the main component determining the density of the solution is sodium nitrate, and according to [10], a linear dependence of the form (4) with a coefficient of volumetric expansion of approximately 0.6 can be used.

Third, the intensity Γ_w of the flow between the mobile and immobile zones should be determined. These formulas may be found, for example, in [11], but we omit them here as they will be excluded from the model later on.

When solving the problem (1), the change in porosity is usually neglected, considering it to be constant, and taking into account only the time derivative of the porosity. Then, similarly to [12–14] for the time derivative in the first equation of system (1) we will use the representation:

$$\frac{\partial \left(S_m \varphi_m \rho_m \right)}{\partial t} \approx \rho_m \frac{\partial \theta_m}{\partial t} + S_m \rho_m s_{stor} \frac{\partial h}{\partial t} + \theta_m \sum_{i=1}^{N_{comp}} \varkappa_{vol,i} \frac{\partial C_i}{\partial t}. \tag{5}$$

Due to the complexity of both the mathematical model (1) and its practical parameterization, let's consider the possible simplifications of system (1) that are relevant for the "Severny" polygon model. An essential feature of this problem has to be focused on: LRW propagation occurs in deep aquifers in a saturated pressure regime. This means that the effect of dual porosity is not significant for the upper layers of the model (1–2 layers) in the zone of variable saturation, since, firstly, pollution does not reach them, and secondly, the moisture exchange between the mobile and immobile zones can be neglected, supposing that the model of flow in a single porosity medium is sufficient to describe the flow process in this region. In turn, for the reservoirs of RAW, flow processes in a medium with variable saturation are not relevant. This factor taken into account in the model it can be assumed that the change in saturation refers only to the mobile zone, and the saturation of the immobile zone does not change.

Indeed, in the fully saturated zone of the GEOPOLIS model, the saturation does not change either in the mobile or in the immobile zones. Besides, in the zone of partial saturation, the presence of an immobile zone with constant saturation reduces the flow problem to the standard flow problem in a single porosity medium, while the effect of the presence of the immobile zone on transport remains. But the transport for the GEOPOLIS model, first, is not significant in the unsaturated zone (pollution does not reach it), and second, this model in any case provides wider possibilities for calculating the transfer in the unsaturated region due to the presence of a mass exchange coefficient ζ. The latter, like other coefficients, can be calibrated. In the upper layers, where it is possible to form a zone of variable saturation, the model reduces to transport in a single porosity vadose zone. Thus, we can take

$$\theta_{im} = \varphi_{im}, \qquad \Gamma_w = 0. \tag{6}$$

Given the assumptions made (5)–(6), the model (1) is converted to the form:

$$
\begin{cases}
\rho_m \frac{\partial \theta_m}{\partial t} + S_m \rho_m s_{\text{stor}} \frac{\partial h}{\partial t} + \theta_m \sum_{i=1}^{N_{\text{comp}}} \varkappa_{\text{vol},i} \frac{\partial C_i}{\partial t} + \nabla \cdot (\rho_m \boldsymbol{u}) = Q \rho_s, \\[2mm]
\frac{\partial \left(\left(\theta_m + \rho_{b_m} k_{d_{i,m}} \right) C_{i,m} \right)}{\partial t} + \lambda \left(\theta_m + \rho_{b_m} k_{d_{i,m}} \right) C_{i,m} + \nabla \left(\boldsymbol{u} C_{i,m} \right) \\[2mm]
\quad - \nabla D_{i,m} \nabla C_{i,m} + \zeta \left(C_{i,m} - C_{i,\text{im}} \right) = Q C_{i,s}, \\[2mm]
\varphi_{\text{im}} R_{\text{im}} \frac{\partial C_{i,\text{im}}}{\partial t} + \lambda \varphi_{\text{im}} R_{\text{im}} C_{i,\text{im}} = \zeta \left(C_{i,m} - C_{i,\text{im}} \right), \quad i = 1, \ldots, N_{\text{comp}}.
\end{cases}
\tag{7}
$$

In (7), the retardation factor in the immobile zone was introduced, [-]:

$$
R_{i,\text{im}} = \left(1 + \frac{\rho_b}{\varphi_{\text{im}}} k_{d_{i,\text{im}}} \right).
\tag{8}
$$

The index "m" in the flow equation will be omitted here, since in this model there are no corresponding variables for the immobile zone. It is necessary to supply the system with dependences $\theta(h), K(h)$. To simulate the flow in unconfined regime, we will use a pseudo-saturated formulation, a similar approach is proposed in [12]. It is more convenient to write out the dependence for the moisture content in terms of the pressure head, and then use formula for the transition from ψ to h:

$$
h = \psi + z.
\tag{9}
$$

It is assumed that $\theta(\psi)$ has two intervals of linear dependence on the pressure head ψ:

$$
\theta(\psi) = \begin{cases}
\varphi, & \psi > 0, \\[2mm]
\varphi - \frac{\varphi(1 - \alpha_\phi)}{\psi_r}, & \psi_r < \psi < 0, \\[2mm]
\varphi \left(\alpha_\phi - \alpha_\theta \left(\psi_r - \psi \right) \right), & \psi_r > \psi.
\end{cases}
\tag{10}
$$

There are three parameters in the model: $(\alpha_\phi, \alpha_\theta, \psi_r)$. The first two are dimensionless, the latter has dimension [L]. The principles of choosing these parameters should be described. Note that $\theta(\psi_r) = \alpha_\varphi \varphi$ is a quantity that is an analog of the residual moisture content in unsaturated flow models. The parameter α_θ characterizes the slope of $\theta(\psi)$ as ψ decreases below ψ_r. The parameter ψ_r characterizes the value of the pressure head at which the rock is close to dry. We follow the basic principle of flow modeling in unconfined conditions: the presence of a sharp boundary between fully saturated and dry media, and the possibility of fluid flow only in saturated region. In this case, the parameters α_θ and α_ϕ should be chosen sufficiently small ($\ll 1$), and the parameter ψ_r should be equal to the characteristic vertical dimension of the grid cell, when the groundwater level in the cell fell by a given amount, it could be considered drained, containing only residual moisture. We also note that when choosing the parameter α_θ, it is necessary to ensure the nonnegativity of $\theta(\psi)$, that is, with the maximum possible drop in the pressure head, the moisture content should be at least zero. In the case of the "Severny" polygon model, the following parameters were selected, thus providing the above properties of the model:

$$\alpha_\phi = 0.1, \alpha_\theta = 0.001, \psi_r = -30 \text{ m}. \tag{11}$$

For the $K(h)$ dependence, the following expression is accepted:

$$K(h) = K_r(h)K_s, \tag{12}$$

where K_s is the saturated hydraulic conductivity and the relative permeability $K_r(h)$ is identically equal to the saturation of $S(h)$:

$$K_r(h) = S(h). \tag{13}$$

Thus, the improved mathematical model of flow and transport at the "Severny" polygon which allows for the unconfined regime of the flow and the presence of weakly permeable interlayers in aquifers (dual porosity), consists of a system of equations (7), the constitutive equations (3), (4), (10), (13), the relation (9) between the variables h and ψ.

3 Discretization of the Model in Time and Space

For the diffusion operator, including the flow problem, the finite volume method (FV) is used with a linear two-point flow approximation scheme. To discretize the convection operator we use the FV explicit TVD-scheme with piecewise linear reconstruction of the concentration on the computational grid cells. Since the calculations of the model take very long time, and the use of the explicit scheme imposes a significant limitation on the time step (about 1 day), a transition to the use of an implicit FVM scheme of the first order of accuracy for approximation of the convective term is possible. The latter will allow one to remove restrictions on the time step and significantly accelerate the calculations especially when carrying out long-term forecasts. To discretize the last equation in (7) we will use an implicit scheme. In general, to solve the transport problem (the second and third equations (7)) we use splitting by physical processes. This will make it possible to separate the solution of the mass exchange problem from the transport and flow problems. To discretize the solution of the coupled flow-transport problem (7) we use the sequential iterative scheme. We represent the discrete system of equations corresponding to the system of differential equations (7) in the following form:

$$\begin{cases} F\left(h^{n+1}, C_m^{n+1}\right) = 0, \\ T\left(h^{n+1}, C_m^{n+1}, C_{im}^{n+1}\right) = 0, \\ E\left(C_m^{n+1}, C_{im}^{n+1}\right) = 0. \end{cases} \tag{14}$$

Here $h^{n+1}, C_{\mathrm{m}}^{n+1}, C_{\mathrm{im}}^{n+1}$ are the vectors of equivalent freshwater heads and concentrations in the mobile and immobile zones in cells of the computational grid at the next, $(n+1)$-th time step. The system (14) consists of three subsystems of nonlinear equations. The first one, $F\left(h^{n+1} C_{\mathrm{m}}^{n+1}\right) = 0$, corresponds to the flow equation, that is, to the first equation in (7). The second, $T\left(h^{n+1}, C_{\mathrm{m}}^{n+1}, C_{\mathrm{im}}^{n+1}\right) = 0$, consists of discrete transport equations corresponding to the second equation of (7). The third subsystem corresponds to the last equation of the system (7) and describes the mass transfer between the mobile and immobile zones. To solve the nonlinear system (14), the fixed point iteration method is used. In this case, the time step of the scheme is constructed as follows:

1. As the initial guess, the values from the previous time step are selected: $h^{n+1,0} = h^n$, $C_{\mathrm{m}}^{n+1,0} = C_{\mathrm{m}}^n$. The iteration counter $k=0$ is set.
2. The flow problem with density and concentration from the previous iteration is calculated: $F\left(h^{n+1,k+1}, C_{\mathrm{m}}^{n+1,k}\right) = 0$. The head $h^{n+1,k+1}$ and the corresponding flow $\boldsymbol{u}^{n+1,k+1}$, the moisture content $\theta_{\mathrm{m}}^{n+1,k+1}$ are calculated.
3. The transport problem with a known flux $\boldsymbol{u}^{n+1,k+1}$ is calculated:

$$
\begin{cases}
T\left(h^{n+1,k+1}, C_{\mathrm{m}}^{n+1,k+1}, C_{\mathrm{im}}^{n+1,k+1}\right) = 0, \\
E\left(C_{\mathrm{m}}^{n+1,k+1}, C_{\mathrm{im}}^{n+1,k+1}\right) = 0.
\end{cases}
\tag{15}
$$

From here there are concentrations on the new iteration $C_{\mathrm{m}}^{n+1,k+1}$, $C_{\mathrm{im}}^{n+1,k+1}$.
4. The iteration count is updated: $k = k+1$.
5. The criterion for the convergence of the iterative process is checked. If the criterion is satisfied, we set $h^{n+1} = h^{n+1,k}$, $C_{\mathrm{m}}^{n+1} = C_{\mathrm{m}}^{n+1,k}$, $C_{\mathrm{im}}^{n+1} = C_{\mathrm{im}}^{n+1,k}$. Otherwise, go back to step 2.

The solution of the transport problem (15) has to be explained. The system of transport equations which is solved at each time step, has the form:

$$
\begin{cases}
\dfrac{\left(\theta_{\mathrm{m}}^{n+1,k+1}+\rho_b k_{d_{i,\mathrm{m}}}\right) C_{i,\mathrm{m}}^{n+1,k+1} - \left(\theta_{\mathrm{m}}^n + \rho_b k_{d_{i,\mathrm{m}}}\right) C_{i,\mathrm{m}}^n}{\Delta t} + \lambda \left(\theta_{\mathrm{m}}^{n+1,k+1}+\rho_b k_{d_{i,\mathrm{m}}}\right) C_{i,\mathrm{m}}^{n+1,k+1} \\
\quad + \nabla\left(\boldsymbol{u}^{n+1,k+1} C_{i,\mathrm{m}}^n\right) - \nabla D_{i,\mathrm{m}} \nabla C_{i,\mathrm{m}}^{n+1,k+1} \\
\quad + \zeta \left(C_{i,\mathrm{m}}^{n+1,k+1} - C_{i,\mathrm{im}}^{n+1,k+1}\right) = Q C_{i,z}^{n+1,k+1}, \\
\varphi_{\mathrm{im}} R_{\mathrm{im}} \dfrac{C_{i,\mathrm{im}}^{n+1,k+1} - C_{i,\mathrm{im}}^n}{\Delta t} + \lambda \varphi_{\mathrm{im}} R_{\mathrm{im}} C_{i,\mathrm{im}}^{n+1,k+1} = \zeta \left(C_{i,\mathrm{m}}^{n+1,k+1} - C_{i,\mathrm{im}}^{n+1,k+1}\right), \\
i = 1, \dots, N_{\mathrm{comp}}.
\end{cases}
\tag{16}
$$

Here we use the explicit approximation in time of the convection operator originally used in GEOPOLIS, and the implicit ones for all other operators. The possibility of an implicit approximation of convection will be explained later. We introduce the intermediate values of the concentration in the mobile zone $\hat{C}_{i,\mathrm{m}}^{n+1,k+1}, \tilde{C}_{i,\mathrm{m}}^{n+1,k+1}$ in order to use the splitting according to the physical processes to solve Eq. (16). We rewrite the time derivative of the first equation of (16) using the intermediate concentration values:

$$\frac{\left(\theta_{\mathrm{m}}^{n+1,k+1} + \rho_b k_{d_{i,\mathrm{m}}}\right) C_{i,\mathrm{m}}^{n+1,k+1} - \left(\theta_{\mathrm{m}}^n + \rho_b k_{d_{i,\mathrm{m}}}\right) C_{i,\mathrm{m}}^n}{\Delta t}$$

$$= \frac{\left(\theta_{\mathrm{m}}^{n+1,k+1} + \rho_b k_{d_{i,\mathrm{m}}}\right)\left(C_{i,\mathrm{m}}^{n+1,k+1} - \hat{C}_{i,\mathrm{m}}^{n+1,k+1} + \hat{C}_{i,\mathrm{m}}^{n+1,k+1} - \tilde{C}_{i,\mathrm{m}}^{n+1,k+1} + \tilde{C}_{i,\mathrm{m}}^{n+1,k+1}\right)}{\Delta t}$$

$$- \frac{\left(\theta_{\mathrm{m}}^n + \rho_b k_{d_{i,\mathrm{m}}}\right) C_{i,\mathrm{m}}^n}{\Delta t} \tag{17}$$

$$= \left(\theta_{\mathrm{m}}^{n+1,k+1} + \rho_b k_{d_{i,\mathrm{m}}}\right) \frac{C_{i,\mathrm{m}}^{n+1,k+1} - \tilde{C}_{i,\mathrm{m}}^{n+1,k+1}}{\Delta t} + \left(\theta_{\mathrm{m}}^{n+1,k+1} + \rho_b k_{d_{i,\mathrm{m}}}\right)$$

$$\times \frac{\tilde{C}_{i,\mathrm{m}}^{n+1,k+1} - \hat{C}_{i,\mathrm{m}}^{n+1,k+1}}{\Delta t} + \frac{\left(\theta_{\mathrm{m}}^{n+1,k+1} + \rho_b k_{d_{i,\mathrm{m}}}\right) \hat{C}_{i,\mathrm{m}}^{n+1,k+1} - \left(\theta_{\mathrm{m}}^n + \rho_b k_{d_{i,\mathrm{m}}}\right) C_{i,\mathrm{m}}^n}{\Delta t}.$$

Then the splitting scheme is to be written as follows (step-by-step, for each component $i = 1, \ldots, N_{\mathrm{comp}}$):

I:

$$\frac{\left(\theta_{\mathrm{m}}^{n+1,k+1} + \rho_b k_{d_{i,\mathrm{m}}}\right) \hat{C}_{i,\mathrm{m}}^{n+1,k+1} - \left(\theta_{\mathrm{m}}^n + \rho_b k_{d_{i,\mathrm{m}}}\right) C_{i,\mathrm{m}}^n}{\Delta t} + \nabla\left(u^{n+1,k+1} C_{i,\mathrm{m}}^n\right) = Q C_{i,z}^{n+1},$$

II:

$$\left(\theta_{\mathrm{m}}^{n+1,k+1} + \rho_b k_{d_{i,\mathrm{m}}}\right) \frac{\tilde{C}_{i,\mathrm{m}}^{n+1,k+1} - \hat{C}_{i,\mathrm{m}}^{n+1,k+1}}{\Delta t} + \lambda \left(\theta_{\mathrm{m}}^{n+1,k+1} + \rho_b k_{d_{i,\mathrm{m}}}\right) \tilde{C}_{i,\mathrm{m}}^{n+1,k+1}$$

$$- \nabla D_{i,\mathrm{m}} \nabla C_{i,\mathrm{m}}^{n+1,k+1} = 0, \tag{18}$$

III:

$$\begin{cases} \left(\theta_{\mathrm{m}}^{n+1,k+1} + \rho_b k_{d_{i,\mathrm{m}}}\right) \frac{C_{i,\mathrm{m}}^{n+1,k+1} - \tilde{C}_{i,\mathrm{m}}^{n+1,k+1}}{\Delta t} + \zeta \left(C_{i,\mathrm{m}}^{n+1,k+1} - C_{i,\mathrm{im}}^{n+1,k+1}\right) = 0, \\[2mm] \varphi_{\mathrm{im}} R_{\mathrm{im}} \frac{C_{i,\mathrm{im}}^{n+1,k+1} - C_{i,\mathrm{im}}^n}{\Delta t} + \lambda \varphi_{\mathrm{im}} R_{\mathrm{im}} C_{i,\mathrm{im}}^{n+1,k+1} = \zeta \left(C_{i,\mathrm{m}}^{n+1,k+1} - C_{i,\mathrm{im}}^{n+1,k+1}\right). \end{cases}$$

The first substep of this scheme explicitly calculates the problem of convective transport taking into account the sources and sinks in the right-hand side. Note that in the case of a source, $C_{i,\mathrm{m}}^{n+1}$ is the contaminant concentration in the source averaged over time step, and in the case of a drain $C_{i,\mathrm{m}}^{n+1}$ is approximated explicitly, that is, $C_{i,\mathrm{m}}^{n+1} = C_{i,\mathrm{m}}^n$. On the second substep, the problem of diffusion and radioactive decay is implicitly calculated. On the third substep, again, the problem of mass exchange between the flow and non-current zones is implicitly solved. In case of implicit approximation of the convection problem, substeps I and II are aggregated, and the intermediate concentration $\hat{C}_{i,\mathrm{m}}^{n+1,k+1}$ is eliminated.

We now explain the verification of the convergence criterion for a nonlinear solver. As the criterion of convergence it is proposed to use the following expression. The user initially sets two values: $\varepsilon_{\mathrm{rel}}, \varepsilon_{\mathrm{abs}}$. It is assumed that the convergence is achieved when one of two conditions is satisfied:

$$\max\left(\left\|h^{n+1,k}\right\|, \left\|h^{n+1,k+1}\right\|\right) < \varepsilon_{\mathrm{abs}} \tag{19}$$

or

$$\frac{\left\| h^{n+1,k} - h^{n+1,k+1} \right\|}{\max \left(\left\| h^{n+1,k} \right\|, \left\| h^{n+1,k+1} \right\| \right)} < \varepsilon_{\text{rel}}. \tag{20}$$

The checking of the condition (19) being fulfilled before the condition (20) ensures that the error of division by zero does not occur.

4   Parallel Solution of the Models

We have performed the numerical solution of several models with the GeRa computational code. The INM RAS cluster [15] was used. The configuration of the cluster is the following:

- x6core: Compute Node Asus RS704D-E6, 12 nodes (two 6-node processor Intel Xeon X5650@2.67 GHz), 24 GB RAM;
- x8core: Compute Node Arbyte Alkazar+ R2Q50, 16 nodes (two 8-node processor Intel Xeon E5-2665@2.40 GHz), 64 GB RAM;
- x10core: Compute Node Arbyte Alkazar+ R2Q50, 20 nodes (two 10-node processor Intel Xeon E5-2670v2@2.50), 64 GB RAM;
- x12core: Compute Node Arbyte Alkazar+ R2Q50, 4 nodes (two 12-node processor IntelXeon E5-2670v3@2.30 GHz), 64 GB RAM.

The last 3 types of nodes x8core, x10core, and x12core can be used together as a single queue "e5core".

For the solution of linear systems we have exploited the package PETSc. The BiCGstab iterations were performed for Additive Schwarz preconditioning with overlap size 1 and ILU($k = 1$) as a subdomain solver. The relative tolerance stopping criterion 10^{-9} was applied.

4.1   Model "Lake"

As a sample test we have used the simple model "Lake". The model domain has an impermeable boundary and contains a lake with dense brine. A production well is working forcing the upper cells in the model to unsaturated conditions and the dense brine from the lake to move in the well direction and sink down as well.

The timestep was 10 days, while the total number of timesteps was 37. The total number of cells in a small size model ("Lake-small") was 37100 and the one in a medium size model ("Lake-medium") was 263020.

The computations were performed in the queue "x6core". In all tables below we have used the following notation: "#proc" denotes the number of cores used; "Time" is the total solution time; "Speedup" is the actual speedup with respect to the run on 1 core; S_2 is the relative speedup with respect to the previous run on the 2 times less number of cores.

In Tables 1 and 2 the numerical results for Lake-small and Lake-medium models are presented. The results show quite reasonable speedup, it was equal to 11.92 for 32 cores for a small size model and 41.15 for 128 cores for a medium size model.

Table 1. Lake-small model.

#proc	Time	Speedup	S_2
1	662.99	1.00	—
2	391.35	1.69	1.69
4	216.62	3.06	1.80
8	118.28	5.60	1.83
16	71.76	9.23	1.64
32	55.59	11.92	1.29

Table 2. Lake-medium model.

#proc	Time	Speedup	S_2
1	7118.78	1.00	—
2	4203.46	1.69	1.69
4	2427.53	2.93	1.73
8	1508.87	4.71	1.60
16	901.24	7.89	1.67
32	423.53	16.80	2.12
64	243.23	29.26	1.74
128	172.97	41.15	1.40

4.2   Model "Polygon"

For a real process modelling we have used the model "Polygon", which is actually
a model of the "Severny" LRW injection site. The model contains ten geological
layers. LRW is injected in the 9-th and the 5-th layers (aquifer I and aquifer
II respectively). Dense brines are injected in layer 9, while in the 5-th layer
the density varies only several grams per liter. In this test we show the first
application of the new numerical model to the real site and assess its parallel
efficiency.

The total number of cells in a small size model ("Polygon-small") was 43524
and the one in a large size model ("Polygon-large") was 421033. Figures 1 and 2
demonstrate the mesh distribution over 64 processors for Polygon-large model.
One can see the geological fault line as well as regions with mesh refinement
over numerous wells. Figure 3 shows the computational results with contaminant
concentration distribution for the Polygon-large model.

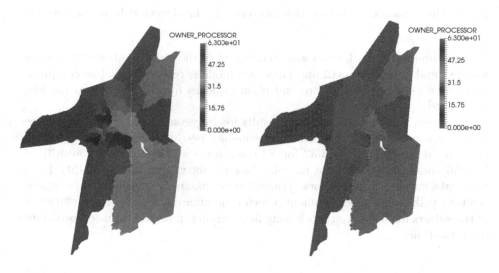

Fig. 1. Distribution of the mesh over 64 processors for Polygon-large model.

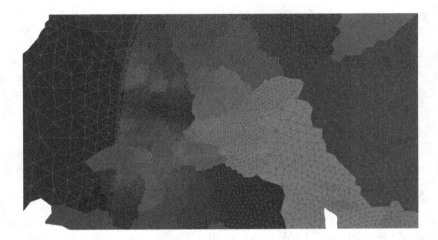

Fig. 2. Fragment of mesh distribution over 64 processors for Polygon-large model.

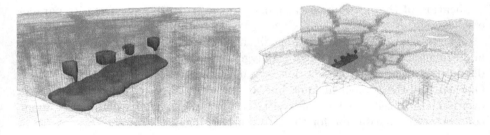

Fig. 3. The contaminant plume position forecast calculated using Polygon-large model.

The timestep in both cases was 10 days, while the total number of timesteps was 183 and 1608, for small and large size models, respectively. The computations were performed in x6core and e5core queues for the above two models, respectively.

In Tables 3 and 4 the numerical results for Polygon-small and Polygon-large models are presented. Results show reasonable speedup for such a complicated problem, it was equal to 13.32 for 32 cores for a small size model and 31.54 for 256 cores for a large size model. Thus we show program scalability for a moderate number of processors. Transition to massively parallel computations is a next task in GeRa development, which is dominated by the parallel efficiency of the solvers for linear systems arising from implicit flow and diffusion problems discretizations.

Table 3. Polygon-small model.

#proc	Time	Speedup	S_2
1	1104.08	1.00	—
2	642.87	1.71	1.71
4	384.39	2.87	1.67
8	218.83	5.04	1.75
16	142.13	7.76	1.53
32	82.84	13.32	1.71

Table 4. Polygon-large model.

#proc	Time	Speedup	S_2
1	67214.51	1.00	—
2	45161.05	1.48	1.48
3	22870.51	2.93	1.97
8	18147.68	3.70	1.26
16	9470.35	7.09	1.91
32	7515.65	8.94	1.26
64	5013.85	13.40	1.49
128	3150.89	21.33	1.59
256	2130.66	31.54	1.47

5 Conclusion

A mathematical model of flow and transport in geological media is proposed, which takes into account the flow in unconfined conditions, dual porosity and variable density of solutions. Based on the analysis of processes applied to the disposal of LRW "Severny", simplifications of the model are proposed. The model is numerically implemented in the GeRa code and tested with respect to the LRW injection polygon. The results of parallel runs show a good scalability of computations up to 256 computational cores. The novel model will be used further for the assessment of the role of density effects on the deep well radioactive waste injection sites.

References

1. Rybalchenko, A.I., Pimenov, M.K., Kostin, P.P., et al.: Deep disposal of liquid radioactive waste, IzdAT, Moscow 256 p. (1994). (in Russian)
2. Compton, K.L., Novikov, V., Parker, F.L.: Deep Well Injection of Liquid Radioactive Waste at Krasnoyarsk-26, Vol. I, RR-00-1, International Institute for Applied Systems Analysis, Laxenburg, Austria (2000)
3. Kapyrin, I.V., Ivanov, V.A., Kopytov, G.V., Utkin, S.S.: Integral code GeRa for the safety substantiation of disposal of radioactive waste. Mountain J. **10**, 44–50 (2015). (in Russian)
4. Kapyrin, I.V., Konshin, I.N., Kopytov, G.V., Nikitin, K.D., Vassilevski, Yu.V.: Hydrogeological modeling in problems of safety substantiation of radioactive waste burials using the GeRa code. In: Supercomputer Days in Russia: Proceedings of the International Conference, Moscow, 28–29 September 2015, pp. 122–132. Publishing House of Moscow State University, Moscow (2015). (in Russian)
5. Plenkin, A.V., Chernyshenko, A.Y., Chugunov, V.N., Kapyrin, I.V.: Methods of constructing adaptive unstructured grids for solving hydrogeological problems. Comput. Methods Program. **16**, 518–533 (2015). (in Russian)
6. Hammond, G.E., Lichtner, P.C., Mills, R.T.: Evaluating the performance of parallel subsurface simulators: an illustrative example with PFLOTRAN. Water Resour. Res. **50**(1), 208–228 (2014)
7. Freedman, V.L., et al.: A high-performance workflow system for subsurface simulation. Environ. Model. Softw. **55**, 176–189 (2014)

8. Zhang, K., Wu, Y.S., Pruess, K.: User's guide for TOUGH2-MP-A massively parallel version of the TOUGH2 code (No. LBNL-315E). Ernest Orlando Lawrence Berkeley National Laboratory, Berkeley, CA (US) (2008)

9. Lekhov, A.V.: Physical-chemical hydrogeodynamics, KDU 2010, 500 p. (2010). (in Russian)

10. Rabinovich, V.A., Khavin, Z.Ya.: Brief Chemical Reference Book, Leningrad, 392 p. (1978). (in Russian)

11. Simunek, J., Van Genuchten, M.T., Sejna, M.: The HYDRUS-1D software package for simulating the movement of water, heat, and multiple solutes in variably saturated media, version 3.0, HYDRUS software series 1. Department of Environmental Sciences, University of California Riverside (2005)

12. Diersch, H.J.: FEFLOW-white papers, vol. I. WASY GmbH Institute for water resources planning and systems research, Berlin, 366 p. (2005)

13. Liu, X.: Parallel modeling of three-dimensional variably saturated ground water flows with unstructured mesh using open source finite volume platform OpenFOAM. Eng. Appl. Comput. Fluid Mech. **7**(2), 223–238 (2013)

14. Paniconi, C., Putti, M.: A comparison of Picard and Newton iteration in the numerical solution of multidimensional variably saturated flow problems. Water Resour. Res. **30**(12), 3357–3374 (1994)

15. INM, RAS Cluster. http://cluster2.inm.ras.ru. Accessed 15 Apr 2018. (in Russian)

New QM/MM Implementation
of the MOPAC2012 in the GROMACS

Arthur O. Zalevsky[1,4(✉)], Roman V. Reshetnikov[2,3,4],
and Andrey V. Golovin[1,3,4]

[1] Faculty of Bioengineering and Bioinformatics,
Lomonosov Moscow State University, 119234 Moscow, Russia
aozalevsky@fbb.msu.ru
[2] Institute of Gene Biology, Russian Academy of Sciences, 119334 Moscow, Russia
[3] Apto-Pharm LLC, Kolomensky ps. 13A, Moscow, Russia
[4] Sechenov First Moscow State Medical University,
Trubetskaya str. 8-2, 119991 Moscow, Russia
http://vsb.fbb.msu.ru/

Abstract. Hybrid QM/MM simulations augmented with enhanced
sampling techniques proved to be advantageous in different usage scenarios, from studies of biological systems to drug and enzyme design.
However, there are several factors that limit the applicability of the approach. First, typical biologically relevant systems are too large and hence
computationally expensive for many QM methods. Second, a majority
of fast non *ab initio* QM methods contain parameters for a very limited
set of elements, which restrains their usage for applications involving
radionuclides and other unusual compounds. Therefore, there is an inces-
sant need for new tools which will expand both type and size of simulated
objects. Here we present a novel combination of widely accepted molecular modelling packages GROMACS and MOPAC2012 and demonstrate
its applicability for design of a catalytic antibody capable of organophos-
phorus compound hydrolysis.

Keywords: QMMM · hpc · Molecular modelling · Rational design

1 Introduction

Hybrid QM/MM simulations were introduced in 1976 [1] and only somewhat
recently began to be actively used in the molecular dynamics (MD) simulations,
becoming a popular tool for studying biomolecular systems. Now the method
allows to gather a ms-scale statistics on protein dynamics, where thermal motion
could significantly contribute to chemical reactivity and conformational space of
the system [2]. An additional momentum to the rise of hybrid QM/MM applications in biosystems' studies was given by Parinello and colleagues who developed
the "metadynamics" method [3–5]. This method allows to scan a conformational
space of biomolecular systems searching for rare events, such as reactions cat-
alyzed by protein enzymes. Generally, a quantum subsystem in metadynamic

© Springer Nature Switzerland AG 2019
V. Voevodin and S. Sobolev (Eds.): RuSCDays 2018, CCIS 965, pp. 279–288, 2019.
https://doi.org/10.1007/978-3-030-05807-4_24

modeling of enzymatic reactions consists of up to 1000 atoms, and to simulate a thermal movement of the system, hundreds of thousands steps of energy gradients and geometry optimization calculations can be made. It requires considerable computing resources even at a low level of QM calculations. Optimization of computational tools for the task is the key factor affecting the progress of enzyme design and an engineering of other biopolymers with desired properties.

A number of software packages capable of hybrid simulations have been developed. The popular MD program package GROMACS [6] has a QM/MM interface to several quantum chemistry software tools. Significant drawback of the original interface implementation is that the data exchange between the programs occurs via the file system, which imposes a significant performance limitation when used on cluster computing systems with distributed storage environment. Here, we propose the modification of the widely accepted MOPAC-2012 [7] semi-empirical QM tool allowing to use it as a GROMACS library. The resulting software features a high speed computing with OpenMP and CUDA acceleration options for both QM and MM hybrid subsystems. An important advantage of the proposed implementation is a wide range of supported chemical elements and a variety of available semi-empirical QM parameters for biological systems.

2   Methods

2.1   ONIOM

The ONIOM approach [8] allows to divide a system into several layers with an independent description of intra-layer interactions. In QM/MM case, force gradients are first evaluated for the isolated QM subsystem using a selected semi-empirical or *ab initio* model. Next, the gradients and the total potential energy of the system are calculated using the corresponding MM force field and added to the ones obtained for the isolated QM subsystem. Finally, in order to avoid duplicated contribution from the QM subsystem a molecular mechanics calculation is performed for the isolated QM region and the result is subtracted from total sum:

$$E_{tot} = E_I^{QM} + E_{I+II}^{MM} - E_I^{MM} \qquad (1)$$

where the subscripts I and II refer to the QM and MM subsystems, respectively. The superscripts indicate at what level of theory the energies are computed. Physical separation between systems is achieved by introduction of linking atoms (LA) which are rendered as hydrogens in quantum part and do not introduce additional interactions into mechanical part [9].

2.2   Implementation of the QM/MM Interface

Using the implementation of the GROMACS/ORCA interface as a reference, we modified corresponding parts of GROMACS (MD engine, input-output module and CMake configuration files) and MOPAC2012 (input-output and parameters verification modules). In the reference implementation the exchange between GROMACS and ORCA packages is performed through text files, which limits precision: data is printed and read with the "%10.7f" pattern, which provides only 7 decimal digits, corresponding to single precision, while internally ORCA uses double precision. Our implementation uses MOPAC2012, which was compiled and assembled into static library and directly linked into GROMACS static binary during compilation allowing direct data transfer between MOPAC2012 and GROMACS with double precision. To optimize the control of QM calculations, the GMX_QM_MOPAC2012_KEYWORDS environment variable was added, which holds MOPAC2012 keywords and is read at the initial simulation step (Fig. 1).

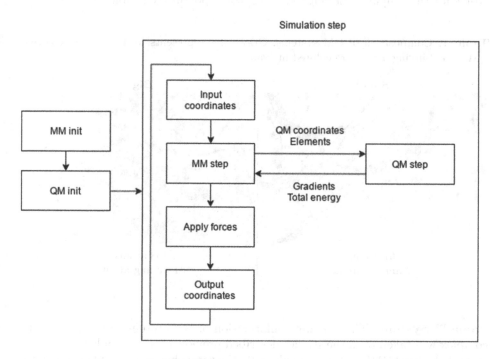

Fig. 1. Scheme of a generic QM/MM interface

2.3 Building Environment

Compilation of MOPAC2012 and GROMACS 5.0.7 into statically linked binaries was performed on local workstation (Intel(R) Core(TM) i7-6700K CPU @ 4.00 GHz, NVIDIA GTX1070, 32 Gb RAM, A-Data XPG SX8000NP 128 Gb SSD, Ubuntu 16.04.4 amd64). Due to the MOPAC dependencies from the Intel®MKL libraries, Intel®Composer suite 2015 update 1 packaged with MKL version 11.2 was used. We also compiled GROMACS versions with GPU support (NVIDIA®CUDA toolkit version 7.5 on the local system and version 8.0 on the "Lomonosov-2" supercomputer). Several GROMACS/ORCA and

GROMACS/MOPAC versions were prepared: single precision, single precision with GPU support and double precision (double precision with GPU is not supported). MPI support was disabled and OpenMP support was enabled in all versions. We used ORCA version 4.2.1 distributed as precompiled binaries.

2.4   Test Systems

Performance test of the QM/MM tools was done on two model systems: "small" - artificial enzyme with paraoxone substrate [10] and "big"—complex of butyril choline esterase with echotiophate. The QM subsystem size was 64 and 408 atoms for the "small" and "big" systems, correspondingly (Table 1).

Table 1. Composition of QM systems. Gray color represents MM part and colored QM part. Linking atoms are colored in cyan.

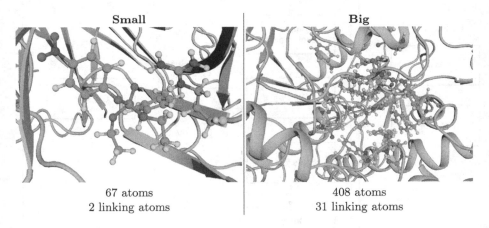

Small	Big
67 atoms	408 atoms
2 linking atoms	31 linking atoms

"Small" System. The starting conformation of the protein-ligand enzymatic complex was taken from the studies described previously [11]. Simulation system was filled with TIP3P water molecules [12], the total charge was neutralized with Na^+ or Cl^- ions. Water and ions were equilibrated around the protein-paraoxon complex with a 100-ps MD simulation with a restrained positions of protein and paraoxon atoms. For the MM subsystem we used the parameters from parm99 force field with corrections [13]. The QM subsystem was described with semi-empirical Hamiltonian PM3 [14] and consisted of paraoxone, Arg35 and Tyr37 side-chain atoms and the two closest water molecules.

"Big" System. Coordinates of esterase were taken from PDB ID 1LXW [15]. Missing residues V377, D378, D379, Q380 and C66 were added using PDB entry 2XMD [16] as a template. Protonation state was reconstructed according to table values with the **pdb2gmx** tool from the GROMACS package.

Molecular docking experiments were performed using Autodock Vina [17] to place the ligand in the active site of the enzyme. Preparation of the input PDBQT files and output processing were done with AutoDock Tools [18]. Initial echotiophate structure with partial Gasteiger charges was created using Avogadro software [19]. Docking cell contained whole protein with a margin of 5 from the edge atoms. The "exhaustiveness" parameter, which affects sampling, was set to 64 because of a moderate number of torsion angles. We performed 20 independent docking runs with freezed protein atoms remained and flexible ligand.

For further calculations one random configuration was selected with the distance between the echotiophate phosphorus atom and catalytic core less than 3.5.

Further equilibration was performed as for the previous system. Final QM system contained backbone parts of Tyr111-Phe115, full residues (with NH or CO parts for the terminal aminoacids) from Gly193-Gly197 loop, Met434-Ile439 loop, side chains of Gln220, Ser221, Asn319, Glu322, Tyr416, whole ligand Ech527 and 3 water molecules in the catalytic area.

Production Run. The prepared systems were subjected to QM/MM simulation with the modified GROMACS/ORCA or GROMACS/MOPAC2012 package [6,11]. The time step used was 0.2 fs. Temperature coupling with Nos-Hoover scheme allowed observation of the behavior of systems at human body temperature, 310 K. The total length of simulations was set to 100 steps and 10 independent replicas were calculated for each system.

QM Parameters. Following parameters were used for QM calculations: for GROMACS (common part for both MOPAC2012 and ORCA)

```
QMMM         = yes
QMMM-grps    = QM
QMMMscheme   = ONIOM
QMmethod     = RHF         ; required but ignored
QMbasis      = STO-3G      ; required but ignored
QMcharge     = XX
QMmult       = 1
```

for MOPAC2012:

```
PM3 1SCF GRADIENTS CHARGE=XX singlet THREADS=YY
```

for ORCA:

```
! RHF PM3 NOFROZENCORE CONV HUECKEL
%rel SOCType  1 end
%elprop Dipole false end
%scf MaxIter 5000 end
%output PrintLevel Nothing end
```

where charge and number of MOPAC2012 threads were altered depending on the test system.

3   Results and Discussions

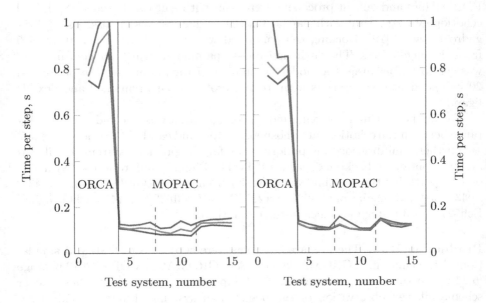

Fig. 2. Performance of QM/MM packages for the "small" system on local workstation (left) and supercomputer "Lomonosov-2" wit Lustre filesystem (right). Red line: maximum time per step value across ten replicas. Green: median time per step value across ten replicas. Blue: minimal time per step value across ten replicas. Description of systems is given in the Table 2 (Color figure online)

Computational Times. Hybrid QM/MM calculation of the "small" system with MOPAC library showed an eight-fold acceleration in comparison with the ORCA binary (Fig. 2, the test systems decoding is given in Table 2). The performance difference was mainly associated with the necessity of the data exchange between GROMACS and ORCA via file system; the QM subsystem energy gradients were calculated with a similar speed by MOPAC and ORCA tools. The use of double precision version of GROMACS slowed the calculation speed on the local workstation approximately 1.5 times, but the difference was completely leveled when the calculations were performed on the supercomputer. Surprisingly, the maximum performance was achieved on the local computer with a fast file system (NVMe 1.2). The use of threads for QM and MM subsystems gave a performance gain of 10%, the most significant effect was observed for the slow SCF (Self-Consistent-Field) steps on the "Lomonosov-2" supercomputer. This observation indicates the effectiveness of Intel libraries parallelization while getting the wavefunction to converge.

In comparison with "small" system we observed non-linear increase in computational time (Fig. 3, the test systems decoding is given in Table 3). While system size expanded in 6.5 times, computational time increased in 30 times

Table 2. Test system descriptions. Threads counts concerns to both MM and QM subsystems. Precision description describe Gromacs precision. MM GPU concerns to Gromacs acceleration of MM subsystem.

System number	Description
1	Orca single precision
2	Orca single precision and MM GPU
3	Orca double precision
4	Mopac single precision; 1 thread
5	Mopac single precision; 2 threads
6	Mopac single precision; 4 threads
7	Mopac single precision; 8 threads
8	Mopac single precision; 1 thread and MM GPU
9	Mopac single precision; 2 threads and MM GPU
10	Mopac single precision; 4 threads and MM GPU
11	Mopac single precision; 8 threads and MM GPU
12	Mopac double precision; 1 thread
13	Mopac double precision; 2 threads
14	Mopac double precision; 4 threads
15	Mopac double precision; 8 threads

Table 3. Test system descriptions. Threads counts concerns to both MM and QM susbsystems. Precision description describe Gromacs precision. MM GPU concerns to Gromacs acceleration of MM subsystem.

System number	Description
1	Orca single precision
2	Orca single precision and MM GPU
3	Orca double precision
4	Mopac single precision; 1 thread
5	Mopac single precision; 1 thread and MM GPU
6	Mopac double precision; 1 thread

for both ORCA and MOPAC2012 versions. Nevertheless internal performance ration between ORCA and MOPAC2012 remained the same.

Reproducibility. Because implementations of computational protocols in different packages can vary it's very important to verify reproducibility of results across different software. To verify that our interface is producing correct results we compared energy values between corresponding runs with MOPAC2012 and

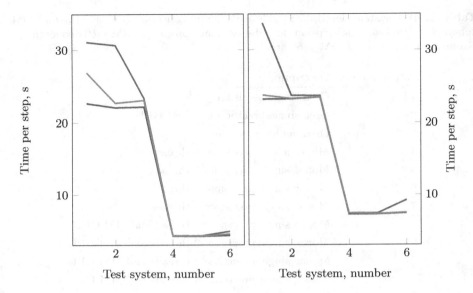

Fig. 3. Performance of test systems for the "big" system on local workstation (left) and supercomputer "Lomonosov-2" wit Lustre filesystem. Red line maximum time value across ten replicas. Green is median time values across ten replicas. Blue is minimal time values across ten replicas. Description of systems are given in the Table 3. (Color figure online)

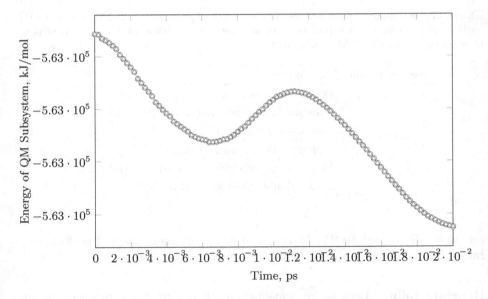

Fig. 4. Comparison of energy values for QM subsystem for MOPAC/Gromacs: red circles and ORCA/Gromacs: blue triangles. (Color figure online)

ORCA. Typical error is less than 0.1% at timescale of 100 steps (Fig. 4) and can be explained by difference in precision of data transfer (single in ORCA vs double for MOPAC2012).

4   Conclusions

We developed a new implementation of QM/MM interface between the MD program package GROMACS and the QM semi-empirical MOPAC tool that performs eight times faster than the original GROMACS interface to ORCA. Linking MOPAC as a GROMACS library allows to use the tool in supercomputer environment with Lustre distributed storage file system without input-output delays. The ability of MOPAC to inexpensively simulate molecular systems of large size allows efficient application of this tool to modern problems in life sciences. Our implementation of QM/MM interface provides a base for even better performance of hybrid simulations considering extensive ongoing development of GROMACS and MOPAC tools.

It pledged the way for combining Gromacs with next MOPAC2016 release with a better support of multithreading as well as GPU acceleration which will allow even to include into QM part even bigger regions which is very important for life science problems.

Acknowledgements. Authors are grateful to Dr. J. Stewart for sources of MOPAC2012 and initial hints on implementation. Computational experiments were carried out using the equipment of the shared research facilities of HPC computing resources at Lomonosov Moscow State University supported by the project RFMEFI62117X0011. The study was supported by the Russian Ministry of Education and Science grant RFMEFI57617X0095.

References

1. Warshel, A., Levitt, M.: Theoretical studies of enzymic reactions: dielectric, electrostatic and steric stabilization of the carbonium ion in the reaction of lysozyme. J. Mol. Biol. **103**(2), 227–249 (1976)
2. Vanatta, D.K., Shukla, D., Lawrenz, M., Pande, V.S.: A network of molecular switches controls the activation of the two-component response regulator NtrC. Nat. Commun. **6**, 7283 (2015)
3. Laio, A., Parrinello, M.: Escaping free-energy minima. Proc. Natl. Acad. Sci. U.S.A. **99**(20), 12562–12566 (2002)
4. Quhe, R., Nava, M., Tiwary, P., Parrinello, M.: Path integral metadynamics. J. Chem. Theory Comput. **11**(4), 1383–1388 (2015)
5. Bussi, G., Laio, A., Parrinello, M.: Equilibrium free energies from nonequilibrium metadynamics. Phys. Rev. Lett. **96**(9), 090601 (2006)
6. Schulz, R., et al.: GROMACS 4.5: a high-throughput and highly parallel open source molecular simulation toolkit. Bioinformatics **29**(7), 845–854 (2013)
7. Stewart, J.J.: Optimization of parameters for semiempirical methods VI: more modifications to the NDDO approximations and re-optimization of parameters. J. Mol. Model. **19**(1), 1–32 (2013)

8. Dapprich, S., Komromi, I., Suzie, B., Morokuma, K., Frisch, M.J.: A new ONIOM implementation in Gaussian98. part i. The calculation of energies, gradients, vibrational frequencies and electric field derivatives1dedicated to professor Keiji Morokuma in celebration of his 65th birthday.1. J. Mol. Struct. THEOCHEM **461-462**, 1–21 (1999)

9. Abraham, M.J., van der Spoel, D., Lindahl, E., Hess, B.: GROMACS User Manual version 5.0.7 (2015)

10. Smirnov, I., et al.: Reactibodies generated by kinetic selection couple chemical reactivity with favorable protein dynamics. Proc. Natl. Acad. Sci. U.S.A. **108**(38), 15954–15959 (2011)

11. Smirnov, I.V., et al.: Robotic QM/MM-driven maturation of antibody combining sites. Sci. Adv. **2**(10), e1501695 (2016)

12. Jorgensen, W.L., Chandrasekhar, J., Madura, J.D., Impey, R.W., Klein, M.L.: Comparison of simple potential functions for simulating liquid water. J. Chem. Phys. **79**(2), 926–935 (1983)

13. Lindorff-Larsen, K., et al.: Improved side-chain torsion potentials for the Amber ff99SB protein force field. Proteins **78**(8), 1950–1958 (2010)

14. Stewart, J.: Optimization of parameters for semiempirical methods i. method. J. Comput. Chem. **10**(2), 221–264 (1989)

15. Nachon, F., Asojo, O.A., Borgstahl, G.E., Masson, P., Lockridge, O.: Role of water in aging of human butyrylcholinesterase inhibited by echothiophate: the crystal structure suggests two alternative mechanisms of aging. Biochemistry **44**(4), 1154–1162 (2005)

16. Nachon, F., et al.: X-ray crystallographic snapshots of reaction intermediates in the G117H mutant of human butyrylcholinesterase, a nerve agent target engineered into a catalytic bioscavenger. Biochem. J. **434**(1), 73–82 (2011)

17. Trott, O., Olson, A.J.: AutoDock Vina: improving the speed and accuracy of docking with a new scoring function, efficient optimization, and multithreading. J. Comput. Chem. **31**(2), 455–461 (2010)

18. Morris, G.M., et al.: AutoDock4 and AutoDockTools4: automated docking with selective receptor flexibility. J. Comput. Chem. **30**(16), 2785–2791 (2009)

19. Hanwell, M.D., Curtis, D.E., Lonie, D.C., Vandermeersch, T., Zurek, E., Hutchison, G.R.: Avogadro: an advanced semantic chemical editor, visualization, and analysis platform. J. Cheminform. **4**(1), 17 (2012)

Orlando Tools: Energy Research Application Development Through Convergence of Grid and Cloud Computing

Alexander Feoktistov[1($\boxtimes$)], Sergei Gorsky[1], Ivan Sidorov[1],
Roman Kostromin[1], Alexei Edelev[2], and Lyudmila Massel[2]

[1] Matrosov Institute for System Dynamics and Control Theory of SB RAS,
Irkutsk, Russia
{agf,gorsky,ivan.sidorov,kostromin}@icc.ru
[2] Melentiev Energy Systems Institute of SB RAS, Irkutsk, Russia
{flower,massel}@isem.sei.irk.ru

Abstract. The paper addresses the relevant problem related to the development of scientific applications (applied software packages) to solve large-scale problems in heterogeneous distributed computing environments that can include various infrastructures (clusters, Grid systems, clouds) and provide their integrated use. We propose a new approach to the development of applications for such environments. It is based on the integration of conceptual and modular programming. The application development is implemented with a special framework named Orlando Tools. In comparison to the known tools, used for the development and execution of distributed applications in the current practice, Orlando Tools provides executing application jobs in the integrated environment of virtual machines that include both the dedicated and non-dedicated resources. The distributed computing efficiency is improved through the multi-agent management. Experiments of solving the large-scale practical problems of energy security research show the effectiveness of the developed application for solving the aforementioned problem in the environment that supports the hybrid computational model including Grid and cloud computing.

Keywords: Scientific application · Grid · Cloud · Energy research

1 Introduction

Nowadays, heterogeneous distributed computing environments often integrate resources of public access computing centers using various Grid computing models and cloud infrastructures. Such an integration causes new challenges for computation management systems in a process of solving large-scale problems in the environments. These challenges related to existing differences in the models of cloud and Grid computing, and conflicts between the preferences of resource owners and the quality criteria for solving environment user problems [1].

The multi-agent approach enables to significantly mitigate the above differences and conflicts through interactions of agents representing the resources of centers and clouds, as well as their owners and users. At the same time, coordinating agent actions

V. Voevodin and S. Sobolev (Eds.): RuSCDays 2018, CCIS 965, pp. 289–300, 2019.
https://doi.org/10.1007/978-3-030-05807-4_25

can significantly improve the quality of the computation management, especially in case market mechanisms for regulating supply and demand of resources are used [2].

Another direction for improving the distributed computing quality is the problem-orientation of the management systems [3]. Its importance is due to the need for the effective integrated use of heterogeneous resources of the environment in the process of solving common problems. We should take into account the problem specifics, matching of preferences of resource owners and the problem-solving criteria, and supporting of automatic decision-making in management systems [4, 5].

The effectiveness of the agents functioning directly depends on the knowledge they use [6]. In the known tools for multi-agent computation management [7], processes of elicitation and application of knowledge by agents remain an actual problem and require their development [8].

We propose a new approach to the creation and use of scalable applications that is based on the multi-agent management in the heterogeneous distributed computing environment integrating Grid and cloud computing models. Advantages of the proposed approach are demonstrated by example of an application for solving important large-scale problems of the energy research on the example of energy security field.

Within the proposed approach, the multi-agent system for the computation management is integrated with the toolkits that are used in the public access computer center "Irkutsk Supercomputer Center of the SB RAS" to create scalable applications based on the paradigms of parallel and distributed computing.

The rest of this paper is organized as follows. In the next Section, we give a brief overview of research results related to the convergence of Grid and cloud computing. Section 3 describes the proposed approach to the development and use of scalable applications in the heterogeneous distributed computing environment. Section 4 provides an example of the scalable application for solving the complex practical problem of vulnerability analysis of energy critical infrastructures in terms of energy security including experimental analysis. The last section concludes the paper.

2   Related Work

The scientific application (applied software package) is a complex of programs (modules) intended for solving the certain class of problems in the concrete subject domain. In such application, a computational process is described by problem-solving scheme (workflow). The rapid advancement of technologies for distributed computing has led to significant changes in the architecture of scientific applications [9]. They retained modular structure, but it became distributed.

For a long time, the well-known tools Globus Toolkit [10], HTCondor [11], BOINC [12] or X-COM [13] are used as middleware for distributed application execution. At the same time, workflow management systems are actively developed and applied for the same purpose. Nowadays, there are the systems Askalon, DAGMan Condor, Grid Ant, Grid, Flow, Karajan, Kepler, Pegasus, Taverna, Triana, etc. [14]. Often, workflow management systems and middleware are used together.

However, the possibilities of the above tools and systems often restrict potential opportunities of the modern scientific applications within the certain computational model [15]. In this regard, the research field related to the integration of various computational models becomes very topical.

Integration of the cloud and Grid computing models leads to the need to solve problems interface unification, application adaptation and their scalability, providing quality of service, mitigating uncertainty of different kinds, intellectualization of resource management system, monitoring heterogeneous resources, etc. A lot of scientific and practical results were provided to solve above problems the last ten years. We brief overview some of them.

Rings et al. [16] propose an approach to the convergence of the Grid and cloud technologies within the next generation network through the design of standards for interfaces and environments that supports multiple realizations of architectural components.

Kim et al. [17] study efficiency of the hybrid platform when changing resource requirements, quality of service and application adaptability.

Mateescu et al. [18] represent a hybrid architecture of high-performance computing infrastructure that provides predictable execution of scientific applications and scalability when a number of resources with different characteristics, owners, policies and geographic locations are changed.

A cloud application platform Aneka is implemented to provide resources of the various infrastructures, including clouds, clusters, Grids, and desktop Grids [19].

An example of the web-oriented platform that supports the use of different computational systems (PC, computational cluster, cloud) is represented in [20]. The subject domain specific of this platform is computational analyses of genomic data.

The Globus Genomics project provides cloud-hosted software service for the rapid analysis of biomedical data [21]. It enables to automate analysis of large genetic sequence datasets and hide the details of the Grid or cloud computing implementation.

Mariotti et al. [22] propose an approach to the integration of Grids and cloud resources using data base management system for deploying virtual machine (VM) images in cloud environments on requests from Grid applications.

Talia [6] analyses cloud computing and multi-agent systems. He shows that many improvements in distributed computing effectiveness can be obtained on the base of their integrated use. Among them, provisioning powerful, reliable, predictable and scalable computing infrastructure for multi-agent based applications, as well as making Cloud computing systems more adaptive, flexible, and autonomic for resource management, service provisioning and executing large-scale applications.

To this end, we provide a special framework named Orlando Tools for developing scalable applications and creating the heterogeneous distributed computing environment that can integrate Grid and cloud computing models. In addition, we apply a multi-agent system to improve the job flow management for the developed applications. Thus, in comparison with aforementioned projects, we ensure developing and using the joint computational environment included virtualized Grid and cloud resources under the multi-agent management.

3 Orlando Tools

Orlando Tools is the special framework for the development and use of scientific applications (applied software packages) in heterogeneous distributed computing environments. It includes the following main components:

- Web-interface supporting user access to other components of Orlando Tools,
- Model designer that is applied to specify knowledge about an application subject domain in both the text and graphic modes (in text mode, a knowledge specification is described in the XML terms),
- Knowledge base with information about the application modular structure (sets of applied and system modules), schemes of a subject domain study (parameters, operations and productions of a subject domain, and their relations), hardware and software infrastructure (characteristics of the nodes, communication channels, network devices, network topology, failures of software and hardware, etc.),
- Executive subsystem providing the problem-solving scheme schedulers and scheme interpreters that use the subject domain specification (computational model) for the distributed computing management at the application level,
- Computation database, which stores parameter values used in the problem-solving processes.

Orlando Tools provides an integration of the developed applications. The model designer enables application developers to use fragments of subject domain descriptions, software modules, input data and computation results of other applications in the process of creating a new application. Therefore, the time needed to develop applications and carry out experiments is reduced. Figure 1 shows an integration scheme of computational infrastructures into the joint environment with Orlando Tools.

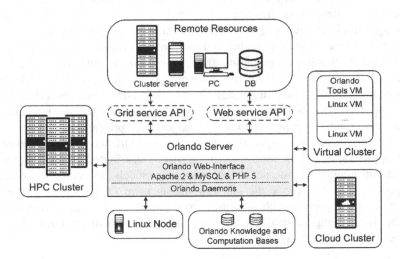

Fig. 1. Integration scheme of computational infrastructures into a heterogeneous distributed computing environment with Orlando Tools.

The integration of computational infrastructures is carried out through Orlando Server. It provides the Web-Interface and Daemons that implement functions of the executive subsystem in the automatic mode. Orlando Server is placed in the dedicated or non-dedicated computational nodes. The Server enables to include the following infrastructures into the integrated environment:

- HPC-clusters with the local resource manager system PBS Torque,
- Linux nodes (individual nodes with operating system Linux) that can be used to include non-portable (located in specialized nodes) software in an application,
- Virtual clusters that are created by using non-cloud resources with the special VMs of Orlando Tools in the images of which is placed PBS Torque (Fig. 2),
- Cloud clusters that are created using cloud resources with the Orlando Tools VMs,
- Remote resources that are included through the Grid or Web service API.

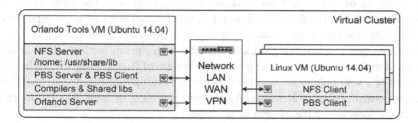

Fig. 2. Architecture of the virtual cluster

The images Orlando Tools VM and Linux VM are preconfigured and packed using Open Virtualization Format – Version 1.0 that provides the VM portability on different hardware and compatibility with various hypervisors.

Nodes have one of the hypervisors (Oracle VM VirtualBox, ESXi, XEN). The virtual cluster is created through placing VMs in the nodes united by LAN, WAN or VPN. NFS Server is used to provide shared access of Linux VM to DLL (/usr/share/lib), application programs and data (/home) on the network. An access to Orlando Tools is carried out by IP-address of Orlando Tools VM.

Applications that are developed in Orlando Tools generate job flows. The job describes the problem-solving process and includes information about the required computational resources, used programs, input and output data, communication network and other necessary data. Job flows are transferred to computational infrastructures that are included in the environment. The Orlando Tools scheduler decomposes job flows between the infrastructures taking into account the performance of their nodes relative to the problem-solving scheme. The node performance evaluation is obtained through the preliminary experiments with application modules. When job flows are transferred to the environment based on the resources of the Irkutsk Supercomputer Center, they are managed by multi-agent system. In this case, the distributed computing efficiency can be significantly improved [23]. The Orlando Tools architecture and model of application subject domains are considered in detail in [24].

4 Energy Research Application

4.1 Problem Formulation

One of the main energy security (ES) aspects is the ensuring conditions for the maximum satisfaction of consumers with energy resources in emergency situations. Investigation of this ES aspect requires the identification of critically important objects (CIO) in the energy sector in general or a particular energy system. CIO is a facility, partial or complete failure of which causes significant damage to the economy and society from the energy sector side.

Today, more than 90% of Russian natural gas is extracted in the Nadym-Pur-Tazovsky district of the Tyumen region. The distance between that district and the main natural gas consumption areas of Russia is 2–2.5 thousand km and the countries that import Russian natural gas are located 2–3 thousand km further. Thus, practically all Russian natural gas is transmitted for long distances through the system of pipelines. It has a number of mutual intersections and bridges, moreover, the pipes of essential gas pipelines are often laid near each other. Some intersections of main pipelines are extremely vital for the normal natural gas supply system operation.

The technique of identifying critical elements in technical infrastructure networks [25] is used to determine CIO of the gas supply system of Russia [26]. The criticality of an element or a set of elements is defined as the vulnerability of the system to failure in a specific element, or set of elements. An element is critical in itself if its failure causes severe consequences for the system as a whole.

Identifying critical elements is usually a simple task when only dealing with single failures. It becomes difficult when considering multiple simultaneous failures. A single element failure or multiple simultaneous element failures are referred to in [25] as failure sets. A failure set is a specific combination of failed elements and is characterized by a size, which is equal the number of elements that fail simultaneously. The investigation of failure sets with synergistic consequences is especially difficult because those consequences cannot be calculated by summarizing the consequences of the individual failures. For example, synergistic consequences of failure sets of size 3 cannot be determined by summarizing the consequences of failure subsets of size 2 and size 1. The number k of possible combinations of failed elements to investigate is defined by the following formula:

$$k = \frac{m!}{(m - n)!n!}, \tag{1}$$

where m is the number of system elements to fail and n is the size of the failure sets.

In formula (1), the number k increases rapidly in case n become greater. Evaluating all possible combinations of failures is practically impossible for a personal computer. Reducing k leads to losing important information about the system's vulnerability to failures. For example, the considering of only combinations of elements that are critical by themselves can miss "hidden" elements. The single failure of these elements causes little consequences for the system. At the same time, combining them with other elements results the failure sets with large synergistic consequences.

Another way to get rid of the rapid growth of k in the real energy systems with big m is applying high-performance computing for the analysis of system's vulnerability to failure sets (also without synergistic effects).

4.2   Application Subject Domain Structure

The process of organization of scientific research groups evolves towards virtual geographically distributed groups working on a project. It is necessary to ensure the availability of information and computing resources of the project for all its participants [27]. An energy research environment consists of computational, information and telecommunication infrastructures. The concept of creating an energy research environment is methodologically justified using the fractal stratified model (FS-model) of information space [28]. The FS-model allows mapping all available domain information into a set of interrelated layers that unite information objects, which have the same set of properties or characteristics. Each layer in turn can be stratified. FS-modeling represents an IT-technology as a set of information layers and their mappings. The IT-technology includes tools to describe information layers and facilities to support mappings from any layer to each.

Graphically, the FS-model of an energy research environment is represented as a set of nested spherical layers defined by the triple $\{S, F, G\}$ (Fig. 3), where S denotes set of layers, F is the set of mappings, and G is the set of invariants.

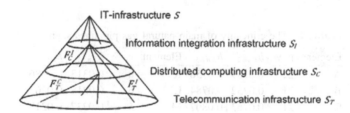

IT-infrastructure S

Information integration infrastructure S_I

Distributed computing infrastructure S_C

Telecommunication infrastructure S_T

Fig. 3. The FS-model of an energy research environment

An energy research environment S according to the FS-methodology is stratified into the information integration infrastructure S_I, distributed computing infrastructure S_C and telecommunications infrastructure S_T. S_I is layered into data and the metadata layers (S_{ID}, S_{IM}), S_C is layered into programs and their meta description layers (S_{CP}, S_{CM}). In the set F, there are the following mappings: $F_C^I : S_I \rightarrow S_C$, $F_T^I : S_I \rightarrow S_T$, $F_T^C : S_C \rightarrow S_T$. The invariants G denote energy research objectives detailed for each layer. The information models, data structure models, and ontologies are used to describe the meta-layers.

Orlando Tools is used for the mapping support when the energy research environment is created. Schemes of its knowledge and databases relate to S_I. The module structure reflects S_C. Information about the hardware and software infrastructure ensures to implement S_T. Relations between objects of the computational model provide F_C^I, F_T^I and F_T^C. Depending of the energy research type, for example, the FS-model for the vulnerability analysis of energy critical infrastructures [26] or the energy sector development investigation [24] can be described. Merging both of them into one constitutes the FS-model of ES research environment. Thus, the researcher creating different FS-models can build the variety of energy research environments on the basis of Orlando Tools using the same set of modules.

4.3 Computational Experiment

The modern gas supply system of Russia model consists of 378 nodes, including: 28 natural gas sources, 64 consumers (Russian Federation regions), 24 underground gas storages, 266 main compressor stations, and 486 arcs which represent main transmission pipelines segments and outlets to distribution networks.

In the first experiment 415 arcs were chosen as elements to fail and the failure sets of size 2 and 3 were considered. The experiment results are represented in Table 1, where v shows the total natural gas shortage in percentages, $h_{1,n=2}$ and $h_{2,n=3}$ express the contribution of a specific element's synergistic consequences to the total synergistic consequences for size n failure sets. According to the second column of Table 1, the single failure of elements from A to E leads to the significant gas shortage from 15 to 21% of the total system demand. Thus, they can be identified as CIO of the gas supply system of Russia. In opposite, elements from N to T are not CIO because the total natural gas shortage due to their failure is less than 5%.

Table 1. The criticality of main natural gas pipelines segments

Element	v, %	$h_{1,n=2}$	$h_{2,n=3}$	Element	v, %	$h_{1,n=2}$	$h_{2,n=3}$
A	21	0.0739	0.0756	K	8	0.0313	0.0386
B	21	0.0752	0.0784	L	6	0.0218	0.0263
C	21	0.0763	0.0807	M	5	0.0237	0.0318
D	15	0.0573	0.0614	N	4	0.0255	0.0348
E	15	0.0573	0.0614	O	4	0.0303	0.0411
F	9	0.0352	0.0424	P	4	0.0302	0.0410
G	9	0.0351	0.0423	Q	4	0.0161	0.0211
H	8	0.0327	0.0401	R	4	0.0161	0.0211
I	8	0.0354	0.0418	S	3	0.0226	0.0325
J	8	0.0369	0.0444	T	3	0.0227	0.0324

The measures $h_{1,n=2}$ and $h_{2,n=3}$ in Table 1 are used to prioritize preemptive efforts to reduce system-wide vulnerability. An element with large synergistic consequences would score higher on the preparedness activity than the other ones with the same total natural gas shortage value. For example, the element C is more preferable than A and B for the implementation of preparedness options. The obtained experimental results allowed forming the new recommendations to correct the importance of CIO that affects the budget of preparedness options and order of their implementation.

The experiment was performed in a heterogeneous distributed computing environment that is created through applying Orlando Tools and based on the resources of the Irkutsk Supercomputer Center [29]. We used the following node pools:

(1) 10 nodes with 2 processors Intel Xeon 5345 EM64T (4 core, 2.3 GHz, 8 GB of RAM) for each (non-dedicated resources),
(2) 10 nodes with 2 processors AMD Opteron 6276 (16 core, 2.3 GHz, 64 GB of RAM) for each (dedicated resources),
(3) 10 nodes with 2 processors Intel Xeon CPU X5670 (18 core, 2.1 GHz, 128 GB of RAM) for each (dedicated and non-dedicated resources).

Pool nodes have also various types of the interconnection (1 GigE, QDR Infiniband) and hard disks (HDD, SSD). All dedicated resources are virtualized.

4.4 Experimental Analysis

Table 2 shows the problem-solving time for $n \in \{2, 3, 4\}$. We demonstrate the following parameters: number l of possible failure sets, time $t_{k=1}$ of solving the problem on PC with the processor Intel Core i5-3450 (4 core, 3.10 GHz, 4 GB of RAM), time $t_{k=400}$ of solving the problem in the pools 1 and 2, and time $t_{k=760}$ of solving the problem in the pools 1–3, where k is the maximum number of available cores.

Table 2. Problem-solving time

n	l	$t_{k=1}$	$t_{k=400}$	$t_{k=760}$
2	85905	16	67	70
3	11826255	656640	7200	2730
4	1218104265	31536000	1555200	622080

We obtained the $t_{k=1}$, $t_{k=400}$ and $t_{k=760}$ values for $n = 2$ and $n = 3$ through the real experiments with Orlando Tools and then evaluated the $t_{k=1}$, $t_{k=400}$ and $t_{k=760}$ values for $n = 4$. It is obvious that the augment of n affects to the rapid rise of l thereby increasing the problem-solving time in many times.

The expediency of computing on PC is shown only for $n = 2$. In this case, the problem-solving time in the pools is the time obtained on PC due to the existence of overheads associated with the transfer of data between pool nodes. For $n = 3$, the problem-solving time in the pools is significantly lower the time shown on PC. When $n = 4$, problem-solving on PC is practically impossible.

Figure 4a and b show a number of the used cores and slots for solving the problems with $n = 3$ and $k = 400$ or $k = 760$. A job flow decomposition between the pools taking into account the performance of their nodes ensures the problem-solving time decrease about 6%.

In additional, we apply the special hypervisor shell for multi-agent management of jobs when dedicated and non-dedicated resources were used [25]. The management improvement is achieved by using the problem-oriented knowledge and information about software and hardware of the environment that are elicited in the process of the job classification, and parameter adjustment of agents. The hypervisor shell ensures to launch VMs in both the dedicated and non-dedicated resources within the framework of the joint virtualized environment. It also allows to use free slots in schedules of local resource manager systems of non-dedicated resources. We evaluate the problem-solving time decrease over 9% due to the hypervisor shell applying. Thus, the job flows decomposition and hypervisor shell use provide the significant problem-solving time decrease over 15%.

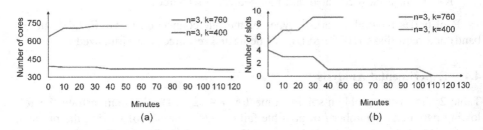

Fig. 4. Number of the used cores (a) and slots (b)

Figure 5a demonstrates the high average CPU load. It is a bit less in the pools 1 and 3 due to overheads of virtualization. At the same time, the virtualization overheads in nodes of all pools were less than 5%. Figure 5b shows the improvement (decrease) in the problem-solving time with $n = 3$ in both $k = 400$ and $k = 760$.

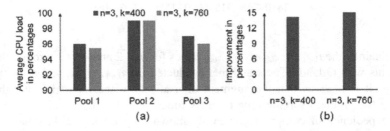

Fig. 5. Average CPU load in pools (a) and the improvement of problem-solving time (b)

5   Conclusions

Nowadays, Grid technologies continue to play important role in the development of scientific computing environments. At the same time, clouds are quickly evolving. The area of their applying constantly expands and often intersects with the field of Grid computing. In this regard, there is a want to use benefits both Grid and cloud computing.

To this end, we propose a new approach to the development and use of scalable applications, and create the special framework named Orlando Tools to support it.

Orlando Tools provides the opportunity to not only develop applications, but also include various computational infrastructures (individual nodes, clusters, Grids and clouds) in the heterogeneous distributed computing environment and share their possibilities and advantages in the problem-solving process. Thus, the capabilities of developed applications can be supported by the needed part of the environment.

We demonstrate the benefits of the proposed approach by example of an application for solving important large-scale problems of research on energy security. A problem solution enables to clarify the previous study results. The experimental analysis shows the advantages and drawback in the use of different infrastructures, and improvement of distributed computing efficiency under the multi-agent management.

Acknowledgements. The work was partially supported by Russian Foundation for Basic Research (RFBR), projects no. 16-07-00931, and Presidium RAS, program no. 27, project "Methods and tools for solving hard-search problems with supercomputers". Part of the work was supported by the basic research program of the SB RAS, project no. III.17.5.1.

References

1. Toporkov, V., Yemelyanov, D., Toporkova, A.: Anticipation scheduling in grid with stakeholders preferences. Commun. Comput. Inf. Sci. **793**, 482–493 (2017)
2. Sokolinsky, L.B., Shamakina, A.V.: Methods of resource management in problem-oriented computing environment. Program. Comput. Soft. **42**(1), 17–26 (2016)
3. Singh, A., Juneja, D., Malhotra, M.: A novel agent based autonomous and service composition framework for cost optimization of resource provisioning in cloud computing. J. King Saud Univ. Comput. Inf. Sci. **29**(1), 19–28 (2015)
4. Singh, S., Chana, I.: QoS-aware autonomic resource management in cloud computing: a systematic review. ACM Comput. Surv. **48**(3) (2016). Article no. 42
5. Qin, H., Zhu, L.: Subject oriented autonomic cloud data center networks model. J. Data Anal. Inform. Process. **5**(3), 87–95 (2017)
6. Talia, D.: Cloud computing and software agents: towards cloud intelligent services. In: Proceedings of the 12th Workshop on Objects and Agents. CEUR Workshop Proceedings, vol. 741, pp. 2–6 (2011)
7. Madni, S.H.H., Latiff, M.S.A., Coulibaly, Y.: Recent advancements in resource allocation techniques for cloud computing environment: a systematic review. Cluster Comput. **20**(3), 2489–2533 (2017)
8. Farahani, A., Nazemi, E., Cabri, G., Capodieci, N.: Enabling autonomic computing support for the JADE agent platform. Scalable Comput. Pract. Exp. **18**(1), 91–103 (2017)
9. Ilin, V.P., Skopin, I.N.: Computational programming technologies. Program. Comput. Soft. **37**(4), 210–222 (2011)
10. Foster, I., Kesselman, C.: Globus: a metacomputing infrastructure toolkit. Int. J. High Perform. Comput. Appl. **11**(2), 115–128 (1997)
11. Couvares, P., Kosar, T., Roy, A., Weber, J., Wenger, K.: Workflow management in condor. In: Taylor, I.J., Deelman, E., Gannon, D.B., Shields, M. (eds.) Workflows for e-Science, pp. 357–375. Springer, London (2007). https://doi.org/10.1007/978-1-84628-757-2_22
12. Anderson, D.: Boinc: a system for public-resource computing and storage. In: Buyya, R. (ed.) Proceedings of the 5th IEEE/ACM International Workshop on Grid Computing, pp. 4–10. IEEE (2004)

13. Voevodin, V.V.: The solution of large problems in distributed computational media. Automat. Remote Control **68**(5), 773–786 (2007)
14. Talia, D.: Workflow systems for science: concepts and tools. ISRN Soft. Eng. **2013** (2013). Article ID 404525
15. Tao, J., Kolodziej, J., Ranjan, R., Jayaraman, P., Buyya, R.: A note on new trends in data-aware scheduling and resource provisioning in modern HPC systems. Future Gener. Comput. Syst. **51**(C), 45–46 (2015)
16. Rings, T., et al.: Grid and cloud computing: opportunities for integration with the next generation network. J. Grid Comput. **7**(3) (2009). Article no. 375
17. Kim, H., el-Khamra, Y., Jha, S., Parashar, M.: Exploring application and infrastructure adaptation on hybrid grid-cloud infrastructure. In: Proceedings of the 19th ACM International Symposium on High Performance Distributed Computing, pp. 402–412. ACM (2010)
18. Mateescu, G., Gentzsch, W., Ribbens, C.J.: Hybrid computing – where HPC meets grid and cloud computing. Future Gener. Comput. Syst. **27**(5), 440–453 (2011)
19. Vecchiola, C., Calheiros, R.N., Karunamoorthy, D., Buyya, R.: Deadline-driven provisioning of resources for scientific applications in hybrid clouds with Aneka. Future Gener. Comput. Syst. **28**(1), 58–65 (2012)
20. Goecks, J., Nekrutenko, A., Taylor, J.: Galaxy: a comprehensive approach for supporting accessible, reproducible, and transparent computational research in the life sciences. Genome Biol. **11**(8), 1–13 (2010)
21. Allen, B., et al.: Globus: a case study in software as a service for scientists. In: Proceedings of the 8th Workshop on Scientific Cloud Computing, pp. 25–32. ACM (2017)
22. Mariotti, M., Gervasi, O., Vella, F., Cuzzocrea, A., Costantini, A.: Strategies and systems towards grids and clouds integration: a DBMS-based solution. Future Gener. Comput. Syst. (2017). https://doi.org/10.1016/j.future.2017.02.047
23. Feoktistov, A., Sidorov, I., Sergeev, V., Kostromin, R., Bogdanova, V.: Virtualization of heterogeneous HPC-clusters based on OpenStack platform. Bull. S. Ural State Univ. Ser. Comput. Math. Soft. Eng. **6**(2), 37–48 (2017)
24. Edelev, A., Zorkaltsev, V., Gorsky, S., Doan, V.B., Nguyen, H.N.: The combinatorial modelling approach to study sustainable energy development of Vietnam. Commun. Comput. Inf. Sci. **793**, 207–218 (2017)
25. Jonsson, H., Johansson, J., Johansson, H.: Identifying critical components in technical infrastructure networks. Proc. Inst. Mech. Eng. O J. Risk Reliab. **222**(2), 235–243 (2008)
26. Vorobev, S., Edelev, A.: Analysis of the importance of critical objects of the gas industry with the method of determining critical elements in networks of technical infrastructures. In: 10th International Conference on Management of Large-Scale System Development (MLSD). IEEE (2017). https://doi.org/10.1109/mlsd.2017.8109707
27. Massel, L.V., Kopaygorodsky, A.N., Chernousov, A.V.: IT-infrastructure of research activities realized for the power engineering system researches. In: 10th International Conference on Computer Science and Information Technologies, pp. 106–111. Ufa State Aviation Technical University (2008)
28. Massel, L.V., Arshinsky, V.L., Massel, A.G.: Intelligent computing on the basis of cognitive and event modeling and its application in energy security studies. Int. J. Energy Optim. Eng. **3**(1), 83–91 (2014)
29. Irkutsk Supercomputer Center of SB RAS. http://hpc.icc.ru. Accessed 13 Apr 2018

Parallel FDTD Solver with Static and Dynamic Load Balancing

Gleb Balykov[✉]

Lomonosov Moscow State University, Moscow 119991, Russia
balykov.gleb@yandex.ru

Abstract. Finite-difference time-domain method (FDTD) is widely used for modeling of computational electrodynamics by numerically solving Maxwell's equations and finding approximate solution at each time step. Overall computational time of FDTD solvers could become significant when large numerical grids are used. Parallel FDTD solvers usually help with reduction of overall computational time, however, the problem of load balancing arises on parallel computational systems. Load balancing of FDTD algorithm for homogeneous computational systems could be performed statically, before computations. In this article static and dynamic load balancing of FDTD algorithm for heterogeneous computational systems is described. Dynamic load balancing allows to redistribute grid points between computational nodes and effectively manage computational resources during process of computations for arbitrary computational system. Dynamic load balancing could be turned into static, if data required for balancing was gathered during previous computations. Measurements for presented algorithms are provided for IBM Blue Gene/P supercomputer and Tesla CMC server. Further directions for optimizations are also discussed.

Keywords: Computational electrodynamics · FDTD
Parallel FDTD · MPI

1 Overview

The FDTD method, originated in 1966 [1], is widely used in electrodynamics solvers. Since then, sequential FDTD solvers evolved to high-performance parallel solvers, which incorporate different parallelization techniques.

Load balancing in parallel FDTD algorithms could significantly impact overall performance, as it allows to efficiently distribute load across all computational nodes of computational system based on their performance and current parameters of computation [2,3]. For homogeneous computational systems load balancing could be performed before the start of computation, which is described in our previous work [4]. In general, for heterogeneous computational systems, characteristics of each computational node should be taken into account, because all computational nodes may have different (in general, arbitrary) performance and share time between each other.

© Springer Nature Switzerland AG 2019
V. Voevodin and S. Sobolev (Eds.): RuSCDays 2018, CCIS 965, pp. 301–313, 2019.
https://doi.org/10.1007/978-3-030-05807-4_26

Different load distributions across computational nodes of the computational system could be chosen as the most efficient based on these performance parameters. For example, it could be more efficient to assign different number of grid points to different computational nodes, or it could be more efficient to disable some computational nodes in order to exclude heavy share operations. One example of such computational systems could be the computational system with GPU only on a single computational node, performance of which is higher than the performance of other computational nodes.

This also applies to the current parameters of computations, as they could also lead to different load distributions. One example could be the case, when with specified number of grid points it is more efficient to perform computations sequentially on a single computational node rather than spread computations across all computational nodes.

One group of approaches for dynamic load balancing for heterogeneous computational systems is based on measurements of execution time of each computational node, after which migration of tasks from one node to another is performed [5,6]. This approach should be updated with specifics of algorithm it is applied to in order to achieve the best balancing properties. Also, migration could be a heavy operation, especially when data is transferred between computational nodes with low bandwidth.

Another popular approach for dynamic load balancing is "work stealing" [7,8]. In the core of this and similar approaches lies queue of tasks, which are to be performed. Each computational node takes element of this queue, performs computations on it, and then takes the next element from queue if it's not empty. This will balance computations across all computational nodes on granularity of the single element of queue, because faster node will finish its computations faster than slower one, i.e. faster node will take the next element of queue. Queues could exist for each computational node separately, then, faster computational node can "steal" element from queue of the slower one.

Work stealing approach should be used cautiously, because choice of queue element could significantly impact overhead and balancing abilities of algorithm. Even for systems with shared memory incorrectly chosen granularity could lead to non-optimal balancing, for example, when there are less queue elements than overall number of computational nodes. For systems with distributed memory this problem is combined with the problem of data locality, because for many numerical algorithms data from neighboring grid points is required. This means that faster computational node may be unable to steal anything from the slower node without performance degradation.

In this article, parallel FDTD solver with static and dynamic load balancing for heterogeneous computational systems is introduced [9], which incorporates dynamic balancing of computations between computational nodes for different dimensions and different virtual topologies. Characteristics of each computational node are identified dynamically, during computations, and load balancing is performed based on them. This dynamic data could be saved to disk and then used for static balancing on this same computational system even for different computation parameters. Presented algorithm allows to preserve good data locality over all computational nodes, because it takes into account properties of FDTD algorithm. Besides, dynamic balancing overhead can be removed fully if only static part of balancing is performed.

2 Parallel Algorithm Description

Electrodynamics modeling could be performed in one-dimensional (1D), two-dimensional (2D) or three-dimensional (3D) modes. For all dimensions Cartesian computational grid is introduced: Ox axis for $1D$ mode, Ox and Oy axes for $2D$ mode, Ox, Oy and Oz axes for $3D$ mode. Yee grid [10] for field components is then set, and all points of Yee grid are spread between all computational nodes, so that each point of Yee grid is assigned to some computational node. For sequential solver all Yee grid points remain on the single computational node.

Let N be the number of computational nodes used in computations. Let's consider it being specified by user of the solver at start of computations. Let S be the total number of Yee grid points and let T be the total number of time steps, which are also specified at start of computations by user.

Yee grid points are assigned to different computational nodes in a natural way: Yee grid is divided in rectangular chunks, which are then assigned to different computational nodes. All N computational nodes are considered to have buffers, which store data from the neighboring computational nodes. This computational nodes' layout could be mapped directly on MPI virtual topology with single MPI process launched on every computational node.

Each time step each computational node performs computations for Yee grid points from the chunk assigned to it and then performs share operations with the neighboring computational nodes. Let's consider only buffers of size 1 by the axis for which each buffer is defined, then, no additional computations are performed for buffer points.

Overall computational time τ_{total} is sum of computational time and share time for each time step τ_{total}^t, and only the maximum time for each time step is taken into account. As described in [4], share operations are performed in all available directions sequentially, one after another (for example, for 2D case there are 8 directions). τ_{calc}^t is the computational time, τ_{share}^t is the sum of maximum share times for all directions on time step t.

$$\tau_{total} = \sum_{t=0}^{T-1} \tau_{total}^t = \sum_{t=0}^{T-1} (\tau_{calc}^t + \tau_{share}^t) \tag{1}$$

$$\tau_{total} = \sum_{t=0}^{T-1} (\max_{i=0}^{N-1} \tau_{calc_i}^t + \sum_{dir} \max_{i,j \in dir} \tau_{share_{i,j}}^t) \tag{2}$$

As also stated in [4], there are several possible virtual topologies for each dimension. Each topology has its own set of directions, and both computational time and share time should be minimized in order to minimize total execution time τ_{total}.

Let S_i^t be the number of Yee grid points assigned to i computational node at time step t, $i = 0, .., N - 1$, $t \geq 0$. Number of grid points assigned to i computational node may change as computations proceed. Process of load balancing, or grid balancing, distributes computations between computational nodes based on their performance parameters.

$$S = \sum_{i=0}^{N-1} S_i^t, \forall t \geq 0 \qquad (3)$$

Let $U_{i,j}^t$ be the number of grid points that are shared between i and j computational nodes at time step t, $i, j = 0, .., N - 1$, $t \geq 0$. If computational nodes i and j do not performed any share operations between each other at time step t, $U_{i,j}^t = 0$. Besides, $S_i^t = 0$ for nodes which are disabled on time step t.

Let $state_i^t$ be the state of i computational node on time step t. In case computational node i is disabled on time step t, $state_i^t = 0$, otherwise $state_i^t = 1$. There should exist at least one computational node j for each time step t, for which $state_j^t = 1$.

Let $perf_i^t$ be the performance of i computational node right before the time step t, in other words, number of grid points, on which operations were performed, divided by elapsed time. In order to increase accuracy of balancing, let's calculate $perf_i^t$ as average values for all previous time steps. In this case, the longer computations are running, the more accurate values of performances are obtained.

Let T_{perf} be the number of time steps, after which performance values are updated. Then, right before time step $t = l * T_{perf}$, $l = 0, 1, .., T/T_{perf}$:

$$perf_i^{l*T_{perf}} = \frac{T_{perf} * \sum_{v=0}^{l-1} S_i^{v*T_{perf}}}{\sum_{v=0}^{l*T_{perf}-1} \tau_{calc_i}^v} \qquad (4)$$

$$S_i^{v*T_{perf}} = S_i^{v*T_{perf}+k}, U_{i,j}^{v*T_{perf}} = U_{i,j}^{v*T_{perf}+k} \qquad (5)$$

where $k = 1, .., T_{perf} - 1$, $v = 0, .., l - 1$. $\tau_{calc_i}^v = 0$ if computational node i is disabled on time step v. If no previously computed data is loaded from file, $perf_i^0 = 0$, and computations are spread evenly between all computational nodes. Computational overhead of balancing could be minimized by tuning values of parameters T_{perf} based on the values of T and S.

2.1   1D Case

Let $A > 0$ be the size of Yee grid by Ox axis, with a_i^t grid points assigned to i computational node on time step t: $S = A$, $S_i^t = a_i^t$, $i = 0, .., n - 1$, $t = 0, .., T$.

Let n be the number of computational nodes, between which Ox axis is spread. For one-dimensional case $N = n$.

Let $x_{L_i}^t$ and $x_{R_i}^t$ be the start and end coordinates of chunk assigned to i computational node on time step t, $x_{L_0}^t = 0$, $x_{R_i}^t = x_{L_{i+1}}^t$, $x_{R_{n-1}}^t = A$, $x_{R_i}^t - x_{L_i}^t = a_i^t$.

In one-dimensional case all computational nodes have same buffer sizes buf, which means that no matter how computations are spread between computational nodes, buffer sizes remain the same. Then, if computational nodes i and j do not perform share operations at time step t, $U_{i,j}^t = 0$, otherwise, $U_{i,j}^t = buf$.

Share time between i and j computational nodes could be estimated using Hockney [11] communication model:

$$\tau_h(i, j, x) = latency_{i,j} + \frac{x}{bandwidth_{i,j}} \qquad (6)$$

where x - is the number of grid points, which are shared between i and j computational nodes, $latency_{i,j}$ is the time required to setup the sharing procedure between i and j computational nodes, $bandwidth_{i,j}$ is the speed of communication between i and j computational nodes.

Let $\tau_{s_{i,j}}^t(x)$ be the measured average share time between i and j computational nodes for x grid points for time steps $t - T_{perf}, .., t - 1$. For each number of grid points x there could be different amount of performed share iterations $D_{i,j}^t(x)$, by which the average value is taken. For in-process measurements $D_{i,j}^t(U_{i,j}^{t-T_{perf}}) = T_{perf}$, for additional measurements $D_{i,j}^t(x)$ could vary based on the overhead of such measurements.

Then, for in-process measurements, i.e. for measurements, which were taken during computations:

$$\tau_{s_{i,j}}^t(U_{i,j}^{t-T_{perf}}) = \frac{\sum_{v=t-T_{perf}}^{t-1} \tau_{share_{i,j}}^v}{D_{i,j}^t(U_{i,j}^{t-T_{perf}})} \qquad (7)$$

Latency could be obtained by performing additional measurements of share time $\tau_{s_{i,j}}^t(0)$ for empty messages or share time $\tau_{s_{i,j}}^t(x)$ for sizes $x \neq U_{i,j}^{t-T_{perf}}$. Another option is to choose $\tau_{s_{i,j}}^t(1)$ as latency, as $1 << bandwidth_{i,j}$.

Linear regression for all the measurements $\tau_{s_{i,j}}^t(x)$ is used to determine latency and bandwidth. Also, in order to increase accuracy, latency and bandwidth are calculated as average values for all time steps. Share time $\tau_{share_{i,j}}^t$ for the next time step t then could be estimated using Hockney model.

Spreading Computations. In case $perf_i^0$, $\forall i$ are equal to 0, computations are spread evenly between all computational nodes. In case $perf_i^0$, $\forall i$ are somehow defined, grid points are spread proportionally to performances of computational nodes. Let α_i^t be the ideal number of grid points, which could be assigned to i computational node on t time step, based on performance of computational node, $\alpha_i^t \in R$, $\alpha_i^t \geq 0$. α_i^t could be non-integer, in this case integer value to assign to a_i^t has to be calculated.

Let $I_{L_i}^t$ and $I_{R_i}^t$ be the start and end coordinates of ideal chunk, which could be assigned to i computational node on time step t, $I_{L_0}^t = 0$, $I_{R_i}^t = I_{L_{i+1}}^t$, $I_{R_{n-1}}^t = A$, $I_{L_i}^t, I_{R_i}^t \in R$. Ideal number of grid points α_i^t is calculated in the next manner:

$$\alpha_i^t = A * \frac{perf_i^t * state_i^t}{\sum_{v=0}^{n-1} perf_v^t * state_v^t}, i = 0, .., n-1, t \geq 0 \tag{8}$$

The goal is to choose a_i^t having the lowest deviations from ideal values α_i^t. For computational node $i = 0$ left border is always the same. As calculation time on i computational node is proportional to the number of grid points assigned to it, there are only two candidates for $x_{R_0}^t$: $[I_{R_0}^t]$ and $[I_{R_0}^t] + 1$, because they have the lowest deviation from the ideal value $I_{R_0}^t$.

Let $\delta_{L_i}^t$ be the deviation of $x_{L_i}^t$ from the ideal left border value $I_{L_i}^t$, same for $\delta_{R_i}^t$ and $I_{R_i}^t$: $x_{g_i}^t = I_{g_i}^t - \delta_{g_i}^t, g \in \{L, R\}, i = 0, .., n-1$.

Also let $\delta_i^t = I_{R_i}^t - [I_{R_i}^t]$. Algorithm consists of the following steps:

1. Update $\delta_{L_i}^t$: $\delta_{L_0}^t = 0$, for other computational nodes: $\delta_{L_i}^t = \delta_{R_{i-1}}^t$.
2. Update $\delta_{R_i}^t$ for computational nodes $i = 0, .., n-1$: $\delta_{R_{n-1}}^t = 0$
   (a) $\delta_{R_i}^t = \delta_i^t$, if $|\delta_{L_i}^t - \delta_i^t| \leq |\delta_{L_i}^t + 1 - \delta_i^t|$
   (b) $\delta_{R_i}^t = \delta_i^t - 1$, otherwise
3. Update $x_{L_i}^t$, $x_{R_i}^t$, a_i^t for computational nodes $i = 0, .., n-1$ according to the formulas above. i computational node is disabled if $a_i^t = 0$.

Both directions (from start to end and from end to start) should be checked and the minimum between them should be chosen. This algorithm allows us to choose integer values of a_i^t, which are the closest to the ideal values α_i^t, and these values lead to the lowest τ_{calc}^t.

Disabling Computational Nodes. In some cases it could be more efficient to disable some computational nodes at time step t in order to decrease overall computational time by decreasing τ_{share}^t. For 1D case there are two possible directions (positive and negative across Ox axis), and total share time is next:

$$\tau_{share}^t = 2 * \max_{i=0}^{N-2} \tau_{share_{i,i+1}}^t \tag{9}$$

Let's consider connections between computational nodes across Ox axis, starting from the connection with the highest share time between i and $i+1$ computational nodes. $perf_{all}^t$ is performance of all enabled computational nodes at time step t, $perf_R^t = perf_{all}^t - perf_L^t$ and

$$perf_L^t = \sum_{j=0}^{i} perf_j^t \tag{10}$$

There are next possibilities.

1. Disable all computational nodes to the left or to the right from the border between i and $i+1$ computational nodes, if computational time on this reduced set of computational nodes is lower. The next condition checks if total computational time on computational nodes $0, .., i$ is less than total computational time on $0, .., N-1$ (i.e. condition for disabling nodes $i+1, .., N-1$):

$$\tau_{calc_L}^t + \tau_{share_L}^t < \tau_{calc_{all}}^t + \tau_{share_{all}}^t - \varepsilon \tag{11}$$

As the border with the highest share time is considered, $\tau_{share_{all}}^t > \tau_{share_L}^t$. Similar condition could be written for disabling computational nodes $0, .., i$. Accuracy ε is the parameter, which helps to deal with inaccuracies of computations of performance parameters. In case Yee grid points are spread between all computational nodes proportionally to performance, computational and share time could be estimated in the next manner:

$$\tau_{calc_L}^t \approx \frac{S}{per f_L^t} \tag{12}$$

$$\tau_{share_L}^t = 2 * \max_{j=0}^{i-1} \tau_{share_{j,j+1}}^t \approx 2 * \max_{j=0}^{i-1}(latency_{j,j+1}^t + \frac{U_{j,j+1}^t}{bandwidth_{j,j+1}^t}) \tag{13}$$

2. Disable K computational nodes to the left or to the right from the border between i and $i+1$ computational nodes, in case computational time on these reduced sets of computational nodes is lower.

For this case additional measurements have to be performed for share operations, which have not been performed yet, otherwise, share time can't be estimated.

Let's check the case of disabling K computational nodes to the left from the border between i and $i+1$ computational nodes. There are $i+1$ cases for K from 1 to $i+1$, i.e. cases for disabling computational nodes from $i-K+1$ to i. From all values of K the one with the smallest computational time is chosen.

$$per f_1^t = \sum_{j=0}^{i-K} per f_j^t + \sum_{j=i+1}^{N-1} per f_j^t \tag{14}$$

The condition for disabling computational nodes from $i-K+1$ to i is next:

$$\tau_{calc_1}^t + \tau_{share_1}^t < \tau_{calc_{all}}^t + \tau_{share_{all}}^t - \varepsilon \tag{15}$$

Computational and share time could be estimated as:

$$\tau_{calc_1}^t \approx \frac{S}{per f_1^t} \tag{16}$$

$$\tau_{share_1}^t = 2 * \max(\max_{j=0}^{i-K} \tau_{share_{j,j+1}}^t, \max_{j=i+1}^{N-1} \tau_{share_{j,j-1}}^t) \tag{17}$$

3. Disable all computational nodes, except the one w with the highest performance $perf_w^t$. This case could be checked once per rebalance. The disabling condition is similar to previous cases considering 0 share time.

After all cases are checked, the resulting balancing is chosen, so that the minimum computational time for all the cases is reached:

$$\tau_{total}^t = \min(\min_K \tau_{total_1}^t, \min_K \tau_{total_2}^t, \tau_{total_L}^t, \tau_{total_R}^t, \tau_{total_w}^t) \qquad (18)$$

This process could be continued for the next highest share time between i and $i+1$ computational nodes, until the conditions from all cases are no longer satisfied or until all the pairs of neighbors are checked, or there is nothing else to disable.

Enabling Computational Nodes. As performance parameters of computational nodes are measured dynamically, inaccuracies arise. This leads to the case, when some computational nodes were disabled, performance parameters were updated, and now it is more efficient to enable them. Partially, ε should help with this. Other thing that should help to reduce inaccuracies is the accumulated average values of performance parameters, so the further computations continue, the more accurate values are obtained.

The conditions for enabling computational nodes are similar to the condition for disabling, except that they are reversed:

$$\tau_{total_L}^t > \tau_{total_{all}}^t + \varepsilon \qquad (19)$$

Other conditions could be written in the same way. It would be inefficient to check all combinations of computational nodes for each joint set of disabled nodes. Solution is to enable computational nodes in the same sets, as they were disabled, which will significantly reduce the number of possible combinations.

So, each T_{perf} time steps conditions for disabling and enabling computational nodes should be checked.

2.2   2D and 3D Cases

Same logic could be applied to higher dimensions in case only one axis is spread between computational nodes, i.e. for $2D - X$, $2D - Y$, $3D - X$, $3D - Y$, $3D - Z$ virtual topologies. All changes in formulas are related to the size of chunks assigned to computational nodes, as they are two- or three-dimensional for $2D$ and $3D$ modes correspondingly.

For $2D - XY$, $3D - XY$, $3D - YZ$, $3D - XZ$ and $3D - XYZ$ virtual topologies similar algorithms with minor changes could be applied. The most significant change is that nodes can't be disabled or enabled separately from the line or plane for $2D$ and $3D$ modes correspondingly.

Let's consider $2D$ case with $2D - XY$ virtual topology. Let $S = A * B$, where A and B are the sizes by Ox and Oy axes correspondingly, $N = n * m$, where n

and m are the sizes of nodes' grid by Ox and Oy axes correspondingly. Both A and B are defined same to $1D$ mode.

$$S = \sum_{k=0}^{N-1} S_k^t = \sum_{i=0}^{n-1} \sum_{j=0}^{m-1} S_{(i,j)}^t = \sum_{i=0}^{n-1} \sum_{j=0}^{m-1} a_i^t * b_j^t, \forall t \geq 0 \tag{20}$$

In the same way to $1D$ case $U_{(i,j),(k,l)}^t$ and performance parameters $perf_{(i,j)}^t$, $latency_{(i,j),(k,l)}^t$, $bandwidth_{(i,j),(k,l)}^t$ are setup.

Computations could be spread for each axis independently by the same algorithm as discussed in the previous sections, except that $perf_i^t$ now is the performance of the whole line, for example:

$$perf_i^t = \sum_{j=0}^{m-1} perf_{(i,j)}^t \tag{21}$$

Another difference arises in the disabling and enabling conditions of computational nodes. Let the latencies for share operations between i and $i+1$ columns be the next:

$$latency_{(i,all),(i+1,all)}^t = \frac{\sum_{j=0}^{m-1} latency_{(i,j),(i+1,j)}}{m} \tag{22}$$

$$latency_{(i,all),(i+1,all+1)}^t = \frac{\sum_{j=0}^{m-2} latency_{(i,j),(i+1,j+1)}}{m-1} \tag{23}$$

$$latency_{(i,all),(i+1,all-1)}^t = \frac{\sum_{j=1}^{m-1} latency_{(i,j),(i+1,j-1)}}{m-1} \tag{24}$$

In the same way bandwidth is setup. Share time between i and $i+1$ columns could be estimated using Hockney model:

$$U_{(i,all),(i+1,all)}^t = \sum_{j=0}^{m-1} U_{(i,j),(i+1,j)} \tag{25}$$

$$\tau_{share(i,all),(i+1,all)}^t \approx latency_{(i,all),(i+1,all)}^t + \frac{U_{(i,all),(i+1,all)}^t}{bandwidth_{(i,all),(i+1,all)}^t} \tag{26}$$

All connections across Ox and Oy axes are considered in sorted order, descending by the values of share time across that axis and across diagonals. For border between i and $i+1$ columns across Ox axis this value is next:

$$\tau_{share(i,all),(i+1,all)}^t + \tau_{share(i,all),(i+1,all+1)}^t + \tau_{share(i,all),(i+1,all-1)}^t \tag{27}$$

Disabling conditions are similar for the ones from $1D$ case, except that whole lines are considered, and there are 8 possible directions for communications, and same for enabling conditions.

2.3 Saving Profiling Data

Gathered dynamic data could be saved to disk for further re-usage. Specifically, this includes performance values $perf_i^T$, $latency_{i,j}^T$, $bandwidth_{i,j}^T$. In order to maintain average values of these parameters, they should be saved in the special manner. At time step T all parameters could be described as: $perf_i^T = \frac{Q_{perf_i}}{P_{perf_i}}$, $latency_{i,j}^T = \frac{Q_{latency_{i,j}}}{P_{latency_{i,j}}}$, $bandwidth_{i,j}^T = \frac{Q_{bandwidth_{i,j}}}{P_{bandwidth_{i,j}}}$. Q_{perf_i} is the total number of grid points, on which computations were performed on i computational node, P_{perf_i} is the total time, which was spent on computations on i computational node, similar for latency and bandwidth.

The values of Q and P are then saved to disk. When file with Q and P is loaded, performance parameters right before time step t could be calculated in the next manner:

$$perf_i^t = \frac{Q_{perf_i} + T_{perf} * \sum_{v=0}^{l-1} S_i^{v*T_{perf}}}{P_{perf_i} + \sum_{v=0}^{t-1} \tau_{calc_i}^v} \tag{28}$$

$$latency_{i,j}^t = \frac{Q_{latency_{i,j}} + \sum_{v=1}^{l} lat_{i,j}^{v*T_{perf}}}{P_{latency_{i,j}} + l} \tag{29}$$

where $l = t/T_{perf}$. Bandwidth is calculated similar to latency. Balancing could now be performed before the start of computations, as $perf_i^0$, $latency_i^0$, $bandwidth_i^0$ are now defined. If balancing is performed only before the start of computations, balancing overhead is fully removed, and balancing is static. Also, this approach allows to improve accuracy of performance parameters by storing just 6 values on disk.

2.4 Further Work

Performance parameters, measurements of which were discussed in previous sections, describe the performance of the computational node. However, this implied that all computational nodes perform the same amount of measurements for each grid point. This could be incorrect in case some electromagnetic sources are setup, for example point source or plane wave source, or some additional measurements have to be performed only for part of grid points, for example near-to-far field computations. In such cases additional computations have to be considered in load balancing algorithm, which will be discussed in further work.

3 Measurements

All measurements were performed on IBM Blue Gene/P supercomputer and Tesla CMC server. IBM Blue Gene/P is a massively parallel computational system. It contains 8192 calculation cores (2048 calculation nodes, 4 core each) with peak performance at 27.9 tflops. Single calculation core is a PowerPC 450 with frequency at 850 MHz having 4 GB of RAM. Communicational network is a three-dimensional torus and unites all the nodes: single node has 6 bidirectional connections with 6 neighbors. Tesla CMC [12] is a computational system of Moscow State University with two Intel Xeon E5620 CPUs and set of Nvidia GPUs, including Nvidia Tesla K20c with 5 GB memory.

Basic FDTD computation was chosen as a benchmark (no PML, no TF/SF, point wave source for each computational node). In each computation virtual topology was mapped on computational nodes of Blue Gene/P in such a way that virtual topology matches physical topology, so, computational nodes, which are neighbors in virtual topology, will be neighbors in physical topology too, and no additional share expenses arise.

In order to demonstrate dynamic balancing, computational nodes with different performances were emulated on IBM Blue Gene/P nodes for $3D$ Yee grid of size $S = 100 * 100 * 100$. Computations were performed on 8 nodes, 4 of which were emulated to be slower by specific delay at each time step. Delay varies from large delay of 2 s, to small delay of 0.02 s, for which performance of slower computational node was 80% of performance of normal computational node for each time step.

The measurements in Table 1 show, that load balancing could significantly improve total execution time, especially for long running tasks, even when computational nodes' delay is fairly small (i.e. computational nodes have nearly similar performances).

On Tesla CMC server computations were performed on both GPU and CPUs. For cases, when computation data could be fully located in GPU memory, usually, it is more efficient to perform all computations just on GPU, because this will remove overhead of data copying to/from GPU.

However, this is not the case for most FDTD modeling, especially for three-dimensional cases, as large numerical grids are required for good accuracy. In such cases there is no way to remove data copying to/from GPU at each time step, as data can not be fully located in GPU memory. Additional share time with GPU could be taken into account in dynamic load balancing algorithm described in previous sections.

Measurements on Tesla CMC were performed on 2 computational nodes, one with Tesla K20c, for grid of size $768 * 512 * 512$, which can't be fully located in GPU memory. Balanced computation took on 15% less execution time than computations on a single node with GPU, and on 50% less execution time than on a equal spread between two computational nodes.

Table 1. Measurements for 3D mode for 8 computational nodes for Yee grid with size $S = 100 * 100 * 100$.

Nodes delay, seconds	Time steps	Execution time decrease with rebalance
2	100	61.3%
2	1000	Above 90%
0.2	1000	14.4%
0.2	10000	28.2%
0.02	10000	3.6%
0.02	20000	5.0%

4   Conclusion

Developed FDTD solver provides features for dynamic load distribution between computational nodes. Measurements prove that described dynamic load balancing algorithm could significantly improve overall computational time on heterogeneous computational systems. This approach could be useful for computational systems with different CPUs on different computational nodes. Besides, this approach could also be used on computational systems with different GPUs on some computational nodes, where performance of GPU and additional communication time with GPU are taken into account. In further work dynamic load balancing would be improved for cases where different amount of computations is performed for different Yee grid points.

References

1. Yee, K.S.: Numerical solution of initial boundary value problems involving maxwell's equations in isotropic media. IEEE Trans. Antennas Propag. **14**(3), 303–307 (1966)
2. Franek, O.: A simple method for static load balancing of parallel FDTD codes. In: Proceedings of the International Conference on Electromagnetics in Advanced Applications (ICEAA 2016), pp. 587–590 (2016)
3. Shams, R., Sadeghi, P.: On optimization of finite-difference time-domain (FDTD) computation on heterogeneous and GPU clusters. J. Parallel Distrib. Comput. **71**(4), 584–593 (2011)
4. Balykov, G.: Parallel FDTD solver with optimal topology and dynamic balancing. In: Voevodin, V., Sobolev, S. (eds.) Supercomputing. RuSCDays 2017. CCIS, vol. 793, pp. 337–348. Springer, Cham (2017). https://doi.org/10.1007/978-3-319-71255-0_27
5. Sharma, R., Priyesh, K.: Dynamic load balancing algorithm for heterogeneous multi-core processors cluster. In: Communication Systems and Network Technologies (CSNT), pp. 288–292. IEEE (2014)
6. De Grande, R., Azzedine, B.: Dynamic balancing of communication and computation load for HLA-based simulations on large-scale distributed systems. J. Parallel Distrib. Comput. **71**(1), 40–52 (2011)

7. Ravichandran, K., Lee, S., Pande, S.: Work stealing for multi-core HPC clusters. In: Jeannot, E., Namyst, R., Roman, J. (eds.) Euro-Par 2011. LNCS, vol. 6852, pp. 205–217. Springer, Heidelberg (2011). https://doi.org/10.1007/978-3-642-23400-2_20

8. Cederman, D., Tsigas, P.: Dynamic load balancing using work-stealing. In: GPU Computing Gems. Elsevier (2012)

9. Parallel FDTD solver. https://github.com/zer011b/fdtd3d

10. Taflove, A., Hagness, S.C.: Computational Electrodynamics: The Finite-difference Timedomain Method, 3rd edn. Artech House, Norwood (2000)

11. Hockney, R.: The communication challenge for MPP: Intel Paragon and Meiko CS-2. Parallel Comput. **20**(3), 389–398 (1994)

12. Tesla CMC Server of Moscow State University. http://hpc.cmc.msu.ru/tesla

Parallel Supercomputer Docking Program of the New Generation: Finding Low Energy Minima Spectrum

Alexey Sulimov[1,2], Danil Kutov[1,2], and Vladimir Sulimov[1,2(✉)]

[1] Dimonta, Ltd., Moscow 117186, Russia
{as, dk}@dimonta.com, vladimir.sulimov@gmail.com
[2] Research Computer Center, Lomonosov Moscow State University,
Moscow 119992, Russia

Abstract. The results of studies of the energy surfaces of the protein-ligand complexes carried out with the help of the FLM docking program belonging to the new generation of gridless docking programs are presented. It is demonstrated that the ability of the FLM docking program to find the global energy minimum is much higher than one of the "classical" SOL docking program using the genetic algorithm and the preliminary calculated grid of potentials of ligand atoms interactions with the target protein. The optimal number of FLM local optimization reliable finding of the global energy minimum and all local minima with energies in the 2 kcal/mol interval above the energy of the global minimum is found. This number is 250 thousand. For complexes with the ligand containing more than 60 atoms and having more than 12 torsions and with more than protein 4500 protein atoms the number of FLM local optimizations should be noticeably increased. There are several unique energy minima in this energy interval and for most complexes these minima are located near (RMSD < 3 Å) the global minimum. However, there a complexes where such minima are located far from the global minimum with RMSD (on all ligand atoms) > 5 Å.

Keywords: Generalized docking · Local optimization · Global minimum
Low-energy local minima spectrum · High-performance computing
Molecular modeling · Drug design

1 Introduction

Discovery of new inhibitors of the protein associated with a given disease is the initial and most important stage of the whole process of the new pharmaceutical substances discovery [1, 2]. Computer-aided molecular modeling can considerably increase effectiveness of the new inhibitors design. Protein-ligand binding free energy calculation is one of the key problems of this molecular modeling. Docking is the popular molecular modeling method based on the search for the ligand binding pose in the target protein and the subsequent estimation of the protein-ligand binding free energy [3]. There are a lot of docking programs now [4–6]. However, accuracy of the binding energy prediction is still not high enough for the lead inhibitors optimization on the base of such calculations [7]. There are a lot of sources of errors decreasing the

© Springer Nature Switzerland AG 2019
V. Voevodin and S. Sobolev (Eds.): RuSCDays 2018, CCIS 965, pp. 314–330, 2019.
https://doi.org/10.1007/978-3-030-05807-4_27

accuracy of such calculations [8]: imperfections of the force field used, using too simplified solvent models or neglect solvent at all, an incomplete search of the best ligand poses, inadequate approximations made in construction of 3D atomistic models of the target protein and ligands, and, finally, simplifications which are used to accelerate the docking performance. The latter is the most harmful trend which over-simplifies many aspects of the complicated docking problem. Certainly this trend was essential and unavoidable at the dawn of docking programs development, 20–30 years ago. However now, when large supercomputer resources are available, we can return to the rigorous docking task formulation and try to solve this problem accurately.

One approximation is widely used in many docking programs and limits their accuracy strongly. To accelerate docking the preliminary calculated grid of potentials is used – the grid approximation. This grid contains usually in its nodes the potentials of non-bonded interactions (Coulomb and Van der Waals) of all possible types of ligand atoms with the protein. When docking, the energy of the ligand in any position in the active site of the protein is calculated as the sum of the grid potentials over all atoms of this ligand. This approach gives a large acceleration because all resource-intensive operations are performed at the grid calculation stage before the docking proper. This approach is used in AutoDock [9], ICM [10], Dock [11], SOL [12], and possibly in many other docking programs. However this approximation leads to several limitations resulting in docking inaccuracy [8]. First, it is impossible to perform accurately a local optimization of the energy of the protein-ligand complex varying either coordinates of only ligand atoms or ligand and protein atoms both. Second, implicit solvent models that describe nonlocal interactions of atom charges of a solute molecule with polar-ization charges on the solvent excluded surface (SES) cannot be sufficiently accurately reduced to predetermined local potentials at grid nodes; hence the contribution of the interaction with water into the protein-ligand energy cannot be described with the grid approximation accurately. Third, there are different fitting parameters (in addition to fixed force field parameters) in existing docking programs. These parameters help to adjust the docking results to crystallized ligand poses in target proteins and to repro-duce in calculations binding constants obtained in experiments for a training set of protein-ligand complexes. Fitting parameters serve to demonstrate a semblance of high accuracy of calculations. However, utilization of fitting parameters obscure the reasons of bad docking accuracy for new protein-ligand complexes which are different from ones in the training set.

So, docking programs of new generation possessing a heightened accuracy should not use the preliminary calculated grid of potentials and any fitting parameters; in the process of ligand positioning the energy of the protein-ligand complex should be calculated in the frame of a given force field or a quantum-chemical method without simplifications and fitting parameters.

Another problem of the accurate docking is connected with the ligand positioning procedure. Until now in some popular docking programs the procedure "lock-and-key" is used for ligand positioning. In the frame of this procedure the ligand should be embedded into the protein active site in a manner as a key is inserted into the lock: the ligand molecular surface must complement the active site surface. In another ligand positioning procedures some prepositioning points in the active site are used for placing near them ligand atoms of definite types, e.g. for formation hydrogen bonds. However, the most general ligand positioning procedure is based on the so-called docking-paradigm [13–15]. This paradigm assumes that the ligand binding pose in the active site of the target protein corresponds to the global minimum of the protein-ligand energy function or is near it. In accordance with this paradigm the docking problem is reduced to the search of the global minimum on the multi-dimensional protein-ligand energy surface the dimensionality of which is defined by the number of protein-ligand system degrees of freedom.

Several supercomputer docking programs of this type (no preliminary calculated grid of potentials, no fitting parameters, and the docking algorithm is based on the global energy minimum search) were developed recently: FLM [13] and SOL-T [14] for docking flexible ligands into rigid proteins and SOL-P [16, 17] which is able to dock flexible ligands into the protein with moveable atoms as well as into the rigid protein. FLM performs the massive multiprocessor exhaustive search for low energy minima of the protein-ligand complex in the rigid protein model. The protein-ligand system energy is calculated in the frame of the MMFF94 force filed either in vacuum or with the rigorous implicit solvent model. Each local energy optimization is carried out from the random position of the ligand in the specified region of the active site of the protein. SOL-T and SOL-P are much faster than FLM due to the new TT-docking algorithm based on the tensor train (TT) decomposition of multi-dimensional tensors and the TT-Cross approximation and the respective global optimization method. However, if enough supercomputing resources are available, FLM will perform the complete low energy minima search thoroughly. The reliability of the global minimum finding and the fullness of the low energy minima identification, especially those belonging to the narrow energy interval of several kcal/mol above the global minimum, define high docking accuracy. Moreover, such a carefully found low energy minima can be used for the validation of other docking algorithms. Experience of practical use of the FLM program and comparison of energy minima found by this program with minima found by other programs, such as SOL-T and SOL-P, compel us to investigate more accurate performance of the FLM program. Also the conception of the quasi-docking approach [18, 19] has been introduced recently, and it is essentially based on the completeness of the whole low energy minima spectrum found by the FLM program. In this work we present the results of studies of the energy surfaces of the protein-ligand complexes carried out with the help of the FLM docking program. The employment of supercomputing resources of Lomonosov Moscow State University made it possible to conduct unprecedentedly detailed studies of the low-energy minima spectra of the test set of protein-ligand complexes containing various proteins and different ligands. Optimal computing resources are determined which are necessary for reliable finding of the global minimum and local energy minima with their energy values near to the energy of the global minimum. Realizability of the docking paradigm and the optimal choice of the size of low energy minima spectrum for the quasi-docking procedure are investigated. Comparison of FLM performance with one of the classical SOL docking program which uses a preliminary calculated grid of potentials is made and the advantage of the FLM program is demonstrated.

2 Materials and Methods

Due to thermal motion in the thermodynamic equilibrium state the ligand continuously jumps from one binding pose to another, and for the binding energy estimation we should find not only the global minimum of the energy of the protein-ligand system where the ligand spends most time but also the low-energy part of the whole local energy minima spectrum. The landscape of the multi-dimensional protein-ligand energy surface is very complicated containing hundreds and thousands of local minima even when positions of the ligand center are limited within the spatially restricted area of the protein active site. Parallel multi-processor performance of the minima search program and available supercomputer resources make it possible to solve practically the complicated problem of determination of all low energy minima on this complicated energy surface.

2.1 FLM Program

The FLM program is the MPI-based program which was developed to find low-energy minima of the ligand-protein system [13]. During the minima search, the protein is considered as rigid and the ligand is fully flexible. The name of the program is the abbreviation of its main function: Find Local Minima. We describe and use here the version FLM-0.05 in which the MMFF94 force field is implemented, and the energy of any configuration of the protein-ligand system is calculated in the frame of this force field in vacuum without simplifications and approximations. There are two different options of the FLM performance: (i) the search for low energy minima of the protein-ligand system and (ii) the search for low energy minima of the free (unbound) ligand.

FLM performs the massive multi-processor search for low energy minima of the energy function of a protein-ligand complex. Each local energy optimization is carried out from the random position of the ligand in the specified region of the active site of the protein. These random positions are obtained by random throws of the ligand with continuous deformations of the ligand when changing its torsion angles (describing the internal rotation around a single acyclic bond of the ligand) and translations and rotations of the ligand as a whole rigid body.

- The ligand geometrical center (the center of gravity when all atomic masses are equal) is moved to a random point in the search area. The search area is defined as the sphere of a given radius R_{in} with the center at the ligand native position geometrical center. The ligand native position is the position of the ligand in the crystallized protein-ligand complex structure. The present investigations are conducted when the radius is equal to $R_{in} = 8$ Å. This sphere covers active sites of all test protein-ligand complexes. There is also a larger sphere with the radius R_{grid} which center is in the same point as the smaller sphere. The whole ligand should be inside this larger sphere during the docking process. In the present study $R_{grid} = 24$ Å.
- The ligand is rotated as a whole by a random angle from the interval [−,] around a random axis passing through the ligand center.
- The ligand torsions are rotated by random angles from the interval [−,].

After each random throw not all random system conformations are further optimized. At first, atom-atom distances are checked: atoms from each ligand-ligand or protein-ligand atom pair must be separated by more than 0.5 Å. Otherwise this random system conformation is rejected. For the acceleration of the checkup of the existence of such protein-ligand atoms clashes a special 3D array is constructed covering the active site region with sufficiently fine grid (the grid step is 0.1 Å). The special region where this grid is created is the sphere with the radius R_{grid}. Each cell of the grid contains an indication of the presence or absence of a protein atom. So, any random initial pose of the ligand in the active site (inside the sphere with the radius R_{grid}) is accepted for the further local optimization if no atom of the ligand finds itself in the cell with an indicator of the presence of the protein atom.

Local optimization is performed using the L-BFGS gradient algorithm [20, 21] without any restrictions on the positions of the ligand atoms in the search area. All Cartesian coordinates of ligand atoms move during optimization. Each local optimization stops when the maximal component of the gradient of the optimized energy function decreases down to the value 10^{-5} kcal/mol/Å. The gradient of the energy function is calculated numerically using 6 points and the step of numerical differentiation is equal to 10^{-8} Å. If the ligand center moves out of the search area after the optimization (out from the sphere of the radius R_{in}), the respective local minimum is rejected. We call each accepted local optimization the trial or test optimization.

The local minima search is parallelized: independent local optimizations from different initial ligand conformations are continuously performed in parallel by different MPI processes. The optimization results are collected in the master process to form the low-energy minima set. The current collected minima set is repeatedly sent back from the master process to other processes, so other processes can select only promising minima to send. This results in the good scalability of the program with an increase in the number of computing cores. The program works for a specified time on the specified number of computing cores. If we continue such FLM calculations sufficiently long time at sufficiently large number of computing cores, we'll certainly find the global energy minimum and also all low energy minima above the global one in the given energy interval or a given number of lowest in energy minima. One of the questions we address in this study: how many trial optimizations should be done for the reliable finding the global minimum, and/or the given number of lowest energy minima? The answer to this question is presented below in the section Results.

A set of found unique local minima with the lowest potential energies is being kept in operative memory during FLM calculations. A new computed local minimum is included into the set, if it differs from any minimum of the set, and the minimum with the highest energy is excluded from this set. Two minima are different if RMSD between them exceeds a given value, the uniqueness parameter, e.g. 0.1 or 0.2 Å. The RMSD is calculated over the ligand heavy atoms without taking into account possible chemical symmetry. Obviously, the larger the uniqueness parameter the lesser number of the unique minima will be collected in the low energy minima set.

FLM can save different numbers of the unique low energy minima. In the present study 8192 (2^{13}) unique low energy minima are saved for each protein-ligand complex from the respective test set (see below).

After finishing the FLM program performance the FLM-PP postprocessing programs starts. FLM-PP conducts the more accurate analysis of the uniqueness of found minima taking into account ligand chemical symmetry. It calculates RMSD values between all minima in respect with all ligand atoms. As a result of FLM-PP performance the only unique minima from the whole pool of the minima found by the FLM program are kept. The minima are considered different from each other when the RMSD value is larger than a given distance, for example 0.2 Å, and the difference of their energies is larger than a given value, for example 1 kcal/mol. FLM-PP can also perform local energy optimization with the L-BFGS method and calculate the system energy in solvent – in one of the two implicit solvent models: PCM and SGB in the points corresponding to the minima.

The explanation of saving so large number of low energy minima is closely related to the quasi-docking procedure [8, 18, 19] which is as follows. Suppose that all low energy minima are found for a given protein-ligand complex in the frame of the given force field. In respect with the docking paradigm the global energy minimum should be found near the crystallized native ligand pose. However our analysis reveals [13] that this is not true for many complexes in the frame of the MMFF94 force field in vacuum. This force field does not describe adequately the energy of many protein-ligand complexes and the docking paradigm is not satisfied for these complexes. So, we should use better force field or quantum chemical methods for the energy calculation. However, is it possible to avoid the time-consuming search for the low energy minima spectrum with a new energy function? Yes, it is possible, and the quasi-docking procedure is proposed [18, 19]. All low energy minima found in a given force field are recalculated in another force field or quantum-chemical methods. Each minimum is used as the initial configuration for local energy optimization. Positions of the minima (poses of the ligand) change insignificantly but the energies of the minima can change very strongly and the minimum with high energy in the initial force field can become the global minimum in the new method of energy calculations. This quasi-docking procedure has already demonstrated that for docking the CHARMM force field is much better than MMFF94, and the PM7 quantum-chemical method is better than CHARMM [18, 19]. All these methods should be used with the respective implicit solvent models.

2.2   Test Sets of Protein-Ligand Complexes

The test set of 16 protein-ligand complexes was chosen from the Protein Data Bank (PDB) [22]. These structures have been chosen due to good resolution and the broad range of ligands different size and flexibility (Table 1). Protein structures were prepared by the elimination of all "HETATM" records, i.e. the records corresponding to water molecules, atoms, ions, and molecules which are not part of the protein structure, from the PDB files of the complexes, and then hydrogen atoms were added to the protein structures by the original APLITE program [12]. Although some complexes have been crystallized in low acidic conditions (pH = 5.2–6.5), all test proteins are active at the neutral conditions, and the APLITE program adds hydrogen atoms according to the standard amino acid protonation states at pH = 7.4. The histidine protonation state is chosen by comparing of electrochemical potentials for hydrogen atoms at "HD1" and

"HE2" positions. Optimization of hydrogen atoms positions is performed with MMFF94 force field after the hydrogen atoms pre-placement. During this optimization, all rotation variants of torsionally moveable hydrogen atoms, e.g., a hydroxyl hydrogen atom from tyrosine, are tested. Ligands were also taken from the PDB files. Hydrogen atoms were added to the ligands by the Avogadro program [22].

Table 1. Test set of 16 protein ligand complexes. PDB ID is the identifier which is assigned to the respective structure of the protein-ligand complex in the Protein Data Bank [22]. The table contains information about the numbers of ligand atoms and torsions, ligand charges and numbers of protein atoms. Numbers of atoms include hydrogen atoms. uPA – urokinase-type plasminogen activator, CHK1 – checkpoint kinase 1, ERK2 – extracellular signal-regulated kinase 2.

Protein name	PDB ID	Number of ligand atoms	Number of ligand torsions	Ligand charges	Number of protein atoms
uPA	1C5Y	20	2	1	3869
	1F5L	24	6	1	3823
	1O3P	46	6	1	3839
	1SQO	34	4	1	3823
	1VJ9	74	19	1	3859
	1VJA	61	17	1	3858
thrombin	1DWC	71	12	0	4494
	1TOM	64	10	2	4455
Factor Xa	2P94	60	7	0	3676
	3CEN	50	7	0	3676
CHK1	4FSW	26	0	0	4342
	4FT0	42	3	−1	4255
	4FT9	32	5	0	4394
	4FTA	35	6	−1	4336
ERK2	4FV5	52	8	0	5414
	4FV6	57	12	0	5449

Also, the locally optimized ligand native position has RMSD from the original native pose less than 1.5 1.5 Å for all 16 test complexes, both for the optimization with MMFF94 in vacuum and for the optimization with the PM7 method in vacuum.

2.3   Minima Indexes

All local energy minima of a given protein-ligand complex for a given energy function can be sorted by their energies in the ascending order; that is, every minimum gets its own index equal to its number in this sorted list of minima. The lowest energy minimum has index equal to 1. We introduce a special index [13, 14, 19] to analyze the docking positioning accuracy and the feasibility of the docking paradigm as follows. The list of minima can include some minima corresponding to ligand positions located near the nonoptimized native (crystallized) ligand pose in the given crystallized protein-ligand complex structure taken from the Protein Data Bank [22]. By our definition the ligand is near the nonoptimized native ligand position if the RMSD, the root-mean-square deviation between equivalent atoms of the ligand in the two positions, is less than 2 Å. Let us designate the index of such minimum which is close to the native (crystallized) ligand position as INN. It is the abbreviation of the term "Index of Near Native". If there are several such minima, we attribute INN to the minimum with the lowest energy among all minima which are close to the native ligand pose.

How can this index be used to analyze the docking positioning accuracy? The docking program performs on the base of the docking paradigm: the best position of the ligand found in the docking procedure is the global energy minimum. So, if INN is equal to 1, then the best ligand position corresponding to the global energy minimum is situated near the experimentally defined ligand pose. Therefore the positioning accuracy of docking with different methods of protein-ligand energy calculation can be compared by the simple comparison of INN index: the closer INN to 1 the better positioning accuracy is observed.

3 Results

3.1 Effectiveness of the Global Energy Minimum Search by SOL, FLM, SOL–P Programs

The comparison of the performance of different docking programs is a non-trivial task because there are several features reflecting quality of the performance: the positioning accuracy, the accuracy of the binding energy calculation, effectiveness of the global energy minimum finding, effectiveness of all low energy minima finding in a given energy interval above the global minimum, etc. The first two features are closely connected with experimentally measured values and they depend on models of proteins and ligands, and on the method of the protein-ligand energy calculation – the choice of the force field or the quantum-chemical method. The third feature reflects better the performance of the docking algorithm. So, we compare the ability of the docking programs to find the global energy minimum in the frame of the given force field. The comparison of the docking programs of the new generation (FLM and SOL-P) and the "classical" SOL docking program is made for 16 testing protein-ligand complexes. SOL does not use the local energy optimization. That is why the additional treatment of the ligand poses found by SOL is made as follows. For each protein-ligand complex all 99 poses, which are found by SOL in 99 independent runs of the genetic algorithm, are locally optimized. The protein atoms are kept fixed and the energy of the protein-ligand complex is optimized in the frame of MMFF94 force field in vacuum with variations of Cartesian coordinates of all ligand atoms. The local optimization method is the same as one which is implemented in FLM and SOL-P programs (L-BFGS), and the same accuracy of the optimization is taken for all these cases. All local energy minima, which are found in these optimizations, are subjected to filtering of unique minima by the FLM-PP program with the same uniqueness 0.2. Difference ΔE_{GM} between the energy of the global minimum found by this procedure and the energy of the global minimum found by programs of the new generation (FLM and SOL-P) is presented in Table 2 for all 16 test protein ligand complexes.

Table 2. Energies of the global minima found by FLM and SOL-P programs for 16 test complexes. The energies are presented relatively the energy of the global minimum found in the procedure involving docking with the "classical" SOL docking program.

PDB ID	ΔE_{GM} FLM, kcal/mol	ΔE_{GM} SOL-P, kcal/mol	PDB ID	ΔE_{GM} FLM, kcal/mol	ΔE_{GM} SOL-P, kcal/mol
1C5Y	0.00	−0.08	2P94	−2.75	−0.92
1DWC	−67.56	−67.56	3CEN	−2.19	5.59
1F5L	−1.16	−1.16	4FSW	−12.04	−12.04
1O3P	−5.13	−5.11	4FT0	−24.80	−26.57
1SQO	−0.11	−0.11	4FT9	−18.31	−18.31
1TOM	−5.18	−1.13	4FTA	−17.53	−17.53
1VJ9	−5.55	−3.04	4FV5	−36.65	−21.03
1VJA	−5.33	−2.93	4FV6	−16.93	−7.39

We can see in Table 2 that the energy of the global minimum which is found by FLM is lower than or equal to the energy of the global minimum found in the procedure with SOL usage for all test protein-ligand complexes. Only for one complex (1C5Y) energies of SOL and FLM global minima are the same. For all other 15 complexes the SOL related procedure cannot find global energy minima and it is obvious that the effectiveness of finding of the global minimum by the SOL related procedure is much lower than one demonstrated by the FLM program. We see in Table 2 that SOL-P is worse than FLM in finding of the global minimum but the former is also better than the performance of the SOL related procedure.

3.2   Ligand Internal Strain Energy

The FLM program can work not only in the search mode of the low energy minima spectrum of the protein-ligand complex with a rigid protein, but also it can find the spectrum of low energy (actually all) minima of the free ligand. The latter is necessary in calculating the binding energy of the protein-ligand complex in the multiwell approximation [8, 13]. In this case, an important contribution in the binding energy is the ligand internal strain energy, which is calculated as the difference between the energy of the free ligand in its conformation in the bound state in the protein and the energy of the free ligand in its the conformation corresponding to the global minimum of the free (unbound) ligand. The physical meaning of this quantity is simple – in order for the ligand to "get into" the active center of the target protein, it often needs to change its configuration in comparison with its configuration in the unbound state, i.e. when the ligand conformation corresponds to the global energy minimum of the free ligand. This change of the ligand conformation requires additional energy cost which is the ligand internal strain energy. The strain energy works against protein-ligand binding. The energy of the ligand internal strain is not taken into account in most of docking programs. However, this energy can be quite large, as can be seen from Table 3, in which the internal strains of native ligands are given for 16 test complexes.

Table 3. The internal strain energy E_{strain} of native ligands for 16 test complexes.

PDB ID	E_{strain}, kcal/mol	PDB ID	E_{strain}, kcal/mol
1C5Y	4.59	2P94	23.34
1DWC	94.83	3CEN	20.26
1F5L	5.30	4FSW	1.75
1O3P	19.85	4FT0	9.74
1SQO	11.54	4FT9	18.62
1TOM	21.10	4FTA	10.39
1VJ9	47.53	4FV5	30.02
1VJA	47.18	4FV6	25.77

As we can see in Table 3 that the ligand internal strain energy can be sufficiently large: from several units to several dozen of kcal/mol, and it should be taken into account when the protein-ligand binding energy is calculated. This means that taking into account the energy of the ligand in its global minimum in the unbound state is extremely important, and FLM does this in the MMFF94 force field.

3.3 Finding the Global Minimum and Local Ones Above It in the 2 Kcal/Mol Interval

Good quality of low energy minima finding means that not only the global minimum is found but also all minima with energies in the given energy interval are found as well. It is reasonably to choose the value of this interval equal to 2 kcal/mol because only these minima will be occupied at room temperature. It is found that for the most part of the test complexes the last update of the energy of the global minimum occurs quickly – after about first 5 thousand local optimizations. The largest number of local optimization (97 thousand) is needed for finding the global energy minimum if the 1VJA complex which contains a sufficiently large and flexible ligand: 61 atoms and 17 torsions (see Table 1).

And how many local optimizations are needed for the reliable finding of all local minima with energies in the interval 2 kcal/mol above the energy of the global minimum? To answer this question we investigated the dependence of such unique minima as a function of the number of FLM performed local optimizations. The results show that for most of test complexes it is sufficient to perform 250 thousand local optimization for finding of all minima with energies in 2 kcal/mol above the energy of the global minimum (see Fig. 1). Additional minima appear in this energy interval after more than 250 thousand local optimizations only in 2 sufficiently larger protein-ligand complexes: 1DWC and 1VJA (see Fig. 1).

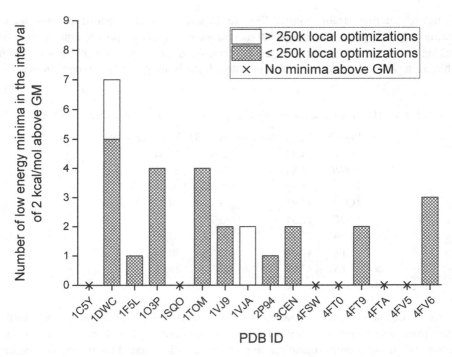

Fig. 1. Number of low energy minima in the interval of 2 kcal/mol above the global minimum depending on the number local optimizations: less than 250 thousand of local optimizations or more than 250 thousand of local optimizations. The symbol "×" on the horizontal axis marks the complexes with no energy minima in the 2 kcal/interval above the energy of the global minimum.

It should be noted that there are no energy minima in the 2 kcal/mol interval above the global one in several (in 6 of 16) complexes. Such complexes are marked by symbol "×" on the horizontal axis in Fig. 1. When calculating the protein-ligand binding energy in such complexes in the multiharmonic approximation, it is sufficient to take into account the characteristics of only the global minimum.

And how these low energy minima are located in space in respect to the position of the global energy minimum? This is shown in Fig. 2 where for each complex the value of the random mean square deviation (RMSD) between poses of ligand corresponding to these minima and the ligand pose corresponding to the global energy minimum. RMSD is calculated on positions of respective ligand atoms in these poses.

As we see, despite the fact that a significant number of minima are relatively closely located in space from the global minimum (RMSD < 3 Å), in some complexes ligand poses in the low energy minima (in the 2 kcal/mol interval above the global minimum) can be strongly different on RMSD from the ligand pose corresponding to the global minimum: RMSD can be larger than 5 or 10 Å.

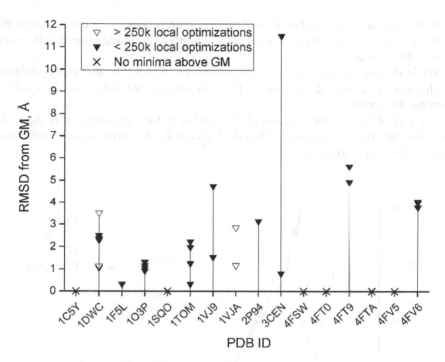

Fig. 2. The deviation RMSD of the ligand poses corresponding to the low energy minima with energies in the interval of 2 kcal/mol above the global minimum from the ligand pose in the global energy minimum.

3.4 Optimizing the FLM Performance Time

The FLM program will carry out infinitely many local optimizations if of course unrestricted supercomputer resources are available. FLM stops in respect with the specified the computing time parameter. The longer FLM works the more local optimizations are performed and a larger number of unique minima can be found in the pool of saved low energy minima. To define the optimum time of FLM calculations it is necessary to find the number of fulfilled local optimization after which the pool of saved minima is near the saturation for all test complexes. Let the function $f(x)$ is the dependence of the number of the pool updates on the number x of performed local optimizations. Our observations show that for larger size of the pool of saved low energy minima the number of pool updates is larger. So, it is convenient to normalize the number of pool updates on the size of the pool. The derivative of the normalized number of pool updates by the number of local optimizations displays the rate $D(x)$ of updates of saved minima in the pool:

$$D(x) = \frac{f'(x)}{N} \approx \frac{1}{N} \lim_{\Delta x \to 1} \frac{f(x_0 + \Delta x) - f(x_0)}{\Delta x} \tag{1}$$

Here x_0 is the given number of local energy optimizations, Δx is the change of the number of local energy optimizations, N is the number of the saved minima in the pool, i.e. it is the pool size.

While the rate of pool updates $D(x)$ is sufficiently high, the low energy minima search must be conducted. As soon as the pool updates rate falls down it is safe to complete the search.

The plot of $D(x)$ as the function of the number of local optimization is shown in Fig. 3 for three typical complexes. The calculation of the derivative is made with a step of $\Delta x = 10^3$ local optimizations.

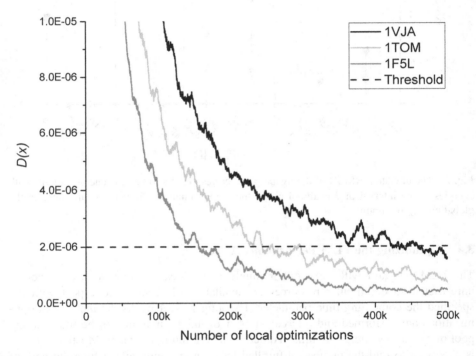

Fig. 3. The normalized derivative $D(x)$ of the number of updates of the low energy minima pool by the number of local optimizations as a function of the number of local optimizations x for three protein-ligand complexes with PDB ID: 1VJA, 1TOM, and 1F5L.

Curves of the rate of pool updates $D(x)$ as functions of the number of conducted local optimizations x for most of test protein-ligand complexes are between the curves for 1F5L and 1TOM. As it is shown above (see Sect. 3.3), for these complexes it is sufficient to perform 250 thousand of local optimizations for finding of all low energy minima in the energy interval 2 kcal/mol above the energy of the global minimum. A further search does not result in finding of new minima with energies in this 2 kcal/mol energy interval above the energy of the global minimum. This means that the pool of low energy minima is saturated, and further search will not practically result in finding of additional minima. Thus, it is possible to find the threshold of the saturation of the pool of low energy minima. It is practically reasonable to finish the search below this threshold. The threshold of 2×10^{-6} is chosen using curves in Fig. 3.

Curves of the rate of pool updates $D(x)$ for all four complicated complexes (1DWC, 1VJ9, 1VJA и 4FV6) are near the curve for 1VJA. To reach the saturation of the pool of saved minima for the complicated complexes it is necessary to spend many more local optimizations, about 450 thousand.

The time expenses of the FLM performance for each of 16 test complexes are shown in Fig. 4. The grey histogram bar show the time of the global minimum search. If there is no a grey histogram bar for a complex, this means than less than 150 CPU·hours (about 5000 local optimizations) are spent for the global minimum search. The shaded histogram bars show the time of conducting local optimizations while the rate of pool updates is larger than the threshold (2×10^{-6}). Unshaded histogram bars show the time which is spent during the presented supercomputer investigations.

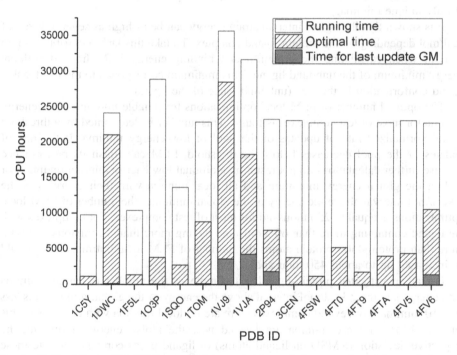

Fig. 4. Optimal time of the FLM run which is needed to reach the saturation of the pool of saved low energy minima.

We can see in Fig. 4, that the time of FLM performance is enough to reach the pool saturation for all test complexes. Also the time expended to find the global minimum is much less than the time to find all low energy minima in the pool of specified size. The longest time is needed to reach the saturation for the "difficult" complexes: 1DWC,

1VJ9, 1VJA и 4FV6. These complexes contain either a larger and flexible ligand (1DWC, 1VJ9, 1VJA) or a flexible ligand and a protein with large number of atoms (4FV6) as it can be seen in Table 1.

4 Conclusions

The performance of the docking program of new generation, the FLM program, is investigated. Some results of studies of the energy surfaces of the protein-ligand complexes with this program are presented. Comparison of performance of docking programs of new generation, FLM and SOL-P, with the classical SOL docking program is made. It is demonstrated for all test complexes that these docking programs of the new generation find the global minimum with considerably lower energy than the energy of the global minimum which is found by the classical SOL program in the same conditions: the same energy function, the same local optimization method, the same accuracy of the optimization, and the same uniqueness parameter for the filtering of only unique minima.

It is shown that the ligand internal strain energy can be as high as several dozen of kcal/mol depending on the protein-ligand complex. To take this large contribution into account when calculating the protein-ligand binding energy FLM finds the global energy minimum of the unbound ligand. This minimum corresponds to the most stable ligand conformation in the free (unbound) state of the ligand.

The optimal number of FLM local optimizations for reliable finding of low energy minima for most of test complexes is found. This number is determined by a threshold of the normalized rate of updates of the pool of low energy minima. If the rate of updates of the pool decreases below the threshold, FLM can finish its performance because further calculations do not result in additional low energy minima finding. For finding the global energy minimum and all local minima with their energies in the 2 kcal/mol interval above the energy of the global minimum the number of FLM local optimizations is equal to 250 thousand for most of test complexes. For complexes with the ligand containing more than 60 atoms and having more than 12 torsions and with more than protein 4500 protein atoms the number of FLM local optimizations should be increased up to about 450 thousand.

For all investigated protein-ligand test complexes the number of unique minima with energies in the 2 kcal/mol interval above the energy of the global minimum is less than 10 minima. For several complexes there are no local minima with energies in this interval. Most of these minima are located near the global energy minimum: the respective deviation (RMSD on ligand atoms) of ligand poses corresponding to these minima from the ligand pose of the global energy minimum is less than RMSD < 3 Å. However, some of the low energy minima can correspond to ligand poses far from the pose of the global minimum (RMSD > 5 Å).

It is shown that for all complexes the time of the global energy minimum finding is considerably smaller than the time of finding of the whole low energy minima spectrum. Supercomputer resources needed for the reliable determination of all low energy minima for most of test complexes are less than 10000 CPU · hours.

Acknowledgements. The work was financially supported by the Russian Science Foundation, Agreement no. 15-11-00025-П. The research is carried out using the equipment of the shared research facilities of HPC computing resources at Lomonosov Moscow State University, including the Lomonosov supercomputer [24].

References

1. Sliwoski, G., Kothiwale, S., Meiler, J., Lowe, E.W.: Computational methods in drug discovery. Pharmacol. Rev. **66**, 334–395 (2014). https://doi.org/10.1124/pr.112.007336
2. Sadovnichii, V.A., Sulimov, V.B.: Supercomputing technologies in medicine. In: Voevodin, V.V., Sadovnichii, V.A., Savin, G.I. (eds.) Supercomputing Technologies in Science, Education, and Industry, pp. 16–23. Moscow University Publishing (2009). (in Russian)
3. Sulimov, V.B., Sulimov, A.V.: Docking: molecular modeling for drug discovery. AINTELL, Moscow, 348 p. (2017). (in Russian), ISBN 978-5-98956-025-7
4. Chen, Y.C.: Beware of docking! Trends Pharmacol. Sci. **36**(2), 78–95 (2015)
5. Yuriev, E., Holien, J., Ramsland, P.A.: Improvements, trends, and new ideas in molecular docking: 2012–2013 in review. J. Mol. Recognit. **28**, 581–604 (2015)
6. Pagadala, N.S., Syed, K., Tuszynski, J.: Software for molecular docking: a review. Biophys. Rev. **9**(2), 91–102 (2017)
7. Mobley, D.L., Dill, K.A.: Binding of small-molecule ligands to proteins: "what you see" is not always "what you get". Structure **17**(4), 489–498 (2009). https://doi.org/10.1016/j.str.2009.02.010
8. Sulimov, A.V., Kutov, D.C., Katkova, E.V., Kondakova, O.A., Sulimov, V.B.: Search for approaches to improving the calculation accuracy of the protein—ligand binding energy by docking. Russ. Chem. Bull. Int. Ed. **66**(10), 1913–1924 (2017)
9. Forli, S., Huey, R., Pique, M.E., Sanner, M.F., Goodsell, D.S., Olson, A.J.: Computational protein-ligand docking and virtual drug screening with the AutoDock suite. Nat. Protoc. **11**(5), 905–919 (2016)
10. Neves, M.A., Totrov, M., Abagyan, R.: Docking and scoring with ICM: the benchmarking results and strategies for improvement. J. Comput. Aided Mol. Des. **26**(6), 675–686 (2012)
11. Allen, W.J., et al.: DOCK 6: impact of new features and current docking performance. J. Comput. Chem. **36**(15), 1132–1156 (2015)
12. Sulimov, A.V., Kutov, D.C., Oferkin, I.V., Katkova, E.V., Sulimov, V.B.: Application of the docking program SOL for CSAR benchmark. J. Chem. Inf. Model. **53**(8), 1946–1956 (2013)
13. Oferkin, I.V., et al.: Evaluation of docking target functions by the comprehensive investigation of protein-ligand energy minima. Adv. Bioinf. **2015**, 12 (2015). https://doi.org/10.1155/2015/126858. Article ID 126858
14. Oferkin, I.V., Zheltkov, D.A., Tyrtyshnikov, E.E., Sulimov, A.V., Kutov, D.C.: Evaluation of the docking algorithm based on tensor train global optimization. Bull. South Ural State Univ. Ser. Math. Model. Program. Comput. Softw. **8**(4), 83–99 (2015). https://doi.org/10.14529/mmp150407
15. Sulimov, A.V., Kutov, D.C., Katkova, E.V., Sulimov, V.B.: Combined docking with classical force field and quantum chemical semiempirical method PM7. Adv. Bioinf. **2017**, 6 (2017). https://doi.org/10.1155/2017/7167691. Article ID 7167691
16. Sulimov, A.V., et al.: Evaluation of the novel algorithm of flexible ligand docking with moveable target protein atoms. Comput. Struct. Biotechnol. J. **15**, 275–285 (2017). https://doi.org/10.1016/j.csbj.2017.02.004

17. Sulimov, A.V., et al.: Tensor train global optimization: application to docking in the configuration space with a large number of dimensions. In: Voevodin, V., Sobolev, S. (eds.) RuSCDays 2017. CCIS, vol. 793, pp. 151–167. Springer, Cham (2017). https://doi.org/10. 1007/978-3-319-71255-0_12

18. Sulimov, A.V., Kutov, D.C., Katkova, E.V., Sulimov, V.B.: Combined docking with classical force field and quantum chemical semiempirical method PM7. Adv. Bioinform. **2017**, 6 p. https://doi.org/10.1155/2017/7167691, Article ID 7167691, Accepted 22 Dec 2016

19. Sulimov, A.V., Kutov, D.C., Katkova, E.V., Ilin, I.S., Sulimov, V.B.: New generation of docking programs: Supercomputer validation of force fields and quantum-chemical methods for docking. J. Mol. Graph. Model. **78**, 139–147 (2017)

20. Byrd, R.H., Lu, P., Nocedal, J., Zhu, C.: A limited memory algorithm for bound constrained optimization. SIAM J. Sci. Comput. **16**(5), 1190–1208 (1995). https://doi.org/10.1137/ 0916069

21. Zhu, C., Byrd, R.H., Lu, P., Nocedal, J.: Algorithm 778: L-BFGS-B: fortran subroutines for large-scale bound-constrained optimization. ACM Trans. Math. Softw. **23**(4), 550–560 (1997). https://doi.org/10.1145/279232.279236

22. Berman, H.M., Westbrook, J., Feng, Z., et al.: The protein data bank. NucleicAcids Res. **28** (1), 235–242 (2000). https://doi.org/10.1093/nar/28.1.235

23. Avogadro: an Open-Source Molecular Builder and Visualization Tool. Version1. XX (2017). Accessed 26 Apr 2018. https://avogadro.cc/

24. Sadovnichy, V., Tikhonravov, A., Voevodin, V.l., Opanasenko, V.: "Lomonosov": supercomputing at Moscow State University. In: Contemporary High Performance Computing: From Petascale Toward Exascale, pp. 283–307. CRC Press (2013)

Parallelization Strategy for Wavefield Simulation with an Elastic Iterative Solver

Mikhail Belonosov[1], Vladimir Cheverda[2(✉)], Victor Kostin[2],
and Dmitry Neklyudov[2]

[1] Aramco Research Center - Delft, Aramco Overseas Company B.V.,
Delft, The Netherlands
mikhail.belonosov@aramcooverseas.com

[2] Institute of Petroleum Geology and Geophysics SB RAS, Novosibirsk, Russia
{cheverdava,kostinvi,neklyudovda}@ipgg.sbras.ru

Abstract. We present a parallelization strategy for our novel iterative method to simulate elastic waves in 3D land inhomogeneous isotropic media via MPI and OpenMP. The unique features of the solver are the preconditioner developed to assure fast convergence of the Krylov-type iteration method at low time frequencies and the way to calculate how the forward modeling operator acts on a vector. We successfully benchmark the accuracy of our solver against the exact solution and compare it to another iterative solver. The quality of the parallelization is justified by weak and strong scaling analysis. Our modification allows simulation in big models including a modified 2.5D Marmousi model comprising 90 million cells.

Keywords: Elastic equation · MPI · OpenMP · Preconditioner
Krylov iterations

1 Introduction

Advances in supercomputing technology make it feasible to solve big data problems. Large simulations seemed impossible in the recent past but have become commonplace today. In geosciences, supercomputers have been opening new horizons for understanding subsurface structures by allowing 3D imaging and velocity model estimation at high fidelity for fine-scale reservoir characterization. To keep up to date with cutting edge technologies, oil and gas companies are spending huge budgets to buy, lease and upgrade/maintain supercomputers (e.g., [1]). Dealing with high channel count field data processing and velocity estimation, the associated forward modeling and inversion algorithms may require petabytes of RAM and petaflops of computing power. The large datasets and high computation requirements are why modern algorithms have to be parallel.

Velocity reconstruction with frequency-domain full waveform inversion (FWI) [19, 24, 26] has been actively developing in the last decades. These days, even 3D elastic inversion, that may bring the most valuable information to interpreters, seems to be feasible. The most time consuming part of this process is forward modeling performed

© Springer Nature Switzerland AG 2019
V. Voevodin and S. Sobolev (Eds.): RuSCDays 2018, CCIS 965, pp. 331–342, 2019.
https://doi.org/10.1007/978-3-030-05807-4_28

several times at each iteration. For macro velocity reconstruction, only a few low frequency (up to 10 Hz) monochromatic components of the seismic wavefield are used.

The common practice is to perform simulation in the time domain [12, 16, 22], extracting the needed frequencies via Fourier Transform applied on the fly with a Discrete Fourier Transform. Time-domain simulation is usually parallelized over sources via MPI, so that each MPI process is independent of the others. Then, within each MPI process, parallelization is carried out via OpenMP [9], or MPI through domain decomposition of the target area, involving exchanges between adjacent subdomains or groups of subdomains (e.g., [11] and [18]).

The alternative, 3D frequency-domain modeling becomes feasible with the appearance of computing technology able to operate with big data. Here, we model the wavefield in the frequency domain only for needed frequencies. There are certain theoretical advantages of frequency-domain modeling that might improve the overall efficiency. Two different approaches are distinguished, including: direct [17] and iterative approaches based on a Krylov-type iteration method [21]. The main bottleneck of the first one is the necessity to store LU factors of the forward modeling matrix, requiring hundreds of gigabytes of RAM in a 3D case even for the acoustic case. This process can be parallelized with MPI via Domain Decomposition (e.g., [4]), decreasing the memory requirements per MPI process. However, in big models this involves vast computational resources, making the method computationally too expensive. Other attempts to resolve this issue are based on applying data compression techniques using Hierarchically Semi-Separable formats for storing data and Low Rank approximation of matrices [14, 27].

The memory requirements of the iterative approach are much more modest, because matrix factorization is not performed. However, its indefiniteness in seismic applications leads to very slow convergence or even divergence of the Krylov-type iteration method. Attempts to overcome this issue involve an appropriate preconditioner. In case of a 3D elastic simulation, only a few preconditioners have been developed so far [7, 15, 20].

In this paper, we use our own preconditioner that we briefly introduced in [13]. It is a modification for the elastic case of the preconditioner presented in [5] that was designed for simulation of low frequency monochromatic components of a wavefield in a 3D acoustic medium. Here we do not illustrate that our method is superior to other aforementioned approaches. This is a feasibility study showing that the method is capable of performing simulation in 3D elastic models of big size at low frequencies needed for macro velocity reconstruction with FWI. This is currently only possible using parallelization that we developed using MPI and OpenMP.

2   Iterative Method to Solve a 3D Elastic Equation

2.1   Statement of the Problem

We solve the following elastic equation written in the velocity-stress form describing propagation of a monochromatic component of the wave in a 3D isotropic heterogeneous medium

$$\left[i\omega\begin{pmatrix} \rho I_{3\times3} & 0 \\ 0 & S_{6\times6} \end{pmatrix} - \begin{pmatrix} 0 & \hat{P} \\ \hat{P}^T & 0 \end{pmatrix}\frac{\partial}{\partial x} - \begin{pmatrix} 0 & \hat{Q} \\ \hat{Q}^T & 0 \end{pmatrix}\frac{\partial}{\partial y} - \gamma(z)\begin{pmatrix} 0 & \hat{R} \\ \hat{R}^T & 0 \end{pmatrix}\frac{\partial}{\partial z}\right]v = f, \quad (1)$$

where vector of unknowns v comprises nine components. These components include the displacement velocities (v_x, v_y, v_z) and components of the stress tensor $(\sigma_{xx}, \sigma_{yy}, \sigma_{zz}, \sigma_{yz}, \sigma_{xz}, \sigma_{xy})$. ω is the real time frequency, $\rho(x, y, z)$ is the density, $I_{3\times3}$ is 3 by 3 identity matrix, $S_{6\times6}(x, y, z) = \begin{pmatrix} A & 0 \\ 0 & C \end{pmatrix}$ is 6 by 6 compliance matrix and

$$A = \begin{pmatrix} a & -b & -b \\ -b & a & -b \\ -b & -b & a \end{pmatrix}, C = \begin{pmatrix} c & 0 & 0 \\ 0 & c & 0 \\ 0 & 0 & c \end{pmatrix}, \hat{P} = \begin{pmatrix} 1 & 0 & 0 & 0 & 0 & 0 \\ 0 & 0 & 0 & 0 & 0 & 1 \\ 0 & 0 & 0 & 0 & 1 & 0 \end{pmatrix},$$

$$\hat{Q} = \begin{pmatrix} 0 & 0 & 0 & 0 & 0 & 1 \\ 0 & 1 & 0 & 0 & 0 & 1 \\ 0 & 0 & 0 & 1 & 0 & 0 \end{pmatrix}, \hat{R} = \begin{pmatrix} 0 & 0 & 0 & 0 & 1 & 0 \\ 0 & 0 & 0 & 1 & 0 & 0 \\ 0 & 0 & 1 & 0 & 0 & 0 \end{pmatrix}. \quad (2)$$

Coefficients $a(x, y, z)$, $b(x, y, z)$ and $c(x, y, z)$ are related to the Lame parameters λ and μ as follows $a = \frac{\lambda+\mu}{\mu(2\mu+3\lambda)}$, $b = \frac{\lambda}{2\mu(2\mu+3\lambda)}$, $c = \mu^{-1}$. f is the right-hand side representing the seismic source. In our experiments, we consider either a volumetric or vertical point-force source. $\gamma(z)$ may be either unity or the damping along z representing the Perfectly Matched Layer (PML) [6].

Equation (1) is solved in a cuboid domain of $N_x \times N_y \times N_z$ points. It is assumed that this domain includes sponge layers [5] on the horizontal and PML on the vertical boundaries (top and bottom) imitating an elastic radiation condition at infinity. The top boundary can be also the free surface. Usage of the sponge layers assures the coefficients of the partial derivatives by x and y are constant matrices. We use this property in the next section, applying the Fourier Transform over these coordinates to construct our preconditioner. Equation (1) along with the boundary conditions on the outer boundaries of the absorbing layers ($v_z = \sigma_{yz} = \sigma_{xz} = 0$ on the horizontal and periodic conditions on the vertical faces) produce the boundary value problem that we solve by means of the Krylov iterations. Their straightforward application does not guarantee convergence. This is why in the next section we develop a special preconditioner.

2.2 Preconditioned Iterative Method

Denote by L the right-hand side operator in Eq. (1) that is subject to the same boundary conditions as the boundary value problem we solve. Let L_0 be the same operator as L, but with $\rho(x, y, z) = \rho_0(z)$, $S_{6\times6}(x, y, z) = (1+i\beta) \cdot S_0(z)$, where

$$S_0(z) = \begin{pmatrix} A_0 & 0 \\ 0 & C_0 \end{pmatrix}, A_0 = \begin{pmatrix} a_0 & -b_0 & -b_0 \\ -b_0 & a_0 & -b_0 \\ -b_0 & -b_0 & a_0 \end{pmatrix}, C_0 = \begin{pmatrix} c_0 & 0 & 0 \\ 0 & c_0 & 0 \\ 0 & 0 & c_0 \end{pmatrix} \quad (3)$$

and a real positive number $\beta < 1$ introduces a complex shift by analogy with the shifted Laplacian [8]. Functions $\rho_0(z) > 0$, $a_0(z)$, $b_0(z)$ and $c_0(z)$ are the averaging of their 3D counterparts, providing proximity of operators L and L_0 to each other.

We use operator L_0 as the preconditioner and search for solution v of the original boundary value problem by solving the 2nd kind Fredholm integral equation

$$LL_0^{-1}\tilde{v} = f \tag{4}$$

with the same boundary conditions as for Eq. (1). Finally, we compute unknown v by formula $v = L_0^{-1}\tilde{v}$. Denoting $\delta L = L - L_0$ and substituting it into Eq. (4) we arrive at

$$(I - \delta L L_0^{-1})\tilde{v} = f, \tag{5}$$

where δL is the zero-order operator – pointwise multiplication by a matrix. This is valid because we consider Eq. (1) with the compliance matrix.

We solve Eq. (5) via a Krylov-type iterative method. From the variety of them, we choose the biconjugate gradient stabilized method (BiCGSTAB) [25] because of its moderate memory requirements. In principle, other methods of the same type are also applicable, for instance IDR [23]. This assumes computing several times per iteration (depending on a method) the product of the left-hand side operator of Eq. (5) by a particular vector w, i.e. computing $[w - \delta L L_0^{-1}w]$. This process breaks down into three computational steps:

1. first, computing $q_1 = L_0^{-1}w$ by solving boundary value problem $L_0 q_1 = w$;
2. then, computing $q_2 = \delta L q_1$, that in the discrete case is a pointwise multiplication of a tridiagonal matrix by a vector;
3. finally, subtracting the two vectors $[w - q_2]$.

To solve $L_0 q_1 = w$ we assume that function $w(x, y, z)$ is expanded into a Fourier series with respect to the horizontal coordinates with coefficients $\hat{w}(k_x, k_y, z)$, where k_x and k_y are the respective spatial frequencies. These coefficients are solutions to the boundary value problems for ordinary differential equations (ODEs)

$$\left[i\omega \begin{pmatrix} \rho_0 I_{3\times3} & 0 \\ 0 & S_0 \end{pmatrix} - ik_x \begin{pmatrix} 0 & \hat{P} \\ \hat{P}^T & 0 \end{pmatrix} - ik_y \begin{pmatrix} 0 & \hat{Q} \\ \hat{Q}^T & 0 \end{pmatrix} - \gamma(z) \begin{pmatrix} 0 & \hat{R} \\ \hat{R}^T & 0 \end{pmatrix} \frac{\partial}{\partial z} \right] \hat{v} = \hat{w}, \tag{6}$$

with the same boundary conditions as for Eq. (1) in the z-direction. We solve it numerically, applying a finite-difference approximation, that results in a system of linear algebraic equations (SLAEs) with a banded matrix, whose bandwidth depends on the order of the finite-difference scheme. In this case, computation of $\hat{w}(k_x, k_y, z)$ can be performed via the 2D Fast Fourier Transform (FFT) and after $\hat{v}(k_x, k_y, z)$ are found, $L_0^{-1}w$ can be computed via the inverse 2D FFT.

It is worth mentioning, that here we assumed, that solutions to the boundary value problems (6) exist. This assumption is partly justified by numerous successful numerical tests that we've carried out.

3 Parallelization

The FWI for macro velocity reconstruction involves simulations for different seismic sources at different low frequencies. This means, that in fact, many boundary value problems for Eq. (1) are solved at the same time, each having its own right-hand side f. Since they are solved independently of each other, we solve each one with a separate MPI process, assigned to a single node or a group of cluster nodes. This is the highest level of our parallelization strategy. There are no communications between these MPI processes. Assuming that all computational nodes have similar performance, this parallel process scales very well. This is why we do not mention this level of parallelization in subsequent tests and consider the case of one seismic source and one frequency only.

Four computational processes including Krylov iteration method, the 2D forward and inverse FFTs and solving the boundary value problem for Eq. (6), mainly drive our solver. We decompose the computational domain along one of the horizontal coordinates and parallelize these processes via MPI. The main exchanges between the MPI processes are while performing FFTs. For computing them, we use the Intel MKL library [10] supporting the decomposition along one direction only. In principle, the decomposition along the second horizontal dimension may be also applied with minor corrections of the code using a 2D FFT realization, supporting this functionality. Decomposition along the z-direction is not that obvious, since this involves solving each boundary value problems for Eq. (6) in parallel.

Following this strategy, each MPI process would independently solve its own set of $N_x \cdot N_y / N$ (N – the number of MPI processes) problems. We solve them in a loop, parallelized via OpenMP. Schematically, our parallelization strategy is presented in Fig. 1.

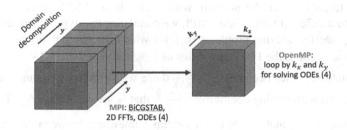

Fig. 1. Parallelization scheme.

Below, we present the results of scaling analysis for both MPI and OpenMP. All results presented here have been computed on a HPC cluster comprising nodes with two Intel® Xeon® E5-2680v4 @ 2400 MHz CPUs and interconnected with 56 Gb FDR InfiniBand HCA. Double precision floating point format has been used in the computations. This is necessary, when dealing with vectors of huge dimensions, for instance, for computing their dot product. As a stopping criterion for the BiCGSTAB, we used a 10^{-3} threshold for the relative residual of the L_2-norm providing enough accuracy for FWI applications. We assume a vertical point-force source as the source type unless explicitly stated to be volumetric.

For our tests, we construct a 2.5D land model (Fig. 2) from the open source 2D Marmousi model. It is discretized with a uniform grid of $551 \times 700 \times 235$ points. The horizontal cell sizes are 16 m and the vertical cell size is 8 m, corresponding to 5 and 10 points per minimal wavelength at frequency of 10 Hz respectively. Denser vertical sampling is required, because the finite-difference approximation along z-axis is only 4th order.

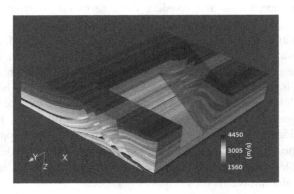

Fig. 2. 2.5D P-velocity model of $8.8 \times 11.2 \times 1.88$ km, constructed from the Marmousi model.

3.1 MPI Strong Scaling Analysis

MPI strong scalability of the solver is defined as ratio t_M/t_N, where t_M and t_N are elapsed run times to solve the problem with N and $M > N$ MPI processes each corresponding to a different CPU. Using MPI, we parallelize two types of processes. First, those scaling ideally (solving problems (6)), for which the computational time with N processes is $\frac{T}{N}$. Second, the FFT, that scales as $\frac{T_{FFT}}{\alpha(N)}$, with coefficient $1 < \alpha(N) < N$. The total computational time becomes $\frac{T}{N} + \frac{T_{FFT}}{\alpha(N)}$ (here we simplify, assuming no need of synchronization) with scaling coefficient $\frac{T + T_{FFT}}{\frac{T}{N} + \frac{T_{FFT}}{\alpha(N)}}$, that is greater than $\alpha(N)$. This is why, we expect very good scalability of the algorithm, somewhere between the scalability of the FFT and the ideal scalability. We did not take into account OpenMP, which can be switched on for extra speed-up. It is worth noting, that we can not use MPI instead of OpenMp here, since then the scaling would degrade. MPI may have worked well if $T \gg T_{FFT}$, but this is not the case.

We estimate the strong scaling for modeling at 5 Hz in two different models. Each model comprises $200 \times 600 \times 155$ points. The first one is a subset of the model depicted in Fig. 2 with $V_{p_{max}}/V_{p_{min}} = 2.85$, $V_{s_{min}} = 867$ m/s and the second one is part of the SEG/EAGE overthrust model [2]: $V_{p_{max}}/V_{p_{min}} = 2.75$, $V_{s_{min}} = 1258$ m/s. Grid cells are 30 m. From Fig. 3 we conclude that our solver scales very well up to 64 MPI processes.

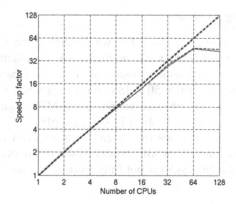

Fig. 3. Strong MPI scaling of our solver: blue dashed line is the result for the Marmousi model, the red line - for the SEG/EAGE overthrust model. The dashed grey line is the ideal scalability. (Color figure online)

3.2 MPI Weak Scaling Analysis

For weak scaling estimation, we assign the computational domain to one MPI process and then extend the size of the computational domain along the y-direction, while increasing the number of MPI processes. Here, we use one MPI process per CPU. The load per CPU is fixed. For the weak scaling, we use function $f_{weak}(N) = \frac{T(N)}{T(1)}$, where $T(N)$ is the average computational runtime per iteration with N MPI processes. The ideal weak scalability corresponds to $f_{weak}(N) = 1$.

To estimate it in our case, we considered a part of the model presented in Fig. 2 of size $200 \times 25 \times 200$ points with a decreased 4 m step along the y-coordinate. After extending the model in the y-direction 64 times, we arrive at a model of size $200 \times 1600 \times 150$ points. Figure 4 demonstrates that for up to 64 MPI processes, weak scaling of our solver has small variations around the ideal weak scalability.

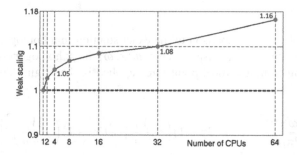

Fig. 4. Weak scaling measurements: the blue line is the result of the iterative solver and the dashed grey line is the ideal weak scaling. (Color figure online)

3.3 OpenMp Scaling Analysis

As already explained above, with OpenMP we parallelize the loop over spatial frequencies for solving the boundary value problems (6). To estimate the scalability of this part of our solver, we performed simulations in a small part of the SEG/EAGE overthrust model comprising $660 \times 50 \times 155$ points on a single CPU having 14 cores with hyper-threading switched off and without using MPI. Figure 5 shows that our solver scales well for all threads involved in this example.

It is worth mentioning, that we use OpenMP as an extra option applied when further increasing of the number of MPI processes doesn't improve performance any more, but the computational system is not fully loaded, i.e., there are free cores.

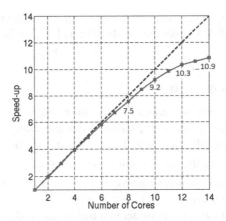

Fig. 5. Strong scalability analysis on one CPU of the part parallelized via OpenMP: the dashed blue line is the ideal scalability and the red line is the iterative solver scalability. (Color figure online)

4 Numerical Experiments

4.1 Benchmarking

We verify our solver by comparison to the exact solution in a homogeneous medium. A frequency-domain vertical displacement $u_z(x, \omega)$ in a homogeneous unbounded medium, resulting from a vertical point-force applied in the origin, has the analytical representation [3]:

$$
\begin{aligned}
u_z(x, \omega) = &\frac{S(\omega)e^{i\omega r/V_p}}{4\pi\rho V_p^2 r}\left[\gamma + (3\gamma - 1)\left(-\frac{V_p}{i\omega r}\right) + (3\gamma - 1)\left(-\frac{V_p}{i\omega r}\right)^2\right] \\
&- \frac{S(\omega)e^{i\omega r/V_s}}{4\pi\rho V_s^2 r}\left[\gamma - 1 + (3\gamma - 1)\left(-\frac{V_s}{i\omega r}\right) + (3\gamma - 1)\left(-\frac{V_s}{i\omega r}\right)^2\right],
\end{aligned}
\tag{7}
$$

where $S(\omega)$ is the monochromatic component of the source function, $r = \sqrt{x^2 + y^2 + z^2}$ is the radius vector, V_p and V_s are P and S-wave velocities respectively and $\gamma = \left(\frac{z}{r}\right)^2$. To get the frequency-domain vertical velocity, we multiply the displacement by $i\omega$ in the Fourier domain (for the derivative):

$$v_z(\boldsymbol{x}, \omega) = i\omega \cdot u_z(\boldsymbol{x}, \omega). \tag{8}$$

Our example comprises a volume of size $12 \times 12 \times 4.5$ km filled with constant elastic properties: $V_p = 2600$ m/s, $V_s = 1500$ m/s and $\rho = 2210$ kg/m³. This volume is discretized using a uniform grid of 60 m in the horizontal directions and 15 m in the vertical direction. An analytical solution for vertical velocity is computed for a frequency of 10 Hz using formula (8), assuming that the origin is the middle of the volume. To obtain our solver solution, we surround the computational area with absorbing layers. In Fig. 6 the corresponding solutions along the vertical line $x = 6400$ m and $y = 3200$ m are given, showing good agreement between the results. We compute the root mean square (RMS) error in percent

$$\text{RMS} = 100 \cdot \frac{\|u_{solver} - u_{exact}\|_{L_2}}{\|u_{exact}\|_{L_2}}, \tag{9}$$

with u_{exact} and u_{solver} being the exact and solver solutions respectively. In our example, the RMS difference is 0.88%. This small error can be reduced by using denser vertical sampling or using a higher order finite-difference approximation for solving the boundary value problems (6).

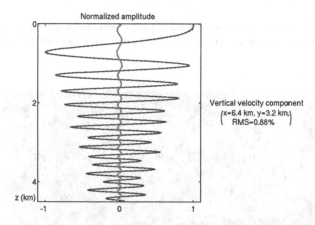

Fig. 6. 1D vertical profiles (real part) of V_z computed in the homogeneous model at receiver location x = 6400 m, y = 3200 m. The exact solution (dashed blue line) overlies our solver solution (solid red line) and the residuals (solid green line) are multiplied by 5. (Color figure online)

We compare the convergence rate of our solver to another 3D elastic iterative solver – CARP-CG [15]. Generating the solution at 7.5 Hz using the homogeneous model our solver converges in 39 iterations in 794 s, whereas other solver converges in 1402 iterations in 1244.42 s (these data were taken from Table 9 in [15]). The number of cores involved in the computations were the same for both solvers. It is worth mentioning, that the computational time comparison is a bit unfair, since the hardware for running these solvers were slightly different.

4.2 Convergence Analysis in the Marmoussi Model

To understand how the convergence rate varies with increasing frequency we consider the full 2.5D model depicted in Fig. 2 containing more than 90 million cells. Simulations were performed at different low frequencies from 2 to 10 Hz, considered a sufficient range for macro velocity reconstruction with FWI. From Fig. 7 we infer, that

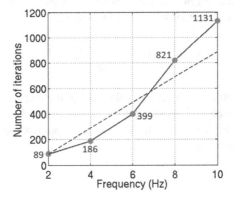

Fig. 7. Convergence versus frequency in the 2.5D Marmousi model: the blue line is the iterative solver convergence curve and the dashed grey line is the linear increase with slope 10. (Color figure online)

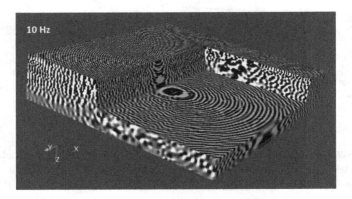

Fig. 8. A 3D view of the real part of a frequency-domain wavefield of the V_z component computed for the model depicted in Fig. 2.

for the higher frequencies the convergence is slower. In this particular case, the convergence curve varies around the linear increase with slope 10. It is worth mentioning, that pure BiCGSTAB would diverge in such a model (number of iterations > 10000) at any of those frequencies. The 10 Hz monochromatic component of the computed wavefield is presented in Fig. 8. Using 9 nodes with 7 MPI processes per node and 4 cores per process, the total computation time is 348 min.

5 Conclusions

We present a parallel iterative solver capable of modeling wavefields in 3D elastic land models of big size at low frequencies. The solver includes both MPI and OpenMP to reduce the computation time and shows good scalability. Further improvement of MPI scaling may be achieved by incorporating domain decomposition along the two horizontal directions into the current MPI parallelization scheme. Another strategy, that may be also considered, is parallelization using domain decomposition along the vertical direction for solving the boundary value problems for Eq. 4.

Acknowledgments. We are grateful to Vincent Etienne and Michael Jervis for reviewing of our manuscript. Special thanks to Maxim Dmitriev for useful discussions and advice on this topic. Two of the authors (Victor Kostin and Vladimir Tcheverda) have been sponsored by the Russian Science Foundation grant 17-17-01128.

References

1. Albanese, C., Gilblom, K.: This oil major has a supercomputer the size of a soccer field. Bloomberg, 18 January 2018
2. Aminzadeh, F., Brac, J., Kuntz, T.: 3-D Salt and Overthrust Models. SEG/EAGE Modelling Series, no. 1. SEG Book Series, Tulsa, Oklahoma (1997)
3. Aki, K., Richards, P.G.: Quantitative Seismology, Theory and Methods, vol. 1. W.H. Freeman and Co., San Francisco (1980)
4. Belonosov, M.A., Kostov, C., Reshetova, G.V., Soloviev, S.A., Tcheverda, V.A.: Parallel numerical simulation of seismic waves propagation with intel math kernel library. In: Manninen, P., Öster, P. (eds.) PARA 2012. LNCS, vol. 7782, pp. 153–167. Springer, Heidelberg (2013). https://doi.org/10.1007/978-3-642-36803-5_11
5. Belonosov, M., Dmitriev, M., Kostin, V., Neklyudov, D., Tcheverda, V.: An iterative solver for the 3D Helmholtz equation. J. Comput. Phys. **345**, 330–344 (2017)
6. Berenger, J.P.: Three-dimensional perfectly matched layer for the absorption of electromagnetic waves. J. Comput. Phys. **127**, 363–379 (1996)
7. Darbas, M., Louer, F.: Analytic preconditioners for the iterative solution of elastic scattering problems. HAL, hal-00839653, pp. 1–32 (2013)
8. Erlangga, Y.A., Nabben, R.: On a multilevel Krylov method for the Helmholtz equation preconditioned by shifted Laplacian. Electron. Trans. Numer. Anal. **31**, 403–424 (2008)
9. Etienne, V., Tonellot, T., Thierry, P., Berthoumieux, V., Andreolli, C.: Optimization of the seismic modeling with the time-domain finite-difference method. In: 84th Annual International Meeting, SEG, Expanded Abstracts, pp. 3536–3540 (2014)

10. Intel, 2018, Intel®Math Kernel Library (Intel®MKL). https://software.intel.com/en-us/intel-mkl
11. Khajdukov, V.G., et al.: Modelling of seismic waves propagation for 2D media (direct and inverse problems). In: Malyshkin, V. (ed.) PaCT 1997. LNCS, vol. 1277, pp. 350–357. Springer, Heidelberg (1997). https://doi.org/10.1007/3-540-63371-5_36
12. Kostin, V., Lisitsa, V., Reshetova, G., Tcheverda, V.: Local time-space mesh refinement for simulation of elastic wave propagation in multi-scale media. J. Comput. Phys. **281**, 669–689 (2015)
13. Kostin, V., Neklyudov, D., Tcheverda, V., Belonosov, M., Dmitriev, M.: 3D elastic frequency-domain iterative solver for full-waveform inversion. In: 86th Annual International Meeting, SEG, Expanded Abstracts, pp. 3825–3829 (2016)
14. Kostin, V., Solovyev, S., Liu, H., Bakulin, A.: HSS cluster-based direct solver for acoustic wave equation. In: 87th Annual International Meeting, SEG, Expanded Abstracts, pp. 4017–4021 (2017)
15. Li, Y., Metivier, L., Brossier, R., Han, B., Virieux, J.: 2D and 3D frequency-domain elastic wave modeling in complex media with a parallel iterative solver. Geophysics **80**, T101–T118 (2015)
16. Lisitsa, V., Tcheverda, V., Botter, C.: Combination of discontinuous Galerkin method with finite differences for simulation of elastic wave. J. Comput. Phys. **311**, 142–157 (2016)
17. Operto, S., Virieux, J., Amestoy, P., L'Excellent, J., Giraud, L., Hadj, H.: 3D finite-difference frequency-domain modeling of visco-acoustic wave propagation using a massively parallel direct solver: a feasibility study. Geophysics **72**, SM195–SM211 (2007)
18. Pissarenko, D., Reshetova, G., Tcheverda, V.: 3D finite-difference synthetic acoustic log in cylindrical coordinates: parallel implementation. J. Comput. Appl. Math. **234**(6), 1766–1772 (2010)
19. Pratt, R.G.: Seismic waveform inversion in the frequency domain, Part 1: theory and verification in a physical scale model. Geophysics **64**, 888–901 (1999)
20. Rizzuti, G., Mulder, W.A.: A multigrid-based iterative solver for the frequency-domain elastic wave equation. In: 77th EAGE Conference and Exhibition, Expanded Abstracts, pp. 1–4 (2015)
21. Saad, Y.: Iterative Methods for Sparse Linear Systems, 2nd edn. SIAM, Philadelphia (2003)
22. Sirgue, L., Etgen, J., Albertin, U., Brandsberg-Dahl, S.: System and method for 3D frequency domain waveform inversion based on 3D time-domain forward modeling. U. S. Patent, 11/756,384 (2007)
23. Sonneveld, P., van Gijzen, M.B.: IDR(s): a family of simple and fast algorithms for solving large nonsymmetric systems of linear equations. SIAM J. Sci. Comput. **31**, 1035–1062 (2008)
24. Symes, W.W.: Migration velocity analysis and waveform inversion. Geophys. Prospect. **56**(6), 765–790 (2008)
25. Van Der Vorst, H.A.: BI-CGSTAB: a fast and smoothly converging variant of BI-CG for the solution of nonsymmetric linear systems. SIAM J. Sci. Stat. Comput. **13**(2), 631–644 (1992)
26. Virieux, J.: Seismic wave modeling for seismic imaging. Lead. Edge **28**, 538–544 (2009)
27. Wang, S.V., de Hoop, M., Xia, J., Li, X.S.: Massively parallel structured multifrontal solver for time-harmonic elastic waves in 3-D anisotropic media. Geophys. J. Int. **191**, 346–366 (2012)

Performance of Time and Frequency Domain Cluster Solvers Compared to Geophysical Applications

Victor Kostin[1], Sergey Solovyev[1(✉)], Andrey Bakulin[2], and Maxim Dmitriev[2]

[1] Institute of Petroleum Geology and Geophysics SB RAS, Novosibirsk, Russia
{kostinvi,solovevsa}@ipgg.sbras.ru
[2] Geophysics Technology, EXPEC ARC, Saudi Aramco, Dhahran, Saudi Arabia
{andrey.bakulin,maxim.dmitriev}@aramco.com

Abstract. In the framework of frequency-domain full waveform inversion (FWI), we compare the performance of two MPI-based acoustic solvers. One of the solvers is the time-domain solver developed by the SEISCOPE consortium. The other solver is a frequency-domain multifrontal direct solver developed by us. For the high-contrast 3D velocity model, we perform the series of experiments for varying numbers of cluster nodes and shots, and conclude that in FWI applications the solvers complement each other in terms of performance. Theoretically, the conclusion follows from considerations of structures of the solvers and their scalabilities. Relations between the number of cluster nodes, the size of the geophysical model and the number of shots define which solver would be preferable in terms of performance.

Keywords: Geophysical problem · 3D acoustic solvers
Frequency-domain · Time-domain · Modeling · Sparse matrix
Low-rank approximation

1 Introduction

Numerical simulation of acoustic wavefields is an important part of many algorithms developed to solve problems arising in exploration geophysics. In particular, it serves as an engine for full-waveform inversion (FWI) that is typically done in the frequency domain using a hierarchical multiscale strategy (e.g., [8,10,12] and references cited therein). For macro velocity reconstruction, repeated simulations are performed for a number of (usually) low frequencies (up to 10 Hz) at each iteration of the process. In simulations, the pressure wavefield is excited by a point source working as a harmonic oscillator at a particular frequency. In industrial applications, the number of shots can be tens of thousands or more, whereas receivers are even larger in quantity.

Any approach for frequency-domain acoustic wavefield simulation has both advantages and drawbacks that may be crucial in a particular situation. Here, we consider two representative solvers and compare their accuracy and performance.

© Springer Nature Switzerland AG 2019
V. Voevodin and S. Sobolev (Eds.): RuSCDays 2018, CCIS 965, pp. 343–353, 2019.
https://doi.org/10.1007/978-3-030-05807-4_29

One conventional solver is based on time-domain simulation followed by Fourier transform. Alternative approaches directly tackle the Helmholtz equation. In 3D, it requires solving a system of linear equations with a huge sparse coefficient matrix. Iterative solvers are one notable member of this family [2,9]. Comparatively new ideas of intermediate data compression [5,13,14] applied to solution algorithms for linear equation systems with sparse coefficient matrices make using direct solvers possible. Our second solver for comparison utilizes these ideas.

2 Method

The wave equation

$$\frac{\partial^2 p}{\partial t^2} - c^2(x, y, z)\Delta p = f(t)\delta(x - x_s, y - y_s, z - z_s), \tag{1}$$

can be solved in the time-domain for a particular point source position (x_s, y_s, z_s) and we can then apply the Fourier transform

$$\hat{p}(\omega, x, y, z) = \int e^{-i\omega t} p(t, x, y, z) dt \tag{2}$$

to obtain the frequency-domain solution. Here $c(x, y, z)$ denotes velocity at point (x, y, z), $f(t)$ is the source function, δ denotes the Dirac delta function and $p(t, x, y, z)$ is the pressure at point (x, y, z) and time t. Here we use a time domain solver developed by the SEISCOPE consortium (https://seiscope2.osug.fr). We refer to it as the time-domain finite-difference (TD) solver. Perfectly matched layers (PML) [3] are used to decrease the influence of the boundaries on the computational domain.

Provided the Fourier transform $\hat{f}(\omega)$ of the source function is not zero at the frequencies of interest, the solutions for all frequencies become available at the same time. For different source point positions, computations are done independently of each other. This explains the ideal MPI scalability of the TD solution with respect to the number of shots when one shot is assigned per one cluster node.

In the frequency-domain approach, the Helmholtz equation

$$\Delta u + \frac{\omega^2}{c^2(x, y, z)} u = \delta(x - x_s, y - y_s, z - z_s) \tag{3}$$

is solved in the domain of interest. Supplying the equation with PMLs and zero boundary conditions one comes to a boundary value problem. The finite-difference approximation of the boundary value problem for Eq. (3) transforms this to a system of linear equations

$$Au = f. \tag{4}$$

Coefficient matrix A of this system is sparse, complex-valued (due to use of PMLs) and symmetric. Its non-zero elements depend on the values of velocity $c(x, y, z)$ in the computational domain and on frequency ω. The right-hand side vector f is defined by the source point position.

The multifrontal [4] direct approach to solve the system (4) consists of factorization of a permuted (P stands for some permutation) matrix

$$\hat{A} = PAP^t = LDL^t. \tag{5}$$

into a product of triangular matrices L, L^t and a diagonal matrix D, followed by solving a system with triangular coefficient matrices

$$y = L^{-1}P^t f;$$
$$z = D^{-1}y; \tag{6}$$
$$u = PL^{-t}z.$$

Factorization (5) is done once and can be used for all right-hand sides. This is the main advantage of the direct solution of boundary value problems for the Helmholtz equation with many source points. We refer to this approach as the frequency-domain finite-difference (FD) direct solver.

The direct approach under consideration faces a challenge called a *fill-in phenomenon* – the triangular factor L has many more non-zero elements than matrix A that prevents an unlimited increase of the number of grid points. To overcome this difficulty, we introduce intermediate data compression [6,11] in the solver. In this way, non-zero ($m_{ij} \times n_{ij}$) blocks L_{ij} of triangular matrix L are approximated by products

$$L_{ij} \cong U_{ij}V_{ij}^t \tag{7}$$

where U_{ij} is a $m_{ij} \times r$ matrix, V_{ij} is a $n_{ij} \times r$ matrix with small value of rank r. Instead of storing block L_{ij} factors, U_{ij} and V_{ij} are stored. This trick helps to reduce memory requirements by 5–6 times and also reduces the computational effort. To find matrices U_{ij} and V_{ij} approach based on randomized sampling [7] is used. This data compressed technique is similar to hierarchically semi separable (HSS) methods [5,13,14] and the respective solver is referred to below as the HSS solver.

The HSS technique is not sufficient by itself to solve huge problems. To scale this for larger systems, the factorization in the solver is implemented on distributed memory systems with data distributed among multiple cluster nodes.

3   Algorithms of MPI-Parallelization

The time-domain (TD) solver is easily parallelized and scales ideally with respect to number of right hand sides (N_{shots}). If the number of cluster nodes $N_{nodes} >$

N_{shots} one shot can be solved on each node without any data communications between nodes. Therefore, the computational time of the TD solver is

$$T_{TD}(N_{proc}, N_{shots}) = T_{TD}(1,1)max\left(\frac{N_{shots}}{N_{proc}}, 1\right) \tag{8}$$

and can be estimated by solution time of 1 shot on 1 node.

The frequency-domain (FD) solver uses another type of parallelization because of the large memory and computation requirements of LDL^t factorization. This factorization requires much more memory than the TD solver, so for each problem there is a minimum number of cluster nodes required to solve it.

Another obstacle of FD parallelization is the large number of sequential operations in the factorization process that prevents achieving ideal parallelization.

The idea of parallelizing the factorization step is based on the special structure of the A- and L-factors which results from the permutation process in Eq. (5) [4]. This structure can be associated with the elimination tree [6]. The details of such parallelization, including the scaling measurements are shown in paper [6]. The scalability of this approach is good for a small number of cluster nodes and becomes worse for many cluster nodes.

Performance can be improved in the case of many shots (i.e., when the solver time is more than the factorization time) by using a modification of MPI-parallelization for the solver. The details of this improvement are below.

HSS computational time is the sum of T_{FCT} (reordering+factorization) and T_{SLV} (solve time, i.e. inversion LDL^t factors):

$$T_{HSS}(N_{proc}, N_{shots}) = T_{FCT}(N_{proc}) + T_{SLV}(N_{proc}, 1)N_{shots} \tag{9}$$

Starting from some number of nodes, T_{HSS} doesn't decrease with increasing N_{proc}. The reason is the poor scalability both of factorization and solve steps for many cluster nodes. Tests show $T_{FCT}(N_{proc}) \approx T_{FCT}(\hat{N})$ and $T_{SLV}(N_{proc}, 1) \approx T_{SLV}(\hat{N}, 1)$ for a constant $\hat{N}$.

One idea for improving scalability is separating all shots into groups and solve these groups in parallel on different sets of cluster nodes. The $M = N_{proc}/\hat{N}$ groups (sets) have been proposed:

set #1: Shots $[1 \ldots \frac{N_{shots}}{M}]$ are solved on the $[1 \ldots \hat{N}]$ nodes;
set #2: Shots $[\frac{N_{shots}}{M} + 1 \ldots 2*\frac{N_{shots}}{M}]$ are solved on the $[\hat{N}+1 \ldots 2*\hat{N}]$ nodes; $\ldots$
set #M: Shots $[(M-1)*\frac{N_{shots}}{M} + 1 \ldots M*\frac{N_{shots}}{M}]$ are solved on the $[(M-1)\hat{N} + 1 \ldots N_{proc}]$ nodes.

The computational time can be written as:

$$T_{HSS}(N_{proc}, N_{shots}) = \begin{cases} T_{FCT}(N_{proc}) + T_{SLV}(N_{proc}, 1)N_{shots}, & \text{if } N_{proc} <= \hat{N} \\ T_{FCT}(\hat{N}) + T_{SLV}(\hat{N}, 1)\hat{N}max\left(\frac{N_{shots}}{N_{proc}}\right), & \text{if } N_{proc} > \hat{N}. \end{cases} \tag{10}$$

Selecting $\hat{N}$ can be done through trial and error. The final scalability and computation time both for time-domain and frequency domain solvers are presented below.

4   Numerical Experiments

Performance and accuracy tests were run on the Shaheen II supercomputer (https://www.hpc.kaust.edu.sa/content/shaheen-ii) with the following hardware characteristics: 2×Intel®Xeon®CPU E5-2698 v3 @2.3 GHz per cluster node, 128 GB RAM/per node. Theoretical Peak (Rpeak) performance of one node is 1,2TFlop/s. The main data storage solution is a Lustre Parallel file system based on Cray Sonexion 2000 with a usable storage capacity of 17.2 PB delivering around 500 GB/s of I/O throughput.

For the purpose of numerical comparison, we use a Overthrust (referred below as OT) model (see [1]) with dimensions of $9 \times 9 \times 4.5$ km. In this model, velocity varies between 2 300 m/s and 6 000 m/s. Example model cross-sections are shown in Fig. 1.

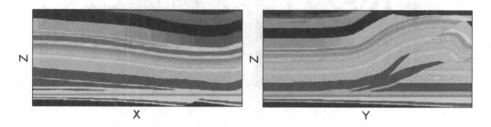

Fig. 1. The OT velocity model showing cross-sections in the X-Z plane (left), and the Y-Z (right).

Table 1 lists the parameters of the numerical experiments used to evaluate the solvers in the OT model. To provide approximately the same number of points per wavelength (ppw) for frequencies 5, 7 and 15, we computed solutions using different grid steps, h, in meters. The corresponding numbers of cells along each axis are shown in columns Nx, Ny and Nz along with the total numbers of grid points (N). The time discretization step used for TD is shown in the last column. The TD solver computations were performed using a time interval $[0, 10\,\text{s}]$.

Table 1. Parameters of numerical experiments.

$\nu(Hz)$	$h(m)$	ppw	Nx	Ny	Nz	N	dt(s)
5	90	5.1	110	110	52	$0.6 \cdot 10^6$	0.00742
7	60	5.4	165	165	78	$2.1 \cdot 10^6$	0.00495
15	30	5.1	330	330	155	$17 \cdot 10^6$	0.00247

The sources were placed at a depth of one grid point at the center of the domain. Receivers were placed throughout the X-Y plane at the same depth. To demonstrate that the HSS solver is working correctly, we provide 2D snapshots of the real part of the computed wavefield taken at the depths of the receiver plane (Fig. 2). Figures 3 and 4 show TD and HSS solutions along selected profiles from Fig. 2, as well as the magnified difference between two solutions.

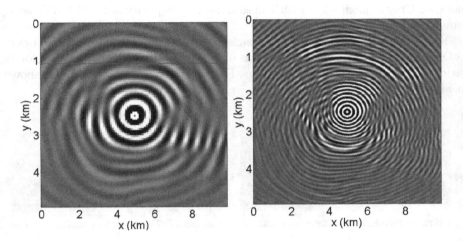

Fig. 2. Snapshots of the solutions (real part) computed at the plane of surface receivers using a HSS solver. The left picture is for 5 Hz, and the right is for 15 Hz. (Color figure online)

Good agreement can be observed between curves in Fig. 3 for the 5 Hz and in Fig. 4 for the 15 Hz solutions.

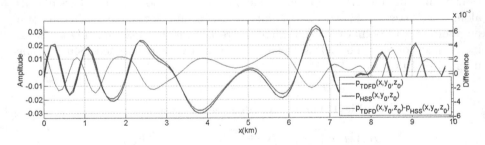

Fig. 3. Real part of the frequency-domain solution at 5 Hz shown along the red profile from Fig. 2. Red and blue lines correspond to TDFD and HSS solutions, respectively; black line represents the magnified (x5) difference between them. (Color figure online)

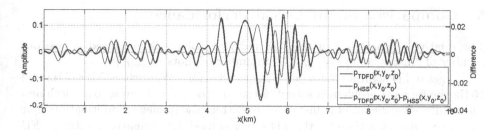

Fig. 4. Real part of the frequency-domain solution at 15 Hz shown along the red profile from Fig. 2. Red and blue lines correspond to TDFD and HSS solutions, respectively; black line represents the magnified (x5) difference between them. (Color figure online)

Four metrics were computed to quantify the difference between solutions. Three of them are simple relative differences of two grid functions u and v computed using classical functional norms:

$$\beta_k(u,v) = \frac{||u-v||_k}{||u||_k}, \quad k = 1, 2, \infty \tag{11}$$

If computations of the norms in formula (11) are done for spheres of radius r and centered at the source point, one gets function $\beta_k(r)$. The wavefields u,v in formula (11) have singularities at the source locations. Therefore, we exclude a sphere of small radius r_0 around the source from the volume of interest. The last (fourth) metric is computed in a similar way using:

$$\gamma(u,v) = \left|1 - \left|\frac{(u,v)}{||u||||v||}\right|\right|. \tag{12}$$

The function $\gamma(r)$ is getting by the similar way as $\beta_k(r)$. Graphs of functions $\beta_k(r)$ and $\gamma(r)$ are shown in Fig. 5. The radius r_0 of a small excluded sphere was taken to be equal to $10h/3$. We clearly observe that the discrepancy between the two solvers generally increases with the volume radius and frequency, as is generally expected. A somewhat larger jump between 5 and 15 Hz may suggest insufficient grid size for the complex OT model with high acoustic contrasts.

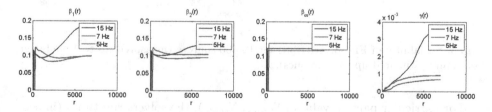

Fig. 5. Difference between solutions obtained with Seiscope and HSS solvers measured as a function of the domain size.

5   Comparison of Numerical Performance

The relative performance of the TD and HSS solvers is measured on various
number of cluster nodes (N_{nodes}) and number of shots (N_{shots}).

To get a TD solution for specific frequencies for FWI such as 5 Hz, 7 Hz and
15 Hz the forward problem is solved in the time domain (1) on a 30 m mesh and
the result forward Fourier transformed to extract the frequency information. The
total computational time for the TD solver is used for comparison with the FD
solver.

To get a solution using the FD solver for these frequencies, the Helmholtz
problem (3) is solved for each frequency 5 Hz, 7 Hz and 15 Hz using parameters
from Table 1. The sum of these three FD times are presented below. The computational times for coarse (90 m) and medium (60 m) meshes are significantly
less than times for a fine mesh (30 m). Table 2 shows the ratio of the TD and
FD solver computation times. The sum of times for three meshes is used for the
T_{HSS} values.

The scaling is shown in the Fig. 6.

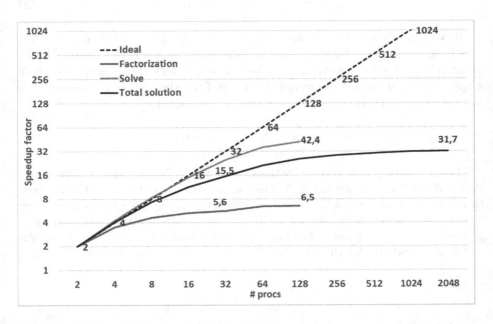

Fig. 6. Scalability of FD solver (HSS) for 12 800 sources (measured up to 128 cluster
nodes and estimated up to 2048 ones).

For particular pairs of values (N_{nodes}, N_{shots}), the solvers run times (in seconds) are provided in Table 2 in the form of fractions, where run time T_{TD} is
put as the numerator and T_{HSS} as the denominator (Fig. 7).

The TD solver scales ideally with respect to N_{shots}. Measured run time for
one shot on one node is 161 s. 128 shots can be computed on 32 nodes for

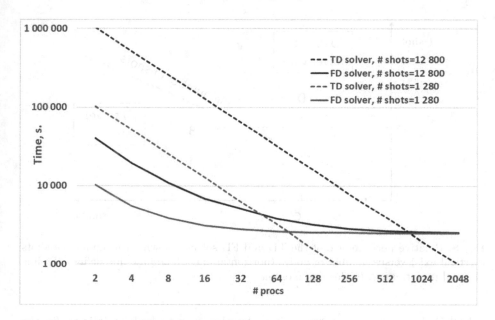

Fig. 7. Timing of TD and FD solvers (measured up to 128 cluster nodes and estimated up to 2048 ones).

Table 2. Timing results for two solvers represented as ratios with T_{TD}/T_{HSS}. Note that for ratios >1.0 the HSS solver becomes relatively more efficient.

N_shots	1	128	1280	12800
N_nodes				
32	161/2945	644/2978	6440/3275	64400/6245
64	161/2625	322/2648	3220/2857	32200/4945
128	161/2555	161/2575	1610/2757	16100/4570

644 s, on 128 nodes for 161 s, but further increases in N_{nodes} do not lead to run time reduction. Scalability of the HSS solver for one shot is determined by the scalability of factorization and decays with increasing N_{nodes}. Increasing N_{shots} gives additional opportunity to improve scalability.

Figure 8 illustrates how the FD and TD solvers complement each other when comparing N_{shots} and N_{nodes}.

The blue line in Fig. 8 defines the line of equal performance of the two solvers. For a given number of nodes (N_{nodes}), this line defines the number of shots that is "big enough" to fully reap the benefits of the HSS solver and reach the numerical performance of the TD solver. For example, for a problem with a comparatively small number of shots which has to be solved on a particular cluster (a point close to point C of line segment CD), the fastest way would be using the TD solver. If the number of shots increases (the point moves to D along CD) then

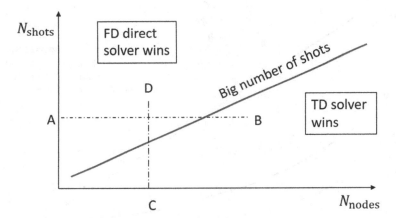

Fig. 8. Relative performance of the TD and FD solvers shown with number of shots (vertical axis) versus number of nodes (horizontal axis). The blue line defines the line of equal performance. (Color figure online)

a FDFD direct solver becomes faster. Another way to look at it is to fix the number of shots, such as defined by horizontal line AB. Then moving from point A towards B along horizontal segment AB means solving the same problem on clusters with increasing number of nodes. For a fixed number of shots, a FDFD direct solver would usually be faster for comparatively small clusters, but on larger clusters TD is more efficient.

6   Conclusions

We compare two solvers for the acoustic wave equation for different cluster sizes and variable numbers of shots. The time-domain finite-difference (TD) solver scales perfectly, with effort linearly increasing with number of shots. In contrast, a direct frequency-domain HSS solver obtains the global solution for the entire domain and all the shots. HSS computational effort is non-linearly dependent on available resources, and the benefit/cost ratio increases sharply with increasing number of shots. We have performed a series of numerical experiments with real-world scenarios using the Overthrust velocity model and the Shaheen II supercomputer. These experiments demonstrate the existence of the line of equal performance, comparing number of shots versus number of available nodes. The TD solver usually wins in the segment of comparatively large values of the ratio N_{nodes}/N_{shots} and the HSS solver wins where the ratio is lower.

While generally larger numbers of nodes become available with time (as computing power becomes cheaper), this is offset by the rapidly increasing trace density of seismic acquisition systems that are constantly growing in number of shots and receivers per square kilometer. These two concurrent trends would likely maintain a need for both types of solvers depending on exact survey geometry, available computing power and number of shots for a particular FWI problem. Therefore, an optimal FWI toolbox should contain both solvers so that the most efficient one can be used for a specific scenario based on the line of equal performance revealed in this study.

Acknowledgments. We also appreciate KAUST for providing access to Shaheen II supercomputer.

References

1. Aminzadeh, F., Brac, J., Kuntz, T.: 3-D salt and overthrust models. In: SEG/EAGE Modelling Series. SEG Book Series (1997)
2. Belonosov, M., Dmitriev, M., Kostin, V., Neklyudov, D., Tcheverda, V.: An iterative solver for the 3D Helmholtz equation. J. Comput. Phys. **345**, 330–344 (2017). https://doi.org/10.1016/j.jcp.2017.05.026
3. Collino, F., Tsogka, C.: Application of the perfectly matched layer absorbing layer model to the linear elastodynamic problem in anisotropic heterogeneous media. Geophysics **66**, 294–307 (2001)
4. Duff, I.S., Reid, J.K.: The multifrontal solution of indefinite sparse symmetric linear systems. ACM Trans. Math. Softw. (TOMS) **9**(3), 302–325 (1983). https://doi.org/10.1145/356044.356047
5. Hackbusch, W.: A sparse matrix arithmetic based on H-Matrices. Part I: introduction to H-matrices. Computing **62**(2), 89–108 (1999). https://doi.org/10.1007/s006070050015
6. Kostin, V., Solovyev, S., Liu, H., Bakulin, A.: HSS cluster-based direct solver for acoustic wave equation. In: 87th Annual International Meeting, SEG Technical Program Expanded Abstracts, pp. 4017–4021 (2017)
7. Martinsson, P.G.: A fast randomized algorithm for computing a hierarchically semiseparable representation of a matrix. SIAM J. Matrix Anal. Appl. **32**(4), 1251–1274 (2011)
8. Mulder, W.A., Plessix, R.E.: How to choose a subset of frequencies in frequency-domain finite-difference migration. Geophys. J. Int. **158**(3), 801–812 (2004)
9. Plessix, R.E.: A Helmholtz iterative solver for 3D seismic-imaging problems. Geophysics **72**(5), SM185–SM194 (2007). https://doi.org/10.1190/1.2738849
10. Shin, C., Cha, Y.H.: Waveform inversion in the Laplace domain. Geophys. J. Int. **173**(3), 922–931 (2008)
11. Solovyev, S., Vishnevsky, D., Liu, H.: Multifrontal hierarchically semi-separable solver for 3D Helmholtz problem using optimal 27-point finite-difference scheme. In: 77th EAGE Conference and Exhibition Expanded Abstracts (2015)
12. Virieux, J., et al.: Seismic wave modeling for seismic imaging. Lead. Edge **28**(5), 538–544 (2009)
13. Wang, S., de Hoop, M., Xia, J.: On 3D modeling of seismic wave propagation via a structured parallel multifrontal direct Helmholtz solver. Geophys. Prospect. **59**, 857–873 (2011). https://doi.org/10.1111/j.1365-2478.2011.00982.x
14. Xia, J.: Efficient structured multifrontal factorization for large sparse matrices. SIAM J. Sci. Comput. **35**, A832–A860 (2013). https://doi.org/10.1137/1208670

Population Annealing and Large Scale Simulations in Statistical Mechanics

Lev Shchur[1,2]([☒]) [iD], Lev Barash[1,3][iD], Martin Weigel[4][iD], and Wolfhard Janke[5][iD]

[1] Landau Institute for Theoretical Physics, 142432 Chernogolovka, Russia
`levshchur@gmail.com`
[2] National Research University Higher School of Economics, 101000 Moscow, Russia
[3] Science Center in Chernogolovka, 142432 Chernogolovka, Russia
[4] Applied Mathematics Research Centre, Coventry University,
Coventry CV1 5FB, UK
[5] Institut für Theoretische Physik, Universität Leipzig,
Postfach 100 920, 04009 Leipzig, Germany

Abstract. Population annealing is a novel Monte Carlo algorithm designed for simulations of systems of statistical mechanics with rugged free-energy landscapes. We discuss a realization of the algorithm for the use on a hybrid computing architecture combining CPUs and GPGPUs. The particular advantage of this approach is that it is fully scalable up to many thousands of threads. We report on applications of the developed realization to several interesting problems, in particular the Ising and Potts models, and review applications of population annealing to further systems.

Keywords: Parallel algorithms · Scalability
Statistical mechanics · Population annealing
Markov Chain Monte Carlo · Sequential Monte Carlo
Hybrid computing architecture CPU+GPGPU

1 Introduction

Over the last decade or so, the development of computational hardware has followed two main directions. The first is based on the parallelization of devices based on traditional silicon technology. This approach splits in turn into two main groups. The first concerns the further development of CPUs featuring a growing number of cores placed onto a single silicon die. This direction is sometimes also known as the heavy-core approach, expressing the fact that these cores are as complex as traditional full CPUs themselves. Currently, the typical maximum number of such cores is of the order of 30. The second group concerns the development of auxiliary computing devices, which in turn is represented by mainly two subgroups. The first type is based on Intel's Xeon Phi architecture, featuring more than 70 relatively light, but otherwise fully developed x86 cores on a single chip. The second and much larger group comprises graphics processing units (GPUs) featuring several thousands of light cores in one device. The

ⓒ Springer Nature Switzerland AG 2019
V. Voevodin and S. Sobolev (Eds.): RuSCDays 2018, CCIS 965, pp. 354–366, 2019.
https://doi.org/10.1007/978-3-030-05807-4_30

use of GPUs in general purpose and scientific computing is sometimes known as GPGPU – general purpose GPU, used outside of graphics applications. It is mostly due to the progress of development within this stream that Moore's famous law still holds today, even at the level of supercomputers.

The second main direction of recent development of computational hardware concerns the construction of computers based on quantum bits (qubits). This development also splits into (at least) two main streams. Quantum annealers or analog quantum computers calculate the ground state for some class of Hamiltonians. The most famous examples in this class are the D-Wave machines with currently up to about 2000 qubits annealed in parallel. Digital quantum computers, on the other hand, are based on logic realizing quantum operators. Examples of this approach include IBM's 16 qubit universal quantum computing processor as well as the recently announced Google 72 qubit quantum processor Bristlecone.

The big challenge for scientific computing in this landscape is to develop algorithms and computational frameworks that use the available hardware efficiently. Despite the great attention that quantum computers attract worldwide, their use for actual calculations is still quite restricted. In the case of the D-Wave machines, this shortcoming is connected to the fixed Hamiltonian simulated in the annealing that features a rather special topology of connections between qubits. Another open issue relates to the problem of decoherence, where progress is being made by further lowering the operating temperature and minimizing other noise in the system. At present, quantum annealers have not been able to demonstrate a fundamental speed-up against classical machines in the sense of applications to problems that are exponentially hard on classical machines and only polynomial on the quantum computer. Digital or universal quantum computers are at present mostly restricted by the limited size of the available realizations, but more generally there is only a rather limited number of known quantum algorithms showing the profound speed-up against classical methods mentioned above, the most prominent example being Shor's factorization algorithm. Another difficulty concerns error correction, with one of the possible solutions being Kitaev's topological quantum computations based on anyons [1].

Currently available universal supercomputers are still based on the traditional silicon multicore architecture. The motivation for building supercomputers are particularly challenging problems in scientific computing that require such extensive resources. It is very demanding, however, to efficiently use all the power of a supercomputer in a single run. Such a program should be able to run on potentially millions of cores. Hence the computation must be divided into millions of tasks to be scheduled on individual cores. It is thus of crucial importance to develop new, fully scalable algorithms, new programming techniques, and new methods to build programs which can efficiently use the power of supercomputers.

The problem is even more complicated when running code on hybrid supercomputers, with auxiliary accelerated computing devices in addition to the conventional CPUs. In this case, one needs to take into account the specific architecture of the auxiliary devices, too. This includes specific arithmetic/logical operations, a specific memory organization, and a specific input/output layout.

In this paper we present a mini-review of a recently developed parallel algorithm, dubbed "population annealing", for simulations of spin and particle systems. The idea of this approach goes back to the beginning of the century [2,3], and the algorithm as we use it now was presented about a decade later in Ref. [4]. The most promising feature of this algorithm is its near ideal scalability both on parallel and hybrid architectures.

2   Population Annealing

The population annealing algorithm consists of two alternating steps. The first is a cooling step [2], where equilibrium is maintained by differential reproduction (resampling) of replicas. This part is a realization of a sequential Monte Carlo algorithm. At each temperature step the resampling propagates the population from inverse temperature $\beta_i = 1/k_B T_i$ to the target distribution at inverse temperature $\beta_{i+1} = \beta_i + \Delta\beta_i$. The second step is an equilibration of replicas. It is performed for each replica independently, and any Markov Chain Monte Carlo (MCMC) algorithm can be used. The initial configurations at β_0 should be chosen at equilibrium. Typically $\beta_0 = 0$, where equilibrium can be easily guaranteed.

In total the algorithm comprises the following steps:

1. Set up an equilibrium population of $R_0 = R$ independent copies (replicas) at inverse temperature β_0.
2. Propagate the population to the inverse temperature β_i ($i = 1, 2, \ldots$): Resample replicas $j = 1, \ldots, R_{i-1}$ with their normalized Boltzmann weights $\hat{\tau}_i(E_j) = (R/R_{i-1}) \exp\left[-(\beta_i - \beta_{i-1})E_j\right]/Q_i$, where

$$Q_i = \sum_{j=1}^{R_{i-1}} \frac{e^{-(\beta_i - \beta_{i-1})E_j}}{R_{i-1}}. \tag{1}$$

3. Update each replica by θ sweeps of the chosen MCMC algorithm at inverse temperature β_i.
4. Calculate estimates for observables $\mathcal{O}$ as population averages $\sum_j \mathcal{O}_j/R_i$.
5. Goto step 2 unless the target temperature β_f has been reached.

In addition to the basic algorithm, it was pointed out in Ref. [4] that both statistical errors as well as bias can be reduced by combining the results of several independent population annealing simulations. Bias is minimized if the combination is performed as a weighted average of the results from M runs, where the weight of the m-th simulation should be chosen as

$$\omega_m(\beta_i) = \frac{e^{-\beta_i \hat{F}_m(\beta_i)}}{\sum_{m=1}^{M} e^{-\beta_i \hat{F}_m(\beta_i)}}. \tag{2}$$

Here, $\hat{F}_m$ corresponds to an estimate of the free energy from the m-th run that can be deduced from the normalization factors Q_i in Eq. (1), namely [4]

$$- \beta_i \hat{F}(\beta_i) = \ln Z_{\beta_0} + \sum_{k=1}^{i} \ln Q_k, \tag{3}$$

where Z_{β_0} is the partition function at the initial inverse temperature β_0. At least for $\beta_0 = 0$ this can often be determined analytically. It is clear that statistical errors decrease with the size R of the population as $1/\sqrt{R}$, and it was proposed in Ref. [4] that, asymptotically, the bias decays as $1/R$.

It follows that statistical as well as systematic errors can be arbitrarily reduced by adding additional parallel resources used to simulate a larger population. Consequently, population annealing (PA) by construction is an ideally scalable algorithm. With a well-designed implementation, the approach should allow one to efficiently utilize nearly arbitrarily large parallel resources (supercomputers). In particular, the method is well suited for GPUs which are known to be most effective if the number of parallel threads significantly exceeds the number of available cores and latency hiding works well [5]. For GPUs with a few thousand cores, efficient operation is guaranteed with a total of several ten to hundred thousand threads, where each thread simulates a single replica [6]. As a consequence, it is possible to efficiently simulate hundreds of millions of replicas on compute clusters with hybrid CPU+GPU architecture.

3   Algorithmic Improvements

A number of algorithmic improvements to the algorithm as discussed in the previous section have been attempted [6–8]. From the description of the algorithm one identifies a number of tunable parameters: the temperature step $\Delta\beta_i = \beta_i - \beta_{i-1}$, the number of MCMC rounds θ, and the size R of the population. For the approach using weighted averages, one can additionally vary the number M of independent runs. Optimal values of these parameters will depend on the model simulated and on the prescribed accuracy. In Sect. 5 we present some examples of simulations of different models, but first we discuss some considerations regarding the parameter choice.

We first turn to the choice of temperature step. For too large a step the resampling factors will be dominated by rare events, leading to very low diversity in the population after resampling. Very small temperature steps, on the other hand, lead to resampling factors that are all very close to unity, and so the resampling just leaves the population invariant. How can a close to optimal value of the temperature step be found? In Ref. [6] we have proposed an adaptive scheme for choosing temperature steps to achieve this. It calculates the overlap of energy histograms between the neighbouring temperatures, and either increases or decreases the inverse temperature step in a way to keep the overlap of neighbouring energy histograms fixed. We find that values of histogram overlap between 50% and 90% usually provide a good compromise between sufficient histogram overlap without an unnecessary proliferation of temperature steps. If the simulation procedure needs to use averaging over independent runs, obtaining the sequence of temperatures via calculations of histogram overlap is carried out only once, during the first run, and the rest of the runs use the same annealing schedule.

Regarding θ and R, it turns out that θ needs to have a (model dependent) minimal value to ensure that the anneal remains in equilibrium, especially if the energy is a slowly relaxing variable. Beyond that, increasing R reduces statistical as well as systematic errors. A detailed discussion of the dependence of performance on these parameters can be found in Ref. [7].

Another algorithmic improvement is given by the multi-histogram reweighting method, also referred to as Weighted Histogram Analysis Method (WHAM) [9–11]. Assume that we have performed a PA simulation with inverse temperature points $\beta_1, \beta_2, ..., \beta_K$ and population sizes R_i, where $i = 1, ..., K$. The multi-histogram reweighting technique is based on reweighting the measurements from all temperatures to a chosen reference point and combining them by calculating error weighted averages. If we have histograms $P_{\beta_i}(E)$ at the inverse temperatures β_i (normalized such that $\sum_E P_{\beta_i}(E) = R_i$), we obtain from the error-weighted combined histogram at the common reference point [6]

$$\Omega(E) = \frac{\sum\limits_{i=1}^{K} P_{\beta_i}(E)}{\sum\limits_{i=1}^{K} R_i \exp[\beta_i(F_i - E)]}, \tag{4}$$

which is an estimate of the density of states (DOS). As we have the estimates (3) from a PA run at hand, we can immediately evaluate (4) and in many cases find good results without further iterations (which, nevertheless, can always be used to further improve the accuracy). Note that if one refrains from using iterations, one needs to store and update at each temperature step only the single partially summed histogram $\sum_{i=1}^{K} P_{\beta_i}(E)$ for each energy. Therefore, memory requirements are quite moderate and, for example, for the 2D Ising model only $\sim L^2$ values have to be stored in total.

4   GPU Accelerated PA Algorithm

As explained above, the PA algorithm appears to be ideally suited for implementations on massively parallel hardware. We presented an implementation of PA for the Ising model on Nvidia GPUs in Ref. [6]. The individual GPU kernels are [6, 12]:

(1) initialization of the population of replicas (kernel `ReplicaInit`)
(2) equilibrating MCMC process (kernel `checkKerALL`)
(3) calculation of energy and magnetization for each replica (kernel `energyKer`)
(4) calculation of $Q(\beta, \beta')$ (kernel `QKer`)
(5) calculation of the number of copies n_i of each replica i (kernel `CalcTauKer`)

(6) calculation of the partial sums $\sum_{i=1}^{j} n_i$, which identify the positions of replicas in the new population (kernel `CalcParSum`)
(7) copying of replicas (kernel `resampleKer`)
(8) calculation of observables via averaging over the population (kernel `CalcAverages`)
(9) calculation of histogram overlap (kernel `HistogramOverlap`)
(10) updating the sum of energy histograms $\sum_{i=1}^{K} P_{\beta_i}(E)$ for the multi-histogram reweighting (kernel `UpdateShistE`)

Some details for the most important GPU kernels (1)–(4) are as follows:

(1) *Equilibration process.* To create sufficient parallel work for the GPU devices, it turns out to be useful to combine the replica-level parallelism with an additional domain decomposition, thus exploiting also spin-level parallelism. The basic step consists of a checkerboard decomposition of the lattice which allows for independent updates of all spins of one sub-lattice [13,14]. The code works with thread blocks of size `EQthreads`, which should be a power of two. The optimal value of `EQthreads` is 128 for $L < 128$, and it is sometimes beneficial to increase it to 256, 512 or 1024 in the case of large system size $L \geq 128$. Each block of threads works on a single replica of the population, using its threads to update tiles of size $2 \times$ `EQthreads` spins. To this end it flips spins on one checkerboard sub-lattice first, moving the tiles over the lattice until it is covered, synchronizes and then updates the other sub-lattice. An important (if well known) trick for optimization is that the value of the exponential function for the Metropolis criterion should not be calculated at each spin flip, since the exponential is not a fast single-cycle operation. However, there are only a few possible values of ΔE. For example, for the 2D Ising model, $\Delta E = 2Js_i \sum_{\text{neighb}} s_j + 2Hs_i$. A lookup table containing 10 possible values of $\min[1, \exp(-\beta \Delta E)]$ is placed in the fast *texture* memory of the GPU for an optimal performance.

(2) *Calculation of energies and other observables.* The GPU parallel reduction algorithm is employed. First, each of the $N = 2^n$ threads calculates one particular summand. All calculated N summands $s_1, \ldots, s_N$ are placed in shared memory. Then, the first $N/2^i$ threads perform $s_i := s_i + s_{i+N/2^i}$ for $i = 1, 2, 3, \ldots$. This scheme can be visualized as a binary tree [15]. For the case of modern devices with CUDA compute capability ≥ 3.0, we use the "shuffle" operations that allow threads to access registers from different threads in the same warp. The latter approach is faster, but it is not supported by older devices, where we use the former method and store partial results in shared memory.

(3) *Calculation of the normalization factor* $Q_i = \frac{1}{R_{i-1}} \sum_{j=1}^{R_{i-1}} \exp[-(\beta_i - \beta_{i-1})E_j]$. Since Q_i is a large sum, each of the summands can be independently calculated with one GPU thread. Then, the same reduction method as in (2) can be used.

(4) *Calculation of numbers of copies* n_i. For each replica, the values τ_i and n_i are calculated by a separate GPU thread.

In the present program we use the random number generator PHILOX available also in the CURAND library [16,17]. Alternatively, one might employ the generators previously developed by some of us in Refs. [18–21]. These libraries allow to use up to 10^{19} uncorrelated parallel streams of random numbers.

5   Applications

In this Section we discuss a number of applications of the method, including the Ising model that undergoes a continuous phase transition, the Potts model with an additional first-order regime, models with disorder and frustration as well as simulations of off-lattice systems.

5.1   Ising Model and Second-Order Phase Transitions

The Ising model plays the same role in statistical mechanics as the fruit fly in biology. New algorithms and other new ideas are usually tested on the Ising model first. It is the first model of statistical mechanics which was proven to have singularities near the spontaneous transition from the disordered to the ordered phase. The Ising model is the simplest model of a ferromagnet, for example for a piece of iron that develops a spontaneous magnetic moment below the Curie temperature T_c, while it is paramagnetic above T_c. The ferromagnetic Ising model in zero external magnetic field and on a square lattice of linear size L is given by the Hamiltonian

$$H = -J \sum_{\langle ij \rangle} s_i s_j, \tag{5}$$

where $J > 0$. The summation is restricted to nearest neighbours and the variables s_i can take one of two values, $s_i = +1$ or $s_i = -1$. Periodic boundary conditions are assumed. The ground state is achieved when all variables are equal, i.e., either all of them take the value $+1$ or all of them are -1. Hence the ground state is two-fold degenerate. The interpretation is that the variables s_i can be viewed as magnetic moments, pointing only in two directions, up for the $+1$ value of s_i, or down for -1. In other words, all spins are aligned in the ground state, and the total magnetic moment $|M|$ of the system is equal to the maximal possible value L^2 and the energy E is equal to $-2L^2 J$. This ground state is reached at zero temperature. For high temperatures the spins will take random orientations and, consequently, the average magnetic moment will be zero. A phase transition occurs at inverse temperature $\beta_c = J/k_B T_c = \ln(1 + \sqrt{2})/2$ [22]. The model admits an exact solution, which was first derived by Onsager [23]. The resulting detailed knowledge about its non-trivial behavior turns this model into an ideal testing ground for new ideas.

Simulating the square-lattice Ising model with population annealing as implemented on GPU and a reference CPU implementation, we show in Table 1 the performance of the two codes in terms of time per spin flip in nanoseconds, for

Table 1. Peak performance of the CPU and GPU PA implementations in units of the total run time divided by the total number of spin flips performed, for different system sizes. The best GPU performance is achieved for large θ, and here $\theta = 500$ was chosen for a population of $R = 50\,000$ replicas. GPU performance data are for the Tesla K80 (Kepler card) and GeForce GTX 1080 (Pascal card). The sequential CPU code was benchmarked on a single core of Intel Xeon E5-2683 v4 CPU running at 2.1 GHz.

	Time per spin flip (ns)		
	CPU	GPU (Kepler card)	GPU (Pascal card)
$L = 16$	23.1	0.094	0.038
$L = 32$	22.9	0.092	0.034
$L = 64$	22.6	0.092	0.036
$L = 128$	22.6	0.097	0.039

$\theta = 500$ and a population size $R = 50\,000$ for different system sizes. The GPU code is found to be about 620 times faster than the sequential CPU program. The additional application of a multi-spin coding technique yields a further speedup of up to 10 times for the GPU implementation, for details see Ref. [6].

Figure 1 shows the temperature steps generated by our adaptive stepping algorithm applied to the two-dimensional Ising model with $J = 1$, while imposing a histogram overlap of approximately 85% (see also Sect. 3). One can see that the closest inverse temperature spacing is found close to the transition point, as expected. The adaptive algorithm automatically suggests the optimal annealing schedule and often saves a significant amount of computing time.

5.2 Potts Model and First-Order Phase Transitions

A natural generalization of the Ising model is the q-state Potts model with Hamiltonian

$$H = -J \sum_{\langle ij \rangle} \delta_{s_i s_j}, \tag{6}$$

where the spins s_i can take q values, $s_i = 1, 2, ..., q$, and $J > 0$ is the ferromagnetic coupling constant. We study the model on the square lattice with periodic boundary conditions, where the sum in Eq. (6) goes over all nearest-neighbour pairs $\langle ij \rangle$. Below we set $J = 1$ to fix units. The transition temperature of the model follows from self-duality and is given [24,25] by the relation $\beta_t = J/k_B T_t = \ln(1+\sqrt{q})$. The phase transition is continuous for a number of states $q \leq 4$, with additional logarithmic corrections at $q = 4$. For $q > 4$ the phase transition is discontinuous with a jump in the internal energy, i.e., a first-order phase transition. The distinctive features of first-order phase transitions are phase coexistence and metastability, which can be observed experimentally in the process of heating and cooling the system [26]. These effects are also pronounced in simulations with canonical Monte Carlo methods, leading to hysteresis of magnetization and energy profiles in a heating/cooling cycle [27]. The

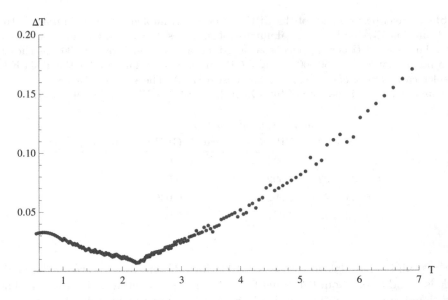

Fig. 1. Schedule of temperature steps for the adaptive temperature-step extension of the PA algorithm applied to the Ising model on an $L = 64$ square lattice with parameters $\theta = 100$, $R = 5000$, and overlap ≈ 0.85.

strength of the transition increases with q. The correlation length ξ does not diverge at the transition temperature T_t, i.e., it is finite, although the value can be extremely large depending on the number of states q. For $q = 20$ it is as small as 2.7 lattice spacings and for $q = 5$ states it is extremely large, $\xi \approx 2552$ [28].

Like standard canonical simulations, PA as described above also leads to hysteresis effects for the Potts model in the first-order regime and as such it is not immediately well suited for simulating systems undergoing first-order transitions. Using the inherent free-energy estimate provided by Eq. (3) leads to a possible way of determining the transition temperature, however. To this end, one extends the free-energy branches of the ordered and disordered phases and locates the transition at the point where the two curves cross [27]. For the case of the cooling temperature schedule, we start with the initial inverse temperature $\beta_0 = 0$ at which the partition function value is given by $Z(\beta_0) = q^N$, where N is the number of lattice sites. Additionally, we also use a heating temperature schedule, where the initial free energy at zero temperature $\beta \to \infty$ can be calculated as [29]

$$- \beta F_{\beta \to \infty} = \ln q - \beta E_0, \tag{7}$$

with the ground-state energy $E_0 = -2N$.

Figure 2 shows the resulting metastable free-energy branches calculated during cooling from infinite temperature and heating from zero temperature for the Potts model with a number of spin components of $q = 6$ and $q = 10$, respectively.

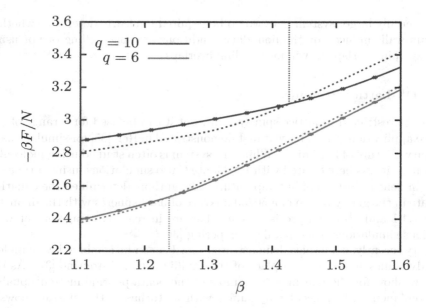

Fig. 2. Metastable free energy for the q-state Potts model with $q = 6$ (green) and $q = 10$ (magenta) for a cooling (solid lines) and a heating (dashed lines) cycle. The vertical dotted lines indicate the locations of the transition points β_t. The lattice size is $L = 32$, and the PA parameters are $\theta = 10$, $R = 10000$, and $\Delta\beta = 0.01$. Taken from Ref. [29]. (Color figure online)

The cooling and heating curves intersect close to the transition temperature, our results being compatible with the presence of small deviations of the order of $1/L^2$ as predicted in Ref. [27].

5.3   Frustrated Models and Spin Glasses

Similar to the parallel tempering heuristic, population annealing is an approach particularly suited for the simulation of systems with complex free-energy land-scapes. A number of applications to frustrated and disordered systems have been reported in the literature. Population annealing has been successfully used for the relaxation to the ground state of a frustrated Ising antiferromagnet on the stacked triangular lattice with a ferromagnetic interlayer coupling [30]. Previous simulations using conventional, canonical Monte Carlo algorithms were not able to reach the ground state. More generally, the properties of PA for ground-state searches of spin systems were investigated in Ref. [31]. Apart from that, PA has been used in a series of simulational studies of spin glasses [32–34]. Among the questions addressed there is that of the number of thermodynamic states in the low-temperature phase. The results appear to be fully compatible with a single pair of pure states such as in the droplet-scaling picture. However, the results for whether or not domain walls induced by changing boundary conditions are space filling are also compatible with scenarios having many thermodynamic states, such as the chaotic pairs picture and the replica-symmetry breaking scheme.

Substantially larger system sizes would be required to clearly determine whether domain walls induced by the boundary conditions are space filling or not using average spin overlaps in windows or link overlaps.

5.4   Off-Lattice Systems

PA is not restricted to lattice spin systems, but it can be used for a range of off-lattice applications too. This was first demonstrated in Ref. [35] for simulations of a binary mixture of hard disks. While this system is often studied using molecular dynamics, it was here treated with PA Monte Carlo simulations, using a range of steps in density instead of in temperature. The authors determined the equation of state in the glassy region of a 50/50 mixture of hard spheres with the diameter ratio 1.4:1 and obtained precise results that are in reasonable agreement with previous simulations using parallel tempering [36].

Very recently, it was also demonstrated that PA can be combined with molecular dynamics simulations in place of the equilibrating sub-routine [37]. As the authors show for the benchmark example of the penta-peptide met-enkephalin, PA combined with molecular dynamics with a stochastic thermostat shows a performance similar to that of the more established parallel tempering method for this problem while providing far superior parallel scaling properties.

6   Conclusion

We have given a brief review of recent developments relating to the population annealing algorithm. The method is attractive especially since it is a highly parallel approach and efficient realizations can be quite easily developed for any known parallel architecture. In particular, we discussed here an implementation of the algorithm with CUDA, and demonstrated an acceleration of the simulations as compared with a conventional CPU realization by several orders of magnitude. The corresponding simulations work well for systems with continuous transitions such as the Ising model. For first-order transitions as present in the Potts model with a sufficiently large number of states, the approach suffers from metastability, but the natural free-energy estimate at least allows for a determination of the transition point by thermodynamic integration. Population annealing in the micro- or multicanonical ensembles promises to provide potential improvements in this respect. There is further scope for improvement in the implementation of the algorithm: a realization within the MPI-CUDA paradigm is currently being developed.

Acknowledgment. This work was partially supported by the grant 14-21-00158 from the Russian Science Foundation and by the Landau Institute for Theoretical Physics in the framework of the tasks from the Federal Agency of Scientific Organizations. The authors acknowledge support from the European Commission through the IRSES network DIONICOS under Contract No. PIRSES-GA-2013-612707.

References

1. Kitaev, A.Yu.: Fault-tolerant quantum computation by anyons. Annals Phys. **303**, 2–30 (2003)
2. Iba, Y.: Population Monte Carlo algorithms. Trans. Jpn. Soc. Artif. Intell. **16**, 279–286 (2001)
3. Hukushima, K., Iba, Y.: Population annealing and its application to a spin glass. In: AIP Conference Proceedings, vol. 690, pp. 200–206 (2003)
4. Machta, J.: Population annealing with weighted averages: a Monte Carlo method for rough free-energy landscapes. Phys. Rev. E **82**, 026704 (2010)
5. Weigel, M.: Monte Carlo methods for massively parallel computers. In: Holovatch, Yu. (ed.) Order, Disorder and Criticality, vol. 5, pp. 271–340. World Scientific, Singapore (2018)
6. Barash, L.Yu., Weigel, M., Borovský, M., Janke, W., Shchur, L.N.: GPU accelerated population annealing algorithm. Comp. Phys. Comm. **220**, 341–350 (2017)
7. Weigel, M., Barash, L.Yu., Shchur, L.N., Janke, W.: Understanding population annealing Monte Carlo simulations (in preparation)
8. Amey, C., Machta, J.: Analysis and optimization of population annealing. Phys. Rev. E **97**, 033301 (2018)
9. Ferrenberg, A.M., Swendsen, R.H.: Optimized Monte Carlo data analysis. Phys. Rev. Lett. **63**, 1195–1198 (1989)
10. Kumar, S., Bouzida, D., Swendsen, R.H., Kollman, P.A., Rosenberg, J.M.: The weighted histogram analysis method for free-energy calculations on biomolecules. I. The method. J. Comp. Chem. **13**, 1011–1021 (1992)
11. Kumar, S., Rosenberg, J.M., Bouzida, D., Swendsen, R.H., Kollman, P.A.: Multidimensional free-energy calculations using the weighted histogram analysis method. J. Comp. Chem. **16**, 1339–1350 (1995)
12. Code repository for the GPU accelerated PA algorithm is located at: https://github.com/LevBarash/PAising
13. Weigel, M.: Performance potential for simulating spin models on GPU. J. Comput. Phys. **231**, 3064–3082 (2012)
14. Yavors'kii, T., Weigel, M.: Optimized GPU simulation of continuous-spin glass models. Eur. Phys. J. Special Topics **210**, 159–173 (2012)
15. McCool, M., Reinders, J., Robison, A.: Structured Parallel Programming: Patterns for Efficient Computation. Morgan Kaufman, Waltham (2012)
16. Salmon, J.K., Moraes, M.A., Dror, R.O., Shaw, D.E.: Parallel random numbers: as easy as 1, 2, 3. In: Proceedings of 2011 International Conference for High Performance Computing, Networking, Storage and Analysis, SC 2011, article no. 16. ACM, New York (2011)
17. Manssen, M., Weigel, M., Hartmann, A.K.: Random number generators for massively parallel simulations on GPU. Eur. Phys. J. Special Topics **210**, 53–71 (2012)
18. Barash, L.Yu., Shchur, L.N.: RNGSSELIB: program library for random number generation, SSE2 realization. Comp. Phys. Comm. **182**, 1518–1526 (2011)
19. Barash, L.Yu., Shchur, L.N.: RNGSSELIB: program library for random number generation. More generators, parallel streams of random numbers and Fortran compatibility. Comp. Phys. Comm. **184**, 2367–2369 (2013)
20. Guskova, M.S., Barash, L.Yu., Shchur, L.N.: RNGAVXLIB: program library for random number generation, AVX realization. Comp. Phys. Comm. **200**, 402–405 (2016)

21. Barash, L.Yu., Shchur, L.N.: PRAND: GPU accelerated parallel random number generation library: using most reliable algorithms and applying parallelism of modern GPUs and CPUs. Comp. Phys. Comm. **185**, 1343–1353 (2014)

22. Kramers, H.A., Wannier, G.H.: Statistics of the two-dimensional ferromagnet. Part I. Phys. Rev. **60**, 252–262 (1941)

23. Onsager, L.: Crystal statistics. I. A two-dimensional model with an order-disorder transition. Phys. Rev. **65**, 117–149 (1944)

24. Baxter, R.J.: Potts model at the critical temperature. J. Phys. C Solid State Phys. **6**, L445–L448 (1973)

25. Wu, F.Y.: The Potts model. Rev. Mod. Phys. **54**, 235–268 (1982). ibid **55**, 315 (1983). Erratum

26. Binder, K., Heermann, D.: Monte Carlo Simulation in Statistical Physics. Springer, Heidelberg (2010). https://doi.org/10.1007/978-3-642-03163-2

27. Janke, W.: First-order phase transitions. In: Dünweg, B., Landau, D.P., Milchev, A.I. (eds.) Computer Simulations of Surfaces and Interfaces, NATO Science Series, II. Mathematics, Physics and Chemistry, vol. 114, pp. 111–135. Kluwer, Dordrecht (2003)

28. Borgs, C., Janke, W.: An explicit formula for the interface tension of the 2D Potts model. J. Physique I **2**, 2011–2018 (1992)

29. Barash, L.Yu., Weigel, M., Shchur, L.N., Janke, W.: Exploring first-order phase transitions with population annealing. Eur. Phys. J. Special Topics **226**, 595–604 (2017)

30. Borovský, M., Weigel, M., Barash, L.Yu., Žukovič, M.: GPU-accelerated population annealing algorithm: frustrated Ising antiferromagnet on the stacked triangular lattice. In: EPJ Web of Conferences, vol. 108, p. 02016 (2016)

31. Wang, W., Machta, J., Katzgraber, H.G.: Comparing Monte Carlo methods for finding ground states of Ising spin glasses: population annealing, simulated annealing, and parallel tempering. Phys. Rev. E **92**, 013303 (2015)

32. Wang, W., Machta, J., Katzgraber, H.G.: Evidence against a mean-field description of short-range spin glasses revealed through thermal boundary conditions. Phys. Rev. B **90**, 184412 (2014)

33. Wang, W., Machta, J., Katzgraber, H.G.: Chaos in spin glasses revealed through thermal boundary conditions. Phys. Rev. B **92**, 094410 (2015)

34. Wang, W., Machta, J., Munoz-Bauza, H., Katzgraber, H.G.: Number of thermodynamic states in the three-dimensional Edwards-Anderson spin glass. Phys. Rev. B **96**, 184417 (2017)

35. Callaham, J., Machta, J.: Population annealing simulations of a binary hard-sphere mixture. Phys. Rev. E **95**, 063315 (2017)

36. Odriozola, G., Berthier, L.: Equilibrium equation of state of a hard sphere binary mixture at very large densities using replica exchange Monte Carlo simulations. J. Chem. Phys. **134**, 054504 (2011)

37. Christiansen, H., Weigel, M., Janke, W.: Population annealing for molecular dynamics simulations of biopolymers. Preprint arXiv:1806.06016

Simulation and Optimization of Aircraft Assembly Process Using Supercomputer Technologies

Tatiana Pogarskaia$^{(\boxtimes)}$ [ID], Maria Churilova [ID],
Margarita Petukhova [ID], and Evgeniy Petukhov [ID]

Peter the Great St.Petersburg Polytechnic University,
195251 Saint Petersburg, Russia
pogarskaya.t@gmail.com, m_churilova@mail.ru,
{margarita, eugene}@lamm.spbstu.ru

Abstract. Airframe assembly is mainly based on the riveting of large-scale aircraft parts, and manufacturers are highly concerned about acceleration of this process. Simulation of riveting emerges the necessity for contact problem solving in order to prevent the penetration of parts under the loads from fastening elements (fasteners). Specialized methodology is elaborated that allows reducing the dimension and transforming the original problem into quadratic programming one with input data provided by disposition of fasteners and initial gap field between considered parts.

While optimization of a manufacturing process the detailed analysis of the assembly has to be done. This leads to series of similar computations that differ only in input data sets provided by the variations of gap and fastener locations. Thus, task parallelism can be exploited, and the problem can be efficiently solved by means of supercomputer.

The paper is devoted to the cluster version of software complex developed for aircraft assembly simulation in the terms of the joint project between Peter the Great St.Petersburg Polytechnic University and Airbus SAS. The main features of the complex are described, and application cases are considered.

Keywords: Aircraft assembly · Optimization · Supercomputing
Task parallelism · Quadratic programming

1 Introduction

During the assembly process, it is important to control both gaps between joined parts and stresses caused by installed fastening elements. On the one hand, tight contact between parts should be achieved; and on the other hand, engineers should avoid cracks, composite layer delamination, and part damage.

V. Voevodin and S. Sobolev (Eds.): RuSCDays 2018, CCIS 965, pp. 367–378, 2019.
https://doi.org/10.1007/978-3-030-05807-4_31

The main goal of the presented work is to develop a special tool that allows performing simulations in order to evaluate displacements and stresses of aircraft parts on the assembly line. For this purpose, specialized software complex ASRP (Assembly Simulation of Riveting Process) is developed for contact problem solving. As a result, we determine the deformed stress state of the assembly loaded by the forces from fastening elements.

This contact problem has following peculiar properties to be taken into account in order to derive efficient algorithm:

1. The contact may occur only in junction area that is known a priori. Thus, there is no need to implement complicated procedures for detection the zone of possible contact.
2. The installed fasteners and rivets restrict relative tangential displacements of assembled parts in the junction area. Therefore, the relative tangential displacements in junction area are negligible in comparison with normal ones. This special feature of the problem justifies implementation of node-to-node contact model that is much simpler than general surface-to-surface model.
3. Loads from fastening elements are applied inside junction area.
4. Only the stationary solution of the problem is of interest.
5. Friction forces between assembled parts in the contact zone do not play significant role due to small relative tangential displacements. So the friction can be omitted from consideration.
6. Stress state of each part in the assembly is described by the linear theory of elasticity.

Solving of considered contact problem comes down to the variation simulation that is used to predict the final assembly state taking into account the part variations. These variations arise from the manufacturing tolerances and can be provided by measurement data or statistical models (see [1, 2]).

The simplest approach is rigid variation simulation when the part deformations are excluded from consideration as in [3]. Consequently, the results are far from the reality. If the mechanical behavior of assembled parts is involved in simulation, then there is a need for finite element analysis (FEA). FEA is used in number of studies and is implemented in specialized commercial software for tolerance analysis [4–6].

Direct application of FEA in variation simulation is inefficient, as even one FEA run may take considerable time for real aircraft models. To overcome this problem, the Method of Influence Coefficients (MIC) is introduced in [1]. The MIC approach establishes linear relationship between a part variation and corresponding assembly variation via sensitivity matrix calculated by FEA. However, possible contact interaction of parts is neglected. Authors of [7] combined MIC with contact modeling for variation simulation in automotive industry.

This paper presents the approach developed in [8] that is similar to MIC to some extent. Reduced stiffness matrix is computed (like sensitivity matrix of influence coefficients), and contact problem is transformed into Quadratic Programming Problem (QPP). Then efficient algorithms are derived for QPP solving. Studies [9, 10] suggest the analogous ideas.

2   The Basics of Numerical Algorithm

We would give here only the main idea of the method, details of numerical algorithms, as well as validation tests, are described in [8].

Let us consider artificial finite element model of the upper wing-to-fuselage junction that is shown in Figs. 1 and 2 (fragment). There are two parts in the assembly: the first part is the wing (light blue) and the second one is the fragment of center wing box (yellow).

Green points in Fig. 2 mark the nodes of possible contact (the nodes in junction area). We denote these nodes as **computational** ones. The set of all computational nodes is referred as **calculation net**.

Using the standard finite element modeling technique, we formulate the contact problem in discrete variation form [11]:

$$\min_{U \in S_h} \left(\frac{1}{2} U^T \cdot K \cdot U - F^T \cdot U \right) \tag{1}$$

Here U is the displacement vector of finite element nodes, K is the stiffness matrix of finite element system, F is the vector of applied loads, S_h is the admissible set that is determined with regard to boundary and non-penetration conditions.

Typically, number of nodes in a finite element model of airframe junction (e.g. wing-to-fuselage junction, as it is shown in the figures above) is much bigger than number of nodes in the junction area. Therefore, the elimination of displacements outside the junction area reduces the problem dimension dramatically. However, we have to divide all the finite element nodes into two groups containing nodes in junction area (green points in Fig. 2) and all the rest. Then the displacement vector can be written as follows

$$U = \begin{pmatrix} U_C \\ U_R \end{pmatrix},$$

where U_C is the vector of node displacements in junction area and U_R is the vector of displacements in other finite element nodes. The same procedure can be implemented for matrix $K = \begin{pmatrix} K_{CC} & K_{CR} \\ K_{CR}^T & K_{RR} \end{pmatrix}$ and vector $F = \begin{pmatrix} F_C \\ F_R \end{pmatrix}$.

Calculating Schur complement of K_{RR}, we get reduced stiffness matrix K_C from the formula $K_C = K_{CC} - K_{CR} \cdot K_{RR}^{-1} \cdot K_{CR}^T$. Now it is possible to derive the reduced QPP:

$$\min_{N \cdot U_C \leq G} \left(\frac{1}{2} U_C^T \cdot K_C \cdot U_C - F_C \cdot U_C \right) \tag{2}$$

where N is the linear operator which defines normal to contact surface, G is the initial gap vector in the junction area.

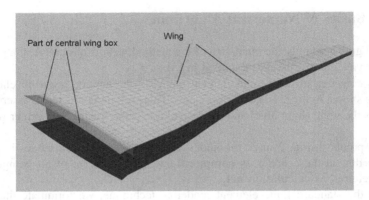

Fig. 1. Finite element model of an artificial wing-to-fuselage junction. (Color figure online)

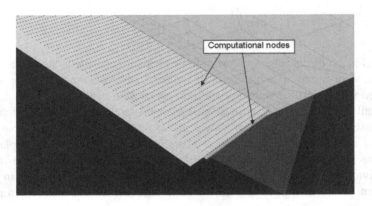

Fig. 2. The nodes of junction area. (Color figure online)

In addition, we mention that U_C contains only the normal components of node displacements. This simplification is possible due to the smallness of tangential displacements regarding normal ones.

Thus, we reduce the initial contact problem to the quadratic programming problem. Moreover, the dimension of the reduced problem is much smaller (e.g. the finite element model depictured in Figs. 1 and 2 has around 130 000 degrees of freedom and the reduced problem (2) for this model has only 16 000 unknowns). This approach is known as substructuring in finite element modeling.

Similar approaches to contact problem solving can be found in [9, 10, 12].

3   Software Overview

3.1   ASRP Desktop Version

The desktop version of the software for assembly simulation is divided into three modules: Preprocessor, Simulator and Postprocessor in order to fully separate the data preparation from the assembly simulation process and subsequent detailed stress analysis, see [13].

ASRP Preprocessor

Preprocessor is designed to prepare models for Simulator on the base of imported finite element model of the assembly in MSC Nastran format.

Preprocessor generates all data structures required for Simulator:

- Reduced stiffness matrix describing mechanical properties of assembled parts. The matrix is computed using MSC Nastran as external finite element solver;
- Geometry for visualization created from finite element mesh;
- Positions of points used for determination of initial gap in ASRP Simulator;
- Positions and diameters of holes for fastening elements in every part.

ASRP Simulator

Simulator is the central part of ASRP software complex. It is designed for the riveting process simulation. This tool permits calculating gaps between assembled parts, absolute displacements, reaction forces caused by contact in junction area, loads in fastening elements needed to achieve contact. In addition, Simulator provides great variety of extra tools for simulation of riveting process and optimization of assembly technology:

- Capabilities for statistical analysis using sets (clouds) of random gaps. For example, user can compute the percentage of examined points with resulting gap within given range (e.g. less than 0.2 mm) for predefined arrangement of fasteners;
- Automatic positioning of fastening elements in order to minimize gap by given number of fasteners. In doing so user can consider either determined initial gap or the cloud of random gaps with given roughness and deviation;
- Powerful tools for editing and visualization of fastening elements (including work with groups of fastening elements, a special library of standard fastening elements etc.);
- Different options for visualization of simulation results;
- Automation of simulation process using script files.

The Simulator is a standalone application that does not need any external software (like MSC Nastran) but can exchange data with other ASRP parts and third party software (e.g. import of measured initial gap or export the instructions for fitting machining).

ASRP Postprocessor

ASRP Postprocessor is aimed at computing the stresses caused by the riveting process.

The main purpose of this module is to evaluate the stresses arising during the assembly process of aircraft junction without solving the contact problem by standard means of finite element analysis but using the results of ASRP simulation instead. Thus, the input data for Postprocessor are the finite element model of junction appropriate for stress computations (in MSC Nastran format) and the file with computation results exported from ASRP Simulator. Postprocessor makes it possible to apply the results imported from ASRP Simulator to the finite element model as the boundary conditions for subsequent static stress analysis.

3.2 ASRP Cluster Version

In order to obtain robust and reliable results, the assembly should be thoroughly analyzed over the wide range of input data that may include the initial gap measurements from the final assembly line or the information about geometric tolerances. Even if the input (i.e. initial gap) is undefined during manufacturing stage, it can be generated in ASRP Simulator using statistical methods and certain gaps properties [14].

According to the assembly technology, the aircraft parts are temporary connected with fasteners installed in about 50% of all holes. These fasteners are called temporary ones and their main objective is to provide contact between the parts for further technological operations. The challenge we face is to find the best temporary fastener positions (fastener pattern) using minimum possible number of elements that still provides sufficient quality of the assembly or to rearrange existing fasteners to minimize all the gaps.

Thus, we have to deal with the clouds of initial data that may account several hundreds of entities that obviously cause significant increase of computations. To overcome this difficulty, the cluster version of ASRP Simulator is developed.

ASRP Cluster Version (CV) is a console application written in C++ that uses MPI for process communication. It does not require any external libraries for executing. Figure 3 illustrates the flowchart of ASRP CV.

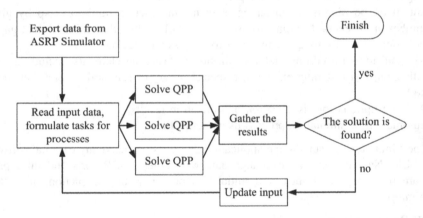

Fig. 3. ASRP CV flowchart

The computations with different input data are so-called task parallel as they can be done independently. The aim is to state the correct QPP problem for each parallel process and then to gather the resulting data at the root process.

The core function in the flowchart is solving QPP (2). Three solvers that are based on the most common methods for such kind of problems are implemented in ASRP CV:

1. Active Set method [15, 16] and its adaptation to the features of given problem [8];
2. Interior Point method [17];
3. Projected Gradient method [18] and its adaptation.

Depending on the considered model user may choose the most appropriate and the fastest solver.

Let us consider the application issues of ASRP CV.

Verification of Current Fastener Configuration

The gap values computed under the loads from the fastener configuration are checked against some predefined value for each initial gap from the cloud. The percentage of "closed" (relatively small) gaps is calculated.

Each process receives its own initial gap field, QPP is solved, and then the root process gathers the statistics for computed gap values. Verification is done in one pass, no update input in Fig. 3 is needed.

Fastener Initial Positioning

Sometimes it is necessary to install a fixed number of new fastening elements. This process is iterative and starts having no new fasteners installed. Each process solves the QPP with the specific gap and identifies the hole for the next fastener according to some criteria. Then the root process gathers data from all the processes, chooses the most suitable hole for fastener installation, and broadcasts its index to all the processes. The algorithm continues until the required number of installed fasteners is reached.

Optimization of Fastener Positions

Due to the time-consuming calculation of the objective functions (calculating resulting gaps with each pattern modification for hundreds of initial gaps) and impossibility to calculate its derivatives, the local variations' method is applied. The optimization procedure is an iterative exhaustive search of optimal position for each fastener one-by-one among predefined holes.

The local variations' algorithm for minimizing function $F(P)$, where P is a vector of hole numbers where fasteners are installed and P_0 is the initial pattern, is as follows:

> **Initialization**: $P := P_0$, Iteration := 1.
> **Repeat**
>   Set Progress := false;
>   **For** each hole i with installed fastener
>   **For** each empty hole j;
>   Obtain pattern $P*$ by moving fastener from hole i to hole j;
>   **For** each initial gap of the cloud
>   Calculate the resulting gap with fastener pattern $P*$;
>   **End for**
>   Evaluate $\Delta F = F(P*) - F(P)$;
>   If $\Delta F < 0$, keep the new pattern $P = P*$ and set Progress := true;
>   **End for**
>   **End for**
>   Iteration := Iteration +1;
> **Until** Progress = false (no $F(P)$ improvement in one iteration).

The local variations' algorithm is implemented in ASRP CV. Iterations continue until the algorithm has converged to some local optimum. The benefit of this approach is that only fastener patterns that improve the goal function are accepted and therefore the optimization algorithm can be stopped at any moment.

4   Application Example

As an application example of the described methodology, we consider a problem of a real aircraft[1] when it is necessary to improve current temporary fastener pattern for a wing-to-fuselage junction. The gap between the parts should be reduced to a given critical value by rearranging the constant number of fasteners.

While assembly, the wing is positioned slightly below the Central Wing Box (CWB), and initial gap between the parts is measured in several predefined points along the junction area (see Fig. 4). The optimized temporary fastener pattern has to be suitable for any similar parts of one aircraft series what makes the procedure difficult as the number of measured initial gaps is very limited. In order to avoid this problem and guarantee fastener pattern quality, the optimization is performed over a set of artificial initial gaps modeled on available real measurements.

For the considered model, the set of 209 measured initial gaps was provided (initial gap cloud). The wing-to-fuselage junction model consists of two independent junction areas. The first one includes the upper wing panel and the part of CWB with 7308 computational nodes in junction area and the second one includes a lower wing panel and the part of CWB with 6452 computational nodes and has more complex geometry.

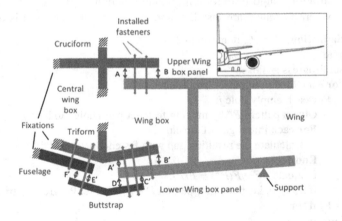

Fig. 4. Mechanical scheme of wing-to-fuselage junction

---

[1] Due to confidentiality reasons, the model details could not be provided in the paper.

Temporary fasteners on these two junctions have to be rearranged. Thus, the objective function is chosen as the percent of computational nodes where the gap between connected parts exceeds the desired value for all initial gaps in the cloud.

Results for the Upper Panel

The optimization steps are described in Table 1. After three iterations, the optimization procedure stops due to no further improvement. The total number of gap computations is about 430 000 what would take nearly 3.5 years of computations on a personal computer without parallelization.

The Fig. 5 illustrates the results of a new (optimized) pattern validation for four different gap clouds. The gap cloud on which the optimization is done is denoted as No. 3. The different gap clouds were obtained by adding local roughness to the measurements in order to simulate part variations. The methodology of initial gap generation is described in detail in [14].

The percent of computational nodes with gap less than X mm is plotted on the vertical axis in the Fig. 5. Thus, the percent of nodes where the gap is less than 0.2 mm for all gaps in this cloud is around 97% for gap cloud No. 4 and the initial fastener pattern. For all gap clouds with the optimized pattern, the plot lines (dashed) are located above the lines, corresponding to the initial pattern (solid) which means that the resulting gaps are decreased after fastener rearrangement.

Results for the Lower Panel

The optimization steps for a lower panel are described in Table 2. One gap computation for this model is almost two times longer than for the previous one because of more complex geometry. The total number of gap computations is about 554 000 what would take nearly 7 years of computations on a personal computer without parallelization.

The Fig. 6 illustrates the results of a new (optimized) pattern validation for three different gap clouds. Optimization is done for gap cloud No. 2. The figure shows that the initial temporary fastener pattern eliminates gaps almost everywhere. Therefore, the optimization provides only slight improvement.

Table 1. Optimization steps for upper panel junction.

Operation	Computational time	Goal function, %
Computation of the goal function value	4.6 min	0.466 (7127 nodes)
Optimization, 1st iteration	53.2 h	0.263 (4013 nodes)
Optimization, 2nd iteration	52.1 h	0.244 (3727 nodes)
Optimization, 3rd iteration	52.1 h	0.242 (3694 nodes)
Total computational time	6.6 days	

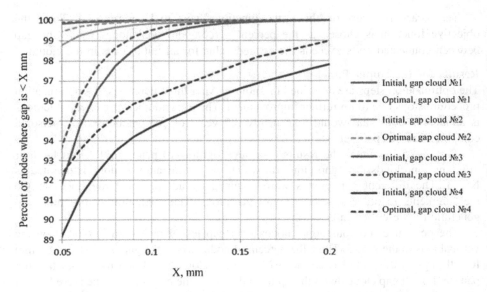

Fig. 5. Validation of optimized fastener pattern for upper panel

Table 2. Optimization steps for lower panel junction.

Operation	Computational time	Goal function, %
Computation of the goal function value	7.6 min	0.706 (9529 nodes)
Optimization, 1st iteration	115.1 h	0.567 (7647 nodes)
Optimization, 2nd iteration	105.7 h	0.563 (7597 nodes)
Optimization, 3rd iteration	106.2 h	0.563(7593 nodes)
Total computational time	13.6 days	

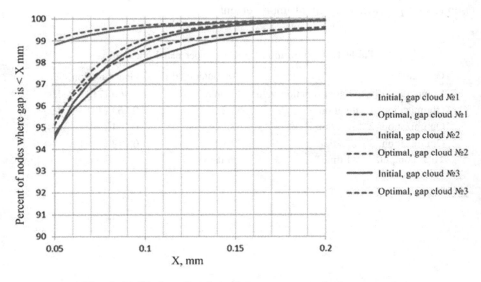

Fig. 6. Validation of optimized fastener pattern for lower panel

5 Conclusion

ASRP complex is developed for diverse but very specific engineering challenges. Some of these problems can be solved using commercial software, such as MSC Nastran, ANSYS etc., but ASRP application results in gain of time and a better quality. It can be explained by the fact that specialized algorithms and data saving strategies are implemented for solving of the narrow contact problem class. The key feature of ASRP optimization and verification methodology is assembly analysis over a cloud of initial gaps that involves series of similar computations for different initial gaps. This makes possible to parallelize the main optimization procedure and to use high-performance computers for executing simultaneous processes.

This software is successfully applied both to modification of existing fastener pattern for wing-to-fuselage junction and to the validation of new pattern against the old one. According to the obtained results, assembly engineers can update the technology at the final assembly line.

We expect the further investigations will be aimed at improvement of ASRP CV, investigation of parallel computing technologies for better performance.

The results of the work are obtained using computational resources of Peter the Great St.Petersburg Polytechnic University Supercomputing Center (www.spbstu.ru).

References

1. Liu, S.C., Hu, S.J.: Variation simulation for deformable sheet metal assemblies using finite element methods. ASME J. Manuf. Sci. Eng. **119**(3), 368–374 (1997)
2. Dahlström, S.: Variation simulation of sheet metal assemblies for geometrical quality: parameter modeling and analysis. Ph.D. thesis. Chalmers University of Technology, Göteborg, Sweden (2005)
3. Marguet, B., Chevassus, N., Falgarone, H., Bourdet, P.: Geometrical behavior laws for computer aided tolerancing: AnaTole a tool for structural assembly tolerance analysis. In: 8th CIRP Seminar on Computer-Aided Tolerancing, Charlotte, pp. 124–131 (2003)
4. Warmefjord, K., Söderberg, R., Lindau, B., Lindkvist, L., Lorin, S.: Joining in nonrigid variation simulation. Computer-aided Technologies, Intech (2016)
5. Wang, H., Ding, X.: Identifying sources of variation in horizontal stabilizer assembly using finite element analysis and principal component analysis. Assembly Autom. **13**(1), 86–96 (2013)
6. Dimensional Control System (DCS) Homepage. http://www.3dcs.com. Accessed 21 May 2017
7. Dahlström, S., Lindkvist, L.: Variation simulation of sheet metal assemblies using the method of influence coefficients with contact modeling. J. Manuf. Sci. Eng. **129**(3), 615–622 (2007)
8. Petukhova, M., Lupuleac, S., Shinder, Y., Smirnov, A., Yakunin, S., Bretagnol, B.: Numerical approach for airframe assembly simulation. J. Math. Ind. **4**(8), 1–12 (2014)
9. Lindau, B., Lorin, S., Lindkvist, L., Söderberg, R.: Efficient contact modeling in nonrigid variation simulation. J. Comput. Inf. Sci. Eng. **16**(1), 011002 (2016)
10. Yang, D., Qu, W., Ke, Y.: Evaluation of residual clearance after pre-joining and pre-joining scheme optimization in aircraft panel assembly. Assembly Autom. **36**(4), 376–387 (2016)

11. Wriggers, P.: Computational Contact Mechanics, 2nd edn. Springer, Berlin (2006). https://doi.org/10.1007/978-3-211-77298-0
12. Pedersen, P.: A direct analysis of elastic contact using super elements. Comput. Mech. **37**(3), 221–231 (2006)
13. Lupuleac, S., Petukhova, M., Shinder, J., Smirnov, A., et al.: Software complex for simulation of riveting process: concept and applications. In: SAE Technical Papers, 2016-01-2090 (2016)
14. Lupuleac, S., Zaitseva, N., Petukhova, M., Shinder, J., et al.: Combination of experimental and computational approaches to A320 wing assembly. In: SAE Technical Papers, 2017-01-2085 (2017)
15. Goldfarb, D., Idnani, A.: A numerically stable dual method for solving strictly quadratic programs. Math. Program. **27**(1), 1–33 (1983)
16. Powell, M.J.D.: On the quadratic programming algorithm of Goldfarb and Idnani. In: Cottle, R.W. (ed.) Mathematical Programming Essays in Honor of George B. Dantzig, Part II, Mathematical Programming Studies Book Series, vol. 25, pp. 46–61. Springer, Heidelberg (1985)
17. Stefanova, M., Yakunin, S., Petukhova, M., Lupuleac, S., Kokkolaras, M.: An interior-point method-based solver for simulation of aircraft parts riveting. Eng. Optim. **50**(5), 781–796 (2017)
18. Bertsekas, D.: Projected Newton methods for optimization problems with simple constraints. SIAM J. Control Optim. **20**(2), 221–246 (1982)

SL-AV Model: Numerical Weather Prediction at Extra-Massively Parallel Supercomputer

Mikhail Tolstykh[1,2,3]($\boxtimes$), Gordey Goyman[1,2], Rostislav Fadeev[1,2,3],
Vladimir Shashkin[1,2,3], and Sergei Lubov[4]

[1] Marchuk Institute of Numerical Mathematics Russian Academy of Sciences,
Moscow, Russia
mtolstykh@mail.ru, gordeygoyman@gmail.com,
rost.fadeev@gmail.com, vvshashkin@gmail.com
[2] Hydrometcentre of Russia, Moscow, Russia
[3] Moscow Institute of Physics and Technology, Dolgoprudny, Russia
[4] Main Computer Center of Federal Service for Hydrometeorology
and Environmental Monitoring, Moscow, Russia
s.lubov@meteorf.ru

Abstract. The SL-AV global atmosphere model is used for operational medium-range and long-range forecasts at Hydrometcentre of Russia. The program complex uses the combination of MPI and OpenMP technologies. Currently, a new version of the model with the horizontal resolution about 10 km is being developed. In 2017, preliminary experiments have shown the scalability of the SL-AV model program complex up to 9000 processor cores with the efficiency of about 45% for grid dimensions of $3024 \times 1513 \times 51$. The profiling analysis for these experiments revealed bottlenecks of the code: non-optimal memory access in OpenMP threads in some parts of the code, time losses in the MPI data exchanges in the dynamical core, and the necessity to replace some numerical algorithms. The review of model code improvements targeting the increase of its parallel efficiency is presented. The new code is tested at the new Cray XC40 supercomputer installed at Roshydromet Main Computer Center.

Keywords: Global atmosphere model · Numerical weather prediction
Interannual predictability of atmosphere · Massively parallel computations
Combination of MPI and OpenMP technologies

1 Introduction

The common ways to improve the quality of numerical weather prediction and fidelity of the atmosphere model 'climate' are the increase of the atmospheric model resolution and advancements in parameterized description of unresolved subgrid-scale processes. Both ways imply the increase in computational complexity of the atmospheric models. Operational numerical weather prediction requires the forecast to be computed rapidly, usually in less than 10 min per forecast day, while the climate modelling requires many multi-year runs to be completed in reasonable time. The resolution of the atmospheric models grows permanently, so these models should be able to use tens of thousands processor cores efficiently. Currently, the typical horizontal resolution of the global

© Springer Nature Switzerland AG 2019
V. Voevodin and S. Sobolev (Eds.): RuSCDays 2018, CCIS 965, pp. 379–387, 2019.
https://doi.org/10.1007/978-3-030-05807-4_32

medium-range numerical weather prediction models is 9–25 km with about 100 vertical levels [1]. Thus, the approximate number of grid points for these models is about 10^8–10^9. Most numerical weather prediction centers plan to increase the resolution of their models [1]. Many supercomputers of weather services and climate research centers have peak performance about 5–10 Pflops [1] and they are in the first hundred of Top500 list [2]. For example, UK MetOffice and Hadley Climate Centre supercomputer currently has the peak performance of 8.1 Pflops and is at 15th place of Top500 list (as of November 2017).

It is essential that parallel efficiency of an atmospheric model be considered together with its computational efficiency, i.e. ability of the model to compute the forecast of a given accuracy combined with a minimum wall-clock time for a given number of processors. Sophisticated numerical methods in the dynamical core of atmospheric models usually allow longer time steps but scale worse than simple explicit time-stepping algorithms so the balance between complexity of the applied numerical methods and their scalability should be found. Computational efficiency is under permanent evaluation in many world leading weather prediction centers and is an important criterion in selecting their development strategies [3, 4].

SL-AV is the global atmosphere model applied for the operational medium-range weather forecast at Hydrometeorological center of Russia and as a component of the long-range probabilistic forecast system. It is also used in experiments on interannual predictability and is an atmospheric component of the coupled atmosphere-ocean-sea-ice model [5]. SL-AV [6] is the model acronym (semi-Lagrangian, based on Absolute-Vorticity equation). It is developed at Marchuk Institute of Numerical Mathematics, Russian Academy of Sciences (INM RAS) in cooperation with the Hydrometeorological centre of Russia (HMCR). The dynamical core of this model uses the semi-implicit semi-Lagrangian time-integration algorithm [7]. The most part of subgrid-scale processes parameterizations algorithms are developed by ALADIN/LACE consortium [8, 9]; however, the model includes CLIRAD SW [10] and RRTMG LW [11] for parameterization of shortwave and longwave radiation respectively. The multilayer soil model developed at INM RAS [12] is also included. The parallel implementation of SL-AV model uses the combination of one-dimensional MPI decomposition and OpenMP loop parallelization [7]. The model code is also adapted to run at Intel Xeon Phi processors [13]. The code is written in Fortran language and consists of several hundred thousands lines.

The parallel structure of the model code is as follows: one-dimensional MPI decomposition is used along latitude or Fourier-space wave number. MPI-processes perform computations in the bands of grid latitudes during the first phase of the time-step, while OpenMP threads are used to parallelize loops along longitude or vertical coordinate. In the second phase of SL-AV time-step, each MPI-process performs computations for the set of longitude Fourier coefficients from pole to pole, and OpenMP parallelization for loops in vertical is applied.

The first tests of the model with grid dimensions $3024 \times 1513 \times 51$ showed that it scales up to 9072 cores with an efficiency of about 45%. However, these results were obtained without possibility of profiling, tuning and running the model on a system with such a number of processor cores. In this paper, we present recent works on improving scalability of the SL-AV program complex. Section 2 describes the changes in interprocessor MPI communications, and Sect. 3 gives an overview of OpenMP optimizations in the model code. We then have tested the modified code at Cray XC40 system using up to 27208 processor cores. The results of these works are presented in Sect. 4.

2 Parallel Communications Optimization

2.1 Semi-Lagrangian Algorithm Optimization

The semi-Lagrangian advection algorithm [14] consists of two main blocks: the calculation of the backward in time trajectories (position of air particles that arrive to the grid points at the next time step) and the spatial interpolation of advected variables to the departure points of these trajectories. Parallel implementation of this algorithm requires halo exchanges (exchanges of latitudinal bands adjacent to the process boundaries) with the width determined by the position of the furthest departure point and interpolation stencil. The width of the exchanges (in terms of grid point distance) in the original version of the model is calculated using predefined wind speed (estimate of the global maximum wind speed) and the model time step. This a priori estimate is the upper limit for data amount that may be required for calculations. Thus, the semi-Lagrangian advection block in the standard version of the model requires the exchange of values for wind speed components and advected variables of a fixed predefined width. Such an estimate for the region of parallel dependence is rough, and the actual one can be significantly lower, especially in regions with a small wind speed. When using one-dimensional MPI decomposition, the size of messages does not decrease with the increase in the number of computational cores, moreover, the number of neighbor processes increases, which negatively affects the parallel efficiency of the model. To reduce the volume of halo exchanges, another approach has been implemented in this block. First, the calculation of backward trajectories with predefined exchanges size of wind speed components is performed. Knowing the coordinates of the trajectories departure points, the width of exchanges necessary for each processor is computed and data exchanges for the values of advected variables are carried out for their further interpolation. Application of this approach allows to reduce significantly the average size of messages and the number of MPI-processes involved in these exchanges. The schematic of this algorithm is shown in Fig. 1.

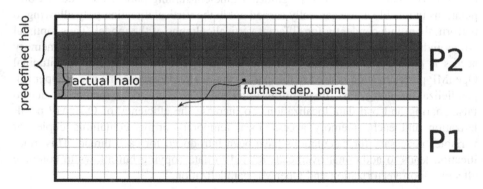

Fig. 1. Schematic of the semi-Lagrangian advection halo optimization.

2.2   Reducing Number of Global Communications

In the SL-AV model, the fast Fourier transforms are used to convert systems of linear algebraic equations arising from discretization of elliptic problems (Helmholtz equation, hyper diffusion equation, wind velocity reconstruction, see [7] for details) allowing to reduce two-dimensional problems to a set of one-dimensional ones, which are then solved by a direct algorithm. The parallel implementation of this method includes data transpositions, i.e. global redistribution of data between processes. The use of data transpositions limits parallel efficiency and should be avoided whenever possible. Initially, the SL-AV model used four transpositions per time step. The model code modifications and rearrangements have been implemented to reduce this number to two per time step. One can note that the work is underway to introduce new solvers that allows abandoning the use of data transpositions [13].

3   OpenMP Optimizations

3.1   OpenMP Loop Parallelization

Initially, OpenMP technology was used in the SL-AV model to parallelize loops along the same direction as MPI decomposition. This approach limited the maximum number of processor cores used, so most of the code in the model was modified in a way to parallelize the loops along the additional direction, thereby forming a quasi-two-dimensional domain decomposition. However, due to the low computational cost and the laboriousness of code modification, some of the code parts remained unchanged. The available MPI-parallelism is exhausted when the number of processor cores is higher than 1512 (for horizontal grid dimensions of 3024×1513), and then these code sections become sequential in terms of OpenMP parallelization. This loss of parallel efficiency can be noticeable at extra-parallel scales. So the abovementioned modifications of the remaining code sections have been implemented.

3.2   Memory Access Optimization

The block computing right-hand sides of prognostic equations describing parameterized subgrid-scale processes is a significant time-consuming part of the model. Computations in this block are generally carried out in the vertical direction only allowing to perform them independently for different vertical columns of the model. Thus, optimal arrays indices arrangement for this block in terms of loop vectorization and memory access is *(horizontal dimension, vertical dimension)*. At the same time, longitudinal OpenMP parallelization is used in this part of the model. However, the use of OpenMP parallelization along 'fast' first Fortran array index is likely to be inefficient, due to false sharing and bad data localization. To increase the efficiency of OpenMP parallelization and cache memory access, local temporary arrays containing copies of variables necessary for calculations have been introduced for each thread. This modification leads to additional overheads related to data copying but allows to combine effective loop vectorization and OpenMP parallelization.

4 Numerical Experiments

4.1 Model Setup and System Configuration

We have tested two versions of the SL-AV model [6, 7]. Both have the same horizontal resolution of 0.119° (approximately 13 km at the equator), the first one has 51 vertical levels, and the second one has 126 vertical levels. The grid dimensions are $3024 \times 1513 \times 51$ and $3024 \times 1513 \times 126$ respectively. The version with 51 vertical levels has the same resolution as was used for preliminary tests of the code before the modifications described above.

All the experiments were carried out at the Cray XC40 system [15] installed at Roshydromet. This system consists of 936 nodes having two Intel Xeon E2697v4 18-core CPUs and 128 GB memory. All the nodes are connected with Cray ARIES interconnect. The peak performance is 1.2 PFlops.

4.2 Results

First, we compare the parallel efficiency of the modified SL-AV model code with respect to the previous version using up to 9072 processor cores. The results are shown in Fig. 2.

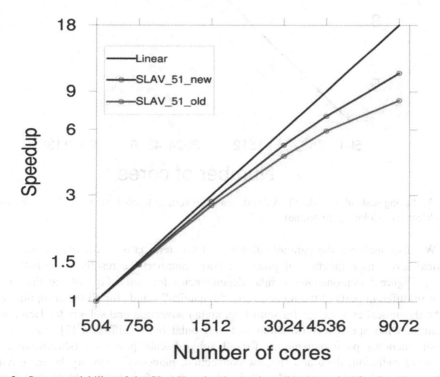

Fig. 2. Strong scalability of the SL-AV code (the version with 51 vertical levels): the version before modifications (red curve), the new version (blue curve), linear speedup (black curve). (Color figure online)

Note that the parallel efficiency of the code (the ration of achieved speedup to linear one) has increased by approximately 15% while using 9072 cores, from 45.5 to 60%.

We have studied strong scalability of the same code but having 126 vertical levels. The results are presented in Fig. 3. We are able to launch the code at 27216 cores, however, there is just 20% acceleration with respect to 13608 cores. The SL-AV code has the parallel efficiency of about 53% while running at 13608 processor cores.

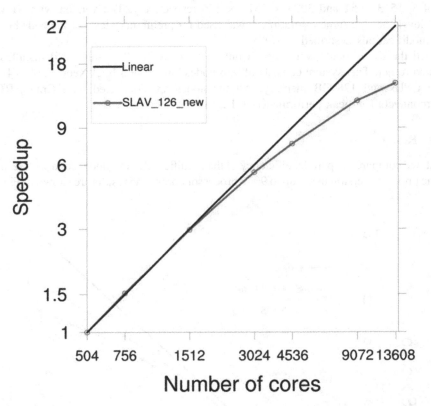

Fig. 3. Strong scalability of the SL-AV code with 126 vertical levels (red curve); linear speedup (black curve). (Color figure online)

We also analyzed the parallel efficiency of different parts of the model with 126 vertical levels as a function of processor cores number; the results are depicted in Fig. 4. Figure 5 demonstrates similar dependencies for percentage of the time step spent in different parts of the model. Here "dynamics" stands for all the computations in the dynamical core except for semi-Lagrangian advection and solvers for Hemholtz equation, wind speed reconstruction and horizontal hyper diffusion [7], "subgrid_-param" denotes parameterizations for all subgrid-scale processes (shortwave and longwave radiation, deep and shallow convection, planetary boundary layer, gravity wave drag, microphysics et al.), "SL_advection" refers to the block of semi-Lagrangian advection described in Subsect. 2.1., and "elliptic solver" corresponds to the abovementioned solvers in Fourier longitudinal space.

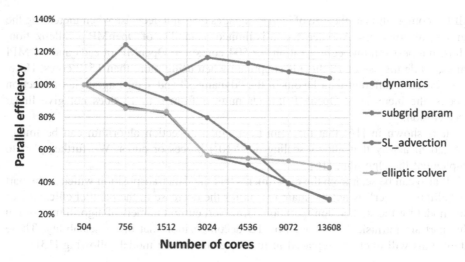

Fig. 4. Parallel efficiency for different parts of the model code as a function of processor core number. See text for details.

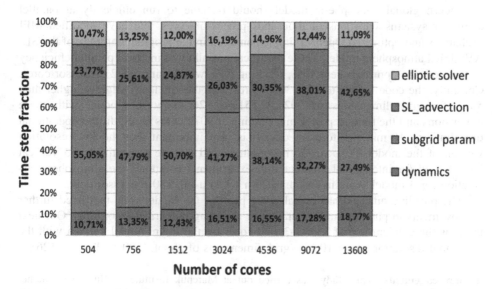

Fig. 5. Percentage of the time step occupied by different parts of the model code as a function of processor cores number.

One can see that parameterizations computations are well parallelized. That could be expected, as most of these computations involve independent computations in vertical columns only. For most of processor cores numbers, the parallel efficiency of this part is higher than for 504 cores. This can be explained by better use of cache memory in these cases. For other parts of the model, the parallel efficiency gradually decays.

The "SL_advection" and "dynamics" blocks seem to be bottlenecks of current model implementation. This is, on one hand, related to the nature of the one-dimensional

MPI decomposition as the size of halo exchanges does not decrease when increasing the number of processes. Another reason is limited scalability of OpenMP parallelization. Thus, the best runtime configuration for 504 cores is 6 OpenMP threads and 84 MPI ranks, while the use of 18 threads is optimal when using more than 1512 cores. However, because different data layout in the "dynamics" and semi-Lagrangian advection blocks, the increase of OpenMP threads number from 6 to 18 does not give linear acceleration.

It is shown in [16] that the semi-Lagrangian advection algorithm can be implemented in a way that allows scaling at $O(10^4)$ processor cores. We further plan to implement the algorithm [16].

It also can be seen from the experiments that the data transposition which is the part of "elliptic solver" is not the main reason for the decrease in the parallel efficiency of the model for the studied configuration. However, current numerical algorithms used in this part are intrinsically bounded by 1D decomposition that limits scalability. These algorithms will likely be replaced in future version of the model following [13].

4.3 Conclusions

A modern global atmospheric model should be able to run efficiently at parallel computer systems with tens of thousands processor cores. We have modified MPI exchanges and optimized OpenMP implementation in the program complex of the SL-AV global atmosphere model. These modifications allow to increase parallel efficiency of this code by approximately 15%, reaching 63% while using 9072 processor cores. Currently, the code is able to use 13608 cores with the efficiency slightly higher than 50%, for grid dimensions of $3024 \times 1513 \times 126$. Now the model with these dimensions and the time-step of 4 min running at 9072 cores would fit the operational time limit of 10 min per forecast day. It is also important that the low-resolution version of the model ($0.9 \times 0.72°$ in longitude and latitude respectively, 85 vertical levels) used for interannual predictability experiments computes the atmosphere circulation for 3 model years in less than 20 h while using 180 processor cores.

The profiling analysis has revealed the parts of the model code that need further improvements in parallel implementation or replacing numerical algorithms. Our next target is the efficient use of 25000–35000 cores for the future model version with the horizontal resolution about 10 km (grid dimensions of about $4500 \times 2250 \times 126$).

Acknowledgements. This study was carried out at Marchuk Institute of Numerical Mathematics, Russian Academy of Sciences. The study presented in Sects. 2 and 3 was supported with the Russian Science Foundation grant No. 14-27-00126P, the work described in Sect. 4 was supported with the Russian Academy of Sciences Program for Basic Researches No. I.26P.

References

1. WGNE Overview of plans at NWP Centres with Global Forecasting Systems. http://wgne.meteoinfo.ru/nwp-systems-wgne-table/
2. TOP500 Supercomputer sites. https://www.top500.org/

3. Wedi, N.P., et al.: The modelling infrastructure of the integrated forecasting system: recent advances and future challenges. Technical Memorandum 760, ECMWF (2015)
4. Dynamical Core Evaluation Test Report for NOAA's Next Generation Global Prediction System (NGGPS). https://www.weather.gov/media/sti/nggps/NGGPS%20Dycore%20Phase%202%202%20Test%20Report%20website.pdf
5. Tolstykh, M.A., et al.: Development of the multiscale version of the SL-AV global atmosphere model. Russ. Meteor. Hydrol. **40**, 374–382 (2015). https://doi.org/10.3103/s1068373915060035
6. Fadeev, R.Yu., Ushakov, K.V., Kalmykov, V.V., Tolstykh, M.A., Ibrayev, R.A.: Coupled atmosphere–ocean model SLAV–INMIO: implementation and first results. Russ. J. Num. An. Math. Mod. **31**, 329–337 (2016). https://doi.org/10.1515/rnam-2016-0031
7. Tolstykh, M., Shashkin, V., Fadeev, R., Goyman, G.: Vorticity-divergence semi-Lagrangian global atmospheric model SL-AV20: dynamical core. Geosci. Model Dev. **10**, 1961–1983 (2017). https://doi.org/10.5194/gmd-10-1961-2017
8. Geleyn, J.-F., et al.: Atmospheric parameterization schemes in Meteo-France's ARPEGE N.W.P. model. In: Parameterization of Subgrid-Scale Physical Processes, ECMWF Seminar Proceedings, pp. 385–402. ECMWF, Reading, UK (1994)
9. Gerard, L., Piriou, J.-M., Brožková, R., Geleyn, J.-F., Banciu, D.: Cloud and precipitation parameterization in a Meso-Gamma-Scale operational weather prediction model. Mon. Weather Rev. **137**, 3960–3977 (2009). https://doi.org/10.1175/2009MWR2750
10. Tarasova, T., Fomin, B.: The use of new parameterizations for gaseous absorption in the CLIRAD-SW solar radiation code for models. J. Atmos. Ocean. Technol. **24**, 1157–1162 (2007). https://doi.org/10.1175/JTECH2023.1
11. Mlawer, E.J., Taubman, S.J., Brown, P.D.: RRTM, a validated correlated-k model for the longwave. J. Geophys. Res. **102**, 16663–16682 (1997). https://doi.org/10.1029/97jd00237
12. Volodin, E.M., Lykossov, V.N.: Parametrization of heat and moisture transfer in the soil-vegetation system for use in atmospheric general circulation models: 1. Formulation and simulations based on local observational data. Izvestiya Atmos. Ocean. Phys. **34**, 402–416 (1998)
13. Tolstykh, M., Fadeev, R., Goyman, G., Shashkin, V.: Further development of the parallel program complex of SL-AV atmosphere model. In: Voevodin, V., Sobolev, S. (eds.) RuSCDays 2017. CCIS, vol. 793, pp. 290–298. Springer, Cham (2017). https://doi.org/10.1007/978-3-319-71255-0_23
14. Staniforth, A., Cote, J.: Semi-Lagrangian integration schemes for atmospheric models – a review. Mon. Weather Rev. **119**, 2206–2233 (1991)
15. Cray XC40 specifications. https://www.cray.com/sites/default/files/resources/cray_xc40_specifications.pdf
16. White III, J., Dongarra, J.: High-performance high-resolution tracer transport on a sphere. J. Comput. Phys. **230**, 6778–6799 (2011). https://doi.org/10.1016/j.jcp.2011.05.008

Supercomputer Simulation Study of the Convergence of Iterative Methods for Solving Inverse Problems of 3D Acoustic Tomography with the Data on a Cylindrical Surface

Sergey Romanov$^{(\boxtimes)}$ [iD]

Lomonosov Moscow State University, Moscow, Russia
romanov60@gmail.com

Abstract. This paper is dedicated to developing effective methods of 3D acoustic tomography. The inverse problem of acoustic tomography is formulated as a coefficient inverse problem for a hyperbolic equation where the speed of sound and the absorption factor in three-dimensional space are unknown. Substantial difficulties in solving this inverse problem are due to its nonlinear nature. A method which uses short sounding pulses of two different central frequencies is proposed. The method employs an iterative parallel gradient-based minimization algorithm at the higher frequency with the initial approximation of unknown coefficients obtained by solving the inverse problem at the lower frequency. The efficiency of the proposed method is illustrated via a model problem. In the model problem an easy to implement 3D tomographic scheme is used with the data specified at a cylindrical surface. The developed algorithms can be efficiently parallelized using GPU clusters. Computer simulations show that a GPU cluster capable of performing 3D image reconstruction within reasonable time.

Keywords: Ultrasound tomography · Medical imaging
Inverse problem · Gradient method

1 Introduction

This paper is dedicated to methods of acoustic tomography, or, to be more specific, to ultrasound tomography used for imaging of soft tissues in medicine [12,22,23]. Differential diagnosis of breast cancer is one of the most important problems in modern medicine. Ultrasonic tomography could be the most promising method for regular mammographic screening.

However, currently existing ultrasonic devices are not tomographic. The reflectivity images obtained by these devices reveal only the contours of tissue irregularities and do not make it possible to characterize the tissues with

© Springer Nature Switzerland AG 2019
V. Voevodin and S. Sobolev (Eds.): RuSCDays 2018, CCIS 965, pp. 388–400, 2019.
https://doi.org/10.1007/978-3-030-05807-4_33

sufficient resolution. Hence developing high-resolution ultrasound tomographs for imaging of soft tissues is very important task.

The most interesting results in ultrasound tomography are associated with the development of methods for solving problems of 3D wave tomography in terms of mathematical models incorporating both diffraction and absorption effects. Breakthrough results in mathematical methods for solving inverse problems are due to the possibility of directly computing the gradient of the residual functional between the computed and experimental data at the detectors [2,8,11,13,16,17]. Detectors are placed at a cylindrical surface surrounding the object studied. This is the underlying approach of the methods used in this paper.

One approach to ultrasound tomography involves using simplified linearized models [20,22,24]. However, a linearized model can provide only a rough characterization of tissues [21].

The development of numerical methods for solving direct and inverse problems of wave tomography is a challenging computational task [3]. We solve the direct problem using finite difference approximation of hyperbolic-type differential equations. To solve inverse problems, we use iterative parallel algorithms based on direct computation of the gradient of the residual functional. To compute the gradient, we solve the conjugate problem in reverse time. The amount of input data and the number of unknowns in the inverse 3D problem exceed 1 Gb and 100 Mb, respectively. The algorithms are implemented on a supercomputer.

Because of the nonlinearity of the inverse problems of wave tomography the residual functional is not convex, and this presents one of the main mathematical challenges. As a consequence, gradient-based methods that minimize the residual functional converge to some local minimum rather than the global minimum of the functional. There are various approaches to find the global minimum of a non-convex functional. Attempts were made to construct "global" methods [2,13,14] to solve inverse problems.

In this paper, the dual-frequency method is proposed for finding approximate solutions of 3D coefficient inverse problems in acoustic tomography. The dual-frequency method extends the domain of convergence of gradient-based algorithms. This method can be applied primarily to ultrasound tomography. The efficiency of the proposed method is illustrated via model problems. The developed algorithms are easily parallelized using supercomputers and GPU-clusters [7,18]. Computer simulations show that a GPU cluster capable of performing 3D image reconstruction within reasonable time.

2 Formulation and Solution Methods of the Inverse Problem of 3D Acoustic Tomography

The aim of acoustic tomography is to reconstruct the internal structure of the object using measurements of the acoustic pressure $u(r, t)$ obtained on some surface surrounding the object. Figure 1 shows the scheme of a 3D acoustic tomographic examination where the measurements are taken on a cylindrical surface. The emitters of sounding pulses are located on the same cylindrical surface. This tomographic scheme can be used for ultrasonic mammography.

The formulation of the inverse problem of ultrasound tomography with the data specified at a cylindrical surface is a novel approach. The peculiarity of the algorithms developed for solving the inverse problem considered is that numerical methods use Cartesian coordinate system, making it necessary to minimize errors of the interpolation of the computed wave field onto the cylindrical surface. The tomographic scheme with the data specified at a cylindrical surface is easy to implement in physical experiments.

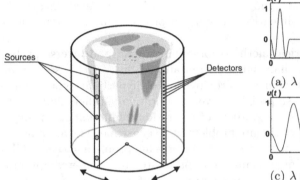

Fig. 1. The scheme of a tomographic examination.

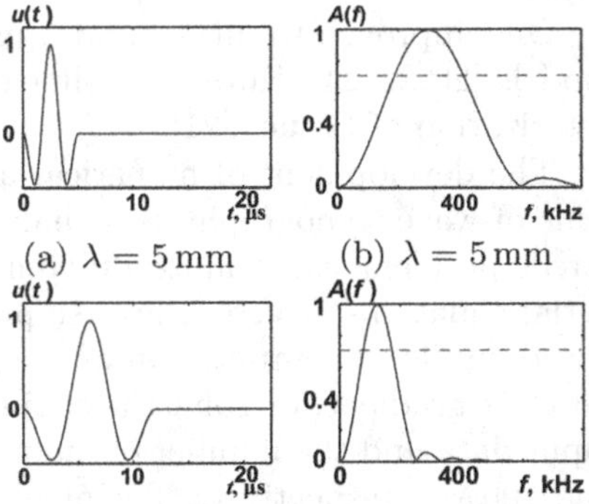

Fig. 2. Sounding pulse for different wavelengths: waveform (a, c); frequency spectrum (b,d).

The scalar wave model is used to describe the wave propagation process. This model also takes into account such important factor as ultrasound absorption in the medium. The initial pulse emitted by a point source was calculated using the formula $u(x) = \sin(2\pi x/(3\lambda)) \cdot \sin(6\pi x/(3\lambda)), x \leq 1.5\lambda$. Figure 2a,b shows the waveform and the frequency spectrum of a pulse with a wavelength of 5 mm, Fig. 2c,d shows the waveform and the frequency spectrum of a pulse with a wavelength of 12 mm. The bandwidth of these pulses at -3 dB level (indicated by a dotted line) is approximately 65% of the central frequency.

The simplest absorption model is used in this paper [9]. The inverse problem of ultrasound tomography can then be formulated as a coefficient inverse problem of reconstructing the unknown coefficients $c(\boldsymbol{r})$ and $a(\boldsymbol{r})$ in the wave equation, given the measurements of the acoustic pressure on the surface S made with different positions $\boldsymbol{q}$ of the sources:

$$c(\boldsymbol{r})u_{tt}(\boldsymbol{r},t) + a(\boldsymbol{r})u_t(\boldsymbol{r},t) - \Delta u(\boldsymbol{r},t) = \delta(\boldsymbol{r}-\boldsymbol{q}) \cdot f(t); \tag{1}$$

$$u(\boldsymbol{r},t)|_{t=0} = 0, \ u_t(\boldsymbol{r},t)|_{t=0} = 0, \ \partial_n u(\boldsymbol{r},t)|_{ST} = p(\boldsymbol{r},t). \tag{2}$$

Here $c(r) = 1/v^2(r)$, where $v(r)$ is the speed of sound in the medium; $r \in \mathbb{R}^N (N = 3)$, the position of the point in space; u, the acoustic pressure; Δ, the Laplace operator with respect to the variable r. The function $f(t)$ describes the sounding pulse generated by a point source at q; $\partial_n u(r,t)|_{ST}$ is the derivative along the normal to the surface S in the range $(r,t) \in S \times (0,T)$, where T is the duration of the measurement. The function $p(r,t)$ is known, and $a(r)$ is the absorption factor. Formula (2) represent the initial conditions and the Neumann conditions at the boundary of the computational domain.

It is assumed that inhomogeneity of the medium is caused by variations of the sound speed and absorption factor. Outside of the region studied the absorption factor is equal to zero, $a(r) = 0$, and the speed of sound is known and equal to $v_0 = const$. This simple model of wave propagation with absorption (1) can be used to describe ultrasound waves in soft tissues.

The inverse problem is formulated as the problem of minimizing the residual functional $\Phi(c(r), a(r))$ with respect to its argument (c, a):

$$\Phi(u(c,a)) = \frac{1}{2} \int_0^T \int_S (u(s,t) - U(s,t))^2 \, ds \, dt. \tag{3}$$

Here $U(s,t)$ is the acoustic pressure measured at the boundary S for the duration $(0,T)$, and $u(s,t)$ is the solution of the direct problem (1, 2) for the given $c(r) = 1/v^2(r)$ and $a(r)$. For multiple ultrasound sources the total value of the residual functional is the sum of the residuals (3) for each source.

Formulas for the gradient $\Phi'(c,a)$ of the residual functional for two- and three-dimensional inverse problems in various formulations were derived in the works [2,16,17]. A strict mathematical derivation of the gradient for the inverse problem (1, 2) using a model that accounts for the diffraction and absorption effects was presented in the papers [6,8,10].

The gradient $\Phi'(u(c,a)) = \{\Phi'_c(u), \Phi'_a(u)\}$ of the functional (3) with respect to the variation $\{dc, da\}$ of the sound speed and absorption factor has the following form:

$$\Phi'_c(u(c)) = \int_0^T w_t(r,t)u_t(r,t) \, dt, \quad \Phi'_a(u(a)) = \int_0^T w_t(r,t)u(r,t) \, dt. \tag{4}$$

Here $u(r,t)$ is the solution of the main problem (1, 2), and $w(r,t)$ is the solution of the "conjugate" problem for the given $c(r)$, $a(r)$, and $u(r,t)$:

$$w_{tt}(r,t) - a(r)w_t(r,t) - \Delta w(r,t) = 0; \tag{5}$$

$$w(r,t=T) = 0, \ w_t(r,t=T) = 0, \ \partial_n w|_{ST} = u|_{ST} - U. \tag{6}$$

The boundary condition $\partial_n w|_{ST} = 0$ is applied at the part of the boundary S where there are no detectors. To compute the gradient (4), the direct problem (1, 2) and the "conjugate" problem (5, 6) must be solved.

Given the representation of the gradient (4), various iterative algorithms can be used to minimize the residual functional. One of the simplest algorithms is the steepest descent method.

The finite difference time-domain (FDTD) method on a uniform grid is used to solve the three-dimensional Eq. (1, 2) and (5, 6). The grid step h and time step τ are related by the Courant stability condition $\sqrt{3} \cdot c^{-0.5}\tau < h$, where $c^{-0.5} = v$ is the speed of sound. The following second-order finite difference scheme is used to approximate Eq. (1):

$$c_{ijl}\frac{u_{ijl}^{k+1} - 2u_{ijl}^{k} + u_{ijl}^{k-1}}{\tau^2} + a_{ijl}\frac{u_{ijl}^{k+1} - u_{ijl}^{k-1}}{\tau} - \frac{\Delta u_{ijl}^{k}}{h^2} = 0 \tag{7}$$

Here $u_{ijl}^{k} = u(x_i, y_j, z_l, t_k)$ is the value of $u(\boldsymbol{r}, t)$ at the point (i, j, l) at the time step k; c_{ijl} and a_{ijl} are the values of $c(\boldsymbol{r})$ and $a(\boldsymbol{r})$ at (i, j, l). Δ is the discrete Laplacian, which is computed using the formula:

$\Delta u_{i_0, j_0, l_0}^{k} = \sum\limits_{i=i_0-1}^{i_0+1} \sum\limits_{j=j_0-1}^{j_0+1} \sum\limits_{k=k_0-1}^{k_0+1} b_{ijl} u_{ijl}^{k}$. The coefficients b_{ijl} are provided,

for example, in the paper [15]. A similar finite difference scheme is used to solve Eq. (5, 6) in reverse time. The iterative process for solving the inverse problem numerically is described in [4].

3   GPU-Implementation of the Explicit Finite-Difference Algorithm for 3D Ultrasound Tomography

The algorithm for solving the inverse problem of ultrasound tomography is highly data parallelizable [1,5]. The values at all the grid points are computed at all time steps by the same formula both in the "direct" (1, 2) and "conjugate" (5, 6) problem and are independent of each other. Such algorithms can be efficiently parallelized on SIMD/SPMD-architectures.

A typical size of the 3D problem is $\approx 400^3$, while the number of parallel threads supported by a GPU device is on the order of 10000. Therefore, the algorithm processes the 3D data array sequentially along the Z-axis (z-marching method), as shown in Fig. 3. Sequential memory accesses are efficiently and automatically cached by modern GPUs and can be rapidly loaded into GPU registers.

In the Z-marching FDTD implementation, at each step along the Z-axis the next layer ($z = z_0 + 1$) is loaded into the registers, and the results for the current computed layer ($z = z_0$) are saved to the global memory. The process is then repeated for the next value of $z = z + 1$. The data for the current horizontal layer reside in the cache and can be read by all threads with no performance penalty. Each thread computes the results for points (x, y, z), where x and y are fixed, and the Z range includes several dozen points. The highest performance on the devices tested is achieved if each 32×4-thread block processes a $32 \times 4 \times 32$-point volume of data.

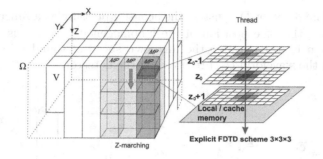

Fig. 3. GPU-implementation of the explicit 3D algorithm.

Nearly constant speed of sound, which differs from the speed of sound in water by no more than 10%, is a specific feature of soft tissue tomography. Taking into account this feature, the algorithm dynamically adjusts the computational domain and processes only the volume where the waves emitted by the source can be present. This volume is designated as V in Fig. 3 and is the intersection of the cubic computational domain Ω and the sphere of radius $Vmax \cdot t$ centered at the source. Here $Vmax$ is the maximum sound speed in soft tissues, and t is the current simulation time. The set of blocks that intersect with the domain V is precomputed, and only these blocks are launched at a given simulation time t. Dynamic computational domain adjustment increases the performance by 30–40%.

4   Convergence of the Iterative Algorithms Used to Solve the Inverse Problem

The inverse problem of wave tomography in the proposed formulation is a non-linear coefficient inverse problem. The residual functional of a nonlinear problem is typically non-convex, and thus it has local minima. In this paper, we investigate how the behavior of the residual functional depends on physical parameters, such as the wavelengths of sounding pulses. The ultimate goal of this work is to develop algorithms that make it possible to obtain an approximate solution of the coefficient inverse problem using some initial approximation.

A simple one-dimensional model problem is presented to illustrate how the wavelength of sounding pulses affects the convergence of gradient minimization algorithms. The ultrasound pulse propagates through one-dimensional medium with the speed of sound defined by the relations $c(x) = \bar{c}$ for $|x| \leq r$, $c(x) = c_0$ for $|X| > r$. Figure 4 shows the scheme of propagation of sounding pulses emitted by the source S. The sounding pulse 1 in Fig. 4a propagates from the source S at a speed of c_0. In the inhomogeneous region $|x| < r$, the pulse propagates at a speed of $\bar{c}$. Should there be no inhomogeneity, the sounding pulse in Fig. 4a would arrive at position 2 at some time T. The difference between $\bar{c}$ and c_0 in the inhomogeneous region causes the pulse to shift and arrive at position 3 at the time T. The detector D registers the acoustic pressure as a function of time $U(t)$. Assuming that the position of the inhomogeneity $|x| < r$ is known, the inverse problem is to determine the unknown speed of sound $\bar{c}$ in the region $|x| < r$, given the waveform $U(t)$ received by the detector.

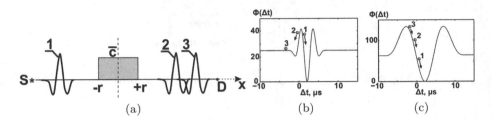

(a)   (b)   (c)

Fig. 4. Positions of sounding pulses with the inhomogeneity present and with the inhomogeneity absent (a). Plots of the residual functional for $\lambda = 5\,\text{mm}$ (b) and for $\lambda = 12\,\text{mm}$ (c).

We introduce the residual functional $\Phi(c) = \|U(t) - u(c,t)\|^2$, where $u(c,t)$ is the numerically simulated pulse at the detector position D computed assuming that the speed of sound in the inhomogeneous region is equal to c. The point of the global minimum of functional $\Phi(c)$ is the exact solution of the inverse problem. The value of the functional at this point is 0.

Let us denote the difference in pulse arrival times caused by the difference between c and c_0 as $\Delta t(c) = 2(r/c_0 - r/c)$. Figure 4b,c shows the plots of the residual functional $\Phi(\Delta t) = \|U(t) - u(c(\Delta t), t)\|^2$ as a function of the pulse arrival time difference Δt for different wavelengths. Here $\|U(t) - u(c(\Delta t), t)\|^2 = \int_{t=0}^{T} (U(t) - u(c(\Delta t), t))^2 \, dt$.

Figure 4b shows that for a shorter wavelength the iterative process of minimizing the residual functional converges to the global minimum from the initial approximation 1 and does not converge from initial approximations 2 and 3. For a longer wavelength the iterative process converges to the global minimum from any initial approximation 1,2 or 3, as shown in Fig. 4c.

Hence the use of sounding pulses of at least two different central wavelengths λ_1 and λ_2, $\lambda_1 > \lambda_2$ seems to be a promising approach for expanding the domain of convergence of the iterative method. First, a number of iterations of the gradient method is performed using the longer wavelength λ_1. The resulting approximate solution falls into the domain of convergence of the iterative process for the shorter wavelength λ_2. Then the residual functional is minimized via the gradient method using the wavelength λ_2. This idea forms the basis of the proposed dual-frequency method used to solve the problem of acoustic tomography.

In reality, the inverse problem is three-dimensional. Outside of the object, the speed of sound is $c(\boldsymbol{r}) = const = c_0$, $\boldsymbol{r} \in \mathbb{R}^3$. The object is insonified using the pulses emitted by sources S. Detectors D register the acoustic pressure $U(t)$ as a function of time. It turns out that the properties of the residual functional in the three-dimensional case are quite similar to the one-dimensional example described above.

Figure 5 shows the positions of wave fronts at some time T. The central wavelengths of the pulses are $\lambda = 5$ mm and $\lambda = 12$ mm for Fig. 5a and b, respectively. The dotted line 1 in Figs. 5a,b shows the position of the wave front at time T in a homogeneous medium, assuming that the object is absent. It corresponds to the pressure field $u(r, T)$ computed at the initial iteration, according to the formulation of the inverse problem (3). The dotted line 2 shows the position of the wave front that has passed through the object. It corresponds to the measured pressure field $U(r, T)$. The gray area in Figs. 5a,b corresponds to the pulse width. A cross-section of the pressure field along the A–A line would be similar to the one-dimensional case.

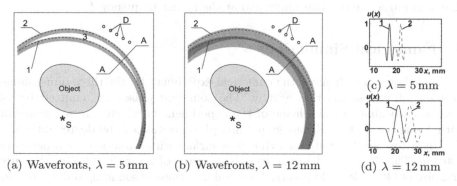

(a) Wavefronts, $\lambda = 5$ mm   (b) Wavefronts, $\lambda = 12$ mm   (c) $\lambda = 5$ mm   (d) $\lambda = 12$ mm

Fig. 5. Wavefront of the sounding waves (a,b) and cross-sections of the pressure field along the A–A line (c,d) for different wavelengths.

Figure 5c,d shows the pressure fields along the A–A line with the object present and with the object absent, for $\lambda = 5$ mm (Fig. 5c) and for $\lambda = 12$ mm (Fig. 5d) respectively. The number "1" denotes the simulated wave $u(r, T)$ propagating in a homogeneous medium at a velocity of c_0. The number "2" denotes the wave that has passed through the object.

The pulses with $\lambda = 5$ mm do not overlap, and there exists a region 3 between the wave fronts, where $u \approx 0$ and $U \approx 0$. Similarly to the one-dimensional case, the gradient of the residual functional between the measured wave 2 and the simulated wave 1 in this case is equal to zero. Therefore, the iterative gradient method used to minimize the residual functional does not converge. If the wavelength is increased to $\lambda = 12$ mm, the waves 1 and 2 overlap. Then the initial approximation c_0 falls into the domain of convergence of the iterative process.

Actual acoustic tomography experiments involve dozens of source positions and thousands of detector positions. The residual functional (3) is the sum of the residuals computed for every source-detector pair.

These examples show that choosing an initial approximation is an important issue, which affects the convergence of the iterative gradient-based method.

If there is no prior information about the structure of the inhomogeneity, the known sound speed c_0 of the surrounding medium can be used as an initial approximation. This approach is natural in ultrasonic mammography, where the sound speed difference between soft tissues and water is less than 10%.

These examples lead to the following conclusions. The following dual-frequency method to find an approximate solution of the inverse problem can be used to extend the domain of convergence of the gradient method. The measurements are performed using sounding pulses with two different central frequencies f_1 and f_2, $f_1 < f_2$. The frequencies f_1 and f_2 should differ by a factor of 2 to 3. An approximate solution is found by minimizing the residual functional at the lower frequency f_1. The frequency f_1 is chosen low enough so that an initial approximation of c_0 is sufficient for the gradient descent method to converge. The obtained solution is used as an initial approximation for the iterative process that minimizes the residual functional at the higher frequency f_2.

5   Numerical Simulations

Figure 1 shows the scheme of the numerical experiment for the three-dimensional problem of ultrasound tomography. The sounding pulses are emitted by 24 sources according to the scheme of the experiment in which a rotating mount with 4 ultrasound transducers steps through 6 positions in 60-degree intervals. The detectors are located on a cylindrical surface with a diameter and height of 130 mm. The simulated detector array has a pitch of 2 mm. The finite difference grid size is $448 \times 448 \times 448$ points. The total durations of sounding pulses are $5\,\mu s$ and $12\,\mu s$, which corresponds to the central wavelengths of 5 mm and 12 mm.

The numerical experiment consisted of solving the direct problem of wave propagation and computing the acoustic pressure $U(\boldsymbol{s}, t)$ at the detectors located at points $\boldsymbol{s}$ of a cylindrical surface, and then using the data obtained, $U(\boldsymbol{s}, t)$, to solve the inverse problem and reconstruct the speed of sound $c(\boldsymbol{r})$ and the absorption factor $a(\boldsymbol{r})$.

Figure 6 shows the cross-sections of the sound speed $c(\boldsymbol{r})$ and absorption factor $a(\boldsymbol{r})$ in the simulated phantom, for which the direct problem was solved. The acoustic parameters of the phantom were chosen to match typical parameters of soft tissues: the speed of sound ranges from 1400 to $1600\,\mathrm{m \cdot s^{-1}}$, the absorption factor ranges from 0 (in water) to $1.2\,\mathrm{dB/cm}$. The ambient sound speed c_0 is $1500\,\mathrm{m \cdot s^{-1}}$. The wave propagation model (1, 2) assumes the frequency-independent absorption law.

The phantom contains inclusions ranging in size from 2 to 10 mm with various sound speeds c and absorption factors a, and an area filled with an anisotropic texture with spatial frequencies ranging from 0.5 to $3\,\mathrm{mm^{-1}}$. The exact values of the sound speed and absorption factor of the phantom are denoted as $\{\bar{c}, \bar{a}\}$.

Figures 7, 8, and 9 show the reconstructed images of the sound speed $c(\boldsymbol{r})$ and absorption factor $a(\boldsymbol{r})$ obtained via numerical simulations. In the first numerical simulation we attempt to reconstruct the coefficients $c(\boldsymbol{r})$ and $a(\boldsymbol{r})$ inside the simulated object using the gradient method with an initial approximation of

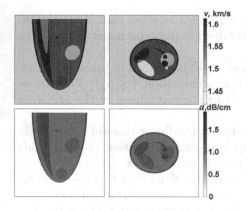

Fig. 6. Simulated phantom: speed of sound $c(r)$ and absorption factor $a(r)$.

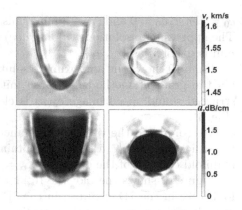

Fig. 7. Images $\{c, a\}_{loc}$ reconstructed using $\lambda = 5\,\text{mm}$ and the initial approximation $c_0 = const$, $a_0 = 0$.

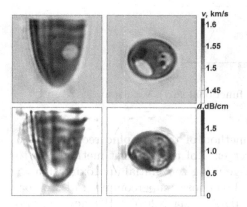

Fig. 8. Images $\{c_1, a_1\}$ reconstructed using $\lambda = 12\,\text{mm}$ and the initial approximation $c_0 = const$, $a_0 = 0$.

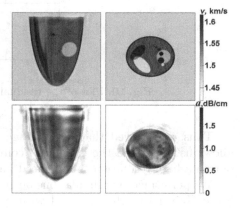

Fig. 9. Images $\{c_2, a_2\}$ reconstructed using $\lambda = 5\,\text{mm}$ and the initial approximation $\{c_1, a_1\}$ shown in Fig. 8.

$c_0 = const$, $a_0 = 0$. The central wavelength of the sounding pulses was set to 5 mm. Figure 7 shows the result of this numerical simulation. The images in Fig. 7 were obtained at the 100th iteration of the gradient method, after which the iterative process stopped at the local minimum of the residual functional. The obtained sound speed and absorption factor are denoted as $\{c, a\}_{loc}$. The global minimum of the functional corresponds to the exact image (Fig. 6) and the coefficients $\{\bar{c}, \bar{a}\}$. The value of the functional $\Phi(\bar{c}, \bar{a}) = 0$.

In the second numerical experiment the wavelength of the sounding pulses was increased to 12 mm for the initial approximation c_0 to become close enough to the global minimum of the residual functional. Figure 8 shows the $c(r)$ and $a(r)$ images obtained via the gradient method with an initial approximation of

$c_0 = const, a_0 = 0$ and $\lambda = 12$ mm. This image has a very low spatial resolution. The approximate solution $c(\boldsymbol{r})$ and $a(\boldsymbol{r})$ obtained in the experiment with $\lambda = 12$ mm is denoted as $\{c_1, a_1\}$.

Figure 9 shows the approximate solution $\{c_2, a_2\}$ obtained in this experiment via the gradient method using the initial approximation $\{c_1, a_1\}$. The reconstructed sound speed image $c_2(\boldsymbol{r})$ is close to the exact image of the phantom (Fig. 6).

We computed the values of the functional on the linear manifold containing the exact solution $\{\bar{c}, \bar{a}\}$ and the obtained approximate solution $\{c, a\}_{loc}$. This manifold consists of the elements $X_\alpha = (1 - \alpha) \cdot \{c, a\}_{loc} + \alpha \cdot \{\bar{c}, \bar{a}\}$.

Figure 10 shows the plot the residual functional $\Phi(\alpha)$ over this linear manifold. Like in the 1D example, in the 3D case the plot of the residual functional resembles the waveform of the sounding pulse. The approximate solution $\{c, a\}_{loc}$ is at a local minimum of the functional ($\alpha = 0$).

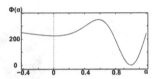

Fig. 10. Plot of the residual functional $\Phi(\alpha), \lambda = 5$ mm.

Thus, we showed that a two-stage method of tomographic reconstruction allows us to expand the domain of convergence of the gradient method and to obtain a high-resolution sound speed image using zero initial approximation.

We performed our simulations on the Lomonosov supercomputer of Moscow State University [19] equipped with NVIDIA Tesla X2070 GPU devices, two devices per computing node, and QDR Infiniband 4x (40 GBit/s) interconnect network. Single-precision floating point arithmetic was used in 3D computations. The number of GPU devices employed was equal to that of ultrasound sources (24). The 3D image was reconstructed in approximately two hours.

6 Conclusions

This paper presents a method for ultrasound tomography imaging of soft tissues for medical and biological research. Efficient algorithms have been developed for solving inverse problems of low-frequency tomography on a supercomputer. The most important issue in ultrasound tomography data interpretation is the nonlinearity of the coefficient inverse problem.

Our study showed that the lower is the sounding frequency the broader is the convergence domain of iterative processes. A method is proposed that involves the use of multiple frequency bands. The method employs an iterative gradient-based minimization algorithm at the higher frequency with the initial approximation of unknown coefficients obtained by solving the inverse problem at the lower frequency.

The numerical simulations were performed using the setup where acoustic pressure is measured on a cylindrical surface. Such a setup can be easily implemented using rotating vertical transducer arrays.

The methods developed can be used to design tomographic devices for differential diagnosis of breast cancer. The iterative algorithms used to solve the inverse problems of wave tomography can be efficiently parallelized using GPU clusters. The increasing performance of modern GPU clusters makes them a suitable computing device for ultrasound tomographs currently being developed.

Acknowledgement. This work was supported by Russian Science Foundation [grant number 17-11-01065]. The research is carried out at Lomonosov Moscow State University. The research is carried out using the equipment of the shared research facilities of HPC computing resources at Lomonosov Moscow State University.

References

1. Bazulin, E.G., Goncharsky, A.V., Romanov, S.Y., Seryozhnikov, S.Y.: Parallel CPU- and GPU-algorithms for inverse problems in nondestructive testing. Lobachevskii J. Math. **39**(4), 486–493 (2018). https://doi.org/10.1134/S1995080218040030

2. Beilina, L., Klibanov, M.V., Kokurin, M.Y.: Adaptivity with relaxation for ill-posed problems and global convergence for a coefficient inverse problem. J. Math. Sci. **167**(3), 279–325 (2010). https://doi.org/10.1007/s10958-010-9921-1

3. Boehm, C., Martiartu, N.K., Vinard, N., Balic, I.J., Fichtner, A.: Time-domain spectral-element ultrasound waveform tomography using a stochastic quasi-newton method. In: Proceedings of SPIE 10580, Medical Imaging 2018: Ultrasonic Imaging and Tomography p. 105800H, March 2018. https://doi.org/10.1117/12.2293299

4. Goncharsky, A.V., Romanov, S.Y., Seryozhnikov, S.Y.: A computer simulation study of soft tissue characterization using low-frequency ultrasonic tomography. Ultrasonics **67**, 136–150 (2016). https://doi.org/10.1016/j.ultras.2016.01.008

5. Goncharsky, A.V., Romanov, S.Y., Seryozhnikov, S.Y.: Supercomputer technologies in tomographic imaging applications. Supercomput. Front. Innov. **3**(1), 41–66 (2016). https://doi.org/10.14529/jsfi160103

6. Goncharsky, A.V., Romanov, S.Y., Seryozhnikov, S.Y.: Low-frequency three-dimensional ultrasonic tomography. Doklady Phys. **61**(5), 211–214 (2016). https://doi.org/10.1134/s1028335816050086

7. Goncharsky, A.V., Seryozhnikov, S.Y.: The architecture of specialized GPU clusters used for solving the inverse problems of 3D low-frequency ultrasonic tomography. In: Voevodin, V., Sobolev, S. (eds.) RuSCDays 2017. Communications in Computer and Information Science, vol. 793, pp. 363–395. Springer, Cham. (2017). https://doi.org/10.1007/978-3-319-71255-0_29

8. Goncharsky, A.V., Romanov, S.Y.: Supercomputer technologies in inverse problems of ultrasound tomography. Inverse Probl. **29**(7), 075004 (2013). https://doi.org/10.1088/0266-5611/29/7/075004

9. Goncharsky, A.V., Romanov, S.Y.: Inverse problems of ultrasound tomography in models with attenuation. Phys. Med. Biol. **59**(8), 1979–2004 (2014). https://doi.org/10.1088/0031-9155/59/8/1979

10. Goncharsky, A.V., Romanov, S.Y.: Iterative methods for solving coefficient inverse problems of wave tomography in models with attenuation. Inverse Probl. **33**(2), 025003 (2017). https://doi.org/10.1088/1361-6420/33/2/025003

11. Goncharsky, A., Romanov, S., Seryozhnikov, S.: Inverse problems of 3D ultrasonic tomography with complete and incomplete range data. Wave Motion **51**(3), 389–404 (2014). https://doi.org/10.1016/j.wavemoti.2013.10.001

12. Jirik, R., et al.: Sound-speed image reconstruction in sparse-aperture 3-D ultrasound transmission tomography. IEEE T. Ultrason. Ferr. **59**(2), 254–264 (2012)

13. Klibanov, M.V., Timonov, A.A.: Carleman Estimates for Coefficient Inverse Problems and Numerical Applications. Walter de Gruyter GmbH, january 2004

14. Kuzhuget, A.V., Beilina, L., Klibanov, M.V., Sullivan, A., Nguyen, L., Fiddy, M.A.: Blind backscattering experimental data collected in the field and an approximately globally convergent inverse algorithm. Inverse Probl. **28**(9), 095007 (2012). https://doi.org/10.1088/0266-5611/28/9/095007

15. Mu, S.Y., Chang, H.W.: Dispersion and local-error analysis of compact LFE-27 formulas for obtaining sixth-order accurate numerical solutions of 3D Helmholz equation. Pr. Electromagn. Res. S. **143**, 285–314 (2013)

16. Natterer, F.: Possibilities and limitations of time domain wave equation imaging. In: AMS Vol. 559: Tomography and Inverse Transport Theory, pp. 151–162. American Mathematical Society (2011). https://doi.org/10.1090/conm/559

17. Natterer, F.: Sonic imaging. In: Scherzer, O. (ed.) Handbook of Mathematical Methods in Imaging, pp. 1–23. Springer Nature, New York (2014). https://doi.org/10.1007/978-3-642-27795-5_37-2

18. Romanov, S.: Optimization of numerical algorithms for solving inverse problems of ultrasonic tomography on a supercomputer. In: Voevodin, V., Sobolev, S. (eds.) RuSCDays 2017. Communications in Computer and Information Science, vol. 793, pp. 67–79. Springer, Cham (2017). https://doi.org/10.1007/978-3-319-71255-0_6

19. Sadovnichy, V., Tikhonravov, A., Voevodin, V., Opanasenko, V.: "Lomonosov": supercomputing at Moscow State University. In: Vetter, J. (ed.) Contemporary High Performance Computing: From Petascale toward Exascale, pp. 283–307. Chapman & Hall/CRC Computational Science, CRC Press, Boca Raton (2013)

20. Saha, R.K., Sharma, S.K.: Validity of a modified Born approximation for a pulsed plane wave in acoustic scattering problems. Phys. Med. Biol. **50**(12), 2823 (2005)

21. Sak, M., et al.: Using speed of sound imaging to characterize breast density. Ultrasound Med. Biol. **43**(1), 91–103 (2017)

22. Schmidt, S., Duric, N., Li, C., Roy, O., Huang, Z.F.: Modification of Kirchhoff migration with variable sound speed and attenuation for acoustic imaging of media and application to tomographic imaging of the breast. Med. Phys. **38**(2), 998–1007 (2011). https://doi.org/10.1118/1.3539552

23. Wiskin, J.W., Borup, D.T., Iuanow, E., Klock, J., Lenox, M.W.: 3-D nonlinear acoustic inverse scattering: Algorithm and quantitative results. EEE Trans. Ultrason. Ferroelectr. Freq. Control **64**(8), 1161–1174 (2017)

24. Zeqiri, B., Baker, C., Alosa, G., Wells, P.N.T., Liang, H.D.: Quantitative ultrasonic computed tomography using phase-insensitive pyroelectric detectors. Phys. Med. Biol. **58**(15), 5237 (2013). https://doi.org/10.1088/0031-9155/58/15/5237

Supercomputer Technology for Ultrasound Tomographic Image Reconstruction: Mathematical Methods and Experimental Results

Alexander Goncharsky and Sergey Seryozhnikov[(⊠)]

Lomonosov Moscow State University, Moscow, Russia
gonchar@srcc.msu.ru, s2110sj@gmail.com

Abstract. This paper is concerned with layer-by-layer ultrasound tomographic imaging methods for differential diagnosis of breast cancer. The inverse problem of ultrasound tomography is formulated as a coefficient inverse problem for a hyperbolic differential equation. The scalar mathematical model takes into account wave phenomena, such as diffraction, refraction, multiple scattering, and absorption of ultrasound. The algorithms were tested on real data obtained in experiments on a test bench for ultrasound tomography studies. Low-frequency ultrasound in the 100–500 kHz band was used for sounding. An important result of this study is an experimental confirmation of the adequacy of the underlying mathematical model. The ultrasound tomographic imaging methods developed have a spatial resolution of 2 mm, which is acceptable for medical diagnostics. The experiments were carried out using phantoms with parameters close to the acoustical properties of human soft tissues. The image reconstruction algorithms are designed for graphics processors. Architecture of the GPU cluster for ultrasound tomographic imaging is proposed, which can be employed as a computing device in a tomographic complex.

Keywords: Ultrasound tomography · Coefficient inverse problem
Medical imaging · GPU cluster

1 Introduction

Supercomputer technologies open up new possibilities in medical diagnostics. One such example is ultrasound tomography. This technology is developed extensively in the USA, Germany, and Russia [1–4]. Medical imaging, especially the differential diagnosis of breast cancer, is a promising application of ultrasound tomography. Wave tomography methods can also be applied to nondestructive imaging of solids [5].

The inverse problem of wave tomography is nonlinear. Breakthrough results in the field of solving inverse problems of wave tomography have been obtained

© Springer Nature Switzerland AG 2019
V. Voevodin and S. Sobolev (Eds.): RuSCDays 2018, CCIS 965, pp. 401–413, 2019.
https://doi.org/10.1007/978-3-030-05807-4_34

in recent years. Explicit formulas for the gradient of the residual functional between the measured and the numerically simulated wave fields in various formulations were derived in the works [6–9]. In our earlier papers [10–12] we derived a representation for the gradient of the residual functional in terms of a scalar mathematical model that accounts for diffraction and absorption effects.

Iterative supercomputer-oriented algorithms for reconstructing ultrasound tomographic images have been developed whose efficiency was confirmed via solving numerous model problems [13]. Testing algorithms by applying them to model problems is necessary for design of ultrasound tomography devices. However, numerical simulations can not assess the adequacy of the underlying mathematical model and only physical experiments can answer this question.

In this study, algorithms for solving the inverse problems of ultrasound tomography in the layer-by-layer formulation are tested on experimental data. For this purpose, a test bench for ultrasound tomography studies was constructed. One of the key results of this study is that it demonstrates that the mathematical model agrees well with real physical processes. Solving inverse problems of wave tomography using experimental data showed that tomographic methods can reconstruct inhomogeneities approximately 2 mm in size. Such spatial resolution is quite acceptable for medical imaging. The parameters of the phantoms used in experiments were close to the acoustic properties of human soft tissues.

The inverse problem of wave tomography is a nonlinear inverse problem with millions of unknowns. The algorithms developed are designed for GPU clusters and we used GPU nodes of the "Lomonosov-2" supercomputer at the Moscow State University [14] to process experimental data. In this paper, we discuss the architecture of a GPU cluster which can be used as a computing device in a tomographic complex.

2   Formulation of the Inverse Problem and Its Solution Methods

Figure 1a shows the scheme of a layer-by-layer tomographic examination used in this study. Object G is submerged in water L with known acoustical properties. Acoustic sounding pulses are emitted by transducers at positions 1. Figure 1b shows a typical waveform of a sounding pulse. The scattered ultrasound waves are measured at positions 2. At each position, the signal is recorded as a function of time $u(t)$. Figure 1c shows a typical received signal. This scheme of the experiment allows us to measure both transmitted and reflected waves. The measurement points lie at surface Γ, which in this case is circular.

The objective is to reconstruct the internal structure of the object using these measurements. This formulation of the problem is called the time-domain formulation. The measurements are repeated for multiple horizontal planes, thus obtaining a layered representation of a 3D object. The inverse problem is solved independently for each of the horizontal planes $z = const$.

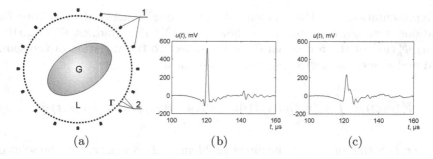

Fig. 1. Layer-by-layer tomographic examination: (a) placement of emitters and detectors; (b) waveform of a sounding pulse; (c) waveform of a received pulse

In this study, we use a scalar wave model, which accounts for all the wave phenomena, such as diffraction, refraction, multiple scattering and absorption of ultrasound. In the scalar model, the acoustic pressure $u(\boldsymbol{r},t)$ is described by a hyperbolic equation. The inverse problem of ultrasound tomography is a coefficient inverse problem, in which we use the measurements of the wave field taken at some surface Γ to obtain the coefficients $c(\boldsymbol{r})$ and $a(\boldsymbol{r})$ of the wave equation:

$$c(\boldsymbol{r})u_{tt}(\boldsymbol{r},t) + a(\boldsymbol{r})u_t(\boldsymbol{r},t) - \Delta u(\boldsymbol{r},t) = 0; \tag{1}$$

$$u(\boldsymbol{r},t)|_{t=0} = F_0(\boldsymbol{r}), \quad u_t(\boldsymbol{r},t)|_{t=0} = F_1(\boldsymbol{r}). \tag{2}$$

Here, $c(\boldsymbol{r}) = 1/v^2(\boldsymbol{r})$, where $v(\boldsymbol{r})$ is the speed of sound; $a(\boldsymbol{r})$ is the absorption factor in the medium; $\boldsymbol{r} = \{x,y\}$ is the position of the point inside the reconstructed 2D plane, and Δ is Laplacian with respect to $\boldsymbol{r}$.

This formulation assumes a transparent (non-reflecting) boundary of the computational domain. A non-reflecting boundary condition [15] in the form $\partial u/\partial n = -c^{-0.5}\partial u/\partial t$ is applied at the boundary. Here, $\boldsymbol{n}$ is a vector pointing towards the ultrasound emitter. The initial conditions $F_0(\boldsymbol{r})$ and $F_1(\boldsymbol{r})$ represent the wave field radiated from the emitter at the initial time of the numerical simulation. We obtain them via the time-reversal method [16] by solving Eq. (1) in reverse time with boundary condition $u(\boldsymbol{s},t)|_{s\in\Gamma} = U_0(\boldsymbol{s},t)$, where $U_0(\boldsymbol{s},t)$ are the data measured at surface Γ in a homogeneous medium without the object.

The coefficient inverse problem considered is ill-posed. The methods for solving such problems were developed in [17–19]. We formulate the inverse problem as that of minimizing the residual functional

$$\Phi(u(c,a)) = \frac{1}{2} \int_0^T \int_S (u(\boldsymbol{s},t) - U(\boldsymbol{s},t))^2 \, d\boldsymbol{s}\, dt \tag{3}$$

for its argument (c,a). Here $U(\boldsymbol{s},t)$ are the data measured at surface Γ for the time period $(0,T)$, $u(\boldsymbol{s},t)$ is the solution of the direct problem (1, 2) for the given $c(\boldsymbol{r}) = 1/v^2(\boldsymbol{r})$ and $a(\boldsymbol{r})$. The residual functional is the sum of the residuals (3) obtained for each position of the emitter.

Representations for the gradient of the residual functional in various formulations were obtained by the authors in [9,13]. The gradient $\Phi'(u(c,a)) = \{\Phi'_c(u), \Phi'_a(u)\}$ of the functional (3) with respect to the variation of the sound speed and absorption factor $\{dc, da\}$ has the form:

$$\Phi'_c(u(c)) = \int_0^T w_t(\mathbf{r},t)u_t(\mathbf{r},t)\,\mathrm{d}t, \quad \Phi'_a(u(a)) = \int_0^T w_t(\mathbf{r},t)u(\mathbf{r},t)\,\mathrm{d}t. \quad (4)$$

Here $u(\mathbf{r},t)$ is the solution of the direct problem (1, 2), and $w(\mathbf{r},t)$ is the solution of the "conjugate" problem with the given $c(\mathbf{r})$, $a(\mathbf{r})$, and $u(\mathbf{r},t)$:

$$c(\mathbf{r})w_{tt}(\mathbf{r},t) - a(\mathbf{r})w_t(\mathbf{r},t) - \Delta w(\mathbf{r},t) = E(\mathbf{r},t); \quad (5)$$

$$w(\mathbf{r},t=T) = 0, \quad w_t(\mathbf{r},t=T) = 0; \quad (6)$$

$$E(\mathbf{r},t) = \begin{cases} u(\mathbf{r},t) - U(\mathbf{r},t), \text{ where } \mathbf{r} \in \Gamma \text{ and } U(\mathbf{r},t) \text{ is known;} \\ 0, \text{ otherwise.} \end{cases} \quad (7)$$

Non-reflecting boundary condition [15] is applied at the boundary of the computational domain. Thus, to compute the gradient (4) we need to solve the direct problem (1, 2) and the "conjugate" problem (5)–(7).

Given a formula for the gradient (4), we can construct various iterative algorithms that minimize the residual functional, such as the steepest descent method. Let us assume that the coefficients $c^{(m)}$ and $a^{(m)}$ for m-th iteration have been determined. To construct the next iterative approximation, we compute the gradient $\{\Phi'_c(u), \Phi'_a(u)\}$ at point $\{c^{(m)}, a^{(m)}\}$ and solve the problem of minimizing a one-dimensional functional along the direction of the gradient. As the next iterative approximation, we choose the point $\{c^{(m+1)}, a^{(m+1)}\} = \arg\min_{\alpha>0} \Phi(c^{(m)} - \alpha\Phi'_c, a^{(m)} - \alpha\Phi'_a)$, and so on.

Minimization methods based on the explicit formula for the gradient allow us to propose efficient numerical algorithms for the approximate solution of inverse problems of wave tomography. The gradient method has regularizing properties and stops when the value of the residual functional becomes equal to the error of the input data [18].

To solve Eqs. (1, 2) and (5)–(7), we use the finite-difference time-domain method (FDTD) on uniform grids. We introduce a uniform discrete grid with a space step of h and a time step of τ:

$$x_i = ih, \ 0 \leq i < n; \quad y_j = jh, \ 0 \leq j < n; \quad t_k = k\tau, \ 0 \leq k < m.$$

The parameters h and τ are related by the Courant stability condition $\sqrt{2}c^{-0.5}\tau < h$, where $c^{-0.5} = v$ is the speed of sound. To approximate the Eq. (1) we use the following second-order finite difference scheme:

$$c_{ij}\frac{u_{ij}^{k+1} - 2u_{ij}^k + u_{ij}^{k-1}}{\tau^2} + a_{ij}\frac{u_{ij}^{k+1} - u_{ij}^{k-1}}{\tau} - \frac{\Delta u_{ij}^k}{h^2} = 0. \quad (8)$$

Here $u_{ij}^k = u(x_i, y_j, t_k)$ are the values of $u(\mathbf{r},t)$ at point (i,j) at the time step k; c_{ij} and a_{ij} are the values of $c(\mathbf{r})$ and $a(\mathbf{r})$ at point (i,j). The first term approximates $c(\boldsymbol{r})u_{tt}(\boldsymbol{r},t)$, the second — $a(\boldsymbol{r})u_t(\boldsymbol{r},t)$. Discrete Laplacian Δ is computed using a fourth-order optimized finite difference scheme [20]. Collecting the terms with u_{ij}^{k+1} for $(k+1)$-th time step, we obtain the explicit formula for simulating the ultrasound wave sequentially in time. A similar scheme is used to solve the Eqs. (5)–(7) for $w(\boldsymbol{r},t)$ in reverse time.

The gradient of the residual functional is computed by formula (4). An explicit formula allows us to use gradient-based mininization methods, such as the steepest descent method [21].

3   GPU Implementation of the Layer-by-layer Tomographic Image Reconstruction Algorithm

The numerical method was implemented on graphics processors using OpenCL^{TM} technology. An iterative gradient method is used for image reconstruction. At the first iteration, we use the initial approximation for the unknown coefficients $c^{(0)}(\boldsymbol{r}) = 1500\,\text{m}\cdot\text{s}^{-1}$, $a^{(0)}(\boldsymbol{r}) = 0$. The following steps are performed at each gradient descent iteration (m):

1. The direct problem (1, 2) is solved and the wave field at the boundary $u(\boldsymbol{s},t)$, $\boldsymbol{s} \in \Gamma$ is stored in memory.
2. The "conjugate" problem (5)–(7) is solved in reverse time. To compute $u(\boldsymbol{r},t)$ we use the boundary data $u(\boldsymbol{s},t)$ previously stored, and to compute $w(\boldsymbol{r},t)$ we use the measured data $U(\boldsymbol{s},t), \boldsymbol{s} \in \Gamma$. The gradient Φ'_c, Φ'_a is computed by formula (4), summed over all the emitter positions and accumulated over time.
3. The gradient descent step γ is determined. Its value is adjusted automatically: if the residual decreases $(\Phi^{(m)} < \Phi^{(m-1)})$, then γ is increased, and vice versa.
4. The current approximation is updated: $c^{(m+1)} = c^{(m)} + \gamma\Phi'_c$, $a^{(m+1)} = a^{(m)} + \gamma\Phi'_a$.

The most computationally expensive tasks are solving the "conjugate" problem (5)–(7) and computing the gradient. Figure 2 illustrates the GPU implementation of the algorithm. To adapt the algorithm for GPU, the task is split into independent thread blocks. The gradient is summed over all the emitter positions, and therefore an efficient solution is to compute the gradient sequentially for each emitter position, accumulating the partial sum in the shared memory of the GPU.

Each thread block processes an $LX \times LY$ area of the $L \times L$ image, where LX is equal to the number of threads in a block, and LY is determined by the amount of available shared memory: $LY = SharedMemorySize/(LX \times 8)$. The number of threads per block LX is chosen to optimize the GPU workload. For tested devices, the optimum value of LX was found to be equal to the maximum supported number of threads (512). Therefore, $LY \leq 12$ for a typical NVidia device with 48 KB of shared memory. We used $LX = L$ for image sizes $L \leq 512$ and $LX = L/2$ for image sizes $512 < L \leq 1024$.

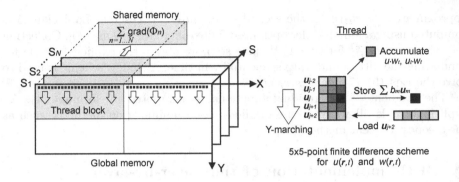

Fig. 2. GPU-optimized algorithm of computing the gradient of the residual functional

Inside each $LX \times LY$ data block the computations are carried out sequentially along the Y-axis (Y-marching method) and sequentially for each emitter position S_n, $n = 1, ..., N$. The Y-marching method ensures high automatic caching efficiency due to the sequential memory access pattern and allows increasing the stencil size with no significant performance loss. At each step along the Y-axis, the value of u_{ij}^{k-1} and a vector $\boldsymbol{u}_{j+2}$ containing the acoustic pressure values for 5 consecutive points along the X-axis are loaded into the registers. Since the discrete Laplacian coefficient matrix is symmetric [20], only 3 components per row are stored in the registers for a 5×5-point finite difference scheme.

Acoustic pressure u_{ij}^{k+1} at the next time step is computed by formula (8), which can be expressed as a scalar product with some coefficients b_m. The coefficients are constant for all points of the stencil except the center point (i, j). The wave field w_{ij}^{k+1} is computed using a similar formula. The gradient $\{u \cdot w_t, u_t \cdot w_t\}$ is then computed and added to the partial sum over emitter positions, which is accumulated in the shared memory. The process is repeated for all emitter positions.

The data are stored in a 3D array $X \times Y \times S$, where S corresponds to the emitter position. A 2D subarray for one position S_n contains the values of $u(\boldsymbol{r}, t)$ and $w(\boldsymbol{r}, t)$ for three time steps. Computations with experimental data showed that 32-bit floating point type is sufficient for data representation. Thus, the amount of memory required to store all the data is $24 \times L^2 \times N$ bytes, where L is the grid size along one dimension, and N is the number of emitter positions. For 24 positions and $L = 768$, the volume of data amounts to 350 MB. Additionally, $32 \times N \times T \approx 100$ MB of global memory are used to store the values of $u(\boldsymbol{r}, t)$ and $w(\boldsymbol{r}, t)$ at the measurement surface Γ. Here T is the number of time steps in the simulation, which is on the order of L. Thus, modern graphics cards with at least 1 GB of memory can host all the data in the global memory, eliminating the need to access the system RAM throughout the simulation.

The algorithm developed was tested on NVidia GeForce GTX Titan and Tesla K40s devices. The program profiling statistics are listed in Table 1.

Table 1. Program profiling statistics (GTX Titan, 768 × 768 grid, 10 iterations)

Runs	Time, μs	Min, μs	Max, μs	Total	%	Function
5040	1924	1866	2425	9699.17 ms	21.16%	ForwardWave
5020	6340	6303	7441	31.83 s	69.44%	BackwardWave
10060	271	181	766	2733.53 ms	5.96%	BoundaryCond
5020	31	25	41	156.72 ms	0.34%	LoadBound
5040	32	24	678	161.89 ms	0.35%	SaveBound
5020	57	53	70	287.69 ms	0.63%	LoadExData
5020	36	35	86	183.14 ms	0.40%	SaveExData
264	2313	1	29224	610.87 ms	1.33%	*Data transfers*

We used OpenCL profiling events to obtain start and end times of each computational kernel and data transfer operation. The statistics were accumulated for 10 gradient descent iterations on a GTX Titan device.

The most time-consuming functions are wave simulations that solve the direct (ForwardWave) and the conjugate (BackwardWave) problems. The time spent in other functions amounted to less than 10%, which means near-linear scaling with respect to the problem size.

Table 2 lists the memory requirements for different problem sizes and the computation times for tested devices. The number of emitter positions was $N = 24$, and 40 gradient descent iterations were performed in each test run. These parameters correspond to the values used to reconstruct the images from experimental data.

Table 2. Memory requirements and computational times for different problem sizes

Grid size (L)	512	768	1024
GPU memory used, MB	220	450	780
Time, GTX Titan	114 s	258 s	477 s
Time, Tesla K40s	130 s	294 s	562 s

The total number of operations per problem is proportional to $L^3 \cdot N$, and the amount of memory is proportional to $L^2 \cdot N$. The experimental reconstructions were performed using a grid size of $L = 768$. For this problem size, reconstructing one horizontal layer using one GPU device takes approximately 5 min.

One of the aims of this study is to assess the parameters of a GPU cluster that can be used as a computing device for medical imaging. The benchmarking results showed that to reconstruct the images for multiple layers in practically feasible time, which is 15–30 min, the number of graphics processors in the cluster should amount to 1/3–1/4 of the number of layers. The computations for each layer are independent. To obtain the images for 30–40 layers, a GPU cluster containing 8–16 devices would be required. Since the memory requirements for layer-by-layer image reconstruction are quite low, the cluster can include 4–8 dual-GPU boards such as GeForce GTX 690. Devices equipped with High Bandwidth Memory (HBM) have approximately 3 times higher performance than GDDR5-equipped devices, so a cluster of 4 such devices can perform the image reconstruction in a reasonable time. Such GPU-cluster would be compact and inexpensive, and can be incorporated as a computing device in a diagnostic facility.

4 The Test Bench for Ultrasound Tomography Studies

The experimental measurements were carried out using a custom-made test bench for ultrasound tomography studies, shown in Fig. 3. The test bench is designed for ultrasound tomographic imaging of 3D objects with parameters close to the acoustical properties of soft tissues. The ultrasound emitter 1 and the receiver 2 can be rotated using the motors 3 around the object 4 and moved in the vertical direction. The maximum size of the inspected object is approximately 12×12 cm.

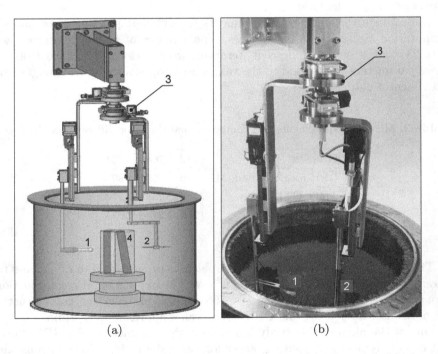

(a)   (b)

Fig. 3. The test bench for ultrasound tomography studies: (a) 3D model, (b) photo of the mechanical assembly

A broadband ultrasound transducer with a central frequency of 400 kHz was used as an emitter. The frequency spectrum of the sounding pulses covers the 100–500 kHz band. The duration of the sounding pulses was approximately $2\,\mu s$. Teledyne Reson TC4038 hydrophone was used as a receiver.

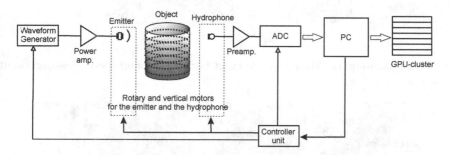

Fig. 4. The functional scheme of the test bench

Figure 4 shows the functional scheme of the test bench. Electrical pulses are generated by the arbitrary waveform generator and amplified by the power amplifier. The amplitude of the pulses at the transducer is $\pm 80\,V$. Acoustic signals are received by the hydrophone, amplified by the preamplifier and digitized using the ADC module. The digitized data are collected by a PC. The motors, the waveform generator, and the ADC module are synchronized by a controller unit.

5   Ultrasound Tomographic Image Reconstruction Using Experimental Data

For ultrasound tomographic image reconstruction experiments we used a cylindrical phantom made of soft silicone, 56 mm in diameter and 130 mm in height. Figure 5 shows the 3D model of the phantom. Phantom 1 contains a tilted 10 mm cylindrical hole 2, and a 15 mm vertical cylindrical hole 3. The phantom is affixed to the mount with three metal pins 1 mm in diameter.

The speed of sound in the silicone is $\approx 1400\,\mathrm{m{\cdot}s^{-1}}$; in water outside of the phantom and inside the holes—1500 m·s^{-1}. Figure 5b shows the horizontal sound speed cross-section of the phantom in the $z=30$ mm plane. Figure 6 shows the reconstructed sound speed images of the phantom in different horizontal cross-sections. The reconstructed speed of sound inside the holes is equal to the speed of sound in water ($1500\,\mathrm{m \cdot s^{-1}}$). The 7 mm rod with a higher speed of sound is clearly resolved in all cross-sections. The reconstructed speed of sound inside the rod is $1800\,\mathrm{m \cdot s^{-1}}$. The tilted hole changes its position from image to image, as the vertical position of the emitter and the receiver changes. The 1 mm metal pins are clearly visible in all cross-sections.

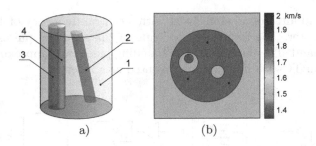

a)   (b)

Fig. 5. The phantom: (a) 3D model, (b) exact horizontal sound speed cross-section at z = 30 mm

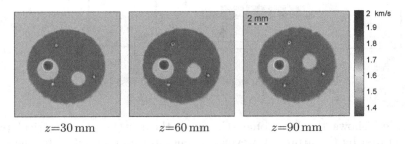

z=30 mm   z=60 mm   z=90 mm

Fig. 6. Reconstructed sound speed cross-sections of the phantom

As is evident from the reconstructed images in Fig. 6, the spatial resolution is approximately 2 mm, while the sounding wavelength is approximately 4 mm. The sound speed difference between the silicone material and water amounted to ≈5%. Yet, the algorithms developed reconstruct the sound speed image with high precision even for a low-contrast object.

The size of the data obtained in a tomographic examination amounted to 35 MB per horizontal layer. The data for 24 emitter positions and 500 receiver positions recorded over a time period of 200 μs with 14-bit precision at a sampling rate of 5 MSPS were used to reconstruct the image in each layer. The grid step was 0.4 mm and the number of gradient descent iterations required to obtain the images amounted to 30–40. Image reconstruction was performed on GPU nodes of the "Lomonosov-2" supercomputer at the Moscow State University.

One of the drawbacks of layer-by-layer ultrasound tomography is that the spatial resolution along the Z-axis is lower than the spatial resolution in X–Y planes. This is due to the fact that in the layered model only waves propagating along the reconstructed plane are taken into account. In order to increase the resolution, out-of-plane wave sources must be used as well. This scheme of tomographic examination requires solving a three-dimensional inverse problem, which is much more computationally expensive than a layer-by-layer problem. The numerical simulations presented in [13,22] showed that the 3D scheme can be implemented in practice and can provide higher image reconstruction quality than the layered scheme. However, a 3D setup requires much more complex measuring equipment.

6   Conclusion

The main result of this study is a successful application of the ultrasound tomography algorithms developed to real experimental data. Numerous experiments performed on a test bench for experimental tomographic studies showed that the scalar wave model based on a second-order hyperbolic equation can be used to reconstruct images of real objects. The tomographic methods developed have high resolution, which is quite acceptable for medical imaging. The experiments were carried out using phantoms with parameters close to the acoustic properties of human soft tissues.

The use of GPU clusters is the most practically feasible option for implementing the iterative solution algorithms for 2D inverse problems. A finite-difference method can be efficiently parallelized using data-parallel architectures such as SIMD and GPU. Experiments showed that single-precision floating-point arithmetic is sufficient for solving direct and inverse problems. Thus, widely available GPU devices can be used for image reconstruction.

The algorithms and the scheme of experiments are designed for implementation on a dedicated GPU cluster The experimental setup assumes detailed recording of the signal on a cylindrical surface for multiple emitter positions. The number of measurement points in each horizontal layer amounts to 500. At the same time, the number of emitter positions is small—approximately 20 in each layer. Thus, an image of a single layer can be reconstructed by a single GPU device in a short time. We propose the architecture of a GPU cluster with a relatively small number of devices proportional to the number of reconstructed layers of the 3D image. The properties of such a GPU cluster, such as the size, power consumption, and cost, allow it to be used in the new medical ultrasound tomographic facilities being developed.

Acknowledgements. This work was supported by Russian Science Foundation [grant number 17-11-01065]. The research is carried out at Lomonosov Moscow State University. The research is carried out using the equipment of the shared research facilities of HPC computing resources at Lomonosov Moscow State University.

References

1. Sak, M., et al.: Using speed of sound imaging to characterize breast density. Ultrasound Med. Biol. **43**(1), 91–103 (2017). https://doi.org/10.1016/j.ultrasmedbio.2016.08.021
2. Jifik, R., et al.: Sound-speed image reconstruction in sparse-aperture 3-D ultrasound transmission tomography. IEEE Trans. Ultrason. Ferr. **59**(2), 254–264 (2012). https://doi.org/10.1109/tuffc.2012.2185
3. Wiskin, J., et al.: Three-dimensional nonlinear inverse scattering: quantitative transmission algorithms, refraction corrected reflection, scanner design, and clinical results. J. Acoust. Soc. Am. **133**(5), 3229–3229 (2013). https://doi.org/10.1121/1.4805138

4. Burov, V.A., Zotov, D.I., Rumyantseva, O.D.: Reconstruction of the sound velocity and absorption spatial distributions in soft biological tissue phantoms from experimental ultrasound tomography data. Acoust. Phys. **61**(2), 231–248 (2015). https://doi.org/10.1134/s1063771015020013

5. Bazulin, E.G., Goncharsky, A.V., Romanov, S.Y., Seryozhnikov, S.Y.: Parallel CPU- and GPU-algorithms for inverse problems in nondestructive testing. Lobachevskii J. Math. **39**(4), 486–493 (2018). https://doi.org/10.1134/S1995080218040030

6. Natterer, F.: Sonic imaging. In: Scherzer, O. (ed.) Handbook of Mathematical Methods in Imaging, pp. 1–23. Springer, New York (2014). https://doi.org/10.1007/978-3-642-27795-5_37-2

7. Klibanov, M.V., Timonov, A.A.: Carleman Estimates for Coefficient Inverse Problems and Numerical Applications, Walter de Gruyter GmbH (2004). https://doi.org/10.1515/9783110915549

8. Goncharsky, A., Romanov, S., Seryozhnikov, S.: Inverse problems of 3D ultrasonic tomography with complete and incomplete range data. Wave Motion **51**(3), 389–404 (2014). https://doi.org/10.1016/j.wavemoti.2013.10.001

9. Goncharsky, A.V., Romanov, S.Y.: Supercomputer technologies in inverse problems of ultrasound tomography. Inverse Probl. **29**(7), 075004 (2013). https://doi.org/10.1088/0266-5611/29/7/075004

10. Goncharsky, A.V., Romanov, S.Y.: Iterative methods for solving coefficient inverse problems of wave tomography in models with attenuation. Inverse Probl. **33**(2), 025003 (2017). https://doi.org/10.1088/1361-6420/33/2/025003

11. Goncharsky, A.V., Romanov, S.Y., Seryozhnikov, S.Y.: Low-frequency three-dimensional ultrasonic tomography. Dokl. Phys. **468**(3), 268–271 (2016). https://doi.org/10.7868/S0869565216150093

12. Goncharsky, A.V., Romanov, S.Y.: Inverse problems of ultrasound tomography in models with attenuation. Phys. Med. Biol. **59**(8), 1979–2004 (2014)

13. Goncharsky, A., Romanov, S., Seryozhnikov, S.: A computer simulation study of soft tissue characterization using low-frequency ultrasonic tomography. Ultrasonics **67**, 136–150 (2016). https://doi.org/10.1016/j.ultras.2016.01.008

14. Voevodin, V., et al.: Practice of "Lomonosov" supercomputer. Open Syst. J. **7**, 36–39 (2012). http://www.osp.ru/os/2012/07/13017641/

15. Engquist, B., Majda, A.: Absorbing boundary conditions for the numerical simulation of waves. Math. Comput. **31**(139), 629–629 (1977). https://doi.org/10.1090/s0025-5718-1977-0436612-4

16. Fink, M.: Time reversal in acoustics. Contemp. Phys. **37**(2), 95–109 (1996). https://doi.org/10.1080/00107519608230338

17. Tikhonov, A.N.: Solution of incorrectly formulated problems and the regularization method. Soviet Math. Dokl. **4**, 1035–1038 (1963)

18. Bakushinsky, A., Goncharsky, A.: Ill-Posed Problems: Theory and Applications. Springer, Netherlands (1994). https://doi.org/10.1007/978-94-011-1026-6

19. Tikhonov, A.N., Goncharsky, A.V., Stepanov, V.V., Yagola, A.G.: Numerical Methods for the Solution of Ill-Posed Problems. Mathematics and Its Applications. Springer, Netherlands (1995). https://doi.org/10.1007/978-94-015-8480-7

20. Hamilton, B., Bilbao, S.: Fourth-order and optimised finite difference schemes for the 2-D wave equation. In: Proceedings of the 16th International Conference on Digital Audio Effects (DAFx-13), pp. 363–395. Springer (2013)

21. Bakushinsky, A., Goncharsky, A.: Iterative Methods for Solving ill-Posed Problems. Nauka, Moscow (1989)
22. Goncharsky, A.V., Seryozhnikov, S.Y.: The architecture of specialized GPU clusters used for solving the inverse problems of 3D low-frequency ultrasonic tomography. In: Voevodin, V., Sobolev, S. (eds.) RuSCDays2017. Communications in Computer and Information Science, vol. 793, pp. 363–395. Springer, Cham (2017). https://doi.org/10.1007/978-3-319-71255-0_29

The Parallel Hydrodynamic Code for Astrophysical Flow with Stellar Equations of State

Igor Kulikov[1]([✉]), Igor Chernykh[1], Vitaly Vshivkov[1], Vladimir Prigarin[2], Vladimir Mironov[3], and Alexander Tutukov[4]

[1] Institute of Computational Mathematics and Mathematical Geophysics SB RAS, Novosibirsk, Russia
{kulikov,vsh}@ssd.sscc.ru, chernykh@parbz.sscc.ru
[2] Novosibirsk State Technical University, Novosibirsk, Russia
vovkaprigarin@gmail.com
[3] Lomonosov Moscow State University, Moscow, Russia
vmironov@lcc.chem.msu.ru
[4] Institute of Astronomy RAS, Moscow, Russia
atutukov@inasan.ru

Abstract. In this paper, a new calculation method for numerical simulation of astrophysical flow at the supercomputers is described. The co-design of parallel numerical algorithms for astrophysical simulations is described in detail. The hydrodynamical numerical model with stellar equations of state (EOS), numerical methods for solving the hyperbolic equations and a short description of the parallel implementation of the code are described. For problems using large amounts of RAM, for example, the collapse of a molecular cloud core, our code upgraded for Intel Memory Drive Technology (IMDT) support. In this paper, we present the results of some IMDT performance tests based on Siberian Supercomputer Center facilities equipped with Intel Optane Memory. The results of numerical experiments of hydrodynamical simulations of the model stellar explosion are presented.

Keywords: Computational astrophysics · Intel Xeon Phi
Numerical methods

1 Introduction

The Type Ia supernova progenitor problem is one of the most exciting problems in astrophysics, requiring detailed numerical modeling to complement observations of these explosions. One possible progenitor is the white dwarf merger scenario, which has the potential to naturally explain many of the observed characteristics of SN Ia [1].

During the last three decades two main approaches have been used to solve astrophysical problems: the Eulerian adaptive mesh refinement (AMR) methods, and the Lagrangian smoothed particle hydrodynamics method (SPH).

© Springer Nature Switzerland AG 2019
V. Voevodin and S. Sobolev (Eds.): RuSCDays 2018, CCIS 965, pp. 414–426, 2019.
https://doi.org/10.1007/978-3-030-05807-4_35

The Lagrangian approach (SPH method) is used in Hydra [2], Gasoline [3], GrapeSPH [4], GADGET [5] packages. The Eulerian approach (including AMR) is used in NIRVANA [6], FLASH [7], ZEUS [8], ENZO [9], RAMSES [10], ART [11], Athena [12], Pencil Code [13], Heracles [14], Orion [15], Pluto [16], CASTRO [17] codes. Eulerian approach with use of AMR was for the first time used on the hybrid supercomputers equipped with graphic accelerators in a GAMER code [18]. BETHE-Hydro [19], AREPO [20], CHIMERA [21], GIZMO [22] codes are based on combination of Lagrangian and Eulerian approaches. The advantages and disadvantages of the method are described in detail in papers [23,24], we provide below a brief comparison. In the last decade Lagrangian-Eulerian methods have been actively used to solve astrophysical problems. These methods unite advantages of both the Lagrangian, and Eulerian approaches. During the past years, us has developed the hybrid Eulerian-Lagrangian approach based on operator splitting method for solving the astrophysical problems [23–29]. In this paper we describe the novel approach to the co-design of the numerical model for astrophysical flows with stellar equations of state, that can be implemented on the overscalable Peta- and Exascale supercomputer complex (Table 1).

Table 1. Advantages and disadvantages of methods

SPH advantages:	AMR advantages:
Robustness of the algorithm	Approved numerical methods
Galilean-invariant solution	No artificial viscosity
Simplicity of implementation	Higher order shock waves
Flexible geometries of problems	Resolution of discontinuities
High accurate gravity solvers	No suppression of instabilities
SPH disadvantages:	AMR disadvantages:
Artificial viscosity is parameterize	The complexity of implementation
Variations of the smoothing length	The effects of mesh
The problem of shock wave and discontinuous solutions	Problem of the minimal mesh resolution
The problem of discontinuous solutions	Problem of the minimal mesh resolution
Instabilities suppressed	Not galilean-invariant solution
The method is not scalable	The method is not scalable

2 Numerical Method

The method is based on the idea of solving the hydrodynamics equations in two stages. At the first, Eulerian, stage the system is solved without advection terms, at the second, Lagrangian, stage the advective transportation is taken into account. The division into stages allows us to solve problems of Galilean non-invariance and other mesh effects efficiently, and to simulate discontinuous solutions and shock waves correctly. The use of regular meshes does not cause new mesh problems and at the same time allows us to use various Cartesian topology of communication for supercomputers and for accelerators. The equations are written on entropy formulation.

In a couple of works in field of numerical hydrodynamics the entropy equation is used instead of nonconservative equation for internal energy written in conservative form [27]. The undeniable advantage of such approach is guarantee of non-decreasing entropy and effect of negative pressure, and ability to correctly describe supersonic flows. But there are some restrictions in usage of this approach (see the discussion in [30]). The undeniable advantage of this notation of equations of hydrodynamics is in their representation in vector conservative form.

$$\frac{\partial}{\partial t}\begin{pmatrix} \rho \\ \rho u \\ \rho S \\ \varepsilon + \rho\frac{u^2}{2} \end{pmatrix} + \nabla \cdot \begin{pmatrix} \rho u \\ \rho u \otimes u + p \\ \rho S u \\ \left(\varepsilon + \rho\frac{u^2}{2} + p\right) u \end{pmatrix} = \begin{pmatrix} 0 \\ \rho \nabla \Phi \\ 0 \\ (\nabla\Phi, \rho u) \end{pmatrix} \tag{1}$$

$$\triangle \Phi = 4\pi G \rho \tag{2}$$

where ρ is density of the gas, u is velocity, S is entropy, $p = p(\rho, S)$ is pressure, ε is internal energy, γ is adiabatic index, Φ is gravity, G is gravity constant. For numerical simulation in this papers we should use next form of equation of state:

$$p = (\gamma - 1)\varepsilon = S\rho^\gamma \tag{3}$$

of course, Eq. (1) allow to take into account any kind of EOS. On Fig. (1) block diagram of the numerical scheme was shown. The details of numerical method were described in [30].

3   Parallel Implementation

3.1   Domain Decomposition

Using a uniform mesh in Cartesian coordinates to solve the equations of hydrodynamics makes it possible to use an arbitrary Cartesian topology for decomposition of the computational domain. Such organization of computing has potentially infinity scalability. In this paper, we use the geometric decomposition of the computational domain. On one coordinate, the external one-dimensional cutting takes place using the MPI technology by means FFTW library. (see Fig. 2). This decomposition is related to the topology and architecture of all supercomputers that were used for computational experiments.

3.2   Parallel Algorithm

In this section, we will consider the parallel implementation of each box in the computational scheme on Fig. (1).

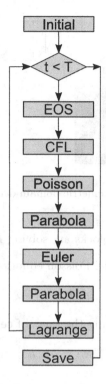

Fig. 1. The block diagram of numerical method

EOS Procedure. The blocks are computing in everyone cells an equation of state, and velocity vector:

1. computing pressure by means equation of state (for ideal gas in this a paper)

$$p = p(\rho, S) = S\rho^\gamma \tag{4}$$

2. computing velocity vector by means median value of the density on 27-point local stencil

$$u = \frac{\rho u}{[\rho]} \tag{5}$$

CFL Procedure. The blocks are computing in everyone cells a time step by means Courant–Friedrichs–Lewy condition:

1. computing local time step for everyone cell by means of equation:

$$\tau = \sqrt{\gamma p/\rho} + \|u\| \tag{6}$$

2. computing global time step for mesh by means MPI_Allreduce function.

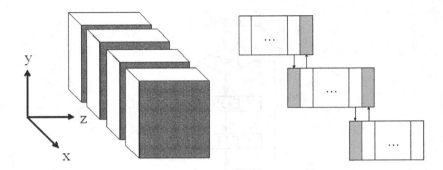

Fig. 2. The geometric domain decomposition

POISSON Procedure. The blocks are solving Poisson equation for gravity:

1. formation the right part of the Poisson equation

$$\triangle \Phi = 4\pi G \rho \tag{7}$$

2. on everyone cells are computing the mass of cell:

$$m = \rho \times h^3 \tag{8}$$

where h is size of cells.
3. computing mass of the gas in computational domain by means MPI_Allreduce function.
4. on boundary cell are computing boundary condition by means fundamental solution:

$$\Phi_{boundary} = -\frac{m}{r} \tag{9}$$

where r is a distance from boundary to center of domain.
5. computing σ is the forward Fast Fourier Transform for the density by means fftwnd_mpi FFTW function (on base MPI_Alltoallv function).
6. solving Poisson equation in harmonic space by means equations:

$$\phi_{jmn} = \frac{\frac{2}{3}\pi h^2 \sigma_{jmn}}{1 - \left(1 - \frac{2sin^2 \frac{\pi j}{J}}{3}\right)\left(1 - \frac{2sin^2 \frac{\pi m}{K}}{3}\right)\left(1 - \frac{2sin^2 \frac{\pi n}{L}}{3}\right)} \tag{10}$$

7. computing Φ is the inverse Fast Fourier Transform for the gravity in harmonic space ϕ by means fftwnd_mpi FFTW function (on base MPI_Alltoallv function).

PARABOLA Procedure. The blocks are construct parabolas for numerical scheme. We construct the piecewise-parabolic function $q(x)$ on regular mesh with step size is h, on interval $[x_{i-1/2}, x_{i+1/2}]$. In general form the parabola can be written on next equation:

$$q(x) = q_i^L + \xi \left(\triangle q_i + q_i^{(6)}(1 - \xi) \right) \tag{11}$$

where q_i is the value in center of cell, $\xi = (x - x_{i-1/2})h^{-1}$, $\triangle q_i = q_i^L - q_i^R$ and $q_i^{(6)} = 6(q_i - 1/2(q_i^L + q_i^R))$ by means conservation laws:

$$q_i = h^{-1} \int_{x_{i-1/2}}^{x_{i+1/2}} q(x)dx \tag{12}$$

to construct $q_i^R = q_{i+1}^L = q_{i+1/2}$ we should use the interpolation function of 4-th order of accuracy:

$$q_{i+1/2} = 1/2(q_i + q_{i+1}) - 1/6(\delta q_{i+1} - \delta q_i) \tag{13}$$

where $\delta q_i = 1/2(q_{i+1} - q_{i-1})$. Input value for construct of the parabola is q_i. Output procedure are all parameters of parabola on each interval $[x_{i-1/2}, x_{i+1/2}]$.

1. We construct $\delta q_i = 1/2(q_{i+1} - q_{i-1})$, and no extreme regularization:

$$\delta_m q_i = \begin{cases} min(|\delta q_i|, 2|q_{i+1} - q_i|, 2|q_i - q_{i-1}|)sign(\delta q_i), \\ (q_{i+1} - q_i)(q_i - q_{i-1}) > 0 \\ 0, (q_{i+1} - q_i)(q_i - q_{i-1}) \leq 0 \end{cases} \tag{14}$$

2. We do the exchange of overlapping domain by means MPI_Send/MPI_Recv functions (details of implementation You can found in appendix).
3. Computing boundary values for parabola:

$$q_i^R = q_{i+1}^L = q_{i+1/2} = 1/2(q_i + q_{i+1}) - 1/6(\delta_m q_{i+1} - \delta_m q_i) \tag{15}$$

4. Reconstruct of the parabola by means equations:

$$\triangle q_i = q_i^L - q_i^R \tag{16}$$

$$q_i^{(6)} = 6(q_i - 1/2(q_i^L + q_i^R)) \tag{17}$$

for the monotone of parabola we should use for boundary values q_i^L, q_i^R next equations:

$$q_i^L = q_i, q_i^R = q_i, (q_i^L - q_i)(q_i - q_i^R) \leq 0 \tag{18}$$

$$q_i^L = 3q_i - 2q_i^R, \triangle q_i q_i^{(6)} > (\triangle q_i)^2 \tag{19}$$

$$q_i^R = 3q_i - 2q_i^L, \triangle q_i q_i^{(6)} < -(\triangle q_i)^2 \tag{20}$$

5. Make a finally upgrade of parameters of parabola

$$\triangle q_i = q_i^L - q_i^R \tag{21}$$

$$q_i^{(6)} = 6(q_i - 1/2(q_i^L + q_i^R)) \tag{22}$$

EULER Procedure. The blocks are solve equations on Euler stage:

$$\frac{\partial}{\partial t}\begin{pmatrix} \rho \\ \rho u \\ \rho S \\ \varepsilon + \rho \frac{u^2}{2} \end{pmatrix} = \begin{pmatrix} 0 \\ \rho \nabla \Phi \\ 0 \\ (\nabla \Phi, \rho u) \end{pmatrix} \tag{23}$$

For approximation of pressure and velocity we should use next equations:

$$U = \frac{u_L(-\lambda t) + u_R(\lambda t)}{2} + \frac{p_L(-\lambda t) - p_R(\lambda t)}{2}\sqrt{\frac{(\sqrt{\rho_L} + \sqrt{\rho_R})^2}{\frac{\gamma_L + \gamma_R}{2}(\rho_L^{3/2} + \rho_R^{3/2})(p_L\sqrt{\rho_L} + p_R\sqrt{\rho_R})}} \tag{24}$$

$$P = \frac{p_L(-\lambda t) + p_R(\lambda t)}{2} + \frac{u_L(-\lambda t) - u_R(\lambda t)}{2}\sqrt{\frac{\frac{\gamma_L + \gamma_R}{2}(\rho_L^{3/2} + \rho_R^{3/2})(p_L\sqrt{\rho_L} + p_R\sqrt{\rho_R})}{(\sqrt{\rho_L} + \sqrt{\rho_R})^2}}, \tag{25}$$

where

$$\lambda = \sqrt{\frac{\frac{\gamma_L + \gamma_R}{2}(p_L\sqrt{\rho_L} + p_R\sqrt{\rho_R})}{\rho_L^{3/2} + \rho_R^{3/2}}} \tag{26}$$

$$q_L(-\nu t) = q_i^R - \frac{\nu t}{2h}\left(\triangle q_i - q_i^6\left(1 - \frac{2\nu t}{3h}\right)\right) \tag{27}$$

$$q_R(\nu t) = q_i^L + \frac{\nu t}{2h}\left(\triangle q_i + q_i^6\left(1 - \frac{2\nu t}{3h}\right)\right) \tag{28}$$

After computing of Euler stage we do the exchange of overlapping domain by means MPI_Send/MPI_Recv functions.

LAGRANGE Procedure. The blocks are solve equations on Lagrange stage:

$$\frac{\partial f}{\partial t} + \nabla \cdot (f v) = 0 \tag{29}$$

where f is the some conservation variable (density, moment of impulse, entropy or total energy). For computing of the flux $F = f v$ by means $\lambda = |v|$ using equations:

$$F = v \times \begin{cases} f_L(-\lambda t), v \geq 0 \\ f_R(\lambda t), v < 0 \end{cases} \tag{30}$$

where $f_L(-\lambda t)$ and $f_R(\lambda t)$ is piecewise-parabolic function for f and velocity cells interface are computing by means Roe average method:

$$v = \frac{v_L\sqrt{\rho_L} + v_R\sqrt{\rho_R}}{\sqrt{\rho_L} + \sqrt{\rho_R}} \tag{31}$$

After computing of Lagrange stage we do the exchange of overlapping domain by means MPI_Send/MPI_Recv functions (details of implementation You can found in appendix).

3.3 Parallel Code Efficiency

To understand the behavior of the paralleled code, we provide the results of scalability tests. In our case, we did weak scalability tests due to a strong memory bounded problem. Typical arithmetic intensity for PDE (Partially Differential Equations) solvers is ranged from 0.1-1 FLOPS per byte [31–33]. For our tests, we use all Intel Xeon X5670 CPU based nodes of Siberian Supercomputer Center's NKS-30T cluster. Each core of CPU was loaded by 256^3 mesh. Figure 3 shows the efficiency T on a different cores number. The efficiency T is

$$T = \frac{Total_1}{Total_p} \tag{32}$$

where $Total_1$ is computations time on one core when using one core, $Total_p$ is computations time on one core when using p cores. The 93% efficiency was achieved on 768 computational cores (see Fig. 3). In numerical experiment the mesh with size 256^3 was computed at each core. As we said before, our problem is the memory bounded problem. Modern CPUs such as Intel Xeon Scalable processors have up to 28 cores with up to 8 sockets on a platform.

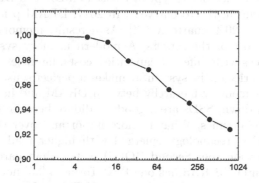

Fig. 3. Scalability of parallel code on the cluster NKS-30T SSCC.

From our point of view, the main challenge for all kinds of PDE solvers is to take full computational power from all CPU's cores. In our case, it will be reasonable to put 2048^3 mesh into the RAM of each computational node with 2x28 cores CPU. We need at least 3TB memory for each computational node for holding this size of mesh in DRAM. It is hard to find this kind of supercomputer's nodes, but we hope that the evolution of Intel Memory Drive Technology (IMDT) [36] and Intel Optane technologies can help to build this kind of systems. In the next chapter, we will present the first results of our solver tests on Intel Optane memory.

3.4 Intel®Optane™Performance Tests

In the recent years the capacity of system memory for high-performance computing (HPC) systems has not been kept with the pace of the increased CPU

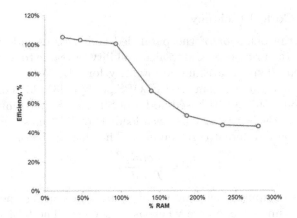

Fig. 4. An efficiency of the Lagrangian stage using Intel Optane SSDs.

power. The amount of system memory often limits the size of problems that can be solved. System memory is typically based on DRAM. DRAM prices have significantly grown up in the recent year. In 2017, DRAM prices were growing up approximately 10–20% quarterly [34]. As a result, memory can contribute up to 90% to the cost of the servers. A modern memory system is a hierarchy of storage devices with different capacities, costs, latencies, and bandwidths intended to reduce price of the system. It makes a perfect sense to introduce yet another level in the memory hierarchy between DRAM and hard disks to drive price of the system down. SSDs are a good candidate because they are cheaper than DRAM up to 10 times. What is more important, over the last 10 years, SSD based on NAND technology emerged with higher read/write speed and Input/Ouput Operations per Second (IOPS) metric than hard disks. Siberian Supercomputer Center SB RAS has one RSC Tornado [35] node equipped with Intel Optane memory (2xPCI-E 375GB P4800X SSDs). Intel Optane SSDs are working as an expansion of 128 GB node's DRAM. IMDT is working as a driver, and transparently integrates the SSDs into the memory subsystem and makes it appear like DRAM to the OS and applications. On the Fig. 4 we presented the efficiency plot of Lagrangian step of the gas dynamic simulation, which is the most time-consuming step (> 90% compute time). This step describes the convective transport of the gas quantities with the scheme velocity for our problem. The number of FLOP/byte is not very high and the efficiency plot follows the trend we described above. We observe a slow decrease in efficiency down to ≈50% when the data does not fit into DRAM cache of IMDT. Otherwise, the efficiency is close to 100%. We achieved this results on an unoptimized code. One of the approaches for improving the performance of IMDT based systems is the more effective usage of Intel Optane's cache which is held in DRAM.

4   Numerical Simulation for Explosion Model

As a model problem of a supernova explosion, let us consider a hydrostatically equilibrium density profile:

$$\rho(r) = \begin{cases} 2r^3 - 3r^2 + 1, & r \leq 1, \\ 0, & r > 1. \end{cases} \tag{33}$$

and pressure profile:

$$p(r) = \begin{cases} -\pi r^8/3 + 44\pi r^7/35 - 6\pi r^6/5 - 4\pi r^5/5 \\ \qquad +8\pi r^4/5 - 2\pi r^2/3 + \pi/7, & r \leq 1, \\ 0, & r > 1. \end{cases} \tag{34}$$

In center of domain injected pressure is equal to $P = 500$. For the velocity the Gauss distribution was used. The results of the simulation are shown in the Fig. (5). The result of numerical experiments have a dense ring (like a Sedov test). A small velocity perturbation leads to unevenness of the ring and has a potential for rupture.

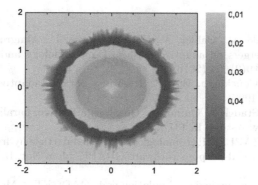

Fig. 5. The nondimensional density in time is equal to t = 0.01

5   Conclusion

In this paper, a novel computation technique for numerical simulations of astrophysical flow at the supercomputers was described. The co-design of parallel numerical algorithms for astrophysical simulations was described in detail. The hydrodynamical numerical model with stellar equations of state, numerical methods for solving the hyperbolic equations and a brief description of the parallel implementation of the code was described. This code is a classical memory bounded partially differential equations solver. We achieved more than 93% weak scalability for 1024 CPU cores. For detailed numerical simulation of our problem, we need to use a large amount of RAM (more than 1TB) on each node.

At this moment, it is hard to find supercomputing facilities with large RAM computational nodes. The evolution of Intel Optane SSDs as well as Intel Memory Drive Technology giving expectations that future systems will be equipped with DRAM and Intel Optane SSDs which are working as an extension of DRAM. First tests of Intel Optanes shows that unoptimized PDE solver working twice slower on SSDs than a DRAM, but it is possible to optimize memory operations for improving performance. One of the approaches for improving the performance of IMDT based systems is more effective usage of Intel Optane's cache which is held in DRAM. This optimization will be done in our future work.

This work is a part of the common joint "Hydrodynamical Numerical Modelling of Astrophysical Flow at the Peta- and Exascale", developed by our team at the Siberian Supercomputer Center ICMMG SB RAS in collaboration with the researchers from the Institute of Astronomy RAS, the Institute of Astronomy of Vienna University and the Southern Federal University, supported by RSC Group.

Acknowledgements. The research work was supported by the Grant of the Russian Science Foundation (project 18-11-00044).

References

1. Katz, M., Zingale, M., Calder, A., Douglas Swesty, F., Almgren, A., Zhang, W.: White dwarf mergers on adaptive meshes. I. methodology and code verification. Astrophys. J. **819**(2), 94 (2016)
2. Pearcea, F.R., Couchman, H.M.P.: Hydra: a parallel adaptive grid code. New Astron. **2**, 411 (1997)
3. Wadsley, J.W., Stadel, J., Quinn, T.: Gasoline: a flexible, parallel implementation of TreeSPH. New Astron. **9**, 137 (2004)
4. Matthias, S.: GRAPESPH: cosmological smoothed particle hydrodynamics simulations with the special-purpose hardware GRAPE. Mon. Not. R. Astron. Soc. **278**, 1005 (1996)
5. Springel, V.: The cosmological simulation code GADGET-2. Mon. Not. R. Astron. Soc. **364**, 1105 (2005)
6. Ziegler, U.: Self-gravitational adaptive mesh magnetohydrodynamics with the NIRVANA code. Astron. Astrophys. **435**, 385 (2005)
7. Mignone, A., Plewa, T., Bodo, G.: The piecewise parabolic method for multidimensional relativistic fluid dynamics. Astrophys. J. **160**, 199 (2005)
8. Hayes, J., Norman, M., Fiedler, R., et al.: Simulating radiating and magnetized flows in multiple dimensions with ZEUS-MP. Astrophys. J. Suppl. Ser. **165**, 188 (2006)
9. O'Shea, B., et al.: Adaptive mesh refinement - theory and applications. Lect. Not. Comput. Sci. Eng. **41**, 341 (2005)
10. Teyssier, R.: Cosmological hydrodynamics with adaptive mesh refinement-A new high resolution code called RAMSES. Astron. Astrophys. **385**, 337 (2002)
11. Kravtsov, A., Klypin, A., Hoffman, Y.: Constrained simulations of the real universe. II. Observational signatures of intergalactic gas in the local supercluster region. Astrophys. J. **571**, 563 (2002)

12. Stone, J., Gardiner, T., Teuben, P., Hawley, J., Simon, J.: Athena: a new code for astrophysical MHD. Astrophys. J. Suppl. Ser. **178**, 137 (2008)
13. Brandenburg, A., Dobler, W.: Hydromagnetic turbulence in computer simulations. Comput. Phys. Commun. **147**, 471 (2002)
14. Gonzalez, M., Audit, E., Huynh, P.: HERACLES: a three-dimensional radiation hydrodynamics code. Astron. Astrophys. **464**, 429 (2007)
15. Krumholz, M.R., Klein, R.I., McKee, C.F., Bolstad, J.: Equations and algorithms for mixed-frame flux-limited diffusion radiation hydrodynamics. Astrophys. J. **667**, 626 (2007)
16. Mignone, A., et al.: PLUTO: a numerical code for computational astrophysics. Astrophys. J. Suppl. Ser. **170**, 228 (2007)
17. Almgren, A., Beckner, V., Bell, J., et al.: CASTRO: a new compressible astrophysical solver. I. Hydrodynamics and self-gravity. Astrophys. J. **715**, 1221 (2010)
18. Schive, H., Tsai, Y., Chiueh, T.: GAMER: a GPU-accelerated adaptive-mesh-refinement code for astrophysics. Astrophys. J. **186**, 457 (2010)
19. Murphy, J., Burrows, A.: BETHE-Hydro: an arbitrary Lagrangian-Eulerian multidimensional hydrodynamics code for astrophysical simulations. Astrophys. J. Suppl. Ser. **179**, 209 (2008)
20. Springel, V.: E pur si muove: Galilean-invariant cosmological hydrodynamical simulations on a moving mesh. Mon. Not. R. Astron. Soc. **401**, 791 (2010)
21. Bruenn, S., Mezzacappa, A., Hix, W., et al.: 2D and 3D core-collapse supernovae simulation results obtained with the CHIMERA code. J. Phys. **180**, 012018 (2009)
22. Hopkins, P.: A new class of accurate, mesh-free hydrodynamic simulation methods. Mon. Not. R. Astron. Soc. **450**, 53 (2015)
23. Kulikov, I.: GPUPEGAS: A new GPU-accelerated hydrodynamic code for numerical simulations of interacting galaxies. Astrophys. J. Suppl. Ser. **214**(1), 12 (2014)
24. Kulikov, I.M., Chernykh, I.G., Snytnikov, A.V., Glinskiy, B.M., Tutukov, A.V.: AstroPhi: a code for complex simulation of dynamics of astrophysical objects using hybrid supercomputers. Comput. Phys. Commun. **186**, 71–80 (2015)
25. Tutukov, A., Lazareva, G., Kulikov, I.: Gas dynamics of a central collision of two galaxies: merger, disruption, passage, and the formation of a new galaxy. Astron. Rep. **55**(9), 770–783 (2011)
26. Vshivkov, V., Lazareva, G., Snytnikov, A., Kulikov, I.: Supercomputer simulation of an astrophysical object collapse by the fluids-in-cell method. Lect. Not. Comput. Sci. **5698**, 414–422 (2009)
27. Godunov, S., Kulikov, I.: Computation of discontinuous solutions of fluid dynamics equations with entropy nondecrease guarantee. Comput. Math. Math. Phys. **54**, 1012–1024 (2014)
28. Vshivkov, V., Lazareva, G., Snytnikov, A., Kulikov, I., Tutukov, A.: Hydrodynamical code for numerical simulation of the gas components of colliding galaxies. Astrophys. J. Suppl. Ser. **194**(47), 1–12 (2011)
29. Vshivkov, V., Lazareva, G., Snytnikov, A., Kulikov, I., Tutukov, A.: Computational methods for ill-posed problems of gravitational gasodynamics. J. Inverse Ill-posed Prob. **19**(1), 151–166 (2011)
30. Kulikov, I., Vorobyov, E.: Using the PPML approach for constructing a low-dissipation, operator-splitting scheme for numerical simulations of hydrodynamic flows. J. Comput. Phys. **317**, 318–346 (2016)
31. Roofline Performance Model. https://crd.lbl.gov/departments/computer-science/PAR/research/roofline/

32. Glinskiy, B., Kulikov, I., Chernykh, I.: Improving the performance of an AstroPhi code for massively parallel supercomputers using roofline analysis. Commun. Comput. Inf. Sci. **793**, 400–406 (2017)

33. Kulikov, I., Chernykh, I., Glinskiy, B., Protasov, V.: An efficient optimization of Hll method for the second generation of Intel Xeon Phi Processor. Lobachevskii J. Math. **39**(4), 543–550 (2018)

34. Markets analytics. https://www.trendforce.com/

35. RSC Tornado architecture. http://www.rscgroup.ru/en/our-solutions

36. Intel Memory Drive Technology. https://www.intel.ru/content/www/ru/ru/solid-state-drives/optane-ssd-dc-p4800x-mdt-brief.html

Three-Dimensional Simulation of Stokes Flow Around a Rigid Structure Using FMM/GPU Accelerated BEM

Olga A. Abramova[1(✉)], Yulia A. Pityuk[1], Nail A. Gumerov[1,2],
and Iskander Sh. Akhatov[3]

[1] Center for Micro and Nanoscale Dynamics of Dispersed Systems,
Bashkir State University, Ufa, Russia
abramovacmndds@gmail.com
[2] Institute for Advanced Computer Studies, University of Maryland,
College Park, USA
[3] Skolkovo Institute of Science and Engineering (Skoltech), Moscow, Russia

Abstract. Composite materials play an important role in aircraft, space and automotive industries, wind power industry. One of the most commonly used methods for the manufacture of composite materials is the impregnation of dry textiles by a viscous liquid binder. During the process, cavities (voids) of various sizes can be formed and then move in a liquid resin flows in the complex system of channels formed by textile fibers. The presence of such cavities results in a substantial deterioration of the mechanical properties of the composites. As a result, the development and effective implementation of the numerical methods and approaches for the effective 3D simulation of the viscous liquid flow around a rigid structure of different configuration. In the present study, the mathematical model and its effective numerical implementation for the study of hydrodynamic processes around fixed structure at low Reynolds numbers is considered. The developed approach is based on the boundary element method for 3D problems accelerated both via an advanced scalable algorithm (FMM), and via utilization of a heterogeneous computing architecture (multicore CPUs and graphics processors). This enables direct large scale simulations on a personal workstation, which is confirmed by test and demo computations. The simulation results and details of the method and accuracy/performance of the algorithm are discussed. The results of the research may be used for the solution of problems related to microfluidic device construction, theory of the composite materials production, and are of interest for computational hydrodynamics as a whole.

Keywords: Stokes flow · Boundary element method
Fast multipole method · High-performance computing · GPUs

1 Introduction

Composite materials are man-made combinations of two or more different materials that produce new materials having unique properties, such as improved

© Springer Nature Switzerland AG 2019
V. Voevodin and S. Sobolev (Eds.): RuSCDays 2018, CCIS 965, pp. 427–438, 2019.
https://doi.org/10.1007/978-3-030-05807-4_36

stiffness, or tailored thermal or electrical conductivity. A main driver for the development of composites has been the increasing need for materials that are at the same time stiff and light, to be used in aircraft, space industries. Other rapidly expanding fields of applications encompass energy generation (wind energy notably), infrastructure and architecture (bridges and buildings), transportation including automotive, marine and rail, as well as biomedical engineering. One of the most commonly used methods for the manufacture of composite materials is the impregnation of dry textiles by a viscous liquid binder. During the impregnation, a liquid resin flows in the complex system of channels formed by textile fibers, which can be considered as a porous medium. In the process of filling the reinforcing structure by the liquid, cavities (voids) of various sizes can be formed. The presence of voids significantly reduces the mechanical properties of composites and increasing design risk. Therefore, the reduction of void content is critical. A common point of view is that the formation of such voids is due to the heterogeneity of the porous structure, which for textiles has two characteristic scales. The microscale is associated with a diameter and packing of individual fibers in the filament, while the mesoscale is associated with a location of filaments in the structure of the reinforcing fabric and with a distance between the layers in multilayer structures. Currently, there are proven algorithms for solving problems of the composites impregnation at the macro level that allows us to describe the overall picture of filling of the preform by binder. Methods of simulation of the impregnation that are based on the solution of the Darcy's equation and given local permeability of the medium are well known and formalized in software packages, that are widely used in research organizations and industry for the design of the impregnation process. However, these methods do not allow to predict the voids volume fraction, their size and shape, since the voids formation is due to the physics of the processes on a smaller scale and cannot be described by Darcy's law. Wide range of work is dedicated to the experimental [5,10] and numerical [4,6,15,16,20] studying of the processes, which associated with the complicated flows that occur while liquid composite moulding (LCM). Since manufacturing of a composite part is in general performed by infiltration of the liquid matrix into the thin porous medium made from the fibre assembly, the direct tree-dimensional numerical simulation of viscous liquid flow around the complicated structure is of interest. Nowadays, modern computational methods and powerful computer resources enable fast large-scale microfluid dynamics simulations, which makes them a valuable research tool. The present work is devoted to the application of efficient computational techniques for the 3D modeling of viscous fluid flow in complex domains containing rigid structures, corresponding to filaments assembly, at low Reynolds numbers. The numerical approach is based on the boundary element method (BEM) accelerated both via advanced scalable algorithms, particularly, the fast multipole method (FMM), and via utilization of advanced hardware, particularly, graphics processors (GPUs) and multicore CPUs. This technique we developed and applied for research the 3D emulsion dynamics and viscous fluid flow in various domains [1–3] and bubble dynamics at low [13] and moderate [14] Reynolds numbers. In our previous works example computations were successfully conducted for 3D dynamics of systems of tens of thousands of deformable drops and systems of several droplets, with very high discretization of the interface in a shear flow [1].

The BEM has been successfully applied for simulation of Stokes flows ([23]; see general overview and details of the BEM in the monograph of Pozrikidis [17]), but its application to simulation of large non-periodic systems is very limited. Thus, the key here is the application of fast algorithms for BEM acceleration. The FMM was first introduced by Rokhlin and Greengard [7] as a method for fast summation of electrostatic or gravitational potentials in two and three dimensions. The first application of the FMM for the solution of Stokes equations in the case of spherical rigid particles was reported by Sangani and Mo [19]. In the work [22], the authors achieved substantial accelerations of droplet dynamics simulations via the use of multipole expansions and translation operators, which is very much in the spirit of the FMM, and can be considered as one- and two-level FMM. However, the $O(N)$ scalability of the FMM can be achieved only on hierarchical (multilevel) data structures, which were not implemented there. Note that the FMM can be efficiently parallelized. The first implementation of the FMM on graphics processors was reported by Gumerov and Duraiswami [8], who showed that the use of a GPU for the FMM for the Laplace's equation in 3D can produce 30 to 60-fold accelerations, and achieved a time of 1 s for the MVP in a case of size 1 million on a single GPU. This approach was developed further and papers by Hu *et al.* [11,12] present scalable FMM algorithms and implementations of heterogeneous computing architectures that may contain many distributed nodes, each of them consisting of one multicore CPU and several GPUs. They performed the MVP for a system of size 1 billion achieving a time of about 10 s on a 32-node cluster, and also for systems of size 1 million in a time of the order of several tenths of a second on a single heterogeneous CPU/GPU node. In the present study, we use a single heterogeneous node with one GPU and apply CPU/GPU parallelism. This realization enables computations of 3D fluid flow at low Reynolds numbers around the structures complicated configurations with tens of thousands triangular elements discretizing the boundary.

2   Problem Statement

In this paper, we consider the viscous fluid flow around a non-deformable fixed object in Cartesian coordinates. It is assumed that the processes are isothermal and slowly enough that the viscosity forces are much more significant than the inertia forces. In this case, the steady flow of incompressible liquids is described by the Stokes equations [9] (1) with the corresponding boundary conditions (2).

$$\nabla \cdot \boldsymbol{\sigma} = -\nabla p + \mu \nabla^2 \mathbf{u} = \mathbf{0}, \quad \nabla \cdot \mathbf{u} = 0, \tag{1}$$

where $\mathbf{u}$ and $\boldsymbol{\sigma}$ are the velocity and the stress tensors, μ is the dynamic viscosity, and p is the pressure, which includes the hydrostatic component.

At the surface of non-deformable fixed structures S

$$\mathbf{u}(\mathbf{x}) = 0, \quad \mathbf{x} \in S, \tag{2}$$

where $\mathbf{x}$ is the radius vector of the current point.

For the carrier fluid the condition $\mathbf{u}(\mathbf{x}) \to \mathbf{u}_\infty(\mathbf{x})$ is imposed, where $\mathbf{u}_\infty(\mathbf{x})$ is a solution of the Stokes equations.

3   Numerical Implementation

The problem is solved using the boundary element method, which is based on the integral equations for the determination of the velocity distribution over the boundary. There is no need to cover all computational domain by mesh as in the case of finite-difference and finite-element methods, but only the boundary of the considered objects. As a result, the BEM is very efficient for solving 3D problems with complicated geometries or in infinite domains.

The boundary integral equations for the volume of fluid occupying domain V bounded by surface S can be written in the form [17]

$$
\begin{aligned}
\mathbf{u}(\mathbf{y}) - \int_S \mathbf{K}(\mathbf{y},\mathbf{x}) \cdot \mathbf{u}(\mathbf{x})\,dS(\mathbf{x}) &= \frac{1}{\mu} \int_S \mathbf{G}(\mathbf{y},\mathbf{x}) \cdot \mathbf{f}(\mathbf{x})\,dS(\mathbf{x}), \quad \mathbf{y} \in V, \\
\frac{1}{2}\mathbf{u}(\mathbf{y}) - \int_S \mathbf{K}(\mathbf{y},\mathbf{x}) \cdot \mathbf{u}(\mathbf{x})\,dS(\mathbf{x}) &= \frac{1}{\mu} \int_S \mathbf{G}(\mathbf{y},\mathbf{x}) \cdot \mathbf{f}(\mathbf{x})\,dS(\mathbf{x}), \quad \mathbf{y} \in S, \quad (3) \\
- \int_S \mathbf{K}(\mathbf{y},\mathbf{x}) \cdot \mathbf{u}(\mathbf{x})\,dS(\mathbf{x}) &= \frac{1}{\mu} \int_S \mathbf{G}(\mathbf{y},\mathbf{x}) \cdot \mathbf{f}(\mathbf{x})\,dS(\mathbf{x}), \quad \mathbf{y} \notin S, V,
\end{aligned}
$$

where the fundamental solutions (Stokeslet and stresslet) are given by the second and third rank tensors

$$
\begin{aligned}
\mathbf{G}(\mathbf{y},\mathbf{x}) &= \frac{1}{8\pi}\left(\frac{\mathbf{I}}{r} + \frac{\mathbf{r}\mathbf{r}}{r^3}\right), \quad \mathbf{T}(\mathbf{y},\mathbf{x}) = -\frac{3}{4\pi}\frac{\mathbf{r}\mathbf{r}\mathbf{r}}{r^5}, \\
\mathbf{K}(\mathbf{y},\mathbf{x}) &= \mathbf{T}(\mathbf{y},\mathbf{x}) \cdot \mathbf{n}(\mathbf{x}), \quad \mathbf{r} = \mathbf{y} - \mathbf{x}, \quad r = |\mathbf{r}|,
\end{aligned} \quad (4)
$$

and $\mathbf{I}$ is the identity tensor, $\mathbf{n}$ is the normal to S. Thus, if $\mathbf{u}$ and $\mathbf{f}$ are known on the boundaries, the velocity field $\mathbf{u}(\mathbf{y})$ can be determined at the any spatial point.

The problem describes a flow around the structures in an unbounded domain. In this case, $\mathbf{u}_\infty$ is prescribed (e.g., a uniform constant flow $\mathbf{u}_\infty = (U,0,0)$), and the boundary integral equation (3) with taking into account the boundary conditions (2) and that the normal pointed into fluid, can be written in the form

$$
\left.\begin{aligned}
\mathbf{y} \in V, \quad & \mathbf{u}(\mathbf{y}) - 2 \cdot \mathbf{u}_\infty(\mathbf{y}) \\
\mathbf{y} \in S, \quad & \frac{1}{2}\mathbf{u}(\mathbf{y}) - \mathbf{u}_\infty(\mathbf{y})
\end{aligned}\right\} = \int_S \left\{-\frac{1}{\mu}\mathbf{G}(\mathbf{y},\mathbf{x}) \cdot \mathbf{f}(\mathbf{x})\right\} dS(\mathbf{x}). \quad (5)
$$

The numerical method is based on the discretization of object surfaces by triangular meshes. The regular integrals over the patches were computed using second order accuracy formulae (trapezoidal rules). The collocation points for a rigid structure surface were located at the center of triangular elements. The computation of singular integrals was performed based on the integral identities for the Stokeslet and stresslet integrals (see [17,23]), which allow to express these integrals via sums of regular integrals over the rest of the surface. More details of the present implementation can be found in [1,2].

The boundary integral equations combined with the boundary conditions in discrete form result in a system of linear algebraic equations (SLAE)

$$\mathbf{AX} = \mathbf{b}, \tag{6}$$

where $\mathbf{A}$ is the system matrix, $\mathbf{X}$ is the solution vector, and $\mathbf{b}$ is the right-hand-side vector.

4 Acceleration of Calculations

Numerical modeling of 3D Stokes flows in regions of non-trivial geometry is a computationally complex and resource-intensive process. Note that calculations using a standard BEM take significant time even for a relatively small mesh resolution of the object surfaces. This is due to the fact that BEM reduces to solving a dense SLAE concerning $3N$ of unknowns (6). Furthermore, the memory required for computations increases proportionally to the square of the number of mesh points. Solving such a large problem requires development and application of efficient numerical methods and techniques.

The direct methods of solution of algebraic systems, having computational complexity $O\left(N^3\right)$, become impractical when N reaches a value of the order of millions. The use of efficient iterative methods reduces this complexity to $O\left(N_{\text{iter}}N^2\right)$, where $N_{\text{iter}} \ll N$ is the number of iterations, and $O\left(N^2\right)$ is the cost of a single matrix-vector product (MVP). In this paper, we used the unpreconditioned general minimal residual method (GMRES) to solve the system [18]. The main computational complexity of the problem is in the calculation of MVP at each iteration. That is why in this study the program module of MVP is accelerated in order to implement the iterative method effectively. Acceleration of MVP is possible due to: (1) application of modern highly efficient methods and algorithms to reduce computational complexity; (2) using of high-performance hardware.

First, in order to speedup the calculations and increasing the problem size we developed program code for MVP without matrix storage in the memory of computational system ("MV Product on the fly"). Each element of matrices $\mathbf{G}$ is calculated using the formulas (4). This program was implemented on GPUs using CUDA technology. Performing calculations on the graphics processors shows excellent results in algorithms that involve parallel processing of data (applying the same sequence of mathematical operations to a set of data), that is why we use GPUs in our study. Herewith the best results are achieved if the ratio of the number of arithmetic instructions to the number of accesses to memory is large enough. The parallelization of the module "MV Product on the fly" is based on the partitioning of the matrices by horizontal bands into m parts such that $M = m \times L$, where M is the dimension of the matrix, L is the number of rows of the matrix in the block, m is the number of threads on the GPUs. At each iteration, each of the m threads computes its part of the solution vector. Each thread stores a part of the resulting vector in the local memory, which is copied to the global memory at the end of the calculation to obtain a complete solution vector.

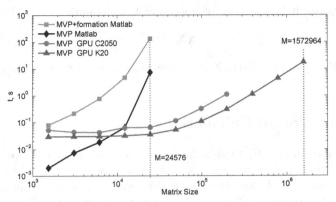

Fig. 1. Runtime for MVP module

The tests were performed on workstations of two configurations: (1) CPU Intel Xeon 5660 and GPUs NVIDIA Tesla K20, 5 GB of global memory (Kepler architecture); (2) CPU Intel Xeon 5660 and GPUs NVIDIA Tesla C2050, 3 GB of global memory (Fermi architecture).

Figure 1 shows a comparison of the MVP execution time on the GPUs and the built-in Matlab function including (MVP + formation) and without (Matlab MVP) matrix formation depending on matrix size ($M = 3 \cdot N$). It can be seen from the curvs that starting with a relatively small $M \approx 2.5 \cdot 10^4$ there is a shortage of the memory of the computer system and further calculations becomes impossible. However the module implemented on the GPUs allows one to solve problems of larger computational complexity. Numerical experiments on the NVIDIA Tesla K20 show the possibility of solving boundary problems for Stokes equations up to $3 \cdot 10^5$ calculation points on one personal workstation in a reasonable time.

But for larger N, this is not fast enough even when using high-performance computing, since the computational cost increases proportionally to the square of N. Thus, the key here is than the MVP specific for the present problem can be computed using the FMM [7]. The main advantage of this method is that the complexity of the MVP involving a certain type of dense matrices can be reduced from $O(N^2)$ to $O(N \log(N))$, or even $O(N)$. In this paper, we used the implementation technique of the FMM for the summation of the Stokeslets and stresslets (Eq. (4)) proposed in [21]. This approach is based on the summation of the fundamental solutions of the three-dimensional Laplace's equation. In the present approach, GPUs acceleration is used in the so-called heterogeneous FMM algorithm, where the system matrix can be decomposed as sum of the two parts: dense and sparse. Subroutines for dense part produce approximate dense MVP on the CPU using OpenMP, and subroutines for sparse part produce direct sparse multiplication on the GPUs using CUDA. A careful tuning of the algorithm and

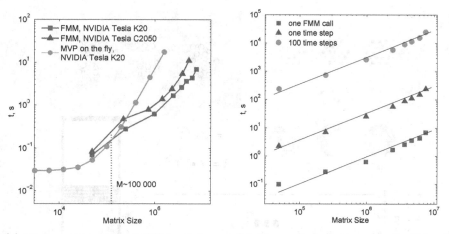

(a) The runtime of the MVP module and (b) The wall-clock time on the PC with
FMM routine   NVIDIA Tesla K20 [1]

Fig. 2. Runtime for MVP of different implementations

the data structure octree depth is based on the work load balance between the
CPU and GPUs. Using such an heterogeneous algorithm on a system with 8 to
12 CPU cores and one GPUs, the overall acceleration can rise by approximately
100 times compared to its value for a single-core CPU implementation. The
features of the algorithm and implementations are presented in [2,8].

It is seen from the Fig. 2(left) that starting with a certain size of the matrix
$M \approx 1 \cdot 10^5$ the using of heterogeneous FMM is preferable for numerical experiments. Numerical experiments show (Figs. 1, 2) that the calculations on the
NVIDIA Tesla K20 graphics card with the Kepler architecture take less time
than ones on the NVIDIA Tesla C2050 GPUs with the Fermi architecture, for
any matrix size. In Fig. 2(right), the computational wall-clock times for one FMM
call, for one time step and for one hundred time steps are shown as functions
of the matrix size. In our work [1] it was reported that the run time for one
FMM call for the system with $M \approx 7.5 \cdot 10^6$ was about 7 s; for one time step, it
was 4 min, and for 100 time steps, about 7 h on one personal workstation. It can
be seen that these functions are close to linear. This allows one to estimate the
computational times for larger scale problems. The use of the FMM for MVP
reduces the computational complexity of the overall problem to $O\left(N_{\text{iter}}N\right)$ per
time step, and potentially can handle direct large scale simulations with millions
of boundary elements.

5   Numerical Results and Discussion

Numerical tests are performed on a workstation equipped with two Intel Xeon
5660, and one NVIDIA Tesla K20. Several algorithm implementations were done,
including CPU and CPU/GPUs versions of the iterative algorithm with the

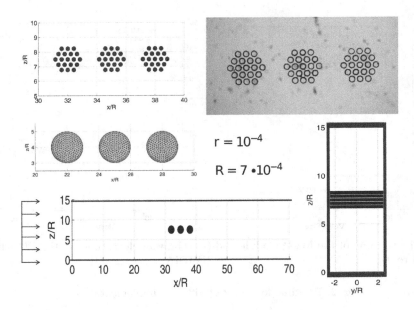

Fig. 3. Geometry of the considered domain

FMM accelerated MVP and a conventional BEM in which the BEM matrices were computed and stored. The latter implementation was developed for verification and validation purposes to ensure that the algorithms produce consistent results.

The implemented methods were tested for the case of one fixed rigid sphere in an unbounded flow. The obtained results were compared against the analytical solution [9]. The relative error in the L_∞-norm was 0.08–0.1% for the velocity component around the sphere, and for traction on the sphere surface the relative error was about 1.8%, for $N_\Delta = 1380$ triangular elements on the surface. The error decreases as the number of mesh points increases.

We conducted simulations of the viscous fluid flow around the fixed rigid structures formed by cylindrical elements with a radius R by an infinite flow with the constant velocity $\mathbf{u}_\infty(\mathbf{x}) = (U, 0, 0)$ at infinity. Represented numerical research are motivated by the study of processes occurring while the viscous binding fluid flow around reinforcement fibers in the manufacture of composite materials. Further comparison of the simulation results of the flow pattern will be carried out with experimental data that will be obtained in the experimental laboratory of our Center. A series of experiments is planned to study the flow pattern around individual filaments and assembly of filaments. Several microdevices are already made for the study of flows in samples with double porosity (Fig. 3, right top corner).

The geometry used in the computational examples for simulation of the single and double porosity is shown in Fig. 3. Structure of the filaments is bounded by the flat planes, which geometry characteristics correspond to the experimental

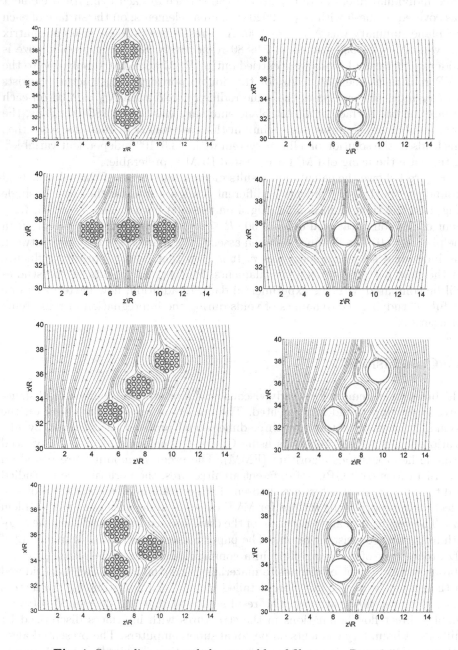

Fig. 4. Streamlines around the assembly of filaments, $Re = 0.5$.

data: $L_x \gg R$, $L_y = 5 \cdot R$, $L_z = 5 \cdot R$. In the case of single porosity there was three individual fibers of radius R and the surface of each cylindrical element was covered by mesh with $N_\Delta = 2920$ triangular elements, on the surface of each flat plane summary was $N_\Delta = 234200$. As a result the problem size and matrix size were $N = 242960$ and $M = 728880$ respectively. Since the problem size is moderate, the calculations were carried out by BEM accelerated only due to the MVP routine at the GPUs. As for the double porosity, the structure consists of 57 cylindrical elements of the same radius $r = \frac{1}{7}R$ with $N_\Delta = 2504$ on each surface, on the surface of all flat plane summary was $N_\Delta = 886400$. Thus, the total number of computational points in this case was $N = 1029128$. Note that this leads to the solution of a linear system with $\sim 3.5 \cdot 10^6$ independent variables. In this case the using of FMM accelerated BEM is preferable.

Figure 4 shows the calculation results of the components of the velocity field around the fixed rigid filaments in different configurations for the cases of single (Fig. 4 on the right) and double (Fig. 4 on the left) porosity, at $Re = 0.5$, fragment of the computational domain $30 \cdot R \le x \le 40 \cdot R, 0.5 \cdot R \le z \le 14.5 \cdot R$ in the plane $y = 0$. The distance between assembly of filaments was R and between the individual small filaments was r. It is seen that double porosity influence on the flow pattern between the filaments. Further, the results of the studies will be compared with the experimental data. In addition, flow patterns can be useful in studying the dynamics of voids during the impregnation of reinforcing structures.

6 Conclusions

The problem statement for steady viscous fluid flow around the fixed rigid structures is developed and implemented. The numerical approach is based on the boundary element method for three-dimensional problems accelerated by utilization of a graphics processors using CUDA technology and via an advanced scalable heterogeneous algorithm (FMM). The program modules are tested on a workstations with GPUs of different architectures, the performance is studied and the optimal parameters are choosen. The software implementation allows to select an appropriate algorithms for MVP calculation depending on the problem size. The verification and validation of the developed codes showed that the algorithms produce consistent results. The paper presents the results of modeling of 3D Stokes flow of a viscous fluid in a complex geometry, corresponding to the fibrous reinforcement of composite materials. Nevertheless the cases considered in the present work require more detailed studies for understanding the factors that affect the flow patterns. The results showed that the developed software enables direct flow simulations in the structures with interfaces discretized by millions of boundary elements on personal supercomputers. The presented algorithms can be mapped onto heterogeneous computing clusters [12], which should both accelerate computations and allow for the treatment of larger systems. The developed software can become a valuable research tool for investigation the influence of geometry features of the reinforcing structure on the flow of viscous liquid and liquid with bubbles to study the formation and dynamics of voids during the manufacturing of composite materials.

Acknowledgements. This study is supported in part by the Skoltech Partnership Program (developing and testing of the accelerated BEM version, conducting of the calculations), part of the study devoted to the problem formulation of the flow near rigid structures is carried out with the support of the grant RFBR 18-31-00074, and FMM routine is provided by Fantalgo, LLC (Maryland, USA).

References

1. Abramova, O.A., Pityuk, Y.A., Gumerov, N.A., Akhatov, I.S.: High-performance BEM simulation of 3D emulsion flow. In: Sokolinsky, L., Zymbler, M. (eds.) PCT 2017. CCIS, vol. 753, pp. 317–330. Springer, Cham (2017). https://doi.org/10.1007/978-3-319-67035-5_23
2. Abramova, O.A., Itkulova, Y.A., Gumerov, N.A., Akhatov, I.S.: Three-dimensional simulation of the dynamics of deformable droplets of an emulsion using the boundary element method and the fast multipole method on heterogeneous computing systems. Vychislitelínye Metody i Programmirovanie **14**, 438–450 (2013)
3. Abramova O.A., Itkulova Y.A., Gumerov N.A.: FMM/GPU accelerated BEM simulations of emulsion flows in microchannels. In: ASME 2013 IMECE (2013). https://doi.org/10.1115/imece2013-63193
4. Arcila, I.P., Power, H., Londono, C.N., Escobar, W.F.: Boundary element method for the dynamic evolution of intra-tow voids in dual-scale fibrous reinforcements using a stokes-darcy formulation. Eng. Anal. Bound. Elem. **87**, 133–152 (2018). https://doi.org/10.1017/s0022112069000759
5. Gangloff Jr., J.J., Daniel, C., Advani, S.G.: A model of a two-phase resin and void flow during composites processing. Int. J. Multiph. Flow. **65**, 51–60 (2014). https://doi.org/10.1016/j.ijmultiphaseflow.2014.05.015
6. Gascon, L., Garcia, J.A., Lebel, F., Ruiz, E., Trochu, F.: Numerical prediction of saturation in dual scale fibrous reinforcements during liquid composite molding. Compos. Part A- Appl. Sci. Manuf. **77**, 275–284 (2015). https://doi.org/10.1016/j.compositesa.2015.05.019
7. Greengard, L., Rokhlin, V.: A fast algorithm for particle simulations. J. Comp. Phys. **73**, 325–348 (1987). https://doi.org/10.1006/jcph.1997.5706
8. Gumerov, N.A., Duraiswami, R.: Fast multipole methods on graphics processors. J. Comput. Phys. **227**(18), 8290–8313 (2008). https://doi.org/10.1016/j.jcp.2008.05.023
9. Happel, J., Brenner, H.: Low Reynolds Number Hydrodynamics. Springer, Netherlands (1981). https://doi.org/10.1007/978-94-009-8352-6
10. Hamidi, Y.K., Aktas, L., Altan, M.C.: Formation of microscopic voids in resin transfer molded composites. J. Eng. Mater. Technol. **126**(4), 420–426 (2004). https://doi.org/10.1115/1.1789958
11. Hu, Q., Gumerov, N.A., Duraiswami, R.: Scalable fast multipole methods on distributed heterogeneous architectures. In: Proceedings of Supercomputing 2011, Seattle, Washington (2011). https://doi.org/10.1145/2063384.2063432
12. Hu, Q., Gumerov, N.A., Duraiswami, R.: Scalable distributed fast multipole methods. In: Proceedings of 2012 IEEE 14th International Conference on High Performance Computing and Communications, UK, Liverpool, pp. 270–279 (2012). https://doi.org/10.1109/ipdpsw.2014.110

13. Itkulova (Pityuk) Y.A., Abramova O.A., Gumerov N.A.: Boundary element simulations of compressible bubble dynamics in stokes flow. In: ASME 2013. IMECE2013-63200, P. V07BT08A010 (2013). https://doi.org/10.1115/imece2013-63200

14. Itkulova (Pityuk) Y.A., Abramova O.A., Gumerov N.A., Akhatov I.S.: Boundary element simulations of free and forced bubble oscillations in potential flow. In: Proceedings of the ASME2014 International Mechanical Engineering Congress and Exposition, vol. 7. IMECE2014-36972 (2014). https://doi.org/10.1115/IMECE2014-36972

15. LeBel, F., Fanaei, A.E., Ruiz, E., Trochu, F.: Prediction of optimal flow front velocity to minimize void formation in dual scale fibrous reinforcements. Int. J. Mater. Form. **7**(1), 93–116 (2014). https://doi.org/10.1007/s12289-012-1111-x

16. Michaud, V.: A review of non-saturated resin flow in liquid composite moulding processes. Transp. Porous Med. **115**, 581–601 (2016). https://doi.org/10.1007/s11242-016-0629-7

17. Pozrikidis, C.: Boundary Integral and Singularity Methods for Linearized Viscous Flow, p. 259. Cambridge University Press, Cambridge (1992). https://doi.org/10.1017/cbo9780511624124

18. Saad, Y.: Iterative Methods for Sparse Linear System, 2nd edn. SIAM, Philadelphia (2000). https://doi.org/10.1137/1.9780898718003

19. Sangani, A.S., Mo, G.: An O(N) algorithm for Stokes and Laplace interactions of particles. Phys. Fluids **8**(8), 1990–2010 (1996). https://doi.org/10.1063/1.869003

20. Tan, H., Pillai, K.M.: Multiscale modeling of unsaturated flow in dual-scale fiber performs of liquid composite molding I: isothermal flows. Compos. Part A-Appl. Sci. Manuf. **43**(1), 1–13 (2012). https://doi.org/10.1016/j.compositesa.2010.12.013

21. Tornberg, A.K., Greengard, L.: A fast multipole method for the three-dimensional Stokes equations. J. Comput. Phys. **227**(3), 1613–1619 (2008). https://doi.org/10.1016/j.jcp.2007.06.029

22. Zinchenko, A.Z., Davis, R.H.: A multipole-accelerated algorithm for close interaction of slightly deformable drops. J. Comp. Phys. **207**, 695–735 (2005). https://doi.org/10.1016/j.jcp.2005.01.026

23. Zinchenko, A.Z., Rother, M.A., Davis, R.H.: A novel boundary-integral algorithm for viscous interaction of deformable drops. Phys. Fluids **9**(6), 1493–1511 (1997). https://doi.org/10.1063/1.869275

Using of Hybrid Cluster Systems for Modeling of a Satellite and Plasma Interaction by the Molecular Dynamics Method

Leonid Zinin[(✉)] [ID], Alexander Sharamet[(✉)] [ID], and Sergey Ishanov [ID]

Immanuel Kant Baltic Federal University, Kaliningrad, Russia
leonid.zinin@gmail.com, alexsharamet@gmail.com,
sishanov@kantiana.ru

Abstract. This article deals with a model of interaction between a positively charged microsatellite and thermal space plasma. The model is based on the method of molecular dynamics (MMD). The minimum possible number of particles necessary for modeling in the simplest geometric problem formulation for a microsatellite in the form of a sphere 10 cm in diameter is ten million. This value is determined by the plasma parameters, including the value of the Debye radius, which is the basis for estimating the spatial dimensions of the modeling domain.

For the solution, MPI and CUDA technologies are used in the version of one MPI process per node. An intermediate layer in the form of multithreading, implemented on the basis of the C++ library of threads, is also used, this provides more flexible control over the management of all kinds of node memory (video memory, RAM), which provides a performance boost. The GPU optimizes the use of shared memory, records the allocation of registers between threads and the features associated with calculating trigonometric functions.

The results of numerical simulation for a single-ion thermal plasma showed significant changes in the spatial distribution of the concentration around the satellite, which depends on three main parameters - the temperature of the plasma components, the velocity of the satellite relative to the plasma and the potential of the spacecraft. The presence of a region of reduced ion concentration behind the satellite and the region of condensation in front of it is shown.

Keywords: Parallel computing · Thermal space plasma · Charged satellite Molecular dynamics method

1 Introduction

The presence of an electric charge of the satellite is an important factor affecting the experimental changes in thermal near-Earth space plasma. The influence of this interaction on the results and processing of measurements has been widely discussed and continues to be discussed in literature (see, for example, the analysis of the measurement results from the Auroral probe [1, 2]). Modern measurement models take into account the effect of temperature anisotropy [3–5]. The analysis of experimental measurements of thermal plasma and their simulation unambiguously indicate that the positive potential of the satellite significantly affects the spatial distribution of the ion flux values. In addition, the temperature anisotropy affects the angles of ion arrival [5].

© Springer Nature Switzerland AG 2019
V. Voevodin and S. Sobolev (Eds.): RuSCDays 2018, CCIS 965, pp. 439–449, 2019.
https://doi.org/10.1007/978-3-030-05807-4_37

These features of measurements can significantly complicate the processing of experimental data in determining the direction of the ion flux, and consequently, the magnitude of the ion velocity components.

At present, there are a number of successful interaction models of a charged satellite and plasma. Separately note the NASA model (NASCAP-2 K [6]) and ESA (SPIS [7, 8]), see also the article [9], where the results of calculations and model codes are compared.

There are enough examples of successful simulation of the electric field distribution around the satellites real form. Thus, for example, [10] present calculations based on the variants of NASCAP and SPIS models have already mentioned for the SCATHA satellite and the modern telecommunication satellite in the geostationary orbit.

Most modern models are based on the Particle-in-Cell method, which has proven its effectiveness. At the same time, this method does not allow us to obtain the trajectories of individual plasma particles, which it is necessary in some cases. The method of molecular dynamics (MMD) can be applied for any parameters of thermal plasma, regardless of the magnitude of the Debye radius, in spite of its computational complexity.

In this article, we consider an interaction model of thermal plasma with a charged microsatellite, which has the simplest spherical shape. The model is based on the molecular dynamics method.

2   Description of the Model

Now the method of classical molecular dynamics is being widely used for modeling physical, chemical and biological systems.

The research of Alder and Wainwright [11, 12] can be considered as the first attempt to apply MMD. (A review of the application of MMD in physics and chemistry is given in the article [13]). In classical MMD, quantum and relativistic effects are neglected, and the particles move according to the laws of classical mechanics.

Let us consider the 3D simulation of a charged microsatellite and thermal Maxwellian plasma in detail [14–16]. Let the plasma consist of protons and electrons. The force of Lorentz acts on every particle in the process of its motion:

$$m\mathbf{a} = q\mathbf{E} + q[\mathbf{v} \times \mathbf{B}], \tag{1}$$

here m is the mass of the ion or electron, q, $\mathbf{a}$, $\mathbf{v}$ is its charge, acceleration and velocity, respectively, $\mathbf{E}$ is the electric field strength, $\mathbf{B}$ is the magnetic field.

The electric field strength at any point in the simulation area is calculated according to the Coulomb law. For its calculation, all the electrons and ions which are in the computational domain are used. The electric field from the satellite is also taken into account.

$$\mathbf{E} = \sum_{i=1}^{n} \frac{q_i}{4\pi\varepsilon_0 r_i^2} \frac{\mathbf{r}_i}{r_i} + \mathbf{E}_{sat} \tag{2}$$

Thus, for each particle one can find the value and direction of force in the right side of the Eq. (1) to obtain the positions and velocities of the particles in the new time level.

The particles at the initial time of simulation are evenly distributed in space and have a Maxwell velocity distribution.

$$F(v) = \left(\frac{m}{2\pi kT}\right)^{3/2} exp\left(-\frac{mv^2}{2kT}\right) \tag{3}$$

where v - velocity of the particle, m - its mass, k - Boltzmann constant, T - temperature.

The number of particles N in the speed range [v, v + dv] for the density n is calculated as

$$N = \int_v^{v+dv} 4\pi n v^2 F(v) dv \tag{4}$$

The model domain is a cube with edge of 1 m. In its center there is a microsatellite which is a ball of 5 cm radius.

The relative velocity of the plasma and the satellite is directed along the OX axis. The magnetic field in these calculations was not taken into account. The temperature of ions and electrons is the same and it is equal to 5000 K, which corresponds to the conditions in the Earth's magnetosphere. The total number of particles in the modeling area was 2×10^7. Thus, the unperturbed concentration of ions and electrons was n = 10 cm^{-3}. The plasma velocities relative to the satellite were assumed to be equal to 10 km/s and 20 km/s. The potential of the satellite U_{sat} was positive and its values were taken equal to +5 V and +10 V. The time step was 10^{-8} s. Such a small amount is due to the need to accurately calculate the trajectories of electrons which thermal velocities are about 500 km/sec. The spatial step for the selected time step value does not exceed 5 mm. At each time step, some of the particles leave the simulation area, while others are injected into this region in accordance with the specified unperturbed Maxwell distribution function. The charge of electrons that hit the satellite surface is compensated by photoelectron emission.

3 Method of Calculations

Algorithm Analysis and Parallel Algorithm Scheme
The model has a good margin of scaling on large computing systems, but, nevertheless, it requires huge computing resources. To calculate the particle's coordinates at the next time step, it is necessary to calculate the electric field strength for this particle, which includes the sum of the stress vectors from all other plasma particles and the electric field strength from the satellite.

The convergence of this method requires from 3000 to 7000 time iterations. With a large number of equations, influenced by the size of the region and the concentration of particles in cm^{-3}, amount of calculations is significantly increased. To solve this problem a huge computing power, concentrated on each of the nodes, is required. So, for example, the computational complexity of the algorithm for the calculated region of 1 m^3 and the concentration of 10 cm^{-3} is 10^{17}.

To solve the problem we use cluster, with MPI technology to communicate between nodes. On each node we have two accelerators and use CUDA to calculate on them. To optimize GPU communication inside node and reduce communication between nodes we don't use a classical scheme "MPI process per GPU". Our scheme assumes that one MPI thread contains three CPU threads (one for communication, other to GPU). To create these threads we use "C++11 thread" library.

For MPI communication we use a classical master-slave scheme. Master reads information about the satellite, a field size, a magnetic field and other. On the second step it generates particles or reads their state from the file. It is very important to have an option to load data from some iteration, because the calculation may take a long time. Later master smashes data into pieces and sends them to slaves. To send data we don't use MPI_Bcast/Mpi_Ibcast, instead we use a cycle with MPI_ISend and synchronization after. This is faster on Intel mpi compilers, and we can control the size of any portion of data. The same way we collect data after iteration. MPI_ISend has another advantage, it doesn't use buffer. Master process gathers data with MPI_Recv and we add the correct address for any pies, so data isn't copied at all in an MPI subsystem.

MPI process consists of two or more threads. These threads have master-slave architecture too. At the beginning the master thread asks CUDA driver and creates as many threads as devices in node. Another thread function is the synchronization between controlling GPU threads and MPI communications. For synchronization we use barriers. The barrier is created on the basis of the "Singleton" pattern and uses mutable exclusion with conditional variables.

So the master thread receives (mpi-slave) or sends (mpi-master) tasks. This task is divided into portions and distributed to slave threads, followed by waiting for the result with barrier synchronization. After calculation some particles leave field or pierce satellite, and disappear. On their places the master thread generates new particles. We don't do it in parallel because rand depends on started seed, and if we try to use time as this seed we should de-synchronize time of starting command. Of course it can be found another way to solve this problem, but the task is small now and the current method is working well. The last master step is saving results to the file (only in some iteration) and sending input data to MPI-slaves.

CUDA Calculation Model and Optimization

The calculation is based on starting some consistent kernels. Kernel is started in CPU function, but calculated at GPU. The main kernel is an electric tension vector calculating for the electric field, because this part of the program contains the main complexity of the algorithm. After it, the new positions of the particles are calculated, later consistent on the CPU calculates an intersection with the satellite and leaving of the space. The only feature of the recalculation particle position is to save results to random access memory (non GPU) with the zero-copy technology, because we need further global exchange and current step data is done once.

Let's review the electric tension vector calculation. The first idea is to give one particle and its electric tension vector one GPU thread, taking into account that we have all particles, but recalculate only some parts of them which are defined at planning iteration in an MPI layer and later in cpu-thread layer.

Besides this condition, amount of threads can be large and greatly exceed the limit of GPU thread, which can be calculated on GPU at one time, so scheduler should be started and exchange calculation blocks. To avoid scheduler loading we give each thread some of the particles to calculate them. Obviously, all discussed above will be effective only with a multiple excess of the number of particles over the number of started threads.

Common model of calculation for one thread is look like:

(1) Load data from DRAM (particles data for which the vector is calculated).
(2) Cooperative load by block threads portion of particles in shared memory, synchronization
(3) Cycle of calculation of the partial vector of electric tension, synchronization
(4) Return to step 2, while all data will not be read.
(5) Choose another particle and Return to step 1

The first level of optimization is coalesced requests to DRAM, and shared memory hasn't bank conflicts. Further, three parameters must be considered to achieve 100% theoretical and practical performance.

(1) Loading of streaming multiprocessors (SM)
(2) Size of shared memory
(3) Count of registers per SM

Due to the fact that in common situations we have the number of particles significantly exceeding the capabilities of computing devices, it will be better to start from the original data about optimal parameters from the driver and execute an appropriate number of blocks and threads in block.

On the current device per SM 1536 and a max block size is 1024, so for full loading we use size of block 512 or less.

Amount of shared memory per SM consists of 48 Kb, and if we have a block size equal to 512 it will split into three blocks. This size is enough to load all necessary data.

The last optimization parameter is registers. We have no direct influence on it, but we can get it count in use by different debugger tools, and take into account that it is affected by the size and the type of variables in kernel and compiler optimization.

For this parameter optimization we move some parameters from an electric tension formula, to multiply them later, thereby reducing the amount of used registers and allowing more threads to be started.

Also one more GPU optimization is the optimization of the algorithm with hard to device trigonometric functions (sin, cos, tg,...), we change these parts of algorithm to their math analogs with the similar coefficient, which reduce the number of operations to the whole calculation.

In order to evaluate the scaling of the algorithm, we estimated the speed of the algorithm for one and two GPUs as shown in Table 1.

Table 1. Algorithm scalability

Numbers of particles	Time in one GPU (sec)	Time in two GPU (sec) (in one node)	Performance increase
10^4	0.621	0.567	9%
10^5	4.549	2.923	55%
10^6	44.40	28.69	54%
10^7	398.69	272.00	46%

Based on the sibling table, it is obvious that the task scales well. With a particle count of 10^4, the performance gain is not large, because the amount of data is not sufficient to load two GPU at once. As the number of particles increases, the performance begins to decrease, because the effects associated with data transfer start to play the more important role.

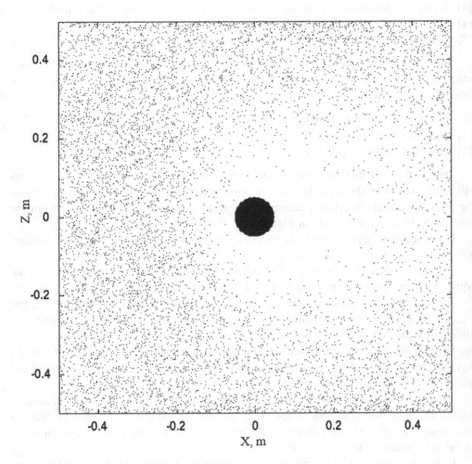

Fig. 1. The spatial distribution of the concentration of hydrogen ions. The axes are the distance from the center of the microsatellite in meters. The plasma velocity relative to the satellite is directed along the horizontal axis OX. V = 10 km / s. U_{sat} is electric potential of satellite equal to +5 Volts

4 Results of Numerical Modeling

Figures 1, 2, 3 and 4 show the results of numerical modeling of the interaction of a charged satellite and thermal Maxwellian plasma. The spatial distribution of the concentration of hydrogen ions in the plane XOZ (y = 0) after 3500 time steps beginning from the numerical experiment, which represents $3.5 \ 10^{-5}$ s from the time the initial modeling is presented. Each point represents a proton, crossed the XOZ plane of the two time steps. Initial conditions were only a spatially uniform distribution of protons and electrons with isotropic velocities Maxwellian distribution corresponding to the temperature 5000 K. The plasma velocity relative to the satellite is V and consisted of 10 km/sec (Figs. 1 and 3) and 20 km/sec (Figs. 2 and 4) along the horizontal axis OX. For all calculations, the plasma velocity relative to the satellite is directed along the horizontal axis OX (in all figures from left to right). U_{sat} satellite potential of +5 Volts (Figs. 1 and 2) and +10 Volts (Figs. 3 and 4), was adopted.

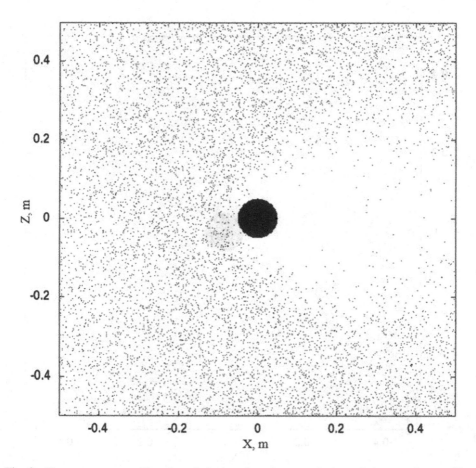

Fig. 2. The same as on Figs. 1 and 2, but for plasma velocity relative to the satellite V = 20 km/s

In all the figures there is a strongly marked ion shadow of microsatellite. Spatial characteristics of the shadows become more clear with increasing relative velocity of the plasma and the satellite and the satellite's positive potential. Thus, for a plasma speed of 10 km/s, and the electric potential of the satellite 5 Volts (Fig. 1), the characteristic dimensions of the shadow of the satellite ion about 20 cm. Moremore, before a satellite there is also a region of a few cm greatly reduced ion concentration. By increasing the relative speed of 20 km/s at the same value of the satellite electric potential (Fig. 2) ion shadow increases up to half of meter (maybe further, but we are limited in spatial scales of modeling the field).

Consider the effect of increasing the positive potential of the satellite. Figures 3 and 4 show the results of modeling the spatial distribution of protons near a satellite with an electrical potential of +10 Volts. When plasma speed of 10 km/sec region of reduced concentration increases significantly as a companion to and behind it. Before the satellite, at a distance of about 10 cm, concentration drops almost to zero. Shadow is greatly increased in the transverse dimension and the satellite extends almost to the

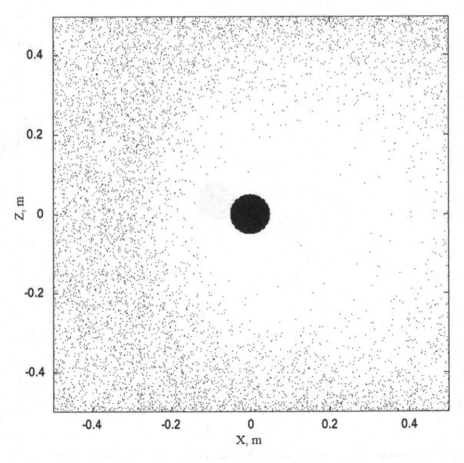

Fig. 3. The same as on Fig. 1, but for satellite electric potential U_{sat} = +10 Volts

boundaries of modeling region. A different picture is observed for twice the relative velocity of the satellite and plasma. Directly in front of the satellite it is observed a thin layer (3–5 cm) lower concentration of protons. But directly in front of this layer the concentration exceeds the background undisturbed conditions. The shadow of the satellite is clear and has about half a meter in diameter. The length of the shade clearly exceeds the spatial domain simulations and probably reaches the meter and more. As expected, pattern is symmetrical along axes OY (not shown), and OZ.

The features of the spatial distribution of the proton concentration near the charged satellite qualitatively correspond to calculations on other models and theoretical concepts. Thus, calculations for the hydrodynamic model for a dense plasma [17] indicate the presence of an ionic shadow behind a satellite with spatial dimensions on the order of the Debye radius. Note that the presence of this shadow was predicted in [18] on the basis of analytical calculations.

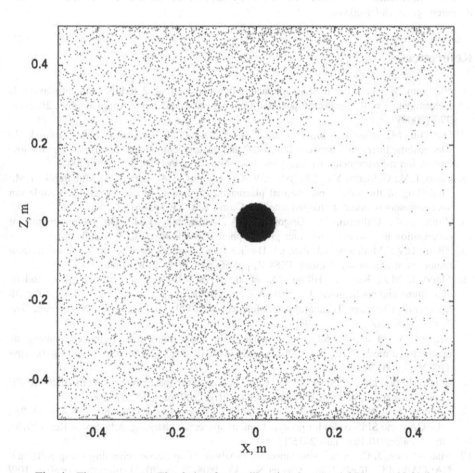

Fig. 4. The same as on Fig. 2, but for satellite electric potential $U_{sat} = +10$ Volts

5   Conclusion

The numerical simulation results show a significant distortion of the spatial distribution of thermal ions in the presence of the positive potential of the satellite, which are expressed in the presence of an area of low concentration to the satellite and ion shadows behind him. These distortions are the more pronounced, the greater the relative velocity of the satellite and the plasma and the magnitude of the positive potential of the satellite.

Thus, the algorithms considered in this paper are easily scaled and can be modified for the case of real microsatellites, such as 1U-3U CubeSats. Modeling the interaction between thermal plasma and larger satellites is currently difficult due to the considerable increase in computational complexity with increasing modeling domain.

Acknowledgements. This work was made by support of Russian Foundation for Basic Research, grant 18-01-00394.

References

1. Bouhram, M., et al.: Electrostatic interaction between Interball-2 and the ambient plasma. 1. Determination of the spacecraft potential from current calculations. Ann. Geophys. **20**, 365–376 (2002)
2. Hamelin, M., Bouhram, M., Dubouloz, N., Malingre, M., Grigoriev, S.A., Zinin, L.V.: Electrostatic interaction between Interball-2 and the ambient plasma. 2. Influence on the low energy ion measurements with Hyperboloid. Ann. Geophys. **20**, 377–390 (2002)
3. Zinin, L.V., Galperin, Y.I., Gladyshev, V.A., Grigoriev, S.A., Girard, L., Muliarchik, T.M.: Modelling of the anisotropic thermal plasma measurements of the energy-mass-angle ion spectrometers onboard a charged satellite. Cosm. Res. **33**, 511–518 (1995)
4. Zinin, L.V., Galperin, Y., Grigoriev, S.A., Mularchik, T.M.: On measurements of polarization jet effects in the outer plasmasphere. Cosmic Res. **36**, 39–48 (1998)
5. Zinin, L.V.: Modelling of thermal H+ ions on charged satellite taking into account temperature anisotropy. Vestnik IKSUR, pp. 56–63 (2009). in Russian
6. Mandell, M.J., Katz, I., Hilton, M., et al.: NASCAP-2K spacecraft charging models: algorithms and applications. In: 2001: A spacecraft charging odyssey. Proceeding of the 7th Spacecraft Charging Technology Conference, pp. 499—507. ESTEC, Noordwijk, The Netherlands (2001)
7. Hilgers, A., et al.: Modeling of plasma probe interactions with a pic code using an unstructured mesh. IEEE Trans. Plasma Sci. **36**, 2319–2323 (2008). https://doi.org/10.1109/TPS.2008.2003360
8. Thiebault, B., et al.: SPIS 5.1: an innovative approach for spacecraft plasma modeling. IEEE Trans. Plasma Sci. **43**, 1 (2015). https://doi.org/10.1109/tps.2015.2425300
9. Novikov, L.S., Makletsov, A.A., Sinolits, V.V.: Comparison of Coulomb-2, NASCAP-2K, MUSCAT and SPIS codes for geosynchronous spacecraft charging. Adv. Space Res. (2015). https://doi.org/10.1016/j.asr.2015.11.003
10. Matéo-Vélez, J.-C., et al.: Simulation and analysis of spacecraft charging using SPIS and NASCAP/GEO. IEEE Trans. Plasma Sci. **43**, 2808–2816 (2015). https://doi.org/10.1109/TPS.2015.2447523

11. Alder, B.J., Wainwright, T.E.: Phase transition for a hard sphere system. J. Chem. Phys. **27**, 1208–1209 (1957)
12. Alder, B.J., Wainwright, T.E.: Molecular dynamics by electronic computers. In: Prigogine, I. (ed.) Transport Processes in Statistical Mechanics, pp. 97–131. Interscience Publishers Inc, New York (1958)
13. Kholmurodov, H.T., Altaisky, M.V., Pusynin, M.V., et al.: Methods of molecular dynamics to simulate the physical and biological processes. Phys. Elementary Part. Atomic Nuclei. **34**, 474–515 (2003)
14. Zinin, L.V., Ishanov, S.A., Sharamet, A.A., Matsievsky, S.V.: Modeling the distribution of charged ions near the satellite method of molecular dynamics. 2-D approximation. Vestnik IKBFU, pp. 53–60 (2012). in Russian
15. Sharamet, A.A., Zinin, L.V., Ishanov, S.A., Matsievsky, S.V.: 2D modelling of a ion shadow behind charged satellite by molecular dynamics method. Vestnik IKBFU, pp. 26–30 (2013). in Russian
16. Zinin, L.V., Sharamet, A.A., Vasileva, A.Y.: Modeling the formation of the ion shadow behind a positively charged microsatellite in an oxygen plasma by the molecular dynamics method. Vestnik IKBFU, pp. 48–52 (2017). in Russian
17. Rylina, I.V., Zinin, L.V., Grigoriev, S.A., Veselov, M.V.: Hydrodynamic approach to modeling the thermal plasma distribution around a moving charged satellite. Cosmic Res. **40**, 367–377 (2002)
18. Alpert, Y.L., Gurevich, A.V., Pitaevskiy, L.P.: Artificial Satellites in Rarefied Plasma. Nauka, Moscow (1964). in Russian

High Performance Architectures, Tools and Technologies

Adaptive Scheduling for Adjusting Retrieval Process in BOINC-Based Virtual Screening

Natalia Nikitina[(✉)] and Evgeny Ivashko

Karelian Research Center of the Russian Academy of Sciences,
Institute of Applied Mathematical Research, Petrozavodsk, Russia
{nikitina,ivashko}@krc.karelia.ru

Abstract. This work describes BOINC-based Desktop Grid implementation of adaptive task scheduling algorithm for virtual drug screening. The algorithm bases on a game-theoretical mathematical model where computing nodes act as players. The model allows to adjust the balance between the results retrieval rate and the search space coverage. We present the developed scheduling algorithm for BOINC-based Desktop Grid and evaluate its performance by simulations. Experimental analysis shows that the proposed scheduling algorithm allows to adjust the results retrieval rate and the search space coverage in a flexible way so as to reach the maximal efficiency of a BOINC-based Desktop Grid.

Keywords: Desktop grid · BOINC · Scheduling
Virtual screening · Game theory · Congestion game

1 Introduction

High-performance computing (HPC) plays a significant role in implementing contemporary fundamental and applied research, developing new materials, new medicines, new types of industrial products. To perform HPC, computational clusters are often used. Should particularly large amounts of resources be required, one can also deploy Grid systems integrating computational clusters. Computing resources can be also provided on-demand using commercial cloud-based services. One more option is the use of Desktop Grids. The term stands for a distributed high-throughput computing system which uses idle time of non-dedicated geographically distributed computing nodes connected over low-speed (as opposed to supercomputer interconnect) regular network. In common case the nodes are either personal computers of volunteers connected over the Internet (volunteer computing) or organization desktop computers connected over local area network (Enterprise Desktop Grid). Desktop Grids can also be integrated into computational clusters or Grid systems (see [1,2] for examples).

As a Desktop Grid is a high-throughput computing tool, it is aimed at processing huge numbers of tasks. Usually, when solving such problems, one does

© Springer Nature Switzerland AG 2019
V. Voevodin and S. Sobolev (Eds.): RuSCDays 2018, CCIS 965, pp. 453–464, 2019.
https://doi.org/10.1007/978-3-030-05807-4_38

not aim to find a particular answer, but rather to select among a large number of prospective solutions candidates for more detailed evaluation by the scientists, for instance, in a laboratory. One of such problems is virtual screening where a HTC tool is used to perform computer modelling of interaction between a target protein and a prospective ligand; molecules with high predicted energy of interaction with the target are studied in detail in laboratories. For such problems, it is important not only to provide high performance when solving them, but also to organize computations in such way that would ensure the balance between the results retrieval rate and the search space coverage.

In this paper, we propose a task scheduling algorithm for adjusting the balance between the results retrieval rate and the search space coverage when performing computations in a BOINC-based Desktop Grid. The basis of the algorithm is a mathematical model which considers the heterogeneous Desktop Grid environment and the limited knowledge about the input dataset structure. The algorithm has been developed for solving the problem of virtual screening, but it can be also used for solving other computationally intensive search problems, where one needs to balance between the results retrieval rate and the search space coverage, probably with limited apriori knowledge about the input dataset.

The rest of the paper has the following structure. In Sect. 2, we provide the motivation for this work. In Sect. 3, we describe the methodology used. In Sect. 4, we provide and analyze the results of the computational experiments. In Sect. 5, we overview the related work. Finally, in Sect. 6, we conclude the paper with result discussion and directions of future work.

2   Motivation

2.1   Desktop Grids

There is a number of approaches to implement a Desktop Grid. There can be peer-to-peer, hierarchical, and other types of Desktop Grid. The diversity of high-level architectures of Desktop Grids has been described, for instance, in [3]. In our work, we consider the Desktop Grid which follows the server-client model, as shown in Fig. 1. The server holds a large number of tasks that are mutually independent pieces of a computationally heavy problem. When a computing node (or a client) is idle, it communicates with the server and requests work. The server replies by sending one or more independent tasks. The node processes them and reports results back to the server. The results are then processed and can be, for instance, stored in the database for further usage. Such architecture has been described in a number of works ([4,5] etc.).

The middleware systems for Desktop Grid operation also vary widely. However, the open source BOINC platform [4] is nowadays considered as *de facto* standard. Since 1990s, BOINC has been a framework for many independent volunteer computing projects. Today, it is the most actively developed Desktop Grid middleware, which supports the widest range of applications.

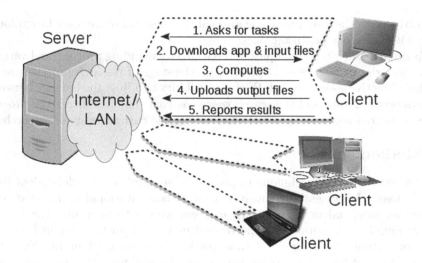

Fig. 1. Desktop Grid

BOINC is based on server-client architecture, where the workflow proceeds as described above. The client part is able to work at an arbitrary number of computers with various hardware and software characteristics.

With a variety of scheduling mechanisms implemented in BOINC, there is still a number of challenges one faces when solving a computationally intensive problem. Such challenges arise due to the heterogeneity and internal uncertainties present in Desktop Grid systems as well as specific requirements imposed by the field of research.

2.2 Virtual Drug Screening

Virtual drug screening [6] (further VS) refers to the creation of new medicines, a time-consuming process with high costs of research and development. It allows to bring *in silico* the first stage of drug development process, namely the identification of a set of chemical compounds called *hits* with predicted desired biochemical activity. Hits are identified among a set of *ligands*, low-molecular compounds able to form a biochemical complex with a protein molecule responsible for disease progression, called *a target*. In the course of VS, one performs computer modeling of the interaction of the candidate ligands with the target and scores the resulting molecular complexes. The ligands with high scores become hits.

In [7], we overview the problem of structure-based VS over large databases and illustrate the need for fast and efficient hits retrieval methods.

At the same time, discovery of novel chemotypes (e.g. essentially novel ligands for the given target) is considered to be one of the major drivers of VS progress for the next years [8]. The problem of finding novel chemotypes is tightly bound with the problem of retrieval of the most chemically diverse hits, fully covering the chemical space [9,10]. In general case, this objective can contradict the fastest hits retrieval, as the least promising areas of the chemical space must be explored along with the most promising ones.

To summarize, VS is a complex and resource-demanding process, and various challenges arise in its course. Depending on the stage of research process, one or another objective steps forward. With a possibility to adjust the balance between the hits retrieval rate and the search space coverage, one can direct the process of VS so as to achieve its maximal efficiency at the current stage of research.

3   Methodology

Being a computational technique to process large numbers of independent fine-grained tasks, VS essentially involves a set of computational nodes that may be seen as independent agents, each of them willing to maximize the reward they receive for computations. The reward may be expressed, for instance, in terms of virtual credit for CPU time (as it is implemented in BOINC), the number of found hits, or any other expression of useful work the node performs. At the same time, decisions of one node may influence the others in case they access limited shared resources, or compete for the fastest results retrieval, etc. Such presentation allows to apply the methods of mathematical game theory for modeling the computational process of VS.

The mathematical model described in this section has been elaborated in [7]. The model is based on a congestion game, which was first proposed by Rosenthal in 1973 [11]. Its important property is the existence of a deterministic Nash equilibrium. The convergence and finite-time convergence of the game iterations are well studied. Existence of a Nash equilibrium is ensured [12] even in the considered case of heterogeneous players and resources. In order to reach the equilibrium situation, we employ the best-response dynamics [12,13].

The idea lying in the basis of a model is as follows. Due to variations in chemical characteristics, molecules have different chances to show high predicted binding affinity. One can expect that these chances are higher for molecules close in topology to a known ligand [14,15]. In contrast, molecules with very large number of atoms are less likely to become hits [16].

Thus, non-overlapping subsets of molecules in the library could be ranked beforehand by their estimated prospectivity for VS. Once specified, the estimated probabilities can be updated in the course of VS according to interim results. At the same time, results originating from the same subset might be redundant. The model is designed so as to explore most prospective subsets first while keeping the desired level of diversity by restricting intensity of subsets exploration.

Consider a computer system with m computational nodes — or players — $C_1, \ldots, C_m$, and a set of computational tasks T. Each node is characterized by its computational performance ops_i, which is the average number of operations performed in a time unit. The input set T is divided into non-overlapping blocks $T = T_1 \bigcup \ldots \bigcup T_n$ such that the estimated portion of VS hits in block T_j is p_j. We define priority of the block T_j as

$$\sigma_j = \frac{p_j}{p_1 + \ldots + p_n}. \tag{1}$$

The blocks with higher priority have to be chosen first for processing. We assume that all tasks in block T_j have the average computational complexity θ_j, i.e., a number of operations to process one task. Each node selects exactly one block.

The nodes make their decisions at time steps $0, \tau, 2\tau, \ldots$. After a node has processed its portion of tasks, it sends the results to the server and is ready for the next portion. Let the utility of node C_i at time step τ express the amount of useful work performed during this step. This amount depends on the number of executed tasks from the chosen block, its computational complexity, priority, and the number of other nodes who have also chosen this block.

The fewer nodes explore block T_j simultaneously, the more valuable their work is. This condition ensures diversification of the interim set of hits. Let n_j be the number of the players who have chosen block T_j at the considered step, and $\delta(n_j)$ be the congestion coefficient for the block, which in the simplest case takes form

$$\delta(n_j) = \frac{1}{n_j + 1} \tag{2}$$

The utility of node C_i that chooses block T_j is

$$U_{ij} = \left(\alpha_i\, \delta(n_j) + (1 - \alpha_i)\, \sigma_j\right) \frac{ops_i}{\theta_j}. \tag{3}$$

Here, $\alpha_i \in [0;1]$ is the parameter to control the balance between block congestion level $\delta(n_j)$ and block prospectivity σ_j. As it tends to zero, the player C_i gains maximal profit of getting the most possible number of hits at a step ("digger"). On the contrary, as α_i tends to one, the player gains maximal profit of selecting the blocks with minimal presence of other players ("explorer").

In such way, different players can have different preferences expressed by the value of parameter α_i. By fixing the preferences, one can direct the computational process as a whole. In Sect. 4, we consider the game where all players have the same preference $\alpha = \alpha_1 = \ldots = \alpha_m$, varying from 0 to 1.

Therefore, at each considered time step, we have a singleton congestion game $G = \langle C, T, U \rangle$, where C is the set of players (computational nodes), T is the set of data blocks of which each node selects exactly one, and U is the set of utility functions. A strategy profile is a *schedule* $s = (s_1, \ldots, s_m)$, where the component $s_i = j$ means that player C_i selects block T_j.

4 Experimental Analysis

In order to perform computational experiments and evaluate the performance of the developed approach, we divide a molecules database into blocks and simulate VS. We prove the efficiency and flexibility of the proposed algorithm by showing the influence of scheduling parameter α on results retrieval rate and search space coverage.

As in [5,7], we use the database GDB-9 of enumerated organic molecules consisting of at most nine atoms of C, N, O, S and Cl (not counting hydrogen). GDB-9 represents about 320 thousand molecules with variety of chemical properties. The chosen database is manageable for performing computational experiments and can be unambiguously divided into several non-overlapping blocks. At the same time, the set of molecules is rich enough to demonstrate the feasibility and practicability of proposed solutions.

For the experiments, we consider three pre-calculated chemical properties of each molecule: the total number of atoms including hydrogens, polar surface area (PSA), and partition coefficient $logP$. Basing on these properties, we divided the database into 16 non-overlapping task blocks. For the sake of time, we do not compute the predicted binding energy as it would be done in a real VS setting. Instead we use the pre-calculated value $logP$. As 0.99% of molecules in GDB-9 have $logP \geq x = 2.7765$, the value $x = 2.7765$ has been taken as a threshold to count a molecule as a hit.

We use two of the considered chemical properties, the total number of atoms and the calculated PSA, for charting the molecules database in two dimensions, and defining blocks of tasks. The resulting decomposition is provided in Table 1. In each cell, the upper number (in bold) stands for the total number of molecules/tasks in the corresponding block. The lower number stands for the hits fraction in this block.

Table 1. Decomposition of GDB-9 database into non-overlapping blocks.

		PSA			
		$[0.0, 24.6)$	$[24.6, 38.05)$	$[38.05, 52.04)$	$[52.04, 118.35]$
Number of atoms	$[4, 17)$	**9308**	**18362**	**24723**	**29411**
		0.00398	0.00408	0.00227	0.00211
	$[17, 19)$	**16969**	**20182**	**20423**	**20333**
		0.00147	0.00357	0.00005	0.00157
	$[19, 21)$	**24661**	**20564**	**19189**	**15695**
		0.01500	0.00146	0.00005	0.00204
	$[21, 31]$	**30871**	**22110**	**15845**	**10732**
		0.06663	0.01072	0.00038	0

We perform computational experiments, simulating virtual screening in a heterogeneous Desktop Grid consisting of 64 computing nodes. The simulations have been implemented within the Center for collective use "High-performance computing center" of Karelian Research Center. The parameters of the simulations are summarized in Table 2.

In Fig. 2, we show the overall process of VS over GDB-9 database for 4 fixed values of parameter α. We observe that the difference in results retrieval rate can be drastic for different values of α, unless the VS process has entered the "tail" phase where all or nearly all hits have been found.

Further, we investigate the influence of the scheduling parameter α value on the characteristics of the computational process.

Table 2. Parameters of the simulations.

Parameter	Value	Description
n	16	Number of task blocks
m	64	Number of computing nodes
ops	15 (nodes C_1–C_{16})	Performance of a computing node (number of conditional operations per time unit)
	20 (nodes C_{17}–C_{32})	
	25 (nodes C_{33}–C_{48})	
	30 (nodes C_{49}–C_{64})	
θ	75 (blocks T_1–T_4)	Complexity of a computational task (number of conditional operations)
	100 (blocks T_5–T_8)	
	125 (blocks T_9–T_{12})	
	150 (blocks T_{13}–T_{16})	
τ	100	Maximal length of a step (number of time units)
x	2.7765	Threshold of $logP$ value for selecting a hit

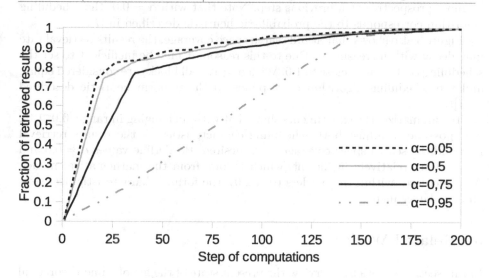

Fig. 2. Results retrieval progress in the course of the simulated virtual screening run over GDB-9 database

Figure 3 depicts the results retrieval rate and search space coverage, both averaged over steps 1–5 (a) and 21–25 (b), as α varies from 0.0 to 1.0. We observe that the search space coverage has a positive correlation with the scheduling coefficient α, while the results retrieval rate has a negative one.

With α varying from 0.0 to 1.0 with step 0.05, we expectedly observe the "scissors" of the two observed characteristics. The diagram shows that the case

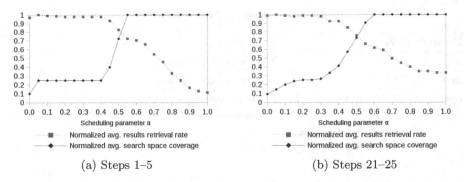

(a) Steps 1–5   (b) Steps 21–25

Fig. 3. Results retrieval rate and search space coverage depending on the scheduling coefficient α (normalized, averaged over 5 steps)

corresponding to zero influence of block congestion ($\alpha = 0.0$) is relatively inefficient both in terms of hits retrieval rate and search space coverage. This is due to the fact that at every step, all players select the same block which appeared to be most prospective at a previous step. Note that with $\alpha = 0.0$, the scheduling algorithm corresponds to the probabilistic heuristic described in [7].

Figure 3 also shows that at full search space coverage, the results retrieval rate may decay with increasing α. Due to this reason, it may be inefficient to set the scheduling coefficient α close to 1.0. With $\alpha = 1.0$, all blocks are considered equal, and the scheduling algorithm corresponds to the uniform heuristic described in [7].

To summarize, the experiments show that with α changing in range $[0.0, 1.0]$, it is possible to adjust both scheduling characteristics — the results retrieval rate and the search space coverage — as desired. Borderline values $\alpha = 0.0$ and $\alpha = 1.0$ are relatively inefficient, which follows from the mathematical model. As α increases within the borders $(0.0, 1.0)$, the former characteristic decreases and the latter increases.

5   Related Work

In this section we briefly overview the present state of the art of game-theoretical approaches to scheduling in Desktop Grids and scheduling for VS.

When developing schedulers for Desktop Grids, game-theoretical methods seem potentially efficient. One can note the work [17]. Its authors describe a game-theoretical approach to the balancing of volunteer computing resources between several projects. Each BOINC client has three parameters: peak performance, resource availability history, and settings of resource sharing between BOINC projects. Each project has its own strategy consisting of the policy of sending tasks, the slack (the value that determines the order of deadlines appointment to the tasks), the intervals between connections to the server, and the replication strategy. The project's payoff function is equal to the cluster equivalence parameter. Based on this model, the authors of the paper find the equilibrium for high-performance computing projects and high-throughput computing projects.

In work [18], co-authored by the authors of the presented paper, an hierarchical game-theoretic model of task scheduling is presented. The model is used to reduce the total server load by creating optimal-sized task parcels instead of sending tasks to clients one by one. Each client decides which size of parcels they will request. The payoff function of the server determines its costs for the formation of task parcels, the time of waiting for the computations results and their processing. The payoff function of the client is the cost of computing and communications. Taking into account the fact that an error in the execution of a single task leads to an error in the entire task parcel, the presented model allows to decrease the number of requests to the server due to the increase in the amount of overhead expenses. The resulting solution can be used in projects with "short" computational tasks; the solution was tested on a VS setting where one task corresponded to a single ligand. In practice, the grouping of ligands is being used in the course of VS, but the optimal parcel size is determined heuristically for specific computing systems (see, for example, [19,20]).

Due to the demand of resources, VS essentially involves high-performance computing tools. A recent review on the latest achievements and current state of VS lists more than a hundred successful examples of ligand discoveries in silico [8]. Among them, 25 of 82 papers published in 2008–2015 explicitly state that VS had been performed using high-performance computing such as computing clusters, supercomputers and grids.

There are works describing fruitful VS using Desktop Grids, as well. Several large-scale volunteer computing projects devoted to drug discovery have gathered and employed significant amounts of "gratis" computational resources due to high public interest to such projects. The most prominent example of such an infrastructure is the World Community Grid [21], which employed the power of over 3.4 million personal computers to support 27 research projects, including drug searches against AIDS, cancer, malaria and other diseases. To name a few, the projects allowed to discover candidate treatments for neuroblastoma [22], tuberculosis [23], leishmaniasis [24] etc.

Thus, high-performance and high-throughput computing tools have proven to be instrumental for implementing VS.

At the same time, there are efforts to automate new VS runs using different types of computational resources [25,26] and to optimize the VS process. For example, the task of fast retrieval of maximally diverse hits is being solved using genetic algorithms and heuristics [9,27].

In work [28], the authors solve the task of fast retrieval of the most prospective ligands on early stages of VS. The method they use is dividing the search space into classes according to molecular properties, and calculating the Bayesian probability of new ligands falling into one or another class. The search space is restricted by well-known Lipinski's rules, which, in general, do not guarantee coverage of all prospective ligands. However, such charting of the database allows to prioritize the ligands apriori, and the experiments prove the efficiency of the approach.

The authors of [20] consider the overall process of VS and investigate its performance in terms of wall-clock time to obtain the results. They divide the database into chunks of ligands and evaluate the performance for different chunk sizes.

The mathematical model described in this paper has been proposed and elaborated in [7]. It was shown that the model allows to boost VS efficiency at early stages, probably at the sacrifice of its productivity at later stages. Its implementation for BOINC platform has been proposed in [5] as a pseudo code. In the present paper we investigate the ability of the model to adjust the results retrieval rate and search space coverage at any stage of VS.

6   Conclusions and Future Work

In this paper, we present an implementation of adaptive scheduling algorithm for virtual screening using BOINC-based Desktop Grid. It is based on the mathematical model of game theory, where task scheduling is considered as a congestion game with computing nodes as players, who choose specific subsets of data blocks for processing. We introduce a scheduling parameter α which expresses the balance between results retrieval rate and search space coverage. We conduct computational experiments and show that by varying α, one is able to set the desired balance value.

The computational experiments to evaluate the performance of the developed algorithm were performed in the Enterprise Desktop Grid based on resources of the Karelian Research Center, Russian Academy of Sciences.

The presented mathematical model and the scheduling algorithm are designed for BOINC-based Desktop Grid. Further study is required to investigate the impact that the overall search process experiences when individual players change their scheduling strategies according to their preferred behavior of either "digger" or "explorer". This is relevant for volunteer computing projects. This will be the subject of future work, as well as assessment of the algorithm performance and effectiveness in multi-objective domains.

Acknowledgements. This work was supported by the Russian Foundation of Basic Research, projects 18-07-00628 and 18-37-00094.

References

1. Afanasiev, A.P., Bychkov, I.V., Manzyuk, M.O., Posypkin, M.A., Semenov, A.A., Zaikin, O.S.: Technology for integrating idle computing cluster resources into volunteer computing projects. In: 5th International Workshop on Computer Science and Engineering: Information Processing and Control Engineering, WCSE 2015-IPCE, pp. 109–114 (2015)
2. Kovács, J., Marosi, A., Visegrádi, Á., Farkas, Z., Kacsuk, P., Lovas, R.: Boosting gLite with cloud augmented volunteer computing. Future Gener. Comput. Syst. **43–44**, 12–23 (2015). https://doi.org/10.1016/j.future.2014.10.005
3. Choi, S., et al.: Characterizing and classifying desktop grid. In: Seventh IEEE International Symposium on Cluster Computing and the Grid, CCGRID 2007, pp. 743–748. IEEE (2007). https://doi.org/10.1109/CCGRID.2007.31
4. Anderson, D.P.: BOINC: a system for public-resource computing and storage. In: The Fifth IEEE/ACM International Workshop on Grid Computing, GRID 2004, pp. 4–10. IEEE CS Press (2004). https://doi.org/10.1109/GRID.2004.14
5. Nikitina, N., Ivashko, E., Tchernykh, A.: Congestion game scheduling implementation for high-throughput virtual drug screening using BOINC-based desktop grid. In: Malyshkin, V. (ed.) PaCT 2017. LNCS, vol. 10421, pp. 480–491. Springer, Cham (2017). https://doi.org/10.1007/978-3-319-62932-2_46
6. Bielska, E., et al.: Virtual screening strategies in drug design – methods and applications. J. Biotech. Comput. Biol. Bionanotechnol. **92**(3), 249–264 (2011). https://doi.org/10.5114/bta.2011.46542
7. Nikitina, N., Ivashko, E., Tchernykh, A.: Congestion game scheduling for virtual drug screening optimization. J. Comput.-Aided Mol. Des. **32**(2), 363–374 (2018). https://doi.org/10.1007/s10822-017-0093-7
8. Irwin, J., Shoichet, B.K.: Docking screens for novel ligands conferring new biology. J. Med. Chem. **59**(9), 4103–4120 (2016). https://doi.org/10.1021/acs.jmedchem.5b0200
9. Rupakheti, C., Virshup, A., Yang, W., Beratan, D.N.: Strategy to discover diverse optimal molecules in the small molecule universe. J. Chem. Inf. Model. **55**(3), 529–537 (2015). https://doi.org/10.1021/ci500749q
10. Harper, G., Pickett, S.D., Green, D.V.: Design of a compound screening collection for use in high throughput screening. Comb. Chem. High Throughput Screening **7**(1), 63–70 (2004). https://doi.org/10.2174/1386207043772884832
11. Rosenthal, R.: A class of games possessing pure-strategy nash equilibria. Int. J. Game Theory **2**(1), 65–67 (1973). https://doi.org/10.1007/BF01737559
12. Milchtaich, I.: Congestion games with player-specific payoff functions. Games Econ. Behav. **13**, 111–124 (1996). https://doi.org/10.1006/game.1996.0027
13. Ieong, S., et al.: Fast and compact: a simple class of congestion games. In: AAAI 2005 Proceedings of the 20th National Conference on Artificial Intelligence, vol. 2, pp. 1–6. AAAI Press (2005)
14. Willet, P., Barnard, J.M., Downs, G.M.: Chemical similarity searching. J. Chem. Inf. Comput. Sci. **38**(6), 983–996 (1998). https://doi.org/10.1021/ci9800211
15. Patterson, D.E., Cramer, R.D., Ferguson, A.M., Clark, R.D., Weinber, L.E.: Neighborhood behavior: a useful concept for validation of "molecular diversity" descriptors. J. Med. Chem. **39**(16), 3049–3059 (1996). https://doi.org/10.1021/jm960290n
16. Hann, M.M., Leach, A.R., Harper, G.: Molecular complexity and its impact on the probability of finding leads for drug discovery. J. Chem. Inf. Comput. Sci. **41**(3), 856–864 (2001). https://doi.org/10.1021/ci000403i

17. Donassolo, B., Legrand, A., Geyer, C.: Non-cooperative scheduling considered harmful in collaborative volunteer computing environments. In: 11th IEEE/ACM International Symposium on Cluster, Cloud and Grid Computing, pp. 144–153 (2011). https://doi.org/10.1109/CCGrid.2011.34

18. Mazalov, V., Nikitina, N., Ivashko, E.: Hierarchical two-level game model for tasks scheduling in a desktop grid. In: Applied Problems in Theory of Probabilities and Mathematical Statistics Related to Modeling of Information Systems, 2014 6th International Congress on Ultra Modern Telecommunications and Control Systems and Workshops (ICUMT), pp. 641–645 (2014). https://doi.org/10.1109/ICUMT. 2014.7002159

19. Jaghoori, M., et al.: A multi-infrastructure gateway for virtual drug screening. Concurrency Computat. Pract. Exper. **27**, 4478–4490 (2015). https://doi.org/10. 1002/cpe.3498

20. Krüger, J., et al.: Performance studies on distributed virtual screening. BioMed Res. Int. 1–7 (2014). https://doi.org/10.1155/2014/624024. Article ID 624024

21. World Community Grid. https://www.worldcommunitygrid.org. Accessed 12 Apr 2018

22. Nakamura, Y., et al.: Identification of novel candidate compounds targeting TrkB to induce apoptosis in neuroblastoma. Cancer Med. **3**(1), 25–35 (2014). https:// doi.org/10.1002/cam4.175

23. Perryman, A.L., et al.: A virtual screen discovers novel, fragment-sized inhibitors of mycobacterium tuberculosis InhA. J. Chem. Inf. Model. **55**(3), 645–659 (2015). https://doi.org/10.1021/ci500672v

24. Ochoa, R., Watowich, S.J., Flórez, A., et al.: Drug search for leishmaniasis: a virtual screening approach by grid computing. J. Comput.-Aided Mol. Des. **30**(7), 541–552 (2016). https://doi.org/10.1007/s10822-016-9921-4

25. Ellingson, S.R., Baudry, J.: High-throughput virtual molecular docking with AutoDockCloud. Concurr. Comput. Pract. Exper. **26**, 907–916 (2014). https:// doi.org/10.1002/cpe.2926

26. Forli, S., et al.: Computational protein-ligand docking and virtual drug screening with the AutoDock suite. Nat. Protoc. **11**(5), 905–919 (2016). https://doi.org/10. 1038/nprot.2016.051

27. Rupakheti, C.R.: Property Biased-Diversity Guided Explorations of Chemical Spaces. Ph.D. dissertation, Duke University, 138 p (2015)

28. Pradeep, P., Struble, C., Neumann, T., Sem, D.S., Merrill, S.J.: A novel scoring based distributed protein docking application to improve enrichment. IEEE/ACM Trans. Comput. Biol. Bioinf. **12**(6), 1464–1469 (2015). https://doi.org/10.1109/ TCBB.2015.2401020

Advanced Vectorization of PPML Method for Intel® Xeon® Scalable Processors

Igor Chernykh[1]([✉]), Igor Kulikov[1], Boris Glinsky[1], Vitaly Vshivkov[1], Lyudmila Vshivkova[1], and Vladimir Prigarin[2]

[1] Institute of Computational Mathematics and Mathematical Geophysics
SB RAS, 630090 Novosibirsk, Russia
{chernykh,kulikov,vsh}@ssd.sscc.ru, gbm@sscc.ru,
lyudmila.vshivkova@parbz.sscc.ru
[2] Novosibirsk State Technical University, 630073 Novosibirsk, Russia
vovkaprigarin@gmail.com

Abstract. Piecewise Parabolic Method on a Local Stencil is very useful for numerical simulation of fluid dynamics, astrophysics. The main idea of the PPML method is the use of a piecewise parabolic numerical solution on the previous time step for computing the Riemann problem solving partial differential equations system (PDE). In this paper, we present the new version of PDE solver which is based on the PPML method optimized for Intel Xeon Scalable processor family. The results of performance comparison between different types of AVX-512 compatible Intel Xeon Scalable processors are presented. Special attention is paid to comparing the performance of Intel Xeon Phi (KNL) and Intel Xeon Scalable processors.

Keywords: Massively parallel supercomputers · Astrophysics
Code vectorization

1 Introduction

For the past decade, the most of research papers in high-performance computing and numerical simulation of different problems using supercomputers are dedicated to parallel algorithms, parallel programming techniques, performance analysis and tests. However, at the same time we can see that the architecture of CPUs evolves not only in number of cores. Modern CPUs have more complex core structure than ten years ago. The most recent processors have many cores/threads and the ability to implement single instructions on an increasingly large data set (SIMD width) [1]. For example, Fig. 1 shows core architecture of Intel Xeon Scalable Processor. From our point of view, the key factor to utilize all computational power of modern CPUs is a FMA +SIMD code optimization as well as using of OpenMP API within a multicore CPU. Intel Xeon Scalable Processor has 2 FMAs per core and AVX-512 support. Also worth noting that base AVX-512 instruction set for Intel Xeon Scalable Processor is supported by Intel Xeon Phi 7290 (KNL architecture) processor. It means that if you have AVX-512 optimized code for Intel Xeon Phi 7290, Intel Xeon Scalable processor will support your optimizations.

V. Voevodin and S. Sobolev (Eds.): RuSCDays 2018, CCIS 965, pp. 465–471, 2019.
https://doi.org/10.1007/978-3-030-05807-4_39

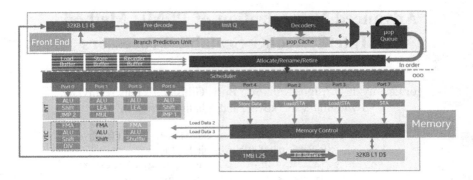

Fig. 1. Intel Xeon Scalable CPU core architecture [2].

Unlike Intel Xeon Phi 7290, Intel Xeon Scalable Processors have cores downclock in case of using AVX-512 instructions. Cores downclock depends on the core's load by AVX-512 instructions. Table 1 shows CPU frequency behavior for three kinds of Intel Xeon Gold Processors which are used for tests in this paper. As you can see from the table, some processors have low base AVX-512 frequency due to TDP value restrictions.

Table 1. Turbo frequencies for Intel Xeon Gold processors. Full load of cores.

Mode	Intel Xeon Gold 6144 (8 cores)	Intel Xeon Gold 6150 (18 cores)	Intel Xeon Gold 6154 (18 cores)
Base (without/with AVX-512) frequency	3.5 GHz/2.2 GHz	2.7 GHz/1.9 GHz	3 GHz/2.1 GHz
Normal turbo	4.1 GHz	3.4 GHz	3.7 GHz
AVX-512 turbo	2.8 GHz	2.5 GHz	2.7 GHz

In the next chapters, we will show the results of performance tests by our solver which is based on PPML method on this three kinds of Intel Xeon Scalable Processors.

2   Mathematical Model and Numerical Method

In our work, we use a multicomponent hydrodynamic model of galaxies considering the chemodynamics of molecular hydrogen and cooling in the following form:

$$\frac{\partial \rho}{\partial t} + \nabla \cdot (\rho \vec{u}) = 0$$

$$\frac{\partial \rho_{H_2}}{\partial t} + \nabla \cdot (\rho_{H_2} \vec{u}) = S(\rho, \rho_H, \rho_{H_2})$$

$$\frac{\partial \rho_H}{\partial t} + \nabla \cdot (\rho_H \vec{u}) = -S(\rho, \rho_H, \rho_{H_2})$$

$$\frac{\partial \rho \vec{u}}{\partial t} + \nabla \cdot (\rho \vec{u} \vec{u}) = -\nabla p - \rho \nabla \Phi \tag{1}$$

$$\frac{\partial \varepsilon}{\partial t} + \nabla \cdot (\varepsilon \vec{u}) = -(\gamma - 1)\varepsilon \nabla \cdot (\vec{u}) - Q$$

$$\frac{\partial E}{\partial t} + \nabla \cdot (E\vec{u}) = -\nabla \cdot (p\vec{u}) - (\rho \nabla \Phi, \vec{u}) - Q$$

$$\Delta \Phi = 4\pi G \rho$$

$$E = \varepsilon + \frac{\rho \vec{u}}{2}$$

$$p = (\gamma - 1)\varepsilon,$$

where ρ is density, ρ_H is atomic hydrogen density, ρ_{H_2} is molecular hydrogen density, $\vec{u}$ is the velocity vector, ε is internal energy, p is pressure, E is total energy, γ is the ratio of specific heats, Φ is gravity, G is the gravitational constant, S is the formation rate of molecular hydrogen, and Q is a cooling function. A detailed description of this model can be found in [3].

The formation of molecular hydrogen is described by an ordinary differential equation [4]:

$$\frac{dn_{H_2}}{dt} = R_{gr}(T)n_H(n_H + 2n_{H_2}) - (\xi_H + \xi_{diss})n_{H_2} \tag{2}$$

where n_H is the concentration of atomic hydrogen, n_{H_2} is the concentration of molecular hydrogen, and T is temperature. Detailed descriptions of the H_2 formation rate R_{gr} and the photodissociation ξ_H, ξ_{diss} of molecular hydrogen, can be found in [5, 6]. Chemical kinetics was don with using of CHEMPAK tool [7, 8].

The original numerical method based on the combination of the Godunov method, operator splitting approach and piecewise-parabolic method on local stencil was used for numerical solution of the hyperbolic equations [9]. The piecewise-parabolic method on local stencil provides the high-precision order. The equation system is solved in two stages: at the Eulerian stage, the equations are solved without advective terms and at the Lagrangian stage, the advection transport is being performed. At the Eulerian stage, the hydrodynamic equations for both components are written in the non-conservative form and the advection terms are excluded. As the result, such a system has an analytical solution on the two-cell interface. This analytical solution is used to evaluate the flux through the two-cell interface. In order to improve the precision order, the piecewise-parabolic method on the local stencil (PPML) is used. The method is the construction of local parabolas inside the cells for each hydrodynamic quantity. The main difference of the PPML from the classical PPM method is the use of the local stencil for computation. It facilitates the parallel implementation by using only one layer for subdomain overlapping. It simplifies the implementation of the boundary conditions and decreases the number of communications thus improving the scalability. The detailed description of this method can be found in [10]. The same approach is used for the Lagrangian stage. Now the Poisson equation solution is based on Fast Fourier Transform method. This is because the Poisson equation solution takes several percents of the total computation time. After the Poisson equation solution, the hydrodynamic equation system solution is corrected. It should be noticed here that the system is over defined. The correction is performed by means of the original procedure for the full energy conservation and the guaranteed entropy nondecrease. The procedure includes the renormalization of the velocity vector length, its direction remaining the same (on boundary gas-vacuum) and the entropy (or internal energy) and dispersion velocity tensor correction. Such a modification of the method keeps the detailed energy balance and guaranteed non-decrease of entropy.

3   Performance Analysis

We used Intel Advisor [11] for performance analysis of our code. Intel Advisor collects different statistics from each cycle of the code. Statistics collection consists of 3 steps: survey collection, trip count collection, visualization and/or extraction of the collected data into a report.

1. Survey collection by command line with advisor:

```
mpirun -n <number of nodes> advixe-cl -collect survey --
trace-mpi -- ./<app_name>
```

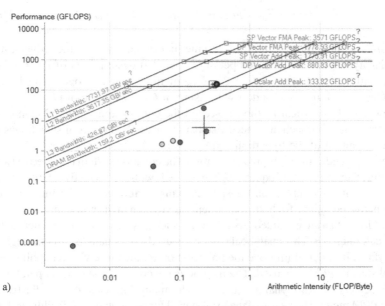

a)

Fig. 2. Roofline chart for our code: (a) 2x Intel Xeon Gold 6144 – 158 GFLOPS; (b) 2x Intel Xeon Gold 6150–125 GFLOPS; (c) 2x Intel Xeon Gold 6154 – 234 GFLOPS.

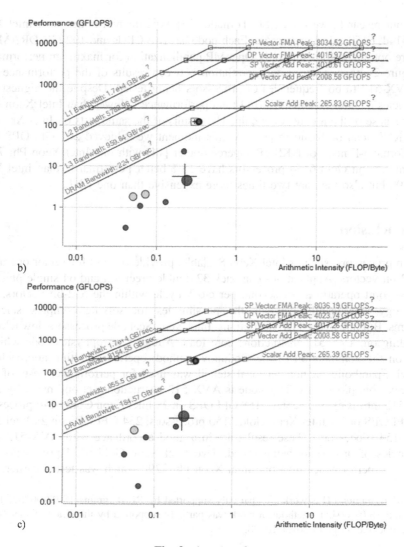

Fig. 2. (*continued*)

2. Trip count collection by command line with advisor:

```
mpirun -n <number of nodes> advixe-cl -collect
trip-counts -flop --trace-mpi -- ./<app_name>
```

3. Extraction of the data in a report:

```
advixe-cl -report survey -show-all-columns --format=text
-- report-output report.txt
```

In our research, we used RSC Tornado [12] experimental nodes with Intel Xeon Gold 6144, 6150, 6154 processors. Each node has two CPUs and 192 GB DRAM. All tests are optimized for using only OpenMP parallelization for maximum performance.

Figure 2 shows a very good correlation between results of the performance tests and AVX-512 turbo frequencies of processors being tested. Despite the highest core frequencies of Intel Xeon Gold 6144 total performance is lower than Intel Xeon Gold 6154 because of the 8 cores on a chip. The results of the same tests on Intel Xeon Phi 7290 (KNL) can be found in [13]. At this moment we achieved 202 GFLOPS from RSC Tornado-F node of NKS-1P supercomputer [14] with one Intel Xeon Phi 7290. Two Intel Xeon Gold 6154 processors have 15% better performance than Intel Xeon Phi 7290 but also they are two times more expensive than one KNL.

4   Conclusion

Modern processors, such as Intel Xeon Scalable, provide support for vector operations on 512-bit vectors. Applications can pack 32 double precision and 64 single precision floating point operations per second per clock cycle within the 512-bit vectors. This technology significantly expands the possibilities for solving complex scientific problems. But due to the TDP restrictions, Intel Xeon Scalable processors downclocked depending on AVX-512 instructions core load. Author's astrophysics code which is based on the combination of the Godunov method, operator splitting approach and piecewise-parabolic method on local stencil was used for performance tests of Intel Xeon Scalable processors. This code is AVX-512 instructions set optimized by using AVX-512 intrinsics. We achieve 158 GFLOPS on 2x Intel Xeon Gold 6144 processors, 125 GFLOPS on 2x Intel Xeon Gold 6150 processors, 234 GFLOPS on 2x Intel Xeon Gold 6154 processors. These results are in a good accordance with AVX-512 turbo frequencies of processors being tested. Two Intel Xeon Gold 6154 processors have 15% better performance than one Intel Xeon Phi 7290 which was tested earlier.

Acknowledgments. This work was partially supported by RFBR grants 18-07-00757, 18-01-00166 and 16-07-00434. Methodical work was partially supported by the Grant of the Russian Science Foundation grant 16-11-10028.

References

1. Vectorization: A Key Tool To Improve Performance On Modern CPUs. https://software.intel.com/en-us/articles/vectorization-a-key-tool-to-improve-performance-on-modern-cpus
2. Intel Xeon Scalable Debuts. https://hothardware.com/reviews/intel-xeon-scalable-processor-family-review?page=2
3. Vshivkov, V.A., Lazareva, G.G., Snytnikov, A.V., Kulikov, I.M., Tutukov, A.V.: Hydrodynamical code for numerical simulation of the gas components of colliding galaxies. Astrophys. J. Suppl. Ser. **194**(47), 1–12 (2011)
4. Bergin, E.A., Hartmann, L.W., Raymond, J.C., Ballesteros-Paredes, J.: Molecular cloud formation behind shock waves. Astrophys. J. **612**, 921–939 (2004)

5. Khoperskov, S.A., Vasiliev, E.O., Sobolev, A.M., Khoperskov, A.V.: The simulation of molecular clouds formation in the Milky Way. Mon. Not. R. Astron. Soc. **428**(3), 2311–2320 (2013)
6. Glover, S., Mac Low, M.-M.: Simulating the formation of molecular clouds. I. Slow formation by gravitational collapse from static initial conditions. Astrophys. J. Suppl. Ser. **169**, 239–268 (2006)
7. Chernykh, I., Stoyanovskaya, O., Zasypkina, O.: ChemPAK software package as an environment for kinetics scheme evaluation. Chem. Prod. Process Model. **4**(4) (2009)
8. Snytnikov, V.N., Mischenko, T.I., Snytnikov, V., Chernykh, I.G.: Physicochemical processes in a flow reactor using laser radiation energy for heating reactants. Chem. Eng. Res. Des. **90**(11), 1918–1922 (2012)
9. Godunov, S.K., Kulikov, I.M.: Computation of discontinuous solutions of fluid dynamics equations with entropy nondecrease guarantee. Comput. Math. Math. Phys. **54**, 1012–1024 (2014)
10. Kulikov, I., Vorobyov, E.: Using the PPML approach for constructing a low-dissipation, operator-splitting scheme for numerical simulations of hydrodynamic flows. J. Comput. Phys. **317**, 316–346 (2016)
11. Intel Advisor. https://software.intel.com/en-us/intel-advisor-xe
12. RSC Tornado. http://www.rscgroup.ru/en/our-technologies/267-rsc-tornado-cluster-architecture
13. Glinskiy, B., Kulikov, I., Chernykh, I.: Improving the performance of an AstroPhi code for massively parallel supercomputers using roofline analysis. Commun. Comput. Inf. Sci. **793**, 400–406 (2017)
14. Siberian Supercomputer Center ICMMG SB RAS. http://www2.sscc.ru

Analysis of Results of the Rating of Volunteer Distributed Computing Projects

Vladimir N. Yakimets[1,2] and Ilya I. Kurochkin[1]

[1] Institute for Information Transmission Problems
of Russian Academy of Sciences, Moscow, Russia
iakimets@mail.ru, kurochkin@iitp.ru
[2] The Russian Presidential Academy of National Economy
and Public Administration, Moscow, Russia

Abstract. Volunteer distributed computing (VDC) is a fairly popular way of conducting large scientific experiments. The organization of computational experiments on a certain subject implies the creation of a project of volunteer distributed computing. In this project, computing resources are provided by volunteers. The community of volunteers is about several million people around the world. To increase the computing power of the volunteer distributed computing project, technical methods for increasing the efficiency of computation can be used. However, no less important are methods of attracting new volunteers and motivating this virtual community to provide computing resources. The organizers of VDC projects, as a rule, are experts in applied fields, but not in the organization of volunteer distributed computing. To assist the organizers of the VDC projects authors conducted a sociological study to determine the motivation of volunteers, created a multiparameter method and rating for evaluating various VDC projects. This article proposes a method for assessing the strengths and weaknesses of VDC projects, based on the approach. The results of multiparameter evaluation and rating of projects can help the organizers of the VDC projects to increase the efficiency of computations, and the community of volunteers to provide a tool for comparing the various VDC projects.

Keywords: Volunteer distributed computing (VDC) · BOINC
The VDC project · Crunchers · Volunteers
Evaluation index of the VDC project
Characteristics for assessing the quality of VDC projects

1 Introduction

The use of distributed computing systems for high-performance computing is an alternative to calculations on supercomputers and other multiprocessor computer systems. Distributed computing systems or grid systems have a number of features, such as heterogeneity of computing nodes, their geographical distance, unstable network topology and high probability of disconnection of a computing node or communication channel. But even with such features, the computing potential of the grid system can be huge because of the large number (hundreds of thousands) of compute nodes. There are

© Springer Nature Switzerland AG 2019
V. Voevodin and S. Sobolev (Eds.): RuSCDays 2018, CCIS 965, pp. 472–486, 2019.
https://doi.org/10.1007/978-3-030-05807-4_40

software platforms for organizing distributed computing, such as HTCondor [1], Legion [2], BOINC [3]. At the moment, the most common platform for organizing distributed computing is BOINC (Berkeley Open Infrastructure for Network Computing) [4]. Public grid systems attract the computing power of volunteers. As a rule, the organizers of public grid systems are scientific or educational organizations. A public computing grid system is called a volunteer distributed computing project.

Volunteer distributed computing projects (VDC projects) are rather heterogeneous, both in terms of topics and in their organization. There are long-lived projects that remain active and successfully conduct experiments for almost 20 years (SETI@home, Folding@home, Einstein@home). There are projects that were created for one large computational experiment or scientific problem, after several years, they stop working (POGS@home, Poem@home). Some projects simultaneously run several independent computing experiments (SZTAKI, LHC@home). Developers of some projects can pay great attention to the design of the project site (POGS@home, Acoustics@home), and other projects are limited to using the standard template of the BOINC platform (Climate@Home, Optima@home [11]). Some projects are considered only on the CPU, others use CPU + GPU.

Interaction with volunteers in projects has general principles [13]:

- Attracting computing resources to help the scientific team;
- Publication of experimental results;
- Publication of scientific and popular scientific publications on the project website;
- Interaction with volunteers;
- Organization of competitions for volunteers within the project;
- Motivation of volunteers with virtual prizes and certificates.

However, for each VDC project there are specific features of interaction with the community of volunteers. Some project is popular, as it was organized by scientists with a global name (LHC@home). Another project has a clear and high goal - the search for a cure for cancer (Rosetta@home). If a media project has been told about a project, this can give an impetus to the project even in the medium term - attracting new volunteers for several months to 1 year. The participation of the project in virtual sports (Formula BOINC, BOINC Pentathlon) can also attract significant computational resources to the project.

The websites of a number of VDC projects based on BOINC platform [3] were studied as well as information about them on the site boinc.ru. The list of VDC projects includes:

- SETI@home – one of the first projects on the basis of which the BOINC platform was developed. This project of a group of scientists from Berkeley University for processing data from radio telescopes.
- Asteroids@home project is organized by the Astronomical Institute of Charles University, Czech Republic. Project aimed at determining the shape and parameters of the rotation of asteroids according to photometric observations In the process of project implementation, the public database DAMIT is filled.
- POGS@home project is organized by the International Centre for Radio Astronomy Research, Australia. This project aimed at building a multispectral atlas (from near

infrared to ultraviolet radiation), as well as at determining the rate of star formation, the stellar mass of galaxies, the distribution of dust and its mass in the galaxy and etc.;

- SAT@home project is organized by the Institute for System Dynamics and Control Theory of Siberian Branch of Russian Academy of Sciences, Russia. Project associated with searching for solutions to such complex problems as inversion of discrete functions, discrete optimization, bioinformatics, and etc., which can be effectively reduced to the problem of the feasibility of Boolean formulas [9];
- Rosetta@home project is organized by Institute for Protein Design, University of Washington, USA. Project associated with solving one of the biggest problems in molecular biology - the calculation of the tertiary structure of proteins from their amino acid sequences;
- MilkyWay@home project is organized by Rensselaer Polytechnic Institute, USA. This project studies the history of our galaxy by analyzing the stars in the Milky Way galaxy's Galactic Halo;
- Some other VDC projects – Folding@home [5], Einstein@home [6], LHC@home [7, 10], Gerasim@home [8] and etc.

It should be noted that in English and Russian literature a majority of publications and articles on Internet resources provide a description of the individual VDC projects [8–10]. Significantly less common are scientific papers which characterize organization of activities and the involvement of crunchers [11, 14]. And it is quite rare to see works which examine various aspects of citizen participation in VDC projects [12, 13]. One of the first attempts of more or less systematic study of the conditions and characteristics of crunchers participating in VDC projects in Russia, as well as their motivations and preferences has become a sociological study among 650 Russian crunchers [15].

2 Description of the Methodology for the Analysis of VDC Projects

The methodology for the index evaluation of VDC projects includes the following main stages:

1. Creating an index model;
2. Collect information to describe the most important elements of VDC projects.
3. Conducting surveys of participants of selected VDC projects to calculate the values of the YaK-index and estimates.
4. Perform calculations and visualize the results.

The YaK-index was developed to assess the quality of the VDC projects implementation, which was applied to a number of the above-mentioned projects in order to identify their "weak" sides, develop proposals to increase their efficiency, increase attractiveness for people interested in the VDC, and providing comparable information for the organizers of VDC projects.

2.1   Model of the YaK-Index of the VDC Projects

We introduce the notation:

$i = \overline{1,n}$ - the ordinal number of the important characteristics of the VDC project (hereinafter referred to as characteristics), it is assumed that n equal to 7–9, that is, from 7 to 9 estimated characteristics of each VDC project will be taken into account;

$s = \overline{1,S}$ - the ordinal number of the VDC project, $S = 34$ (for this paper);

R^s - YaK-index of the VDC project s;

x_i^s - availability of characteristic i of the VDC project s (0 – if not present; 1 – if available);

K^s - number of respondents for the VDC project s;

K - number of respondents for all VDC projects;

α_i^s - the mean weighting factor (significance) of characteristic i of the VDC project s from K^s respondents, $0 \leq \alpha_i^s < 1$, $\sum_{i=1}^{n} \alpha_i^s = 1$;

ρ_i^s - the mean experts assessment of the quality of i-th characteristic of s-th VDC project,

The scale values vary from −2 to 2. A linguistic interpretation of these values is given in the questionnaire. If necessary $\rho_i^s \in \{-2,-1,0,1,2\}$ is converted into the set $\rho_i^s \in \{1,2,3,4,5\}$.

Identically for all n characteristics it is mapped one-to one in a numerical set from the set of possible linguistic estimates. The maximum value of the numeric scale is m. In our case, it is assumed that $m = 5$. The index values vary from 1 to 5. But it is possible to normalize the index values so that they vary from 0 to 1.

There are two possibilities for calculating the value R^s:

1. When the weights of characteristics for a VDC project are individual and independent of what such are such weights for all other VDC projects.
2. When for all VDC projects the same vector of weights is defined.

In the first case, the index R^s is calculated as follows:

$$R_1^s = \frac{\sum_{i=1}^{n} \alpha_i^s \bullet x_i^s \bullet \rho_i^s}{n_1^s m}. \tag{1}$$

Here n_1^s is the number of characteristics for VDC project s, $n_1^s \leq n$.
In the second case

$$R_2^s = \frac{\sum_{i=1}^{n} \alpha_i \bullet x_i^s \bullet \rho_i^s}{n_1^s m}. \tag{2}$$

2.2   Collect Information for Describing the Most Important Elements of VDC Projects

For the survey of volunteers, a preliminary list of characteristics for project evaluation was created. At conferences and round tables with representatives of the volunteer community and experts, discussions were held on this list and some corrections were made. As a result, 9 characteristics were selected to assess the features of VDC projects [16]:

1. The clear concept and vision of the project;
2. Scientific component of the project;
3. The quality of scientific and scientific-popular publications on the topic of the project;
4. Design of the project (site, certificate, screensaver);
5. Informativity of materials on the project site;
6. Visualization of the project results (photo, video, infographic);
7. Organization of feedback (forums, chat rooms, etc.);
8. Stimulation of the cruncher participation in the project (competitions, scoring system, prizes);
9. Simplicity of joining the project (there are no barriers and organizational or technical difficulties).

The assessment of each characteristic was supposed to be 5-point. The questionnaire was initially focused on the Russian-speaking audience, so the following assessments were proposed (with the following linguistic interpretation):

1. "+2" is excellent;
2. "+1" is good.
3. "0" is normal;
4. "−1" - it is necessary to improve;
5. "−2" is bad.

This interpretation of the estimates allows you to get rid of the unnecessary connotation with school grades (from 1 to 5) and makes it possible to use the entire range of estimates.

Due to the heterogeneity of the projects, each characteristic has a different impact on the evaluation of the project. Therefore, it was decided to assess the significance of each characteristic for a particular project. Weight characteristics could vary from "0" - not significant, to "10" - very significant. In addition to evaluating the project, several questions were asked in the questionnaire on the degree of involvement of the respondent in this project: determining the respondent's status, determining the duration of the respondent's interaction with the project.

2.3   Conducting Surveys of Participants of VDC Projects

The questionnaire was implemented using Google Docs in Russian and English [17]. Information about the questionnaire was distributed by the administration of the site BOINC.ru on the profile forums and sites of the community of volunteers.

As a result, estimates were received from 259 respondents for 34 projects. However, for subsequent processing, only those projects for which more than 10 questionnaires were filled were taken (Fig. 1).

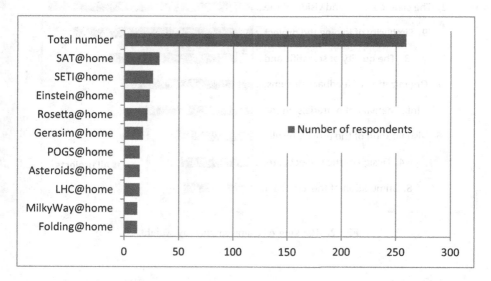

Fig. 1. Distribution of the number of respondents among projects.

The results were obtained in 2016–2017, therefore new projects such as Rake-Search, XANSONS for COD, etc., are not listed in the rating.

2.4   Visualization of Survey Results

To determine the averaged weights of each characteristic, questionnaires were used from all 259 respondents (for all projects). When comparing the average weight of a characteristic for individual projects, the values were not significantly different from the average values for individual projects. As an example, the project characteristics are ranked by weight according to the average values for all projects (Fig. 2).

For the visual presentation of the projects characteristics a radar diagram was used (Fig. 3). The characteristics are arranged in descending order of average weight.

The average estimations for all projects on the radar diagram (Fig. 4) are given, while the ranking of the characteristics in descending order of importance is preserved (Figs. 2 and 3).

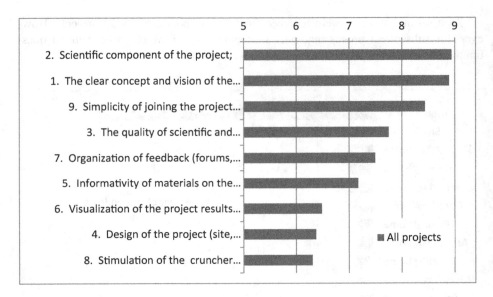

Fig. 2. Ranking of characteristics by weight.

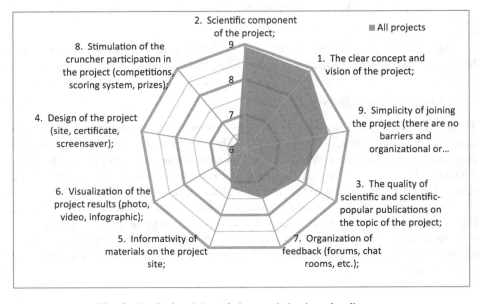

Fig. 3. Ranked weights of characteristics in radar diagram

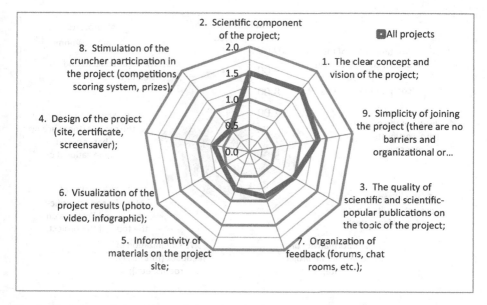

Fig. 4. Estimates of the characteristics of all projects, taking into account their ranking by weight

3   Results for Selected VDC Projects

Let's present estimates of large international VDC projects: Folding@home, Einstein@home, LHC@home and Russian VDC project Gerasim@home.

3.1   Folding@Home

Folding@home is a distributed computing project for disease research that simulates protein folding, computational drug design, and other types of molecular dynamics [10]. The project started in 2000 and is one of the most successful international projects that began to use grid systems from personal computers. The audience of the project is more than 1 million users and continues to grow. For several years, the organizers of Folding@home have maintained compatibility with the BOINC client. This allowed the volunteers, who use a BOINC client, to participate in the project. But, despite this, the community of volunteers is very different from the volunteer community of VDC projects on the BOINC platform. This can be seen from the various estimates of the significance of the characteristics (Fig. 5). The success and attractiveness of the project for volunteers can be seen at Fig. 6, Where project estimates for each characteristic are better than average values for all projects.

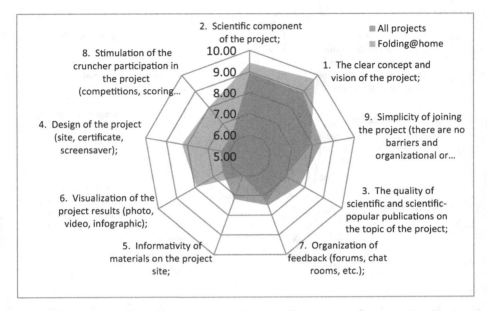

Fig. 5. Weights of characteristics for the project SETI@home

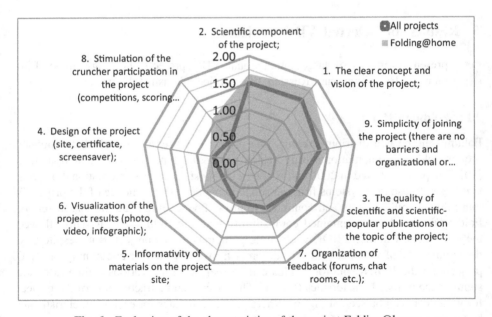

Fig. 6. Evaluation of the characteristics of the project Folding@home

3.2   Einstein@Home

Einstein@home project is supported by the American Physical Society, the US National Science Foundation, the Max Planck Society. Einstein@home organized to search for weak astrophysical signals from spinning neutron stars using data from the LIGO gravitational-wave detectors, the Arecibo radio telescope, and the Fermi gamma-ray satellite [11].

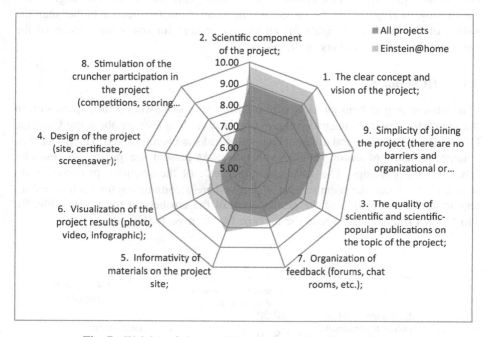

Fig. 7. Weights of characteristics for the project Einstein@home

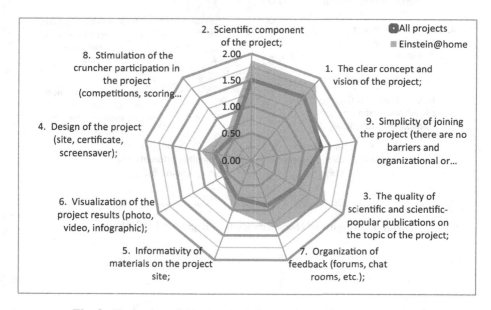

Fig. 8. Evaluation of the characteristics of the project Einstein@home

The project Einstein@home is one of the largest international VDC projects on the BOINC platform. The project is active now. About 500 thousand users were connected to the project. In fact, the Einstein@home project is a standard project, as the evaluation of the significance of the characteristics weights coincides with the average values for all projects (Fig. 7) and the project estimates by characteristics are higher than the values of the averaged weights for all projects except for one - stimulation of the participation of the crunchers in the project (Fig. 8).

3.3 LHC@Home

LHC@home project help physicists from CERN compare theory with experiment, in the search for new fundamental particles and answers to questions about the Universe [12]. In this project, several different computing tasks are being solved. The project has a large audience of about 160,000 volunteers. At present the project successfully functions and develops. The scale and importance of the scientific problems being solved even reduce the requirements of the volunteer community for such a characteristic as the stimulation of the participation of the crushers in the project (Fig. 9). Most of his estimates are above the average for all projects (Fig. 10).

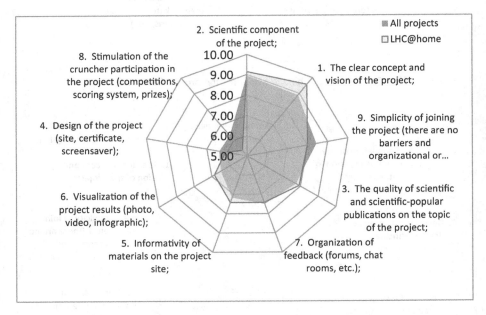

Fig. 9. Weights of characteristics for the project LHC@home

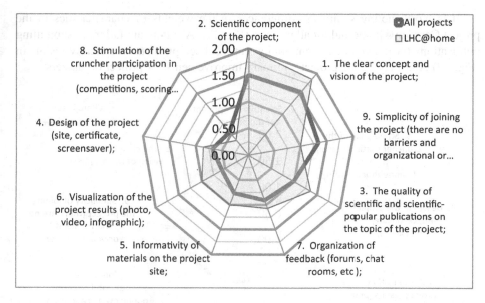

Fig. 10. Evaluation of the characteristics of the project LHC@home

3.4 Gerasim@Home

The Russian project Gerasim@home is notable for the fact that the server part of the project is implemented for the MS Windows operating system, but the standard client part of the BOINC platform is used [13]. Number of users about 4000.

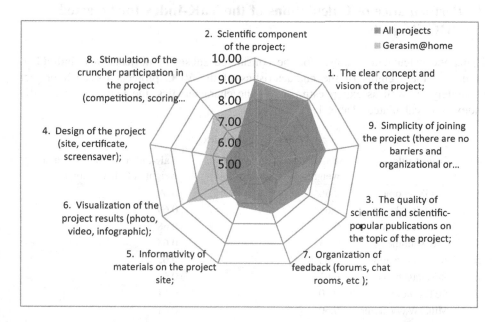

Fig. 11. Weights of characteristics for the project Gerasim@home

It is worth noting significant differences in the weights of characteristics in the project Gerasim@home and for all projects (Fig. 11). A significant failure in submitting publications for the volunteer community and feedback problems are clearly visible in (Fig. 12) And may be the project's primary improvements for project organizers.

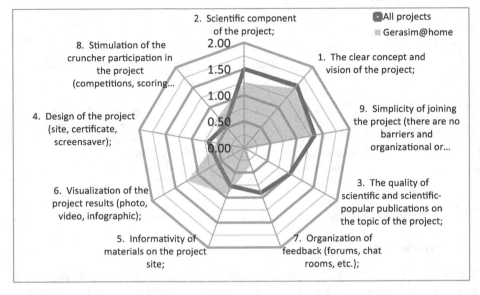

Fig. 12. Evaluation of the characteristics of the project Gerasim@home

4 Performance of Calculations of the YaK-Index for Selected VDC Projects

Using the calculated average (for the considered subset of projects) and individual values of the weights of the characteristics of the VDC projects for each project separately, as well as expert estimates of the characteristics, the values of the YaK-index were calculated (Table 1).

Table 1. YaK-index.

№	Project title	YaK-index (with average weights of characteristics)	YaK-index (with individual weights of characteristics)
1	POGS@home	0.66	0.69
2	Einstein@home	0.65	0.69
3	Folding@home	0.65	0.68
4	LHC@home	0.64	0.64
5	Asteroids@home	0.62	0.62
6	Rosetta@home	0.61	0.63
7	SETI@home	0.60	0.61
8	MilkyWay@home	0.60	0.61
9	Gerasim@home	0.59	0.61
10	SAT@home	0.58	0.57

Judging by the magnitude of the YaK-index values, all 10 VDC projects have a certain capabilities for development. So, in the case of using individual characteristics scales (right column), the highest values of the index (0.69) have both the Einstein@home projects, and the POGS@home project, and the lowest values has the SAT@home project. The teams of the first two projects, referring to the values of the characteristics weights, can determine which characteristics of their projects they should pay attention to in order to increase the values of the YaK-index.

For the project Einstein@home, judging by Figs. 7 and 8, the growth of the values of the YaK-index is possible when a visualization of the project results and a stimulation of the participation of the crunchers in the project are improved.

The lowest values of the YaK-index in both cases (right and second from the right columns of the table) were obtained by the Gerasim@home and SAT@home projects. This is just a bit over half of the maximum possible values of the index. It is clear that this project is "younger" than the rest ones. Nevertheless, by comparing the radar diagrams with weights and estimates of characteristics, we can recommend to the project teams to pay attention to the scientific component and description of the project's concept, to visualize the results and design the project site.

5 Conclusions

The proposed approach to assessing the state of VDC projects through a specially organized survey of project participants and calculation of the values of the developed YaK-index allows teams to obtain in visualized form comparable information on 9 significant characteristics. These materials can be an important tool for project teams to improve work and project management.

The estimated values for a number of well-known VDC projects show the adequacy of the proposed approach, clearly illustrate the advantages of well-organized projects with a solid operating experience, and create certain conditions for monitoring the process of formation and development of newly created and "young" VDC projects.

Acknowledgements. This work was funded by Russian Science Foundation (№16-11-10352).

References

1. Litzkow, M.J., Livny, M., Mutka, M.W.: Condor-a hunter of idle workstations. In: Distributed Computing Systems. IEEE (1988)
2. Grimshaw, A.S., Wulf, W.A.: The Legion vision of a worldwide virtual computer. Commun. ACM **40**(1), 39–45 (1997)
3. Anderson, D.P.: BOINC: a system for public-resource computing and storage. In: Grid Computing. IEEE (2004)
4. The server of statistics of volunteer distributed computing projects on the BOINC platform. http://boincstats.com
5. Homepage of the Folding@home project. https://foldingathome.org/. Accessed 30 Apr 2018
6. Homepage of the Einstein@home project. https://einsteinathome.org/ru/. Accessed 30 Apr 2018

7. Homepage of the LHC@home project. http://lhcathome.web.cern.ch/. Accessed 30 Apr 2018
8. Vatutin, E.I., Titov, V.S.: Volunteer distributed computing for solving discrete combinatorial optimization problems using Gerasim@home project. In: Distributed Computing and Grid-Technologies in Science and Education: Book of Abstracts of the 6th International Conference. JINR, Dubna (2014)
9. Posypkin, M., Semenov, A., Zaikin, O.: Using BOINC desktop grid to solve large scale SAT problems. Comput. Sci. **13**(1), 25 (2012)
10. Lombraña González, D., et al.: LHC@Home: a volunteer computing system for massive numerical simulations of beam dynamics and high energy physics events. In: Conference Proceedings, vol. 1205201, no. IPAC-2012-MOPPD061, pp. 505–507 (2012)
11. Zaikin, O.S., Posypkin, M.A., Semenov, A.A., Khrapov, N.P.: Experience in organizing volunteer computing: a case study of the OPTIMA@home and SAT@home projects. Vestnik of Lobachevsky State University of Nizhni Novgorod, no. 5–2, pp. 340–347 (2012)
12. Tishchenko, V.I., Prochko, A.L.: Russian participants in BOINC-based volunteer computing projects. The activity statistics. Comput. Res. Model. **7**(3), 727–734 (2015)
13. Clary, E.G., et al.: Understanding and assessing the motivations of volunteers: a functional approach. J. Pers. Soc. Psychol. **74**(6), 1516 (1998)
14. Webpage of World Community Grid project. 2013 Member Study: Findings and Next Steps. https://www.worldcommunitygrid.org/about_us/viewNewsArticle.do?articleId=323. Accessed 30 Jan 2018
15. Yakimets, V.N., Kurochkin, I.I..: Voluntary distributed computing in Russia: a sociological analysis. In: Collection of scientific articles of the XVIII Joint Conference "Internet and Contemporary Society" (IMS-2015), St. Petersburg, 23 June 2015, pp. 345–352. ITMO University, St. Petersburg (2015). ISBN 978-5-7577-0502-6
16. Kurochkin, I.I., Yakimets, V.N.: Evaluation of the voluntary distributed computing project SAT@home. National Supercomputer Forum (NSCF-2016), Pereslavl-Zalessky, Russia, 29 November–2 December 2016
17. Yakimets, V.N., Kurochkin, I.I.: Multiparameter and index evaluation of voluntary distributed computing projects. In: Alexandrov, D., Boukhanovsky, A., Chugunov, A., Kabanov, Y., Koltsova, O. (eds.) Digital Transformation and Global Society. DTGS 2018. CCIS, vol 858, pp. 528–542. Springer, Cham (2018). https://doi.org/10.1007/978-3-030-02843-5_44

Application of the LLVM Compiler Infrastructure to the Program Analysis in SAPFOR

Nikita Kataev$^{(\boxtimes)}$ ⓘ

Keldysh Institute of Applied Mathematics RAS, Moscow, Russia
kaniandr@gmail.com

Abstract. The paper proposes an approach to implementation of program analysis in SAPFOR (System FOR Automated Parallelization). This is a software development suit that is focused on cost reduction of manual program parallelization. It was primarily designed to perform source-to-source transformation of a sequential program for execution on parallel architectures with distributed memory. LLVM (Low Level Virtual Machine) compiler infrastructure is used to examine a program. This paper focuses on establishing a correspondence between the properties of the program in the programming language and the properties of its low-level representation.

Keywords: Program analysis · Program parallelization
Source-to-source transformation · LLVM

1 Introduction

The main applications of program analysis are program optimization and program correctness. Program optimization may require a significant transformation of the source code of a program. In this case, it is rather difficult to estimate the quality of the generated code for a particular program in advance. This leads to the fact that compilers are forced to apply fixed sequence of optimizations to all programs, which does not always produce the desired results. Compiling for execution on parallel architectures (multiprocessors, accelerators and distributed memory systems) drastically complicates the situation. The auto-parallelization feature of modern compilers may have the opposite effect and lead to a significant slowdown of the program.

User-guided program transformation that relies on recommendation of some interactive tools is paramount to simplify the mapping of sequential programs to parallel architectures. This approach should be considered as one of the key factors in the development of SAPFOR (System FOR Automated Parallelization) [1]. Unfortunately, we did not managed to find a compiler infrastructure which supports the development of source-to-source transformation passes, provides detailed information about high-level program items (alias analysis,

© Springer Nature Switzerland AG 2019
V. Voevodin and S. Sobolev (Eds.): RuSCDays 2018, CCIS 965, pp. 487–499, 2019.
https://doi.org/10.1007/978-3-030-05807-4_41

data dependency analysis, reduction and induction variables recognition, privatization) and allows us to compile large applications in C and Fortran. This paper is devoted to the use of the capabilities of the LLVM (Low Level Virtual Machine) [2] compiler infrastructure for program analysis in SAPFOR. The considered questions involve the interpretation of information derived from the LLVM intermediate representation (LLVM IR) and its relation to the items of the higher level language. In addition, the possibilities of analysis of the transformed LLVM IR are explored to improve the quality of the source program analysis.

The rest of the paper is organized as follows. Section 2 advocates the necessity of a new compiler architecture for the SAPFOR and determines the directions for the future enhancement of the system. Section 3 discusses open-source compiler infrastructures that most closely match our goals. Section 4 focuses on high-level representation of accessed memory locations based on programming language items. Section 5 presents sequence of LLVMs analysis and transform passes that SAPFOR's analysis uses. Section 6 discusses application of analysis techniques implemented in SAPFOR to explore the C version of the NAS Parallel Benchmarks [3]. Section 7 presents the conclusion and future work.

2   Motivation

SAPFOR (System For Automate Parallelization) [1] is a software development suit that is focused on cost reduction of manual program parallelization. It is developed in Keldysh Institute of Applied Mathematics, Russian Academy of Sciences, with the active participation of graduate students and students of Faculty of Computational Mathematics and Cybernetics of Lomonosov Moscow State University. SAPFOR can be used to produce a parallel version of a program in a semi-automatic way according to DVMH [4,5] model of the parallel programming for heterogeneous computational clusters.

SAPFOR was primarily designed to support Fortran. It was successfully applied to simplify parallelization of different applications including the NAS Parallel Benchmarks [6], programs designed for solving hydrodynamic and geophysics problems and applications from the field of laser material processing [1,7–9].

The system is written in C/C++ and operates with a representation of the program based on abstract syntax tree used in Sage++ [10]. Sage++ is an object-oriented compiler preprocessor toolkit aimed to perform source-to-source program transformation. However, it has not been developed for a long time and the responsibility for the support of modern programming languages lies on the developers of SAPFOR. At the moment, Fortran 95 is only supported.

The system implements static and dynamic analysis techniques and relies on automatic parallelizing compiler. The system is helpful to explore the information structure of programs and to perform automatic parallelization of well-formed sequential programs. This implies that the user must prepare the program himself for parallelization, guided by the hints and results of the program analysis provided to him by SAPFOR. Thus, implicitly parallel programming is applied to the development of parallel programs.

It is important to note that a significant mutation of the program may be required. Sometimes a programmer needs to choose a more parallel but not necessarily completely equivalent algorithm. However, a lot of transformations are essential to reveal hidden parallelism in a potentially parallel code. The work [12] shows the program transformation impact on a parallelization of an application designed for laser material processing [11]. To obtain the implicitly parallel version of a program written in C99, it took about 35 simple transform passes such as variable propagation, loop-invariant code motion, loop unrolling, loop distribution and other. We should clarify that a one-to-one correspondence can be established between the operators of the source and the transformed programs and these two programs are completely equivalent.

These results were collected when we explored different approaches to automate program transformation in SAPFOR. The approach proposed in [12] involves an automatic execution of individual passes in the order specified by the user. Another approach considered in [7] supposes an automatic selection of transform passes depending on the hints identified by the SAPFOR compiler. These hints describe problems which hinder the parallelization.

The desire to reduce the cost of manual parallelization of large-scale computational applications written in C and Fortran encounters problems described in [13]. This dictated the necessity of major improvement of SAPFOR in the following directions: (a) incremental parallelization for heterogeneous clusters [14], (b) automation of program transformations, (c) improving the quality of static and dynamic analysis of programs, (d) the ability to utilize program profiling and code coverage information to determine hot-spots, (e) support for C language.

This, in turn, led to investigation of available compiler infrastructures that have ample opportunities to enable the development of SAPFOR in these areas.

3   Related Works

3.1   Cetus

The Cetus compiler infrastructure [15] is designed for the source-to-source transformation of programs and is being developed at Purdue University since 2004. The system is written in Java and at the moment Cetus can parse programs that follow ANSI C89/ISO C90 standard. ANSI/ISO C99 standard is not fully supported. For example, Cetus parser breaks when it sees a variable declaration within the for statement header.

Cetus is initially designed to support interprocedural analysis across multiple files. This gives it an advantage over standard compilers such as GCC, which compile one source file at a time. Cetus implements the basic types of analysis (data dependence analysis, induction variable recognition and substitution, reduction variable recognition, privatization, points-to analysis, alias analysis) which are necessary for program parallelization. In some situations, Cetus makes too conservative assumptions. For example, the presence of 'goto' statement in loops hinders their analysis. A significant drawback is that only the addition operation is supported for reduction variables.

Cetus does not support standard compiler options, which limits the use of build automation tools to organize code analysis and compilation. Therefore investigation of large software is not straightforward.

Cetus does not support Fortran which is one of the SAPFOR target languages. Further SAPFOR is developed in C/C++ languages and the use of the component written in Java will cause additional difficulties in their integration. The Cetus system is being developed by a small team. Updates do not come out often. The latest version was released in February 2017.

3.2 ROSE

ROSE [16] compiler infrastructure is developed at Lawrence Livermore National Laboratory (LLNL). ROSE is an open source project to build source-to-source program transformation and analysis tools. The main supported languages are C(89/98), C++(98/11), Fortran (77/95/2003), Java, Python, Haskell. The main supported platform in Linux. The project implements a large number of analyses. A software engineering environment is provided for new tools developers.

We have used ROSE to implement a tool for semi-automatic transformation of programs written in C [12]. The transform request is specified in the form of directives placed in a source code. The tool checks preconditions and applies specified transformations.

The experience of using ROSE shows the presence of implementation issues in some libraries. A large size of the project (several million lines of code) makes it difficult to correct them by ourselves. At the same time, the authors of the system give the main preference to the binary analysis subsystem, therefore it is not known when errors may be corrected.

Testing of the system on the NAS Parallel Benchmarks (NPB) [3,6] reveals that ROSE is not capable to process some programs (LU, FT). We attempt to take input source files, build AST (Abstract Syntax Tree), and then unparse the AST back to compilable source code. However, unhandled runtime errors occur.

3.3 OPS

OPS (Optimizing Parallelizing System) [17] is a program tool oriented for development of (a) parallelizing compilers, parallel language optimizing compilers, semi-automatic parallelizing systems; (b) electronic circuits computer-aided design systems; (c) systems of automatic design of hardware based on FPGA.

The main supported language is C, there is a limited support for Fortran. The main research direction of the OPS developers is the automatic mapping of sequential programs to systems with FPGA. Another direction is the user-guided program transformation and the search for new optimizing transformations.

OPS uses Clang [18] to parse programs and build its internal representation. The higher level of the internal representation distinguishes OPS from GCC and LLVM. High-level representation is more convenient to create interactive mode of program optimization into the compiler. Unlike Clang, it allows to perform transformations directly. The system is compatible with Clang 3.3, more recent versions are not supported yet.

Data dependence analysis, reduction variable recognition, privatization and alias analysis are implemented. Reduction variable recognition is only performed for fairly simple patterns. Calculations of maximum and minimum values are not allowed for reduction variables. Dependency analysis assumes that each array subscript must be a combination of surrounding loop indices. Variable substitution solves this problem, but the resulting program may differ significantly from the original version. However, the explicit execution of this transformation in many cases is not required for parallel execution of the program. A large number of different loop transformations are involved, but a source loop must be represented in a canonical form.

3.4   LLVM

LLVM is one of the most robust compiler infrastructures available to the research community. LLVM generates highly-optimized code for a variety of architectures. Its open-source distribution and continuous updates make it attractive. LLVM consists of a number of subprojects, many of them are being used in production. LLVM operates on its own low-level code representation known as the LLVM intermediate representation (LLVM IR). Unlike GCC, LLVM provides a friendly API for designing analysis and transform passes. LLVM is currently written using C++ 11 conforming code. The LLVM libraries are well documented.

One of the most important LLVM features is language-independent type system that can be used to implement data types and operations from high-level languages. LLVM IR does not represent high-level language features directly. Nevertheless, in general it includes enough meta information to utilize analysis results to evaluate a program in a higher level language.

The release in 2017 of the Fortran language front-end Flang [19] as well as the opportunity to obtain directly the LLVM IR for Fortran became an additional motivation for using LLVM to analyze programs in SAPFOR.

4   Representation of Analysis Results

The main interest for us is to describe the memory accesses in the program. It includes data dependence analysis, induction and reduction variable recognition, privatization. Moreover it is necessary to determine whether or not two pointers ever can point to the same object in the memory (alias analysis). All information provided by SAPFOR must be attached to the items of the source program. Lower level of LLVM IR does not directly allow LLVM to be applicable for the description of analysis results.

To achieve this goal, a novel structure is presented. We call it a source-level alias tree. It depicts the structure of accessed memory using source-level debug information. Set of memory locations forms a node of the tree. A memory location specifies a correspondence between the set of IR-level memory locations and some item in higher level programming language.

Each memory location is identified by the address of the start of the location and its size. Note, that sometimes its size can only be known at runtime. LLVM is a load/store architecture. It means that programs transfer values between registers and memory solely via load and store operations using typed pointers [2]. So that, there are no implicit accesses to the memory. Hence, an address can be specified with a sequence of LLVM instructions. At the source-level the distinct elements of arrays are collapsed into one object, if it can not be expressed as a base pointer plus a constant offset. However, members of a structure are distinguished. A special type of memory locations is introduced to summarize the unknown memory accesses when a function is called.

For example, consider directly accesses to a member S.X of a structure S. At the source level it will be represented as a single memory location. From the LLVM point of view the address of the start of the memory location will be represented as the result of the allocation instruction (for example, 'alloca') and as the instruction that calculates the address of a subelement of an aggregate data structure (for example, 'getelementptr'). Different accesses may produce different sequences of instructions. In case of P->X, where P points to S new memory location will be constructed. Regardless the pointer P can refer to different structures at different points in the program, all P->X will be represented by one memory location.

Two memory locations fall into a single node of a source-level alias tree, if they may alias. The pairwise alias analysis information provided by LLVM alias analysis passes is inspected to disambiguate memory references. Each memory location is established only once in the tree and can only refer to one node. A node may have a set of child nodes. The partitioning is done in such a way that the union of all the memory locations from the parent nodes covers the union of the memory locations from a child.

The ability to adjust its structure across the transformation of LLVM IR is a distinct advantage of the source-level alias tree. Figure 1 (a) presents an alias tree fragment for the LLVM IR directly constructed by Clang for the function shown in Listing 1.1. Memory locations which are the dereference of different pointers fall into a common node, and thus may-alias relation is conservatively assumed. This relation is caused not only by mapping of separate low-level memory locations to the item of the source program. This confirmed by the investigation of the information produced by LLVM about alias sets that are active in the function foo.

Figure 1(b) contains the alias tree fragment constructed after memory references have been promoted to be register references. Accesses to structure members are now performed directly through IR-values associated with pointers A and B. Despite that source-level alias-tree also provides information about the items before mutation. The relation between P and IR-values associated with A and B is established so that it disambiguates different accesses to P in a source program. This information will be used to refine results of analyses. Transformed

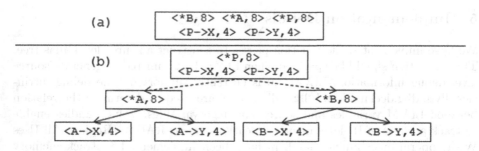

Fig. 1. Fragment of a source-level alias tree before (a) and after (b) transformation of LLVM IR for a function in Listing 1.1.

locations are stored in the alias tree node of a special kind. Dotted lines indicate that the union of memory locations from a given node is not required to cover the union of memory locations from descendant nodes. May-alias relation is assumed instead.

Listing 1.1. Source code for a source-level alias tree in Fig. 1.

```
struct S {float X;  float Y;};
void foo(struct S * restrict A, struct S * restrict B) {
    struct S *P;
    P = A;
    P->X = 0;
    P->Y = 0;
    P = B;
    P->X = 0;
    P->Y = 0;
}
```

The capability to consider a memory location in conjunction with related high-level items is another important feature of the source-level alias tree. All analyses referred to at the beginning of this section are initially performed for memory locations which are explicitly mentioned in a source code. After that the obtained results are propagated to surrounded locations. Suppose A->X is accessed in a loop and is recognized as a private variable. If at the same time A->Y is recognized as a shared variable then the entire structure should be specified as a first private. Alias tree reflects relation between structure and its members and simplifies this investigation.

Another situation arises if several memory locations are explicitly mentioned in a loop and attached to a single node of the alias tree (or there is a path between them in the tree). If these locations correspond to different items in the original program, then a data dependence should be assumed. However, there is no dependency in another loop if only one of these locations is explicitly accessed.

5   Implementation Details

We have implemented the function passes to construct a source-level alias tree. The '-g' option should be specified when LLVM IR is emitted to generate source level debug information. LLVM 4.0 is currently supported. The debug intrinsics 'llvm.dbg.declarae' and 'llvm.dbg.value' are used to determine the relation between LLVM variables and source language variables. Value handles enable to track address of IR-level memory locations across RAUW (Replace All Uses With) operations. Similar handlers have been implemented to track memory locations across rebuilding of the alias tree after IR transformation.

LLVM implements reduction and induction variable recognition technics which are based on the scalar evolution pass. Dependence analysis pass is applied to detect dependences between memory accesses. As noted in the LLVM documentation, it implements the tests described in [20]. Actually these tests satisfy SAPFOR analysis goals. If necessary, the set of tests may be extended in the future. In addition, for scalar privatization a function pass has been written. This pass implements the approach described in [21]. We extended the mentioned approach to enable SAPFOR to analyze statements that do pointer accesses.

The static analysis in SAPFOR uses a sequence of passes discussed below. We divide this sequence into several steps. The first step consists of simple transform passes which simplify the analysis, but they do not require special efforts to maintain the correspondence between LLVM IR and higher level program items. These transformations available in LLVM are implemented by the following passes: unreachable code elimination, removal of declarations of unused functions, elimination of unreachable internal globals, propagation of function attributes and other passes. The next step builds the source-level alias tree and performs variable privatization. Subsequent transform passes may destroy some variables and debugging information. To avoid loss of source-level data dependencies, we perform the preliminary analysis as mentioned above.

After that, the SROA (Scalar Replacement of Aggregates) pass is executed. It breaks up 'alloca' instructions of aggregate type (structure or array) into individual alloca instructions for each member, if possible. Then, if possible, it transforms the individual 'alloca' instructions into SSA (Static Single Assignment) form. Debugging information is modified in accordance with the transformation performed and it allows SAPFOR to determine which variables in the source code correspond to registers.

At the following step a previously constructed source-level alias tree is updated. Then, induction and reduction variable recognition is performed, privatization pass is re-executed and appropriate information is updated. Promotion of memory references realized at the previous step simplifies expressions. Hence, array subscript becomes a combination of surrounding loop indices in many cases. So that data dependencies are discovered and classified.

The next step performs loop rotation transformation. Otherwise some reductions will not be recognized due to the features of the for-loop representation. Finally, reduction recognition is repeated, and previously obtained information is updated. Analysis of other types of dependencies is not repeated, since after the loop rotation, the results of the corresponding analysis may not correspond to the properties of the original program.

6   Evaluation

The implemented analysis techniques were examined on the C versions of the NAS Parallel Benchmarks (NPB) [3]. Each benchmark consists of several files and can be investigated in two modes: (a) file-by-file analysis (b) analysis of preliminary merged files. The last one uses the ability of Clang to merge together several ASTs in order to subsequently generate a single LLVM IR for all files. We have increased the applicability of the merge action to maintain large applications, we have improved the readability of diagnostic messages, and we also have eliminated some implementation errors.

Analysis of preliminary merged files is preferable to the file-by-file analysis for several reasons (a) the possibility of interprocedural analysis (b) the possibility of applying source code transformations affecting the entire project (for example, inline expansion) (c) more detailed debugging information is available. As mentioned above, it is necessary to use debugging information to present analysis results. However, Clang does not generate metadata in some cases, for example, external declarations are ignored. The merge action solves this problem.

The time of the analysis of the merged files, as well as the size of each benchmark in the number of files and lines of code are given in Table 1. The correctness of the merge action was verified by the compilation and execution of the emitted LLVM IR. The analyzer of SAPFOR enables us to use build automation tools, such as Make. Thus the original Makefiles have been easily updated. We have replaced the compilation command with the AST generation command and the linker command with the merge and analysis command.

Table 1. The analysis time (s) of the NAS Parallel Benchmarks (NPB)

Benchmark	BT	CG	DC	EP	FT	IS	LU	MG	SP	UA
Files	20	7	13	6	12	5	23	6	22	16
Lines	4198	1331	3202	587	1333	977	4210	1640	3550	8015
Time (s)	6.59	0.12	N/A	0.05	N/A	0.03	5.98	0.58	4.21	N/A

In some cases, merging is not possible without modifying the original versions of the programs, mainly due to the conflicts between similar names with internal linkage. These benchmarks were not analyzed, and in the time row N/A is set.

Statistics of the analysis results are presented in the Table 2. It includes the total number of loops, the number of loops containing accesses to arrays, and function calls. Reduction, induction and privatizable variables are not considered as loop-carried dependencies in this statistic and they are presented separately.

Table 2. The analysis statistic for the NAS Parallel Benchmarks (NPB)

Benchmark	Number of loops										
	Total	Array	Call	Indep.	Dep.			Priv.		Ind.	Red.
					Total	Call	Only	Total	Indep.		
BT	181	171	51	101	80	50	16	109	65	179	0
CG	47	14	6	17	30	6	2	15	1	46	8
EP	9	6	5	3	6	5	2	3	0	9	2
IS	12	11	3	4	8	3	2	3	0	12	0
LU	187	156	40	96	91	39	27	84	39	171	3
MG	81	37	22	12	69	19	14	38	2	77	1
SP	250	243	48	158	92	47	16	145	87	248	0
Total	767	638	175	391	376	169	79	407	194	742	14

Loops with data dependencies averaged 49% (Dep.) of the total number of loops. Among all loops with dependencies, approximately 45% contain function calls (Dep./Call), but only in 20% of cases the dependency is caused only by the presence of a call (Dep./Only). The analysis is hampered by a large number of global data and inability to determine that parameters of a function do not alias. In general, from the total number of loops only 10% of loops contain dependencies caused only by the function calls. This suggests that the limitation for interprocedural analysis did not have a significant impact on the accuracy of the results obtained. Sometimes, the presence of dependencies is caused by the use of indirect array accesses. But in general, a large number of dependencies is more likely to indicate the need to program transformation for their parallel execution rather than the limitations for analysis capabilities. This confirms that the use of advanced analysis techniques is necessary, but it is not sufficient for program parallelization.

It is important to note that variable privatization is necessary for the parallel execution of about the half of the loops. Moreover, this is also true for loops without data dependencies. To preserve the semantics of the original program, privatizable variables should be classified, first and last private variables should be identified. This advocates the implementation of a separate pass responsible for such kind of analysis. The built-in LLVM dependency analysis at best recognizes output dependency for the privatizable scalar variables. In fact, it can be successfully applied only after memory references have been promoted to be register references. Therefore, it is not applicable for the references that have been promoted. Although, for the compiler, this approach is acceptable, but in the case of source-to-source transformation and parallelization, information should be available on all variables presented in a source program.

7   Conclusion

This paper is devoted to the use of the capabilities of the LLVM compiler infrastructure for program analysis in SAPFOR. Low-level representation is used to obtain information about the original program. Investigation of transformed LLVM IR improves the quality of the source program analysis. We present a structure called the source-level alias tree to restore original program properties after transformation. It depicts the structure of accessed memory using source-level debug information.

Source-level alias tree (a) summarizes IR-level memory locations to higher level items, (b) corresponds to a hierarchical type system of a higher level language, (c) does not directly depend on a programming language and front-end, because it uses metadata, rather than abstract syntax tree, (d) adjust its structure across the transformation of LLVM IR which does not affect the structure of the memory used in the original program, (e) provides investigation of a memory location in conjunction with alias high-level items.

The implemented analysis techniques were examined on the C versions of the NAS Parallel Benchmarks.

Future works involve the usage of scalar evolution analysis in conjunction with source-level alias tree to implement array privatization techniques. We also focused on application of LLVM based analysis to check the correctness of source-level transformations. In our opinion, source-level transformations are vital to provide program parallelization in interaction with a programmer. Implementation of dynamic analysis in order to improve the alias analysis results and the accurateness of alias tree construction is also one of our future goals.

We plan to use the proposed approach in combination with Flang to analyze Fortran programs.

Acknowledgement. The reported study was funded by the Program of the Presidium of RAS 26 "Fundamental basis for creating algorithms and software for perspective ultrahigh-performance computing".

References

1. Klinov, M.S., Krukov, V.A.: Automatic parallelization of fortran programs. Mapping to cluster. Vestnik of Lobachevsky University of Nizhni Novgorod, no. 2, pp. 128–134. Nizhni Novgorod State University Press, Nizhni Novgorod (2009). (in Russian)
2. Lattner, C., Adve, V.: LLVM: a compilation framework for lifelong program analysis & transformation. In: Proceedings of the 2004 International Symposium on Code Generation and Optimization (CGO 2004), Palo Alto, California (2004)
3. Seo, S., Jo, G., Lee, J.: Performance characterization of the nas parallel benchmarks in OpenCL. In: 2011 IEEE International Symposium on Workload Characterization (IISWC), pp. 137–148 (2011)
4. Konovalov, N.A., Krukov, V.A., Mikhajlov, S.N., Pogrebtsov, A.A.: Fortan DVM: a language for portable parallel program development. Program. Comput. Softw. **21**(1), 35–38 (1995)

5. Bakhtin, V.A., et al.: Extension of the DVM-model of parallel programming for clusters with heterogeneous nodes. Bulletin of South Ural State University. Series: Mathematical Modeling, Programming & Computer Software, vol. 18(277), nol. 12, pp. 82–92. Publishing of the South Ural State University, Chelyabinsk (2012). (in Russian)
6. NAS Parallel Benchmarks. https://www.nas.nasa.gov/publications/npb.html. Accessed 14 Apr 2018
7. Kataev, N.A., Bulanov, A.A.: Automated transformation of Fortran programs essential for their efficient parallelization through SAPFOR system. In: Parallel Computational Technologies (PCT 2015): Proceedings of the International Scientific Conference, Ekaterinburg, Russia, 30th March–3rd April 2015, pp. 172–177. Chelyabinsk, Publishing of the South Ural State University (2015). (in Russian)
8. Kataev, N., Kolganov, A., Titov, P.: Automated parallelization of a simulation method of elastic wave propagation in media with complex 3D geometry surface on high-performance heterogeneous clusters. In: Malyshkin, V. (ed.) PaCT 2017. LNCS, vol. 10421, pp. 32–41. Springer, Cham (2017). https://doi.org/10.1007/978-3-319-62932-2_3
9. Bakhtin, V.A., Kataev, N.A., Klinov, M.S., Krukov, V.A., Podderugina, N.V., Pritula, M.N.: Automatic parallelization of Fortran programs to a cluster with graphic accelerators. In: Parallel Computational Technologies (PCT 2012): Proceedings of the International Scientific Conference, Novosibirsk, Russia, 26 March–30 2012, pp. 373–379. Publishing of the South Ural State University, Chelyabinsk (2012). (in Russian)
10. pC++/Sage++. http://www.extreme.indiana.edu/sage/. Accessed 14 Apr 2018
11. Niziev, V.G., Koldoba, A.V., Mirzade, F.H., Panchenko, V.Y., Poveschenko, Y.A., Popov, M.V.: Numerical modeling of melting process of two-component powders in laser agglomeration. Math. Model. **23**(4), 90–102 (2011). (in Russian)
12. Baranov, M.S., Ivanov, D.I., Kataev, N.A., Smirnov, A.A.: Automated parallelization of sequential C-programs on the example of two applications from the field of laser material processing. In: CEUR Workshop Proceedings 1st Russian Conference on Supercomputing Days 2015, vol. 1482, p. 536 (2015)
13. Armstrong, B., Eigenmann, R.: Challenges in the automatic parallelization of large-scale computational applications. In: Proceedings of SPIE 4528, Commercial Applications for High-Performance Computing, 27 July 2001, p. 50 (2001). https://doi.org/10.1117/12.434876
14. Bakhtin, V.A., et al.: Automation of software packages parallelization. In: Scientific Service on the Internet. Proceedings of the International Scientific Conference, Novorossiysk, 19th September–24th 2016, pp. 76–85. Keldysh Institute of Applied Mathematics RAS, Moscow (2016)
15. Lee, S.-I., Johnson, T.A., Eigenmann, R.: Cetus – an extensible compiler infrastructure for source-to-source transformation. In: Rauchwerger, L. (ed.) LCPC 2003. LNCS, vol. 2958, pp. 539–553. Springer, Heidelberg (2004). https://doi.org/10.1007/978-3-540-24644-2_35
16. ROSE compiler infrastructure. http://rosecompiler.org/. Accessed 14 Apr 2018
17. Optimizing parallelizing system. http://ops.rsu.ru/en/about.shtml. Accessed 14 Apr 2018
18. Clang: a C language family frontend for LLVM. https://clang.llvm.org/. Accessed 14 Apr 2018
19. GitHub - flang-compiler/flang. https://github.com/flang-compiler/flang. Accessed 14 Apr 2018

20. Goff, G., Kennedy, K., Tseng, C.-W.: Practical dependence testing. In: Proceedings of the ACM SIGPLAN 1991 Conference on Programming Language Design and Implementation (PLDI 1991), pp. 15–29. ACM, New York (1991)
21. Tu, P., Padua, D.: Automatic array privatization. In: Pande, S., Agrawal, D.P. (eds.) Compiler Optimizations for Scalable Parallel Systems. LNCS, vol. 1808, pp. 247–281. Springer, Heidelberg (2001). https://doi.org/10.1007/3-540-45403-9_8

Batch of Tasks Completion Time Estimation in a Desktop Grid

Evgeny Ivashko[1,2]([✉]) [iD] and Valentina Litovchenko[1,2]

[1] Institute of Applied Mathematical Research, Karelian Research Centre of Russian
Academy of Sciences, Petrozavodsk, Russia
ivashko@krc.karelia.ru
[2] Petrozavodsk State University, Petrozavodsk, Russia
va.lentina97@yandex.ru

Abstract. This paper describes a statistical approach used to estimate batch of tasks completion time in a Desktop Grid. The statistical approach based on Holt model is presented. The results of numerical experiments based on statistics of RakeSearch and LHC@home volunteer computing projects are given.

Keywords: Desktop Grid · BOINC · High-throughput computing
Holt model · Confidence interval · Completion time estimation

1 Introduction and Related Works

Along with computing clusters and Grid systems, Desktop Grid systems keep their valuable place in high-performance computing infrastructure. Desktop Grid is a form of distributed high-throughput computing system which uses idle time of non-dedicated geographically distributed computing nodes connected over a low-speed (Internet or LAN) network. In general, Desktop Grids have a server-client architecture. Such the systems have huge computing potential exceeding 1 ExaFLOPS [20].

Desktop Grids have a number of peculiarities comparing to other computational systems:

- slow data transfer between server and a node (comparing to computing clusters and Grid systems); in general, no direct data transfer between nodes is allowed;
- limited computational capacity of separate nodes;
- high heterogeneity of software, hardware, or both;
- no information on a node state or completion level of a task is available;
- low reliability of computing nodes;
- dynamic set of nodes;
- lack of trust to computing nodes.

V. Voevodin and S. Sobolev (Eds.): RuSCDays 2018, CCIS 965, pp. 500–510, 2019.
https://doi.org/10.1007/978-3-030-05807-4_42

Because of these peculiarities Desktop Grids are significantly differ from computing clusters, Grid systems or cloud computing systems. So, it is necessary to develop special algorithms aimed at solving specific problems of Desktop Grid management. One of these problems is batch of tasks completion time estimation. Many of Desktop Grid computational projects perform separate computational experiments, where each of the experiments consists of a number of tasks (batch of tasks).

Batch of tasks completion time estimation is a complicated and many-sided problem, caused by the intermittent resource availability. This problem relates to a single task or a batch of tasks completion time estimation problem as a part of a more complex scheduling problem.

For example, the paper [5] describes an approach to scheduling thousands of jobs with diverse requirements to heterogeneous grid resources, which include volunteer computers running BOINC software. A key component of this system provides a priori runtime estimation using machine learning with random forests.

Papers [12,13] deal with a novel solution for predicting the runtimes of parameter sweep jobs. In the infrastructure with a dynamic configuration, such as a Desktop Grid, resource uptime and application runtime are critical scheduling factors. The proposed scheduling mechanism maps job runtimes onto resource uptimes. It is based on runtime prediction and its application in the prediction-aware mapping of jobs onto available resources. The aim is to reduce the number of jobs prematurely interrupted due to insufficient resource availability. The parameter sweep prediction framework used to make the predictions is referred to as GIPSy (Grid Information Prediction System). A detailed comparison between the expected results, based on simulation analysis, and the final results is given. By comparing the results for different model building configurations an optimal configuration is found that produces reliable result independent of the chosen job type. Results are presented for a quantum physics problem and two simulated workloads represented by sleepjobs.

In the paper [11] authors try to estimate delays, which are part of the task lifespan, i.e., distribution, in-progress, and validation delays. In the paper, the authors evaluate the accuracy of several probabilistic methods to get the upper time bounds of these delays. An accurate prediction of job lifespan delays allows to make more accurate prediction of a batch of tasks time completion, provide more efficient resources use, higher project throughput, and lower job latency in Desktop Grid projects.

In the paper [15] a dynamic replication approach is proposed to reduce the time needed to complete a batch of tasks; a mathematical model and a strategy of dynamic replication at the tail stage are proposed.

In the paper [19] the authors formulate an analytical model that permits to compare different allocation policies. In particular, the authors study an allocation policy that aims at minimizing the average job completion time; it is shown that the proposed policy can reduce the average completion time by as much as 50% of the completion time required for uniform or linear allocation policies. In the paper [17] a Desktop Grid is dynamically augmented with an Infrastructure as a Service (IaaS) cloud to reduce the average batch of tasks completion time.

Batch of tasks completion time estimation can be performed using various Desktop Grid emulators and simulators. Anderson [3] proposes a BOINC client emulator which is focused on scheduling strategies testing. The proposed emulator allows to trial different scheduling strategies in various usage scenarios, varying different hardware characteristics, defining client availability patterns, etc. The BOINC scheduling policies can be evaluated by several performance metrics. The results of emulations show which scheduling strategies are the most efficient. A part of chapter [8] is devoted to review and description of several Grid and Desktop Grid simulation tools: SimBA, SimBOINC, and EmBOINC software. The paper [6] also consider simulation of various computing systems focusing on the SimGrid simulator. The papers [9,10] by Estrada *et al.* are devoted to Desktop Grid emulation by EmBOINC, also paying attention to several simulation tools. The paper [18] describes the BOINC simulator SimBA.

This paper proposes statistical approach to batch of tasks completion time estimation. It is based on Holt's model and confidence intervals construction. The rest part of the paper is organized as follows. Section 2 briefly describes Desktop Grid and BOINC workflow. Section 3 describes mathematical model including formal problem definition, Holt's statistical model and confidence interval construction. Section 4 presents results of the numerical experiments. Finally, Sect. 5 gives the final discussion.

2   Desktop Grid and BOINC

A Desktop Grid consists of a (large) number of computing nodes and a server which distributes tasks among the nodes. The workflow is as follows. A node asks the server for work; the server replies sending one or more tasks to the node. The node performs calculations and when it finishes it, it sends the result (which is a solution of a task or an error report) back to the server. The server collects the results, assimilates and validates them. The detailed description of the computational process in BOINC-based Desktop Grids is given in [1].

There is a number of middleware systems for Desktop Grid implementation, the paper [14] authors overviews and compares the most popular of them: BOINC, XtremWeb, OurGrid and HTCondor. They review scientific papers devoted to research of the factors that influence the performance of Desktop Grid (from the point of view of both server and the client).

The most popular Desktop Grid platform is BOINC (Berkeley Open Infrastructure for Network Computing). It is actively developing Open Source software with rich functionality; it is a universal platform for scientific project in different domains. Herewith, BOINC is simple in deployment, use, and management [2].

BOINC is based on server-client model. The client software exists for different hardware and software platforms. The server can have arbitrary number of clients. A server can host several different BOINC-projects; each client can be connected to several projects and servers. BOINC-project is identified by its URL. BOINC client can be flexibly tuned up to be unobtrusive to the computer owner.

3   Completion Time Estimation

A statistical approach to batch of tasks completion time estimation is based on studying of time series characteristics, describing the process of results retrieving. A time series is a series of data points listed in time order. More formal, a discrete-time time series is a set of observations x_t, each one being recorded at a specific time t; the set T of times at which observations are made is a discrete set.

3.1   Mathematical Model

We consider a BOINC-based Desktop Grid, consisting of a number of computing nodes. The configuration of the Desktop Grid is dynamic: the set of available nodes can change in time (the nodes can abandon the Desktop Grid and new nodes can appear), the nodes have different characteristics of availability and reliability, tasks can be lost because of missed deadlines or abandoned nodes. A computational experiment of N tasks takes place; we need to construct a forecast on completion time of the computational experiment.

To make a forecast one should determine a functional dependence reflecting to time series. This functional dependence is called a forecast model. The model should minimize absolute difference between forecasted and observed values for a specified horizon (look-ahead period). Based on forecast model, one should find out forecasted values and confidence interval.

Consider a cumulative process of results retrieving. This process is described by a time series

$$Z(t) = Z(t_1), Z(t_2), ..., Z(t_k). \tag{1}$$

The values of the process are observed at discrete time points $t = t_1 < t_2 < ... < t_k$ with non-uniform intervals between them. Note, that the process is steadily increasing, so $Z(t+1) > Z(t)$ $\forall t$. At the point t_k (forecast point) one should estimate a time point t_p, at which observed value $Z(t_p)$ will exceed a specified value A (see Fig. 1).

For convenience, turn to considering a process

$$Y_i = (Z(t))^{-1}, i = 1, ..., k, \tag{2}$$

which describes time points of i-th result receiving. Then at step k (forecast point) one should estimate the value of process Y_i at step A ($A > k$; $p = A - k$ – look-ahead period, see. Fig. 2).

With a view to forecast, take the following assumptions. First, assume that there is a functional dependence between previous and future values of the process:

$$Z(t) = F(Z(t-1), Z(t-2), Z(t-3), ...) + \epsilon_t, \tag{3}$$

here ϵ_t – is a random error with a normal law of distribution. Second, this dependence is piecewise linear with up trend.

From the point of view of Desktop Grid these assumptions mean the following. The observed process describes time points of new results receiving. So, it

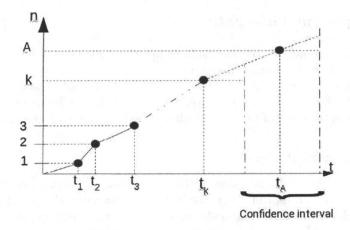

Fig. 1. Time series.

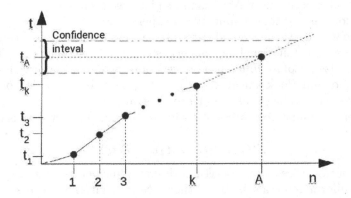

Fig. 2. Forecasted process.

is strictly increasing (two results can not be received at the same moment). An angle of trend line describes performance of the Desktop Grid (the less the angle to x axis, the more performance has the computing system). The performance can vary, according to it changes the trend. We assume that change in performance is linear because non-linear changes usually related to BOINC-project start, computing competitions and so on. Such non-linear effects are limited to a transition period and subside quickly.

In this paper we consider Holt's linear trend method based on exponential smoothing as forecasting statistical model.

3.2 Forecasting Model

Exponential smoothing is a rule of thumb technique for smoothing time series data using the exponential window function with decreasing weights over time.

Holt's forecasting model is extended simple exponential smoothing to allow forecasting of data with a trend. This model involves a forecast equation and two smoothing equations (one for the level and one for the trend) [7]:

$$
\begin{aligned}
Z_t &= L_t + \epsilon_t; \\
L_t &= \alpha Z_{t-1} + (1-\alpha)(L_{t-1} - T_{t-1}); \\
T_t &= \beta(L_{t-1} - L_{t-2}) + (1-\beta)T_{t-1}.
\end{aligned}
\tag{4}
$$

Here

- Z_i – observed level value;
- L_i – smoothed level value;
- T_i – trend value,
- $0 \leq \alpha, \beta \leq 1$ – smoothing coefficients for level and trend accordingly.

Forecast on p steps is constructing with an assumption of keeping the trend by the following formula:

$$
\overline{Z}_{t+p} = L_t + pT_t.
\tag{5}
$$

Forecast is linear and relies on the current trend.

3.3   Confidence Interval

Having right statistical model and keeping the trend, observed values and extrapolated point forecast are mismatching due to

1. inexact parameters of the model;
2. random error ϵ_t.

These errors can be shown as a forecast confidence interval. The formula to calculate a forecast confidence interval is following (see [7]):

$$
\widehat{y}_{k+p} \pm t_\gamma \cdot S_y \cdot \sqrt{\frac{k+1}{2} + \frac{(k+p-\bar{t})^2}{\sum_{t=1}^{k}(t-\bar{t})^2}},
\tag{6}
$$

here

- y_{k+p} – point forecast at the moment $k+p$, where k – number of observed values and p – look-ahead period;
- t_γ – value of Student's t-statistics;
- S_y^2 – mean-square distance between observed and forecasted values;
- $t = 1, 2, \cdots, k$ – process steps;
- $\bar{t} = \frac{k+1}{2}$ – mean step.

Mean-square distance between observed and forecasted values is defined by the following formula:

$$
S_y^2 = \frac{\sum_{t=1}^{k}(y_t - \widehat{y}_t)^2}{k-1},
\tag{7}
$$

where

- y_t – observed values;
- $\widehat{y}_t$ – forecasted values;
- k – number of values.

Therefore, confidence interval width depends on confidence level, look-ahead period, number of values and mean-square distance between observed and forecasted values.

4   Experiments Analysis

We performed set of numerical experiments to assess the approach. We used statistics of RakeSearch [16] and LHC@home [4] BOINC-project as input data. The available data of RakeSearch project contain information on workunit/result/host ids, results create/sent/received times, outcome, elapsed/cpu times; available data of LHC@home project contain information on workunit id and create time, results create/sent/received times, elapsed/cpu times. The summary on the number of records and time periods covering by input data are given in Table 1. We also mention that the records of RakeSearch project data are generated by 1081 computing nodes (we do not have the same information for LHC@home project).

Table 1. Input data characterization.

	RakeSearch	LHC@home
Time period	06/09/2017 - 14/11/2017	13/07/2017 – 30/08/2017
Number of records	117 579	41 797

For the purposes of this research, the input data sets were reduced to timestamps of results receiving. The data sets are depicted on Figs. 3 and 4.

To estimate batch of tasks completion time we considered sequences of tasks with random length (from 100 to 1000 tasks) with random look-ahead periods (from 10 to 50 tasks). The experiments show good covering of real data by confidence intervals. The examples of batch of tasks completion time estimation (based on newer results of RakeSearch project) are given in Figs. 5 and 6.

In the figures the blue line shows results receiving time, solid line is an approximation according to Holt's model, arrow is the point forecast and the red line at the arrow is a confidence interval.

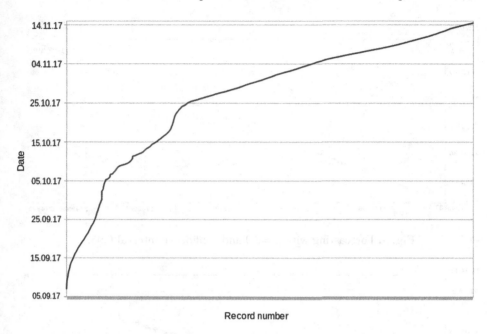

Fig. 3. Results receiving times of RakeSearch project.

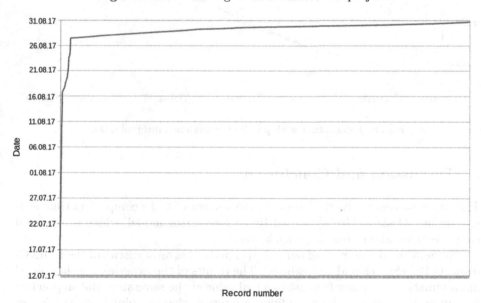

Fig. 4. Results receiving times of LHC@home project.

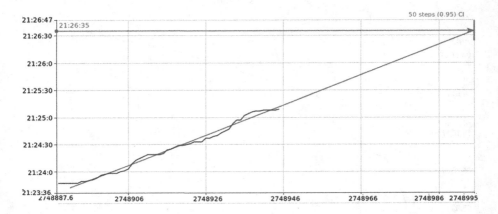

Fig. 5. Forecasting with $p = 50$ and confidence interval 0.95.

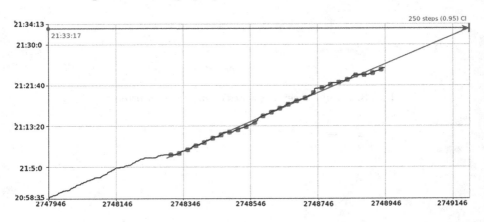

Fig. 6. Forecasting with $p = 250$, confidence interval 0.95.

5   Discussion and Conclusion

The paper presents statistical approach to a batch of tasks completion time estimation in a Desktop Grid. We used Holt's forecasting model, which is extended simple exponential smoothing with a trend.

We performed a number of numerical experiments on statistics of the BOINC-projects RakeSearch and LHC@home. The results of the experiments show good approximation of a point forecast to a real value at the same step. But in practice it is more important to have confidence interval, than a point forecast. So, we use confidence intervals also; the experiments show good covering of real data by confidence intervals. This shows that despite of high heterogeneity and big number of computing nodes (recall, that we used the statistics of RakeSearch project generated by 1081 computing nodes) and different tasks complexity, it is possible to get useful runtime estimation using simple statistical models.

There is a number of aspects that should be taken into account while implementing the described approach. First, approximation accuracy of Holt's model depends on two parameters: smoothing coefficients for level and trend. A procedure to tune-up the parameters is described, for example, in [7].

Second, in some cases it is reasonable to scale up time series when a long look-ahead period is given. Scaling up means summarizing several values of the time series into one single, describing the results received in the same hour, day or week. This allows to reduce computational overhead and increase accuracy while looking ahead long periods (thousands, tens of thousands steps) in large Desktop Grid projects.

Third, not all the values of the time series should be considered to perform forecasting. Only k latest values should be considered. The value k is selected to characterize the current trend and to minimize width of confidence interval, balancing mean-square distance between observed and forecasted values (which is increasing with higher k, in general) and Student's t-statistics (which is decreasing with higher k) according to formula 6.

Finally, fourth, in practice one is needed on-going update of the forecasted values according to new received observations. So, the model should be recalculated with every new observed value.

It also should be noted, that in this work computational complexity (size) and computing nodes performance are not considered. If in a sequence of tasks there is a subsequence in which all the tasks are more (or less) computationally complex than others, the statistical model will have a systematic error until this subsequence will be completed. Also, in practice it should be reasonable to take into account some kind of periodicity (for example, daily or weekly computing nodes periodicity).

All the mentioned above aspects will be taken into account to implement a BOINC-module of statistical batch of tasks completion estimation.

Acknowledgements. This work was supported by the Russian Foundation of Basic Research, projects 18-07-00628, 18-37-00094 and 16-47-100168.

References

1. Anderson, D.P., Christensen, C., Allen, B.: Designing a runtime system for volunteer computing. In: SC 2006 Conference, Proceedings of the ACM/IEEE, p. 33, November 2006. https://doi.org/10.1109/SC.2006.24
2. Anderson, D.P.: BOINC: a system for public-resource computing and storage. In: Proceedings of Fifth IEEE/ACM International Workshop on Grid Computing 2004, pp. 4–10. IEEE (2004)
3. Anderson, D.: Emulating volunteer computing scheduling policies. In: 2011 IEEE International Symposium on Parallel and Distributed Processing Workshops and Phd Forum (IPDPSW), pp. 1839–1846. IEEE (2011)
4. Barranco, J., et al.: Lhc@home: a BOINC-based volunteer computing infrastructure for physics studies at CERN. Open Eng. **7**(1), 379–393 (2017). https://doi.org/10.1515/eng-2017-0042

5. Bazinet, A.L., Cummings, M.P.: Computing the tree of life: leveraging the power of desktop and service grids. In: 2011 IEEE International Symposium on Parallel and Distributed Processing Workshops and Phd Forum, pp. 1896–1902, May 2011. https://doi.org/10.1109/IPDPS.2011.344
6. Beaumont, O., et al.: Towards Scalable, Accurate, and Usable Simulations of Distributed Applications and Systems. Research report rr-7761, Institut National de Recherche en Informatique et en Automatique (2011)
7. Brockwell, P.J., Davis, R.A.: Introduction to Time Series and Forecasting, 2nd edn. Springer, New York (2002). https://doi.org/10.1007/b97391
8. Estrada, T., Taufer, M.: Challenges in designing scheduling policies involunteer computing. In: Cérin, C., Fedak, G. (eds.) Desktop Grid Computing, pp. 167–190. CRC Press (2012)
9. Estrada, T., Taufer, M., Anderson, D.: Performance prediction and analysis of BOINC projects: an empirical study with EmBOINC. J. Grid Comput. **7**(4), 537–554 (2009)
10. Estrada, T., Taufer, M., Reed, K., Anderson, D.: EmBOINC: an emulator for performance analysis of BOINC projects. In: IEEE International Symposium on Parallel & Distributed Processing 2009. IPDPS 2009, pp. 1–8. IEEE (2009)
11. Estrada, T., Taufer, M., Reed, K.: Modeling job lifespan delays in volunteer computing projects. In: Proceedings of the 2009 9th IEEE/ACM International Symposium on Cluster Computing and the Grid, CCGRID 2009, pp. 331–338. IEEE Computer Society, Washington (2009). https://doi.org/10.1109/CCGRID.2009.69
12. Hellinckx, P., Verboven, S., Arickx, F., Broeckhove, J.: Predicting parameter sweep jobs: from simulation to grid implementation. In: 2009 International Conference on Complex, Intelligent and Software Intensive Systems, pp. 402–408, March 2009. https://doi.org/10.1109/CISIS.2009.86
13. Hellinckx, P., Verboven, S., Arickx, F., Broeckhove, J.: Runtime Prediction in Desktop Grid Scheduling, vol. 1, January 2009
14. Khan, M., Mahmood, T., Hyder, S.: Scheduling in desktop grid systems: theoretical evaluation of policies and frameworks. Int. J. Adv. Comput. Sci. Appl. **8**(1), 119–127 (2017)
15. Kolokoltsev, Y., Ivashko, E., Gershenson, C.: Improving "tail" computations in a boinc-based desktop grid. Open Eng. **7**(1), 371–378 (2017)
16. Manzyuk, M., Nikitina, N., Vatutin, E.: Employment of distributed computing to search and explore orthogonal diagonal Latin squares of rank 9. In: Proceedings of the XI All-Russian Research and Practice Conference "Digital Techologies in Education, Science, Society" (2017)
17. Reynolds, C.J., et al.: Scientific workflow makespan reduction through cloud augmented desktop grids. In: 2011 IEEE Third International Conference on Cloud Computing Technology and Science, pp. 18–23, November 2011. https://doi.org/10.1109/CloudCom.2011.13
18. Taufer, M., Kerstens, A., Estrada, T., Flores, D., Teller, P.: SimBA: a discrete event simulator for performance prediction of volunteer computing projects. In: 21st International Workshop on Principles of Advanced and Distributed Simulation, vol. 7, pp. 189–197 (2007)
19. Villela, D.: Minimizing the average completion time for concurrent grid applications. J. Grid Comput. **8**(1), 47–59 (2010). https://doi.org/10.1007/s10723-009-9119-2
20. Wu, W., Chen, G., Kan, W., Anderson, D., Grey, F.: Harness public computing resources for protein structure prediction computing. In: The International Symposium on Grids and Clouds (ISGC), vol. 2013 (2013)

BOINC-Based Branch-and-Bound

Andrei Ignatov[1(✉)] and Mikhail Posypkin[2]

[1] Moscow State University, Moscow, Russia
rayignatov@outlook.com
[2] Federal Research Center Computer Science and Control
of Russian Academy of Sciences, Moscow, Russia
mposypkin@gmail.com

Abstract. The paper proposes an implementation of the Branch-and-Bound method for an enterprise grid based on the BOINC infrastructure. The load distribution strategy and the overall structure of the developed system are described with special attention payed to some specific issues such as incumbent updating and load distribution. The implemented system was experimentally tested on a moderate size enterprise grid. The achieved results demonstrate an adequate efficiency of the proposed approach.

Keywords: BOINC · Branch and Bound · Distributed computing

1 Introduction

Desktop grids is a rapidly growing platform for distributed computing. The most popular software for desktop grids is BOINC (Berkeley Open Infrastructure for Network Computing) [7]. In a nutshell, BOINC is a system that can harness the unused computing power of desktop machines for processing resource demanding applications. Potentially, such grids can collect a tremendous amount of resources. However, efficient usage of this power can be a rather complicated task due to heterogeneity and irregularity of desktop grids.

Traditional approaches developed for standard HPC platforms are based on some assumptions that are not valid in desktop grids. In particular, such approaches assume low-latency fast communications where each parallel processor can serve both as a sender or a receiver. Clearly such approaches are not suitable for BOINC grids where communications are possible only between the server and a client node and are initiated by a client. BOINC communications have high latency as they involve several services and a file system.

Thus, we can conclude that developing new efficient distributed algorithms suitable for desktop grids is an important research direction. In this paper, we consider the implementation of tree-structured computations in desktop grids. Such computational schemes are remarkably popular in global optimization, in particular, with Branch-and-Bound family of methods. The paper is organized as follows. Section 2 discusses existing approaches for solving global optimization

© Springer Nature Switzerland AG 2019
V. Voevodin and S. Sobolev (Eds.): RuSCDays 2018, CCIS 965, pp. 511–522, 2019.
https://doi.org/10.1007/978-3-030-05807-4_43

problems in distributed computing environment. Section 3 outlines the structural, algorithmic and implementation details of the proposed approach. Experimental results are presented in Sect. 4.

2   Related Work

Branch-and-bound is a universal and well-known technique for solving optimization problems. It interprets the input problem as the root of a search tree. Then, two basic operations are recursively executed: branching the problem (node) into several smaller (hopefully easier) problems, and bounding (pruning) the tree node. At any point during the search tree traversal, all subproblems can be processed independently. The only shared resource is the incumbent. Hence, processing the search tree in a distributed way is considered rather natural and has been studied for decades. Since the size and the structure of the branch-and-bound tree are unknown in advance, the even distribution of computations among processors is a challenging task. Load balancing has been comprehensively studied for tightly-coupled multiprocessors. Most efficient schemes use intensive communication among processors to approach uniform distribution. Unfortunately, this approach is not suitable for volunteer desktop grids where direct communications among computing nodes are normally not allowed.

The solution for distributed systems consisting of computational resources connected via wide-area networks (WAN) was proposed in [5,6].

In [20] authors describe the AMPLX toolkit that enables modifying any AMPL script to solve problems by a pool of distributed solvers. The toolkit is based on Everest platform [21] that is used to deploy optimization tools such as web services and run these tools across distributed resources.

The BNB-Grid framework proposed in [4,11] was aimed at running exact (based on Branch-and-Bound) and heuristic search strategies on grids consisting of heterogeneous computing resources. BNB-Grid uses different communication packages on different levels: on the top level, it uses ICE middleware coupled with TCP/IP sockets, and within a single computing element, either MPI or POSIX Thread libraries are used. BNB-Grid was used to solve several challenging problems from various fields, see e.g. [19].

The approach closest to ours was proposed in [8], where authors described a grid enabled implementation of the branch-and-bound method for computational grids based on the Condor [16] middleware. The suggested approach uses a centralized load balancing strategy: the master keeps a queue of sub-problems and periodically sends them to free-working nodes (slaves). When a sub-problem is sent to the slave, it is either completely solved or the resolution process is stopped after a given number of seconds while unprocessed subproblems are sent back to the master. Authors reported successful results for several hard quadratic assignment instances.

In comparison to that research, our major contribution is the implementation of a branch-and-bound optimization problem solver for a conceptually different platform. In addition, in our work, we suggest an effective algorithm of filtering tasks on the server side, which leads to a significant reduction in the computing time.

There have been several attempts to use BOINC desktop grids to run Branch-and-Bound method. Some preliminary experiments on solving knapsack problems on a small test BOINC system were presented in [22]. In [12] authors consider an implementation of a back-track search in volunteer computing paradigm. They present a simple yet efficient parallelization scheme and describe its application to some combinatorial problems: Harmonious Tree and Odd Weird Search, both carried out at the volunteer computing project yoyo@home. The paper also presents a simplified mathematical analysis of the proposed algorithm.

3 Outline of the Approach

Any BOINC application consists of the server and the client parts. The server part running on the BOINC server is responsible for generating tasks and aggregating their results obtained by client applications running on computational nodes. In our case, the client part is a C++ application [3] that solves a global box-constrained optimization problem with the Branch-and-Bound method. Interval analysis [14] is used to compute bounds on the objective. We used our own implementation of interval analysis [2,18] that allows to compute bounds based on the C++ representation of an algebraic expression.

The general outline of our system is presented in Fig. 1. Like all BOINC applications, it contains client and server parts. Server maintains creation, sending, cancellation of tasks, and receiving and processing results. There is also a number of client machines that receive tasks, perform calculations, and send the achieved results back to the server.

The client reads the "state" from input file "in.txt" and writes the produced output to "out.txt". Both files are encoded in JSON format. The input files contain the record (incumbent value) found so far, a set of sub-problems to process and the maximal step limit. The step limit helps to bound the running time to avoid long-running tasks. To make running times of tasks uniform, it is suggested to aggregate several subproblems in one task. Omitting this suggestion may lead to dramatic differences in time elapsed for tasks [15].

Similarly, an output file contains the number of iterations that were performed, the record value, and a set of unprocessed subproblems in case if the upper bound for iterations was exceeded. The workflow of this application is as follows:

1. The pool of subproblems is initialized according to the input file.
2. The first subproblem D is extracted from the pool. Having it, the function value $f(c_D)$ for the centre point c_D of the feasible set X_D of the subproblem D is calculated along with the bound of the function's values in X_D.
3. The discovered value $f(c_D)$ is compared to the current record value f_r. The latter is possibly updated and the new incumbent solution is stored.

4. The optimality test is performed: if the lower bound of the current domain is less than the record function value with the prescribed accuracy ϵ, the feasible set is divided in two across the greatest dimension and the generated subproblems are added to the pool.
5. Steps 2–4 are repeated until the pool is empty or the number of performed steps exceeds the limit.
6. After the computation process is over, the output JSON line is formed and written to "out.txt" file.

This application was compiled for Linux, Windows, and OS X 32- and 64-bit systems.

To run an application in a BOINC project, it is required either to implement the API for server-client communication or to use special wrappers [1,17] available in BOINC public resources. A wrapper is a special program for a particular operating system and architecture that runs the provided functional applications, 'feeds' the tasks to them, and manages all types of communication with the server. For our project, we decided to use the BOINC wrapper [1] for all platforms, which resulted in lesser amounts of code along with the convenience of tracking possible issues as wrappers provide detailed information regarding the running application.

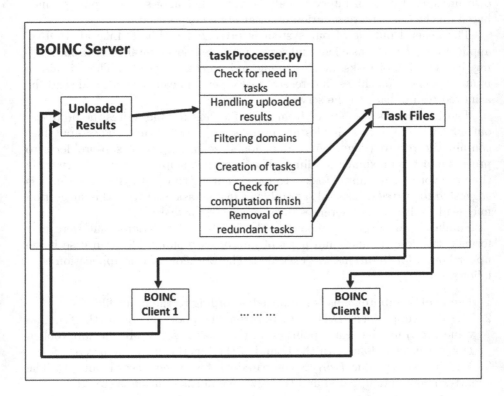

Fig. 1. Basic workflow of the developed system

In the server part, there is a permanently running Python script [3] that finds all uploaded files and processes their contents. Information from each file is extracted and analyzed: the number of performed iterations is added to the global counter, record data is updated, and received subproblems are added to the global pool. After processing, the files are deleted as they are no longer needed, and the formed pool of subproblems is merged with the previous ones saved in the special file containing the pool of subproblems. Subproblems should be also examined at the task formation step in order to avoid sending clearly redundant ones.

Another job performed by the server part is the creation of new workunits. One workunit aggregates a fixed number of subproblems S. It is well known that in some cases the Branch and Bound method may not converge in a short time, so it is worth restricting the total number of iterations for the whole problem with a value denoted below as $N_{totalmax}$. Besides subproblems and the record value, each workunit stores the maximum number of iterations. This number is computed as follows:

$$N = \min \left(\left\lfloor \frac{N_{totalmax} - N_{performed}}{N_{pool}} \right\rfloor , N_{max} \right) , \tag{1}$$

where $N_{performed}$ is the total number of iterations performed so far and N_{max} is the iteration threshold. Initially, it was suggested to generate workunits once per a fixed amount of time, but then it turned out that decrease in this delay time led to creation of a great number of redundant tasks. On the other hand, increasing that amount of time caused a lack of workunits and consequently long idle periods of client PCs. In general, we came to the following conclusion: the later a subproblem is packaged to a workunit, the higher chances that it will discarded by the optimality test are. Therefore, it was decided to implement the 'on-demand' model where new workunits are created when they are really needed. The script checks the number of unsent tasks, and if there are more than T of them, it is concluded that new tasks are not necessary and the script stops generating them.

Regardless of the application's 'on-demand' policy, it is still possible that some submitted or running tasks could be discarded according to the fresh record value. To handle this, all unsent or currently running tasks are examined. It leads to the following rule: if for a task t

$$\min_{s \in t} lb_s \geq f_r - \varepsilon, \tag{2}$$

then the task should be cancelled. Here, s is a subproblem and $lb(s)$ is its lower bound. Indeed, if the inequality is valid than all subproblems in a task are subject to discard and thus should not be processed. In case such redundancy is discovered, the workunit is canceled using the respective BOINC command.

4   Experimental Results and Discussion

For our experiments, we use the following small enterprise [13] BOINC grid system. The server runs on a single-board 64-bit 'Orange Pi 2' machine, which has 1 GB RAM and sustainable Internet access. Clients are collected from the computer class: 12 32-bit machines with Intel Core 2 Duo E4600 CPUs with 2 GB RAM. The server machine uses Armbian OS, while clients run Linux Mint.

Parameters that were used by our load distribution algorithm are summarized in Table 1. These values were experimentally selected.

Table 1. Load distribution algorithm parameters

Parameter	Description	Value
T	The minimal number of workunits in a queue to start generating new tasks	5
S	The (maximal) number of subproblems in a workunit	2
N_{max}	The maximal number of steps performed by a client	10^5
$N_{totalmax}$	The maximal total number of steps to perform	$5 \cdot 10^9$

For testing the system performance, we conducted a number of computational experiments within an enterprise grid consisting of identical machines that have been described above. For testing, we considered benchmark optimization problems Biggs Exp5 and Exp6 function from the test suite [2,18]:

$$\begin{cases} F_{BiggsEXP6}(x_1, x_2, ..., x_N) \to \min \\ x_i \in [-20, 20], \ i \in 1, \ldots, N. \end{cases} \quad (3)$$

The objective function is defined as follows:

$$F_{BiggsEXP6}(x_1, x_2, ..., x_N) = \sum_{n=1}^{13} (x_3 e^{-t_i x_1} - x_4 e^{-t_i x_2} + x_6 e^{-t_i x_5} - y_i)^2, \quad (4)$$

where

$$t_i = 0.1i, \ y_i = e^{-t_i} - 5e^{10t_i} + 3e^{-4t_i} \quad (5)$$

The problem was solved for 5- and 6-dimensional functions. In the 5-dimensional case, Biggs Exp6 function degenerates into Biggs Exp5 one, having 'x_6' replaced with '3' in the described formula. The obtained results as functions of the number of machines are shown in Figs. 2, 3, 4, 5 and 6 presenting the 5-dimensional experiments measurements in blue color and the 6-dimensional ones in orange.

To evaluate efficiency of the implemented system, we use two main metrics, which are the elapsed computational time and the relative time for one task that is calculated in the following way:

$$T_{rel} = \frac{T_{elapsed}}{N_{iterations}} \quad (6)$$

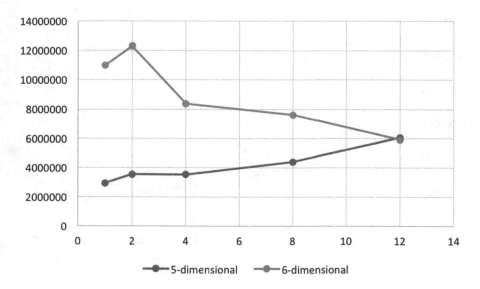

Fig. 2. Number of performed iterations (Color figure online)

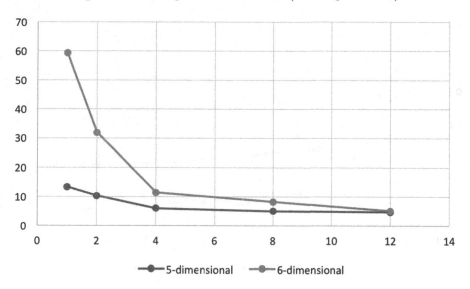

Fig. 3. Elapsed time, min (Color figure online)

The latter metrics is needed because the number of iterations may be affected by the rate of record updating ("search anomaly"). It separates the speedup due to changing the number of iterations and the parallelization.

As seen from the obtained graphs, both metrics decrease when the number of machines grows. However, both graphs turn out to be far from linear as the rate of metrics decrease dramatically falls after more than four machines are

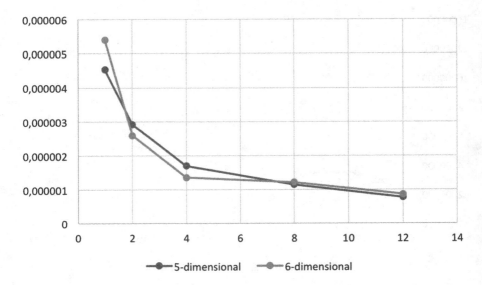

Fig. 4. Average time for one task, min (Color figure online)

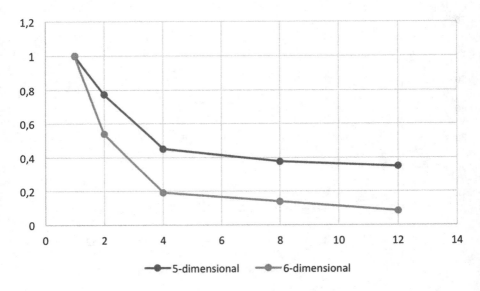

Fig. 5. Elapsed time decrease rate (Color figure online)

involved. The major reason for this is that the number of tasks provided by the server turns out to be less than the overall computational ability of clients, which leads to the situation when a part of clients stay idle due to the lack of workunits.

In order to relieve this issue, we tried to change the criterion of necessity of new tasks so that the number of currently processing tasks would matter

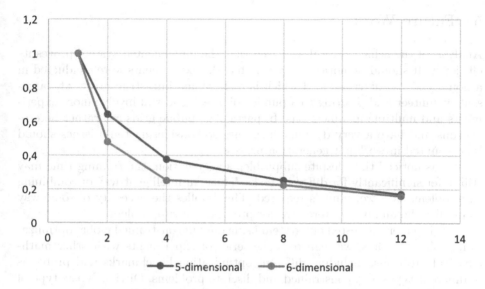

Fig. 6. Time for task decrease rate (Color figure online)

as well, but it led to rise in iterations number of nearly 10% in average and, therefore, to rise in elapsed time, so this strategy was rejected. We conclude that this issue needs to be observed in a more computationally complicated problem where the number of tasks would be considered as much greater than the number of computational units. In case this effect repeats in such an experiment, the strategy of distributing tasks between units should be corrected in a way, otherwise it would be considered as normal behaviour of the system.

Another remarkable result is that trends for the overall number of performed iterations do not match in two experiments. As it can be seen in the first graph in Fig. 2, there is a clear decreasing trend in the 6-dimensional case. It leads to a superlinear speedup in elapsed time between 1 and 4 nodes. Surprisingly the situation with 5-dimensional case is opposite. This issue needs further investigation.

Using the strategy that has been described before, we get the following dependency: the sooner the current record value is updated, the less tasks would need to be processed, and it means the decrease in the number of performed iterations. As using multiple machines speeds up the process of updating record values, the number of iterations is supposed to decrease along with the time elapsed for the whole problem. However special efforts are needed to ensure fast record propagation and to filter the jobs queue before sending to client nodes.

5   Future Work

Analysis of the achieved results showed that the implemented system proves its efficiency. It should be noticed however that the experiments were conducted in a homogeneous enterprise grid which is a simplified infrastructure w.r.t. large-scale volunteer grid. Porting to a public volunteer grid will involve more experiments and multiple modifications. In particular, public grids are rather heterogeneous and have a very dynamic structure so considering these issues should be accounted in workunit generation process.

It was noticed that despite subproblems aggregation, the running time may still differ significantly for different tasks. Thus, an intelligent way of combining subproblems in a workunit is required. This implies the necessity of some way to predict the number of iterations for processing a subproblem.

In this paper, we tested our system using only unconstrained global optimization problems. Thus, we plan to make series of experiments with other mathematical problems, including different optimization benchmarks and problems of different types, e.g. constrained and discrete problems. Obviously the type of the problem can significantly affect the choice of the parallelization strategy. For example, finding the solution of systems of inequalities by Branch-and-Bound method [10] do not involve records exchange and thus can be viewed as a classical back-tracking. Conversely another popular approach to global optimization — branch and cut method assumes passing "cuts" (cutting planes) between processes [9] and thus can significantly increase the network overhead. The mentioned problems may require different load distribution strategies.

6   Conclusion

In this paper, we described a system implementing the Branch and Bound method on a small enterprise grid based on the BOINC infrastructure. Techniques to ensure uniform load distribution were proposed as well as subproblems filtering before sending to working nodes. The series of experiments on two benchmark problems were performed. The experimental results show that this algorithm is rather efficient in terms of total execution time decrease.

The obtained results, however, demonstrate a dramatic drop in time decrease after more than four machines are involved. As we suppose, it may be caused by a lack of workunits for computational nodes, and this assumption should be examined in a more complicated problem.

Acknowledgements. The work was supported by the RAS Presidium program No. 26 "Fundamentals of algorithms and software development for advanced ultra-high-performance computing." Authors are grateful to the head of supercomputer department of Federal Research Center "Computer Science and Control" Vadim Kondrashov and members of this department Ilya Kurochkin and Alexander Albertyan for helping in deploying and maintaining test BOINC grid.

References

1. The BOINC Wrapper (2008). http://boinc.berkeley.edu/trac/wiki/WrapperApp/. Accessed 15 May 2018
2. Box-constrained test problems collection (2017. https://github.com/alusov/mathexplib.git. Accessed 17 June 2018
3. BnB BOINC project codes (2018). https://github.com/AndrewWhyNot/BOINC-Interval-Based-BnB.git. Accessed 17 June 2018
4. Afanasiev, A., Evtushenko, Y., Posypkin, M.: The layered software infrastructure for solving large-scale optimization problems on the grid. Int. J. Comput. Res. **18**(3/4), 307 (2011)
5. Aida, K., Natsume, W., Futakata, Y.: Distributed computing with hierarchical master-worker paradigm forparallel branch and bound algorithm. In: 3rd IEEE/ACM International Symposium on Cluster Computing and the Grid 2003. Proceedings, CCGrid2003, pp. 156–163. IEEE (2003)
6. Alba, E., et al.: MALLBA: a library of skeletons for combinatorial optimisation. In: Monien, B., Feldmann, R. (eds.) Euro-Par 2002. LNCS, vol. 2400, pp. 927–932. Springer, Heidelberg (2002). https://doi.org/10.1007/3-540-45706-2_132
7. Anderson, D.P.: BOINC: a system for public-resource computing and storage. In: Proceedings of the 5th IEEE/ACM International Workshop on Grid Computing, pp. 4–10. IEEE Computer Society (2004)
8. Anstreicher, K., Brixius, N., Goux, J.-P., Linderoth, J.: Solving large quadratic assignment problems on computational grids. Math. Program. **91**(3), 563–588 (2002)
9. Elf, M., Gutwenger, C., Jünger, M., Rinaldi, G.: Branch-and-cut algorithms for combinatorial optimization and their implementation in ABACUS. In: Jünger, M., Naddef, D. (eds.) Computational Combinatorial Optimization. LNCS, vol. 2241, pp. 157–222. Springer, Heidelberg (2001). https://doi.org/10.1007/3-540-45586-8_5
10. Evtushenko, Y., Posypkin, M., Rybak, L., Turkin, A.: Approximating a solution set of nonlinear inequalities. J. Global Optim. **71**, 1–17 (2017)
11. Evtushenko, Y., Posypkin, M., Sigal, I.: A framework for parallel large-scale global optimization. Comput. Sci. Res. Dev. **23**(3–4), 211–215 (2009)
12. Fang, W., Beckert, U.: Parallel tree search in volunteer computing: a case study. J. Grid Comput. **16**, 1–16 (2017)
13. Ivashko, E.E.: Enterprise desktop grids. Programmnye Sistemy: Teoriya i Prilozheniya [Program Systems: Theory and Applications] **1**, 19 (2014)
14. Jaulin, L., Kieffer, M., Didrit, O., Walter, E.: Applied Interval Analysis. With Examples in Parameter and State Estimation, Robust Control and Robotics, vol. 1. Springer, London (2001). https://doi.org/10.1007/978-1-4471-0249-6
15. Samtsevich, A., Posypkin, M., Sukhomlin, V., Khrapov, N., Rozen, V., Oganov, A.: Using virtualization to protect the proprietary material science applications in volunteer computing. Open Eng. **8**(1), 57–60 (2017)
16. Litzkow, M.J., Livny, M., Mutka, M.W.: Condor-a hunter of idle workstations. In: 8th International Conference on Distributed Computing Systems 1988, pp. 104–111. IEEE (1988)
17. Marosi, A.C., Balaton, Z., Kacsuk, P.: Genwrapper: a generic wrapper for running legacy applications ondesktop grids. In: IEEE International Symposium on Parallel & Distributed Processing 2009. IPDPS 2009, pp. 1–6. IEEE (2009)

18. Posypkin, M., Usov, A.: Implementation and verification of global optimization benchmark problems. Open Eng. **7**(1), 470–478 (2017)
19. Semenov, A., Zaikin, O., Bespalov, D., Posypkin, M.: Parallel logical cryptanalysis of the generator A5/1 in BNB-grid system. In: Malyshkin, V. (ed.) PaCT 2011. LNCS, vol. 6873, pp. 473–483. Springer, Heidelberg (2011). https://doi.org/10.1007/978-3-642-23178-0_43
20. Smirnov, S., Voloshinov, V., Sukhoroslov, O.: Distributed optimization on the base of AMPL modeling language and everest platform. Procedia Comput. Sci. **101**, 313–322 (2016)
21. Sukhoroslov, O., Volkov, S., Afanasiev, A.: A web-based platform for publication and distributed execution of computing applications. In: 2015 14th International Symposium on Parallel and Distributed Computing (ISPDC), pp. 175–184. IEEE (2015)
22. Tlan, B., Posypkin, M.: Efficient implementation of branch-and-bound method on desktop grids. Comput. Sci. **15**(3), 239–252 (2014)

Comprehensive Collection of Time-Consuming Problems for Intensive Training on High Performance Computing

Iosif Meyerov[1], Sergei Bastrakov[1,2], Alexander Sysoyev[1],
and Victor Gergel[1(✉)]

[1] Lobachevsky State University of Nizhni Novgorod, Nizhni Novgorod, Russia
{iosif.meerov, sergey.bastrakov,
alexander.sysoyev}@itmm.unn.ru, gergel@unn.ru
[2] Helmholtz-Zentrum Dresden-Rossendorf, Dresden, Germany

Abstract. Training specialists capable of applying models, methods, technologies and tools of parallel computing to solve problems is of great importance for further progress in many areas of modern science and technology. Qualitative training of such engineers requires the development of appropriate curriculum, largely focused on practice. In this paper, we present a new handbook of problems on parallel computing. The book contains methodological materials, problems and examples of their solution. The final section describes the automatic solution verification software. The handbook of problems will be employed to train students of the Lobachevsky University of Nizhni Novgorod.

Keywords: Parallel computing · High performance computing
Education

1 Introduction

Modern computational science is becoming increasingly interdisciplinary and concerns all spheres of human life. The development of computer architectures and related technologies has led to the emergence of new applied fields of science (bioinformatics, computational biomedicine, computer vision, machine learning, big data and so forth). Many existing areas of the natural sciences received further development under the influence of supercomputers. It is widely recognized that supercomputers are driving technological progress in the world.

Further development of the industry requires significant efforts from the community. Among the important areas is the problem of training highly qualified specialists in application of models, methods and technologies of parallel computing to solve scientific and engineering problems using supercomputers. This area is rapidly expanding in many directions, which, on the one hand, enhances its expressive power, but, on the other hand, it places ever higher demands on engineers, their knowledge and skills. Our community needs special training courses on parallel programming, mainly oriented to practice.

V. Voevodin and S. Sobolev (Eds.): RuSCDays 2018, CCIS 965, pp. 523–530, 2019.
https://doi.org/10.1007/978-3-030-05807-4_44

In this paper we briefly review the new handbook of problems on parallel computations developed at the Institute of Information Technology, Mathematics and Mechanics (ITMM) of the Lobachevsky University on the basis of many years of experience in training engineers in supercomputing technologies. The book aims to guide students through a wide range of problem domains, including low level parallel programming, basic parallel algorithms, and parallel numerical methods. We hope that the book can complement the modern lecture courses on parallel programming.

2   Related Work

A significant amount of methodical literature on parallel computations has been prepared during the last decades. Theoretical information on parallel computations and parallel algorithms is presented in the books [1–5]. The development of parallel programs for distributed memory systems using the MPI (including not only a description of the contents and capabilities of the MPI, but also examples of programs) is discussed in detail in the books [6–9] and the in first volume of [10]. Parallel programming for shared memory system based on the OpenMP technology can be studied from the book [11] and the second volume of the book [10]. Development and optimization of parallel programs for Intel Xeon Phi processors are presented in detail in [12, 13]. The books [14, 15] contain various examples of optimizing programs for Xeon Phi from different problem domains. Programming for GPU accelerators with a large number of examples is discussed in detail in [16–18]. Paper [19] should also be noted, where the study of the MPI technology is conducted on the example of solving a number of typical problems of parallel programming – matrix calculations, sorting, processing graphs, etc. The questions of teaching parallel algorithms and technologies are also of considerable interest. In this regard, the activity of the NSF/IEEE-TCPP Curriculum Initiative on Parallel and Distributed Computing [20, 21] should be mentioned. In Russia, significant progress has been made in this direction within the framework of the government program Supercomputing Education [22].

The presented handbook continues a series of publications on parallel programming education. Among these publications are textbooks [5, 10], tutorials and practical classes on various technologies of parallel programming and their application to scientific research in different problem domains [23, 24]. A considerable part of the prepared publications is presented on the web page of the Center for Supercomputer Technologies of Lobachevsky University in Russian (http://hpc-education.unn.ru/ru) and in English (http://hpc-education.unn.ru/en/trainings/collection-of-courses). The training course "Introduction to parallel computing" is available on the website of the e-learning system (http://mc.e-learning.unn.ru/; in Russian).

3   Handbook Overview

Preparing a specialist capable of effectively using modern parallel computing systems is a challenging problem. It may seem that any person who has good theoretical background in mathematics and, for example, physics (chemistry, biology, medicine – depending on a specific problem domain), and also able to implement sequential programs, can easily study parallel programming. However, it is often not enough to read descriptions of OpenMP and MPI and learn how to develop and run software on a supercomputer. The efficiency of parallel software depends highly on specific details, and it turns out that there are quite a lot of these details. For example, it is necessary to have skills of system programming and debugging of parallel code, which is significantly more complicated than the sequential one. In addition, the architecture of modern computing systems has become so complicated that very little manipulation of code and run modes can lead to an order of magnitude difference in computational time, and sometimes even more. Therefore, training of specialists with skills to develop and run parallel programs is still of great importance.

Education in a rapidly changing field cannot be reduced to studying theoretical material and a number of standard use cases. In our universities we need training courses with many practical problems exemplifying typical challenges from different areas of science and engineering. Thus, the development of parallel programs for solving sparse algebra problems requires the study of special data structures and methods for their effective use in parallel computations. Parallelization of Monte Carlo methods requires the understanding of how to use pseudo-random number generators in parallel programs. The experience of running standard performance tests forms extremely useful skills of customization the computational environment and analysis of the experimental data. Each section of the presented handbook illustrates typical data structures, algorithms and approaches in some specific problem domain. We stress that we did not try to write a book on system programming, parallel algorithms or numerical methods. On the contrary, we recommend readers to study the literature that has become classical in this field, giving corresponding links in each section. Theoretical material, which is given in the handbook, is the minimum necessary to understand the problems, and where it can be read about it. Further, we consider data structures and the approaches to parallelization, give problem statements, discuss possible solutions, and formulate the set of problems which should be solved by students. These set of problems can be easily extended to reach the goals of education. Thus, algorithms can be implemented for different types of computing systems with shared and distributed memory using appropriate technologies. By adjusting the input data and, correspondingly, computational load, we can assess the quality of implementation checking correctness, performance, and scaling efficiency. The book also describes a system for automated testing of parallel programs, developed by students. The use of such a system makes it possible to significantly reduce the efforts of teachers in verifying the homework.

4 Handbook Structure and Contents

The book outline includes the following sections:

1. Parallel programming for beginners.
2. Parallel execution of threads and processes.
3. Communication methods.

4. Performance tests.
5. Calculations with matrices.
6. Interpolation and approximation of functions.
7. Numerical integration.
8. Sorting algorithms.
9. Graph algorithms.
10. Numerical solution of systems of ordinary differential Eqs.
11. Numerical solution of systems of partial differential Eqs.
12. Methods of global optimization.
13. Monte Carlo methods.
14. Computational geometry algorithms.
15. Computer graphics and image processing algorithms.
16. Automatic solution verification system SoftGrader.

Most of the sections are organized as follows. A section begins with a brief introduction to the problem area and provides a rationale for the importance of discussed tasks and methods in HPC applications. The following material describes ways of representing and storing data, standard for typical problems in the considered area. Then we give a brief overview of parallelization schemes. The section continues with detailed statements of one or two typical problems. Parallel methods for their solution are outlined and discussed. Then, we formulate tasks for students. An example of a problem statement accompanied by a demonstration of program code finalizes the section.

Let us overview the book contents. In the *Preface* we briefly outline our motivation to write the book. The first section presents the computational problems for beginners and helps to form an introductory laboratory practical work on various academic disciplines in the field of parallel computing (for example, to study the OpenMP and MPI technologies).

The next three sections relate to the area of system programming. In *section 2* we consider the principles of threads execution planning, synchronization mechanisms and standard problems, offer tasks for the use of typical mechanisms of threads interaction. *Section 3* proposes tasks for the development of data transfer programs between processors on distributed memory. When carrying out homework, it is necessary to evaluate the expected time for executing communication operations, implement sample data transfer programs, and compare the results obtained with theoretical estimates. To evaluate the effectiveness of the implemented software, it is necessary to run computational experiments comparing the developed programs with standard tools (for example, with functions of the similar purpose in an MPI library). *Section 4* deals with standard tests that identify the performance parameters of computational clusters. Relevant work with such tests allows an engineer to make decisions about the configuration of the future computing system, calculate the standard performance parameters of ready to use systems, and identify their bottlenecks.

The following three sections concern numerical methods. In *section 5* we consider the following basic linear algebra algorithms: matrix-vector (dense and sparse) and matrix-matrix multiplication. In this regard, we discuss data structures, especially for sparse matrices, and different schemes of parallelization. In *section 6* we discuss interpolation and approximation methods, give examples and propose several tasks for students. In *section 7* we consider one of the simplest formulas of numerical integration (the rectangle rule) and its application to the problem of calculating an integral of a function of two variables. When performing each of the tasks, it is necessary to check correctness, construct theoretical estimates of the run time of an algorithm, perform computational experiments and compare the results with theoretical estimates.

Sections 8 and 9 consider two classical areas of Algorithms and data structures: sorting algorithms and graph algorithms. In *section 8* we discuss different parallel merging schemes which can be used as a part of many sorting algorithms. In *section 9* we consider methods for storing graphs, the properties of graphs algorithms, general approaches to their parallelization, and consider some classical graph processing methods: breadth-first search, minimal spanning tree search, and single-source shortest path paths search.

In *sections 10 and 11* we consider parallel algorithms of solving systems of ordinary differential equations and partial differential equations, correspondingly. In *section 10* we discuss the Euler and Runge-Cutta methods for an ODE system. In *section 11* we consider the numerical solution of the Dirichlet problem for the Poisson equation.

In *section 12* we briefly overview global optimization methods, discuss the applicability of general parallelization approaches to global optimization and present a parallel version of the global search algorithm.

Section 13 concerns Monte Carlo methods. We outline methods of pseudorandom and quasirandom number generation, consider approaches for simulation of distributions with given properties, and discuss how to use random number generators in parallel computations correctly. Next, we introduce the problem of developing a generator of uniformly distributed numbers and demonstrate one of the methods for solving it. Then, we consider an example from financial mathematics: finding the fair price of an option of the European type using the parallel Monte Carlo method.

Section 14 is devoted to approaches to the parallel solution of computational geometry problems. The ideas are demonstrated by the example of constructing a convex hull of points on a plane. In *section 15*, computer graphics and image processing problems are presented. The methods of parallelization of image processing operations are considered.

In *section 16*, the SoftGrader automated testing system is described. We demonstrate main features of the software and emphasize the scheme for preparing the training course and its practical problems for their automated testing in SoftGrader.

5 Section Example

This section of the paper presents a more detailed description of the one of the book sections – Monte Carlo method. We start the section with a brief overview of the problem area and give references to classical books and tutorials concerning the Monte Carlo method and its applications to different problems of modern computational science.

Then, we discuss the general scheme of the Monte Carlo method. First, we outline a problem of generating random numbers in a computer. Then, we give a description of the linear congruential generator and depict its advantages and disadvantages. After that, we briefly describe the Mersenne Twister algorithm for generating random numbers and give relevant references. The description of the Sobol sequence continues the section. As before, we present the advantages and disadvantages of the approach, and compare it with the algorithms mentioned above. At the end of the subsection we put references to the description of the pseudorandom number sampling methods mapping uniformly distributed numbers to other distributions.

In the next subsection we discuss parallelization of the Monte Carlo method. Although this method is known to be embarrassingly parallel, there are important details concerning usage of random number generators in parallel computations. In this regard, we consider the master-worker, leapfrog, and skipahead methods. We also describe some typical errors which are often made during the implementation and their consequences. The discussion regarding the performance of the Monte Carlo method implementation finalizes the subsection. Thus, we pay attention to performance of the random number generation algorithm, including cache efficiency and utilization of the SIMD instructions. We highly recommend using carefully tested high performance implementations given by mathematical software libraries in real-world applications (i.e. the Intel Math Kernel library).

To better understand the nature of random number generation algorithms, we recommend students to implement and test correctness and performance of the MCG59 algorithm. The relevant problem statement and discussion are given in the following subsection of the book. We give the algorithm and implementation overview, formulate the main requirements and recommend using the Pearson's chi-squared test and the Kolmogorov–Smirnov test to check correctness. Then, we ask students implementing a parallel algorithm to generate a chunk of uniformly distributed numbers and analyze its performance and scalability depending of the chunk size. Different parallel programming technologies, computer architectures and random number generation algorithms can be involved.

In the next subsection, we show an example of a financial mathematics problem solved by the Monte Carlo method. In this regard, we introduce the Black-Scholes financial market model, give the main definitions including the European option and fair price concepts. We consider the case where the option price can be found analytically using the Black-Scholes formula to simplify testing correctness of the Monte Carlo method implementation. Then, the general scheme of the Monte-Carlo method for this problem including the parallel algorithm is also discussed. As in the previous subsection, we formulate general requirements and tasks for students.

According to our approach, every problem statement for students mentioned in the previous two subsections of the book, could be widen significantly to make it possible to check correctness and performance automatically. Thus, we need giving a mathematical problem statement and formulate strict requirements for the further implementation including parameters, data structures, algorithm, programming language, technologies, input/output formats, correctness and performance testing criterions. The appropriate example of one task has been done in the end of the Monte Carlo method section.

6 Conclusion

In this paper we presented a new handbook on parallel computing. The main idea of the book is to form the general principles of organizing a practical work on parallel programming with automated testing of developed implementations for correctness and performance. To this end, we have formed a pool of tasks in various areas of computer science and computational sciences, and also presented a way of formalizing the requirements for solving these problems, illustrating it with examples. The authors hope that the book could be useful in teaching students to solve problems using parallel algorithms.

References

1. Andrews, G.R.: Foundations of Parallel and Distributed Programming. Addison-Wesley Longman Publishing Co., Inc., Boston (1999)
2. Kumar, V., Grama, A., Gupta, A., Karypis, G.: Introduction to Parallel Computing, 612 p. Pearson Education, Harlow (2003)
3. Wilkinson, B., Allen, M.: Parallel Programming: Techniques and Applications Using Networked Workstations and Parallel Computers. Prentice Hall, Upper Saddle River (1999)
4. Voevodin V.V., Voevodin V.V.: Parallel Computations. BHV-Petersburg, Saint-Petersburg (2002, in Russian)
5. Gergel, V.P.: Theory and Practice of Parallel Computations. Binom, Moscow (2007). (in Russian)
6. Pacheco, P.: Parallel Programming with MPI. Morgan Kaufmann, San Francisco (1996)
7. Gropp, W., Lusk, E., Skjellum, A.: Using MPI – 2nd Edition: Portable Parallel Programming with the Message Passing Interface (Scientific and Engineering Computation). MIT Press, Cambridge (1999a)
8. Gropp, W., Lusk, E., Thakur, R.: Using MPI-3: Advanced Features of the Message Passing Interface (Scientific and Engineering Computation). MIT Press, Cambridge (1999b)
9. Nemnyugin, S., Stecik, O.: Parallel Programming for Multiprocessor Computing Systems. BHV-Petersburg, Saint-Petersburg (2002)
10. Gergel, V., et al.: Parallel Numerical Methods and Technologies. UNN Press (2013, in Russian)
11. Chandra, R., et al.: Parallel Programming in OpenMP. Morgan Kaufmann Publishers, Burlington (2000)
12. Jeffers, J., Reinders, J.: Intel Xeon Phi Coprocessor High Performance Programming. Newnes, Oxford (2013)
13. Jeffers, J., Reinders, J., Sodani, A.: Intel Xeon Phi Processor High Performance Programming: Knights Landing Edition. Morgan Kaufmann, Boston (2016)
14. Jeffers, J., Reinders, J.: High Performance Parallelism Pearls Volume One: Multicore and Many-Core Programming Approaches. Morgan Kaufmann, San Francisco (2014)
15. Jeffers, J., Reinders, J.: High Performance Parallelism Pearls Volume Two: Multicore and Many-core Programming Approaches. Morgan Kaufmann, San Francisco (2015)
16. Sanders, J., Kandrot, E.: CUDA By Example: An Introduction to General-Purpose GPU Programming. Addison-Wesley Professional, Boston (2010)
17. Pharr, M., Fernando, R.: GPU Gems 2: programming Techniques for High-Performance Graphics and General-Purpose Computation. Addison-Wesley Professional, Reading (2005)

18. Nguyen, H.: GPU Gems 3. Addison-Wesley Professional, Reading (2007)
19. Quinn, M.J.: Parallel Programming in C with MPI and OpenMP. McGraw-Hill, New York (2004)
20. Prasad, S.K., et al.: NSF/IEEE-TCPP curriculum initiative on parallel and distributed computing – core topics for undergraduates, Version I (2012). http://www.cs.gsu.edu/~tcpp/curriculum
21. Prasad, S.K., Gupta, A., Rosenberg, A.L., Sussman, A., Weems, C.C. (eds.): Topics in Parallel and Distributed Computing: Introducing Concurrency in Undergraduate Courses. Morgan Kaufmann, San Francisco (2015)
22. Voevodin, V., Gergel, V., Popova, N.: Challenges of a systematic approach to parallel computing and supercomputing education. In: Hunold, S., et al. (eds.) Euro-Par 2015. LNCS, vol. 9523, pp. 90–101. Springer, Cham (2015). https://doi.org/10.1007/978-3-319-27308-2_8
23. Gergel, V., Liniov, A., Meyerov, I., Sysoyev, A.: NSF/IEEE-TCPP curriculum implementation at University of Nizhni Novgorod. In: Proceedings of Fourth NSF/TCPP Workshop on Parallel and Distributed Computing Education, pp. 1079–1084 (2014)
24. Gergel, V., Kozinov, E., Linev, A., Shtanyk, A.: Educational and research systems for evaluating the efficiency of parallel computations. In: Carretero, J., et al. (eds.) ICA3PP 2016. LNCS, vol. 10049, pp. 278–290. Springer, Cham (2016). https://doi.org/10.1007/978-3-319-49956-7_22

Dependable and Coordinated Resources Allocation Algorithms for Distributed Computing

Victor Toporkov[(⊠)] and Dmitry Yemelyanov

National Research University "MPEI", Moscow, Russia
{ToporkovVV,YemelyanovDM}@mpei.ru

Abstract. In this work, we introduce slot selection and co-allocation algorithms for parallel jobs in distributed computing with non-dedicated and heterogeneous resources. A single slot is a time span that can be assigned to a task, which is a part of a parallel job. The job launch requires a co-allocation of a specified number of slots starting and finishing synchronously. Some existing resource co-allocation algorithms assign a job to the first set of slots matching the resource request without any optimization (the first fit type), while other algorithms are based on an exhaustive search. In this paper, algorithms for efficient, dependable and coordinated slot selection are studied and compared with known approaches. The novelty of the proposed approach is in a general algorithm efficiently selecting a set of slots according to the specified criterion.

Keywords: Distributed computing · Grid · Dependability
Coordinated scheduling · Resource management · Slot · Job · Allocation
Optimization

1 Introduction

Modern high-performance distributed computing systems (HPCS), including Grid, cloud and hybrid infrastructures provide access to large amounts of resources [1, 2]. These resources are typically required to execute parallel jobs submitted by HPCS users and include computing nodes, data storages, network channels, software, etc. These resources are usually partly utilized or reserved by high-priority jobs and jobs coming from the resource owners. Thus, the available resources are represented with a set of time intervals (slots) during which the individual computational nodes are capable to execute parts of independent users' parallel jobs. These slots generally have different start and finish times and vary in performance level. The presence of a set of heterogeneous slots impedes the problem of resources allocation necessary to execute the job flow from HPCS users. Resource fragmentation also results in a decrease of the total computing environment utilization level [1, 2].

There are different approaches for a job-flow scheduling problem in distributed computing environments. Multi-agent application level scheduling [3] actually performs individual jobs execution optimization and, as a rule, does not imply any global resource sharing or allocation policy. Such approach with an unrestricted competition

© Springer Nature Switzerland AG 2019
V. Voevodin and S. Sobolev (Eds.): RuSCDays 2018, CCIS 965, pp. 531–542, 2019.
https://doi.org/10.1007/978-3-030-05807-4_45

for the available computing resources may result in an inefficient and unbalanced resources usage and hence poor overall job-flow execution efficiency.

Job flow scheduling in virtual organizations (VO) [4, 5] suggests uniform rules of resource sharing and consumption, in particular based on economic models. This approach allows improving the job-flow level scheduling and resource distribution efficiency. VO formation and performance largely depends on mutually beneficial collaboration between all the related stakeholders. However, users' preferences and owners' and administrators' preferences may conflict with each other. Users are likely to be interested in the fastest possible running time for their jobs with least possible costs whereas VO preferences are usually tuned for available resources load balancing or node owners' profit boosting. Thus, VO policies in general should respect all members and the most important aspect of rules suggested by VO is their fairness. At the same time VO scheduling policies usually limit individual jobs optimization opportunities, may violate queue order and possess disadvantages common for centralized scheduling structures [6]. In order to implement any of the described job-flow scheduling schemes and policies, first, one needs an algorithm for selecting sets of simultaneously available slots required for each job execution. Further, we shall call such set of simultaneously available slots with the same start and finish times as execution *window*.

In this paper, we study algorithms for optimal or near-optimal heterogeneous resources selection by a given criterion with the restriction to a total cost. Additionally we consider practical implementations for a dependable resources allocation problem.

The rest of the paper is organized as follows. Section 2 presents related works. Section 3 introduces a general scheme for searching slot sets efficient by the specified criterion. Then several implementations are proposed and considered. Section 4 contains simulation results for comparison of proposed and known algorithms. Section 5 summarizes the paper and describes further research topics.

2 Related Works

The scheduling problem in Grid is *NP*-hard due to its combinatorial nature and many heuristic-based solutions have been proposed. In [7] heuristic algorithms for slot selection, based on user-defined utility functions, are introduced. NWIRE system [7] performs a slot window allocation based on the user defined efficiency criterion under the maximum total execution cost constraint. However, the optimization occurs only on the stage of the best found offer selection. First fit slot selection algorithms (backtrack [8] and NorduGrid [9] approaches) assign any job to the first set of slots matching the resource request conditions, while other algorithms use an exhaustive search [10–12] and some of them are based on a linear integer programming (IP) [10] or mixed-integer programming (MIP) model [11]. Moab scheduler [13] implements the backfilling algorithm and during a slot window search does not take into account any additive constraints such as the minimum required storage volume or the maximum allowed total allocation cost.

Modern distributed and cloud computing simulators GridSim and CloudSim [14, 15] provide tools for jobs execution and co-allocation of simultaneously available computing resources. Base simulator distributions perform First Fit allocation algorithms without any specific optimization. CloudAuction extension [15] of CloudSim implements a double auction to distribute datacenters' resources between a job flow with a fair allocation policy. All these algorithms consider price constraints on individual nodes and not on a total window allocation cost. However, as we showed in [16], algorithms with a total cost constraint are able to perform the search among a wider set of resources and increase the overall scheduling efficiency.

GrAS [17] is a Grid job-flow management system built over Maui scheduler [13]. The resources co-allocation algorithm retrieves a set of simultaneously available slots with the same start and finish times even in heterogeneous environments. However, the algorithm stops after finding the first suitable window and, thus, doesn't perform any optimization except for window start time minimization.

Algorithm [18] performs job's response and finish time minimization and doesn't take into account constraint on a total allocation budget. [19] performs window search on a list of slots sorted by their start time, implements algorithms for window shifting and finish time minimization, doesn't support other optimization criteria and the overall job execution cost constraint.

AEP algorithm [20] performs window search with constraint on a total resources allocation cost, implements optimization according to a number of criteria, but doesn't support a general case optimization. Besides AEP doesn't guarantee same finish time for the window slots in heterogeneous environments and, thus, has limited practical applicability.

Main contribution of this paper is a window co-allocation algorithm performing resources selection according to the user requirements and restrictions. The novelty of the proposed approach consists in implementing a dynamic programming scheme in order to optimize heterogeneous resources selection according to the scheduling policy.

3 Resource Selection Algorithm

3.1 General Problem Statement

We consider a set R of heterogeneous computing nodes with different performance p_i and price c_i characteristics. Each node has a local utilization schedule known in advance for a considered scheduling horizon time L. A node may be turned off or on by the provider, transferred to a maintenance state, reserved to perform computational jobs. Thus, it's convenient to represent all available resources as a set of slots. Each slot corresponds to one computing node on which it's allocated and may be characterized by its performance and price.

In order to execute a parallel job one needs to allocate the specified number of simultaneously idle nodes ensuring user requirements from the resource request. The resource request specifies number n of nodes required simultaneously, their minimum applicable performance p, job's computational volume V and a maximum available resources allocation budget C. The required window length is defined based on a slot with the minimum performance. For example, if a window consists of slots with performances $p \in \{p_i, p_j\}$ and $p_i < p_j$, then we need to allocate all the slots for a time

$T = \frac{V}{p_i}$. In this way V really defines a computational volume for each single job subtask. Common start and finish times ensure the possibility of inter-node communications during the whole job execution. The total cost of a window allocation is then calculated as $C_W = \sum_{i=1}^{n} T * c_i$.

These parameters constitute a formal generalization for resource requests common among distributed computing systems and simulators. The overall problem statement lacks specific features of internal nodes processing as well as network communication aspects that are important for a job execution performance in data-intensive HPC systems. However, the problem remains independent from specific HPCS configurations and, thus, the problem solutions may be evaluated against general case criteria. For this purpose we introduce criterion f representing a user preference for the particular job execution during the scheduling horizon L. f can take a form of any additive function and as an example, one may want to allocate suitable resources with the maximum possible total data storage available before the specified deadline.

3.2 General Window Search Procedure

For a general window search procedure for the problem statement presented in Sect. 3.1, we combined core ideas and solutions from algorithm AEP [20] and systems [17, 19]. Both related algorithms perform window search procedure based on a list of slots retrieved from a heterogeneous computing environment.

Following is the general square window search algorithm. It allocates a set of n simultaneously available slots with performance $p_i > p$, for a time, required to compute V instructions on each node, with a restriction C on a total allocation cost and performs optimization according to criterion f. It takes a list of available slots ordered by their non-decreasing start time as input.

1. Initializing variables for the best criterion value and corresponding best window: $f_{max} = 0$, $w_{max} = \{\}$.
2. From the slots available we select different groups by node performance p_i. For example, group P_k contains resources allocated on nodes with performance $p_i \geq P_k$. Thus, one slot may be included in several groups.
3. Next is a cycle for all retrieved groups P_i starting from the max performance P_{max}. All the sub-items represent a cycle body.
   a. The resources reservation time required to compute V instructions on a node with performance P_i is $T_i = \frac{V}{p_i}$.
   b. Initializing variable for a window candidates list $S_W = \{\}$.
   c. Next is a cycle for all slots s_i in group P_i starting from the slot with the minimum start time. The slots of group P_i should be ordered by their non-decreasing start time. All the sub-items represent a cycle body.
   (1) If slot s_i doesn't satisfy user requirements (hardware, software, etc.) then continue to the next slot (3c).
   (2) If slot length $l(s_i) < T_i$ then continue to the next slot (3c).
   (3) Set the new window start time $W_i.start = s_i.start$.
   (4) Add slot s_i to the current window slot list S_W
   (5) Next a cycle to check all slots s_j inside S_W

    i. If there are no slots in S_W with performance $P(s_j) = P_i$ then continue to the next slot (3c), as current slots combination in S_W was already considered for previous group P_{i-1}.

    ii. If $W_i.start + T_i > s_j.end$ then remove slot s_j from S_W as it can't consist in a window with the new start time $W_i.start$.

(6) If S_W size is greater or equal to n, then allocate from S_W a window W_i (a subset of n slots with start time $W_i.start$ and length T_i) with a maximum criterion value f_i and a total cost $C_i < C$. If $f_i > f_{max}$ then reassign $f_{max} = f_i$ and $W_{max} = W_i$.

4. End of algorithm. At the output variable W_{max} contains the resulting window with the maximum criterion value f_{max}.

3.3   Optimal Slot Subset Allocation

Let us discuss in more details the procedure which allocates an optimal (according to a criterion f) subset of n slots out of S_W list (algorithm step 3c(6)). For some particular criterion function f a straightforward subset allocation solution may be offered. For example for a window finish time minimization it is reasonable to return at step 3c(6) the first n cheapest slots of S_W provided that they satisfy the restriction on the total cost. These n slots will provide $W_i.finish = W_i.start + T_i$, so we need to set $f_i = -(W_i.start + T_i)$ to *minimize* the finish time at the end of the algorithm.

However in a general case we should consider a subset allocation problem with some additive criterion: $Z = \sum_{i=1}^{n} c_z(s_i)$, where $c_z(s_i) = z_i$ is a target optimization characteristic value provided by a single slot s_i of W_i. In this way we can state the following problem of an optimal n - size window subset allocation out of m slots stored in S_W:

$$Z = x_1 z_1 + x_2 z_2 + \ldots + x_m z_m, \tag{1}$$

with the following restrictions:

$$x_1 c_1 + x_2 c_2 + \ldots + x_m c_m \leq C,$$
$$x_1 + x_2 + \ldots + x_m = n,$$
$$x_i \in \{0, 1\}, \ i = 1..m,$$

where z_i is a target characteristic value provided by slot s_i, c_i is total cost required to allocate slot s_i for a time T_i, x_i - is a decision variable determining whether to allocate slot s_i ($x_i = 1$) or not ($x_i = 0$) for the current window.

This problem relates to the class of integer linear programming problems and we used 0–1 knapsack problem as a base for our implementation. The classical 0-1 knapsack problem with a total weight C and items-slots with weights c_i and values z_i have the same formal model (1) except for extra restriction on the number of items required: $x_1 + x_2 + \ldots + x_m = n$. To take this into account we implemented the following dynamic programming recurrent scheme:

$$f_i(C_j, n_k) = \max\{f_{i-1}(C_j, n_k), f_{i-1}(C_j - c_i, n_k - 1) + z_i\},$$
$$i = 1, .., m, \ C_j = 1, .., C, \ n_k = 1, .., n, \tag{2}$$

where $f_i(C_j, n_k)$ defines the maximum Z criterion value for n_k-size window allocated out of first i slots from S_W for a budget C_j. After the forward induction procedure (2) is finished the maximum value $Z_{max} = f_m(C, n)$. x_i values are then obtained by a backward induction procedure.

For the actual implementation we initialized $f_i(C_j, 0) = 0$, meaning $Z = 0$ when we have no items in the knapsack. Then we perform forward propagation and calculate $f_1(C_j, n_k)$ values for all C_j and n_k based on the first item and the initialized values. Then $f_2(C_j, n_k)$ is calculated taking into account second item and $f_1(C_j, n_k)$ and so on. So after the forward propagation procedure (2) is finished the maximum value $Z_{max} = f_m(C, n)$. Corresponding values for variables x_i are then obtained by a backward propagation procedure.

An estimated computational complexity of the presented recurrent scheme is $O(m * n * C)$, which is n times harder compared to the original knapsack problem ($O(m * C)$). On the one hand, in practical job resources allocation cases this overhead doesn't look very large as we may assume that $n < < m$ and $n < < C$. On the other hand, this subset allocation procedure (2) may be called multiple times during the general square window search algorithm (step 3c(6)).

3.4 Dependable and Coordinated Resources Allocation

As a practical implementation for a general optimization scheme we propose to study a resources allocation placement problem. Figure 1 shows Gantt chart of 4 slots co-allocation (hollow rectangles) in a computing environment with resources pre-utilized with local and high-priority tasks (filled rectangles).

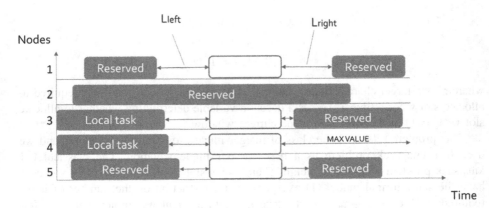

Fig. 1. Dependable window co-allocation metrics.

As can be seen from Fig. 1, even using the same computing nodes (1, 3, 4, 5 on Fig. 1) there are usually multiple window placement options with respect to the slots start time. The window placement generally may affect such job execution properties as cost, finish time, computing energy efficiency, etc. Besides, slots *proximity* to neighboring tasks reserved on the same computing nodes may affect a probability of the job execution delay or failure. For example, a slot reserved too close to the previous task on the same node may be delayed or cancelled by an unexpected delay of the latter. Thus, dependable resources allocation may require reserving resources with some reasonable distance to the neighboring tasks.

As presented in Fig. 1, for each window slot we can estimate times to the previous task finish time: L_{left} and to the next task start time: L_{right}. Using these values the following criterion for the window allocation represents average time distance to the nearest neighboring tasks: $L_{min \ \Sigma} = \frac{1}{n} \sum_{i=1}^{n} min(L_{left \ i}, L_{right \ i})$, where n is a total number of slots in the window. So when implementing a dependable job scheduling policy we are interested in maximizing $L_{min \ \Sigma}$ value.

On the other hand such *selfish* and individual job-centric resources allocation policy may result in an additional resources fragmentation and, hence, inefficient resources usage. Indeed, when $L_{min \ \Sigma}$ is maximized the jobs will try to start at the maximum distance from each other, eventually leaving truncated slots between them. Thus, the subsequent jobs may be delayed in the queue due to insufficient remaining resources.

For a coordinated job-flow scheduling and resources load balancing we propose the following window allocation criterion representing average time distance to the farthest neighboring tasks: $L_{max \ \Sigma} = \frac{1}{n} \sum_{i=1}^{n} max(L_{left \ i}, L_{right \ i})$, where n is a total number of slots in the window. By minimizing $L_{max \ \Sigma}$ our motivation is to find a set of available resources best suited for the particular job configuration and duration. This *coordinated* approach opposes selfish resources allocation and is more relevant for a virtual organization job-flow scheduling procedure.

4 Simulation Study

4.1 Simulation Environment Setup

An experiment was prepared as follows using a custom distributed environment simulator [2, 16, 20]. For our purpose, it implements a heterogeneous resource domain model: nodes have different usage costs and performance levels. A space-shared resources allocation policy simulates a local queuing system (like in GridSim or CloudSim [14]) and, thus, each node can process only one task at any given simulation time. The execution cost of each task depends on its execution time, which is proportional to the dedicated node's performance level. The execution of a single job requires parallel execution of all its tasks.

During the experiment series we performed a window search operation for a job requesting $n = 7$ nodes with performance level $p_i \geq 1$, computational volume $V = 800$ and a maximum budget allowed is $C = 644$. During each experiment a new instance for the computing environment was automatically generated with the following properties. The resource pool includes 100 heterogeneous computational nodes.

Each node performance level is given as a uniformly distributed random value in the interval [2, 10]. So the required window length may vary from 400 to 80 time units. The scheduling interval length is 1200 time quanta which is enough to run the job on nodes with the minimum performance. However, we introduce the initial resource load with advanced reservations and local jobs to complicate conditions for the search operation. This additional load is distributed hyper-geometrically and results in up to 30% utilization for each node (Fig. 2).

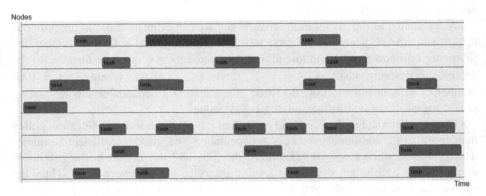

Fig. 2. Initial resources utilization example.

Additionally an independent value $q_i \in [0; 10]$ is randomly generated for each computing node i to compare algorithms against $Q = \sum_{i=1}^{n} q_i$ window allocation criterion.

4.2 General Algorithms Comparison

Firstly we intend to study the proposed resources allocation algorithm against an abstract general-case criterion Q. For this purpose we implemented the following window search algorithms based on the general window search procedure introduced in Sect. 3.2.

- *FirstFit* performs a square window allocation in accordance with a general scheme described in Sect. 3.2. Returns first suitable and affordable window found. In fact, performs window start time minimization and represents algorithm from [17, 19].
- *MultipleBest* algorithm searches for multiple non-intersecting alternative windows using *FirstFit* algorithm. When all possible window allocations are retrieved the algorithm searches among them for alternatives with the maximum Q value. In this way *MultipleBest* is similar to [7] approach.
- *MaxQ* implements a general square window search procedure with an optimal slots subset allocation (2) to return a window with maximum total Q value.
- *MaxQ Lite* follows the general square window search procedure but doesn't implement slots subset allocation (2) procedure. Instead at step 3c(6) it returns the first n cheapest slots of S_W. The total Q value of these n slots is returned as a target

criterion, which is then maximized during the search procedure. Thus, *MaxQ Lite* has much less computational complexity compared to *MaxQ* but doesn't guarantee an accurate solution [20].

Figure 3 shows average $Q = \sum_{i=1}^{n} q_i$ value obtained during the simulation. Parameter q_i was generated randomly on a [0; 10] interval and is independent from other node's characteristics. Thus, for a single window of 7 slots we have the following practical limits specific for our experiment: $Q \in [0; 70]$.

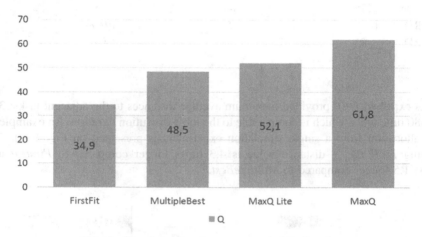

Fig. 3. Simulation results: average window Q value.

As can be seen from Fig. 3, *MaxQ* is indeed provided the maximum average criterion value $Q = 61.8$, which is quite close to the practical maximum, especially compared to other algorithms. The advantage over *MultipleBest* and *MaxQ Lite* is almost 20%. *MaxQ Lite* implements a simple heuristic but still is able to provide a better solution compared to the best of 50 different alternative executions retrieved by *MultipleBest*. *First Fit* provided average Q value exactly in the middle of [0; 70] which is 44% less compared to *MaxQ*.

4.3 Dependable Resources Allocation

For the window placement problem along with *FirstFit* and *MultipleBest* we introduce two pairs of algorithms based on *MaxQ* and *MaxQ Lite* approaches.

- *Dependable (DEP)* and *DEP Lite* perform $L_{min\ \Sigma}$ maximization, i.e. maximize the distance to the nearest running or reserved tasks.
- *Coordinated (COORD)* and *COORD Lite* minimize $L_{max\ \Sigma}$: average distance to the farthest neighboring tasks.

So, by setting $L_{min\ \Sigma}$ and $L_{max\ \Sigma}$ as target optimization criteria we performed scheduling simulation with the same settings described in Sect. 4.1. The results of 2000 independent scheduling cycles are compiled in Table 1.

Table 1. Window placement simulation results

Algorithm	Distance to the nearest task $L_{min\Sigma}$	Distance to the farthest task $L_{max\Sigma}$	Average operational time, ms
Multiple best	253	*159*	*103*
First fit	85	*342*	**4.2**
DEP	**369**	**480**	*1695*
DEP Lite	275	*440*	*4.5*
COORD	**9**	**52**	*1694*
COORD lite	31	*148*	*4.5*

As expected *DEP* provided maximum average distances to the adjacent tasks: 369 and 480 time units, which is comparable to the job's execution duration. An example of such allocation from a single simulation experiment is presented on Fig. 4 (a). The resulting *DEP* $L_{min\ \Sigma}$ distance value is 4.3 times longer compared to *FirstFit* and almost 1.5 longer compared to *MultipleBest*.

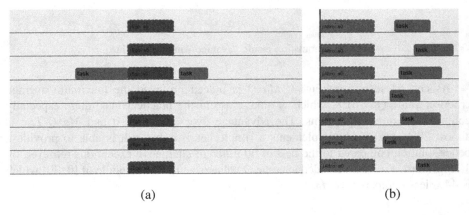

(a)   (b)

Fig. 4. Simulation examples for dependable (a) and coordinated (b) resources allocation for the same job.

Similarly, *COORD* provided minimum values for the considered criteria: 9 and 52 time units. Example allocation is presented on Fig. 4 (b) where left edge represents the scheduling interval start time. As can be seen from the figure the allocated slots are highly coincident with the job's configuration and duration. Here the resulting average distance to the farthest task is three times smaller compared to *MultipleBest* and 9 times smaller when compared with *DEP* solution.

However due to a higher computational complexity it took *DEP* and *COORD* almost 1.7 s to find the 7-slots allocation over 100 available computing nodes, which is

17 times longer compared to *Multiple Best*. At the same time simplified *Lite* implementations provided better scheduling results compared to *Multiple Best* for even less operational time: 4.5 ms. *FirstFit* doesn't perform any target criteria optimization and, thus, provides average L_{min} Σ and L_{max} Σ distances with the same operational time as *Lite* algorithms.

MultipleBest in Table 1 has average distance to the farthest task smaller than to the nearest task because different alternatives were selected to match the criteria: L_{min} Σ maximization and L_{max} Σ minimization. Totally almost 50 different resource allocation alternatives were retrieved and considered by *MultipleBest* during each experiment.

5 Conclusion and Future Work

In this work, we address the problems of dependable and coordinated slot selection and co-allocation for parallel jobs in distributed computing with non-dedicated resources. For this purpose a general window allocation algorithm was proposed along with two practical implementations: for dependable and coordinated resources allocation policies.

A simulation study was carried out to prove the algorithm's optimization efficiency according to the target criteria. As a result, the advantage of the proposed general scheme over traditional scheduling algorithms reaches 20% against an abstract general case criterion and more than 50% when we consider window placement problem.

As a drawback, the general case algorithm has a relatively high computational complexity, especially compared to First Fit approach. In our further work, we will refine a general resource co-allocation scheme in order to decrease its computational complexity.

Acknowledgments. This work was partially supported by the Council on Grants of the President of the Russian Federation for State Support of Young Scientists (YPhD-2297.2017.9), RFBR (grants 18-07-00456 and 18-07-00534) and by the Ministry on Education and Science of the Russian Federation (project no. 2.9606.2017/8.9).

References

1. Dimitriadou, S.K., Karatza, H.D.: Job scheduling in a distributed system using backfilling with inaccurate runtime computations. In: Proceedings of 2010 International Conference on Complex, Intelligent and Software Intensive Systems, pp. 329–336 (2010)
2. Toporkov, V., Toporkova, A., Tselishchev, A., Yemelyanov, D., Potekhin, P.: Heuristic strategies for preference-based scheduling in virtual organizations of utility grids. J. Ambient Intell. Hum. Comput. **6**(6), 733–740 (2015)
3. Buyya, R., Abramson, D., Giddy, J.: Economic models for resource management and scheduling in grid computing. J. Concurr. Comput. Pract. Exp. **5**(14), 1507–1542 (2002)
4. Foster, I., Kesselman, C., Tuecke, S.: The anatomy of the grid: enabling scalable virtual organizations. Int. J. High Perform. Comput. Appl. **15**(3), 200–222 (2001)
5. Carroll, T., Grosu, D.: Formation of virtual organizations in grids: a game-theoretic approach. Econ. Mod. Algorithms Distrib. Syst. **22**(14), 63–81 (2009)

6. Yang, R., Xu, J.: Computing at massive scale: scalability and dependability challenges. In: 2016 IEEE Symposium on Service-Oriented System Engineering (SOSE), pp. 386–397 (2016)

7. Ernemann, C., Hamscher, V., Yahyapour, R.: Economic scheduling in grid computing. In: Feitelson, Dror G., Rudolph, L., Schwiegelshohn, U. (eds.) JSSPP 2002. LNCS, vol. 2537, pp. 128–152. Springer, Heidelberg (2002). https://doi.org/10.1007/3-540-36180-4_8

8. Aida, K., Casanova, H.: Scheduling mixed-parallel applications with advance reservations. In: 17th IEEE International Symposium on HPDC, pp. 65–74. IEEE CS Press, New York (2008)

9. Elmroth, E., Tordsson, J.: A standards-based grid resource brokering service supporting advance reservations, co-allocation and cross-grid interoperability. J. Concurr. Comput. Pract. Exp. **25**(18), 2298–2335 (2009)

10. Takefusa, A., Nakada, H., Kudoh, T., Tanaka, Y.: An advance reservation-based co-allocation algorithm for distributed computers and network bandwidth on QoS-guaranteed grids. In: Frachtenberg, E., Schwiegelshohn, U. (eds.) JSSPP 2010. LNCS, vol. 6253, pp. 16–34. Springer, Heidelberg (2010). https://doi.org/10.1007/978-3-642-16505-4_2

11. Blanco, H., Guirado, F., Lérida, J.L., Albornoz, V.M.: MIP model scheduling for multi-clusters. In: Caragiannis, I., et al. (eds.) Euro-Par 2012. LNCS, vol. 7640, pp. 196–206. Springer, Heidelberg (2013). https://doi.org/10.1007/978-3-642-36949-0_22

12. Garg, S.K., Konugurthi, P., Buyya, R.: A linear programming-driven genetic algorithm for meta-scheduling on utility grids. Int. J. Parallel Emergent Distrib. Syst. **26**, 493–517 (2011)

13. Moab Adaptive Computing. http://www.adaptivecomputing.com. Accessed 12 Apr 2018

14. Calheiros, R.N., Ranjan, R., Beloglazov, A., De Rose, C.A.F., Buyya, R.: CloudSim: a toolkit for modeling and simulation of cloud computing environments and evaluation of resource provisioning algorithms. J. Softw. Pract. Exp. **41**(1), 23–50 (2011)

15. Samimi, P., Teimouri, Y., Mukhtar, M.: A combinatorial double auction resource allocation model in cloud computing. J. Inf. Sci. 357 **C**, 201–216 (2016)

16. Toporkov, V., Toporkova, A., Bobchenkov, A., Yemelyanov, D.: Resource selection algorithms for economic scheduling in distributed systems. In: Proceedings of International Conference on Computational Science, ICCS 2011, 1–3 June 2011, Singapore, Procedia Computer Science, vol. 4, pp. 2267–2276. Elsevier (2011)

17. Kovalenko, V.N., Koryagin, D.A.: The grid: analysis of basic principles and ways of application. J. Programm. Comput. Softw. **35**(1), 18–34 (2009)

18. Makhlouf, S., Yagoubi, B.: Resources co-allocation strategies in grid computing. In: CIIA, CEUR Workshop Proceedings, vol. 825 (2011)

19. Netto, M.A.S., Buyya, R.: A flexible resource co-allocation model based on advance reservations with rescheduling support. In: Technical Report, GRIDSTR-2007-17, Grid Computing and Distributed Systems Laboratory, The University of Melbourne, Australia, 9 October 2007

20. Toporkov, V., Toporkova, A., Tselishchev, A., Yemelyanov, D.: Slot selection algorithms in distributed computing. J. Supercomput. **69**(1), 53–60 (2014)

Deploying Elbrus VLIW CPU Ecosystem for Materials Science Calculations: Performance and Problems

Vladimir Stegailov[1,2,3] and Alexey Timofeev[1,2,3]($\boxtimes$)

[1] Joint Institute for High Temperatures of the Russian
Academy of Sciences, Moscow, Russia
[2] National Research University Higher School of Economics, Moscow, Russia
[3] Moscow Institute of Physics and Technology (State University),
Dolgoprudny, Russia
timofeevalvl@gmail.com

Abstract. Modern Elbrus-4S and Elbrus-8S processors show floating point performance comparable to the popular Intel processors in the field of high-performance computing. Tasks oriented to take advantage of the VLIW architecture show even greater efficiency on Elbrus processors. In this paper the efficiency of the most popular materials science codes in the field of classical molecular dynamics and quantum-mechanical calculations is considered. A comparative analysis of the performance of these codes on Elbrus processor and other modern processors is carried out.

Keywords: Elbrus architecture · VASP · LAMMPS · FFT

1 Introduction

A large part of HPC resources installed during the last decade is based on Intel CPUs. However, the situation is gradually changing. In March 2017, AMD released the first processors based on the novel x86_64 architecture called Zen. In November 2017, Cavium has presented server grade 64-bit ThunderX2 ARMv8 CPUs that are to be deployed in new Cray supercomputers. The Elbrus microprocessors stand among emerging types of high performance CPU architectures [1,2].

The diversity of CPU types complicates significantly the choice of the best variant for a particular HPC system. The main criterion is certainly the time-to-solution of a given computational task or a set of different tasks that represents an envisaged workload of a system under development.

Computational materials science provides an essential part of the deployment time for high performance computing (HPC) resources worldwide. The VASP code [3–6] is among the most popular programs for electronic structure calculations that gives the possibility to calculate materials properties using the non-empirical (so called *ab initio*) methods. Ab initio calculation methods based

© Springer Nature Switzerland AG 2019
V. Voevodin and S. Sobolev (Eds.): RuSCDays 2018, CCIS 965, pp. 543–553, 2019.
https://doi.org/10.1007/978-3-030-05807-4_46

on quantum mechanics are important modern scientific tools (e.g., see [7–11]). According to the recent estimates, VASP alone consumes up to 15–20 percent of the world's supercomputing power [12,13]. Such unprecedented popularity justifies the special attention to the optimization of VASP for both existing and novel computer architectures (e.g. see [14]).

Significant part of calculation time of such software packages for computational materials science is the execution time of the Fourier transform. One of the most time consuming components in VASP is 3D-FFT [15]. FFT libraries are tested on the Elbrus processor in order to determine the most optimal tool for performing a fast Fourier transform. The EML (Elbrus Multimedia Library), developed by Elbrus processor manufacturer, and the most popular FFTW library are under consideration.

In this work we present the efficiency analysis of Elbrus CPUs in comparison with Intel Xeon Haswell CPUs using a typical VASP workload example. The results of the FFT libraries testing on Elbrus processors are presented.

2   Related Work

HPC systems are notorious for operating at a small fraction of their peak performance and the deployment of multi-core and multi-socket compute nodes further complicates performance optimization. Many attempts have been made to develop a more or less universal framework for algorithms optimization that takes into account essential properties of the hardware (see e.g. [16,17]). The recent work of Stanisic et al. [18] emphasizes many pitfalls encountered while trying to characterize both the network and the memory performance of modern machines.

A fast Fourier transform is used in computational modeling programs for calculations related to quantum computations, Coulomb systems, etc. and takes a very substantial part of the program's running time [19], especially for VASP [15]. The detailed optimization of the computation of 3D-FFT in VASP in order to prepare the code for an efficient execution on multi- and many-core CPUs like Intel's Xeon Phi is considered in the article [15]. In this article the threading performance of widely used FFTW (Cray LibSci) and Intel's MKL on a current Cray-XC40 with Intel Haswell CPUs and a Cray-XC30 Xeon Phi (Knights Corner, KNC) system is evaluated.

At the moment, Elbrus processors are ready for use [1,2], so we decided to benchmark them using one of the main HPC tools used for material science studies (VASP) and the library that determines the performance of this code (FFT). The architecture of the Elbrus processors [1,2] allows us to expect that the butterfly cross-linking during the execution of the FFT performs during a smaller number of cycles than for CPUs like Intel's Xeon Phi.

3   Methods and Software Implementation

3.1   Test Model in VASP

VASP 5.4.1 is compiled for Intel systems using Intel Fortran, Intel MPI and linked with Intel MKL for BLAS, LAPACK and FFT calls. For the Elbrus-8S system, lfortran compatible with gfortran ver.4.8 is used together with MPICH, EML BLAS, Netlib LAPACK and FFTW libraries.

Our test model in VASP represents the liquid Si system consisting of 48 atoms in the supercell. The Perdew-Burke-Ernzerhof model for xc-functional is used. The calculation protocol corresponds to molecular dynamics. We use the time for the first iteration of electron density optimization τ_{iter} as a target parameter of the performance metric.

The τ_{iter} values considered in this work are about 5–50 s and correspond to performance of a single CPU. At the first glance, these are not very long times to be accelerated. However, *ab initio* molecular dynamics requires usually $10^4 - 10^5$ time steps and larger system sizes. That is why the decrease of τ_{iter} by several orders of magnitude is an actual problem for modern HPC systems targeted at materials science computing.

The choice of a particular test model has a certain influence on the benchmarking results. However, our preliminary tests of other VASP models show that the main conclusions of this study do not depend significantly on a particular model.

3.2   Fast Fourier Transform

FFTW 3.3.6 is compiled using lcc, the analogue of gcc for Elbrus-4S and Elbrus-8S systems. As an input array for the Fourier transforms, a sinusoidal signal, white, pink and brown noise are used. In this article the results are presented for white noise.

The usual pattern when calling FFT (or MKL through its FFTW interface) is as follows:

1. Preparation stage: create plans for FFT computations, e.g., for FFTW via fftw_plan p=fftw_plan_dft(..), for EML via eml_Signal_FFTInit(...).
2. Execution stage: perform FFT computation using that plan, e.g., for FFTW via fftw_execute_dft(p,in,out), for EML via eml_Signal_FFTFwd(...).
3. Clean up.

We consider the work of the first two stages, since they are the most time consuming. Preparation takes the main time when you start the Fourier transform once for a fixed size of the input array. When the Fourier transform is repeatedly started, the running time of the program can determine the execution time of the Fourier transform itself.

So for these two stages we compare libraries FFTW and EML on processors Elbrus-4S and Elbrus 8S. The library EML has fewer useful functions than in the library FFTW for now. In particular, the size of the input array can only be a power of two, so the preparation stage has to be partially implemented by users. The number of functions in the library EML is much smaller than in the library FFTW.

Plan creation with FFTW can happen with differently expensive planner schemes: FFTW_ESTIMATE (cheap), FFTW_MEASURE (expensive), FFTW_PATIENT (more expensive) and FFTW_EXHAUSTIVE (most expensive). Except for FFTW_ESTIMATE plan creation involves testing different FFT algorithms together with runtime measurements to achieve best performance on the target platform. On servers with processors Elbrus-4S and Elbrus-8S due to the lack of libraries authors managed to compile FFTW only for use in the mode FFTW_ESTIMATE, in which the preparation time is short and the execution time is long.

To average the resulting values of the operating time and to obtain a spread of results, the calculations were repeated from 30 to 1000 times. The spread of the results was within 1%, and sometimes did not exceed 0.001%.

4   Results and Discussion

4.1   VASP Benchmark on Elbrus-8S and Xeon Haswell CPUs

VASP is known to be both a memory-bound and a compute-bound code [14]. Figure 1a shows the results of the liquid Si model test runs. Benchmarks with Intel Haswell CPUs are performed with RAM working at 2133 MHz frequency and C612 chipset.

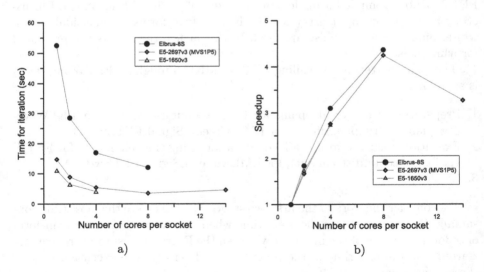

Fig. 1. The dependence of the (a) time and (b) speedup of the first iteration of the liquid Si model test on the number of cores per socket

Figure 1b presents the same data as shown on Fig. 1a but in the coordinates of speedup and number of cores. The considered Intel processors and processor Elbrus show a similar dynamics of speedup. The Elbrus processor shows slightly bigger speedup. It is seen that for the E5-2697v3 processor the optimal number of cores is between 4 and 14, approximately 8. The form of the dependence suggests that a greater speedup value can be achieved for the Elbrus processor.

4.2   Fast Fourier Transform on Elbrus CPUs: EML vs FFTW

We divide the process the Fourier transformation into two stages: the preparation of the algorithm (Figs. 2, 3, 4 and 5), and the execution of the transformation (Figs. 6, 7 and 8). Preparation takes the main time when you start the Fourier transform once. The algorithm execution time can determine the total running time of Fourier transform for the situations when the Fourier transform is started many times for a fixed size of the input array.

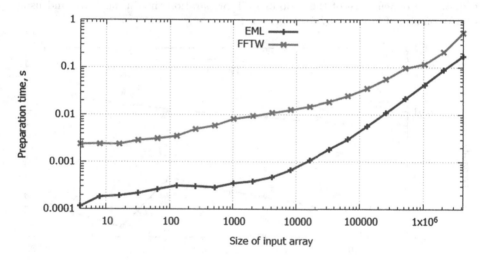

Fig. 2. The dependence of the time for FFT preparation on size of input array for Elbrus-4S

The preparation time of the algorithm FFT for Elbrus-4S appears to be an order of magnitude smaller using EML library in comparison with FFTW for array size smaller than 2^{15} (Figs. 2 and 3). And for large sizes of the array preparation time using EML is only 2–3 smaller than using FFTW. All points have an error less than 1%.

Figures 4 and 5 show that for the Elbrus-8S, the difference in preparation time is even greater. For arrays smaller than 2^{15} the preparation time using EML is 10–20 times less than using FFTW. And for large arrays up to 2^{17} the preparation time using EML is 50–90 times less than the one using FFTW.

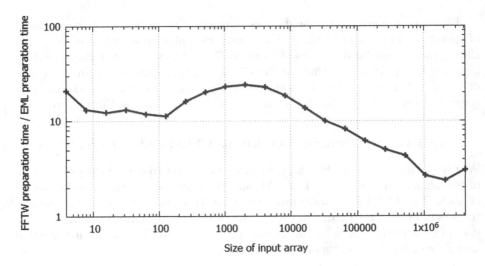

Fig. 3. The dependence of the ratio of FFT preparation time using EML and using FFTW on size of input array for Elbrus-4S

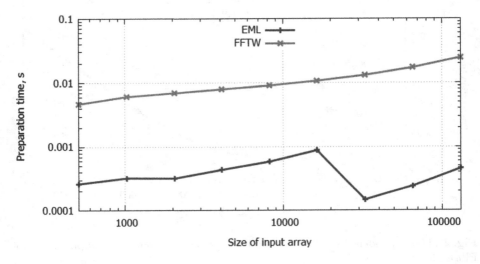

Fig. 4. The dependence of the time for FFT preparation on size of input array for Elbrus-8S

Thus, we can make an interim summary, that single launches of the FFT on Elbrus-4S and Elbrus-8S are more efficient using the EML library, because the preparation of the algorithm FFT using EML is 2–20 times for Elbrus-4S and 10–90 times faster for Elbrus-8S than the ones using FFTW.

And here we consider the second stage of FFT implementation, namely, the execution of the algorithm. The stage of execution of the algorithm takes from one to several orders of magnitude less time than the stage of the algorithm

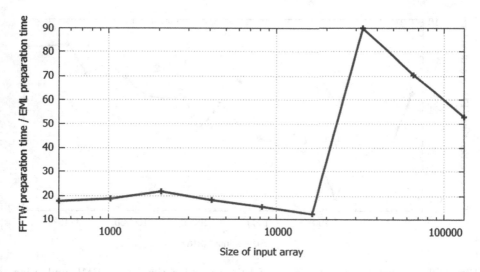

Fig. 5. The dependence of the ratio of FFT preparation time using EML and using FFTW on size of input array for Elbrus-8S

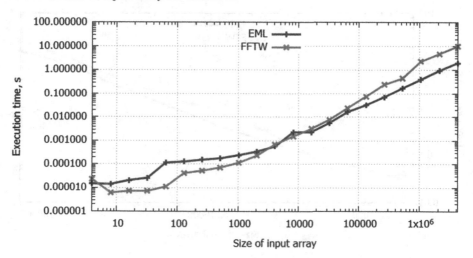

Fig. 6. The dependence of the FFT execution time on size of input array for Elbrus-4S

preparation, so it will have a significant effect only if the algorithm is run multiple times after a single preparation. This often happens when we need to perform an FFT on a set of arrays of the same size.

The execution time of the FFT algorithm using EML turns out to be from 1 to 10 times greater than the execution time of the algorithm using FTTW for array sizes less than 2^{11} (Fig. 8). And for large array sizes the situation reverses and the ratio of execution time using FFTW to the one using EML increases from 1 to 6 for the array of sizes $2^{14} - 2^{22}$.

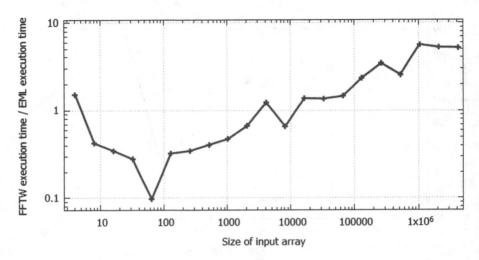

Fig. 7. The dependence of the ratio of FFT execution time using EML and using FFTW on size of input array for Elbrus-4S

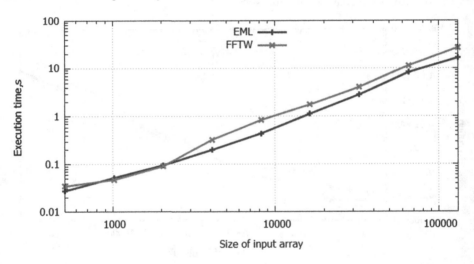

Fig. 8. The dependence of the FFT execution time on size of input array for Elbrus-8S

Figure 8 shows that for the Elbrus-8S, the difference in preparation time is smaller than for the Elbrus-4S. For arrays smaller than 2^{12} the execution time using EML is close to the one using FFTW. And for large arrays up to 2^{18} the ratio of execution time using FFTW to the one using EML is in the range from 1.4 to 1.9.

On Elbrus-4S multiple starts (more than 1000) of FFT using FFTW more efficient for small arrays (less than 2^{11}) than using EML. Execution time using FFTW is 1 to 10 times faster than using a library EML for Elbrus-4S. FFT of array of almost all sizes on Elbrus-8S is more efficient to run using the EML library, but the ration of execution time for FFTW and EML is less than 2.

4.3 Fast Fourier Transform Using FFTW on Elbrus CPUs vs Intel Xeon

Let's compare the work of FFTW on the Elbrus-8S and Intel Xeon E5-2660 processors. One can see that initialization is performed on the intel processor three times faster than on the Elbrus processor for a wide range of sizes of the input array (Fig. 9a). However, this acceleration does not compensate for the difference in time between performing FFT using EML and FFTW libraries on the Elbrus processor. Thus, we can conclude that the phase of preparing the Fourier transform performs on the Elbrus processor using the EML library several times faster than the same phase is performed using the FFTW library on the Intel Xeon E5-2660 processor. The phases of execution of the algorithm on two processors take almost the same time (Fig. 9b). This is quite surprising given the fact that the frequency of Intel processor 2.2 GHz, and the frequency of the Elbrus processor is 1.3 GHz. It is also worth noting that FFTW is significantly optimized by itself and it is especially strongly optimized for Intel processors and it is absolutely not optimized for Elbrus processors.

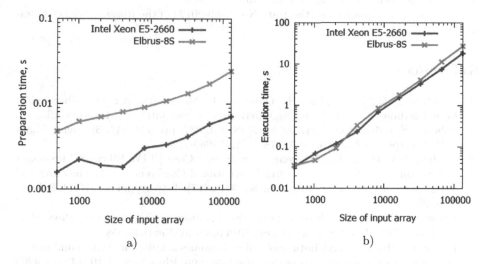

Fig. 9. The dependence of the FFTW preparation time (a) and execution time (b) on the Elbrus-8S and Intel Xeon E5-2660 processors

5   Conclusions

We have performed test calculations for the VASP model on Intel Xeon Haswell and Elbrus-8S CPUs with the best choice of mathematical libraries available. Elbrus-8S shows larger time-to-solution values, however there is no large gap between performance of Elbrus-8S and Xeon Haswell CPUs. The major target for optimization that could significantly speed-up VASP on Elbrus-8S is the FFT library.

We have performed test of native EML library and unoptimised FFTW library on processors Elbrus-4S and Elbrus-8S. Single launches of the FFT on Elbrus-4S and Elbrus-8S are more efficient using the EML library. Multiple starts (more than 10000) of FFT using FFTW is more efficient for small arrays (less than 4000) than using EML. FFT of any array using Elbrus-8S is more efficient to run using the EML library. FFTW performance on Elbrus-8S is competitive with Intel Xeon Broadwell CPUs. EML on Elbrus-8S (1.3 GHz) appears to be close or even more effective than FFTW on Intel Xeon E5-2660v4 (2.2 GHz).

Acknowledgments. The authors acknowledge Joint Supercomputer Centre of Russian Academy of Sciences (http://www.jscc.ru) for the access to the supercomputer MVS1P5. The authors acknowledge JSC MCST (http://www.msct.ru) for the access to the servers with Elbrus CPUs. The authors are grateful to Vyacheslav Vecher for the help with calculations based on hardware counters.

The work was supported by the grant No. 14-50-00124 of the Russian Science Foundation.

References

1. Tyutlyaeva, E., Konyukhov, S., Odintsov, I., Moskovsky, A.: The elbrus platform feasibility assessment for high-performance computations. In: Voevodin, V., Sobolev, S. (eds.) RuSCDays 2016. CCIS, vol. 687, pp. 333–344. Springer, Cham (2016). https://doi.org/10.1007/978-3-319-55669-7_26
2. Kozhin, A.S., et al.: The 5th generation 28nm 8-Core VLIW Elbrus-8C processor architecture. In: Proceedings - 2016 International Conference on Engineering and Telecommunication, EnT 2016, pp. 86–90 (2017). https://doi.org/10.1109/EnT.2016.25
3. Kresse, G., Hafner, J.: Ab initio molecular dynamics for liquid metals. Phys. Rev. B **47**, 558–561 (1993). https://doi.org/10.1103/PhysRevB.47.558
4. Kresse, G., Hafner, J.: Ab initio molecular-dynamics simulation of the liquid-metal-amorphous-semiconductor transition in germanium. Phys. Rev. B **49**, 14251–14269 (1994). https://doi.org/10.1103/PhysRevB.49.14251
5. Kresse, G., Furthmuller, J.: Efficiency of ab-initio total energy calculations for metals and semiconductors using a plane-wave basis set. Comput. Mater. Sci. **6**(1), 15–50 (1996). https://doi.org/10.1016/0927-0256(96)00008-0
6. Kresse, G., Furthmüller, J.: Efficient iterative schemes for ab initio total-energy calculations using a plane-wave basis set. Phys. Rev. B **54**, 11169–11186 (1996). https://doi.org/10.1103/PhysRevB.54.11169

7. Stegailov, V.V., Orekhov, N.D., Smirnov, G.S.: HPC hardware efficiency for quantum and classical molecular dynamics. In: Malyshkin, V. (ed.) PaCT 2015. LNCS, vol. 9251, pp. 469–473. Springer, Cham (2015). https://doi.org/10.1007/978-3-319-21909-7_45

8. Aristova, N.M., Belov, G.V.: Refining the thermodynamic functions of scandium triflouride scf3 in the condensed state. Russ. J. Phys. Chem. A **90**(3), 700–703 (2016). https://doi.org/10.1134/S0036024416030031

9. Kochikov, I.V., Kovtun, D.M., Tarasov, Y.I.: Electron diffraction analysis for the molecules with degenerate large amplitude motions: intramolecular dynamics in arsenic pentafluoride. J. Mol. Struct. **1132**, 139–148 (2017). https://doi.org/10.1016/j.molstruc.2016.09.064

10. Minakov, D.V., Levashov, P.R.: Melting curves of metals with excited electrons in the quasiharmonic approximation. Phys. Rev. B **92**, 224,102 (2015). https://doi.org/10.1103/PhysRevB.92.224102

11. Minakov, D., Levashov, P.: Thermodynamic properties of lid under compression with different pseudopotentials for lithium. Comput. Mater. Sci. **114**, 128–134 (2016). https://doi.org/10.1016/j.commatsci.2015.12.008

12. Bethune, I.: Ab initio molecular dynamics. Introduction to Molecular Dynamics on ARCHER (2015)

13. Hutchinson, M.: Vasp on gpus. when and how. GPU technology theater, SC15 (2015)

14. Zhao, Z., Marsman, M.: Estimating the performance impact of the MCDRAM on KNL using dual-socket Ivy Bridge nodes on Cray XC30. In: Proceedings of the Cray User Group — 2016 (2016)

15. Wende, F., Marsman, M., Steinke, T.: On enhancing 3D-FFT performance in VASP. In: CUG Proceedings, p. 9 (2016)

16. Burtscher, M., Kim, B.D., Diamond, J., McCalpin, J., Koesterke, L., Browne, J.: Perfexpert: an easy-to-use performance diagnosis tool for HPC applications. In: Proceedings of the 2010 ACM/IEEE International Conference for High Performance Computing, Networking, Storage and Analysis, SC 2010, pp. 1–11. IEEE Computer Society, Washington, DC (2010). https://doi.org/10.1109/SC.2010.41

17. Rane, A., Browne, J.: Enhancing performance optimization of multicore/multichip nodes with data structure metrics. ACM Trans. Parallel Comput. **1**(1), 3:1–3:20 (2014). https://doi.org/10.1145/2588788

18. Stanisic, L., Mello Schnorr, L.C., Degomme, A., Heinrich, F.C., Legrand, A., Videau, B.: Characterizing the performance of modern architectures through opaque benchmarks: pitfalls learned the hard way. In: IPDPS 2017–31st IEEE International Parallel & Distributed Processing Symposium (RepPar Workshop), Orlando, US, pp. 1588–1597 (2017)

19. Baker, M.: A study of improving the parallel performance of VASP. Ph.D. thesis, East Tennessee State University (2010)

Design Technology for Reconfigurable Computer Systems with Immersion Cooling

Ilya Levin, Alexey Dordopulo$^{(\boxtimes)}$, Alexander Fedorov,
and Yuriy Doronchenko

Scientific Research Centre of Supercomputers and Neurocomputers (LLC),
Taganrog, Russia
{levin,dordopulo,doronchenko}@superevm.ru,
ss24@mail.ru

Abstract. In this paper, we consider the implementation of reconfigurable computer systems based on advanced Xilinx UltraScale and UltraScale+FPGAs and a design method of immersion cooling systems for computers containing 96–128 chips. We propose the selection criteria of key technical solutions for creation of high-performance computer systems with liquid cooling. The construction of the computational block prototype and the results of its experimental thermal testing are presented. The results demonstrate high energy efficiency of the proposed open cooling system and existence of power reserve for the next-generation FPGAs. Effective cooling of 96–128 FPGAs with the total thermal power of 9.6–12.8 kW in a 3U computational module is the key feature of the considered system. Insensitivity to leakages and their consequences, and compatibility with traditional water cooling systems based on industrial chillers are the advantages of the developed technical solution. These features allow installation of liquid-cooled computer systems with no fundamental change of the computer hall infrastructure.

Keywords: Immersion cooling system · Liquid cooling
Reconfigurable computer systems · FPGA
High-Performance computer systems · Energy efficiency

1 Introduction

A reconfigurable computer system (RCS), which contains an FPGA computational field of large logic capacity [1, 2], is used for implementation of computationally laborious tasks from various domains of science and technique [1–3], because it has a considerable advantage in its real performance and energetic efficiency in comparison with cluster-like multiprocessor computer systems. The RCS provides adaptation of its architecture to the structure of any solving task. In this case a special-purpose computer device is created. It hardwarily implements all computational operations of the information graph of the task with the minimum delays. Here, we have a contradiction between the implementation of the special-purpose device and its general-purpose usage for solving tasks from various problem areas. It is possible to eliminate these contradictions, combining creation of a special-purpose computer device with a wide

© Springer Nature Switzerland AG 2019
V. Voevodin and S. Sobolev (Eds.): RuSCDays 2018, CCIS 965, pp. 554–564, 2019.
https://doi.org/10.1007/978-3-030-05807-4_47

range of solving tasks, within a concept of reconfigurable computer systems based on FPGAs that are used as a principal computational resource [1].

Continuous increasing of the circuit complexity and the clock rate of each new FPGA family leads to considerable growth of power consumption and to growth of the maximal operating temperature on the chip. Practical experience of maintenance of large RCS-based computer complexes proves that air cooling systems have reached their heat limit.

According to the obtained experimental data, further development of FPGA production technologies and conversion to the next FPGA family Virtex UltraScale (power consumption up to 100 W for each chip) will lead to additional growth of FPGA overheat on 20...25 °C. This will shift the range of their operating temperature limit (80...85 °C), which means negative influence on their reliability when chips are filled up to 85–95% of available hardware resource. This circumstance requires a quite different cooling method which provides keeping of performance growth rates of advanced RCS.

2 Reconfigurable Computer Systems with Liquid Cooling

Today liquid cooling systems are the most promising design area for cooling modern high-loaded electronic components of computer systems [4–6], because of heat capacity of liquids which is better than air capacity (from 1500 to 4000 times), and higher heat-transfer coefficient (increasing up to 100 times).

At present the technology of liquid cooling of servers and separate computational modules is developed by many vendors and some of them have achieved success in this direction [7–10]. However, these technologies are intended for cooling computational modules which contain one or two microprocessors. All attempts of its adaptation to cooling computational modules which contain a large number of heat generating components (an FPGA field of 8 chips), have proved a number of shortcomings of liquid cooling of RCS computational modules [5, 6].

The special feature of the RCS produced in Scientific Research Centre of Supercomputers and Neurocomputers is the number of FPGAs, not less than 6-8 chips on one printed circuit board, and high packing density. Usage of an open liquid cooling system is efficient owing to the heat-transfer agent characteristics, and the design and specification of the used FPGA heatsinks, pump equipment, heat-exchangers. The heat-transfer agent must have the best possible electric strength, high heat transfer capacity, the maximum possible heat capacity and low viscosity. The heatsink must provide the maximum possible surface of heat dissipation, must allow circulation of the heat-transfer agent through itself, a turbulent heat-transfer agent flow in itself, manufacturability.

Since 2013 the scientific team of SRC SC and NC has actively developed the domain of creation of next-generation RCS on the base of their original liquid cooling system for computational circuit boards with high packing density and the large number of heat generating electronic components. Design criteria of the computational module (CM) of next-generation RCS with an open loop liquid cooling system are based on the principles as follows:

- the RCS configuration is based on a computational module with the 3U height and the 19" width and with self-contained circulation of the heat-transfer agent;
- one computational module can contain 12–16 computational circuit boards (CCB) with FPGA chips;
- each CCB must contain up to 8 FPGAs with dissipating heat flow of about 100 W from each FPGA;
- a standard water cooling system, based on industrial chillers, must be used for cooling the heat-transfer agent.

The principal element of modular implementation of an open loop immersion liquid cooling system for electronic components of computer systems is a reconfigurable computational module with liquid cooling (see the design in Fig. 1). The CM casing consists of a computational section and a heat exchange section. In the casing, which is the base of the computational section, a hermetic container with heat-transfer agent (dielectric cooling liquid) and electronic components with elements that generate heat during operating, is placed.

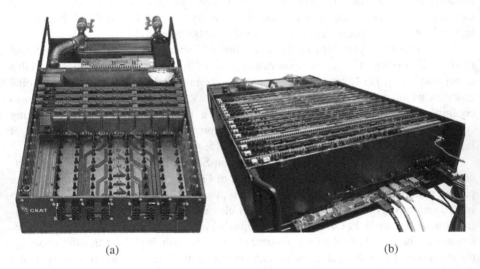

(a)   (b)

Fig. 1. Design of reconfigurable computer system with liquid cooling (a – assembly of CM, b – assembled CM without top cover)

The computational section contains: boards of the computational module (not less than 12–16 (see Fig. 2)), control boards (see Fig. 3), RAM, power supply blocks (see Fig. 4), storage devices, daughter boards, etc. The computational section is closed with a cover. The CCB of the advanced computational module contains 8 Kintex UltraScale XCKU095T FPGAs. Each FPGA has a specially designed thermal interface and a

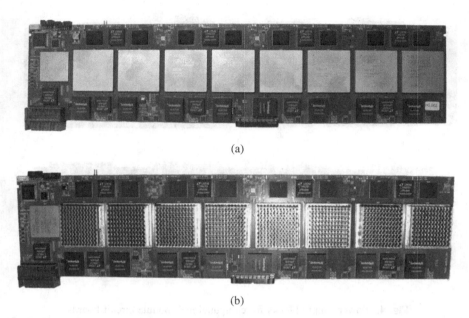

(a)

(b)

Fig. 2. Boards of computational module of reconfigurable computer system with liquid cooling (a – without heatsinks, b – with designed heatsinks)

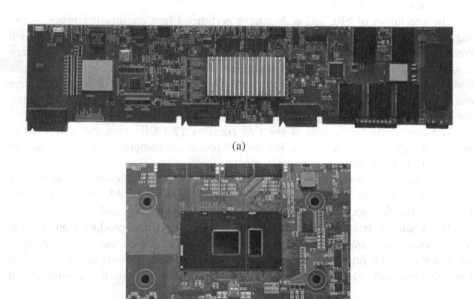

(a)

(b)

Fig. 3. Board of loading and control block (a – without heatsinks, b – with placed heatsinks)

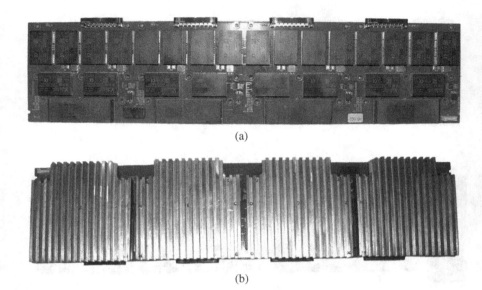

(a)

(b)

Fig. 4. Power supply blocks for computational module circuit boards

small height heatsink for heat dissipation. To cool FPGAs we use heatsinks of an original design (see Fig. 2-b).

The specialists of SRC SC & NC have performed heat engineering research and suggested a fundamentally new design of a heatsink with original solder pins, which create a local turbulent flow of the heat-transfer agent. The used thermal interface cannot be deteriorated or washed out by the heat-transfer agent. Its coefficient of heat conductivity must remain permanently high. The specialists of SRC SC & NC have created an effective thermal interface which fulfills all specified requirements. Besides, the technology of its coating and removal was also perfected.

The computational section of the CM contains 12 CCBs with the power up to 800 W each, 3 power supply units. Besides, all boards are completely immersed into an electrically neutral liquid heat-transfer agent.

For creation of an effective immersion cooling system a dielectric heat-transfer agent was developed. This heat-transfer agent has the best possible electric strength, high heat transfer capacity, the maximum possible heat capacity and low viscosity.

The scientific team of SRC SC & NC has designed and produced an original motherboard based on an Intel Skylake® (Core I5-6300U) processor for the computational block with liquid cooling. The printed circuit board consists of 18 layers and has the minimum size of 490 × 109.7 mm for placement into the computational module.

For the motherboard we have designed an original basic input-output system (**BIOS**), which allows usage of all capabilities of the Intel Skylake® (Core I5-6300U) system-on-chip and external peripheral equipment. Besides, we have designed an immersion power supply block (see Fig. 4), which provides DC/DC 380/12 V transformation with power up to 4 kW for 4 CCBs. Power supply blocks for 4 CCBs are placed into the computational section.

The computational section adjoins to the heat exchange section, which contains a pump and a heat exchanger. The pump provides closed loop circulation of the heat transfer agent in the CM: from the computational module the heated heat-transfer agent passes into the heat exchanger and is cooled there. From the heat exchanger the cooled heat-transfer agent again passes into the computational module and there cools the heated electronic components. As a result of heat dissipation the agent becomes heated and again passes into the heat exchanger, and so on. The heat exchanger is connected to the external heat exchange loop via fittings and is intended for cooling the heat-transfer agent with the help of the secondary cooling liquid. As a heat exchanger it is possible to use a plate heat exchanger in which the first and the second loops are separated. So, as the secondary cooling liquid it is possible to use water, cooled by an industrial chiller. The chiller can be placed outside the server room and can be connected with the reconfigurable computational modules by means of a stationary system of engineering services.

The computational and the heat exchange sections are mechanically interconnected into a single reconfigurable computational module. Maintenance of the reconfigurable computational module requires its connection to the source of the secondary cooling liquid (by means of valves), to the power supply or to the hub (by means of electrical connectors).

In the casing of the computer rack the CMs are placed one over another. Their number is limited by the dimensions of the rack, by technical capabilities of the computer room and by the engineering services.

Each CM of the computer rack is connected to the source of the secondary cooling liquid with the help of supply return collectors through fittings (or balanced valves) and flexible pipes; connection to the power supply and the hub is performed via electric connectors.

Supply of cold secondary cooling liquid and extraction of the heated one into the stationary system of engineering services connected to the rack, is performed via fittings (or balanced valves).

The performance of one next-generation CM is increased in 8.7 times in comparison with the CM "Taygeta". Such qualitative increasing of the system specific performance is provided by more than triple increasing of the system packing density owing to original design solutions, and increasing of the clock frequency and the FPGA logic capacity. Experimental results prove that the complex of the developed solutions concerning the immersion liquid cooling system provide the temperature of the heat-transfer agent not more than 33 °C, the power of 91 W for each FPGA (8736 W for the CM) in the operating mode of the CM. At the same time, the maximum FPGA temperature during heat experiments does not exceed 57 °C. This proves that the designed immersion liquid cooling system has a reserve and can provide effective cooling for the designed RCS based on the advanced Xilinx UltraScale+FPGA family.

On basis of the designed solutions of the advanced computational module we have created a prototype of a computer system Nekkar (see Fig. 5). The RCS Nekkar contains 12 3U computational modules with liquid cooling.

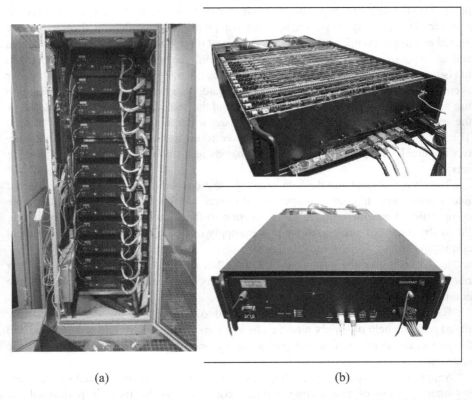

(a)   (b)

Fig. 5. RCS Nekkar based on advanced computational modules (a – without heatsinks, b – with placed heatsinks)

Each computational module contains 12 boards with the power consumption of 800 W each. Each board contains 8 Xilinx Kintex UltraScale XCKU095T FPGAs (100 million equivalent gates in each chip). The total performance of the RCS Nekkar is 1 PFlops, and its power consumption is 124 kW.

The improved design of the computational modules for serial production is shown in Fig. 6.

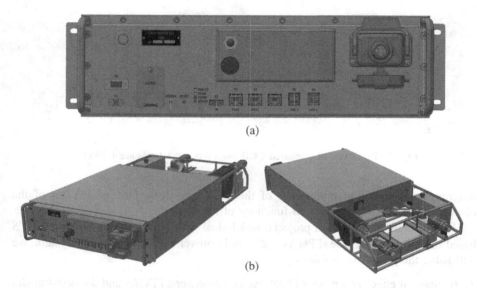

(a)

(b)

Fig. 6. Serial computational module Nekkar (a – front panel, b – top and side view of front and back panels)

3   Advanced Reconfigurable Computer System "Arctur" Based on Xilinx UltraScale+FPGAs

Usage of the UltraScale+FPGAs, which have been implemented on the base of the 16-nm technology 16FinFET Plus and produced by Xilinx since 2017, will provide up to triple ramp of the computational performance owing to increasing of clock frequency and FPGA circuit complexity; the size of the computer system remains unchanged. However, in spite of reduction of relative energetic consumption owing to new technological standards of FPGAs manufacturing, and owing to a certain power reserve of the designed liquid cooling system, it is possible to expect a new approach of FPGA operating temperatures to their critical values.

Besides, the new FPGAs of the UltraScale+family have larger geometric sizes. The size of the FPGAs of the RCS "Nekkar" is 42.5×42.5 mm. The size of the FPGAs, which are going to be placed into the RCS "Arctur", is 45×45 mm. Due to this circumstance it is impossible to use the existing design of the CCB, because the width of the printed circuit board will become larger and therefore will not fit for the standard 19" rack.

In this connection it is necessary to modify the designed open liquid cooling system and the CCB design that will lead to modification of the whole CM. During modification of the CCB design we have created a prototype of an advanced board shown in Fig. 7. The CCB contains 8 UltraScale+FPGAs of high circuit complexity. To provide placement of a new CCB into a 19" rack possible, it is necessary to exclude its CCB controller from its structure. The CCB controller was always implemented as a separate FPGA and provided access to FPGA computational resources of the CCB, FPGA

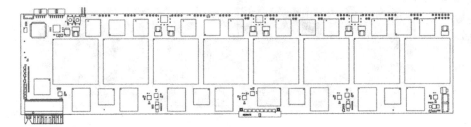

Fig. 7. Prototype of upgraded CCB assembly for UltraScale + FPGAs

programming, condition monitoring of the CCB resources. One of FPGAs of the computation field will perform all functions of the controller.

Within preliminary design projects, which deal with creation of an advanced RCS based on Xilinx UltraScale+FPGAs and with improvement of the cooling system, we will solve the problems as follows:

1. Increase of effective surface of heat-exchange between FPGAs and the heat-transfer agent.
2. Increase of the performance of the heat-transfer agent supply pump.
3. Increase of reliability of the liquid cooling system with the help of immersed pumps.
4. Experimental improvement of the heatsink optimal design.
5. Experimental improvement of the technology of thermal interface coating.

We have designed a prototype of an advanced computational module with a modified immersed cooling system (Fig. 8). The distinctive feature of the new design is immersed pumps and the considerable reliability growth of the CM owing to reduction of the number of components and simplification of the cooling system. According to our plans, the heat exchange section will contain only the heat exchanger. We are working on experimental research of various pump equipment which can operate in the heat-exchange agent.

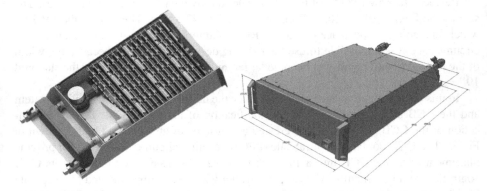

Fig. 8. A prototype of a computational module with a modified immersed cooling system

So, owing to breakthrough technical solutions which we have got during design of the RCS "Nekkar" with the immersed liquid cooling system, we can develop this direction of high-performance RCS design, and after some design improvements we can create a computer system which provides a new level of computational performance.

4 Conclusion

Usage of air cooling systems for the designed supercomputers has practically reached its limit because of reduction of cooling effectiveness with growing of consumed and dissipated power, caused by growth of circuit complexity of microprocessors and other chips. That is why usage of liquid cooling in modern computer systems is a priority direction of cooling systems perfection with wide perspectives of further development. Liquid cooling of RCS computational modules which contain not less than 8 FPGAs of high circuit complexity is specific in comparison with cooling of microprocessors and requires development of a specialized immersion cooling system. The designed original liquid cooling system for a new generation RCS computational module provides high maintenance characteristics such as the maximum FPGA temperature not more than 57 °C and the temperature of the heat-transfer agent not more than 33 °C in the operating mode. Owing to the obtained breakthrough solutions of the immersion liquid cooling system it is possible to place not less than 12 CMs of the new generation with the total performance over 1 PFlops within one 47U computer rack. Power reserve of the liquid cooling system of the new generation CMs provides effective cooling of not only existing but of the developed promising FPGA families Xilinx UltraScale+ and UltraScale 2.

Since FPGAs, as principal components of reconfigurable supercomputers, provide stable ramping of RCS performance, it is possible to get specific performance of RCS, based on Xilinx Virtex UltraScale FPGAs, similar to the one of the world best cluster supercomputers, and to find new perspectives of design of super-high performance supercomputers.

References

1. Tripiccione, R.: Reconfigurable computing for statistical physics. the weird case of JANUS. In: IEEE 23rd International Conference on Application-Specific Systems, Architectures and Processors (ASAP) (2012)
2. Baity-Jesi, M., et al.: The Janus project: boosting spin-glass simulations using FPGAs. In: IFAC Proceedings Volumes, Programmable Devices and Embedded Systems, vol. 12, no. 1 (2013)
3. Shaw, D.E., et al.: Anton, a special-purpose machine for molecular dynamics simulation. Commun. ACM 51(7), 91–97 (2008)
4. Kalyaev, I.A., Levin, I.I., Semernikov, E.A., Shmoilov, V.I.: Reconfigurable multipipeline computing structures. Nova Science Publishers, New York (2012). ISBN 978-1-62081-462-8

5. Dordopulo, A., Kalyaev, I., Levin, I., Slasten, L.: High-performance reconfigurable computer systems. In: Malyshkin, V. (ed.) PaCT 2011. LNCS, vol. 6873, pp. 272–283. Springer, Heidelberg (2011). https://doi.org/10.1007/978-3-642-23178-0_24

6. Dordopulo, A.I., Levin, I.I., Fedorov, A.M., Kalyaev, I.A.: Reconfigurable computer systems: from the first FPGAs towards liquid cooling systems. Supercomputing Front. Innovations, 22–40 (2016). http://superfri.org/superfri/article/view/97, https://doi.org/10.14529/jsfi160102

7. Technology. https://www.coolitsystems.com/technology/. Accessed 25 May 2018

8. Immers 6 R6. http://immers.ru/sys/immers6r6/. Accessed 25 May 2018

9. Eurotech liquid cooling is hot! https://www.eurotech.com/en/hpc/hpc+solutions/liquid+cooling. Accessed 25 Oct 2018

10. RSC. http://www.rscgroup.ru. Accessed 25 May 2018

Designing a Parallel Programs on the Base of the Conception of Q-Determinant

Valentina Aleeva[✉]

South Ural State University (National Research University), Chelyabinsk, Russia
aleevavn@susu.ru

Abstract. The paper describes a design method of parallel programs for numerical algorithms based on their representation in the form of Q-determinant. The result of the method is Q-effective program. It uses the parallelism resource of the algorithm completely. The results of this research can be applied to increase the implementation efficiency of algorithms on parallel computing systems. This should help to improve the performance of parallel computing systems.

Keywords: Q-determinant of algorithm
Algorithm representation as Q-determinant
Q-effective implementation of algorithm
Parallelism resource of algorithm · Parallel computing system
Parallel program · Q-effective program

1 Introduction

The development of parallel computing systems has the history for decades, but the implementation effectiveness of algorithms remains low on those. That problem can be solved by using parallelism resource of algorithms completely. The conception of Q-determinant [1] allows to detect the parallelism resources of numerical algorithms. The basis of the conception is the universal description of algorithms. This is the algorithms representation in the Q-determinant form. All of the algorithm implementation are described by Q-determinant, including the Q-effective one. The Q-effective implementation uses the parallelism resource of the algorithm completely. From a formal point of view that is the most rapid implementation. The description of the method of designing a parallel program for the Q-effective implementation of numerical algorithm is the aim of this paper. The concept of the Q-determinant has been theoretical development and used to study the resource of algorithm parallelism in the papers [2–6]. This conception for the design of parallel program is offered for the first time.

The investigation of parallel structure of algorithms and programs is very important and highly developed for their implementation on parallel computer systems. The basis of research is described in [7,8]. Representations by graphs

© Springer Nature Switzerland AG 2019
V. Voevodin and S. Sobolev (Eds.): RuSCDays 2018, CCIS 965, pp. 565–577, 2019.
https://doi.org/10.1007/978-3-030-05807-4_48

are used for a description of parallel algorithms. The Internet encyclopedia Algo-Wiki [9] is created nowadays. The encyclopedia describes the features, peculiar properties, static and dynamic characteristics of the algorithms. This help to implement algorithms effectively. In the report [10] there is the current state of researches of parallelization algorithms and their implementations on parallel computing systems. The report contains the formulations of some problems. Here are some of these problems. How to represent a potentially infinite graph? How to represent a potentially multi-dimensional graph? How to show dependence graph structure on the size of the tasks? How to express the available parallelism and show affordable way of the parallel execution? The conception of Q-determinant gives answers for these questions. This research contains one of the answers to the last question.

There are many studies, in which the specific nature of algorithms and the architecture of parallel computing systems take into account for the development of parallel programs. Examples of such studies are [11–13]. The efficiency of implementing specific algorithms or implementing algorithms on parallel computing systems of a particular architecture is increased in cases of those studies. However, they do not provide general universal approach. The parallel program synthesis is other approach to creating parallel programs also. The synthesizing parallel programs method is to construct new parallel algorithms using the knowledge base of parallel algorithms for solving more complex problems. The technology of fragmented programming and its implementing language and programming system LuNA are developed on the base of synthesizing parallel programming method. This direction of research is developing [14,15] at present time. The approach is universal, but it does not solve the problem of research and use of the algorithm parallelism resource. The investigation of the parallelism resource of algorithms is provided using their software implementation [16]. If you want to solve the problem of determining the parallelism resource of algorithm, then use of any program implementing of algorithm may be wrong because that program can not contain all implementations of the algorithm. In particular, the Q-effective implementation can be lost under program creation. We can notice that the analysis of the existing approaches of problem solution of studying the parallelism resource of algorithm and its implementation on parallel computing systems shows they are inapplicable, or ineffective, or non-generic. The approach is perspective if it is based on the universal description of the algorithm showing the parallelism resource in full. For example, such approach is an approach on the base of the Q-determinant concept.

2 Q-Determinant of Algorithm

Let $\mathcal{A}$ be an algorithm for solving an algorithmic problems $\bar{y} = F(N, B)$, where N is a set of dimension parameters of the problem, B is a set of input data, $\bar{y} = (y_1, \ldots, y_m)$ is a set of output data, $y_i \notin B$ $(i = 1, \ldots, m)$, m is a computable function parameters N on condition $N \neq \varnothing$ or constant.

The set N satisfies the conditions: $N = \varnothing$, or $N = \{n_1, \ldots, n_k\}$, where $k \geqslant 1$, n_i $(1 \leqslant i \leqslant k)$ are every positive integers. If $N = \{n_1, \ldots, n_k\}$ then we denote vector $(\bar{n}_1, \ldots, \bar{n}_k)$, where $\bar{n}_i$ is some assigned value of parameter n_i $(1 \leqslant i \leqslant k)$ by $\bar{N}$. We denote by $\{\bar{N}\}$ the set $\bar{N}$ of possible vectors.

Let is Q a set of operations those are used by algorithm $\mathcal{A}$. Assume that the operations of Q are 0-ary (constant), unary or binary. An example of a set Q is a set of arithmetic, logic operations and comparison operations. The expressions can be formed by sets B and Q. We call chain an expression that obtained from the n expressions with the help of use of $n-1$ times one of associative operations of Q.

One of the basic notions of Q-determinant conception is Q-term.

Definition of Q-term:

1. If $N = \varnothing$ then unconditional Q-term is called every expression w over B and Q (term of signature Q). Let V be a set of all expressions over B and Q. If $N \neq \varnothing$ then every map $w : \{\bar{N}\} \to V$ is called unconditional Q-term also.
2. Let $N = \varnothing$ and be given an unconditional Q-term w. Let under each interpretation of B the expression w over B and Q have a logical type value. Then unconditional Q-term w is called unconditional logical Q-term. Let $N \neq \varnothing$ and be given an unconditional Q-term $w : \{\bar{N}\} \to V$. Let be an expression $w(\bar{N})$ for every $\bar{N} \in \{\bar{N}\}$ have logical type value under each interpretation of B. Then unconditional Q-term w is called unconditional logical Q-term.
3. Let $u_1, \ldots, u_l$ be an unconditional logical Q-terms. $w_1, \ldots, w_l$ are an unconditional Q-terms. We denote the set of pairs (u_i, w_i) $(i = 1, \ldots, l)$ as $(\bar{u}, \bar{w}) = \{(u_i, w_i)\}_{i=1,\ldots,l}$ and call conditional Q-term of length l.
4. Assume we have a countable set of pairs unconditional Q-terms $(\bar{u}, \bar{w}) = \{(u_i, w_i)\}_{i=1,2,\ldots}$ such that $\{(u_i, w_i)\}_{i=1,\ldots,l}$ is conditional Q-term for each $l < \infty$ then we call it conditional infinite Q-term.
5. If it does not matter whether the Q-term unconditional, conditional or conditional infinite then we call it Q-term.

Q-terms can be calculated.

We mean by the calculation of unconditional Q-term w under each interpretation of B as the calculation of the expression of w if $N = \varnothing$ and the calculation of the expression of $w(\bar{N})$ for some $\bar{N} \in \{\bar{N}\}$ if $N \neq \varnothing$.

We describe the calculation of conditional Q-term $(\bar{u}, \bar{w}) = \{(u_i, w_i)\}_{i=1,\ldots,l}$ under any interpretation of B. If $N = \varnothing$ is necessary to calculate the expressions u_i, w_i $(i = 1, \ldots, l)$. If there are expressions of u_{i_0}, w_{i_0} $(i_0 \leqslant l)$ such that u_{i_0} takes the value **true** and the value of w_{i_0} is determined we will set $(\bar{u}, \bar{w})$ taking value is equal to w_{i_0}. Also we assume the value of $(\bar{u}, \bar{w})$ for this interpretation B is not determined otherwise. If $N \neq \varnothing$ then we set value $\bar{N} \in \{\bar{N}\}$. We obtain the expressions $u_i(\bar{N}), w_i(\bar{N})$ $(i = 1, \ldots, l)$ and calculate them. If there are expressions of $u_{i_0}(\bar{N}), w_{i_0}(\bar{N})$ $(i_0 \leqslant l)$ such that $u_{i_0}(\bar{N})$ takes the **true** and the value of $w_{i_0}(\bar{N})$ is determined we will set $(\bar{u}, \bar{w})$ taking value is equal to $w_{i_0}(\bar{N})$. Also we assume the value of $(\bar{u}, \bar{w})$ is not determined for given $\bar{N}$ and interpretation of B otherwise.

We describe for given interpretation of B the calculation of conditional infinite Q-term $(\bar{u}, \bar{w}) = \{(u_i, w_i)\}_{i=1,2,\ldots}$. If $N = \varnothing$ it is necessary to find u_{i_0}, w_{i_0} such that u_{i_0} is set to **true** and the value of w_{i_0} is determined. Then w_{i_0} is the value of $(\bar{u}, \bar{w})$. If we have no such expressions u_{i_0}, w_{i_0} the value of $(\bar{u}, \bar{w})$ is not determined for this interpretation B. Similarly, we can define calculation of conditional infinite Q-term in the case of $N \neq \varnothing$.

Suppose that I_1, I_2, I_3 are subsets of the set $I = (1, \ldots, m)$, satisfying the following conditions:

1. $I_1 \cup I_2 \cup I_3 = I$;
2. $I_i \cap I_j = \varnothing$ $(i \neq j; i, j = 1, 2, 3)$;
3. One or two subsets I_i $(i = 1, 2, 3)$ may be empty.

We consider the set of Q-terms $\{f_i\}_{i \in I}$ such that:

1. f_{i_1} $(i_1 \in I_1)$ is an unconditional Q-term, $f_{i_1} = w^{i_1}$;
2. f_{i_2} $(i_2 \in I_2)$ is conditional Q-term, $f_{i_2} = \left\{ \left(u_j^{i_2}, w_j^{i_2} \right) \right\}_{j=1,\ldots,l_{i_2}}$, l_{i_2} is either constant or computable function of parameters N for $N \neq \varnothing$;
3. f_{i_3} $(i_3 \in I_3)$ is a conditional infinite Q-term, $f_{i_3} = \left\{ \left(u_j^{i_3}, w_j^{i_3} \right) \right\}_{j=1,2,\ldots}$.

Suppose that the algorithm $\mathcal{A}$ is that Q-term f_i should be computed in order that y_i $(i \in I)$ evaluates. Then the set of Q-terms f_i $(i \in I)$ is called a Q-determinant of algorithm $\mathcal{A}$ and presentation of algorithm in the form $y_i = f_i$ $(i \in I)$ is called a presentation of the algorithm in the form of Q-determinant. Every numerical algorithm can be represented in the form Q-determinant.

3 Q-Effective Implementation of Algorithm

Let $\mathcal{A}$ be an algorithm in the form of Q-determinant $y_i = f_i$ $(i \in I)$. The process of calculating the Q-terms f_i $(i \in I)$ is called an implementation of the algorithm $\mathcal{A}$. If the implementation of the algorithm is such that two or more operations are performed simultaneously, it will be called a parallel implementation. We describe a realization algorithm $\mathcal{A}$ represented in the form of Q-determinant.

Let $N = \varnothing$. We specify the variable interpretation of B. Expressions

$$W = \left\{ w^{i_1}(i_1 \in I_1); u_j^{i_2}, w_j^{i_2}(i_2 \in I_2, j = 1, \ldots, l_{i_2}); \right.$$
$$\left. u_j^{i_3}, w_j^{i_3}(i_3 \in I_3, j = 1, 2, \ldots) \right\} \quad (1)$$

will be calculated at the same time (in parallel). We say that the operation is ready to perform if you have calculated the value of its operands already. In calculating each of the expressions W we perform the operations as soon as they are ready to perform.

If you are ready to perform several operations chain, they are calculated by doubling scheme. For example, the doubling scheme of calculating the chain $a_1 + a_2 + a_3 + a_4$ is the following. First, we calculate $b_1 = a_1 + a_2$ and $b_2 = a_3 + a_4$ simultaneously, after that we calculate $c = b_1 + b_2$.

If we obtain **false** value for some expression u_j^i $(i \in I_2 \cup I_3, j = 1, 2, \ldots)$ the calculation of the corresponding expression of w_j^i comes to an end. If the calculation of some pair of expressions (u_j^i, w_j^i) $(i \in I_2 \cup I_3, j = 1, 2, \ldots)$ implies that the value of one of the expressions is not defined then the calculation of other expression is terminated. If the calculation of some pair of expressions $(u_{j_0}^i, w_{j_0}^i)$ $(i \in I_2 \cup I_3)$ finds that their values define and $u_{j_0}^i$ is set to **true**, the calculation of expressions u_j^i, w_j^i $(i \in I_2 \cup I_3, j = 1, 2, \ldots; j \neq j_0)$ stops. Calculation the identical expressions of u_j^i, w_j^i $(i \in I_3, j = 1, 2, \ldots)$ and their identical subexpressions may not duplicate.

Let N be nonempty. We define $\bar{N} \in \{\bar{N}\}$ and the interpretation of the variables of B. We get the set of expressions

$$W(\bar{N}) = \{ w^{i_1}(\bar{N})(i_1 \in I_1); u_j^{i_2}(\bar{N}), w_j^{i_2}(\bar{N})(i_2 \in I_2, j = 1, \ldots, l_{i_2});$$
$$u_j^{i_3}(\bar{N}), w_j^{i_3}(\bar{N})(i_3 \in I_3, j = 1, 2, \ldots) \}. \quad (2)$$

Expressions $W(\bar{N})$ can be calculated by analogy with the expressions W.

Described implementation of the algorithm $\mathcal{A}$ will be called Q-effective. We say that the implementation of the algorithm $\mathcal{A}$ is realizable, if finite number of operations needed to be performed simultaneously.

4   The Method of Parallel Program Design for the Q-Effective Implementation of Algorithm

Q-determinants can be constructed for any numerical algorithm. Q-determinant allows us to describe Q-effective implementation of the algorithm. If Q-effective implementation of the algorithm is realizable it can be programmed directly. You can develop a sequential program using a flow chart of a numerical algorithm. Also you can develop a parallel program using the Q-determinant of a numerical algorithm. This idea is the basis of the proposed method. The model of the Q-determinant concept allows us to investigate machine-independent properties of algorithms just only. The basic model of the Q-determinant concept is expanded to take into account the features of implementing algorithms on real parallel computing systems. The extended model of the Q-determinant concept obtained by adding model of parallel computing PRAM [17] for shared memory and BSP [18] for distributed memory. We proposed the method of parallel program design for the Q-effective implementation of the algorithm that is based on the extended model of the Q-determinant concept.

The method consists of the following stages.

1. Construction of Q-determinant of algorithm.
2. Description of Q-effective implementation of algorithm.
3. If Q-effective implementation is realizable then a parallel program is developed for it.

The program will be called Q-effective if it is designed with the help of this method. Q-effective program uses the parallelism resource of algorithm completely because it performs a Q-effective implementation of the algorithm. So, it has most high parallelism among the programs implement the algorithm. For this reason the Q-effective program uses the resources of the computing system more efficient than the programs perform other implementations of the algorithm. A Q-effective program can be used parallel computing systems with shared and distributed memory. We show the use of this method for modeling algorithms, which have Q-determinants that consist of Q-terms of various types. First, consider the first two stages of the method.

The Algorithm of Matrix Multiplication

Stage 1. Consider the algorithm of multiplication of matrices

$$A = [a_{ij}]_{i=1,\ldots,n;j=1,\ldots,k} \text{ and } B = [b_{ij}]_{i=1,\ldots,k;j=1,\ldots,m}. \tag{3}$$

The result is matrix $C = [c_{ij}]_{i=1,\ldots,n;j=1,\ldots,m}$, where $c_{ij} = \sum_{s=1}^{k} a_{is}b_{sj}$.
The algorithm of matrix multiplication can be presented in the form of Q-determinant. Q-determinant is composed of nm unconditional Q-terms.

Stage 2. Q-effective implementation of the algorithm for multiplication of the matrices is that all of Q-terms $\sum_{s=1}^{k} a_{is}b_{sj}(i = 1,\ldots,n; j = 1,\ldots,m)$ are calculated simultaneously. First, all multiplication operations should are ready to perform, so they need to be performed simultaneously. The result will nm chains formed by the operation addition. Each chain is calculated by doubling scheme. So, Q-effective implementation of the algorithm for multiplication of matrices is realizable.

Gauss–Jordan Method of Solving of Linear Equation Systems

Stage 1. Gauss-Jordan method of solving linear equation systems $A\bar{x} = \bar{b}$ can be applied to each dimension. For simplicity, we assume that $A = [a_{ij}]$ is a matrix of dimension $n \times n$ with a nonzero determinant. $\bar{x} = (x_1,\ldots,x_n)^T$, $\bar{b} = (a_{1,n+1},\ldots,a_{n,n+1})^T$ has a column-vectors, $\bar{A}$ is augmented matrix of the system. We construct Q-determinant method of Gauss-Jordan.

Gauss-Jordan method consists of n steps.

Step 1.

We select element a_{1j_1} with properties a_{11} if $a_{11} \neq 0$ otherwise $a_{1j} = 0$ for $j < j_1 \leqslant n$ and $a_{1j_1} \neq 0$ as leading element. We get augmented matrix $\bar{A}^{j_1} = [a_{ij}^{j_1}]$, whose elements are calculated by rules

$$a_{1j}^{j_1} = \frac{a_{1j}}{a_{1j_1}}, \; a_{ij}^{j_1} = a_{ij} - \frac{a_{1j}}{a_{1j_1}}a_{ij_1} (i = 2,\ldots,n; j = 1,\ldots,n+1). \tag{4}$$

Step $k(2 \leqslant k \leqslant n)$.

We get augmented matrix $\bar{A}^{j_1\cdots j_{k-1}}$ after we made the step $(k-1)$. We select element $a_{kj_k}^{j_1\cdots j_{k-1}}$ with properties $a_{k1}^{j_1\cdots j_{k-1}}$ if $a_{k1}^{j_1\cdots j_{k-1}} \neq 0$ otherwise $a_{kj}^{j_1\cdots j_{k-1}} = 0$ for $j < j_k \leqslant n$ and $a_{kj_k}^{j_1\cdots j_{k-1}} \neq 0$ as leading element. We get augmented matrix $\bar{A}^{j_1\cdots j_k} = [a_{ij}^{j_1\cdots j_k}]_{i=1,\ldots,n;j=1,\ldots,n+1}$, whose elements are calculated by the rules

$$a_{kj}^{j_1 \cdots j_k} = \frac{a_{kj}^{j_1 \cdots j_{k-1}}}{a_{kj_k}^{j_1 \cdots j_{k-1}}}, \quad a_{ij}^{j_1 \cdots j_k} = a_{ij}^{j_1 \cdots j_{k-1}} - \frac{a_{kj}^{j_1 \cdots j_{k-1}}}{a_{kj_k}^{j_1 \cdots j_{k-1}}} a_{ij_k}^{j_1 \cdots j_{k-1}}$$

$$(i = 1, \ldots, n; i \neq k; j = 1, \ldots, n+1). \quad (5)$$

We get system of equations $A^{j_1 \cdots j_n} \bar{x} = \bar{b}^{j_1 \cdots j_n}$ after step n, where

$$A^{j_1 \cdots j_n} = [a_{ij}^{j_1 \cdots j_n}]_{i=1,\ldots,n; j=1,\ldots,n}, \quad \bar{b}^{j_1 \cdots j_n} = (a_{1,n+1}^{j_1 \cdots j_n}, \ldots, a_{n,n+1}^{j_1 \cdots j_n})^T. \quad (6)$$

We denote by

$$L_{j_1} = \bigwedge_{j=1}^{j_1-1} (a_{1j} = 0) \text{ if } j_1 \neq 1, \ L_{j_1} = \text{true if } j_1 = 1, \quad (7)$$

$$L_{j_l} = \bigwedge_{j=1}^{j_l-1} (a_{lj}^{j_1 \cdots j_{l-1}} = 0) \text{ if } j_l \neq 1, L_{j_l} = \text{true if } j_l = 1 \ (l = 2, \ldots, n). \quad (8)$$

Permutations of elements $(1, \ldots, n)$ may be numbered. Let i be a number of permutation $(j_1, \ldots, j_n)$. Then the terms

$$w_i^{j_l} = a_{l,n+1}^{j_1 \cdots j_n} \ (l = 1, \ldots, n), \quad (9)$$

$$u_i = L_{j_1} \wedge (a_{1j_1} \neq 0) \wedge \left(\bigwedge_{l=2}^{n} \left(L_{j_l} \wedge \left(a_{lj_l}^{j_1 \cdots j_{l-1}} \neq 0 \right) \right) \right) \quad (10)$$

are unconditional Q-terms.

Q-determinant of Gauss–Jordan method consists of n conditional Q-terms and

$$x_j = \{(u_1, w_1^j), \ldots, (u_{n!}, w_{n!}^j)\}(j = 1, \ldots, n) \quad (11)$$

is the representation in the form of Q-determinant.

Stage 2. By the definition of Q-effective implementation all unconditional Q-terms $\{u_i, w_i^j\}(i=1, \ldots, n!; j=1, \ldots, n)$ should be calculated simultaneously. Therefore, two computational process should be carried out at the same time: parallel calculation of matrices $\bar{A}^{j_1}, \bar{A}^{j_1 j_2}, \ldots, \bar{A}^{j_1 j_2 \cdots j_n}$ for all possible values of the numbers $j_1, j_2, \ldots, j_n$, as well as a parallel calculation of the Q-terms $u_i(i=1, \ldots, n!)$. The leading elements of the matrix for each step of the algorithm are determined in the calculation of Q-terms $u_i(i=1, \ldots, n!)$ successively. Calculation of matrices $\bar{A}^{j_1}, \bar{A}^{j_1 j_2}, \ldots, \bar{A}^{j_1 j_2 \cdots j_n}$ stops if they do not correspond to the leading elements. The first cycle of calculations is as follows. We begin to calculate the matrices $\bar{A}^{j_1}$ and Q-terms $u_i(i = 1, \ldots, n!)$ simultaneously. The calculations $u_i(i=1, \ldots, n!)$ start with the subexpressions $L_{j_1} \wedge (a_{1j_1} \neq 0)(j_1=1, \ldots, n)$, because only their operations are ready for execution. Only one of the subexpressions $L_{j_1} \wedge (a_{1j_1} \neq 0)$ will be set to **true**. Let r_1 be a value of j_1. Further we end calculating $\bar{A}^{j_1}(j_1 = 1, \ldots, n)$ and $u_i(i = 1, \ldots, n!)$ under the condition that j_1 is not equal r_1. We compute the matrices $\bar{A}^{r_1j_2}(j_2 = 1,\ldots,n; j_2 \neq r_1)$ and Q-terms $u_i(i = 1,\ldots,n!)$ to $j_1 = r_1$ in the second cycle of calculations simultaneously. In the calculation of $u_i(i = 1,\ldots,n!)$ one should be calculated only subexpression $L_{j_2} \wedge (a_{2j_2}^{r_1} \neq 0)(j_2 = 1,\ldots,n; j_2 \neq r_1)$ so as soon as their operations are ready for execution. Only one of the subexpressions $L_{j_2} \wedge (a_{2j_2}^{r_1} \neq 0)(j_2 = 1,\ldots,n; j_2 \neq r_1)$ will be set to true. Let r_2 be a value of j_2. Further, the calculating $\bar{A}^{r_1j_2}(j_2 = 1,\ldots,n; j_2 \neq r_1)$ and $u_i(i = 1,\ldots,n!)$ to $j_2 \neq r_2$ is stop. The following $n - 3$ cycles of calculations are executed similarly. In conclusion, you need to calculate a single matrix $\bar{A}^{r_1\cdots r_{n-1}j_n}(j_n \neq r_1, r_2,\ldots,r_{n-1})$ as the parameter j_n has single value. Let r_n be a value of j_n. The result is $x_{r_j} = a_{j,n+1}^{r_1\cdots r_n}(j = 1,\ldots,n)$ that is the solution of original system of linear equations. Q-effective implementation of the method of Gauss-Jordan is realizable.

Jacobi Method of Solving a System of Linear Equations

Stage 1. We construct Q-determinant of the Jacobi method of solving a system of linear equations $A\bar{x} = \bar{b}$, where $A = [a_{ij}]_{i,j=1,\ldots,n}$, $a_{ii} \neq 0$ $(i = 1,\ldots,n)$, $\bar{x} = (x_1,\ldots,x_n)^T$, $\bar{b} = (a_{1,n+1},\ldots,a_{n,n+1})^T$. We denote as $c_{ij} = -\frac{a_{ij}}{a_{ii}}$ and $d_i = \frac{b_i}{a_{ii}}$. Let $\bar{x}^0$ be an initial approximation. Then the iteration process can be written as $x_i^{k+1} = \sum_{j=1,\ldots,n; j\neq i} c_{ij}x_j^k + d_i(i = 1,\ldots,n; k = 0,1,\ldots)$. The criterion of the iterative process ending is the condition $||\bar{x}^{k+1} - \bar{x}^k|| < \varepsilon$. There ε is the calculation precision.

Q-determinant of the Jacobi method consists of n conditional infinite Q-terms. Presentation of the Jacobi method in the form of Q-determinant is written as

$$x_i = \{(||\bar{x}^1 - \bar{x}^0|| < \varepsilon, x_i^1),\ldots,(||\bar{x}^k - \bar{x}^{k-1}|| < \varepsilon, x_i^k),\ldots\}(i = 1,\ldots,n). \quad (12)$$

Stage 2. We denote $u^l = ||\bar{x}^l - \bar{x}^{l-1}|| < \varepsilon(l = 1, 2,\ldots)$ that simplifies the description of Q-effective implementation. Then the Q-determinant has the form $x_i = \{(u^1, x_i^1), (u^2, x_i^2),\ldots,(u^k, x_i^k),\ldots\}(i = 1,\ldots,n)$. All unconditional Q-terms $\{u^l, x_i^l\}(i = 1,\ldots,n; l = 1, 2,\ldots)$ by the definition of Q-effective implementation should be calculated simultaneously. At first Q-terms $x_i^1(i = 1,\ldots,n)$ should be calculated simultaneously. Then Q-terms $u^1, x_i^2(i = 1,\ldots,n)$ are calculated simultaneously. If the value of u^1 is true then the calculation should be finished. In this case $x_i = x_i^1(i = 1,\ldots,n)$ is the solution of system of linear equations. If the computation will continue the Q-terms $u^k, x_i^{k+1}(i = 1,\ldots,n)$ are calculated at the same time for any value of $k \geqslant 2$. If the value of u^k is true then the calculation should be finished. In this case $x_i = x_i^k(i = 1,\ldots,n)$ is the solution of system of linear equations. So, Q-effective implementation of the method of Jacobi is realizable.

Stage 3 of our proposed method is a parallel program development for the Q-effective implementation of the algorithm. To make this we should be used for parallel programming tools. We use a description of the Q-effective implementation of the algorithm for developing a Q-effective program for shared memory. The description of the Q-effective implementation of the algorithm should be supplemented with a description of the distribution of computation by computing nodes for development of Q-effective program for distributed memory.

If we use distributed memory computing research is limited by a principle of a "master-slaves". That principle is used on cluster computing systems often. The principle can be described as follows. To compute we use node M (Master) and several computing nodes S (Slave). The computational process is divided into several steps.

Step 0. Initialization.

Step 1. Make a task from the node M to all nodes of S.

Step 2. Make the calculating on the each node of S without exchanges with other nodes.

Step 3. Sending the results of all the nodes S to M node.

Step 4. Merge the results on the node M.

We describe the development features of Q-effective programs using distributed memory.

The Algorithm of Matrix Multiplication

Each Q-term $\sum_{s=1}^{k} a_{is}b_{sj}(i = 1,\dots,n; j = 1,\dots,m)$ is calculated on your compute node of S. If number of nodes of S is less than the number Q-terms then some node of S can be calculated several Q-terms. The result of the calculation of each Q-term is transmitted to the node M.

Gauss–Jordan Method of Solving of Linear Equation Systems

Each matrix $\bar{A}^{j_1}(j_1 = 1,\dots,n)$ and the corresponding Q-term $u_i(i=1,\dots,n!)$ are to be calculated at its node of S. If the number of nodes S is less than n then nodes of S should perform calculations for several values of j_1. Nodes of S receive information from a node M to compute matrices $\bar{A}^{j_1}(j_1 = 1,\dots,n)$ and corresponding Q-terms $u_i(i = 1,\dots,n!)$. Results of calculation r_1 and $\bar{A}^{r_1}$ are transmitted to node M. Nodes of S receive information from a node M to compute matrices $\bar{A}^{r_1j_2}(j_2 = 1,\dots,n; j_2 \neq r_1)$ and corresponding Q-terms $u_i(i = 1,\dots,n!)$. Each matrix $A^{r_1j_2}(j_2 = 1,\dots,n; j_2 \neq r_1)$ and the corresponding Q-term $u_i(i = 1,\dots,n!)$ are to be calculated at its node of S. Results of calculation r_2 and $\bar{A}^{r_1r_2}$ are transmitted to node M. The following $n-2$ cycles of calculations are executed similarly.

Jacobi Method of Solving a System of Linear Equations

Each component of the vector $\bar{x}^k(k = 1,2,\dots)$ is calculated on different compute node S. If the number of nodes of S is less than n then nodes S should perform the calculations for several components of vector $\bar{x}^k(k = 1,2,\dots)$. At first node M sends to the nodes of S the necessary information to calculate the Q-terms $x_i^1(i = 1,\dots,n)$. Results of calculating are transmitted to node M. Node M sends values $x_i^1(i = 1,\dots,n)$ to the nodes S. Node M calculates the value of $||\bar{x}^1 - \bar{x}^0|| < \varepsilon$. At the same time $x_i^2(i = 1,\dots,n)$ are computed on the nodes of S simultaneously. Next iterations are executed similarly.

At the present time Q-effective programs are designed for the considered algorithms. To develop programs we use the programming language C. OpenMP technology is used for shared memory, MPI and OpenMP are used for distributed memory. The research was performed on the supercomputer "Tornado" of South Ural State University. The programs for shared memory were executed on one

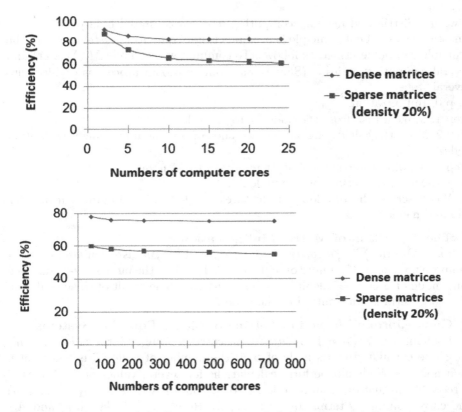

Fig. 1. The efficiency of Q-efficient programs of the matrix multiplication algorithm for shared (top) and distributed memory (bottom)

compute node. Programs for distributed memory used several compute nodes. Dynamic characteristics of the programs were evaluated. The students of South Ural State University Val'kevich [19], Tarasov [20] and Lapteva [21] developed the programs and found the evaluations of their characteristics.

Figure 1 shows the efficiency of Q-effective programs of the matrix multiplication algorithm for shared (top graph) and distributed memory (lower graph) [19].

The study of the Q-effective implementation of the algorithm shows that implementation of principle "master-slaves" for some algorithms impractical because it increases the amount of transfers between computing nodes. An example of such algorithm might be Jacobi method for solution of five-point difference equations. The study also reveals that usage of distributed memory is impractical for some algorithms because it increases execution time compare to usage of shared memory. An example is sweep method which is used for solution of three-point difference equations.

5 Conclusion

This method of designing a parallel program gives the possibility of the development of Q-effective program for numerical algorithms. Q-effective program uses the parallelism resource of the algorithm completely. The method can be used for the development of libraries of parallel programs for different classes of numerical algorithms.

Q-determinant makes the numerical algorithm clearer in terms of structure and implementation. In particular, it makes an opportunity to show the existing parallelism of the algorithm and the possible way of its parallel execution. So we can develop efficient implementations of algorithms for real parallel computing systems.

Acknowledgements. The reported study was funded by RFBR according to the research project No 17-07-00865 à. The work was supported by Act 211 Government of the Russian Federation, contract No 02.A03.21.0011.

References

1. Aleeva, V.N.: Analysis of parallel numerical algorithms: Preprint No. 590. Novosibirsk, Computing Center of the Siberian Branch of the Academy of Sciences of the USSR (1985)
2. Ignatyev, S.V.: Definition of parallelism resource of algorithms on the base of the concept of Q-determinant. In: Scientific Service on the Internet: Supercomputer Centers and Tasks: Proceedings of the International Supercomputer Conference, Novorossijsk, Russia, September 20–25, 2010, pp. 590–595. Publishing of the Moscow State University, Moscow (2010)
3. Svirihin, D.I.: Definition of parallelism resource of algorithm and its effective use for of a finite number of processors. In: Scientific Service on the Internet: the Search for New Solutions: Proceedings of the International Supercomputer Conference, Novorossijsk, Russia, September 17–22, 2012, pp. 257–260. Publishing of the Moscow State University, Moscow (2012)
4. Svirihin, D.I., Aleeva, V.N.: Definition the maximum effective realization of algorithm on the base of the conception of Q-determinant. In: Parallel Computational Technologies (PCT'2013): Proceedings of the International Scientific Conference, Chelyabinsk, Russia, April 1–5, 2013, p. 617. Publishing of the South Ural State University, Chelyabinsk (2013)
5. Aleeva, V.N.: The parallelization of algorithms on the base of the conception of Q-determinant. In: Groups and Graphs, Algorithms and Automata, 2015: Abstracts of the International Conference and PhD Summer School of the 80th Birthday of Professor Vyacheslav A. Belonogov and of the 70th Birthday of Professor Vitaly A. Baransky (Yekaterinburg, Russia, August 9–15, 2015), p. 33. UrFU Publishing house, Yekaterinburg (2015)
6. Aleeva, V.N., Sharabura, I.S., Suleymanov, D.E.: Software system for maximal parallelization of algorithms on the base of the conception of Q-determinant. In: Malyshkin, V. (ed.) PaCT 2015. LNCS, vol. 9251, pp. 3–9. Springer, Cham (2015). https://doi.org/10.1007/978-3-319-21909-7_1

7. Voevodin, V.V., Voevodin, V.V.: The V-ray technology of optimizing programs to parallel computers. In: Vulkov, L., Waśniewski, J., Yalamov, P. (eds.) WNAA 1996. LNCS, vol. 1196, pp. 546–556. Springer, Heidelberg (1997). https://doi.org/10.1007/3-540-62598-4_136

8. Voevodin, V.V., Voevodin, V.V.: Parallel computing. BHV-Petersburg, St. Petersburg (2002). (in Russian)

9. Open Encyclopedia of Parallel Algorithmic Features. http://algowiki-project.org/en/Open_Encyclopedia_of_Parallel_Algorithmic_Features

10. Voevodin, V.V.: Parallel algorithms under the microscope. In: Parallel Computational Technologies (PCT'2016): Report on the International Scientific Conference, Arkhangelsk, Russia, March 28–April 1 2016. http://omega.sp.susu.ru/books/conference/PaVT2016/talks/Voevodin.pdf

11. Gurieva, Y.L., Il'in, V.P.: On parallel computational technologies of augmented domain decomposition methods. In: Malyshkin, V. (ed.) PaCT 2015. LNCS, vol. 9251, pp. 35–46. Springer, Cham (2015). https://doi.org/10.1007/978-3-319-21909-7_4

12. Suplatov, D.A., Voevodin, V.V., Svedas, V.K.: Robust enzyme design: bioinformatic tools for improved protein stability. Biotechnol. J. **10**(3), 344–355 (2015). https://doi.org/10.1002/biot.201400150

13. Venkata, M.G., Shamis, P., Sampath, R., Graham, R.L., Ladd, J.S.: Optimizing blocking and nonblocking reduction operations for multicore systems: hierarchical design and implementation. In: Proceedings of IEEE Cluster, pp. 1–8 (2013)

14. Malyshkin, V.E., Perepelkin, V.A., Schukin, G.A.: Distributed algorithm of data allocation in the fragmented programming system LuNA. In: Malyshkin, V. (ed.) PaCT 2015. LNCS, vol. 9251, pp. 80–85. Springer, Cham (2015). https://doi.org/10.1007/978-3-319-21909-7_8

15. Malyshkin, V.E., Perepelkin, V.A., Tkacheva, A.A.: Control flow usage to improve performance of fragmented programs execution. In: Malyshkin, V. (ed.) PaCT 2015. LNCS, vol. 9251, pp. 86–90. Springer, Cham (2015). https://doi.org/10.1007/978-3-319-21909-7_9

16. Legalov, A.I.: Functional language for creating architecturally independent parallel programs. Comput. Technol. **1**(10), 71–89 (2005). (in Russian)

17. McColl, W.F.: General purpose parallel computing. In: Lectures on Parallel Computation. Cambridge International Series on Parallel Computation, pp. 337–391. Cambridge University Press, Cambridge (1993)

18. Valiant, L.G.: A bridging model for parallel computation. Commun. ACM **33**(8), 103–111 (1990)

19. Val'kevich, N.V.: Q-effective implementation of the algorithm for matrix multiplication on a supercomputer "Tornado SUSU": Graduate qualification work of bachelor in direction "Fundamental informatics and information technology": 02.03.02. 32 s. South Ural State University, Chelyabinsk (2017). http://omega.sp.susu.ru/publications/bachelorthesis/17-Valkevich.pdf

20. Tarasov, D.E.: Q-effective co-design of realization of the Gauss-Jordan method on the supercomputer "Tornado SUSU": Graduate qualification work of master in direction "Fundamental informatics and information technology": 02.04.02. 41 s. South Ural State University, Chelyabinsk (2017). http://omega.sp.susu.ru/publications/masterthesis/17-Tarasov.pdf

21. Lapteva, Yu.S.: *Q*-effective implementation of the Jacobi method for solving SLAE on the supercomputer "Tornado SUSU": Graduate qualification work of bachelor in direction "Fundamental informatics and information technology": 02.03.02. 30 s. South Ural State University, Chelyabinsk (2017). http://omega.sp.susu.ru/publications/bachelorthesis/17-Lapteva.pdf

Enumeration of Isotopy Classes of Diagonal Latin Squares of Small Order Using Volunteer Computing

Eduard Vatutin[1]([✉]), Alexey Belyshev[2], Stepan Kochemazov[3],
Oleg Zaikin[3], and Natalia Nikitina[4]

[1] Southwest State University, Kursk, Russia
evatutin@rambler.ru
[2] BOINC.ru, Moscow, Russia
alexey-bell@yandex.ru
[3] Matrosov Institute for System Dynamics and Control
Theory SB RAS, Irkutsk, Russia
veinamond@gmail.com, zaikin.icc@gmail.com
[4] Institute of Applied Mathematical Research KRC RAS, Petrozavodsk, Russia
nikitina@krc.karelia.ru

Abstract. The paper is devoted to discovering new features of diagonal Latin squares of small order. We present an algorithm, based on a special kind of transformations, that constructs a canonical form of a given diagonal Latin square. Each canonical form corresponds to one isotopy class of diagonal Latin squares. The algorithm was implemented and used to enumerate the isotopy classes of diagonal Latin squares of order at most 8. For order 8 the computational experiment was conducted in a volunteer computing project. The algorithm was also used to estimate how long it would take to enumerate the isotopy classes of diagonal Latin squares of order 9 in the same volunteer computing project.

Keywords: Volunteer computing · Combinatorics · Latin square Diagonal Latin square · Enumeration

1 Introduction

There exist a number of problems in which it is required to enumerate combinatorial objects with specific features. The correspondence between the number of such objects and the problem's dimension can be viewed as an integer sequence. The Online Encyclopedia of Integer Sequences (OEIS) [1] contains about 300 thousands of such sequences. For some of them there exists a formula that allows to easily calculate any member of the sequence. In the cases where such formulas don't exist, the members of the sequences can sometimes be obtained as a result of a computational experiment. Often, such calculations require a large amount of computational resources, that is why high performance computing methods are required to conduct them.

© Springer Nature Switzerland AG 2019
V. Voevodin and S. Sobolev (Eds.): RuSCDays 2018, CCIS 965, pp. 578–586, 2019.
https://doi.org/10.1007/978-3-030-05807-4_49

Latin squares [2] are one of the most well studied combinatorial designs. They have applications both in science and industry. A *Latin square* of order N is a square table with $N \times N$ cells filled with elements from a finite set $\{0, 1, 2, \ldots, N-1\}$ in such a way, that all elements within each row and each column are distinct. If in addition to this both main diagonal and main antidiagonal contain every possible element from $\{0, 1, 2, \ldots, N-1\}$ then such Latin square is called *diagonal Latin square*.

Latin squares have practical applications in various areas, including experiment design, cryptography, error-correcting codes, scheduling. A number of open mathematical problems are associated with Latin squares, for example whether there exists a triple of mutually orthogonal Latin squares of order 10 [3]. Also of interest is the asymptotic behavior of combinatorial characteristics of Latin squares with the increase of dimension N.

There are several examples of applications of high-performance computing in order to search for combinatorial designs based on Latin squares. For example, the hypothesis about the minimal number of clues in Sudoku was first proven on a computing cluster [4].

The present study is aimed at enumeration of isotopy classes of diagonal Latin squares of small order. In other words, the goal is to construct the corresponding new integer sequence, not yet presented in OEIS. To achieve this goal, we develop a new combinatorial algorithm, implement it and employ a volunteer computing project to conduct the computational experiments.

2 Isotopy Classes and Canonical Forms of Diagonal Latin Squares

A number of combinatorial problems related to diagonal Latin squares can be solved significantly faster if only one diagonal Latin square from an equivalence class is considered. Each equivalence class contains diagonal Latin squares with equal features (for instance, the number of orthogonal mates). Such approach can be used to enumerate diagonal Latin squares [5,6] or to construct and enumerate systems of orthogonal diagonal Latin squares and combinatorial designs based on them [7].

In the case of Latin squares, an isotopy class consists of all Latin squares that are equivalent: any Latin square from the class can be produced by any other Latin square from the class by permuting the rows, columns, or the names of the symbols of a Latin square. In the case of diagonal Latin squares, the isotopic classes cannot be constructed in the same manner because some permutations of rows and columns can lead to the violation of the constraint on uniqueness of elements on diagonals. Instead, the isotopy classes of diagonal Latin squares can be constructed using the so-called M-transformations, which were suggested by Chebrakov for magic squares [8] (by definition any diagonal Latin square is a magic square). The M-transformations can be divided into two types. An M-transformation of the first type is a permutation of a pair of columns, which are symmetric with respect to the middle of a square, along with the permutation of rows with the same indexes. An example of an M-transformation of the first type is shown in Fig. 1.

An M-transformation of the second type is a permutation of a pair of columns from the left half of the square, along with the permutation of a pair of columns from the right half of the square, which are symmetric to the columns from the first pair, and the corresponding permutation of two pairs of rows with the same indexes. An example of an M-transformation of the second type is shown in Fig. 2.

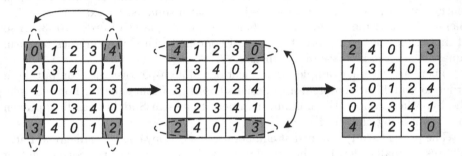

Fig. 1. An example of an M-transformation of the first type.

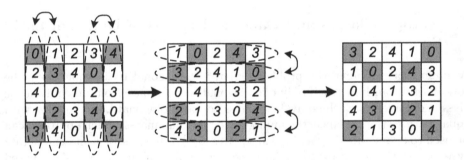

Fig. 2. An example of M-transformation of the second type.

It is easy to see, that several other equivalent transformations can be used for diagonal Latin squares: turning a square by $90 \times h$ degrees, $h \in \{1, 2, 3\}$; transposing it relative to main diagonal or main antidiagonal; mirroring it horizontally or vertically. The examples of all these transformations are shown in Fig. 3.

While all squares in an isotopy class are equal, it is convenient to choose one of them to be a representative of a class. Assume that we consider a diagonal Latin square A of order N. It can be represented as a sequence $s(A)$ with $N \times N$ elements by writing all elements of A one by one from left to right, from top to bottom. Let us refer to $c(A)$ as to *string representation* of A. A *canonical*

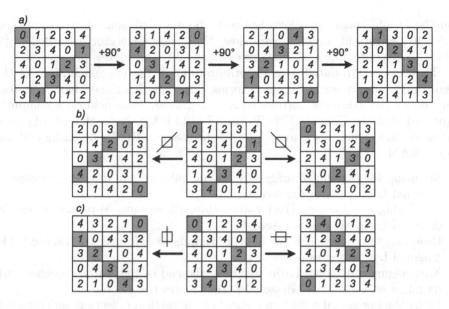

Fig. 3. Examples of equivalence transformations, which are not defined by M-transformations: turns of a square (a); transpose of a square (b); vertical and horizontal reflections (c). One of the square's transversal is marked with grey.

form (CF) $\tilde{A}$ of a diagonal Latin square A is a diagonal Latin square from a corresponding isotopy class $\Theta(A) = \{A_1, A_2, \ldots, A_C\}$, $A \in \Theta(A)$, $\tilde{A} \in \Theta(A)$, which has a lexicographically minimal string representation: $\forall A' \in \Theta(A) \backslash \{\tilde{A}\}$: $s(\tilde{A}) < s(A')$. If for a fixed N we construct a canonical form for every single diagonal Latin square of order N, then the number of unique canonical forms coincides with the number of isotopy classes of diagonal Latin squares of order N (in the sense outlined above), and this fact can be employed in practice.

3   Computational Experiments

The number of isotopy classes of Latin squares of order at most 11 is known [9–11] and is presented in the OEIS (sequence A040082). Meanwhile, the corresponding sequence for diagonal Latin squares has not been known before the present study. The easiest way to determine such number for a particular order N is to check each Latin square of order N whether it is a canonical form or not. To generate all diagonal Latin squares of small order, a fast generator can be used [5]. In order to construct a canonical form for a given Latin square A, the corresponding isotopy class $Q(A) = \{A_1, A_2, \ldots, A_C\}$ is built using M-transformations and other transformations preserving the contents of main diagonal and main antidiagonal of a Latin square (they were mentioned in the previous section). It should be noted, that Latin square A is a canonical form if and only if $A = argmin_{A_i \in \Theta(A)} A_i$. A strict implementation of this approach (on the Delphi language) allows to check about 1–2 diagonal Latin squares of order 10 per second, which is very slow. Note, that generator described in [5] allows to generate about 6.6 millions of diagonal Latin squares per second.

There were introduced several optimizations in the implementation of the algorithm that constructs canonical forms. They are described below, together with estimated gain. The corresponding experiments were held on a computer equipped with the Core i7-4770 CPU and DDR3 RAM. Note, that small size of the used data structures made it possible to use only CPU's L1 cache and not to use RAM.

1. Reducing the amount of usage of dynamic memory and dynamic strings: 31 diagonal Latin squares per second.
2. Employing the Johnson-Trotter algorithm [12] to form M-permutations: 56 diagonal Latin squares per second.
3. Disabling the checkups of the built-in compilator ($R–, ASSERTs, etc.): 118 diagonal Latin squares per second.
4. Normalizing a diagonal Latin square in the end of the search together with its mirrored version: 277 diagonal Latin squares per second.
5. Using the horizontal reflection instead of the vertical reflection: 305 diagonal Latin squares per second.
6. Terminating the processing of a current diagonal Latin square if its string representation is less than that of an original one, high-level optimizations: 642 diagonal Latin squares per second.

It should be noted, that the aforementioned rates were obtained at one of the first sections of the search space (of all possible diagonal Latin squares, in lexicographic order), where the concentration of canonical forms is high and the corresponding checks take quite a lot of time. At the final sections of the search space the canonical forms are sparse, and the performance is several times higher because of early terminations.

Using this implementation (author: Eduard Vatutin) a computational experiment was conducted on a personal computer. As a result, the isotopy classes of diagonal Latin squares of order at most 7 were enumerated, forming the following sequence (for $1 \leq N \leq 7$, where N is an order): 1, 0, 0, 1, 2, 2, 972.

A computational experiment for order 8 was conducted in the volunteer computing project Gerasim@home [13]. The deadline of 7 days, and the quorum of 2 were used. A set of work units (WUs) was formed in such a way, that each WU contains initial values of several elements of a square. Given a particular WU, a client application on a volunteer's personal computer forms all possible values of other elements of a square in order to generate all possible diagonal Latin squares (with fixed values of some elements provided by the WU). The client application enumerates all canonical forms among the generated diagonal Latin squares, and the obtained number is sent to the project server (see Fig. 4).

This experiment was conducted in 2 days. As a result, the isotopy classes of diagonal Latin squares of order 8 were enumerated, it turned out that there are 4 873 096 of them (remind, that each such class corresponds to a canonical form). This result was verified by another program implemented by Alexey Belyshev.

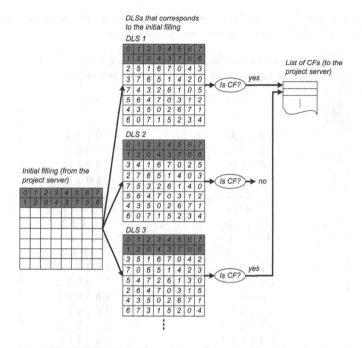

Fig. 4. Schematic representation of the calculation process on a client computer.

His implementation is based on normalization of diagonal Latin squares performed based on the main diagonal. Thus, the following sequence was obtained: '1, 0, 0, 1, 2, 2, 972, 4873096'. It represents the number of isotopy classes of diagonal Latin squares of order at most 8. This sequence was unknown before the present study, it has been reviewed and added to the OEIS with the number A287764.

For order 9 the number of corresponding normalized diagonal Latin squares is 505 699 465 350 758 [5]. Taking into account that an average rate of checking for canonical forms for diagonal Latin squares of order 9 is about 60 thousand squares per second, it means that to enumerate the isotopy classes for order 9 it would take about 267 years of calculations on 1 CPU core, or about 1 year in a volunteer computing project with the real performance of 1 TFLOPS/s.

In the course of the conducted experiments, the minimal and maximal sizes of isotopy classes of diagonal Latin squares of order at most 8 were also found, see Table 1. The corresponding sequences '1, 0, 0, 2, 4, 32, 32, 96' and '1, 0, 0, 2, 4, 96, 192, 1536' are new and have not been represented in the OEIS yet.

First pairs of orthogonal diagonal Latin squares of order 10 were first presented in [14]. The volunteer computing projects Gerasim@home [13] and SAT@home [7] have been used to find systems of mutually orthogonal diagonal Latin squares of order 10. It was done in an attempt to prove the existence or non-existence of a triple of mutually orthogonal diagonal Latin squares (MODLS) of

Table 1. The minimal and maximal sizes of isotopy classes of diagonal Latin squares of order at most 8 and examples of corresponding squares.

N	Minimal size of isotopy class and its representative		Maximal size of isotopy class and its representative	
1	1	-	1	-
2	-	-	-	-
3	-	-	-	-
4	2	0 1 2 3 2 3 0 1 3 2 1 0 1 0 3 2	2	0 1 2 3 2 3 0 1 3 2 1 0 1 0 3 2
5	4	0 1 2 3 4 2 3 4 0 1 4 0 1 2 3 1 2 3 4 0 3 4 0 1 2	4	0 1 2 3 4 2 3 4 0 1 4 0 1 2 3 1 2 3 4 0 3 4 0 1 2
6	32	0 1 2 3 4 5 1 2 0 5 3 4 4 3 5 0 2 1 3 5 1 4 0 2 5 4 3 2 1 0 2 0 4 1 5 3	96	0 1 2 3 4 5 1 2 0 5 3 4 4 3 5 0 2 1 3 0 1 4 5 2 5 4 3 2 1 0 2 5 4 1 0 3
7	32	0 1 2 3 4 5 6 1 2 5 6 0 3 4 4 6 3 0 5 1 2 5 4 6 1 3 2 0 3 5 4 2 6 0 1 6 0 1 5 2 4 3 2 3 0 4 1 6 5	192	0 1 2 3 4 5 6 1 2 5 4 6 0 3 5 0 3 6 2 4 1 2 4 6 1 0 3 5 6 3 4 0 5 1 2 4 5 1 2 3 6 0 3 6 0 5 1 2 4
8	96	0 1 2 3 4 5 6 7 1 2 3 5 7 6 0 4 3 0 1 7 5 4 2 6 5 6 7 4 3 0 1 2 7 3 5 1 6 2 4 0 4 7 6 0 2 3 5 1 6 5 4 2 0 1 7 3 2 4 0 6 1 7 3 5	1 536	0 1 2 3 4 5 6 7 1 2 5 6 7 3 0 4 5 7 1 0 6 2 4 3 7 3 6 4 1 0 5 2 2 6 4 5 3 7 1 0 4 5 3 7 0 6 2 1 3 4 0 2 5 1 7 6 6 0 7 1 2 4 3 5

order 10. Note, that in [15] a triple of diagonal Latin squares of order 10, that is the closest to being a triple of MODLS of order 10 found so far, was provided.

   To exclude the duplication of found systems of ODLS of order 10, for each system the canonical forms of its squares are constructed, and then these canonical forms are added to a special list, in which all elements are unique. This procedure has been performed for all systems of ODLS of order 10 found so far in two mentioned volunteer computing projects. At the present moment (April 2018), this list consists of 309 thousand canonical forms of diagonal Latin squares of order 10 (see Fig. 5). It turned out, that based on these canonical forms, the triple of MODLS of order 10 cannot be constructed, and neither it is possible to construct a pseudotriple which is better than that found in [15].

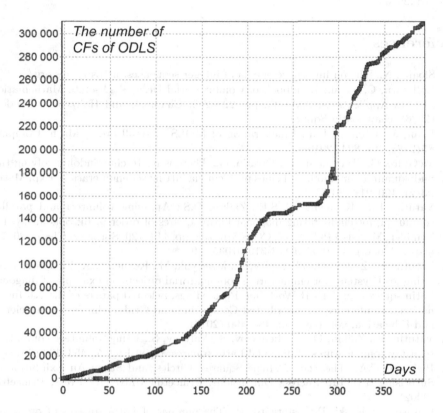

Fig. 5. The number of unique canonical forms met in systems of orthogonal diagonal Latin squares of order 10 found in Gerasim@home for the last year

4   Conclusions

Using a combinatorial algorithm, which allows to construct canonical forms, the isotopy classes of diagonal Latin squares of order at most 8 were enumerated. The experiments for order up to 7 were conducted on a computer, while for order 8 it was conducted in a volunteer computing project. It was also estimated, that isotopy classes for order 9 can be enumerated in reasonable time using the same volunteer computing project. In the nearest future we are planning to launch the corresponding experiment. It was also shown, that the canonical forms can be used to maintain a list of all systems of orthogonal diagonal Latin squares found so far.

Acknowledgements. The research was partially supported by Russian Foundation for Basic Research (grants 16-07-00155-a, 17-07-00317-a, 18-07-00628-a, 18-37-00094-mol-a) and by Council for Grants of the President of the Russian Federation (stipend SP-1829.2016.5).

References

1. Sloane, N.: The on-line encyclopedia of integer sequences. https://oeis.org/
2. Colbourn, C., et al.: Handbook of Combinatorial Designs. Discrete Mathematics and Its Applications, 2nd edn, pp. 224–265. Chapman and Hall/CRC, London (2006). chap. Latin Squares
3. Egan, J., Wanless, I.M.: Enumeration of MOLS of small order. Math. Comput. **85**(298), 799–824 (2016)
4. McGuire, G., Tugemann, B., Civario, G.: There is no 16-clue Sudoku: solving the Sudoku minimum number of clues problem via hitting set enumeration. Exp. Math. **23**(2), 190–217 (2014)
5. Vatutin, E.I., Kochemazov, S.E., Zaikin, O.S.: Applying volunteer and parallel computing for enumerating diagonal Latin squares of order 9. In: Sokolinsky, L., Zymbler, M. (eds.) PCT 2017. CCIS, vol. 753, pp. 114–129. Springer, Cham (2017). https://doi.org/10.1007/978-3-319-67035-5_9
6. Vatutin, E., Zaikin, O., Zhuravlev, A., Manzyuk, M., Kochemazov, S., Titov, V.S.: Using grid systems for enumerating combinatorial objects on example of diagonal Latin squares. In: CEUR Workshop Proceedings, Selected papers of the 7th International Conference on Distributed Computing and Grid-Technologies in Science and Education, vol. 1787, pp. 486–490 (2017)
7. Vatutin, E., Zaikin, O., Kochemazov, S., Valyaev, S.: Using volunteer computing to study some features of diagonal Latin squares. Open Eng. **7**, 453–460 (2017)
8. Pickover, C.A.: The Zen of Magic Squares, Circles, and Stars: An Exhibition of Surprising Structures across Dimensions. Princeton University Press, Princeton (2002)
9. Hulpke, A., Kaski, P., Östergård, P.: The number of Latin squares of order 11. Math. Comput. **80**(274), 1197–1219 (2011)
10. McKay, B.D., Rogoyski, E.: Latin squares of order 10. Electr. J. Comb. **2**(1), 1–4 (1995)
11. McKay, B.D., Wanless, I.M.: On the number of Latin squares. Ann. Comb. **9**(3), 335–344 (2005)
12. Trotter, H.F.: Algorithm 115: perm. Commun. ACM **5**(8), 434–435 (1962)
13. Vatutin, E., Valyaev, S., Titov, V.: Comparison of sequential methods for getting separations of parallel logic control algorithms using volunteer computing. In: Second International Conference BOINC-based High Performance Computing: Fundamental Research and Development (BOINC:FAST 2015), Petrozavodsk, Russia, September 14–18, 2015, vol. 1502, pp. 37–51. CEUR-WS (2015)
14. Brown, J., Cherry, F., Most, L., Parker, E., Wallis, W.: Completion of the spectrum of orthogonal diagonal Latin squares. Lecture Notes in Pure and Applied Mathematics, vol. 139, pp. 43–49 (1992)
15. Zaikin, O., Zhuravlev, A., Kochemazov, S., Vatutin, E.: On the construction of triples of diagonal Latin squares of order 10. Electron. Notes Discrete Math. **54**, 307–312 (2016)

Interactive 3D Representation as a Method of Investigating Information Graph Features

Alexander Antonov[✉] and Nikita Volkov

Lomonosov Moscow State University, Moscow, Russia
asa@parallel.ru, volkovnikita94@gmail.com

Abstract. An algorithm information graph is a structure of wide variety. It can tell a lot about algorithm features, such as computational complexity and resource of parallelism, as well as about sequential operations blocks within an algorithm. Graphs of different algorithms often share similar regular structures — their presence is an indicator of potentially similar algorithm behavior. Convenient, interactive 3D representation of an information graph is a decent method of researching it; it can demonstrate algorithm characteristics listed above and its structural features. In this article we investigate an approach to creating such representations, implement it using our AlgoView system and give examples of using a resulting tool.

Keywords: Information graph · Parallelism · AlgoWiki · AlgoView
Level parallel form

1 Introduction

An information graph of an algorithm is a DAG (directed acyclic graph), where vertices stand for operations within an algorithm and edges stand for data dependencies [1]. Its size and structure can thus depend not only on particular algorithm features, but also on input data size and on an exact execution sample (the latter happens only in case of a nondeterministic algorithm). Information graph can demonstrate algorithm's resource of parallelism, determine primary data flows within it and show parts that require most computation to be executed. Information graphs of different algorithms can possess similar regular structures, demonstrating possible similarities in algorithms' execution processes.

The graph itself can be represented in many forms, which allow to highlight some particular features of its structure. These representations include the level parallel form [1] and projections of a multidimensional graph on a hyperplane with less dimensions. The number of dimensions is an algorithm characteristics which can be defined as the maximum depth of cycle nests within an algorithm. If it's 3 or less, one can easily represent a graph in 3-dimensional space without hiding its regular structure. Otherwise more complex methods may be required

© Springer Nature Switzerland AG 2019
V. Voevodin and S. Sobolev (Eds.): RuSCDays 2018, CCIS 965, pp. 587–598, 2019.
https://doi.org/10.1007/978-3-030-05807-4_50

to display such a graph with all regularities preserved. These include graph projections and macrographs (a generalization of information graphs, where vertices can stand not only for operations, but also for whole information graphs of algorithm's subtasks).

2   Graph Visualization

Representing a graph in a visual form is overall a complex task with many approaches to its solution. The exact approach depends on the primary goal of a given graph visualization task. Most well-developed and widespread techniques aim for automatically built images of really large graphs of, say, social networks or worldwide web parts, with related vertices being grouped into clusters. An example of this approach can be found in [2]. We, however, deal with small graphs that have a feature their bigger brothers don't — a strict, regular structure. Information graphs therefore have much more in common with data flow and control flow graphs used in compilation theory [3]. The limited size of studied graphs comes from their definition — we can take input data size as small as we want, if the graph still retains its regular structure.

3   The AlgoWiki Project

We came up with an idea of the AlgoWiki project a few years ago [4,5]. It's main purpose is to describe algorithms accenting their resource of parallelism, possible implementations for massively parallel systems, researching scalability [7] and efficiency. However, new possible problems to solve appeared during it's development. These include investigating less-known algorithm characteristics, such as data locality [8], building up a classification system for tasks, methods and algorithms, searching for algorithms that could be used as benchmarks and creating supercomputer ratings based on these benchmarks. AlgoWiki is an open web-resource [6] powered by MediaWiki engine. This project is supported by Russian Science Foundation and officially led by a well-known scientist, professor J. Dongarra (University of Tennessee, USA).

4   An Interactive Representation — Features and Problems

A visual representation of an algorithm graph is definitely a powerful instrument for researching its structure. There are, however, so many different information graphs, that a viable and logical form of visual representation simply can't be the same for all of them. Modifications of a visual representation are very useful for displaying certain algorithm features.

Moreover, our AlgoView system used in AlgoWiki to provide visual representations of information graphs [9] generates 3D representations automatically, based on XML algorithms descriptions (which, in their turn, are based on actual program code). This system defines every vertex of an information graph and its position within a generated representation. Edges are defined by pairs of vertices they connect. This information doesn't give a unique representation of vertices and edges, since we don't know, how exactly graph edges should be displayed. This causes certain technical difficulties.

4.1 Level Parallel Form of an Information Graph

A level parallel form is one of many ways to show an information graph. All graph vertices are placed on a certain level, which is defined by two simple rules. A vertex without data dependencies is placed on level one. For any other vertex all sources of data dependencies are assigned to a certain level first; the vertex is then given a level of maximum source level plus one. The main purpose of such a representation is to show algorithm's resource of parallelism. Another one (used in our interactive representations) is to follow operation executions within an algorithm. Our method of displaying an algorithm level parallel form is shown in Fig. 1. The tool we created to show information graphs is available online and interactive, so users should have a certain portion of control over what they see on their screens. In the upper right corner of Fig. 1 a minimized control panel used to manipulate the visualization and provide information can be seen. It consists of several menus which can be maximized separately and allow to easily acquire visualization projections, highlight level-parallel form levels, vertex types, display information like axis names etc., but don't possess scientific value themselves and thus won't be referred to later on.

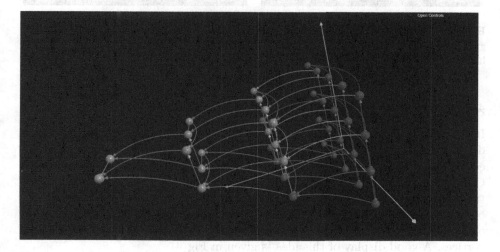

Fig. 1. Possible way to display a level parallel form; vertices from current level are shown in orange, previous levels are dark green, upcoming levels are light green. (Color figure online)

4.2   Information Graph Projections

As it was said before, our visual representation of an information graph is 3-dimensional. However, regular structures within a graph aren't necessarily 3-dimensional themselves. They can have both more and less dimensions. An example of a "1-dimensional" structure is data broadcast. A 3-dimensional representation can be projected on axis or plane to better display such low-dimensional structures. A good projection brings more clearness into information graph structure. For example, in Fig. 2 data dependencies are almost hidden, while in Fig. 3 one can clearly see a basic data broadcast.

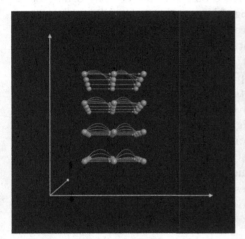
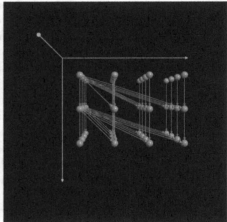

Fig. 2. A projection of an information graph on oYZ plane; broadcast is almost hidden.

Fig. 3. A projection of an information graph on oXZ plane; broadcast can be clearly seen.

4.3   Removing Edges Intersections

Intersections of edges within an information graph visual representation can be divided into several categories: self-intersections, edges intersecting each other, edges passing through vertices that are neither edge beginning nor edge ends. Currently we use only one method to avoid too many edge intersections. If distance between the start and end vertices of an edge surpasses a certain threshold, then Bezier curves are used to layout edges in 3-dimensional space instead of straight lines. Curving extent can also be set to depend on distance between vertices. A good display of this idea is given in Fig. 4.

Sometimes this algorithm doesn't allow to remove all edge intersections, as shown in Fig. 5. However, it at least excludes the possibility of edges completely impositioning each other. In some cases intersection practically don't affect one's ability to distinct different edges. An example of such a graph will be shown later in Fig. 11.

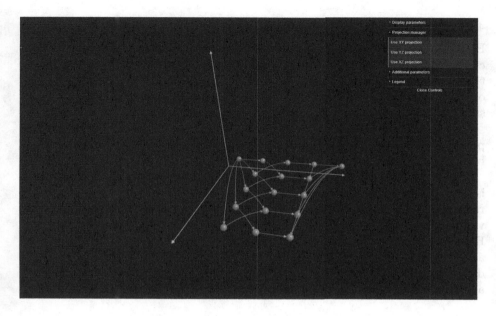

Fig. 4. Broadcasting is displayed with the help of Bezier curves.

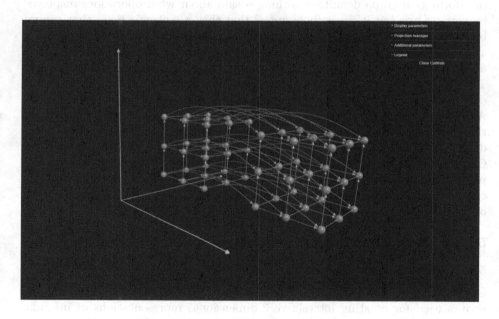

Fig. 5. Edges still intersect, but don't imposition each other.

4.4 Outer Parameters and Point of View

Besides key questions described above, one needs to address lesser tasks when creating a visual representation of an information graph. The most obvious of them is choosing a correct point of view for a given representation. With point of view being a generalization of graph projections, different points of view can simplify or complicate one's understanding of an algorithm structure.

We decided to allow users to choose points of view themselves, while certain points of view (such as projections) are saved, so current point of view can always be reset to one of those.

Information graphs are also different for every given size of input data, which is controlled by so-called outer parameters. These affect not only the size of regular structures within an information graph, but also define whether some vertices and edges are present in a graph or not. A good example of that is Householder algorithm for QR-decomposition of a matrix [10]. There are $N - I$ additional blocks in it, and data is broadcasted from the first block to all of these during every step of an algorithm. If $N - I = 1$, however, there's only one block and the broadcast nature of its data dependency from the first block is hidden.

4.5 Macrographs

In information graph definition nothing is said about what operations displayed by vertices are, but it's usually assumed that they are atomic tasks that cannot be further subdivided into smaller parts. A macrograph is a generalization of an information graph, in which operations displayed by vertices can be pretty much anything, including whole information graphs of an algorithm subtask. Two macrograph vertices have a data dependency between them, if there is a data dependency between any pair of information graph vertices they represent. A concept of macrograph is really useful for representing algorithms and methods that basically consist of several more simple algorithms. Here we give a simple display of a macrograph concept. In Fig. 6, you can see the already familiar Givens method from Fig. 1, except for now vertices as displayed as macros, not just simple operations. In Fig. 7, there's a more thorough structure view of a cyan colored macro vertex from Fig. 6, where green vertices stand for normal operations and red octahedrons stay for input and output data.

5 A Library of Typical Information Graph Structures

5.1 The AlgoView System

Our AlgoView system, described in detail in [9], is a set of applications and scripts used for creating interactive 3-dimensional representations of information graphs. Here we give only a brief overview of it. The AlgoView system includes two main functional tools — a 3D model generator written in C++ and a viewer, which is based on a set of web-pages with JavaScript and WebGL utilities attached to it. It's supported by server-executed node.js code, which brings more functionality to an interactive representation, for example, allowing user to re-generate 3D-models on server if needed.

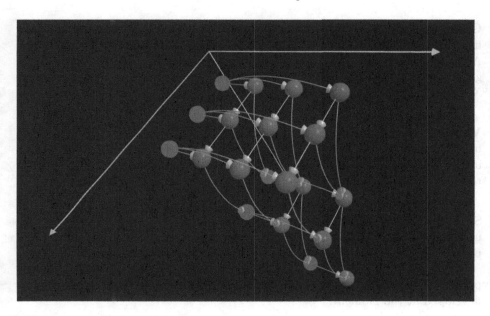

Fig. 6. Givens QR-decomposition method for $N = 3$. At each plane, all vertices but the broadcasting ones display a 2D-vector rotation operation.

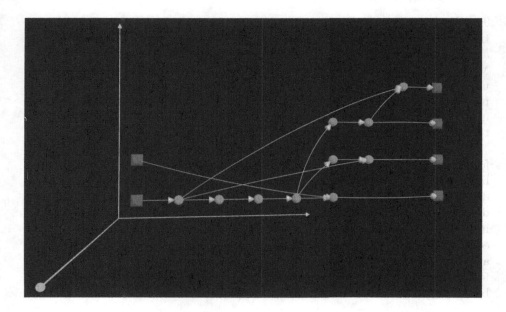

Fig. 7. A more detailed structure of a 2D vector rotation operation from Fig. 6.

Generator. An input for the generator is an XML file of known format. XML files are also built automatically, but XML file builder doesn't belong to the AlgoView system. A set of 3D models representing the information graph is generated. User is asked to provide values for all outer parameters listed in an XML file.

Viewer. Users are supposed to interact with the viewer as follows: prepared 3D models are stored on server. Users open web-pages containing 3D representations in their browsers and JavaScript code is executed on client side. This code is responsible for showing models in browser window and manipulating them.

Server Part. This includes node.js code executed on server, which allows to upload new models to same web-page and to generate 3D models with new parameters on server (where a copy of generator application is stored).

One can read more about AlgoView in the link given above. However, server part isn't described there, since it wasn't even planned at that time. It's being developed at the moment.

5.2   A Library of Typical Information Graph Structures Based on AlgoView

The AlgoView system provides quite a variety for possibilities to display information graphs. These representations are already used to further describe certain algorithms in AlgoWiki [11], but they can also be used for another purposes. Many different algorithms possess similar regular structures. Some of these regular structures are very well known in terms of how deeply their resource of parallelism and other characteristics are investigated. They are also encountered in various algorithms very often and can be called typical. We decided to use the AlgoView system to create a library of such typical regular structures.

The most common ones are 2- and 3-dimensional groups of vertices, which form up a grid or a group of grids. The main difference between these structures lies in the way of how data dependencies are organized. It's convenient to divide dependency patterns into categories not only by addressing them the way it's done in level parallel form, but also by paying attention how much coordinates in n-dimensional space are different in a pair of vertices connected by a graph edge. These categories are good for describing a complicated dependency pattern as a combination of more simple ones. Moreover, they correspond very well to a formal description of an information graph given in XML files, which are used by the AlgoView system. Some of these typical structures are:

- N-dimensional sequential structures, as shown in Fig. 8;
- N-dimensional broadcasts, as shown in Fig. 9;
- Independent sets of sequential structures or broadcasts (Fig. 10);
- Same sets of sequential structures and broadcasts, but with dependencies between them (Fig. 11);

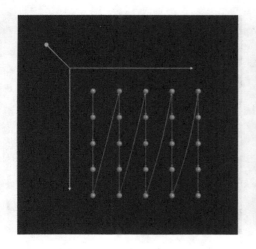

Fig. 8. A 2-dimensional sequential structure.

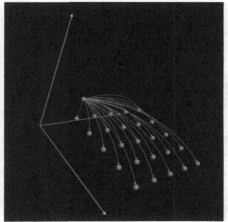

Fig. 9. A 2-dimensional broadcast.

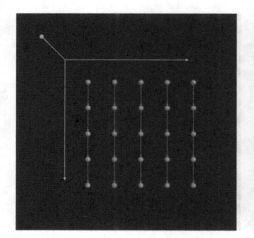

Fig. 10. A set of 1-dimensional, independent, sequential structures.

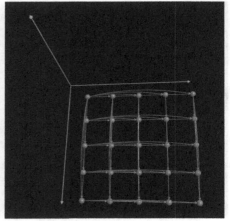

Fig. 11. Two sets of 1-dimensional broadcasts, dependent on each other.

– Schemes with oblique parallelism [10], as shown in Figs. 12 and 13;
– Same as previous category, but with additional broadcasts.

This basic set is already enough to display many structures seen in widely used algorithms, such as [11,12]. More complex structures can be described as a combination of these simple ones. Examples are in Figs. 14 and 15.

In addition to its main purpose, this library can, from our point of view, help in solving other tasks. One of them is testing and further developing the AlgoView system. In case of complex information graphs structures the system build correct, but yet non-aesthetic 3D-models. Since information graph

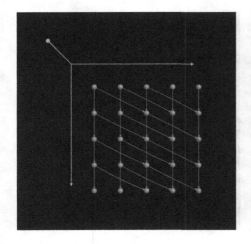

Fig. 12. A parallel scheme.

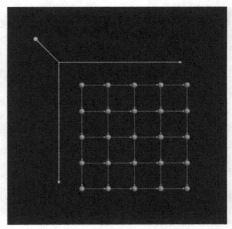

Fig. 13. A scheme with oblique parallelism.

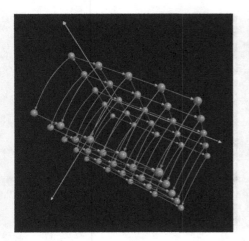

Fig. 14. Two 2-dimensional vertex groups with parallelism, connected to each other.

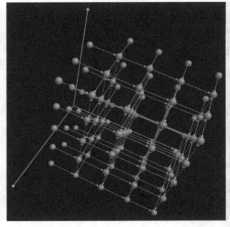

Fig. 15. A 3-dimensional vertex group, being a set of 2-dimensional schemes with oblique parallelism

complexity doesn't have any limits (expect for the maximum number of edges for a graph of given size), we aren't aiming for an ideal 3D-model generation system. However, paying attention to structures common in real algorithms is, in our opinion, definitely worth it.

Our plan is also make this regular structures library public with same intentions as the whole AlgoWiki project. Just like with algorithms, we plan to further classify these structures and create description pages for all of them. Every page will contain an interactive representation of a given structure and examples of algorithms utilizing it.

6   Conclusion

In this paper, we have given a review of various algorithm features that can be represented within an information graph, as well as methods to do so. We have also provided data on our approach to visualizing information graphs considering these features and provided an updated brief description of our AlgoView system. In addition, our two latest ideas were discussed in this paper. The first one is about creating a public library of typical structures found in information graphs with the help of interactive 3D representations. A list of such structures was described and examples of 3D representations were provided. The second idea is all about finding algorithms that could be good benchmarks for massively parallel systems, since we think a good sign of such an algorithm would be numerous typical structures within it. Further research and implementation of these two concepts is therefore our primary concern.

Acknowledgements. The results described in Sects. 1, 2 and 4 were obtained in Lomonosov Moscow State University with the financial support of the Russian Science Foundation (Agreement № 14–11–00190). The research is carried out using the equipment of the shared research facilities of HPC computing resources at Lomonosov Moscow State University supported by the project RFMEFI62117X0011.

References

1. Voevodin, V., Voevodin, Vl.: Parallel Computing. BHV-Petersburg, St. Petersburg (2002)
2. Hu, Y.I., Shi, L.: Visualizing large graphs. WIREs Comput. Stat. **7**(2), 115–136 (2015). https://doi.org/10.1002/wics.1343
3. Gold, R.: Control flow graphs and code coverage. Appl. Math. Comput. Sci. **20**(4), 739–749 (2010). https://doi.org/10.2478/v10006-010-0056-9
4. Voevodin, V., Antonov, A., Dongarra, J.: AlgoWiki: an open encyclopedia of parallel algorithmic features. Supercomput. Front. Innovations **2**(1), 4–18 (2015). https://doi.org/10.14529/jsfi150101
5. Voevodin, V., Antonov, A., Dongarra, J.: Why is it hard to describe properties of algorithms? Procedia Comput. Sci. **101**, 4–7 (2016). https://doi.org/10.1016/j.procs.2016.11.002
6. Open Encyclopedia of Parallel Algorithmic Features. http://algowiki-project.org/en. Accessed 13 Apr 2018
7. Antonov, A., Teplov, A.: Generalized approach to scalability analysis of parallel applications. In: Carretero, J. (ed.) ICA3PP 2016. LNCS, vol. 10049, pp. 291–304. Springer, Cham (2016). https://doi.org/10.1007/978-3-319-49956-7_23
8. Antonov, A., Voevodin, V., Voevodin, Vl., Teplov, A.: A study of the dynamic characteristics of software implementation as an essential part for a universal description of algorithm properties. In: 24th Euromicro International Conference on Parallel, Distributed, and Network-Based Processing Proceedings, pp. 359–363, 17–19 February 2016.https://doi.org/10.1109/PDP.2016.24
9. Antonov, A.S., Volkov, N.I.: An algoview web-visualization system for the AlgoWiki project. In: Sokolinsky, L., Zymbler, M. (eds.) PCT 2017. CCIS, vol. 753, pp. 3–13. Springer, Cham (2017). https://doi.org/10.1007/978-3-319-67035-5_1

10. Householder (reflections) method for the QR decomposition of a square matrix, real point-wise version. http://algowiki-project.org/en/Householder_(reflections)_method_for_the_QR_decomposition_of_a_square_matrix,_real_point-wise_version. Accessed 13 Apr 2018
11. Givens method. http://algowiki-project.org/en/Givens_method. Accessed 13 Apr 2018
12. Dense matrix multiplication (serial version for real matrices). http://algowiki-project.org/en/Dense_matrix_multiplication_(serial_version_for_real_matrices). Accessed 13 Apr 2018

On Sharing Workload in Desktop Grids

Ilya Chernov$^{(\boxtimes)}$ (iD)

Institute of Applied Mathematical Research, Karelian Research Center
of the Russian Academy of Sciences, 185910 Petrozavodsk, Russia
chernov@krc.karelia.ru

Abstract. We consider two optimization problems of trade-off between risk of not getting an answer (due to failures or errors) and precision or accuracy in Desktop Grid computing. Quite simple models are general enough to be applicable for optimizing real systems. We support the made assumptions by statistics collected in a Desktop Grid computing project.

Keywords: Optimal work share · Distributed computing
Desktop Grid

1 Introduction

Desktop Grid is a promising paradigm of high-performance computing because of the quick growth both of amount and power of desktop computers, drop of the cost of computing, development of networks (including the Internet). It is a computing system that consists of desktop computers (called computing nodes or just nodes in the sequel) connected by a LAN or the Internet; usually, the nodes are non-dedicated and their power is used by the Desktop Grid only when they are idle [1]. In Volunteer Computing [2] volunteers grant their resources to a project on a free will and can leave at any time; Enterprise Desktop Grids (e.g., [3,4]) unite resources of an institution and so its structure is more stable.

The classical Desktop Grid architecture is the client-server: the server schedules tasks to nodes and collects the results. In this model, there are no node-to-node connections, even if it is technically possible. Peer-to-peer desktop grids also exist and are studied (e.g., [5,6]); the scheduling problems here are, of course, even more complicated compared to the classical architecture. The challenges of using Desktop Grids for solving calculation-intensive scientific problems include heterogeneity and unpredictable availability of computing nodes, the threat of errors, including sabotage, possible high load of the server, low possibilities for interaction between the nodes.

Heterogeneity [7] means high diversity of computing nodes in terms of both hardware and software type, performance, reliability, availability, predictability, etc. However, the scheduling problem is hard (often NP-hard [8–10]) even under the identical nodes assumption. Nodes grant their idle processor time to desktop grid projects and usually do so without any obligations; therefore they can stop

V. Voevodin and S. Sobolev (Eds.): RuSCDays 2018, CCIS 965, pp. 599–608, 2019.
https://doi.org/10.1007/978-3-030-05807-4_51

working for the project due to switching to something else, switching off, etc. It is impossible to guarantee the availability of a node at some time and so the structure of the desktop grid is changing in a random way. There are many attempts to predict availability (e.g., [11]) or look for stable patterns (e.g., the review [12]) in availability history. Also, it is important to establish correct deadlines and to use replication so that the total performance increases.

In the classical architecture, the nodes ask the server for tasks, solve them, and report the results to the server. If multiple nodes solve rather quick tasks, the server may fail to process all requests in time and so suffer a DoS attack. This may drastically spoil the performance.

Sabotage and other malicious actions are a problem for volunteer computing, mostly. However, unintentional errors are possible too. Often it is reasonable to sacrifice some performance for double-checking the answers. Besides, sometimes calculations can be done with different precision, e.g., resolution in finite-difference schemes. Higher precision is, naturally, more reliable, but also more time-consuming.

The problems best suited for Desktop Grids are "Bags of Tasks"; a bag of tasks is a set of multiple independent relatively easy computational problems. Much attention has been paid to scheduling bags of tasks in Desktop Grids, optimizing various criteria [8, 12–15]. Often it is reasonable to join simple tasks to complex ones called parcels in order to reduce the server load [16], add test tasks to each parcel [17], mutual check, etc. On the other hand, in some cases a task is itself a parcel of simpler tasks; gathering a statistical sample can serve as an example. The reciprocal of making parcels of tasks is work sharing: tasks of a parcel can be distributed among a few nodes in order to reduce some risk. To our knowledge, there are just a few papers on the subject, e.g., [18].

In this paper, we consider two trade-off problems. The first is the trade-off between sharing work and chances to get the work done. Using fewer nodes to complete a joint task (a parcel) means more nodes for other tasks (parcels), but also higher risk of a switch-off, suspending work, failure, etc. It is obvious that in case of absolutely reliable nodes each is given as large parcel as desired, while in the opposite case of unreliable nodes only single tasks can be scheduled in spite of server load and other drawbacks; otherwise, the expected returns would be too few.

The second problem is the trade-off between reducing the risk by either sharing work (and thus reducing the total performance of the system) or using higher precision (so spending more time on each task). In [19] we propose an optimization criterion called cost per a task. It is the average cost of spent time, with possible replication taken into account, and penalty in case of accepting a wrong answer. The computing model is as follows: tasks are solved independently on identical nodes with some known error probability. Cost of solving a task is known (in time units). Each task is solved until ν identical answers are obtained. This number ν is called the quorum. For binary (yes/no) problems with the unit cost of a task, the spent time is between ν and $2\nu - 1$. If a wrong answer is accepted, some penalty is added to the spent time. An interesting question is not only to determine the optimal quorum given the penalty and error risk, but to estimate the penalty that forces the desired replication. We show that under some reasonable assumptions this dependence is logarithmic and, thus, there is no need to know the penalty values precisely. This is a practically useful result because it is not easy to get precise risk values in terms of the cost of a single task.

Here we develop this idea, expanding from replication to work division. We obtain a similar result (the logarithmic dependence of work share on the penalty threat), which makes the proposed model practically useful.

The nodes are assumed to be identical: so we ignore heterogeneity. One can consider a subset of a large Desktop Grid that consists of similar computers. Also, we ignore unintentional errors and malicious actions; this is reasonable for Enterprise Desktop Grids, where properties of nodes are known, connections are reliable, and saboteurs are unlikely. However, the nodes can be unreliable in the sense that they can suspend doing work for the Desktop Grid project or be switched off at any moment. In Sect. 3 we consider a threat in case of an error, which can be either a wrong answer or a missed interesting result.

2   Work Sharing and the Switch-Off Risk

For the sake of definiteness, let us assume that the tasks collect statistical samples evaluating some function in subsets of a multidimensional set. There are restrictions on the trust level, etc., converted to minimal size N of the sample. So, at least N points must be tested by computing nodes; this sets up N individual tasks that can be grouped into parcels. These tasks can be equally distributed over a few (ν) computing nodes in order to reduce the workload of each node to $n = N/\nu$. Let us call a set of n points to process a *task*.

According to the Desktop Grid paradigm, nodes are not absolutely reliable: they can be switched off, loose connection, return wrong results due to errors or malicious actions, etc. We consider only returning an answer in time: if the answer arrives before the deadline, it is for sure correct.

Probability $q_n = q(t_n)$ of returning an answer depends on the time t_n spent on solving the task; assume that $t = \tau n$ (so τ is the average time to solve an individual task, i.e., to process one point).

Denote $p_n = 1 - q_n$ (the fault probability). The function $p(t)$ is the cumulative distribution function of the time t_f up to the first fault: the node fails if its worktime t is less than the time of work before the fault. The only absolutely continuous distribution with no memory (so that the fault sequence is Markovian) is the exponential distribution; so let us assume that

$$q(t) = \mathrm{e}^{-\frac{t}{T}}, \quad T = E(t_f). \tag{1}$$

Here T is the average time of work up to the first fault.

Failed tasks need to be replicated until the necessary size of the sample is gathered, i.e., until exactly ν nodes return answers.

Statement 1. *The optimal workload for each task and the optimal work sharing are*

$$n = \frac{T}{\tau}, \quad \nu = \frac{N}{n}. \tag{2}$$

Proof. The number i of failed tasks can be any, from 0 to infinity; we need exactly ν successful returns, so the distribution is reduced to the negative binomial distribution with parameters ν as the number of successes and q as the success probability in a single try. The negative binomial distribution counts only failures, we need to add all ν successes, so the expectation is

$$E = \sum_0^\infty \binom{\nu+i-1}{\nu-1} q_n^\nu p_n^i \cdot (\nu+i) = \nu + \sum_0^\infty \binom{\nu+i-1}{\nu-1} q_n^\nu p_n^i \cdot i = \nu + \frac{\nu p_n}{q_n} = \frac{\nu}{q_n}. \tag{3}$$

Note that this can be guessed if we consider the scheme as ν geometrical random variables (tries until a success).

Therefore, we need to minimize the function

$$\frac{\nu}{q_n} = \frac{N}{nq_n}, \tag{4}$$

which is equivalent to maximizing $f(n) = nq_n = n\exp(-n\tau/T)$.

Assume that $n \in R$ (not necessarily integer); then the necessary condition for the maximum is $q_n = nq_n\tau/T$, equivalent to

$$n = \frac{T}{\tau}. \tag{5}$$

This can be explained as the number of points processed during the average time before the first failure. If it is less than 1 (highly unreliable nodes), the solution $n = 1$. In case of extremely reliable nodes n can exceed N: in this case, it is reasonable to process more points than the restriction N in order to improve statistical properties of the result.

Statement 2. *The average time spent on gathering the statistics is*

$$\hat{t} = e \cdot N\tau, \tag{6}$$

i.e., e time more than time needed to process N points with no failures, independently of the failure probability p (provided that we have enough computing nodes).

Proof. The average time needed to gather the sample is

$$\frac{\nu}{q_n} \cdot n\tau \tag{7}$$

(the average number of submitted tasks times time per a task). Substitute the optimal values of n and ν to obtain

$$\hat{t} = e \cdot N\tau. \tag{8}$$

In case of optimal $n < 1$ (unreliable nodes) we need to choose $n = 1$ so that

$$\hat{t} = e^{\frac{T}{\tau}} \cdot N\tau. \tag{9}$$

If, additionally, we are limited in a number of nodes (i.e., $\nu \leq M$), then $n \geq N/M$ and the estimation is even worse. However, nodes can process a few tasks sequentially, which may be better than suffering failures. This case, i.e., the optimal $n = 1$, reduces the parcels to individual tasks. In a highly volatile system, it may be reasonable to collect data point by point using scattered periods of idle processor time. However, here one can face challenges with high server load [16].

Let us estimate the error due to the fact that n is integer. If $T/\tau = n - \varepsilon$, $|\varepsilon| \leq 0.5$, then

$$\Delta\hat{t} = e^{1+\varepsilon\frac{\tau}{T}}N\tau - eN\tau. \tag{10}$$

The worst cases are $\approx 107\% N\tau$ (if $\varepsilon = 0.5$, $T/\tau = 1.5$) and $\approx -77\%\ N\tau$ (if $\varepsilon = -0.5$, $T/\tau = 1.5$); in case of small τ compared to T the asymptotic estimation is

$$\Delta\hat{t} \approx NT\varepsilon \left(\frac{\tau}{T}\right)^2 = NT\varepsilon \cdot o\left(\frac{\tau}{T}\right). \tag{11}$$

So, in case of reliable nodes rounding is quite safe. If nodes are unreliable, compare two boundary integer values.

To support the assumption about the exponential distribution of the time up to a failure, we considered statistical data gathered from the *RakeSearch*[1] volunteer computing project [20] at the date of 28 March 2018. In Fig. 1 we show a histogram of the time before a failure due to any reason was recognized (with a 25 h deadline). About 70% of values are below 25 h; we ignore the rest (up to 600 h) because their density is at most 0.5%

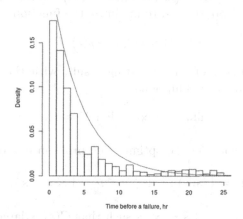

Fig. 1. The histogram of time before failure recognition in the project. The line shows the exponential distribution with $T = \lambda^{-1} = 4.38$ h.

---

[1] http://rake.boincfast.ru/rakesearch/.

3   Precision and Replication

Now let the Desktop Grid search for rare interesting results in a set divided into subsets called *blocks*. The algorithm examines an element and calculates some value; if it is high enough (more than a chosen threshold) the element is interesting and reported to the server.

Often the computational algorithm depends on a parameter that can be called precision: given higher values of this parameter, the algorithm takes more time but is more reliable. This unreliability can have various origins. First, the algorithm itself may produce a wrong answer for some values of its parameters (e.g., if the limit number of terms of a series is insufficient to approximate the sum, or if a descent-type algorithm converges to a false minimum). Second, the task can be to search for an interesting element inside a block; then the number of examined points is the precision parameter and there is always a risk to miss an interesting point unless every point in a block is examined.

Assume that changing some parameter s we are able to reduce the error probability $p(s)$ through more time $C(s) > 0$ spent on solving a task. The functions $C(s)$ and $p(s)$ are continuous. Let $s \geq 1$, $p(1) = p_0 \leq 0.5$, $C(1) = 1$. Again, sharing work between ν identical independent computing nodes is able to reduce the individual load of each node.

Note that this scheme can be used if p is the risk of an error decreased by replication (check by two or more nodes) [19].

The function $p(s)$ is a decreasing function with the lower bound $\bar{p} \geq 0$, which is the unavoidable risk. The cost $C(s)$ increases to infinity as $s \to \infty$.

The probability of finding an interesting answer by at least one computing node is $1 - p(s)^\nu$. If the correct answer is missed, it will be found much later, so that its cost becomes equal to (or even more than) the total cost of the project F. We call F a *threat*. So we need to minimize the function

$$Y(s, \nu) = \nu C(s) + p(s)^\nu F. \tag{12}$$

This is the average cost of an interesting result, with the threat taken into account. Under the made assumptions

$$\lim_{\nu \to \infty} Y = \infty, \quad \lim_{s \to \infty} Y = \infty. \tag{13}$$

So, it is sufficient to consider the optimization problem in a compact set

$$\Omega = \left\{ (s, \nu) : \quad 1 \leq \nu \leq \bar{\nu}, \quad 1 \leq s \leq \hat{s} \right\}, \tag{14}$$

where $\bar{\nu}$ is high enough and $\hat{s} < \infty$ is such that $C(\hat{s})$ is large enough. As Y is continuous in this compact set, it meets its maximal and minimal values. Denote the interior of Ω by Ω_o.

Statement 3. *If there is a solution* $(s^*, \nu^*) \in \Omega_o$, *then* s^* *is a root of the equation*

$$\frac{p'(s^*)}{p(s^*) \ln p(s^*)} = \frac{C'(s^*)}{C(s^*)}, \tag{15}$$

and

$$\nu^* = \frac{\ln C(s^*) - \ln F - \ln \ln p^{-1}(s^*)}{\ln p(s^*)}. \tag{16}$$

Note that s^* depend neither on F nor on the optimal ν^*: (15) shows that for the case $s^* > 1$, while for $s^* = 1$ it is obvious.

Proof. If a solution is inside the domain (i.e., in Ω_o), the derivatives with respect to the variables must vanish:

$$\frac{\partial Y}{\partial \nu} = C(s) + F p(s)^\nu \ln p(s) = 0, \tag{17}$$

$$\frac{\partial Y}{\partial s} = \nu C'(s) + \nu p(s)^{\nu-1} p'(s) F = 0. \tag{18}$$

Divide (18) on (17) to get the convenient condition (15) for the optimal s^* inside the domain.

Also this relation (15) can be rewritten as

$$\frac{d}{ds} \left(\ln \frac{|\ln p|}{C} \right) = 0 \quad \text{or} \quad \frac{d}{ds} \left(\frac{\ln p^{-1}}{C} \right) = 0. \tag{19}$$

So, if the function $\ln p^{-1}(s)/C(s)$ is strictly decreasing, this equation has no roots and, therefore, the only solution is $s^* = 1$: the cost is growing too quickly with respect to the precision. Note that this function can not be strictly increasing because for large s the function $p(s)$ needs to decrease slowly (tending to $\bar{p}$). This gives us a no-root condition:

Statement 4. *The optimal* $s^* = 1$ *if*

$$G(s) = \frac{\ln p^{-1}(s)}{C(s)} \tag{20}$$

is strictly decreasing.

such conditions make exploiting high precision is useless and therefore are practically convenient. Let us derive other sufficient no-root conditions for this equation.

Statement 5. *If*

$$p(s) \geq p_0^{C(s)}, \tag{21}$$

then the optimal $s^* = 1$.

Proof. Consider the differential inequality

$$\frac{p'(s)}{p(s)\ln p(s)} \leq \frac{C'(s)}{C(s)}. \tag{22}$$

It can be transformed to $d\ln|\ln p| \leq d\ln C$ and integrated from 1 to any C:

$$\ln\frac{\ln p}{\ln p_0} \leq \ln C \quad \text{or} \quad \frac{\ln p}{\ln p_0} \leq C. \tag{23}$$

Finally, we get (23).

Note that inequality (23) holds for large s if $\bar{p} > 0$, so solutions, if any, can exist only for reasonably small costs s. Also note that the inequality can be an identity for a special class of functions

$$\tilde{p}(s) = p_0^{C(s)}. \tag{24}$$

In this case the unavoidable risk $\bar{p} = 0$, necessarily. Here lies an important case of $C = s$, $p = 2^{-s}$: linear performance drop with exponential increase of accuracy. For such risk $p(s)$ the cost Y depends, actually, on the joint variable $x = s\nu$. Then either refuse to share work ($\nu^* = 1$) or choose he minimal precision ($s^* = 1$); then, using (16), get

$$x^* = s^*\nu^* = \frac{\ln F + \ln\ln 2}{\ln 2}. \tag{25}$$

In both cases the optimal solution logarithmically depends on the threat F.

Excluding the cases when this equation is an identity and when it has no roots, we can obtain all possible values of s^* which depend only on the properties of the computing system: functions $p(s)$ and $C(s)$.

The right-hand side of equation (16) should be positive; otherwise the optimal $\nu^* = 1$ (the boundary of the domain Ω). So, we see, that the dependence of ν^* on F is logarithmic, i.e., one need not know F precisely: quite coarse estimations are sufficient. In the same way, ν^* is insensible with respect to the cost $C(s)$.

Finally, let us establish the uniqueness in the following sense:

Statement 6. *Let s be given; the optimal ν^* inside Ω_0 is at most one.*

Proof. Consider Eq. (17). Its left-hand side is an increasing function of ν (remember that $p < 1$ and thus $\ln p < 0$). So, it is not able to vanish more than once.

4   Conclusion

In this paper, we consider rather simple yet general mathematical models of work sharing in a computing system of Desktop Grid type. We consider the trade-off between the risk of not getting an answer from some of a few nodes doing a long task and the risk of not getting an answer from some of many nodes doing short tasks. The average time is shown to be constant in case of exponential distribution of availability intervals. Formulae for any distribution are also available. Then we consider a threat of additional cost for missing a valuable result and the trade-off between careful (high-precision) expensive search by a few nodes and cheaper (low-precision) search by multiple nodes under the assumption of independence of the nodes. The main results are the logarithmic dependence of the optimal solution on the penalty values (so that an order of magnitude is sufficient to be known) and the threshold decrease rate of the risk as a function of cost. Unless the risk is exponentially decreasing, using the lowest precision with cross-checking by many nodes looks optimal.

Acknowledgements. The research was supported by the Russian Foundation of Basic Research, projects 18-07-00628, 16-07-00622.

References

1. Foster, I., Kesselman, C., Tuecke, S.: The anatomy of the grid: enabling scalable virtual organizations. Int. J. High Perf. Comput. Appl. **15**(3), 200–222 (2001)
2. Sarmenta, L.F., Hirano, S.: Bayanihan: building and studying web-based volunteer computing systems using Java. Future Gener. Comput. Syst. **15**(5), 675–686 (1999)
3. Kondo, D., Chien, A., Casanova, H.: Scheduling task parallel applications for rapid turnaround on enterprise desktop grids. J. Grid Comput. **5**(4), 379–405 (2007). https://doi.org/10.1007/s10723-007-9063-y
4. Ivashko, E.: Enterprise desktop grids. In: Proceedings of the Second International Conference BOINC-Based High Performance Computing: Fundamental Research and Development, BOINC:FAST 2015, Petrozavodsk, pp. 16–21 (2015)
5. Kwan, S., Muppala, J.: Bag-of-tasks applications scheduling on volunteer desktop grids with adaptive information dissemination. In: IEEE 35th Conference on Local Computer Networks (LCN), pp. 544–551. IEEE (2010)
6. Rius, J., Cores, F., Solsona, F.: Cooperative scheduling mechanism for large-scale peer-to-peer computing systems. J. Netw. Comput. Appl. **36**(6), 1620–1631 (2013). https://doi.org/10.1016/j.jnca.2013.01.002
7. Anderson, D., Reed, K.: Celebrating diversity in volunteer computing. In: 42nd Hawaii International Conference on System Sciences, pp. 1–8. IEEE (2009)
8. Xhafa, F., Abraham, A.: Computational models and heuristic methods for grid scheduling problems. Future Gener. Comput. Syst. **26**(4), 608–621 (2010). https://doi.org/10.1016/j.future.2009.11.005
9. Casanova, H., Dufossé, F., Robert, Y., Vivien, F.: Scheduling parallel iterative applications on volatile resources. In: IEEE International Parallel & Distributed Processing Symposium, pp. 1012–1023. IEEE (2011). https://doi.org/10.1109/IPDPS.2011.97
10. Chmaj, G., Walkowiak, K., Tarnawski, M., Kucharzak, M.: Heuristic algorithms for optimization of task allocation and result distribution in peer-to-peer computing systems. Int. J. Appl. Math. Comput. Sci. **22**(3), 733–748 (2012)
11. Kianpisheh, S., Kargahi, M., Charkari, N.M.: Resource availability prediction in distributed systems: an approach for modeling non-stationary transition probabilities. IEEE Trans. Parallel Distrib. Syst. **28**(8), 2357–2372 (2018). https://doi.org/10.1109/TPDS.2017.2659746

12. Durrani, N., Shamsi, J.: Volunteer computing: requirements, challenges, and solutions. J. Netw. Comput. Appl. **39**, 369–380 (2014). https://doi.org/10.1016/j.jnca. 2013.07.006
13. Choi, S., Kim, H., Byun, E., Hwan, C.: A taxonomy of desktop grid systems focusing on scheduling. Technical report KU-CSE-2006-1120-02, Department of Computer Science and Engineering, Korea University (2006)
14. Estrada, T., Taufer, : M.: Challenges in designing scheduling policies in volunteer computing. In: Cérin, C., Fedak, G. (eds.) Desktop Grid Computing, pp. 167–190. CRC Press (2012)
15. Khan, M., Mahmood, T., Hyder, S.: Scheduling in desktop grid systems: theoretical evaluation of policies and frameworks. Int. J. Adv. Comput. Sci. Appl. **8**(1), 119–127 (2017)
16. Mazalov, V.V., Nikitina, N.N., Ivashko, E.E.: Task scheduling in a desktop grid to minimize the server load. In: Malyshkin, V. (ed.) PaCT 2015. LNCS, vol. 9251, pp. 273–278. Springer, Cham (2015). https://doi.org/10.1007/978-3-319-21909-7_27
17. Yu, J., Wang, X., Luo, Y.: Deceptive detection and security reinforcement in grid computing. In: 5th International Conference on Intelligent Networking and Collaborative Systems, pp. 146–152 (2013)
18. Bazinet, A., Cummings, M.: Subdividing long-running, variable-length analyses into short, fixed-length BOINC workunits. J. Grid Comput. **14**, 429–441 (2016). https://doi.org/10.1007/s10723-015-9348-5
19. Chernov, I., Nikitina, N.: Virtual Screening in a Desktop Grid: Replication and the Optimal Quorum. In: Malyshkin, V. (ed.) PaCT 2015. LNCS, vol. 9251, pp. 258–267. Springer, Cham (2015). https://doi.org/10.1007/978-3-319-21909-7_25
20. Manzyuk, M., Nikitina, N., Vatutin, E.: Employment of distributed computing to search and explore orthogonal diagonal latin squares of rank 9. In: Proceedings of the XI All-Russian Research and Practice Conference "Digital technologies in education, science, society", Petrozavodsk, pp. 97–100 (2017)

On-the-Fly Calculation of Performance Metrics with Adaptive Time Resolution for HPC Compute Jobs

Konstantin Stefanov$^{(\boxtimes)}$ and Vadim Voevodin

M.V. Lomonosov Moscow State University, Moscow 119234, Russia
{cstef,vadim}@parallel.ru

Abstract. Performance monitoring is a method to debug performance issues in different types of applications. It uses various performance metrics obtained from the servers the application runs on, and also may use metrics which are produced by the application itself. The common approach to building performance monitoring systems is to store all the data to a database and then to retrieve the data which correspond to the specific job and perform an analysis using that portion of the data. This approach works well when the data stream is not very large. For large performance monitoring data stream this incurs much IO and imposes high requirements on storage systems which process the data.

In this paper we propose an adaptive on-the-fly approach to performance monitoring of High Performance Computing (HPC) compute jobs which significantly lowers data streams to be written to a storage. We used this approach to implement performance monitoring system for HPC cluster to monitor compute jobs. The output of our performance monitoring system is a time-series graph representing aggregated performance metrics for the job. The time resolution of the resulted graph is adaptive and depends on the duration of the analyzed job.

Keywords: Performance · Performance monitoring
Adaptive performance monitoring · Supercomputer · HPC

1 Introduction

Performance monitoring in High Performance Computing (HPC) is a method to debug performance issues in different types of applications. It uses various performance metrics obtained from the compute nodes the application runs on, and also may use metrics which are produced by the application itself.

The common approach to building performance monitoring systems is to store all the data to a database. After the job is finished the data for the specific job is retrieved from the database. Then per job derived metrics like value average are calculated from the retrieved data. Those data are used to perform an analysis of the finished job. The overall picture is shown in Fig. 1.

© Springer Nature Switzerland AG 2019
V. Voevodin and S. Sobolev (Eds.): RuSCDays 2018, CCIS 965, pp. 609–619, 2019.
https://doi.org/10.1007/978-3-030-05807-4_52

This approach works well when the data stream is not very large. For large performance monitoring data stream this incurs much IO and imposes high requirements on storage systems which process the data.

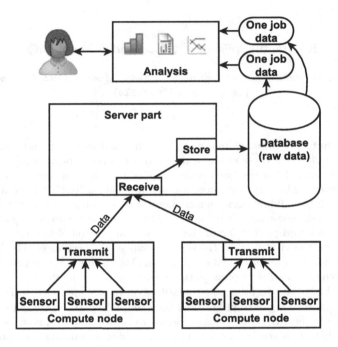

Fig. 1. Common approach to performance monitoring in HPC

Another approach may be to separate data for different applications and store data for different applications separately. This is easy to implement if the allocation of the resources to the applications is static and the rules to separate data flows from different applications can be set up before the monitoring system is started. But for HPC compute jobs the resources are assigned by the resource manager just before the job starts. Therefore we can't setup static separation of data for different jobs.

In this paper we propose an approach to performance monitoring in the case if the distribution of the resources to the applications is dynamic and may change on-the-fly. We used this approach to implement performance monitoring system for HPC cluster to monitor compute jobs. The output of our performance monitoring is a time-series graph representing aggregated performance metrics for the job which we call JobDigest [1]. The time resolution of the resulted graph is adaptive and depends on the duration of the analyzed job. The proposed approach is evaluated on Lomonosov-2, a largest Russian supercomputer.

The contribution of this paper consists of two parts. First, we propose a method for calculating performance monitoring metrics for HPC jobs on-the-fly by dynamically reconfiguring data processing paths when a job is started or finished. The performance data are analyzed without intermediate storing. This method allows reducing the amount of IO operations for processing performance monitoring data. Second, our approach automatically sets the time resolution of the result data from as small as 1 s to greater values. This enables to have the result of approximately equal size for every job without significantly reducing the available information about jobs.

The paper is organized as follows. Section 2 presents common information about performance monitoring in HPC and describes related work in field of performance monitoring. Section 3 gives the details about our proposed approach to calculating performance metrics on-the-fly. Section 4 gives an overview of making time resolution of data for compute job adaptive and depending of job duration. Section 5 describes the test performance monitoring system we implemented to evaluate our approach, and Sect. 6 contains the conclusion.

2   Background and Related Work

In HPC clusters when the compute job is ready to start, it is allocated a set of compute nodes. In most cases a node is assigned exclusively to one job. When the job is finished, the nodes assigned to it become free and will be allocated to other jobs. The main object of the performance monitoring in HPC is a compute job.

Many approaches to performance monitoring have been proposed. Some of them are intended for inspecting only selected jobs. Most often whether the job is to be evaluated is decided before the jobs is started. Such performance monitoring systems [2–7] start agents on nodes which are allocated to the inspected job when the job is starting. In this approach only the data which are needed to evaluate the performance of the selected jobs are collected. But with this approach it is not possible to determine the reasons why a job's performance was worse than it was expected if the job was not set up for performance monitoring before the job was started. Another drawback of this approach is that the data which require superuser privileges to collect are hard to get, as the job and the monitoring agent are run by an ordinary user. And still the data in the database is stored without a label which point to job ID which produced the data. When a job is to be analyzed, the data are retrieved using time period and node list as a key and performance metrics are calculated afterwards.

There also exist approaches [8,9] which run agents on every node of the compute system and store all the data into a database. When a job performance metrics are to be inspected, the job's data are retrieved from the database using time period and node list when and where the job ran. This approach incurs great amount of unnecessary data stored to the database. Indeed, all the data for the jobs which will not be examined later would never be retrieved and are just a waste of all type of resources: CPU to process, net to transmit, IO operations to save, and disk space to store. Moreover, as a database that stores all the data for the compute system may become quite large, the time to retrieve data for the specific job may become quite long, so saving all data creates unnecessary delays when analyzing data.

These performance monitoring systems usually perform data averaging on periods from 1 to 15 min. This period is chosen based on the amount of data which the system can process.

3   Calculating Aggregate Performance Data for Compute Jobs On-the-Fly

Our approach is to separate data flows for different jobs by marking them with a label with job ID. When these separated flows are processed, the data for different jobs are collected separately.

Every compute node runs a monitoring agent which retrieves performance metrics for the OS and the hardware of the compute node. The agent sends the data to a central (server) part of the monitoring system which calculates the aggregated metrics. In our approach the monitoring agent is able to mark data with a label with job ID, and this label can be set when the job is started on the specific node and cleared when the job is finished.

Resource manager is the program component responsible for allocating compute nodes to the compute jobs. Resource manager prologs and epilogs are the programs (scripts) which are executed when a job is about to start or finish. We use them to signal the monitoring agents on the nodes allocated to the job that they should start or cease to label the data.

When the data labeled with a job ID is received in the server part of the monitoring system, they are demultiplexed by the job ID and sent to a collector responsible for collecting information for that specific job. When the job is finished, that object stores the collected data to the disk. Our approach is presented in Fig. 2.

The monitoring agents on compute nodes retrieve data every 1 s and send them to the server part of the monitoring system immediately after all the metrics are collected. In the server part the data are aggregated over a short interval (we use 1 s as an initial value). For every such interval and for every performance metric we collect five aggregated metrics:

- Minimum value of the metrics over the interval
- Maximum value
- Average value
- Minimum of all nodes among average value for one compute node
- Maximum among averages for one compute node.

The last two metrics are calculated as follows: first, we calculate an average of values for the given performance metric for the given compute node. Then we calculate minimum and maximum among all compute nodes running the specific job. These metrics represent a measure of performance imbalance between different nodes running one job.

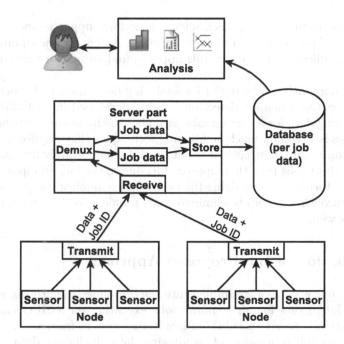

Fig. 2. The proposed approach to calculating per job performance metrics

With this approach we do not have to store large stream of raw data, and then read it from disk thus creating large IO. Instead we collect all the data and calculate derived aggregated metrics in memory. The final result is written to disk only after the job is finished, and the size of the data written is much smaller than the size of the raw data for the job.

4   Making Time Resolution of the Result Adaptive to the Duration of the Compute Job

Another feature of our approach is a variable time period on which we do averaging of raw data. When a job is started, this time period is set to 1 s. Then we calculate predefined number (we use 1000) of samples for derived metrics. When the last sample is calculated, we merge every two adjacent samples into one and double the averaging period. So for jobs shorter than 1000 s we have aggregated metrics for averaging period of 1 s. For longer jobs we have between 500 and 1000 samples for aggregated metrics and each sample is done over 2^n-second period (n is such an integer that job length L satisfies $500 \cdot 2^n < L \leq 1000 \cdot 2^n$).

This allows us to have detailed graphs for relatively short jobs. As the data for every job is visualized in a browser, it is important not to have very large number of data points for every chart, otherwise it will take too long to draw the chart. Our approach gives a reasonable tradeoff between the need in detailed time resolution of the data for every job and the performance of the visualization part. And more, such an approach allows us to have upper bound on memory consumption of the server part of the monitoring system for one running job. As the data are collected in memory, this upper bound on memory consumption is important.

Such an approach can be useful for analyzing performance of short jobs. For example, a tool for anomaly detection is being developed in RCC MSU, which uses monitoring data for detecting jobs with abnormally inefficient behavior [10]. This analysis is based on machine learning techniques that require a significant amount of input data. Currently it is not possible to accurately detect abnormal behavior in short jobs (less than approx. 1 h) due to the lack of input data from existing monitoring systems. Using the performance monitoring system with the adaptive approach will help to eliminate such problem, still allowing to analyze long jobs as well.

5   Evaluation of the Proposed Approach

To evaluate our approach we implemented performance monitoring system for producing JobDigests for all compute jobs executing on a HPC cluster. Our implementation is based on DiMMon monitoring system framework [11].

In DiMMon all processing of monitoring data (including data acquisition from OS or other data sources) is done in monitoring nodes which operate in a monitoring agent. A monitoring agent can run on a compute node or on a dedicated server.

In our implementation every compute node runs a monitoring agent. Server monitoring agent runs on a dedicated server. Compute node monitoring agent contains nodes which acquire data from OS (sensor nodes), a node that prepends a label with job ID to data, a node which transmit data to server monitoring agent for processing, and a resource manager interaction control node which receives notifications about job start and finish. Compute node agent structure is presented in Fig. 3. Server monitoring agent structure is shown in Fig. 2.

We implemented prolog and epilog scripts for SLURM [12] resource manager. The prolog script receives a list of nodes where the job is to be started. The prolog notifies agents on the compute nodes that they should start to send the performance data to server monitoring agent and that data should be marked with job ID ("resource manager interaction" node receives these notifications). It also notifies server monitoring agent that the job with given ID is about to be started. The server agent programs its demultiplexer to create separate data path for that job data. It also creates collector entity which will collect the data for that job.

The epilog script notifies agent running on compute nodes of the finishing job to cease sending data to the server monitoring agent. It also notifies server monitoring agent that the job is finished and the aggregated data for it should be written to disk.

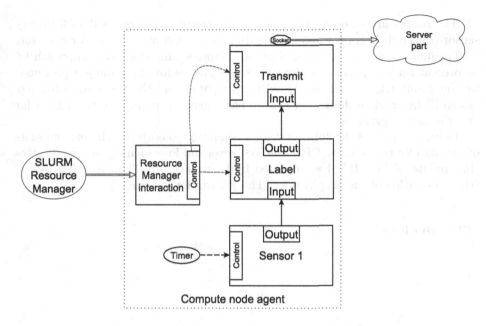

Fig. 3. Compute node monitoring agent structure

For evaluation we collect CPU usage level data and memory (RAM) consumption data, 21 sensors in total. CPU usage level data provide information about what fraction of CPU time is used for executing user mode code, operating system, processing interrupts etc. There are 9 such sensors provided by modern Linux kernels. Memory consumption data give the distribution of different type of memory usage: free memory volume, the amount of memory used for file cache etc. The data from the kernel is retrieved once every 1 s.

We run our performance monitoring system on Lomonosov supercomputer and Lomonosov-2 supercomputers installed in M.V.Lomonosov Moscow State University. Lomonosov is a cluster consisting of more than 6000 compute nodes. Lomonosov-2 is a cluster consisting of more than 1600 compute nodes with peak performance of 2.9 Pflops. These two machines execute about 100–200 parallel jobs simultaneously each.

We run monitoring agents on every compute node and one server monitoring agent. Compute node monitoring agents consume about 1 MB of RAM and less then 0.1% of CPU time each. Server monitoring agent run on a server with 2 4-core Intel Xeon E5450 processors with 4 300 GB SATA disks connected as RAID 10. Our monitoring agent is single-threaded, so it occupies only 1 core. It consumes about 60 MB of RAM and less than 20% of CPU (CPU consumption depends on number of jobs running on the cluster at the moment). All data for jobs executed in 1 week was about 1.6 GB in size. So our new system stores about 11 MB of data per sensor per day.

The results are saved for every job as 5 collected aggregate values for every sensor collected. It is presented to the user as a graph with 5 lines for a sensor.

For illustrating the advantages of our adaptive time resolution approach let us present the comparison of the time series graphs for the same job produced by the monitoring system described in this paper and the previous JobDigest system. The previous JobDigest system has a fixed averaging period of 120 s for every sensor of every job.

In Fig. 4 a part of the JobDigest for a compute job is given. The part consists of the data for two sensors: CPU user load level and Free memory sensor (it gives the amount of Free RAM as reported by Linux kernel). The averaging period (time resolution of the graph) is 1 s. The job ran for 15 min.

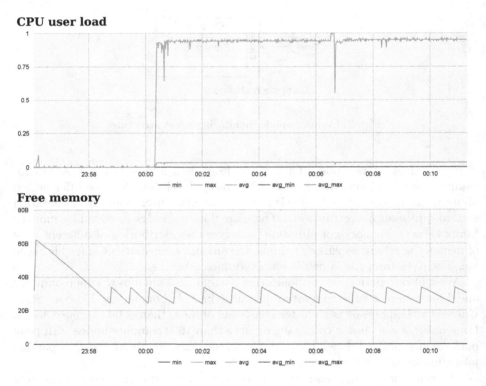

Fig. 4. An example of JobDigest produced by our system

The same graphs produced by the previous JobDigest system are given in Fig. 5 for comparison. Differences in Y-axis scale is caused by the fact that the previous JobDigest system presents CPU usage level in % (the range is 0–100) and our currents system gives the same value in the range 0–1. The memory volume for the previous system is given in kilobytes, and in the current system it is given in bytes, so B (Billion) suffix instead of M (Million) for the previous system is used.

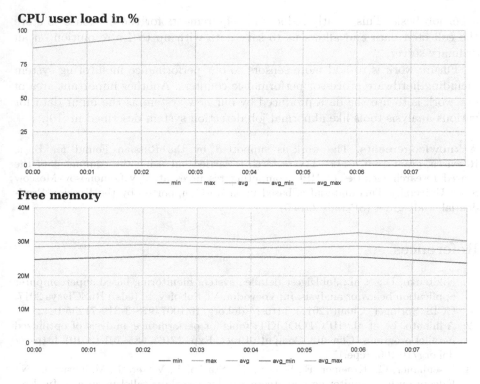

Fig. 5. A JobDigest produced by previous system

We can clearly see the difference in time resolution. On graphs produced by our systems one can identify stages and some other properties of the jobs being evaluated, while the graphs produced by the previous system contain several dots each and cannot reveal any details of the job run time behaviour.

To asses the volume of the data used by the performance monitoring system based on our proposed approach let's compare to other existing system, which is now used for creating JobDigests [1]. That system creates approx. 2.1 GB of data per day and stores min, max and average for every sensor (3 values) per node every 120 s. It stores data of 27 sensors, so it stores approx. 80 MB per sensor per day. Our new system compared to the existing one stores more than 7 times less data while maintaining time resolution up to 120 times better (depends on job length).

6   Conclusion and Future Work

In this paper we present an approach for calculating aggregate metrics for HPC jobs on-the-fly. We label the data for the job by a label based on job ID and thus we may collect the data for separate jobs and store them in a database on a per-job basis. This greatly reduces to requirements for the database to store the collected data and allows us to save data with up to 1-s resolution on an ordinary server.

Future work is to add more sensors to our performance monitoring system including hardware processor performance counters. Another important area of our work is to use the data produced by our new system as the input data for various analysis tools like abnormal job detection system described in [10].

Acknowledgements. The work is supported by the Russian Found for Basic Research, grant 16-07-01121 The research is carried out using the equipment of the shared research facilities of HPC computing resources at M.V.Lomonosov Moscow State University This material is based upon work supported by the Russian Presidential study grant (SP-1981.2016.5).

References

1. Nikitenko, D., et al.: JobDigest detailed system monitoring-based supercomputer application behavior analysis. In: Voevodin, V., Sobolev, S. (eds.) RuSCDays 2017. CCIS. Springer, Cham (2017). https://doi.org/10.1007/978-3-319-71255-0_42

2. Adhianto, L., et al.: HPCTOOLKIT: tools for performance analysis of optimized parallel programs. Concurr. Comput.: Pract. Exp. **22**(6), 685–701 (2010). https://doi.org/10.1002/cpe.1553

3. Eisenhauer, G., Kraemer, E., Schwan, K., Stasko, J., Vetter, J., Mallavarupu, N.: Falcon: on-line monitoring and steering of large-scale parallel programs. In: Proceedings Frontiers 1995. The Fifth Symposium on the Frontiers of Massively Parallel Computation, pp. 422-429. IEEE Computer Society Press, McLean, VA (1995). https://doi.org/10.1109/FMPC.1995.380483

4. Gunter, D., Tierney, B., Jackson, K., Lee, J., Stoufer, M.: Dynamic monitoring of high-performance distributed applications. In: Proceedings 11th IEEE International Symposium on High Performance Distributed Computing, pp. 163–170. IEEE Computer Society (2002). https://doi.org/10.1109/HPDC.2002.1029915

5. Jagode, H., Dongarra, J., Alam, S.R., Vetter, J.S., Spear, W., Malony, A.D.: A holistic approach for performance measurement and analysis for petascale applications. In: Allen, G., Nabrzyski, J., Seidel, E., Albada, G.D., Dongarra, J., Sloot, P.M.A. (eds.) Computational Science ICCS 2009. LNCS, vol. 5545, pp. 686–695. Springer, Heidelberg (2009). https://doi.org/10.1007/978-3-642-01973-9

6. Mellor-Crummey, J., Fowler, R.J., Marin, G., Tallent, N.: HPCView: a tool for top-down analysis of node performance. J. Supercomput. **23**(1), 81–104 (2002)

7. Ries, B., et al.: The paragon performance monitoring environment. In: Supercomputing 1993, Proceedings, pp. 850-859. IEEE (1993). https://doi.org/10.1109/SUPERC.1993.1263542

8. Kluge, M., Hackenberg, D., Nagel, W.E.: Collecting distributed performance data with dataheap: generating and exploiting a holistic system view. Procedia Comput. Sci. **9**, 1969–1978 (2012). https://doi.org/10.1016/j.procs.2012.04.215

9. Mooney, R., Schmidt, K., Studham, R.: NWPerf: a system wide performance monitoring tool for large Linux clusters. In: 2004 IEEE International Conference on Cluster Computing (IEEE Cat. No. 04EX935), pp. 379–389. IEEE (2004). https://doi.org/10.1109/CLUSTR.2004.1392637

10. Shaykhislamov, D., Voevodin, V.: An approach for detecting abnormal parallel applications based on time series analysis methods. In: Wyrzykowski, R., Dongarra, J., Deelman, E., Karczewski, K. (eds.) PPAM 2017. LNCS, vol. 10777, pp. 359–369. Springer, Cham (2018). https://doi.org/10.1007/978-3-319-78024-5_32
11. Stefanov, K., Voevodin, V., Zhumatiy, S., Voevodin, V.: Dynamically reconfigurable distributed modular monitoring system for supercomputers (DiMMon). In: Sloot, P., Boukhanovsky, A., Athanassoulis, G., Klimentov, A. (eds.) 4th International Young Scientist Conference on Computational Science. Procedia Comput. Sci. **66**, 625–634. Elsevier B.V. (2015). https://doi.org/10.1016/j.procs.2015.11.071
12. Slurm workload manager. https://slurm.schedmd.com/

Residue Logarithmic Coprocessor for Mass Arithmetic Computations

Ilya Osinin[✉]

Limited Liability Company Scientific Production Association Real-Time
Software Complexes, 16/2 Krasnoproletarskaya, Moscow 127473, Russia
iposinin@gmail.com

Abstract. The work is aimed at solving the urgent problems of modern high-performance computing. The purpose of the study is to increase the speed, accuracy and reliability of mass arithmetic calculations. To achieve the goal, author's methods of performing operations and transforming data in the prospective residue logarithmic number system are used. This numbering system makes it possible to unite the advantages of non-conventional number systems: a residue number system and a logarithmic number system. The subject of study is a parallel-pipelined coprocessor implementing the proposed calculation methods. The study was carried out using the theory of computer design and systems, methods and means of experimental analysis of computers and systems. As a result of the research and development new scientific and technical solutions are proposed that implement the proposed methods of data computation and coding. The proposed coprocessor has high speed, accuracy and reliability of processing of real operands in comparison with known analogs based on the floating-point positioning system.

Keywords: Logarithmic number system · Residue number system
Residue logarithmic number system · Performance · Accuracy
Reliability

1 Introduction

It is well known that the decimal number system significantly simplified the calculations. This served as a revolutionary impetus for technological progress. After the invention of infinite decimal fractions, it acquired the status of a universal number system. The binary number system is the basis of modern technological progress. The areas of its application are computer facilities and information technology. To date, they are implemented on the basis of nanoelectronics.

The development of technological progress stimulates both the growth of the computational need and the improvement of computations in solving new problems. For example, the development of astronomy and navigation in the 16th century

The work was supported by the Russian Science Foundation, grant N 17-71-10043.

V. Voevodin and S. Sobolev (Eds.): RuSCDays 2018, CCIS 965, pp. 620–630, 2019.
https://doi.org/10.1007/978-3-030-05807-4_53

stimulated the growth of computational needs. As a result, the logarithm appeared, as a means of reducing the complexity of multiplicative operations.

The rapid growth of computing needs in the XX century has aggravated new problems in the field of computing technologies. The key ones are:

- acceleration of calculations due to the parallelism of programs and machine codes, streaming calculations;
- fault tolerance due to self-correction, i.e. detection and correction of errors in the calculation process, including for systems of long-term autonomous existence;
- the accuracy of real calculations in a limited machine bit capacity.

However, their effective solution is almost impossible at the level of computer codes of traditional arithmetic. The reason is the inter-digit relationship of positional numbers. Redundancy is used to improve the reliability of the calculations. Its draw-back is a multiple increase in hardware costs. To work with real numbers, the de facto standard is the floating point [1]

$$x = (-1)^{sign} \cdot 2^{E-bias} \cdot (1 + \frac{M}{2^f}) \tag{1}$$

where *sign* is the sign of the number, 2^{E-bias} is the exponent with bias as 127 for single precision and 1023 for double precision, M is the mantissa in the normalized form, and f is the width of the mantissa. In this case, rounding in the calculation process reduces the accuracy of calculations. As a result, this can lead to an erroneous result of the calculations.

The mentioned problems concern all known universal processors available in the market of high-performance computing. Solutions of leading companies Intel, AMD, NVIDIA and several others are based on positional floating-point arithmetic. Thus, to effectively solve the above problems, the development of a new computer account system is urgent.

One of the applicants is the residue logarithmic number system [2]. It combines two-level coding of numbers. One of the levels is the residue number system (RNS) in the basis of modular arithmetic. Another level is the logarithmic number system (LNS). The subject of study is a parallel-pipelined coprocessor implementing the proposed calculation methods. As a result of the research and development new scientific and technical solutions are proposed that implement the proposed methods of data computation and coding. Let us consider them in more detail.

1.1 Residue Number System

The foundations of modular arithmetic were proposed more than half a century ago. Scientific papers by Akushsky, Yuditsky, Chervyakov, Knyazkov, Garner, Omondi are devoted to this direction [3–6].

The objectives of transition to RNS are:

- to increase the speed of residue operations (addition, multiplication) due to the parallel processing of each digit of a number from the basis of modules $\{p_1, p_2, \ldots, p_n\}$;

– to increase the reliability of calculations due to self-correction of machine codes when the basis is extended by the control modules $\{p_1,p_2,\ldots,p_{n+k}\}$;

where n and k are the numbers of bases of the main and reference ranges, respectively. To date, there are very few completed technical solutions that operate with the use of RNS. One of them is a modular processor, which operates on the basis of artificial neural networks. It was developed in Stavropol, Russia in 2005 [3].

The limiting factor is the slow (sequential) execution of non-residue operations. They are:

– division, and hence, scaling;
– formation of an overflow indication;
– translation from one number system to another.

The previous work was aimed at the research on high-speed devices for the efficient execution of non-residue operations. They are based on integer parallel-pipelined processing of bit-slices of operands [7].

In general, the presence of non-modular operations makes it difficult to use a floating point. There are some works in which the RNS is adapted to frequent scaling and rounding. One of possible approaches is the use of interval positional character-istics [8]. Its disadvantage is the iterative scaling. This has a negative effect on the speed of arithmetic operations.

However, it is possible to fix the point, providing a wide range of representation of numbers. It will be discussed further.

1.2   Logarithmic Number System

In the logarithmic number system (LNS), the real number of the field R is represented by its logarithm along a fixed base. Moreover, the arithmetic operations are isomorphic to the operations of the field R. The interest in the use of LNS in computers first appeared in the late XX century. A fundamental contribution to the study of LSS was made by Coleman, Arnold, Chester, Lewis [9–12]. The goal of transition to the LNS is:

– to increase the speed of multiplicative operations (multiplication, division, raising and root extraction) of the R field;
– to increase the accuracy of calculations by fixing a point separating the integer and the fractional part of the number.

The latter allows us to use LNS as an alternative to floating-point calculations. This ensures a similar range of representation of numbers in the allocated bitmap of the machine number

$$\log_2 x = \log_2 2^E + \log_2\left(1 + \frac{M}{2^f}\right) = E + \log_2\left(1 + \frac{M}{2^f}\right), \tag{2}$$

where E is the exponent, M is the mantissa in the normalized form, and f is the width of the mantissa.

Despite the rather extensive volume of studies performed, there are only a few completed technical implementations to date. One of them is the European Logarithmic Microprocessor [9]. It has a 32-bit RISC architecture.

The reason for this is the slow performance of additive operations. They are expressed through multiplicative operations

$$\log_2(a+b) = \log_2 a + \log_2(1 + 2^{\log_2 b - \log_2 a}),\tag{3}$$

where a and b are operands and a < b. The reason for limiting the single precision is the complexity of the conversion from the traditional number system and back. This fact leads to an exponential increase in the hardware costs of code converters.

The methods of converting numbers, based on the author's research [13], helped to eliminate this drawback by using multilevel interpolation. On the other hand, the use of specialized technical solutions contributed to an increase in the speed of additive operations.

Summing it up, fixing a point in the LNS allows us to work with a real operand the same as with an integer. This allows us to combine its advantages with RNS, thus forming a residue logarithmic number system (RLNS). Let us briefly describe its features.

2   Residue Logarithmic Format and Related Work

In the introduction it was established that the application of RLNS is actual for solving the problems of modern arithmetic devices in terms of increasing the speed, the accuracy and the reliability of real calculations.

In simplified form, the translation of the real positional number into a residue logarithmic format takes place in three stages:

- the base 2 logarithm is calculated from the original real number;
- the point separating the integer and fractional part of the logarithm is discarded;
- the received integer number is converted to the RNS.

An example of converting a double-precision number is shown in Fig. 1.

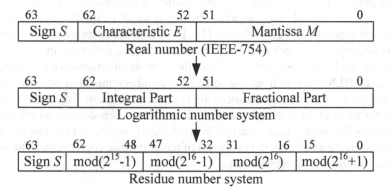

Fig. 1. Converting a double-precision number to RLNS

RLNS integrates implementation advantages both of logarithms:

- multiplicative operations are replaced by additive ones;
- the absence of rounding in the performance of multiplicative operations;

and of the residues of numbers:

- parallel calculation of each digit of a number;
- auto correction of machine codes.

It is obvious that this also accumulates disadvantages:

- the need to convert numbers first in the LNS, then into the RNS;
- the speed of additive operations depends on the speed of the code converters.

Consequently, the classes of computational tasks, where the successes of the RLNS can be achieved are as follows:

- the number of additive operations is commensurable with the number of multiplicative operations, for example, calculation of polynomials, fast Fourier transforms, solution of systems of linear algebraic equations, etc.;
- critical dependence on rounding in ill-conditioned problems, the problems with different-scale coefficients, and also the ones sensitive to the class of equivalent transformations;
- highly reliable real-time systems, for example, missile guidance, management of a nuclear power station, space vehicles operation, etc.

It is worth noting the difference between RLNS and residue logarithmetic [3]. In the latter, the RNS is combined with discrete logarithms. This leads to the preservation of the scaling problem. In contrast, the RLNS allows us to process numbers with a fixed position of a point.

The founder of the RLNS is Arnold. He described the basic capabilities of this number system and proposed a software implementation [1]. However, RLNS did not receive its further development. The accuracy of the numbers was limited to 32 bits. The calculations were adapted to the implementation on general purpose processors. All these negated the advantages of applying known solutions. It is obvious that theoretically basics of computing in the RLNS are currently investigated insufficiently. There are practically no hardware implementations that operate on its base.

So, the previous work was aimed at creating a set of methods describing residue logarithmic computations a general form [14–16]. These methods are aimed at its high-speed parallel-pipelined technical implementation, also mentioned in [14–16]. The proposed methods and devices based on them relate to high-speed execution of operations in RLNS, as well as conversion of numbers into a residue logarithmic format and back. The maximum speed of operations in them (one clock cycle) is achieved by mass arithmetic processing. This condition is necessary to fill the computational pipeline with a data stream. The conducted research and development made it possible to create a basis for the operation of the arithmetic coprocessor. The features of its organization are discussed below.

3 Organization of a Residue Logarithmic Coprocessor

In this section, a device operating on the basis of RLNS is considered. It integrates the entire previous work of the author, allowing us to maximize the benefits of this number system at the hardware level. The use of the device as a coprocessor makes it possible:

- to simplify the internal control device as much as possible;
- to transfer for calculation only a part of the tasks that can be solved most effectively using the residue logarithmic approach.

Figure 2 shows the structural scheme of the relationship with a universal processor based on shared memory. The advantage of this approach is the lack of a mechanism for memory coherency and simplification of the relationship between the processor and the coprocessor.

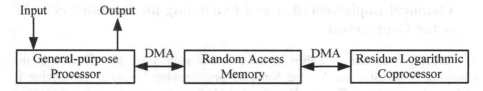

Fig. 2. The structural scheme of the relationship with a universal processor

Figure 3 shows the structural scheme of the coprocessor in its general form. Each computational core of the coprocessor performs independent processing on a particular residue. Arithmetic operations can be performed in both vector and scalar forms.

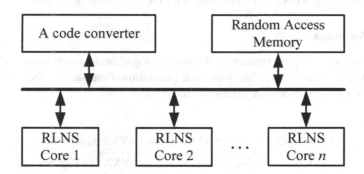

Fig. 3. The structural scheme of the coprocessor in its general form

In general, the number of cores is equal to the number of bases of the main and control range of the RLNS. Their maximum number is limited only by hardware costs. The adder and the multiplier are provided for processing multiplicative operations in the computational core. Their width corresponds to the width of the base on which the processing is performed.

The architecture of the coprocessor is discussed in more detail in [17]. The cores are connected to a parallel-pipeline code converter. The method and the code converter device are given in a general form in [15, 16].

The difference from the arithmetic devices of known universal processors lies in the following as:

- scalar values without pipelining, which increases the speed of calculations;
- noiseless encoding of operands, which allows automatic error correction;
- when performing multiplicative operations there is no rounding operation, and that increases the accuracy of calculations.

Further the coprocessor is compared with the known analogs.

4   Technical Implementation and Evaluating the Parameters of the Coprocessor

The structural schemes described in the previous section are implemented at the functional level using the Verilog hardware programming language. Debugging is performed using Altera Cyclone V programmable logic integrated circuit (FPGA).

At the moment, the residue logarithmic coprocessor is implemented as a stand-alone intellectual property (IP) block. Such blocks are complete components for creating systems-on-a-chip, for example, microprocessors. Hardware costs of the IP-block are about 20 million transistors, that is comparable to the block expansion of Intel Xeon AVX2 processors Broadwell generation [17].

The following are the estimates of the speed, the accuracy, and the reliability of computations using RLNS in comparison with the traditional approach.

4.1   Performance

Table 1 compares the performance of the residue logarithmic coprocessor (RLC) with the Intel AVX2 extension of the Broadwell generation. Precision of the operand corresponds to a double-precision IEEE-754 standard.

Table 1. Comparison of RLC and Intel AVX2 performance.

Vector operation	RLC, clock cycles		AVX2, clock cycles	
	Latency	Throughput	Latency	Throughput
Addition	6	1	3	1
Subtraction	6	1	3	1
Multiplication	1	1	3	0,5
Division	1	1	16–23	16
Exponentiation	1	1	288	270
Square root	1	1	19–35	16–27

The numbers are packed into vectors with a dimension of 8 elements. It is assumed that the conversion to RLNS and back is performed once at the beginning and in the end of the task. This does not introduce significant time delays into the computational process due to pipelining.

Latency and throughput of the arithmetic operation pipeline in Intel AVX2 (Broadwell generation) is taken from Intel 64 and IA-32 Architectures Optimization Reference Manual [18].

The greatest acceleration is achieved in the proposed coprocessor when using multiplicative operations. For example, for raising to the power operations, the difference is three orders of magnitude in favor of the coprocessor, which is explained by the lack of hardware implementation of these operations in Intel coprocessors.

At the same time, slow additive operations are performed with twice the latency, but the same throughput (1 clock cycle per operation). One-step execution of vector operations multiplication, division and exponentiation makes it possible to obtain a significant advantage in speed of execution of a certain class of tasks, where such operations prevail. These include the calculation of polynomials of n-th power, for example, Taylor series expansion, matrix multiplication and a number of others.

4.2   Accuracy of Calculations

One of the options for comparing the accuracy of calculations is the problems of numerical analysis. For example, Laguerre's method is a root-finding algorithm tailored to polynomials. In other words, Laguerre's method can be used to numerically solve the equation $p(x) = 0$ for a given polynomial $p(x)$. The average relative error of calculations is 25% less compared to the traditional type of computation (see Fig. 4).

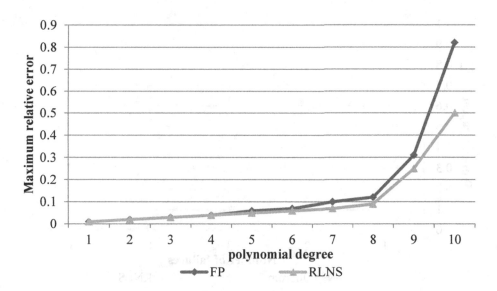

Fig. 4. Accuracy of calculations

This equation has the known analytical solution and, hence, it is possible to calculate a relative error of the solution numerically obtained in the floating-point arithmetic. The experiment is carried out for $n \in [1, 10]$, where n is the degree of the polynomial.

The RLNS error does not exceed half the last significant digit when n = 10. The result will be exact. While for a floating point, the effect of the error will spread to the last digit of the result, which will cause the error to accumulate. This allows us to conclude that the accuracy of the calculations in the RLNS is higher.

4.3 Reliability Assessment

To assess the fault tolerance of RLC, it is advisable to use the availability indicator

$$\delta = \frac{H}{N}, \tag{4}$$

where H is the number of operational states of the system in case of failure and N is the total number of possible states [5].

This indicator allows us to evaluate the effectiveness of the application of corrective codes. Their use makes it possible to increase the coprocessor's fault tolerance in comparison with the quorum - the classic approach of masking failures "2 of 3" [3].

The use of three control bases allows us to correct all double errors with an increase in hardware costs by 75% (4 modules require 3 control modules). The traditional approach has a smaller margin of efficiency (see Fig. 5) with an increase in hardware costs by 300% (quorum need triple hardware costs).

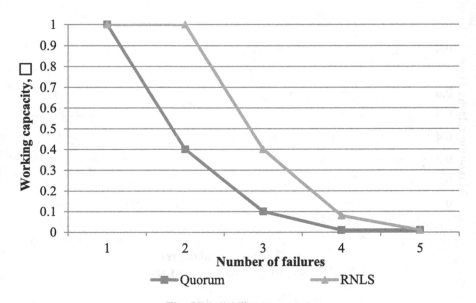

Fig. 5. Reliability assessment

5 Conclusion

The analysis of the problems of modern high-performance computing presented in this article has revealed difficulties in further improving the characteristics without introducing changes on the fundamental level of the principles of computation.

One possible solution is the use of forward-looking residue logarithmic number system.

Its advantages allowed us to improve the technological capabilities of specialized arithmetic devices.

In particular:

- multiplicative operations on each base have the same hardware structure, that is, they consist of the same set of simple blocks;
- the computational element is universal within a set of the same types of bases, which is the basis for building highly reliable schemes;
- the code lag is realizable both at the level of a separate computational element and a modular code in general, which opens prospects for the development of highly reliable computing structures.

The coprocessor considered in this paper functions on the basis of RLNS. To eliminate the disadvantages of non-traditional numbering systems associated with the slow implementation of non-modular (in the part of the RNS) and additive (in the part of the LNS) operations special methods and parallel-pipelined devices based on them were developed.

All this allowed us to increase the speed, the accuracy and the reliability of mass arithmetic calculations at a comparable level of hardware costs.

Future work will be focused on testing the applicability of the coprocessor using for different classes of applied problems. It is possible to provide some evaluated results on synthetic benchmarks and compare it to general purpose processor.

References

1. 754-2008 – IEEE Standard for floating-Point Arithmetic. Revision of ANSI/IEEE Std 754-1985. http://ieeexplore.ieee.org. Accessed 12 Apr 2018
2. Arnold, M.G.: The residue logarithmic number system: theory and implementation. In: 17th IEEE Symposium on Computer Arithmetic, pp. 196–205 (2002)
3. Chervyakov, N.I.: Modular Parallel Computing Structures of Neuroprocessor Systems. Fizmatlit Publishers, Moscow (2003)
4. Knyazkov, V.S., Isupov, K.S.: Parallell multiple-precision arithmetic based on residue number system. Program Syst. Theor. Appl. 7(28), 61–97 (2016)
5. Garner, H.: Number Systems and Arithmetic. In: Advances in Computers, vol. 6, pp. 131–194 (1966)
6. Omondi, A.: Residue Number System: Theory and Implementation. Imperial College Press, London (2007)
7. Osinin, I.P.: The organization of a parallel-pipelined VLSI-processor for direct conversion of numbers based on the arithmetic of the bit-slices. In: News of higher Educational Institutions, Technical Science, The Volga Region, vol. 4, pp. 5–13 (2014)

8. Isupov, K.S.: Algorithms for estimating modular numbers in floating-point arithmetic. In: Science Innovation and Technologies, vol. 4, pp. 44–56 (2016)

9. Coleman, J.N., Chester, E.I., Softley, C.I., Kadlec, J.: Arithmetic on the European logarithmic microprocessor. IEEE Trans. Comput. **49**, 702–715 (2000). https://doi.org/10.1109/12.863040

10. Arnold, M., Bailey, T., Cowles, J., Winkel, M.: Arithmetic co-transformations in the real and complex logarithmic number systems. IEEE Trans. Comput. **47**, 777–786 (1998). https://doi.org/10.1109/12.709377

11. Coleman, J.N.: Simplification of table structure in logarithmic arithmetic. Electron. Lett. **31**, 1905–1906 (1995). https://doi.org/10.1049/el:19951304

12. Lewis, D.M.: An architecture for addition and subtraction of long word length numbers in the logarithmic number system. IEEE Trans. Comput. **39**, 1325–1336 (1990). https://doi.org/10.1109/12.61042

13. Osinin, I.P.: Optimization of the hardware costs of interpolation converters for calculations in the logarithmic number system. In: Proceedings of the International Scientific Conference Parallel Computational Technologies (PCT 2018), pp. 153–164 (2018)

14. Osinin, I.P.: Method and device of parallel-pipelined arithmetical computers in modular-logarithmic system. Eurasian Union Sci. **3**(48), 45–56 (2018)

15. Osinin, I.P.: Method and device of direct parallel-pipelined transformation of numbers with floating point in modular-logarithmic format. Eurasian Union Sci. **3**(48), 56–69 (2018)

16. Osinin, I.P.: Method and device for the backward parallel-pipelined transformation of modular-logarithmic numbers in a format with a floating point. Eurasian Union Sci. **3**(48), 70–83 (2018)

17. Osinin, I.P.: A modular-logarithmic coprocessor concept. In: Proceedings International Conference on High Performance Computing and Simulation (HPCS-2017), pp. 588–595 (2017). https://doi.org/10.1109/hpcs.2017.93

18. Intel 64 and IA-32 Architectures Optimization Reference Manual. http://www.intel.com. Accessed 12 Apr 2018

Supercomputer Efficiency: Complex Approach Inspired by Lomonosov-2 History Evaluation

Sergei Leonenkov[1,2]([⊠]) and Sergey Zhumatiy[1]([⊠])

[1] Research Computing Center,
Lomonosov Moscow State University, Moscow, Russia
{leonenkov, serg}@parallel.ru
[2] Faculty of Computational Mathematics and Cybernetics,
Lomonosov Moscow State University, Moscow, Russia

Abstract. These days the number of supercomputer users and the jobs they execute is rapidly growing, especially for supercomputers, providing computing time to external users. Supercomputers and their computing time are highly expensive, so their efficiency is crucial for both users and owners. There are several ways to increase operational efficiency, however, in most cases it involves a trade-off between efficiency metrics. This brings about a need to define "efficiency" in each specific case. We use the historical data from two largest Russian supercomputers to create a number of metrics in order to provide the definition of resource management "efficiency". The data from both Lomonosov and Lomonosov-2 supercomputers consists of over one year history of job executions. Lomonosov and Lomonosov-2 efficiency in terms of CPU hours utilization is considerably high, nevertheless, our global goal is to offer the way to maintain or improve this metric when maximizing others examined in the paper.

Keywords: High-performance computing · Resource management
Supercomputer job scheduling efficiency

1 Introduction

The average performance of TOP500 List supercomputers is growing every year, along with the number of users and their tasks, which are growing even faster. For instance, the same situation occurs in the Research Computing Center of Moscow State University (RCC MSU), where two largest CIS systems are installed: Lomonosov and Lomonosov-2.

Lomonosov supercomputer, which shows theoretical peak performance of 1.7 Pflop/s and Linpack performance of 0.901 Pflop/s, increases the base of its users every year, since 2014 the number of users has grown by many times [3, 4]. For the last year it maintained an average level of CPU hours utilization at 92.3%. In turn, Lomonosov-2 supercomputer with theoretical peak performance of 4.946 Pflop/s and Linpack performance of 2.478 Pflop/s does not have such a longstanding history of use, but the available usage history data shows that during last year over 1000 users launched over

V. Voevodin and S. Sobolev (Eds.): RuSCDays 2018, CCIS 965, pp. 631–640, 2019.
https://doi.org/10.1007/978-3-030-05807-4_54

200000 tasks, while utilization of CPU hours was 88.7% [8]. Thus, task of increasing CPU hours utilization is very important for RCC MSU.

To secure RCC MSU clusters efficiency we have researched Lomonosov and Lomonosov-2 historical data of using Simple Linux Utility for Resource Management (SLURM) to analyze current status [1]. Within the framework of this research, we were able to develop our definition of supercomputer efficiency based on limitations of RCC MSU and selected list of metrics.

The paper organized as follows. Section 2 is devoted to Lomonosov and Lomonosov-2 supercomputers SLURM statistics review. Section 3 describes main trends of RCC MSU supercomputer usage. Section 4 provides an approach to define supercomputer resource management efficiency and represents list of efficiency metrics used. Our conclusions and future plans are drawn in the final section.

2 Background

Lomonosov and Lomonosov-2 supercomputers are two core high performance computing systems of Moscow State University. Both provide access for various scientific research groups whose number has already exceeded 900. As a rule 1000 jobs are processed every day and requested by 50–70 different users in general. There are approximately 3000 active accounts and this number is still growing every year. To cope with the constantly growing workload a SLURM (Simple Linux Utility for Resource Management) system and its implementation of Backfill scheduling algorithm are used.

2.1 SLURM

SLURM is a highly scalable, fault-tolerant cluster resource manager and job scheduler for big computational systems [2]. SLURM cluster manager is used on many supercomputers specified in Top500 rating, including half of supercomputers stated in the top 10 (as per November 2013). The system is available under GNU GPL V2 license and well-documented. [1].

SLURM is based on the hierarchical model of supercomputer management systems. SLURM is designed for heterogeneous clusters with up to 10 million processors possible. It is successfully used on a supercomputer with more than 98000 nodes. Those who use a supercomputer managed via SLURM can set up to 1000 jobs for execution per second. Manager can perform up to 500 jobs per second (depending on the system configuration and equipment).

System administrator can set logical configuration of computing system, which would be supported by SLURM, flexibly and vary a set of cluster parameters easily by changing a configuration file or using appropriate commands, also external schedulers are allowed [5, 6].

2.2   Lomonosov and Lomonosov-2 Supercomputers

Lomonosov supercomputer consists of 12346 CPUs (mainly Intel Xeon X5570 2.93 GHz), but even this amount is not enough to provide all of MSU research groups with necessary computation power. All CPUs are physically separated into groups of 4 or 6, called nodes. A selection of nodes may be combined into a logical group, which is called partition and includes a queue of incoming jobs. Partitions can be limited by, for example, indicating users who can use them, size of a job or processing time limits. Each partition focuses on the specific needs of users. Current configuration of the Lomonosov supercomputer includes 8 separate sections, each consisting of 1 to 4096 nodes [3, 4].

We have researched Lomonosov supercomputer usage statistics from March 2014 to March 2017. During this time, users submitted overall 828389 jobs. Column 2 of Table 1 shows jobs state distribution gathered by SLURM over considered period. It should be mentioned that statuses cancelled, completed, failed and timeout do not imply CPU hours losses in most cases, it is mainly depends on program architecture. Node_fail job status represents that job terminated due to failure of one or more allocated nodes.

Table 1. Lomonosov supercomputer jobs statistics over researched period.

Status	Number of jobs (Lom)	Number of jobs (Lom-2)
Cancelled jobs	93758	37582
Completed jobs	384990	125648
Failed jobs	267550	33847
Node_fail jobs	4423	2000
Timeout jobs	77668	6201

Lomonosov-2 is top Russian supercomputer built by T-Platforms. Lomonosov-2 is an Intel Xeon/FDR InfiniBand cluster, accelerated with NVIDIA Tesla K40 s and Tesla P100 GPUs [8]. Lomonosov-2 supercomputer consist of 1696 CPUs (Intel Haswell-EP E5-2697v3, 2.6 GHz, 14 cores and Intel Xeon Xeon Gold 6126 2.60 GHz, 12 cores). Current configuration of the Lomonosov-2 supercomputer includes 4 separate partitions: "compute", "low_io", "tesla" and "test", which consists of 1120, 384, 160 and 32 nodes respectively [7]. We have researched Lomonosov-2 supercomputer usage statistics from August 2016 to August 2017. During this time, users submitted overall 205278 jobs. Column 3 of Table 1 shows jobs state distribution gathered by SLURM over considered period. Next section provides more detailed analysis of Lomonosov and Lomonosov-2 supercomputers usage history and examines different efficiency metrics.

3   Usage History Analysis

Data was gathered using Octotron [9] and SLURM sacct plugin. We developed SLURM usage history analysis and visualization tool. All metrics computations and figures in this section was made in this tool.

Figure 1 shows CPU time utilization of Lomonosov-2 supercomputer over one year period from August 2016 to August 2017. It is defined by the ratio of the number of nodes that execute users' jobs (black line) to the number of available nodes at a given moment of time (blue line).

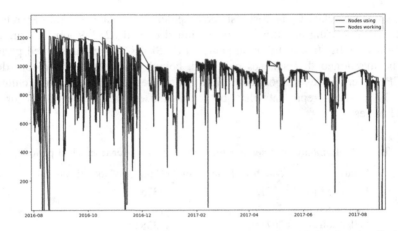

Fig. 1. Lomonosov-2 supercomputer CPU time utilization. (Color figure online)

The overall CPU time utilization rate over this period amounted to 88.7%. The graph illustrates that before the year 2017 CPU time utilization was lower than the average due to the fact that the majority of users got an access to Lomonosov-2 supercomputer in January 2017. The average number of users raised at this time from approximately 3 to more than ten users per time unit. The efficiency of Lomonosov-2 supercomputer scheduling in terms of CPU time utilization starting from January 2017 almost reached the maximum. As a consequence, any further optimization of scheduling techniques must be based on other efficiency metrics that will allow to improve supercomputer users' experience.

For accurate analysis the Lomonosov supercomputer statistics was also examined on the same period (only for queue "regular4"). Note that the maximum number of available nodes for this period was only around 2000 (not 4096) because of maintenance works. CPU time utilization over the period totaled over 92%.

Owing to the fact that Lomonosov supercomputer operates for many years, it has a significantly wider user base than Lomonosov-2 supercomputer. Another valuable observation that at the given moment of time the number of users in the queue is much more than the number of users whose jobs are on execution. This tendency represents the future development of user base and distribution of user jobs on queue and on execution of Lomonosov-2 supercomputer (Fig. 2).

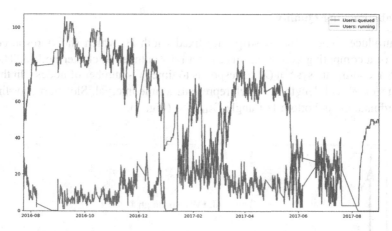

Fig. 2. Lomonosov-2 supercomputer number of jobs queued or running.

In spite of the high utilization of CPU time, RCC MSU supercomputers does not always use the most efficient settings for scheduling algorithms for majority of users. After analyzing historical data, we selected metrics that will help to assess more precisely the user's experience on the supercomputer.

4 Supercomputers' Scheduling Efficiency

Because of such considerable load, RCC MSU has faced a challenge of SLURM's settings scarcity. For instance, one of the serious problems of CPU time scheduling is the fact that each user's waiting time is too long, even when using a job scheduler based on Backfill algorithm. This causes a common problem: each user queues several jobs (often with the purpose of program debug) and waits. When the first job of the user is started, top of the queue is already occupied by this user's jobs, so the majority of other users are waiting for one. This issue cannot be solved by putting a limitation on the number of user's running jobs, as the efficiency of the queue may decrease significantly if it contains many short and/or small jobs.

Due to this limitations we have already tested some optimizations of Backfill algorithm. For example, we added individual CPU hours limit, this decision showed great performance on historical data. This solution has been tested compared to standard SLURM backfill plugin on historical data queue "regular4" Lomonosov supercomputer, and showed good acceleration for first job start time of each user (up to ∼9% compared to standard SLURM backfill plugin and average acceleration is ∼9.5 min per week for first job start time of each user), while the total start time of jobs almost has not been increased (only 0.2% compared to standard SLURM backfill plugin and with an average 10 s start time delay for all users jobs, except first user's jobs) [10].

4.1   Jobs Packing Quality

Let's introduce some terms. A strip with fixed width H, which shows resources utilization of a computing system in time, is set (H - number of cluster nodes). The strip has an XY coordinate system (X corresponds to time, Y - number of nodes). In the strip we set a slot W with length T, which represents a time interval. Slot start coordinate is the coordinate of its bottom left angle (X_0, Y_0) (Fig. 3).

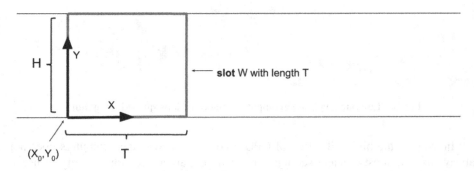

Fig. 3. Lomonosov supercomputer number of users, whose jobs queued or running.

Job is a user's program that has two states: it is either in a queue or is being executed in computing resources.

Job is a set of elements: $Ji = \{X_i, T_i, H_i, R_i, U_i, Q_i\}$ (Fig. 4), where

X_i — execution start time of a task in computing resources;

T_i — time length of job execution in computing resources;

H_i — number of computing nodes required to execute a task;

R_i — non-empty setup of j pairs (y_{ij}, h_{ij}), which describes task allocation in nodes as a rectangle with bottom left angle coordinate (Xi, yij), T_i execution time and h_{ij} number of nodes such that $\sum_j h_{ij} = H_i$ (Figs. 4 and 5);

U_i — identifier of a user associated with a task;

Q_i — queueing time

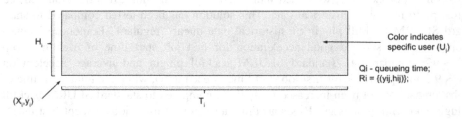

Fig. 4.   Job

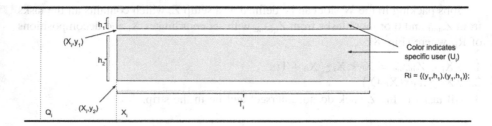

Fig. 5. Job J_i in a strip and R_i decomposition example.

Notations interpretation: a job is represented as a rectangle with set coordinates of a bottom left angle, defined size (H_i and T_i are rectangle's sizes on Y and X axes respectively) and color (corresponds to user identifier), decomposition of R_i among nodes (Fig. 5).

Let's consider two setups of jobs:

1. Z_{start} is a setup of jobs executing in the $X_i = X_0$ moment of time (Fig. 6).
2. Z_{queue} is a setup of queued jobs, for which X_i coordinates and R_i decomposition are not set and $Q_i \leq X_0 + T$ (Figure 7).

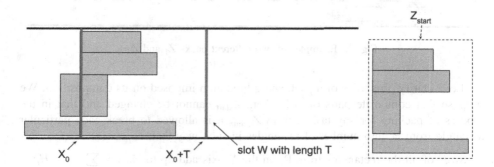

Fig. 6. Z_{start} setup and an example of its position in the W slot.

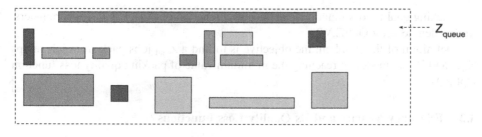

Fig. 7. Z_{queue} setup

Jobs packing in the W slot can be defined as a setup Z, which contains all the tasks from Z_{start} and 0 or more tasks from Z_{queue} with set coordinates X_i and decompositions of R_i, where (Fig. 8):

1. $\forall X_i \in Z_{queue} \Rightarrow X_0 < X_i \le X_0 + T$;
2. $\forall X_i \ge \max(X_0, Q_i)$;
3. All tasks in the Z pack do not intersect and lie in the strip.

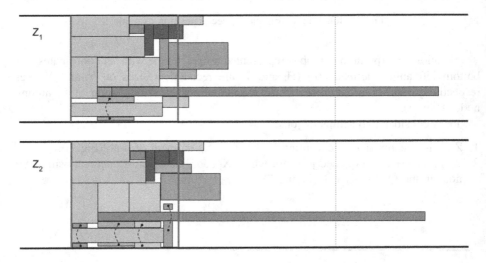

Fig. 8. Examples of two different packs Z_1 and Z_2

Let's clarify properties of a pack and a limitation imposed on its composition. We suppose that composite parts of tasks from Z_{start} cannot be changed and that in the process of packing for any task J_i from Z_{queue} it is allowed to break each particular rectangle from R_i into number of rectangles in such way that:

1. Sizes of all the rectangles from R_i on the Y-axis add up to H_i, i.o. $\sum_j h_{ij} = H_i$;
2. X-coordinate of the left side of all rectangles is equal to X_i;
3. Sizes of all the rectangles on the X-axis is equal to T_i;
4. It is not allowed to rotate rectangles.

Packing quality loss function is defined as a function of Z pack and slot parameters. Let us denote it by Opt(Z,W).

Definition of the problem: the objective is to find a Z_{end} jobs pack from job setups Z_{start} and Z_{queue} in slot W reaching the minimum value of packing quality loss function Opt(Z,W).

4.2   Efficiency Metrics and Its Quality Loss Functions

Basing on the conducted research, we offer a set of metrics, which allows to consider the task of CPU hours scheduling efficiency more comprehensively, and a formula, which provides a means of comparing different settings of any scheduling algorithms. A list of metrics that should be considered when determining the operational efficiency of RCC MSU was formulated based on the historical data from supercomputers Lomonosov and Lomonosov-2 is presented below.

Let's introduce methods for evaluating the quality loss function of a Z pack. Each of proposed metrics represents some quality characteristic of supercomputer's resource management.

Given that

- UserNum(Z) — number of users, whose tasks belong to Z pack,
- UserJob(i) — a set of jobs of ith user in Z pack,
- Class(A,B) — a set of jobs, for which H_i belongs to set half-interval [A, B].

Most widely used resource management quality characteristic is utilization of computing nodes. The main idea is to minimize the number of free resources.

$$\text{Opt}(Z,W) = \text{Utilization}(Z,W) = 1 - \sum_{i=0}^{|Z|} (H_i * (min(T, X_i + T_i) - X_i)/(H * T);$$

Another useful metric is average start time of the first job of users in slot W. The objective is to minimize the average distance from a job of each color, for which $X_i - X_0$ is minimal among jobs of this color, to the beginning of the slot $W - X_0$.

$$\text{Opt}(Z,W) = \text{FUJStartTime}(Z,W)$$
$$= \sum_{user=1}^{UserNum} min_{j \subset UserJobs(user)}(X_j - Q_j)/UserNum(Z);$$

Next two metrics represents number of running jobs (minimizing the number of not running jobs from Z_{queue}) and number of users, whose jobs from Z_{queue} were started in slot W (minimizing the number of users, whose jobs from Z_{queue} were not started in slot W).

$$\text{Opt}(Z,W) = \text{StartedJobs}(Z,W) = (|Z_{start}| + |Z_{queue}| - |Z|)/(|Z_{queue}|);$$

$$\text{Opt}(Z,W) = \text{StartedUsers}(Z,W)$$
$$= \text{UserNum}(Z_{start}) + \text{UserNum}(Z_{queue}) - \text{UserNum}(Z);$$

Last metric shows average start time of jobs belonging to a specific class (minimizing the average distance from each job, which size $H_i \in [A, B)$, to the beginning of the slot $W - X_0$).

$$\text{Opt}(Z,W) = \text{AVGStartTime}(Z,W,\text{Class}) = \sum_{i \subset Class}(X_j - Q_j)/|Class|;$$

Finally, we propose supercomputer's resource management efficiency definition and a related formula, which represents supercomputer scheduling efficiency based on normalized values of proposed metrics. *PriorityCoefficient* variables indicate priorities of each metrics, it should be predefined due to analysis made on supercomputer administration needs. *MetricsValue$_i$* is one of five represented metrics.

$$Efficiency = \sum_{i=1..5} \left(PriorityCoefficient_i * MetricsValue_i \right) \text{ , where}$$
$$\sum_{i=1..5} \left(PriorityCoefficient_i \right) = 1.$$

Proposed formula is based on operational statistics of RCC MSU supercomputers, so in other cases metrics can be chosen in different way (depending on goals of supercomputer centres) and added to this formula. The only limitation is that $MetricsValue_i$ should be normalized on interval $[0,1]$.

5 Future Work

Future work lies in field of designing fast online algorithm for proposed "efficiency" function minimum search. All future research will be conducted on Lomonosov-2 and Lomonosov supercomputers.

Acknowledgments. This material is based upon the work supported by Russian Foundation for Basic Research (Agreement N 17-07-00664 A).

References

1. Yoo, A.B., Jette, M.A., Grondona, M.: SLURM: simple linux utility for resource management. In: Feitelson, D., Rudolph, L., Schwiegelshohn, U. (eds.) JSSPP 2003. LNCS, vol. 2862, pp. 44–60. Springer, Heidelberg (2003). https://doi.org/10.1007/10968987_3
2. Slurm workload manager (2015). http://slurm.schedmd.com/slurm.html
3. Sadovnichy, V., Tikhonravov, A.: LOMONOSOV: supercomputing at moscow state university. In: Contemporary High Performance Computing: From Petascale toward Exascale, pp. 283–307 (2013)
4. Lomonosov—T-Platforms (2015). http://www.top500.org/system/177421
5. Lipari, D.: The SLURM Scheduler Design (2012). http://slurm.schedmd.com/slurm_ug_2012/SUG-2012-Scheduling.pdf
6. Jones, M.: Optimization of resource management using supercomputers SLURM (2012). http://www.ibm.com/developerworks/ru/library/l-slurm-utility/
7. Lomonosov-2 supercomputer configuration (2018). http://users.parallel.ru/wiki/pages/22-config
8. Lomonosov-2 supercomputer on TOP50 list (2018). http://top50.supercomputers.ru/?page=stat&sub=ext&id=593
9. Antonov, A., et al.: An approach for ensuring reliable functioning of a supercomputer based on a formal model. In: Wyrzykowski, R., Deelman, E., Dongarra, J., Karczewski, K., Kitowski, J., Wiatr, K. (eds.) PPAM 2015. LNCS, vol. 9573, pp. 12–22. Springer, Cham (2016). https://doi.org/10.1007/978-3-319-32149-3_2
10. Leonenkov, S., Zhumatiy, S.: Introducing new backfill-based scheduler for SLURM resource manager. Proc. Comput. Sci. **66**, 661–669 (2015)

Supercomputer Real-Time Experimental Data Processing: Technology and Applications

Vladislav A. Shchapov[1,2](✉) ⓘ, Alexander M. Pavlinov[1] ⓘ, Elena N. Popova[1] ⓘ,
Andrei N. Sukhanovskii[1] ⓘ, Stanislav L. Kalyulin[2] ⓘ,
and Vladimir Ya. Modorskii[2] ⓘ

[1] Institute of Continuous Media Mechanics of the Ural Branch of Russian Academy
of Science, 614013 Academician Korolev Street, 1, Perm, Russia
{shchapov,pam,popovadu,san}@icmm.ru
[2] Perm National Research Polytechnic University, Perm, Russia
{ksl,modorsky}@pstu.ru
https://www.icmm.ru/, http://pstu.ru/

Abstract. The study is focused on the technology of remote real-time
processing of intensive data streams from experimental stands using
supercomputers. The structure of distributing data system, software for
data processing, optimized PIV algorithm are presented. Using of real-
time data processing makes possible realization of experiments with feedback when external forcing depends on internal characteristics of the
system. Approbation of this technique is demonstrated on experimental
study of intensive cyclonic vortex formation from localized heat source
in a rotating layer of fluid. In this study the heating intensity depends on
velocity of the flow. The characteristics of the flow obtained by supercomputer real-time processing of PIV images are used as input parameters
for the heating system. The concept of using developed technology in the
experimental stands of aircraft industry is also described.

Keywords: Supercomputer · Experimental data processing
PIV · SciMQ · Laboratory analog of tropical cyclone · Feedback

1 Introduction

At the present significant e-Science projects deal with the processing of large
amounts of data obtained from experimental stands (for example, CERN LHC
in high-energy physics). The growth in the amount of experimental data makes
impossible the data processing using local computer power. Traditionally, grid
computing is used to solve this problem, when the received data are distributed

---

The study was supported by the grants of the Russian Foundation for Basic Research
(RFBR): project 17-45-590846 (V. Shchapov, A. Pavlinov, E. Popova, A. Sukhanovskii,
Sects. 1–3.1, 4) and project 17-47-590017 (S. Kalyulin, V. Modorskii, Sect. 3.2).

V. Voevodin and S. Sobolev (Eds.): RuSCDays 2018, CCIS 965, pp. 641–652, 2019.
https://doi.org/10.1007/978-3-030-05807-4_55

for processing and analysis in many computer centers. However, this scheme is not suitable for the cases when data processing should be performed directly during the experiment. It is necessary for on-line analyses of the results or controlled forcing. For example experimental study of formation of laboratory analog of tropical cyclone [1–4] with a controlled feedback between flow velocity and heat release. Realization of controlled feedback required solution of a number of technical problems such as data acquisition and storage, real-time data processing, integration of PIV (Particle Image Velocimetry) and heating control systems. The common way of application of PIV technique for velocity field reconstruction consist of acquisition of images of tracers and their postprocessing. The main problem toward realization of real-time PIV (RTPIV) measurements is a high computing cost of data postprocessing [5]. There are several ways for realization of RTPIV. Simplified PIV algorithms and small images allow to processed PIV images up to 15 Hz [6]. Another way is using FPGA (field programmable gate array) technology, the strong limitations of which is requirement of using specific hardware language for programming the PIV code. The detailed description of FPGA can found in [5, 7]. Growth of GPU computing power results in RTPIV based on implementation of GPU code for PIV processing [8]. Alternative efficient solution of the described problem is a main goal of the present study. The key idea of our approach is a transfer of resource-intensive data processing to the supercomputer. Integration of experimental measurement system and supercomputer is realized by using of data stream manager SciMQ [9] and supercomputer software for PIV processing.

2   Real-Time Supercomputer Processing of Experimental Data

Remote data processing using external computing system requires organization of data transport network between measuring and computational systems. Another problem is efficient distribution of experimental data on computing nodes and return of processed data back to the main experimental computer [10, 11]. The possibility of distributed processing imposes some restrictions on the structure of the data stream. The data stream from the measurement system must be discrete, that is, consist of independent blocks (messages). The processing of each message by application algorithms must be independent of other messages and the results of their processing. If the data stream from the measurement system meets these requirements, then it can be processed on remote supercomputers in real time by distributing stream messages to accessible computing nodes.

The presence of a discrete stream with independent processing of each message allows us to apply the concept of queues for organizing the processing of experimental data in real time on a supercomputer. The use of queues for data transmission makes possible the separation of the measuring and computing systems, renouncing the synchronization at the level of computational nodes, and isolation of distributing and collecting data from computational nodes in one subsystem – the data stream manager. The architecture of a system for processing experimental data in real time on a supercomputer is shown in Fig. 1.

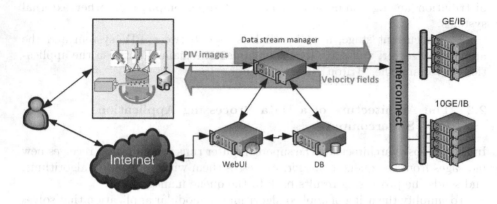

Fig. 1. The architecture of a system for processing experimental data on a supercomputer

The advantage of this application is parallel data transfer from the data stream manager to the supercomputer, which significantly increases the efficiency of data transmission in the case of using a remote long distance supercomputer [11].

Development of our own software stack for controlling of data streams was necessary because existing solutions do not satisfy high requirements for network capacity (for messages large than 10 MB), usability and efficiency for data transfer through extensive communication channels [10].

2.1 Software-Hardware Platform for Processing Experimental Data

A supercomputer "Triton" with a peak performance of 23.1 TFlops, built on the basis of Intel Xeon E5450 processors (Harpertown, SSE 4.1) and Intel Xeon E5-2690v4 (Broadwell, AVX2) is used for experimental data processing. Computing nodes are combined with InfiniBand (Harpertown nodes – 20 Gbps DDR, Broadwell nodes – 56 Gbps FDR) and Ethernet (Harpertown nodes – 1 Gbps, Broadwell nodes – 10 Gbps) interconnects.

The SciMQ data stream manager [9,11] is installed on the HP ProLiant DL360p Gen8 application server (2x Intel Xeon CPU E5-2660, 2.20 GHz, RAM 128 GB), which is connected to the network at a speed of 10 Gbps. SciMQ software consists of the following main components:

1. Data stream manager;
2. Database;
3. WebUI – Web interface.

SciMQ data stream manager is a highly performance queue server, developed by authors which is ready for efficient work with large messages up to tens of megabytes. This condition is necessary for PIV images transport. Manager is responsible for control of the sending messages in queues, intermediate storing, distribution among computational nodes of supercomputer or other external systems.

At the current stage, a 1 Gbps link is used between PIV system and the application server, and a 10 Gbps communication channel between the application server and the "Triton" supercomputer.

2.2 The Architecture of a Data Processing Application for a Supercomputer

In the proposed architecture, the supercomputer runs software that receives new messages from the queue manager, processes them with application algorithms, and sends the processing results back to the queue manager.

To simplify the using of applied algorithms, a modular application that solves problems of receiving and sending messages, parsing the original messages, and formatting the processing results was implemented. The application supports the development of modules of applied algorithms that are implemented in the form of C++ classes using the specified interface.

Main application features:

- Organization of a network connection with the data stream manager with control over its operation and reconnection in the event of a disconnection.
- Receiving of new messages for processing, including prefetch mode, when the following message is downloaded in the background mode. The use of preload allows to increase the efficiency of the use of processor cores when the supercomputer is located far from the data stream manager.
- Sending the results of processing to the data stream manager. Both receiving and sending messages are executed along with processing in the dedicated thread.
- United management of configuration of the system and using algorithms through a single configuration file.
- Integration with the Intel Threading Building Blocks (Intel TBB) library, which allows data processing algorithms to use Intel TBB functions for parallel processing of one message.
- Integration with MPI, which allows to inform the data stream manager about the completion of processing only once (not from each compute node with the application running).
- The organization of storing and distributing by computing nodes of the history of processed data, which will be available for processing subsequent messages. It allows to implement adaptive algorithms that take into account the specifics of initial and earlier processed data.

Figure 2 shows the block diagram of the data processing application for the supercomputer.

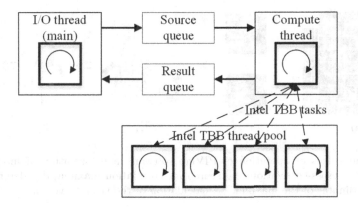

Fig. 2. The block diagram of the data processing application for the supercomputer.

The application works in several threads. The main thread is responsible for input/output (I/O) – receiving messages from the data stream manager and sending processing results. The second thread is computational – it starts the basic functions of applied algorithms. In addition, the Intel TBB library thread pool is used to speed up data processing in application algorithms. The interaction between the computational thread and the I/O thread is done by two internal queues, one of which transmits input data from the I/O thread to the computational thread, and the other transmits processed results from the computational thread to the I/O thread.

2.3   Module of PIV Data Processing

As a basic PIV algorithm we chose the one realized in PIVlab software [12] published by BSD license. Since there is no Matlab software installed on supercomputer "Triton" PIV algorithms from PIVlab initially realized in Matlab were rewritten in C++ and optimized by using Intel libraries (MKL and IPP). Some algorithms were optimized by manual vectorization using SSE and AVX instructions. The MKL library is used in Sequential mode, because MKL functions are called from parallelized code. For manually-optimized functions, when an application is started, an implementation using the processor-supported instruction set is selected.

The stages of operation of the algorithm are shown in Fig. 3.

The processing of each interrogation area is completely independent of the others. This allows parallel processing of one measurement by distributing interrogation areas across different cores of the compute node using the functions of the Intel TBB library.

Implementation of efficient paralleling of processing using all available cores of computational node was successful on all stages except initial unpacking compressed PNG (portable network graphics) images. At this stage only two data streams are used (because of the two frames in one measurement) excluding the

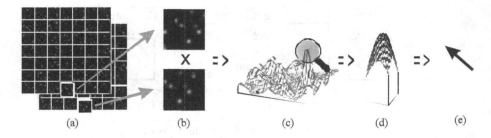

Fig. 3. The stages of operation of PIVlab software: a – formation of interrogation windows, b – cross-correlation, c – search for correlation maxima, d – determination of the coordinates of the maxima, e – calculation of the velocity vector.

interpolation algorithm which are called four times during data processing. The interpolation algorithm uses global mutex and only one thread, since the Intel MKL PARDISO solver used in it is not thread safe in Sequential mode.

3 Application of the Developed System

Developed system is actively used in laboratory study of tropical cyclone formation. Specifically it is applied for modelling of complex process of latent heat release. Feedback between intensity of the flow and the amount of heat flux is realized.

It is also planned to use the developed system for recording and processing of fast processes with feedback during aircraft flights. The main problem is the processes of icing of structural aircraft elements.

3.1 Laboratory Modelling of Tropical Cyclones with Controlled Forcing

Connection between hydrodynamic and thermodynamic processes is important factor in different natural and technological systems. For example velocity or flow topology variation can lead to increasing or decreasing of heat flux in processes of convective heat transfer, combustion or exothermic chemical reactions. Open problem is a link between wind velocity and latent heat release during formation of large-scale atmospheric vortices like tropical cyclones (hurricanes, typhoons). The problem of tropical cyclogenesis attracts great attention because of multiple human losses and vast economical damage. The main problem is long-term reliable forecast. The quality of prediction of tropical cyclone intensity and track of its motion strongly depends on the choice of mathematical models. Up to now capabilities of numerical modeling are restricted. Most of numerical simulations are carried out using spatial resolution of 2–3 km with parametrization of the subgrid processes. Some effects like the influence of secondary flows with characteristic scale of 1–3 km on heat and mass transfer are either parametrized or neglected. Another serious problem for numerical modeling is a large number of parameters (humidity, compressibility, physical properties of media and many others). Taking into account that the time of one full-scale 3D run is one week or more to study the role of all parameters is hardly possible.

Limited capabilities of direct numerical modeling of atmospheric flows increase interest to the laboratory modeling of geophysical processes. On the base of approach described in [1] and using of PIV (Particle Image Velocimetry) system it was shown [2–4] that the structure of laboratory convective vortex is similar to the structure of typical tropical cyclone. Supercomputer data processing can be used for modelling of latent heat release in the process of tropical cyclone formation. Realization of controlled feedback between velocity and heating required solution of a number of technical problems such as data acquisition and storing, real-time data processing, integration of PIV and heating control systems.

Experimental Setup. Experimental model is a cylindrical vessel of diameter $D = 300\,\text{mm}$, and height $H = 40\,\text{mm}$ (Fig. 4(a)). The sides and bottom were made of plexiglass with a thickness 3 mm and 20 mm respectively. There was no cover or additional heat insulation at the sidewalls. The heater is a brass cylindrical plate mounted flush with the bottom. The diameter of the plate d is 104 mm, and its thickness is 10 mm. The brass plate is heated by an electrical coil placed on the lower side of the disc. Massive heater provides uniform heating which is optimal for vortex excitation. Cylindrical vessel was placed on a rotating horizontal table (Fig. 4(b)). Silicon oil PMS-5 (5 cSt at $T = 25°$) is used as working fluid. In all experiments, the depth of the fluid layer h was 30 mm and the surface of the fluid was open. The room temperature was kept constant by air-conditioning system, and cooling of the fluid was provided mainly by the heat exchange with surrounding air on the free surface and some heat losses through sidewalls. Details of experimental setup, structure and characteristics of the laboratory analog of tropical cyclone can be found in [3,4]. The images for PIV processing were obtained with a 2D particle image velocimetry (PIV) system Polis and the software package Actual Flow.

Experiment with a feedback requires real-time data processing. Standard kit of Actual Flow software does not have such option. Also we need to note that PIV technique is resource demanding and even for measurements at relatively low frequency (0.5 Hz) computing resources of personal computer are not sufficient for real-time velocity reconstruction. In order to solve this problem we use supercomputer for PIV images processing. The variation of number of computational nodes allows to achieve necessary rate of image processing.

Actual Flow is a commercial software which does not have documented API for interaction with external systems and realization of required series of measurements. For solution of this problem following procedure was used. During experiment Actual Flow saves for each measurement three files: two images and file with metadata in xml format. Developed data loader using WinAPI analyses catalog of Actual Flow data. For each new xml file availability of images is checked. When all three files are available they are packed and transferred to the data stream manager for sending to supercomputer for processing.

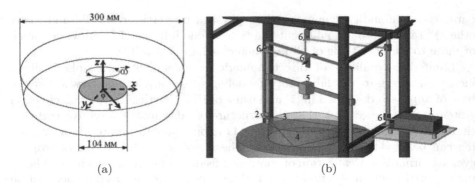

(a)   (b)

Fig. 4. Experimental model, dimensions and location of the coordinate system (a); experimental stand (b): 1 – dual pulsed laser for PIV, 2 – laser sheet system, 3 – laser sheet, 4 – tracers, 5 – CCD camera, 6 – mirrors.

Test Experiments. The developed system for realization of experiments with feedback was checked on test experiments. The test experiment was done as follows. The experimental model was placed on a rotating stand. The solid-state rotation was achieved. It takes about 2 h (for a period of rotation $T = 77$ s). After that the measurement process was started using the PIV system "Polis".

Simultaneously with a measurement process, for the organization of the initial radial flow, a constant power heating (about 30% of the maximum) was switched on for 30 s. After initiation of radial motion the heating was switched to a feedback mode in which the heating power was proportional to the average radial motion velocity in the heating region. The area of measurements and the complex structure of forming convective flows after the switching on the heating is illustrated in Fig. 5.

Figure 6 shows the time dependences of the heating power (proportional to the mean radial velocity) and the kinetic energy of the cyclonic motion. It is clearly seen that for a given relationship between the power and the mean velocity of the radial flow, the transition to the quasi-stationary state occurs fairly quickly (in the time of 5–6 revolutions of the model). The characteristic oscillations of the mean velocity and kinetic energy are a feature of the convective vortex flows in the system under consideration.

Figure 7 shows vector fields for the early and late stages of the vortex development. Test experiments showed the efficiency of the complex integration of measurement system, supercomputer and heating system. The next step will be a detailed study of the effect of functional dependencies between the heating intensity and the flow structure.

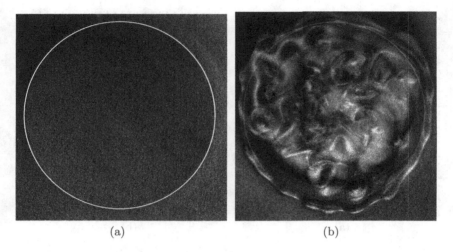

(a)   (b)

Fig. 5. a – field of measurements, thick white line shows the heating area of diameter $D = 104\,\text{mm}$; b – formation of convective structures after switching on the heating, vizualization by aluminum powder.

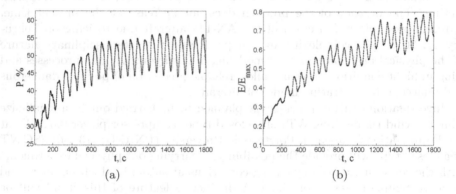

(a)   (b)

Fig. 6. a – temporal variation of the heating power (percentage of the maximum value); b – temporal variation of the kinetic energy of azimuthal flow in the heating area.

3.2 Application of Recording and Processing Techniques of Fast Processes for Icing Research of Structural Aircraft Elements

For another applied problem – modelling of icing processes it is also necessary to use real-time processing of PIV data. The formation of ice on the structural elements of aircraft in flight can lead to a significant deterioration of the aerodynamic characteristics and controllability of the aircraft, the failure of control systems, as well as to the destruction of engine components, which is an actual problem of safety of flights. According to accident statistics ("Army AircraftIcing" data, 2002), between 1985 and 1999, 255 cases of aircraft icing occurred, of which 12% with victims, and total losses amounted to 28 million $.

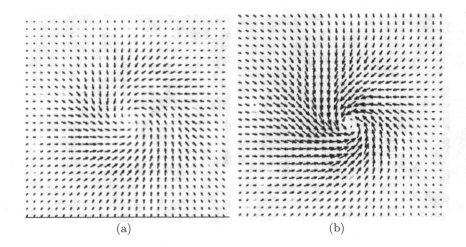

Fig. 7. Vector velocity fileds, left – t = 100 s, right – t = 1500 s.

According to the data of "Aircraft Owners and Pilot Association" (2007) –
202 cases of aircraft icing occurred between 1998 and 2007, of which 21% with
victims. The urgency of the problem does not decrease at the present time.
A recent example is the crash of the AN-148 aircraft due to icing on February 11, 2018. The complexity and, in particular, the interdisciplinary nature
of the physical and mathematical modeling of gas-hydrodynamic processes and
icing in flight conditions does not allow reliable prediction of possible dangerous
and unacceptable operating modes of aircraft.

Investigation of icing processes is planned to be carried out in a small-size
climatic wind tunnel (SSCWT) of a closed type (compressor power 0.3 MW) at
the Perm National Research Polytechnic University (PNRPU) [13,14]. SSCWT
of a closed type will provide the possibility of carrying out physical experiments
with the reproduction of flight icing conditions at subsonic Mach numbers and
negative temperatures of breaking. A distinctive feature of this stand will be
high energy efficiency, due to the closed isolated wind tunnel and the relatively
small size of the working part. The closed working part of SSCWT will ensure a
high degree of uniformity of the watered airflow and the possibility of a detailed
study of the physical processes during icing in flight.

For the real-time measurements in fast processes it is planned to use phase
doppler anemometry (PDA) (Fig. 8) and high-speed cameras or time-resolved
PIV systems.

It is planned that experimental data will be transferred to the high-
performance computer complex PNRPU [15] for processing – calculation of gas-
dynamic flow parameters: velocity fields, temperatures and distributions of droplets, qualitative and quantitative parameters of icing. All data must be recorded
simultaneously in several sections of SSCWT. Control parameters of experiments
will be also recorded.

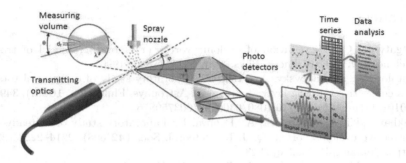

Fig. 8. Optical scheme for analysis of particle dynamics (Particle Dynamics Analysis).

At the same time, physical experiments are planned to be conducted on reduced models. These results will be used for verification of numerical simulations. After that numerical simulations of full-size models will be carried out. For the verification, it is necessary to specify in detail the fields of distribution of gas-dynamic parameters, obtained from the results of the physical experiment, as boundary conditions for carrying out a numerical simulation.

The results of processing of the initial data obtained in real time are planned to be used to refine the course of the experiments, as well as for the parallel execution of the full-scale and computational experiment, where the results of the full-scale experiment are used in the computational experiment.

4   Conclusion

Developed system for real-time experimental data processing is described. The key feature of this system is using supercomputer for computationally intensive processing. The possibility of the transfer of complex calculations from the main experimental computer to the external computing system (supercomputer) is demonstrated. The description of the modular application for writing algorithms for the supercomputer processing is given. The algorithm for processing of PIV measurements using supercomputer is described. The developed system for realization of studies with feedback was checked on test experiments that showed the efficiency of the complex integration of measurement system, supercomputer and heating system. The next step will be a detailed study of the effect of functional dependencies between the heating intensity and the flow structure. The system is ready to use high-speed cameras for obtaining PIV images. Development of adaptive PIV algorithms, taking into account the dynamics of the investigated flow is planned. Supercomputer processing can be used for the studying of icing of aircraft elements in a wind tunnel.

References

1. Bogatyrev, G.P.: Excitation of cyclonic vortex or laboratory model of tropical cyclone. Sov. J. Exp. Theor. Phys. Lett. **51**(11), 630 (1990)
2. Batalov, V., Sukhanovsky, A., Frick, P.: Laboratory study of differential rotation in a convective rotating layer. J. Geophys. Astrophys. Fluid Dyn. **104**(4), 349–368 (2010). https://doi.org/10.1080/03091921003759876
3. Sukhanovskii, A., Evgrafova, A., Popova, E.: Laboratory study of a steady-state convective cyclonic vortex. Q. J. R. Meteorol. Soc. **142**(698), 2214–2223 (2016). https://doi.org/10.1002/qj.2823
4. Sukhanovskii, A., Evgrafova, A., Popova, E.: Non-axisymmetric structure of the boundary layer of intensive cyclonic vortex. Dyn. Atmos. Oceans **80**, 12–28 (2017). https://doi.org/10.1016/j.dynatmoce.2017.08.001
5. Kreizer, M., Ratner, D., Liberzon, A.: Real-time image processing for particle tracking velocimetry. Exp. Fluids **48**(1), 105–110 (2010)
6. Willert, C.E., Munson, M.J., Gharib, M.: Real-time particle image velocimetry for closed-loop flow control applications. In: 15th International Symposium on Applications of Laser Techniques to Fluid Mechanics (2010)
7. Yu, H., Leeser, M., Tadmor, G., Siegel, S.: Real-time particle image velocimetry for feedback loops using FPGA implementation. J. Aerosp. Comput. Inf., Commun. **3**(2), 52–62 (2006)
8. Gautier, N., Aider, J.L.: Real-time planar flow velocity measurements using an optical flow algorithm implemented on GPU. J. Vis. **18**(2), 277–286 (2015)
9. Shchapov, V.A., Masich, A.G., Masich, G.F.: The technology of processing intensive structured dataflow on a supercomputer. J. Syst. Softw. **127**, 258–265 (2017). https://doi.org/10.1016/j.jss.2016.07.003
10. Masich, G., Shchapov, V.: The software platform of transmission of intense data streams on remote supercomputers. In: CEUR Workshop Proceedings, vol. 1482, pp. 720–731 (2015)
11. Shchapov, V., Masich, G., Masich, A.: Platform for parallel processing of intense experimental data flow on remote supercomputers. Procedia Comput. Sci. **66**, 515–524 (2015). https://doi.org/10.1016/j.procs.2015.11.058
12. Thielicke, W., Stamhuis, E.J.: PIVlab - towards user-friendly, affordable and accurate digital particle image velocimetry in MATLAB. J. Open Res. Softw. **2**(1), e30 (2014). https://doi.org/10.5334/jors.bl
13. Kalyulin, S.L., Shavrina, E.V., Modorskii, V.Y., Barkalov, K.A., Gergel, V.P.: Optimization of drop characteristics in a carrier cooled gas stream using ANSYS and globalizer software systems on the PNRPU high-performance cluster. In: Sokolinsky, L., Zymbler, M. (eds.) PCT 2017. CCIS, vol. 753, pp. 331–345. Springer, Cham (2017). https://doi.org/10.1007/978-3-319-67035-5_24
14. Kalyulin, S.L.: Numerical design of the rectifying lattices in a small-sized wind tunnel. In: Kalyulin, S.L., Modorskii, V.Ya., Paduchev, A.P. (ed.) AIP Conference Proceedings - 2016, vol. 1770, Article no. 030110 (2016)
15. Modorskii, V.Ya.: Research of aerohydrodynamic and aeroelastic processes on PNRPU HPC system. In: Modorskii, V.Ya., Shevelev, N.A. (ed.) AIP Conference Proceedings - 2016, vol. 1770, Article no. 020001 (2016)

The Conception, Requirements and Structure of the Integrated Computational Environment

V. P. Il'in[✉]

Institute of Computational Mathematics and Mathematical Geophysics, SBRAS,
Novosibirsk State University, Novosibirsk, Russia
ilin@sscc.ru

Abstract. The general conception, main requirements and functional architecture of the integrated computational environment (ICE) for the high-performance mathematical modeling of a wide class of the multi-physics processes and phenomena on the modern and future post-petaflops supercomputers are considered. The new generation problems to be solved are described by the multi-dimensional direct and inverse statements for the systems of nonlinear differential and/or integral equations, as well as by variational and discrete inequalities. The objective of the ICE is to support all the main technological stages of large-scale computational experiments and to provide a permanent and extendable mathematical innovation structure for wide groups of the users from various fields, based on the advanced software and on integration of the external products. The technical requirements and architecture solutions of the project proposed are discussed.

Keywords: Mathematical modeling
Integrated computational environment · High performance
Interdisciplinary direct inverse problems · Numerical methods
Program technologies

1 Introduction

In the epoch of fantastic growth of the postpetaflops supercomputers, the mathematical modeling of a wide class of multi-physics processes and phenomena has become the third way of the knowledge mining and deep learning, together with theoretical and natural experimental investigations. The huge computational and big data resources, the achievements of theoretical, applied and numerical mathematics, as well as the success in artificial intelligence and information technologies are a challenging chance for innovations in industry, nature, economy, social and human applications. The extremal machine experiments acquire a great importance both for basic research and practical issues.

The work is supported by the Russian Scientific Foundation, grant N 14-11-00485 P., and by Russian Foundation of Basic Reseach, grant N 16-29-15122 ofi-m.

V. Voevodin and S. Sobolev (Eds.): RuSCDays 2018, CCIS 965, pp. 653–665, 2019.
https://doi.org/10.1007/978-3-030-05807-4_56

As was pointed out in [1], in the near future the main problem for the computer science community will consist in the development of large volumes of the new generation software for upcoming supercomputers with millions and hundred millions of nodes, cores, special accelerators, and other units. The great opportunities for simulation are opened based on Data Centers, Grid Technologies and Cloud Computing. In such a situation the unique possibility to overcome the actual world crisis in programming is based on creating a modern software paradigm and on organizing a wide cooperation of various development groups, which should unify the academic expertise, software manufacturing and potential users. It is important to understand that an advanced algorithm can be proposed, investigated and tested efficiently by the operating mathematicians, but the robust implementation and code optimization for the method should be done by a professional programmer under the production requirements. And creative interaction between such specialists is the best way to a fast scientific innovation to industry.

Historically, the scientific simulation software has been evolved in the three main frameworks, that are either free accessible or commercial: applied program packages (ANSYS [2] or FEniCS [3], for example), algorithm libraries (see NETLIB [4] and MKL INTEL [5]) and special tools for particular operations (grid generation–NETGEN [6], visualization–PARAVIEW [7], etc.). The other trend in the past decades was based on creating an integrated computational environment (ICE), from the system standpoint, to support the general issues of mathematical modeling. We can mention such projects as OpenFOAM [8], DUNE (Distibuted Unified Numerical Environment [9]), MATLAB [10], INMOST [11], and Basic System of Modeling (BSM, [12]). Such a kind of software products is considered as instrumental media for automatic construction of the new computational methods and technologies which present the background for a flexible creation of the configurations on the principle of popular intellectual constructor LEGO, for particalur applications. The objective of such intellectual system is to provide such alternative properties as the high efficiency, performance, robustness and the universality of the resulting codes.

In general, the statements of mathematical modeling problems, interdisciplinary direct and inverse ones, are described by the nonlinear systems of differential, integral, and/or discrete equations, or by equivalent, in a sense, variational relations in the corresponding functional spaces. In the applications with real data, the unknown solutions are defined in the computation domains with complicated geometries of multi-scale boundaries and contrast material properties in different domains.

The numerical experiments in various applications are based on a large number of algorithms and software technologies, but the fact is, an enormous set of numerical procedures can be classified and divided into separate stages: geometrical and functional modeling that is responsible for the automatic construction of the models for tasks to be solved, generation of adapted grids, approximation of the original continuous problems by the corresponding discrete relations, solving the obtained algebraic tasks, using methods of optimization for solving the inverse problems, post-processing and visualization of the results obtained, control analysis and decision making, etc. Each of these steps has a rich extendable intellectual functionality and is presented by the library of algorithms. Such subsystems act reciprocally via coordinated data structures, which also provides the interaction with external software products.

These components form the mathematical kernel that can be efficiently used for the automatic construction of algorithms in the framework of applied program configurations for particular practical fields, which are oriented to a wide class of the end users involved in various problems. Let us make one more remark. The numerical problems can be considered based on a hierarchical set of models: coarse, middle and fine, for example. And computer experiments should be made succesively, with increasing the accuracy of approximations, to provide the adequacy of the obtained simulation results. Correspondingly, the computational tasks to be solved can be classified in the three groups: small, middle, and large problems. The conventional division can be chosen as follows: the run times of the first, the second and the third groups of tasks present similar values (seconds, minutes, hours, etc.) on the gigaflops, teraflops and petaflops computers, respectively. Of coarse, the strategies and tactics of the parallelization are different in these situations, and the main topic of our interest is solving the large (and very large) problems.

The considered Integrated Computational Environment (ICE) should provide a high productivity of applied programmers from different groups of developers and conceptually represent a community project. In general, the resulting codes present an open source, but the implemented algorithms and technologies can be used for creating the confidential applications or for the special computing services.

The conception of a global environment for the computer simulation presents a tremendous software project. The described modeling problems can be examined in various coordinates. From the productive branches point of view, the processes and phenomena to be simulated, can be referred to energetics, machinery, biology, geophysics, chemistry, ecology, etc. From another standpoint, the same problems can be considered by means of the applied statements from hydro-gasdynamics, elasticity, electromagnetism, thermo-mass-transfer, and/or in multi-physics formulations. The favorable circumstances are that all these multi-variant particular cases can be presented by a finite set of abstract mathematical relations. So, the mathematical modeling involves in a powerful chain various branches of the fundamental sciences and technologies: theoretical, applied, and numerical mathematics, informatics and programming, intensive data computing, artificial intelligence, and different application fields. But the mathematical modeling is impossible without high-technological software, and the mission of the ICE concludes in unification the people of new mass professionals on the general industrial platform.

The contribution of this paper can be presented as follows: motivation of creating the Integrated Computational Environment (ICE) for a high performance scientific software for supercomputer modeling; description of the architecture of the ICE functional kernel for the instrumental support of the main stages for solving a wide class of the interdisciplinary direct and inverse problems; technical requirements for constructing the ICE and technological principles of implementing big programs; expected results of realizing the ICE for valuable increasing productivity of the development and for the effective wide distribution of the created applied products.

The paper is organized as follows. Section 2 presents the formal statements of the problems to be solved. Section 3 includes a description of technological stages of mathematical modeling, the corresponding data structures, the general description of the numerical algorithms, the issues of the scalable parallelism, and the implementation features of the functional kernel for a proposed scientific software. In Sect. 4, the general technical requirements and some architecture solutions for the integrated computational environment are presented. In conclusion, we discuss several methodological aspects and the conception of the ICE, and, also, the future activity dealing with the proposed project in terms of the fundamental problems of the mathematical modeling.

2   Statements of Problems

In principle, the general mathematical statement of the numerical simulation is described by the interdisciplinary, or multi-physics, direct or inverse problem. In the abstract form, a direct task can be presented by the initial boundary value problem (IBVP) for the operator equation:

$$L\vec{u} = \vec{f}(\vec{x},\, t),\ \vec{x} \in \bar{\Omega},\ 0 < t \leqslant T < \infty, \tag{1}$$

where the unknown solution $\vec{u}$ is the vector function which is defined in the space-time computation domain $(\vec{x},\, t) \in \bar{\Omega} \times [0,\, T]$ and satisfies the boundary and initial conditions

$$l\vec{u} = \vec{g}(\vec{x},\, t),\ \vec{x} \in \Gamma = \Gamma_D \bigcup \Gamma_N,\ \vec{u}(\vec{x},\, 0) = \vec{u}^0(\vec{x}). \tag{2}$$

Here the operator l is responsible for the conditions on the boundary Γ. For example, L in (1) can be presented by the differential operator

$$L = A\frac{\partial}{\partial t} + \nabla B \nabla + C \nabla + D,\ \bar{\Omega} = \Omega \bigcup \Gamma \tag{3}$$

with the matrix coefficients A, B, C, D whose entries can depend on independent space-time variables $\vec{x}, t$ and, in nonlinear cases, on components of the solution $\vec{u}(\vec{x},\, t)$ to be sought for. The examples of the boundary conditions on different parts of the boundaries Γ_D, Γ_N can be described as follows:

$$u = g_D,\ x \in \Gamma_D;\ D_N u + A_N \nabla_n u = g_N,\ x \in \Gamma_N, \tag{4}$$

where D_N and A_N are, in general, some matrices. The computation domain Ω can be presented as a union of the subdomains Ω_j with the corresponding interior and external boundaries Γ_i^j, Γ_j^e:

$$\bar{\Omega} = \bigcup \bar{\Omega}_j,\ \Gamma = \Gamma^e \bigcup \Gamma^i,\ \Gamma^i = \bigcup \Gamma_{j,\,k}^i = \bigcup (\bar{\Omega}_j \bigcap \bar{\Omega}_k). \tag{5}$$

It is important to remark that the coefficients of original equations can have sufficiently contrast values in different subdomains. Moreover, different equations can be solved in different subdomains. It is usually supposed that the input data in the direct problem (1)–(5) provide the existence and uniqueness of the solutions to be sought for some functional spaces. However sometimes we need to make a computational experiment for a given problem even without theoretical knowledge about its properties. The descriptions of various mathematical statements for differential and integral equations, as well as the review of the modern literature, can be found in the recent books [13,14].

In real cases the ultimate goal of the research consists in solving not direct but inverse problems, which means, for example, the identification of the model parameters, the optimization of some processes, etc. The universal optimization approach to solving the inverse problems is formulated as minimization of the objective functional

$$\Phi_0(\vec{u}(\vec{x},\, t,\, \vec{p}_{opt})) = \min_{\vec{p}} \Phi_0(\vec{u}(\vec{x},\, t,\, \vec{p})), \tag{6}$$

which depends on the solution $\vec{u}$ and on some vector parameter $\vec{p}$ which is included in the input data of the direct problem. The constrained optimization is carried out under the linear conditions

$$p_k^{min} \leqslant p_k \leqslant p_k^{max}, \quad k = 1, \ldots, m_1, \tag{7}$$

and/or under the functional inequalities

$$\Phi_l \vec{u}(\vec{x},\, t,\, \vec{p})) \leqslant \delta_l, \quad l = 1, \ldots, m_2. \tag{8}$$

Formally, the direct problem can be considered as the state equation and can be written down as follows:

$$L\vec{u}(\vec{p}) = \vec{f}, \quad \vec{p} = \{p_k\} \in \mathcal{R}^m, \quad m = m_1 + m_2. \tag{9}$$

There are two main kinds of the optimization problems. The first one consists in the local minimization. This means that we look for a single minimum of the objective function in the vicinity of the initial guess $\vec{p}^0 = (p_1^0, \ldots, p_m^0)$. The second problem is more complicated and presents the global minimization, i. e. the search for all extremal points of $\Phi_0(\vec{p})$.

The numerical solution and high-performance implementation with scalable parallelism on the modern multi-processor computational systems (MPS) of the above mathematical problems present a tremendous set of the complicated computational schemes, (see the discussions in [15,16], for example). The efficient scalability of the parallel algorithms is attained in a weak or in a strong sense. The first point of view means that the computing run time is approximately the same if the degree of freedom (d.o.f.) of the mathematical problem to be solved and the number of computational hardware units simultaneously increase. The second case corresponds to the situation, when the run time for a big problem with a fixed d.o.f. decreases in proportion to enlarging the volume of computational equipment. Different strategies and tactics of parallelization are attained by the mapping of algorithmic structures onto computer architectures. Here, the quantitative characteristics are estimated by the speedup S_p and the parallelization efficiency E_p:

$$S_p = T_1/T_p, \ E_p = S_p/P, \ T_p = T_p^a + T_p^c, \tag{10}$$

where T_p is the computer time needed for a given task (or the algorithm) on p processors. The description carefully analyzes a real model of computations at the heterogeneous cluster systems with distributed and hierarchical shared memory presenting a sufficiently complicated problem and will not be the topic of our consideration. We just mention that the conventional parallel technologies are based on creating the MPI (Message Passing Interface) processes on the multi-thread computations, on the vectorization of the operations by AVX special tools, as well as on the intensive computing at the fast graphical accelerators (GPGPU or Intel Phi, for example). It is important to consider the coefficient

$$Q_p = N_p^a/(N_p^a + N_p^c), \tag{11}$$

where N_p^a and N_p^c are the number of arithmetic operations and the volume of communications, respectively, because the interprocessor data transfers not only decelerate the computational process, but are essentially energy consuming. So, the problem of incresing the coefficient Q_p in (11) is an unexpected mathematical consequence of the engineer requirements.

3   Technological Stages of Mathematical Modeling

In total, the large-scale computational experiments can be divided into the following main technological stages. In the framework of the ICE, all computational steps are implemented by the corresponding autonomous subsystems which are connected with each other via specified data structures. In general, a set of such subsystems forms the functional kernel of the integrated operating environment.

The first stage of a computing scenario consists in the geometrical and functional modeling. From the mathematical and technological standpoints, this means the automatic construction and modifications of the original problem. On the one hand, we need to describe the computation domain which includes different types of three dimensional (3-D), 2-D, 1-D and 0-D geometrical objects: the computation domain $\bar{\Omega} = \Omega \bigcup \Gamma$, the subdomains $\bar{\Omega}_j = \Omega_j \bigcup \Gamma_j$, the surface boundary segments (the faces Γ_j), the edges E_p (intersections of the surfaces)and the vertices, or points, P_q. In the dynamic and shape optimization problems, various operations can be defined on these objects: shifts, rotations, scaling, as well as topological and theoretical-set transformations. In the recent decades, many modern mathematical approaches have been developed : analytical metrics, differential calculus, isogeometric analysis (see [17,18], for example).

The second part of the mathematical model includes the description of functional objects, i. e. the equations to be solved, their coefficients, initial values, boundary conditions, objective functionals, required accuracy, and so on. Of course, these data should be connected with geometric information. The end results of the first stage of the computer simulation consist in the geometric and functional data structures GDS and FDS, respectively, which map the whole input information onto a set of integer and real arrays. We will call the considered subsystem of the ICE as VORONOI.

It is important to remark that there is a large world market of the computer-aided design products (CAD, CAE, CAM, PLM), which include the huge intellectual solutions in the geometric and visualization problems. The modern trend consists in the convergence, or integration, of CAD-systems and scientific codes for the mathematical modeling. Of course, the subsystem VORONOI should include convertors from the GDS and the FDS to the conventional information formats of the external computer design products, as well as to the popular visualization tools (PARAVIEW, for example).

Based on geometric and functional data structures we can realize the grid generation, which presents an important and resources-consuming stage of the numerical solution for the multi-dimensional problems. In the world software market, there are many available (free of charge) and expensive commercial codes, but the problem of constructing optimal or even "good" 3-D grids in the complicated computation domain is far enough from its final solution. Moreover, it is not easy to define the concept of an optimal or a good mesh, and there are many different quantitative characteristics of the grid quality.

Formally, the grid data structures are similar to the GDS and include the following objects: the grid computaion domain $\bar{\Omega}^h = \Omega^h \bigcup \Gamma^h$ with the boundary Γ^h, the grid subdomains $\bar{\Omega}_k^h = \Omega_k^h \bigcup \Gamma_k^h$ with the corresponding faces Γ_k^h, the grid edges E_p^h and the nodes P_q^h. The conventional approach to the discretization of the computation domain consists in constructing an adaptive grid. This means that the vertices P_q, the edges E_p and the surfaces Γ_k of the computation domain Ω should coincide with the corresponding grid nodes P_q^h, the grid edges E_p^h and the faces Γ_k^h. All these objects, in contrast to micro-objects, we call macro-objects: finite elements, or volumes, $\bar{T}_r^h = T_r^h \bigcup F_r^h$, with the element faces F_r^h, the mesh ribs R_s^h (intersection of the neighbouring faces F_r^h), and the mesh-points Q_l^h.

There are many kinds of the grids with different types of finite elements with various distributions of the meshsteps h, local refinement and multi-grid approaches included. Also, there are a lot of algorithms for the mesh generation, which are based on the frontal principles, on comformal or quasi-conformal transformations, on the differential geometry and various metrics. Here we consider the quasi-structured grids. This means that the grid structures can be different in different grid subdomains. For example, a grid can be non-structured, which means that the neighbouring mesh-points for every node can be defined by the enumeration only. In general, the quasi-structured grid consists of the grid subdomains which can have different types of the finite volumes, and grids in different subdomains can be constructed by the different algorithms.

In a sense, the grids considered present a two-level hyper-graph, at the macro- and micro-, or mesh, levels. In the technological sense, the final result of the grid generator should be the mesh data structure (MDS) with full mapping of the input geometric and functional data onto the micro (mesh) level. In particular, all inter-connections between grid objects should be strictly defined. Also, the affiliation of each finite volume T_j^h, grid face Γ_k^h into the corresponding subdomains Ω_k and the boundary surface segment Γ_p must be given.

The principles of constructing the library DELAUNAY are described in [19]. In fact, it presents the integrated instrumental media for the considered class of problems, based on original algorithms, as well as on re-using the external codes (there are popular free available mesh generators NETGEN, GMESH, TETGEN, for example). In the world "grid developer community", there are several popular grid formats, and the subsystem DELAUNAY should include the corresponding data convertors with the MDS. At this stage one of important operations includes the decomposition of the grid computation domain into grid subdomains, when the number of mesh points is too large, from 10^8 to 10^{10}, for example. In this case, it is natural to implement such procedures in parallel, and form the corresponding MDS for subdomains, distributed among different processors and MPI-processes.

There are many numerical approaches to construct the qualitive or optimal grids, but, in general, these mathematical questions are open yet. Also, we do not consider here resource-consuming problems of generating the dynamic meshes which are changed during the computational process.

The next stage of the mathematical modeling presents the approximation of the original IBVP, based on the MDS, the GDS and the FDS. In this case, the most popular approaches are finite difference, finite volume, finite element, and discontinuty Galerkin methods (FDM, FVM, FEM, and DGM, see [14, 20, 21], for example). The advanced theoretical and applied mathematical results have profound foundations and technologies for constructing and justification of high order accuracy numerical schemes for complicated IBVPs with real data. The implementation of such algorithms on the non-structured grids is not simple, and the tools for automatic construction of the scheme are very useful for such problems, (see [3], for example). It is important to remark that a very powerful approach here is based on the element technology with computing the local matrices and assembling the global matrix, which provide the "natural" parallelization and easy programming of the algorithms.

The concept of the integrated operating environment for the methods of approximation of the multi-dimensional IBVP is presented in the library CHEBYSHEV [22] based on original algorithms and re-using the external software. The end result of this subsystem consists in the algebraic data structure (ADS) which presents the original problem to be solved at the discrete level. To provide the necessary accuracy, the obtained systems of linear algebraic equations (SLAEs) should have very large dimensions (10^8 and more) and sparse matrices. To save such systems in the memory, the conventional compressed formats are used, Compressed Sparse Row (CSR), for example. Of course, for the large d.o.f., the distributed versions of the CSR are used, i.e. the matrix is divided into block rows, and each one is placed in the corresponding MPI-process.

The most resource-consuming stage of mathematical modeling is a numerical solution of large sparse SLAEs, because the volume of arithmetic operations grows nonlinearly, when the number of unknowns increases. Fortunately, the computational algebra is one of the most progressive parts of numerical mathematics, both in algorithmic and in technological senses, (see [23–26] and the literature, cited therein). In particular, there are many applied software packages and libraries with algebraic solvers which are free accessible. The main approaches to solve large sparse SLAEs are based on the preconditioned iterative methods in the Krylov subspaces. The scalable parallelism is provided in the framework of the two-level iterative domain decomposition methods (DDM) by means of hybrid programming with using MPI tools for the distributed memory of the heterogeneous cluster MPS, multi-thread computing of the shared memory of the multi-core CPUs, vectorization of operations by means of the AVX system, as well as fast computation on the graphic accelerators (GPGPU or Intel Phi).

The grid computation domain is decomposed into subdomains with parametrized overlapping and different interface conditions on the interior boundaries. Algebraically, the external iterative process presents the multi-preconditioned generalized minimal residual (GMRES) or a semi-conjugate residual (SCR) algorithm, based on the parallel block Schwartz-Jacobi method, coarse grid correction, deflation and/or augmentation procedures, and an advanced low-rank approximation of the original matrix. At each external iteration, solving the auxiliary SLAEs in subdomains is implemented synchronously by means of direct or iterative algorithms, with various preconditioning matrices.

The described parallel methods are realized in the framework of the library KRYLOV [25] which presents the integrated algebraic environment, based on the original algorithms and efficient re-using the external products. In particular, the robust matrix-vector operations and other algebraic tools from the library MKL Intel are applied in KRYLOV in a productive manner.

If the original continuous problem is nonlinear, then after its discretisation we will have the system of nonlinear algebraic equations (SNLAEs). In these cases, the quasi-linearization process is applied, based on Newtonian type of iterations, and at each of such steps the linear equations are solved.

The optimization methods for solving inverse problems are presented in the library KANTOROVICH. This broad class of algorithms includes solving the tasks of linear, integer and nonlinear programming, constrained or non-constrained, local or global minimization of functionals. In the recent decades, such a scientific direction has been developed dramatically fast. The advanced approaches include the conjugate equation approaches, modified Lagrangians and regularization, interior point methods, successive quadratic programming, trust region algorithms, kriging technologies, and surrogate optimization (see [27, 28] and the literature therein). The last mentioned approach corresponds to the situation, when the run time for computing one value of the objective function is too expensive and requires several hours or more. In such a case, the design, or planning, of numerical experiments is very important, as well as using special methods for approximation of the investigated functionals by means of the radial basic functions (RBF, see [28], for example).

So, in general, we have multi-step computational process, and on each stage it is necessary to repeat geometric and functional modeling, grid generation, approximation, and so on. When the numerical solution of the problem has been obtained, we have to understand and interpret the digital results which are presented usually by the values of the scalar or/and vector functions defined on the multi-dimensional non-structured grid or by the coefficients of the expansion of these solutions into the series of some basis functions. If we simulate some physical 3-D fields, for example, it is interesting to see the iso-surfaces, force lines, gradients, some extremal points and other characteristics, dynamic behavior included. Such postprocessing and vizualization approaches are very computation-consuming and may constitute the main run time of computer experiments. The usual way to overcome such issues consists in using the graphic accelerators.

We consider mainly the intensive computing stages of mathematical modeling, and such important questions as the general control of large-scale numerical experiments and decision making systems on the simulation results present the special topic for further research.

4   Technical Requirements for Integrated Environment

The unification of the program implementations of the above -described mathematical stages of modeling presents a huge software complex, which consists of the functional kernel of the basic system of modeling [12]. From the system standpoint, it is a method-oriented set of tools for solving a wide class of mathematical problems. In order to provide the resulting success of the project in question, the BSM should be organized as an open source program product with a long lifecycle and professional maintenance, with active participation of different groups of developers and end users. In accord with such a conception, the integrated computational environment should satisfy the following evident requirements.

- The flexible extendability of the content of the models and of the problems to be solved in the framework of the ICE, as well as a manifold of applicable advanced numerical methods and program technologies without limitations on the d.o.f. and on the number of computer nodes, cores and other units. The matter of fact is that theoretical, applied and numerical mathematics, as well as computational and informational technologies are permanently fast developed, and the new generation software should be currently modernized in order to provide the scientific innovations for practical applications.
- Adaptation to the evolution of computer architectures and platforms. Component object technologies (COM, see [29], for example) to provide the automatic concordance of the internal and external interfaces. Of course, such a feature of the applied software is obvious for a wide-spread distribution of the advanced computational technologies, and needs the automatic mapping of algorithm structures onto hardware equipment.

- Compatibility of the flexible and expandible data structures with the conventional formats to provide the efficient re-using the external products. Unification of the interfaces and possibilities for their convertation is a powerful device for the integration of intellectual software properties.
- High performance of the software developed, scalable parallelism based on the hybrid programming tools and code optimization on the heterogeneous multi-processor supercomputers with distributed and hierarchical shared memory on the cluster nodes and many-core CPUs (Central Processor Unit), respectively, as well as using vectorization of operations by means of the AVX system and graphic accelerators (GPGPU or Intel Phi, for example). A special attention should be given to constructing the computational schemes with minimal data communications.
- The polyglot interaction and consistency of various program components, as well as opening the working contacts for different groups of developers. The similar multi-language requirements are necessary to create friendly interfaces for the end users of different professional applications. As it was pointed out in [30], the general problem consists in a software language engineering for creating domain-specific languages. Figuratively speaking, the road-map of artificial intelligence for the computer science community is to move from paleo-informatics to neo-informatics.

In order to satisfy such strict and diverse requirements, the integrated program environment should have a valuable system support. Such intelligent components constitute the infrastructure whose goal is to provide the maintenance, information security, collective exploitation and further development of the ICE. The corresponding instruments should include the following main procedures:

- automatic verification and validation of the codes, as well as testing and comparative analysis of the efficiency of algorithms on representative sets of the model and real life examples;
- construction of the particular program configurations for a specific application by assembling the functional modules from the ICE, multi-version control included;
- preparing the documentation on the project components (manuals, user guides, special descriptions, etc.), different examples for demonstrations included;
- data structures control and transformations, manufacturing the problem-oriented languages and compilers, or convertors, providing the friendly interfaces for developers and users;in particular, such a problem includes the analytical transformations of the analytical expressions, as it is done in the popular systems MAPLE and MATHEMATICS; here we should recall the formula by Niclaus Wirt: "Program = algorithm + data structure".
- a permanently extended knowledge database, which includes the information about mathematical statements, supported by the ICE, employing computational methods and technologies, as well as industrial or natural problems

and other applications, which can be solved; ontology principles and cognitive instruments can help the users to recognize the statement and pecularities of the problem, to choose the corresponding algorithm and/or available code for computational experiments.

5   Conclusion

In the recent years, the widespread discussions on the digital economy and on the post-industrial society require the understanding of the role of mathematical and computer sciences in the product and social life management. Mathematical modeling is becoming a significant device in the business and human evolution. And the high-performance numerical simulation is impossible without integrated program environment which should be a permanent background for the collective development of the advanced algorithmic approaches, for support, maintenance, and promotion of the unique software support to provide the wide practical innovation.

An important consequence of the global mathematical modeling consists in the appearance of new mass professions: developers, distributers and modellers. The latter present the new generation of the end users, from theoretical physicists to designers of building or aircraft industries, whose main working instrument becomes a power computer with intellectual interface.

Creating an integrated computational environment presents an unprecedent large project, based on the unification of the various proffesions. In a sense, this is the way from individual or small group program productions to the industrial conception of the unified community activity in the scientific software business. It is well known that the expenses of the supercomputer hardware and software are comparable and big enough. So, the mathematical modeling is becoming the subject of the digital economy, and organizing solutions should take into acount this side of scientific innovations.

References

1. IESP. www.exascale.org/iesp
2. ANSYS. www.ansys.com
3. FEniCS. http://fenicsproject.org
4. Netlib. http://netlib.org
5. Intel R Mathematical Kernel Library. http://software.intel.com/en-us/intel-mkl
6. Schoberl, J.: Netgen-an advancing front 2D/3D-mesh generator based on abstract rules. Comput. Vizualization Sci. 1(1), 41–52 (1997)
7. PARAVIEW. www.paraview.org
8. OpenFOAM. http://www.openfoam.com
9. DUNE. http://www.dune-project.org
10. MATLAB. https://www.mathworks.com/products/matlab.html
11. INMOST: A toolkit for distributed mathematical modelling. www.inmost.org

12. Il'in, V.P., Gladkih, V.S.: Basic system of modelling (BSM): the conception, architecture and methodology (in Russian). In: Proceedings of International Conference on Modern Problems of Mathematical Modelling, Image Processing and Parallel Computing, MPMMIP&PC-2017, pp. 151–158. DSTU Publ., Rostov-Don (2017)
13. Brugnano, L., Iavernano, F.: Line Integral Methods for Conservative Problems. CRC Press/Taylor & Francis Group, New York (2015)
14. Il'in V.P.: Mathematical Modelling, Part I: Continuous and Discrete Models (in Russian). SBRAS Publ, Novosibirsk (2017)
15. Il'in, V.P.: Fundamental issues of mathematical modeling. Her. Russ. Acad. Sci. **86**(2), 118–126 (2016)
16. Il'in, V.P.: On the parallel strategies in mathematical modeling. In: Sokolinsky, L., Zymbler, M. (eds.) Parallel Computational Technologies. PCT 2017. CCIS, vol. 753, pp. 73–85. Springer, Heidelberg (2017). https://doi.org/10.1007/978-3-319-67035-5_6
17. Delfour, M., Zolesio, J.-P.: Shape and Geometries. Metrics, Analysis Differential Calculus, and Optimization. SIAM Publications, Philadelphia (2011)
18. Cottrell, J., Hughes, T., Bazilevs, Y.: Isogeometric Analysis. Towards Integration of CAD and FEA. Wiley, Singapore (2009)
19. Il'in, V.P.: DELAUNAY: technological environment for grid generation (in Russian). Sib. J. Ind. Math. **16**, 83–97 (2013)
20. Brenner, S.C., Scott, L.R.: The Mathematical Theory of Finite Element Methods. Springer, New York (2008). https://doi.org/10.1007/978-0-387-75934-0
21. Riviere, B.: Discontinuous Galerkin Methods for Solving Elliptic and Parabolic Equations. Theory and Implementation. SIAM, Philadelphia (2008)
22. Butyugin, D.S., Il'in, V.P.: CHEBYSHEV: the principles of automatical constructions of algorithms for grid approximations of initial-boundary value problems (in Russian). In: Proceedings of International Conference, PCT-2014, pp. 42–50. SUSU Publ., Chelyabinsk (2014)
23. Saad, Y.: Iterative Methods for Sparse Linear Systems. PWS Publ., New York (2002)
24. Dolean, V., Jolivet, P., Nataf, F.: An Introduction to Domain Decomposition Methods: Algorithms, Theory and Parallel Implementation. SIAM, Philadelphia (2015)
25. Butyugin, D.S., Gurieva, Y.L., Il'in, V.P., Perevozkin, D.V., Petukhov, A.V.: Functionality and algebraic solvers technologies in Krylov library (in Russian). Vestnik YuUrGU. Ser. Comput. Math. Inform. **2**(3), 92–105 (2013)
26. Il'in, V.P.: Problems of parallel solution of large systems of linear algebraic equations. J. Math. Sci. **216**(6), 795–804 (2016)
27. Il'in, V.P.: On the numerical solution of the direct and inverse electromagnetic problems in geoprospecting (in Russian). Sib. J. Num. Math. **6**(4), 381–394 (2003)
28. Forrester, A., Sobester, A., Keane, A.: Engineering Design via Surrogate Modeling. A Practical Guide. Wiley, New York (2008)
29. Maloney, J.: Distributed COM Application Development Using Visual C++. Prentice Hall, New York (1999)
30. Kleppe, A.: Software Language Engineering: Creating Domain-Specific Language Using Metamodels. Addison-Wesley, New York (2008)

The Elbrus-4C Based Node as Part of Heterogeneous Cluster for Oil and Gas Processing Researches

Ekaterina Tyutlyaeva[1]($\boxtimes$), Igor Odintsov[1], Alexander Moskovsky[1], Sergey Konyukhov[1], Alexander Kalyakin[2], and Murad I. Neiman-zade[2]

[1] ZAO RSC Technologies, Moscow, Russia
{xgl,igor_odintsov,moskov,s.konyuhov}@rsc-tech.ru
[2] MCST/INEUM, Moscow, Russia
{kalyakin,muradnz}@mcst.ru

Abstract. This paper briefly examines the advantages and disadvantages of Elbrus architectures as building blocks for Seismic Processing cluster system. The configuration of a heterogeneous clustered system build for research Oil and Gas Company is examined in more detail. In this system, processing nodes with different architecture (x86, GPU and e2k) are integrated in a single computing cluster through a high performance global networking topologies. Heterogeneous cluster with Elbrus node provides a good opportunity for software cross-architectural migration. To demonstrate the potential of Elbrus nodes usage, the multispectral data analysis application has been optimized for e2k architecture. Paper includes performance results and scalability analysis for implemented module using e2k and x86 nodes. It is anticipated that the heterogeneous cluster with Elbrus node will form an integral part of the preparation process of the domestic supercomputer under development, based on the Elbrus processors. The basic software stack in Seismic Processing will be naturally emerged on the use of Elbrus node as part of the heterogeneous cluster.

Keywords: VLIW · Elbrus in HPC · e2k · Heterogeneous cluster Elbrus-4C Based Node

1 Introduction

HPC facilities have become an important part of scientific researches in all areas of technology. Oil and gas companies use HPC for seismic processing [1], reservoir modeling process, interpretation and visualization. Developing the oil and gas fields can cost hundreds of millions of dollars, especially in geologically, environmentally and operationally challenging areas. Complex innovative techniques for analyzing and visualization of huge seismic and other geophysical datasets using HPC facilities could minimize the costs and risks associated with exploration and drilling in challenging areas.

V. Voevodin and S. Sobolev (Eds.): RuSCDays 2018, CCIS 965, pp. 666–674, 2019.
https://doi.org/10.1007/978-3-030-05807-4_57

Most of legacy algorithms used today were designed with x86 and GPUs in mind, so they can take advantage of the compute power that comes from using target hardware. At the same time, however, domestic hardware platform such as e2k is a reliable environment for processing nationally-important data and developing new competitive approaches.

It is well known that cross-architectural software migration is critical issue and requires a lot of time and efforts. Including the Elbrus node in heterogeneous cluster allows to use more specific allocation of work between nodes with different architectures:

- Complex legacy software libraries build for x86 architectures could be used on x86 nodes
- Complex visualization libraries parallelized with CUDA threads could be used on NVidia node
- New modules algorithms and modules could be designed and developed for Elbrus

The noteworthy feature here is decreasing qualified time for making determined efforts to adaptation of existing software for e2k architecture. Engineers could use Elbrus node only for developing new modules.

In addition, users sometimes runs what are called pre-processing and post-processing applications on the master/login node. The pre-processing applications create the input data for jobs that will run on the cluster (a job is a generic term to mean application launched by a job scheduler). The post-processing applications take the output of the cluster job and perform an analysis on the output data. Some of these operations could be also done on the Elbrus node.

Heterogeneous clustered system is the useful solution for the development and testing of advanced libraries for seismic data processing.

The main contribution of this paper is the example of application module ported to Elbrus node and x86 node, and comparative analysis of the optimizations and performance results.

The rest of the paper is structured as follows. Section 2 provides a detailed specification of the used heterogeneous system. Section 3 reviews basic e2k microarchitecture features. After that, Sect. 4 describes the testing application that processes multispectral image using Fourier transform. Section 5 describes the implementation and architecture-specific optimizations of these algorithms for e2k and x86 nodes. Finally, Sect. 5.4 provides with measured performance results. Afterwards, Sect. 6 concludes the paper with discussion of the obtained results.

2 Heterogeneous Clustered System

The heterogeneous system developed for research Oil and Gas Company consists of three types of nodes:

- The **x86 nodes** are 2-socket, based on Intel Xeon E5-2697A v4 processors with DDR4-2400 DRAM Memory, 8 × 16 GB,
- The **e2k node** is 4-socket, based on Elbrus-4C processors with DDR3-1600 DRAM memory, 12 × 4 GB,
- The **GPU node** is based on 2-socket Intel Xeon E5-2697A v4 processors with NVIDIA GRID Tesla M10 co-processor.

Current installation includes 6 × **x86 nodes**, 1 × **e2k node** and 1 × **GPU node** connected using high-performance Mellanox Infiniband FDR interconnect and service Ethernet interconnect. Simplified scheme of the installation is shown on the Fig. 1.

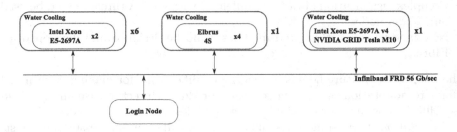

Fig. 1. Heterogeneous HPC system scheme

Programming for heterogeneous cluster is a difficult task. Among others, performance and fault tolerance are probably the most important and challenging issues of heterogeneous parallel and distributed programming.

By the way, this heterogeneous cluster can be divided on three logical parts. The partition based on **x86 nodes** (Intel Xeon E5-2697A v4) have the property of homogeneity. According to the Dongarra and Lastovetsky [2], there is only one way for such a system to be homogeneous: all processors in the system have to be identical and interconnected via a homogeneous communication network, that is, a network providing communication links of the same latency and bandwidth between any pair of processors. But this definition is not complete. One more important restriction has to be satisfied: the system has to be dedicated, that is, at any time it can execute only one application providing all its resources to this application. Theoretical peak performance of this partition is about 10 TFlops, that is reasonable number for basic computational seismology tasks.

Visualization tasks could be performed using **GPU node** (NVIDIA GRID Tesla M10 co-processor).

Moreover, unified liquid cooling provides increased density and energy efficiency of the overall heterogeneous system.

3   Elbrus Node Characteristics

Microprocessor Elbrus architecture e2k is classified as VLIW (Very Large Instruction Word) microprocessor architecture. According to the VLIW principles [3], compiler forms a single instruction (Very large Instruction Word) using a multiple independent operations grouped together for parallel execution. Thus, the microprocessor design allows to take advantage of as much Instruction Level Parallelism (ILP) as possible. Optimizing compiler exempts hardware from their parallelization responsibilities and provide performance improvements using ILP techniques.

Along with effective ILP usage, e2k architecture includes the basis for implementing another levels of parallelism, such as vector parallelism, thread parallelism in shared memory, task parallelism in large distributed memory cluster.

Another basic feature of e2k architecture is effective binary compatibility with x86 architecture.

Elbrus software development strategy focused on growth of compatibility. A roadmap for Elbrus-based systems includes large list of planned software adaptation in different areas. For example, the study [4] published in 2018 includes analysis of the wide range of computing, graphical, network and disk-related software in the field of compatibility with Elbrus.

Theoretically, e2k design allows achieving better performance with reduced power consumption rate. As noted above, compilers for e2k processors are required to package multiple instructions into Very Large Instruction Word, which are later treated by CPU pipeline as single instruction. Conversely, superscalar machines uses hardware for analysis of dependences among instructions at runtime. Using the compiler for the exploiting and scheduling of instructions allows to get rid of energy demanding mechanisms of dynamic instruction scheduling for Elbrus architecture [5].

An important implication of this microarchitecture design is that the way in which the application implemented and further compiler works can have a large effect on code execution performance.

4   Application

The testing application processes multispectral images archive in HDF5 format. All input images are roughly the same size, about 60 Mb of double precision data.

This module repeatedly performs Direct Fourier Transforms and Overdetermined real linear systems solving routines in moving window for each picture. The simplified flow graph of the module is shown on the Fig. 2.

Input and output data details are presented on the Table 1.

The Fourier transform is fundamental to seismic data analysis [6]. It applied to almost all stages of processing.

- The analysis of a seismic trace into its sinusoidal components is achieved by the forward Fourier transform.

Fig. 2. Simplified application flow graph

Table 1. Data specifications

Stage	Data format	Size, MB	Data type	Dimensions H × W
Input	HDF5	60	Double	3072 × 4064
Output	ENVI	96	Double	3072 × 4064

- The synthesis of a seismic trace from the individual sinusoidal components is achieved by the inverse Fourier transform.
- The two-dimensional (2-D) Fourier transform is a way to decompose a seismic wavefield, such as a common-shot gather, into its plane-wave components, each with a certain frequency propagating at a certain angle from the vertical.

For example, the Fourier transform is used to convert the signal into the frequency domain.

Thus, the application example given covers the most frequently used domain in seismic image processing – analysis in moving window using Fourier transform.

Loginov and Ishin [7] proposed the optimized FFT algorithm for Elbrus processor including special optimization for 32-bit floating point data type.

5   Implementation Details

The testing application has been implemented for Intel x86 nodes, based on Intel Xeon E5-2697A v4 processors and for e2k node, based on Elbrus-4C processors.

Basic source code has been implemented using C++ programming language to reach the maximal level of compiler-assisted optimization. The basic C++ version of the code implemented in straight-line manner; all repeatedly performed loops have single entry and single, not data-dependent exit. Compliance with these requirements and conditions is effective for both further implementations.

Furthermore, the two architecture-dependent branches has been implemented.

In summary, virtually all multi-core architectures now requires basic optimization in order to obtain a satisfactory result.

The best solution for mathematical calculation for both platforms is architecture dependent mathematical libraries: EML [9] (Elbrus Mathematical Library) for Elbrus and MKL [10] (Math Kernel Library) for Intel.

Each architecture requires specific code optimization.

Table 2. Architecture dependent implementations

Feature	Intel implementation	Elbrus implementation
MPI compiler	Intel® MPI library version 2018	MPICH-2
C++ compiler	Intel® Compilers 2018	lcc-1.23
Math library	Intel® MKL library (2018 Studio)	EML (Elbrus mathematical library) + lapack-3.8.0 (Elbrus optimized)
I/O library	The HDF5-1.8.20 Technology suite	The HDF5-1.8.20 Technology suite
Specific compilation keys	-ipo -xCORE-AVX2	-mcpu=elbrus-4c

5.1 Optimization for Intel Node

For Intel implementation the auto-vectorization [8] features has been used to improve the vectorizer performance:

- Avoiding, if it possible, dependencies between loop iterations
- Aligning data to 32 byte boundaries
- Using pragmas to give a hint to the compiler to asserts that data within the following loop is aligned, there aren't any data dependencies etc.

It is worst noting, however, that stronger optimizations, such as using Intel intrinsic instructions (C style functions that provide access to many Intel instruction, including AVX2, available on Intel Xeon E5-2697A v4 processors) haven't been used. So there are still rooms for optimization to fully exploit available processor vector capabilities.

5.2 Optimization for Elbrus Node

Optimizations are very important for high performance VLIW, especially for microprocessors with static scheduling. Optimizing compiler [11] have taken up a lot of the work in the program parallelization, including utilization of instruction level parallelism, short vector instruction parallelism, and thread level parallelism. Optimizations for the studied module for Elbrus node included several steps:

- moving some invariant memory allocations/deallocations outside the loops
- replacing frequent invokes of "hypot" function in the innermost loops with vectorized "vhypot" version invoke outside these loops
- minor rearrange of certain local data in order to avoid indirect addressing in hot loops
- collapsing some simple loop nests to reduce overhead of loop software pipelining
- using linearized multidimentional arrays

- building project in -fwhole (-flto) that allowed to inline cross-module invokes of functions; better inline [12] is known to have increased positive effect on VLIW architecture processors
- modifying minor subroutine that contained indirect memory writes in inner-most loop; replacing it with loop nest with indirect memory reads, to make loop software pipelining more effective.

As a result, program execution profile shows >85% of cycles spent in library functions, top of them being fftw/eml functions.

5.3   Data Parallelization

Archive processing structure of the module described in Sect. 4 enables each picture to be computed independently. This enables the algorithm to be implemented in the data parallel way which is very efficient for high performance computing. Each MPI rank reads, processes and writes results for its own images independently, so there are minimum communications between processes.

For Intel x86 architectures, the OpenMP threads also used to exploit the Intel Hyper-Threading Technology [13] that enables effective multiple threads execution on each core. OpenMP threads used on Sharpness Index computation step for processing independent data in different positions of moving window.

5.4   Performance Testing

The performance testing has been conducted using **x86 node** (based on 2 × Intel Xeon E5-2697A v4 Processors) and **Elbrus node** (based on 4 × Elbrus-4C Processors).

Testing procedure included a series of 10 executions per each combination of input data set and input feature sets.

Appropriate preparatory steps has been done prior to each execution, especially removing the results of previous computations and cleaning up the caches and swap.

The Fig. 3 presents the median execution times for all executions. It's important to outline that number of used cores is equal to the number of processed pictures, so the ideal time result should be straight horizontal line. However, as the number of used cores and processed images is increased, performance characteristics deviate from linearity. It can be concluded that the collective work with memory and I/O slightly increases execution time with increased number of used cores and processed pictures.

According to the obtained results, the one Elbus-4C core requires about 93 sec to process one image; one Intel Xeon E5-2697A v4 core takes about 32 sec to process the same image. However, different levels of optimization should be taken into account.

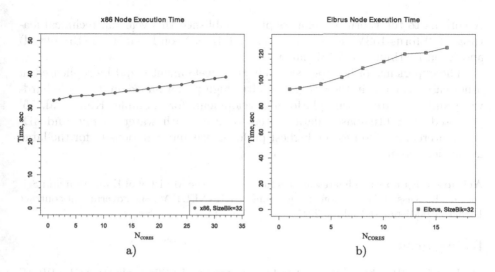

Fig. 3. Median execution times. (a) x86 testbed (b) Elbrus testbed

6   Conclusion

The important process of import substitution industrialization is actual now, in which domestic industries should be favored over imports. Heterogeneous cluster with Elbrus node provides an good opportunity for natural process of an increase the level of qualified technical expertise for e2k architecture usage and creating software stack in seismic processing area. Implementation Elbrus node as a part of heterogeneous cluster also simplifies software development and cross-architectural migration. Software engineer can easily check the correctness and compare performance results. It is hoped that usage of such architectures will identify possibilities for mutual reinforcement between Elbrus nodes/x86 nodes and Nvidia GPU.

As an example, the data parallel algorithm of multispectral images processing using Fourier transform is successfully implemented for both architectures.

While the optimization complexity is different, the experience gained could be replicated in other optimizations. Some important processing routines could be further reused.

The obtained performance results are in line with expectations, based on technical characteristics of studied nodes. Heterogeneous structure of the cluster allows to easily checking correctness and comparing performance characteristic.

The potential for Elbrus microprocessors for domestic HPC is tremendous; our current implementation has used previous generation of processors. However, the Elbrus processors must be further studied to fully exploit it potential. Elbrus-8S, new generation of Elbrus microprocessors is now available. The Elbrus-8S [14] includes eight cores, and is manufactured using a 28 nm process. Elbrus-8S equipped with L3 cache, shared across cores; It has 1300 MHz frequency versus 750 MHz for Elbrus-4C and supports up to 4 DDR3-1600 channels.

According to the Elbrus developers plans, published this year [5], technical features of "Elbrus-16SV" is expected to match Intel Xeon E7-4850 v4 (Broadwell) and Xeon Platinum 8153 (Skylake).

The experience of heterogeneous cluster development could be replicated in other cases, such as in the systems of ultra-high performance on domestic hardware and software Elbrus platform development, for example. Kim et al. [15] proposed an Elbrus-based data centers solution with water cooling and 4D-torus interconnect to provide highest performance and high density for the large domestic system.

Acknowledgments. This research is funded from the Ministry of Education and Science of the Russian Federation (Programme "SKIF-NEDRA", the government contract 14.964.11.0001 dated 17 June 2015).

References

1. Liao, T.: HPC challenges in oil and gas upstream scientific applications. In: PPME Workshop, Portland, OR, USA, pp. 1–30 (2013)
2. Dongarra, J., Lastovetsky, A.: An Overview of Heterogeneous High Performance and Grid Computing. Engineering the Grid: Status and Perspective. American Scientific Publishers, February 2006
3. Kim, A., Perekatov, V., Ermakov S.: Microprocessors and Computing Systems of the Elbrus Family. Piter, Saint-Petersburg, p. 272 (2013). (in Russian) ISBN 978-5-459-01697-0
4. Molchanov, I.A., Bychkov, I.N.: Study of Elbrus computing platform in the field of compatibility including software adaptation. Voprosy radioelektroniki **2**, 14–22 (2018). (in Russian)
5. Kim, A.K., Perekatov, V.I., Feldman, V.M.: On the way to Russian exasistemes: plans of the Elbrus hardware-software platform developers on creation of an exaflops performance supercomputer. Voprosy radioelektroniki **2**, 6–13 (2018). (in Russian)
6. Yilmaz, O.: Seismic Data Analysis: Processing, Inversion, and Interpretation of Seismic Data. Society of Exploration Geophysics, vol. 1 (2001)
7. Loginov, V.E., Ishin, P.A.: 32-bit floating-point fast fourier transform optimization for Elbrus processor. Voprosy radioelektroniki, IVT Series, vol. 3 (2012). (in Russian)
8. Intel Corporation: A Guide to Vectorization with Intel C++ Compilers (2012)
9. EML Mathematical Library Home Page. http://www.mcst.ru/vysokoproizvoditelnye_biblioteki
10. Intel Math Kernel Library Documentation. https://software.intel.com/en-us/mkl/documentation
11. Volkonskiy, V., et al.: Program parallelization methods implemented in optimizing compiler. Voprosy radioelektroniki **4**(3), 63–88 (2012)
12. Ermolitckii, A., Neiman-Zade, M., Chetverina, O., Markin, A., Volkonskii, V.: Aggressive inlining for VLIW. Proc. Inst. Syst. Program. **27**(6), 189–198 (2015)
13. Marr, D., et al.: Hyper-threading technology architecture and microarchitecture. Intel Technol. J. **6**(1), 4–15 (2002)
14. Microprocessor Elbrus-8S. http://mcst.ru/elbrus-8c. (in Russian)
15. Kim, A.K., Perekatov, V.I., Feldman, V.M.: Data centers based on Elbrus servers. Voprosy radioelektroniki **3**, 6–12 (2017). (in Russian)

The Multi-level Adaptive Approach for Efficient Execution of Multi-scale Distributed Applications with Dynamic Workload

Denis Nasonov[1]([✉]), Nikolay Butakov[1], Michael Melnik[1], Alexandr Visheratin[1], Alexey Linev[2], Pavel Shvets[3], Sergey Sobolev[4], and Ksenia Mukhina[1]

[1] ITMO University, Saint-Petersburg, Russia
denis.nasonov@gmail.com, alipoov.nb@gmail.com,
mihail.melnik.ifmo@gmail.com, mukhinaks@gmail.com
[2] Lobachevsky State University of Nizhni Novgorod, Nizhny Novgorod, Russia
alipoov.nb@gmail.com
[3] Research Computing Center of Moscow State University, Moscow, Russia
shvets.pavel.srcc@gmail.com
[4] Moscow State University, Moscow, Russia
sergeys@parallel.ru

Abstract. Today advanced research is based on complex simulations which require a lot of computational resources that usually are organized in a very complicated way from technical part of the view. It means that a scientist from physics, biology or even sociology should struggle with all technical issues on the way of building distributed multi-scale application supported by a stack of specific technologies on high-performance clusters. As the result, created applications have partly implemented logic and are extremely inefficient in execution. In this paper, we present an approach which takes away the user from the necessity to care about an efficient resolving of imbalance of computations being performed in different processes and on different scales of his application. The efficient balance of internal workload in distributed and multi-scale applications may be achieved by introducing: a special multi-level model; a contract (or domain-specific language) to formulate the application in terms of this model; and a scheduler which operates on top of that model. The multi-level model consists of computing routines, computational resources and executed processes, determines a mapping between them and serves as a mean to evaluate the resulting performance of the whole application and its individual parts. The contract corresponds to unification interface of application integration in the proposed framework while the scheduling algorithm optimizes the execution process taking into consideration the main computational environment aspects.

Keywords: Multi-scale applications · Distributed computing · HPC Optimization · Multi-agent modeling · MPI

© Springer Nature Switzerland AG 2019
V. Voevodin and S. Sobolev (Eds.): RuSCDays 2018, CCIS 965, pp. 675–686, 2019.
https://doi.org/10.1007/978-3-030-05807-4_58

1   Introduction

The growth in the performance of computing systems (CS) for scientific computing and the increasing complexity of computer simulation models is one of the
leading trends in the development of information technologies. Currently, the
implementation of the exascale computing by 2020 is being discussed. On the
other hand, this performance increase is mainly associated with the complexity
of the CS architecture. Efficient use of this type of integrated CS in modelling
is a complex engineering task. In addition to that, following challenges emerge:

1. the need to use multi-scale and multi-physical models, various modelling
   methods (grid and drains) in the solution of one applied problem;
2. the use of specialized computation resources (for example, graphics processors);
3. the problem of balanced spatial decomposition due to the complexity of the
   geometry of the domain of definition;
4. dynamic change in the complexity of different parts of the problem: with
   spatial decomposition due to the change in the geometry of the system or
   due to the emergence of areas of high computational complexity (for example,
   clustering of agents in multi-agent systems, slow convergence regions for grid
   methods;
5. the development of cloud computing technologies, in which the CS architecture is hidden from the user, and the need to meet their requirements.

This leads to a significant slowdown in the development of new simulation models, the use of obsolete technologies for parallel problem solving, the low efficiency of resource allocation and, ultimately, the inability to master the exascale
computing system. In these circumstances, the traditional approach in which
the responsibility for the parallel performance of software implementation of the
model is assigned to the developer of the model is not efficient. Within the framework of this project, the approach to the separation of the development process
of a simulation model from solving the problem of the efficient use of computing
resources is given, which should be solved automatically. At the same time, efficient allocation of resources is impossible without knowledge of the internal logic
of models; therefore, a tool should be proposed for its formal description in the
resource allocation system and for providing the system with access to the task
decomposition; dynamic allocation of resources should be ensured. The development of this approach for automatic resource allocation will significantly reduce
the complexity of implementing computationally complex simulation models,
allowing the developer to concentrate entirely on modelling methods, increase
the efficiency of using computing resources. The authors of the proposal are
sure that without the solution of these problems and the development of the
proposed approach, the CS's exascale performance will not be accessible from a
practical point of view. Additionally, simplifying the implementation of models
stimulates interest in more complex modelling methods, for example, using multiscale approaches. In this paper, we present an approach for efficient execution
of multi-scale distributed applications with the dynamic overflowing workload.

This approach includes dynamic (variable) graph model for allocating the structure of a multiscale distributed application as well as the unified framework for constructing this graph model and applying the algorithm of efficient management of executed tasks on dedicated computational resources.

2 Related Works

Since multiscale applications have a graph structure, the task of their planning is generally considered as the task of planning composite applications. To date, there is a wide variety of algorithms and methods for solving this problem. In [1], the authors proposed a coevolutionary genetic algorithm (CGA) for planning scientific composite applications that have execution deadlines. Experimental studies have shown the high efficiency of the developed method for optimizing the value of the resources used. However, this algorithm does not allow to optimize further the execution time, which reduces the possibility of its use. The heuristic IPEFT algorithm was proposed in [2]. The results of the experiments showed a fairly low execution time for small applications, as well as better results compared to the predecessors - HEFT and PEFT. Despite this, the authors' experience with heuristic algorithms [3,4] shows that their ability to find optimal solutions is very limited. An additional direction of research in the field of planning composite applications is the study of the ways to ensure energy efficiency of tasks. So in [5], the authors presented the heuristic EONS algorithm for planning composite computations taking into account the energy consumption of computing resources. But, since in most modern projects energy efficiency is not a key factor for optimization, it must be considered in conjunction with the implementation time and the cost of using resources. In [6] the methods of planning composite applications in the conditions of time constraints and the budget for the computing resources rent in the cloud environment are studied. The ideas of the planning algorithms proposed in the article are based on concrete, well-structured templates of the composite application. This approach is not always efficient because there are strict requirements for the structure of the composite application, which in general will rarely be met. There are works devoted to the development of systems for organizing the implementation and design of composite applications. For example, [7] presents a system for modelling designing and integrating composite applications into a computing environment. Such systems are aimed at simplifying the process of creating and executing applications. [8] presents a platform for organizing the planning process based on the flow of tasks and the dependencies between them. The main idea of the architecture of the platform is to present all the computational tasks in one composite application, which expands due to the tasks entering the platform. In [9], a detailed analysis of the types of multiscale applications, as well as possible ways of their implementation in a distributed environment, is given. The authors identified three types of applications: related, scalable and prioritized applications. For each type, application examples were selected and manual optimization was performed. As shown by the results of experimental studies, the use of knowledge about the nature of applications can significantly accelerate their implementation. However, the authors did not offer automated optimization paths, which is a critical drawback of the work. The cloud platform for analyzing and visualizing multiscale data is presented in [10]. The platform is based on the integration of tools and services for data analysis with services for data storage and composite applications execution. The main objectives of the platform development were to provide a convenient tool for modelling the processing of multiscale data, their implementation with automatic scaling of the computing environment and visualization of the analysis results (for example, climate data). In [11], the authors analyze the capabilities of existing composite application management platforms for efficient work with extreme-scale composite applications. Under extreme-scale composite applications, the authors mean applications that require advanced high-performance computing technologies for highly accurate predictive models based on the analysis of large volumes of multi-scale data. The authors of [12] presented a modified version of the framework for distributed execution of multi-scale MUSCLE-HPC applications. Its main advantage over the previous version of MUSCLE-2 is the more efficient distribution of tasks in high-performance clusters by analyzing the relationships between applications and the location of closely related tasks within a single cluster. Despite the presented advantages, in MUSCLE-HPC, as well as in previous versions, there are no mechanisms for optimizing the applications themselves during the execution.

3   Multilevel Approach

This section describes basic parts of architecture of the proposed approach taking into consideration problem statement aspects.

3.1   Problem Statement

As mentioned before, the main problem lays between complexity of model distributed blocks interconnection and infrastructure appropriate mapping. Consider that application has execution environment that is organized as a computational grid $G <V, E>$, where $V = \{v_j\}$ corresponds to vertexes and $E = \{e_{j1,j2}\}$ represents edges. Upon the environment, grid computational elements $W = \{w_l\}$ form actual load of model logic. This elements may move in the environment from one nodes to another each computational iteration. V nodes are divided between computational resources $R = \{r_m\}$. Let consider $S = \{w_l^j\}$ as current distribution of actual load and define reorganization function

$$f(S_1, S_2) = \Sigma \frac{c_{w_l^j}}{e_{j,j'}} \cdot \delta_{j'}^j,$$

$$\delta_{j'}^j = \begin{cases} 1, & \text{if } j \text{ and } j' \text{ are on different resources} \\ 0, & \text{otherwise;} \end{cases}, \forall j, j' = 1, \dots, J_m,$$

where $c_{w_i^j}$ - is amount of metadata needed to transfer load from v_j to $v_{j'}$; while $e_{j1,j2}$ corresponds to network channel throughput.

$$T(S) = max_m(\Sigma_j^{J_m} \frac{w}{p_m} + \Sigma_j \Sigma_{j'} \frac{w_j}{e_{j,j'}} \cdot \tau_{j'}^j)$$

$$\tau_{j'}^j = \begin{cases} 1, & \text{if there is actual load moving from } j \text{ to } j' \\ 0, & \text{otherwise;} \end{cases}, \forall j, j' = 1, \ldots, J_m,$$

Having these equations we can define the change environment criteria for the optimization algorithm:

$$T(S_{prev}) \cdot \theta > f(S_{prev}, S_{new}) + T(S_{new}) \cdot \theta.$$

Here θ is statistically depending value that corresponds to the rate of changing of actual load through the elements of the environment.

3.2   The Approach

Our approach of efficient execution of distributed applications consists of 3 main parts: a three-layer model of performance; a scheduling algorithm that uses the model to estimate performance of an application in different configurations; the partition-based model of computations that allows user to provide its own routines for computations. The basic concept is presented in Fig. 1.

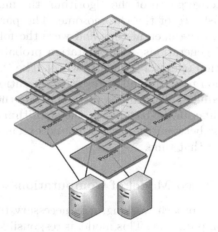

Fig. 1. Three-tier design of architecture

To solve the problem of scaling using the proposed model we developed a genetic-based algorithm to balance the workload on individual processes and thus improve overall performance. The objective of the algorithm is to reconfigure the computing resources based on the current load profiles generated by the computing processes on the resources. Concerning the task of modelling the behaviour of the population in specified urbanized areas (see City-Simulator application in Case Study section), the developed algorithm adaptively manages the allocation of modelling areas to physical computing processes based on the current specific load of these processes.

At the core of the model, there are few logical entities - "process", "agent", as well as the matrix of process contiguity with each other. Based on the fact that individual processes are responsible for modelling individual geographic areas, and moving agents across the city imply moving only between adjacent areas (in the simplest case), the adjacency matrix defines the subsequent area on the agent path at each time point. In this case, if modelling processes on nodes of computing clusters are placed arbitrarily, the absence of excessive network interaction is not guaranteed.

3.3   The Scheduling Algorithm

To perform scheduling and rescheduling of partitions among all processes in the distributed application, we implemented a special version of the genetic algorithm (GA) as a part of our approach. The genetic algorithm was chosen because its generality and ability to search through the whole solution space.

Our version of GA performs a search of optimal mapping between partitions and processes.

The mutation operator is implemented as random choosing of a host process for a random partition. As the crossover operator, the single-point crossover was chosen. To speed up convergence of the algorithm, the mutation may happen more than once per instance of the chromosome. The parameters of this GA stayed the same for all experimental runs and were the following: size of population - 100, count of generations - 300, mutation probability - 0.7, crossover probability - 0.3, selection operator - roulette wheel. The three-layer model was used as a fitness function to estimate the resulting execution time of modelling per iteration. The scheduling algorithm is used as before the start of the execution as during the runtime. In the latter case, the algorithm is being periodically run according to a shift between the last estimated execution time of iteration and the current value of that time.

3.4   The Partition-Based Model of Computations

To make the proposed approach working, it is necessary to introduce a model to describe required computations. This model is responsible for: (a) integrating of user-written computing functions into the framework based on the proposed approach and (b) obtaining required for the three-layer model monitoring and profiling data.

To achieve the stated goals, we propose the following model that can be easily expressed on any high-level programming language.

Let introduce main entities: $a_t^i = <s_t, x_t>$, $p_t^k = <S_p, \{a_t^i\}>$, $e_{t+1}^r = <\{a_{t+1}^j\}, l_{t+1}^a>$, where a_t^i is an agent, p_t^k is a partition, e_t^r is an envelope, all of them in moment t. An agent represents data the computing should happen on (x_t - location of the agent in modeling space, s_t - the rest of data associated with the agent). A partition represents an area in modeling space which has a set of associated agents with it and some static information required for the execution S_p. A partition serves as a unit of scaling. An envelope is a unit of data transferring between two connected partitions.

The user has to supply two functions m_u and g_u to be used for computing new state across all partitions and all agents on each node according to the following transformations.

$$p_{t+1}^k = m_u(p_t^k, \{e_t^r\})$$

$$l_{t+1}^a = g_u(x_{t+1})$$

$$<p_{t+1}^k, \{e_{t+1}^r\}> = f(m, g, p_t^k, \{e_t^r\})$$

where m_u - compute functions that implement the logic of modelling on the set of agents belonging to the area and multiple agent inflows, g_u - a function that determines a partition the agent should reside to according to its new coordinates x_{t+1} in the modelling space. Function f uses these user-defined functions, partitions data and incoming flows of the agent to calculate new states and perform all service functionality including optimization of data exchange between individual processes of the distributed applications.

4   Case Study: A Large Scale Multiagent Simulation of Urban Traffic

4.1   Urban Traffic Simulation

To carry out experimental studies of the performance and scalability of the proposed approach, a test case was developed for the field of multiscale modelling of urban mobility of the population in urbanized areas. It was called City-Simulator. In the developed example, we simulate the daily dynamics of people moving around the city, taking into account the specifics of these movements - from the sleeping areas to the city centre in the morning and from the centre to the sleeping areas in the evening. The city is divided into areas of modelling, the number of areas corresponds to the number of allocated computing resources allocated to the application. Each area and agents inside it are modelled in interrelation with other areas because, during the simulation, agents move between regions. This application assumes the dynamism of the computational load on the processes associated with the movement of agents by location, each of which is processed at the designated node for it.

Data on the mobility of the population on a city scale are simulated through the software package sim_city_package, after which the output array is transferred to the sim_district_package software package, responsible for multiagent modeling of individuals' behavior on the scales of individual regions (in units), the interaction between which does not involve large transmitted data and, as a consequence, is not a determining factor when planning placement on the nodes. The microscale modelling of the behaviour of agents within a set of small areas of the district (counted in hundreds, thousands and tens of thousands) is carried out by multiple copies of the package sim_object_package. The amount of communication between the instances of this package is most significant across the entire application, as a result of which it is the determining one when planning the placement of processes on the nodes, as provided by the extreme scaling (ES) pattern [13]. An example of dividing a city into regions (the city is divided into eight regions) and the general scheme of interaction taking into account intra- and inter-district communication are presented on Fig. 2).

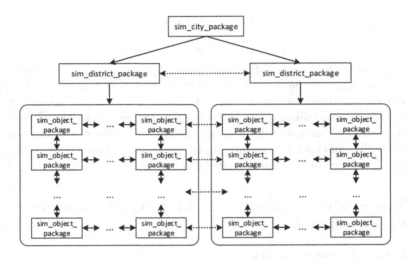

Fig. 2. Example of computations structure in distributed application City-Simulator

To configure the parametric performance models, the City-Simulator application is profiled depending on the size of the simulated area and the number of simulated agents. The experiments were conducted on the resources of the computing cluster of the University ITMO. The number of simulated agents ranged from 100 to 500 thousand in increments of 100 thousand. The width of the modelling area varied from 5 to 15 km in increments of 2.5 km. The dependence of the total calculation time and the time of data transfer between the modelling blocks on the number of processes on which the sim_object_package packet is placed, on one iteration is shown Fig. 3.

Figure 3 shows that the most intensive decrease in computation time is observed in the range of 4 to 32 processes. Also, there is a significant increase in the time spent on communication overhead, as the number of computing processes increases. The initial reduction in communication time in each of the scenarios (2–8 processes) is due to the placement of processes on one node from the calculation of 8 processes to one physical computing node, which does not cause network interaction between processes.

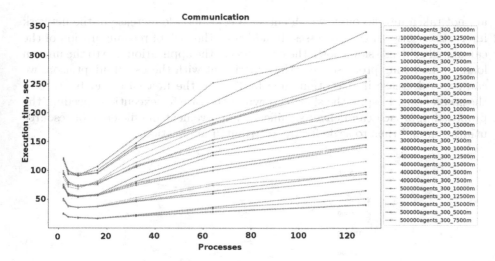

Fig. 3. City-Simulator execution time dependence on number of processes

The obtained results of performance profiling clearly show that the distributed application build with static partitioning of modelled areas doesn't scale well and eventually can't fully utilize available resources. With the growth of the simulation scale and preserving the same or close patterns of agents dynamic - e.g. increases in the number of agents and/or a number of areas - the problem with efficient resources utilization is getting worse. The latter makes the user wait more and thus slows down research speed. Using the proposed approach is possible to ease the pain caused by this problem by improving scaling capabilities of the application exploiting patterns in its dynamic.

For this test case, the application execution scheduling component can provide planning optimization through the methods described in The approach section - accounting for agent movements between modelling areas when scheduling tasks and load balancing of compute nodes by reconfiguring the grid.

4.2   Experimental Results

The experiments were carried out - with and without application of adaptations of the computational template to the application. Activity modelling was carried out for 6 million agents to produce simulations on a city scale both regarding the size of the calculation area and regarding the size of the population of agents. The application was planned for 100, 200, 500 and 1000 cores. The results of the experiments are shown in Fig. 4. As can be seen from the graph, for experiments without the use of adaptations, the total simulation time with changing in the number of processes from 100 to 200 decreases (by 9%), but with a further increase in the number of computing cores, the total execution time increases (by 7% at 500 cores and 33% for 1000 cores). Such an effect, first of all, is due to the fact that when planning tasks, the adjacency matrices of the movements of agents are not taken into account, which entails a considerable increase in the network interaction between the processes. In addition, the lack of reconfiguration of the calculated grid also slows down the execution of the application due to the uneven load on modelling processes. For the experiment with the use of adaptation, we can see a decrease in execution time by 32% in the first case and by 18% in the second one. Despite the slowing down of the rate for execution speedup, the proposed approach can deliver significant improvement in efficiency of resource utilization and thus to scale for the distributed application.

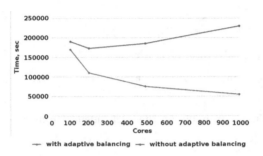

Fig. 4. Performance comparison of classical static partitioning approach and the proposed approach in case of City-Simulator distributed application

The results of the experiments in the scenario using the modelling template adaptations demonstrate a stable decrease in execution time with an increase in the number of computational cores. However, it is worth noting that the change in execution time slows down as the number of cores increases. This is due to the fact that even with the use of adaptations of the computational template with a large number of modelling processes, the contribution of network interaction inevitably increases, which is not capable of completely levelling even the InfiniBand data transmission channel.

To further analyze the results obtained, changes in the computational load of modelling processes over time were investigated. Since the computational load is expressed in terms of the number of simulated agents in a certain process, for the simulation experiment with 1000 computational cores, data was collected on the number of agents at each time point and the dynamics of the change in the number of agents for scenarios with and without reconfiguration of the grid was analyzed. Figures 5 and 6 show the graphs of three processes in which the characteristic effects of the optimization performed by the PC EO are visible

For the process, the graphs of which are represented in Fig. 5, by reducing the cell size it is possible to reduce the computational load on average from 2500 to 1000 agents per iteration. In the process, the graphs of which are shown in Fig. 6, the situation was the reverse - at a certain stage of modeling the load dropped and fell almost to zero. By distributing the load from other processes it was possible to increase the workload of the process, thereby reducing the probability of downtime.

Fig. 5. Workload decreasing for a processes in City-Simulator

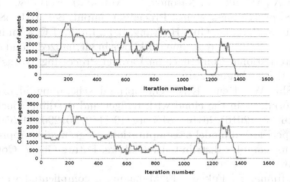

Fig. 6. Workload increasing for a processes in City-Simulator

Analysis of simulation results in individual processes showed that the reconfiguration of the computational grid allows relatively equalizing the computational load on the processes that perform agent modeling of the movement of people in the urban environment. This factor, coupled with the contiguity matrices in the planning process, allows you to significantly optimize the execution of the application and ensure its scalability on a large number of compute nodes.

5   Conclusion

The experimental case study demonstrated an improvement in execution time with the growth of exploited cores for: up to 32% in case of 200 cores, up to 18% in case of 500 cores thus showing that the proposed approach is able to deliver significant improvement in efficiency of scaling for distributed applications. This research is financially supported by The Russian Science Foundation, Agreement #14-11-00823.

References

1. Liu, L., Zhang, M., Buyya, R., Fan, Q.: Deadline-constrained coevolutionary genetic algorithm for scientific workflow scheduling in cloud computing. Concurr. Comput. Pract. Exp. **29**(5) (2017)
2. Zhou, N., Qi, D., Wang, X., Zheng, Z., Lin, W.: A list scheduling algorithm for heterogeneous systems based on a critical node cost table and pessimistic cost table. Concurr. Comput. Pract. Exp. **29**(5) (2017)
3. Visheratin, A.A., Melnik, M., Nasonov, D.: Dynamic resources configuration for coevolutionary scheduling of scientific workflows in cloud environment. In: Pérez García, H., Alfonso-Cendón, J., Sánchez González, L., Quintián, H., Corchado, E. (eds.) SOCO/CISIS/ICEUTE -2017. AISC, vol. 649, pp. 13–23. Springer, Cham (2018). https://doi.org/10.1007/978-3-319-67180-2_2
4. Visheratin, A.A., Melnik, M., Nasonov, D.: Automatic workflow scheduling tuning for distributed processing systems. Procedia Comput. Sci. **101**, 388–397 (2016)
5. Chen, H., Zhu, X., Qiu, D., Guo, H., Yang, L.T., Lu, P.: EONS: minimizing energy consumption for executing real-time workflows in virtualized cloud data centers. In: 2016 45th International Conference on Parallel Processing Workshops (ICPPW), pp. 385–392. IEEE, August 2016
6. Wang, Y., Shi, W., Kent, K.B.: On optimal scheduling algorithms for well-structured workflows in the cloud with budget and deadline constraints. Parallel Proc. Lett. **26**(02) (2016). https://doi.org/10.1142/S0129626416500092
7. Balis, B.: HyperFlow: a model of computation, programming approach and enactment engine for complex distributed workflows. Future Gener. Comput. Syst. **55**, 147–162 (2016)
8. Zenmyo, T., Iijima, S., Fukuda, I.: Managing a complicated workflow based on dataflow-based workflow scheduler. In: 2016 IEEE International Conference on Big Data (Big Data), pp. 1658–1663. IEEE, December 2016
9. Borgdorff, J., et al.: Performance of distributed multiscale simulations. Phil. Trans. R. Soc. A **372**(2021) (2014). https://doi.org/10.1098/rsta.2013.0407
10. Lu, S., et al.: A framework for cloud-based large-scale data analytics and visualization: case study on multiscale climate data. In: 2011 IEEE Third International Conference on Cloud Computing Technology and Science (CloudCom), pp. 618–622. IEEE, November 2011
11. da Silva, R.F., Filgueira, R., Pietri, I., Jiang, M., Sakellariou, R., Deelman, E.: A characterization of workflow management systems for extreme-scale applications. Future Gener. Comput. Syst. **75**, 228–238 (2017)
12. Belgacem, M.B., Chopard, B.: MUSCLE-HPC: a new high performance API to couple multiscale parallel applications. Future Gener. Comput. Syst. **67**, 72–82 (2017)
13. Alowayyed, S., Groen, D., Coveney, P.V., Hoekstra, A.G.: Multiscale computing in the exascale era. J. Comput. Sci. **22**, 15–25 (2017)

Using Resources of Supercomputing Centers with Everest Platform

Sergey Smirnov, Oleg Sukhoroslov$^{(\boxtimes)}$, and Vladimir Voloshinov

Institute for Information Transmission Problems of the Russian Academy of Sciences,
Moscow, Russia
sasmir@gmail.com, {sukhoroslov,vv_voloshinov}@iitp.ru

Abstract. High-performance computing plays an increasingly important role in modern science and technology. However, the lack of convenient interfaces and automation tools greatly complicates the widespread use of HPC resources among scientists. The paper presents an approach to solving these problems relying on Everest, a web-based distributed computing platform. The platform enables convenient access to HPC resources by means of domain-specific computational web services, development and execution of many-task applications, and pooling of multiple resources for running distributed computations. The paper describes the improvements that have been made to the platform based on the experience of integration with resources of supercomputing centers. The use of HPC resources via Everest is demonstrated on the example of loosely coupled many-task application for solving global optimization problems.

Keywords: High-performance computing · Clusters
Many-task applications · Distributed computing · Web services
Global optimization

1 Introduction

Computational methods are now widely used for solving complex scientific and engineering problems. These methods often require a large amount of computations and the use of high-performance computing (HPC) resources. Such resources can be provided by clusters operated on-premises, supercomputing centers, distributed computing infrastructures or clouds. Supercomputing centers represent an important source of HPC resources. In contrast to on-premises clusters, supercomputers have significantly more resources. However these resources are usually shared among many users and projects, which can lead to queues and high wait times. In contrast to distributed computing infrastructures, supercomputers support efficient execution of tightly coupled parallel applications. In comparison to public clouds, supercomputers use specialized hardware and are generally free for scientific projects.

The wide use of HPC resources among scientists is complicated due to a number of problems. There is a lack of convenient interfaces for running computations on such resources. Typically this procedure involves copying of input

© Springer Nature Switzerland AG 2019
V. Voevodin and S. Sobolev (Eds.): RuSCDays 2018, CCIS 965, pp. 687–698, 2019.
https://doi.org/10.1007/978-3-030-05807-4_59

data, compilation of a program, preparing a job script, submission of a job via a batch system, checking the job state and collecting the job results. All these steps should be performed via a command line environment. Such environments and batch systems are unfamiliar and too low-level for many researchers, requiring a considerable effort to master and use them instead of focusing on a problem being solved. This can demotivate researchers with less technical background. At the same time, there is a lack of tools for automation of routine activities that can be useful even for advanced users. Such tools are necessary for execution of single jobs, many-task applications such as parameter sweeps or complex workflows involving multiple jobs with dependencies. Also there is a need for tools supporting reliable execution of long-running computations or massive computations spanning multiple resources. In order to be efficient, such tools should take into account various characteristics and policies of HPC resources.

In this paper, we present an approach for solving the aforementioned problems using Everest [1,18], a web-based distributed computing platform. A distinguishing feature of Everest is the ability to serve multiple distinct groups of users and projects by implementing the Platform as a Service (PaaS) cloud computing model. The platform is not tied to a single computing resource and allows the users to attach their resources and bind them to the applications hosted by the platform. These features make it possible to use the publicly available platform instance without having to install it on-premises. The use of Everest enables convenient access to HPC resources by means of domain-specific computational web services. It also allows one to seamlessly combine multiple resources of different types for running massive distributed computations.

The paper is organized as follows. Section 2 discusses the related work and compares it with the presented approach. Section 3 provides an overview of the Everest platform and its relevant features. Section 4 describes the integration of HPC resources with Everest including improvements of platform components. Section 5 demonstrates the use of Everest for running a loosely coupled many-task application for solving global optimization problems on HPC resources. Section 6 concludes the paper and discusses the future work.

2   Related Work

The use of web technologies for building convenient interfaces to HPC systems has been exploited since the emergence of the World Wide Web. For example [7] describes a user-friendly web interface for reconstruction of tomography data in a clinical environment backed by the Cray T3D massively parallel computer. In [6] authors describe a Web/Java based framework for remote submission of parallel applications on HPC clusters.

The emergence of grid computing [8] and the web portal technologies enabled the development of grid portals [21] and science gateways [5] facilitating the access to distributed computing resources and offering additional services such as collaborative capabilities. One of the early examples is the NPACI HotPage portal [20] consisted of a set of services for accessing the grid and individual

HPC resources via a web browser. In [14] authors describe Gateway, a computational web portal system implementing a set of generic core services such as user management, security, job submission, job monitoring, file transfer, etc., that were used to build web interfaces for running applications from different domains on HPC systems. A number of similar frameworks for development of computational portals were proposed [26], which facilitated the development of many specialized portals in different domains of computational science.

While providing convenient web interfaces to HPC systems or specific computational packages, the first generation systems had the following drawbacks. First, it was difficult to combine multiple packages, run multi-step workflows and other many-task applications [15] typically found in science. This has lead to development of web based environments supporting the description and execution of user-defined workflows [3,4,10]. Second, the extension of portal functionality, e.g. publishing new applications, was difficult and limited to administrators only. While more recent systems [4,12] introduced standard tools for application development, the deployment is still limited to privileged users. Third, classic portals were tightly coupled with particular HPC systems or a limited set of resources configured by administrators. Current science gateways support different types of resources ranging from supercomputers to clouds, but still not allow users to attach and access their own resources and accounts via the gateway.

In this paper we use Everest [18], a web-based distributed computing platform that addresses the mentioned drawbacks.

3   Everest Platform

Everest [1,18] is a web-based distributed computing platform. It provides users with tools to publish and share computing applications as web services. The platform also manages the execution of applications on external computing resources attached by the users. In contrast to traditional distributed computing platforms, Everest implements the PaaS model by providing its functionality via remote interfaces. A single instance of the platform can be accessed by many users in order to create, run and share applications with each other.

Everest supports the development and execution of applications following a common model. An application has a number of *inputs* that constitute a valid request to the application and a number of *outputs* that constitute a result of computation corresponding to some request. Upon each request Everest creates a new *job* consisting of one or more *tasks* generated by the application from the job inputs. The tasks are executed by the platform on computing resources specified by a user. The dependencies between tasks are currently managed internally by the applications. The results of completed tasks are passed to the application and are used to produce the job outputs or new tasks if needed. The job is completed when all its tasks are completed. The described model is generic enough to support a wide range of computing applications.

Everest users can publish arbitrary applications with a command-line interface via the platform's web interface. Upon the invocation of the application the platform produces a single task corresponding to a specific command run. It is possible to dynamically add new tasks or invoke other applications from a running application via the Everest API. This allows users to create and publish complex many-task applications with dependencies between tasks, such as workflows. Everest also includes a general-purpose application for running a large number of independent parametrized tasks such as parameter sweeps [23]. Each application is automatically published as a RESTful web service. This enables programmatic access to applications, integration with third-party tools and composition of applications into workflows. The platform also implements a web interface for running the applications via a web browser. The application owner can manage the list of users allowed to run the application.

Instead of using a dedicated computing infrastructure, Everest performs the execution of applications on external resources attached by users. The platform implements integration with standalone machines and clusters through a developed *agent* [16]. The agent runs on a resource and acts as a mediator between it and Everest enabling the platform to submit and manage tasks on the resource. The agent performs routine actions related to staging of input files, submitting a task, monitoring a task state and collecting the task results. The platform also supports integration with grid infrastructures [16], desktop grids [19] and clouds [24]. Everest users can flexibly bind the attached resources to applications. In particular, a user can specify multiple resources, possibly of different type, for running an application [16]. In this case the platform performs dynamic scheduling of application tasks across the specified resource pool.

4   Integration of HPC Resources with Everest

In this section, we describe the improvements that have been made in Everest to support the efficient use of HPC resources via the platform. These developments are based on the experience of integration with several supercomputing centers in Russia including the Data Processing Center of NRC Kurchatov Institute (NRC KI) and the Supercomputer Simulation Laboratory of South Ural State University (SSL SUSU).

4.1   Agent Improvements

The integration of computing resource with Everest is achieved by means of the Everest agent [16]. The main functions of the agent are execution of tasks on the resource and provision of information about characteristics and current status of the resource. The agent supports interaction with various types of resource managers through the extensible adapter mechanism. The adapter receives generic resource requests and translates these requests into commands specific to a particular resource manager. There are three adapters have been developed for HPC clusters that support integration with TORQUE, Slurm and Sun Grid Engine.

In case of a cluster, the agent should by deployed by a user on the submission node. Since the agent has minimal dependencies and system requirements, it can be quickly deployed by the users without special skills and superuser privileges. We have not found any serious problems when launching the agent on the used supercomputing centers. Basically, the agent was able to operate in environments with old Python versions and hard network restrictions.

However, while setting up the agent on the HPC4 supercomputer at NRC KI we have found that the agent can not submit more than 64 concurrent jobs. It was due to the limit on the maximum number of jobs per user imposed by the administrators via the Slurm manager. Thus the basic Slurm adapter running a single job per Everest task was not able to fully utilize the user quota on this system. In case of loosely coupled many-task applications with single-core tasks, such as described in Sect. 5, it was possible to utilize only 64 cores at once.

Two possible approaches to solving this problem were considered that relied on advanced features of Slurm: job arrays and complex many-task job scripts. Job arrays is a mechanism for managing collections of similar jobs in Slurm. It allows to submit and manage such jobs faster. However every job in the array is still treated by Slurm as a single job. So it does not allow to overcome the imposed jobs per user limit. Another option is to create complex Slurm jobs consisting of multiple tasks started with srun command inside a job script submitted with sbatch. Using this feature a set of Everest tasks can be submitted as a single Slurm job requesting resources needed to run all these tasks. In the job script a new Slurm task is created for every Everest task by calling the srun command.

The second approach was implemented in the new advanced Slurm adapter. The tasks received by the adapter from the agent are grouped by their Everest job ID and are accumulated. If a specified number of tasks belonging to the same job has accumulated, or no new tasks have arrived within the specified time, a complex job containing accumulated tasks is submitted to Slurm.

Below is an example of a job script generated by the adapter for two tasks:

```
#!/bin/bash
#SBATCH -D TASK_DIR/job1-2
#SBATCH -e srstderr
#SBATCH -o srstdout
#SBATCH -p hpc4-3d
#SBATCH -c 6
#SBATCH -n 2
( srun --exclusive -D TASK_DIR/job1-2 --output TASK_DIR/job1-2/stdout \
    --error TASK_DIR/job1-2/stderr -n1 -N1 bash TASK_DIR/job1-2/jobfile
  _ECODE=$?
  echo errorcode 0 $(ps -o etime= -p "$$") $_ECODE
  exit $_ECODE ) &
( srun --exclusive -D TASK_DIR/job1-1 --output TASK_DIR/job1-1/stdout \
    --error TASK_DIR/job1-1/stderr -n1 -N1 bash TASK_DIR/job1-1/jobfile
  _ECODE=$?
  echo errorcode 1 $(ps -o etime= -p "$$") $_ECODE
  exit $_ECODE ) &
wait
```

The -`exclusive` switch provides a uniform distribution of task processes to cluster nodes that have been allocated by Slurm for the job. To track the completion of a task, immediately after the `srun` call is completed, its return code is printed along with the sequence number of the task. The `srun` call is grouped with these operations using the parentheses operator, so that the group of commands is executed in a separate bash subshell. The ampersand operator allows tasks to be started concurrently while the `wait` call blocks further execution until all `srun` calls are completed.

Checking the state of tasks running inside a complex Slurm job is carried out in two stages. First, the adapter obtains the status of the job via the `scontrol` command. If the job is running, the state of individual tasks is checked by reading the `srstdout` file, where the job script prints the return codes of completed tasks. Thus the agent can process completed tasks before the whole job has completed.

While Everest supports the cancellation of jobs and tasks, in the described implementation it is difficult to stop the individual tasks. Therefore the cancellation is done for the entire Slurm job containing the tasks being canceled.

There are a number of other improvements that has been made in the agent and the Slurm adapter, such as automatic tasks directory cleanup, handling of job submission failures and propagation of environment variables.

The work on integrating the Tornado supercomputer [11] at SSL SUSU is ongoing right now. Because this system is quite overloaded, the calls to Slurm command line utilities can fail with a timeout. This is a new challenge because some rework of the agent's internals is needed to support such timeouts.

4.2   Supporting Advanced Resource Requirements

A significant limitation of Everest related to running parallel applications on HPC resources was the lack of explicit support for resource requirements. By default, all Everest tasks submitted via the agent on a resource were run as single-core jobs in a predefined queue. It was possible to change the number of CPU cores per task, but only on the level of the agent and for all tasks submitted via it. The execution of a parallel application with specific requirements for the number of nodes, processes per node and system resources required the development of an auxiliary script, which complicated publishing of such applications in Everest.

The aforementioned limitation was removed by enabling the Everest users to specify the application resource requirements. A number of new parameters were added to the configuration of Everest application such as the number of CPU cores per node (default is 1), the maximum number of CPU cores per node, the number of nodes (default is 1), the maximum number of nodes, the total amount of memory, the amount of memory per core and the wall clock time limit.

The application developer can specify the values of these parameters and optionally allow the users to override some of them when running the application. This approach is flexible enough to support the various use cases. The default values correspond to the previously supported case of a single-core task. Since for many applications the resource requirements depend on the input data, it is desirable to allow the users to tune these parameters according to their problems. A possible future improvement would be to allow application developers to provide performance models that can be used to automatically compute runtime settings for a given problem. In cases where the problems have a fixed size or resources are limited, it makes sense to use only the predefined runtime settings.

The platform and the agent were modified to take into account the described resource requirements during the task scheduling and execution. The requirements are embedded in each task and are examined during the resource selection. In order to do this, the agent was modified to provide the additional information about the resource, such as number of nodes, cores and memory per node, available queues with their limits and states. Everest scheduler matches this information with the task requirements in order to select a suitable resource and a queue for the task execution. After the resource is selected, the final resource requirements of the task are computed and passed along with the task to the remote agent. The agent was modified to take into account these requirements during the submission of the task to a local resource manager. Currently, this functionality is implemented in the advanced Slurm adapter, which translates the task requirements to the corresponding Slurm options.

5   Experimental Evaluation Using Global Optimization

In this section, an application use case is presented which demonstrates the use of HPC resources via Everest for running loosely coupled many-task applications.

For many years the branch-and-bound (B&B) algorithm for discrete and global optimization has been considered as a challenge for parallel computing [2]. Most of known implementations for distributed computing environments follow the so called fine-grained approach and low-level parallelization. Here the parallel search tree traversal is coordinated by a master process delegating searching in the subtrees to a number of slave processes. To achieve a good load balancing the master performs dynamic redistribution of the work among slaves which implies intensive two-way data exchanges between the processes. The master also keeps track of a global incumbent value. This approach has drawbacks: the complexity of implementation; the need for low level communication between B&B solvers; usually it requires a homogeneous computing environment, e.g. cluster.

DDBNB (Domain Decomposition B&B) is an Everest application which implements an alternative approach based on a coarse-grained parallelism and preliminary decomposition of a feasible domain of the problem by some heuristic rules. Subproblems obtained via decomposition are solved by a pool of standalone B&B solvers running in parallel. The incumbent values found by each solver are intercepted and delivered to other solvers in order to speed up the traversal of B&B search tree. Current implementation of DDBNB[1] is based on SCIP and CBC open source solvers. Incumbent values exchange is based on a special messaging service implemented in Everest. More details can be found in [17,25]. DDBNB is an example of loosely coupled many-task applications [15].

---

[1] https://github.com/distcomp/ddbnb.

Initially DDBNB has been tested by solving a well known Travelling Salesman Problem formulated as a mixed-integer linear problem [17, 25]. The experiments conducted in a heterogeneous distributed environment with standalone servers and virtual machines (up to 28 CPU cores in total) have demonstrated promising results. The described integration with HPC resources allowed to try DDBNB for solving hard global optimization problems requiring more computing resources.

Two well known problems in combinatorial geometry, which may be treated as global optimization problems, have been considered: the so called Tammes[2] and Thomson[3] problems. Both problems in their original form concern the arrangement of N points $(x_i, i=1{:}N)$ on a unit sphere in $\mathbb{R}^3$:

(Tammes) to maximize the minimal distance between any pair of points x_i, x_j

$$y \to \max_{y \in \mathbb{R}, x_i \in \mathbb{R}^3, \; i=1{:}N} \quad s.t.:$$
$$y \leqslant \|x_i - x_j\|, (1 \leqslant i < j \leqslant N), \|x_i\|=1 \; (i=1{:}N);$$

(Thomson) to minimize electrostatic Coulomb energy of unit charges put in x_i:

$$\sum_{1 \leqslant i < j \leqslant N} \frac{1}{\|x_i - x_j\|} \to \min_{x_i \in \mathbb{R}^3, \; i=1{:}N} \quad s.t.: \; \|x_i\|=1 \; (i=1{:}N).$$

Both problems are the well known challenges for computer science. It is hard to proof the global optimality of a given set of points $\{x_i, \; i=1{:}N\}$ even for small values of N, e.g. Tammes problem has computer-assisted proof for $N \leqslant 14$ [13]. This proof substantially relies on the problem's specifics and is based on an enumeration of millions of the so called irreducible contact graphs.

DDBNB application enables one to try another, rather general approach. It is based on the implementation of the B&B algorithm for mathematical programming problems with polynomial in constraints and objective function in the SCIP solver [9, 22]. Both problems may be reduced to the proper form:

Tammes - minimize the maximum of scalar products $x_i^\mathsf{T} x_j$ for any $i, j{:}1 \leqslant i < j \leqslant N$:

$$z \to \min_{x_i, z} \quad s.t.:$$
$$x_i^\mathsf{T} x_j = \sum_{d=1:3} x_{i,d} x_{j,d} \leqslant z, (1 \leqslant i < j \leqslant N); \tag{1}$$
$$\|x_i\|^2 = \sum_{d=1:3} (x_{i,d})^2 = 1, z \in \mathbb{R}, x_i \in \mathbb{R}^3, \; i=1{:}N.$$

Thomson - via the auxiliary variables for the values of Coulomb energy:

$$\sum_{i,j=1, i<j}^{N} z_{ij} \to \min_{x_i, z_{ij}} \quad s.t.:$$
$$z_{ij}^2 (x_i - x_j)^\mathsf{T} (x_i - x_j) = 1 \; (1 \leqslant i < j \leqslant N); \tag{2}$$
$$x_i^\mathsf{T} x_i = 1 \; (i=1{:}N), \; x_i \in \mathbb{R}^3, z_{ij} \in \mathbb{R}.$$

---

[2] http://neilsloane.com/packings/.
[3] http://www-wales.ch.cam.ac.uk/~wales/CCD/Thomson/table.html.

It is worth to mention that in addition to major constraints, some auxiliary constraints have been added to reduce the number of redundant solutions that might be obtained by 3D rotations, mirror transformations and renumbering of points. For example, the first point is always fixed as $x_1 = (0, 0, 1)$.

The first attempts to solve the Tammes problem by a standalone single-threaded SCIP process were failed even for $N = 8$. The solver running with default settings quickly occupies a lot of memory (29 GB in 45 min) to store the search tree data. By that time, the difference between the lower bound and the incumbent value was more than 90%, i.e. the B&B was far from completion.

The further optimizations are based on the understanding of how SCIP handles non-convex polynomial constraints [9,22]. Because any polynomials may be converted to a sum of bilinear summands by introducing additional variables and constraints, it is enough to explain how SCIP handles a bilinear function in constraints. SCIP uses the so called McCormik envelopes which give convex lower bound and concave upper bound (both are piecewise linear) for a bilinear function on a rectangle (Fig. 1).

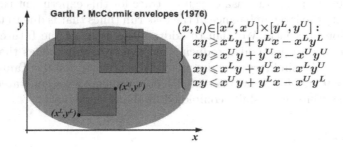

Garth P. McCormik envelopes (1976)

$$(x, y) \in [x^L, x^U] \times [y^L, y^U]:$$
$$\begin{cases} xy \geqslant x^L y + y^L x - x^L y^L \\ xy \geqslant x^U y + y^U x - x^U y^U \\ xy \leqslant x^L y + y^U x - x^L y^U \\ xy \leqslant x^U y + y^L x - x^U y^L \end{cases}$$

Fig. 1. McCormik convex lower bound and concave upper bound for bilinear function.

The smaller the diameter of the rectangle the more is the accuracy of convex approximations and the better is the lower bound for global optimum value given by solving of relaxed convex subproblems. Thus the greater the number of "small" rectangles, the higher the accuracy of the B&B algorithm. As a rough approximation, assume that the memory consumption is proportional to the current number of rectangles overlapping the feasible domain in problems (1) and (2). Assume also that the number of rectangles is proportional to the multi-dimension volume of the domain. Let's divide the domain, e.g. into two equal subdomains with twice as less volume (Fig. 2), and solve these subproblems in parallel by using different SCIP processes and exchanging incumbents between the solvers. One can hope that in this case each solver process will require twice as less memory. This approach allows to solve the problem in parallel while reducing the system requirements of individual tasks.

Consider a possible simple decomposition into subdomains with equal volumes. Let's $K \leqslant (N-1)$, take vectors x_k, $k = 2 : K+1$ (beginning from 2, because x_1 is fixed to $(0, 0, 1)$) and perform decomposition by their first coordinate's sign.

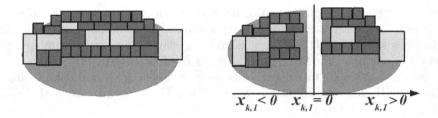

$$x_{k,1} < 0 \quad x_{k,1} = 0 \qquad x_{k,1} > 0$$

Fig. 2. Domain decomposition idea for global optimization.

Formally, for any K-set of parameters $\{p_k = \pm 1, k = 1{:}K\}$ the following inequality constraints should be added to problems (1) or (2): $p_k \cdot x_{k+1,1} \leqslant 0, (k = 1{:}K)$. Thus the original problem can be subdivided into 2^K subproblems.

We have run several experiments for solving the Tammes problem using the DDBNB application and the HPC4 cluster at NRC KI. In the experiment with $N=8$ and $K=7$ the problem has been decomposed into 128 tasks that were completed within 100 min. The task execution trace for this experiment is presented on Fig. 3. Note the large imbalance in the task run times that is due to the chosen decomposition approach. We plan to address this issue in the future by using other decomposition methods producing much more subproblems that may be more balanced. Preliminary experiments were conducted for the Tammes problem with $N=9$ and $K=8$. In this case 256 subproblems were generated and the computations were successfully completed in about three days.

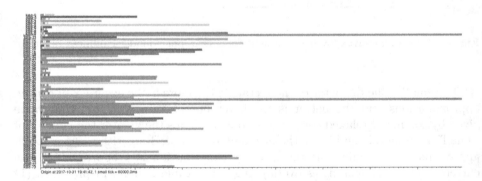

Fig. 3. Trace of solving the Tammes problem $N=8$ at HPC4 cluster at NRC KI.

6   Conclusion and Future Work

In this paper we have presented the integration of Everest platform with HPC resources of supercomputing centers. The described implementation has been tested by solving hard global optimization problems with the DDBNB Everest application. The use of HPC resources with DDBNB has provided the significant increment in computing power available for our experiments. A number of improvements has been made in Everest and its agent in order to support the efficient use of HPC resources and overcome the limitations found during the early experiments. The presented approach enables convenient access to HPC resources, execution of many-task applications, and pooling of multiple resources for running distributed computations.

Future work will focus on integration with other supercomputing centers and improving the described implementation. We also plan to conduct the large-scale experiments involving resources of multiple centers via Everest. Regarding the DDBNB application, we plan to investigate the use of alternative decomposition rules producing thousands of subtasks for solving the considered problems.

Acknowledgments. This work is supported by the Russian Science Foundation (project No. 16-11-10352). This work has been carried out using computing resources of the federal collective usage center Complex for Simulation and Data Processing for Mega-science Facilities at NRC "Kurchatov Institute", http://ckp.nrcki.ru/.

References

1. Everest. http://everest.distcomp.org/
2. Talbi, E.G. (ed.): Parallel Combinatorial Optimization, vol. 58. Wiley, Hoboken (2006)
3. Afanasiev, A., Sukhoroslov, O., Voloshinov, V.: MathCloud: publication and reuse of scientific applications as RESTful web services. In: Malyshkin, V. (ed.) PaCT 2013. LNCS, vol. 7979, pp. 394–408. Springer, Heidelberg (2013). https://doi.org/10.1007/978-3-642-39958-9_36
4. Afgan, E., et al.: Galaxy: a gateway to tools in e-science. In: Yang, X., Wang, L., Jie, W. (eds.) Guide to e-Science. CCN. Springer, London (2011). https://doi.org/10.1007/978-0-85729-439-5_6
5. Allan, R.N.: Virtual Research Environments: From Portals to Science Gateways. Elsevier, Amsterdam (2009)
6. Chen, Z., Maly, K., Mehrotra, P., Vangala, P.K., Zubair, M.: Web-based framework for distributed computing. Concurrency Pract. Exp. **9**(11), 1175–1180 (1997)
7. Formiconi, A., et al.: World wide web interface for advanced spect reconstruction algorithms implemented on a remote massively parallel computer. Int. J. Med. Inform. **47**(1–2), 125–138 (1997)
8. Foster, I., Kesselman, C.: The Grid 2: Blueprint for a New Computing Infrastructure. Elsevier, Amsterdam (2003)
9. Gleixner, A., et al.: The SCIPoptimization suite 5.0. Technical report 17-61, ZIB, Takustr.7, 14195 Berlin (2017)
10. Kacsuk, P.: P-grade portal family for grid infrastructures. Concurrency Computat. Pract. Exp. **23**(3), 235–245 (2011)
11. Kostenetskiy, P., Safonov, A.: SUSU supercomputer resources. In: Proceedings of the 10th Annual International Scientific Conference on Parallel Computing Technologies (PCT 2016), Arkhangelsk, Russia, vol. 1576, pp. 561–573 (2016)
12. McLennan, M., Kennell, R.: HUBzero: a platform for dissemination and collaboration in computational science and engineering. Comput. Sci. Eng. **12**(2), 48–53 (2010)

13. Musin, O.R., Tarasov, A.S.: The Tammes problem for $N = 14$. Exp. Math. **24**(4), 460–468 (2015)
14. Pierce, M.E., Youn, C., Fox, G.C.: The gateway computational web portal. Concurrency Comput. Pract. Exp. **14**(13–15), 1411–1426 (2002)
15. Raicu, I., et al.: Toward loosely coupled programming on petascale systems. In: Proceedings of the 2008 ACM/IEEE Conference on Supercomputing, p. 22. IEEE Press (2008)
16. Smirnov, S., Sukhoroslov, O., Volkov, S.: Integration and combined use of distributed computing resources with everest. Procedia Comput. Sci. **101**, 359–368 (2016)
17. Smirnov, S., Voloshinov, V.: Implementation of concurrent parallelization of branch-and-bound algorithm in everest distributed environment. Procedia Comput. Sci. **119**, 83–89 (2017)
18. Sukhoroslov, O., Volkov, S., Afanasiev, A.: A web-based platform for publication and distributed execution of computing applications. In: 2015 14th International Symposium on Parallel and Distributed Computing (ISPDC), pp. 175–184, June 2015
19. Sukhoroslov, O.: Integration of Everest platform with BOINC-based desktop grids (2017)
20. Thomas, M., Mock, S., Boisseau, J.: Development of web toolkits for computational science portals: the NPACI HotPage. In: The Ninth International Symposium on High-Performance Distributed Computing, 2000. Proceedings, pp. 308–309. IEEE (2000)
21. Thomas, M., et al.: Grid portal architectures for scientific applications. In: Journal of Physics: Conference Series, vol. 16, p. 596. IOP Publishing (2005)
22. Vigerske, S., Gleixner, A.: SCIP: global optimization of mixed-integer nonlinear programs in a branch-and-cut framework. Optim. Methods Softw. **33**, 1–31 (2017)
23. Volkov, S., Sukhoroslov, O.: A generic web service for running parameter sweep experiments in distributed computing environment. Procedia Comput. Sci. **66**, 477–486 (2015)
24. Volkov, S., Sukhoroslov, O.: Simplifying the use of clouds for scientific computing with everest. Procedia Comput. Sci. **119**, 112–120 (2017)
25. Voloshinov, V., Smirnov, S., Sukhoroslov, O.: Implementation and use of coarse-grained parallel branch-and-bound in everest distributed environment. Procedia Comput. Sci. **108**, 1532–1541 (2017)
26. Yang, X., Martin, T., Mark, H., Mark, C., Ligang, H., Peter, M.: Survey of major tools and technologies for grid-enabled portal development. In: Proceedings of the UK e-Science All Hands Meeting (NeSC 2006). University of Cambridge Press (2006)

Author Index

Abramova, Olga A. 427
Akhatov, Iskander Sh. 427
Aleeva, Valentina 565
Antonov, Alexander 587

Bakulin, Andrey 343
Balykov, Gleb 301
Barash, Lev 354
Barkalov, Konstantin 50
Bastrakov, Sergei 523
Belonosov, Mikhail 331
Belozerov, Ivan 208
Belyshev, Alexey 578
Berezovsky, Vladimir 208
Bibin, Vladimir 159
Bochenina, Klavdiya 136
Butakov, Nikolay 675
Bykov, Andrei 242

Chernov, Ilya 599
Chernykh, I. G. 29
Chernykh, Igor 414, 465
Cheverda, Vladimir 3, 331
Chigerev, Eugenii 76
Churilova, Maria 367

Dmitriev, Maxim 343
Dolganina, Natalia Yu. 185
Dordopulo, Alexey 554
Doronchenko, Yuriy 554

Edelev, Alexei 289

Fadeev, Rostislav 379
Fedorov, Alexander 554
Feoktistov, Alexander 289

Gergel, Victor 88, 523
Glinskiy, B. M. 29
Glinsky, Boris 465
Golovin, Andrey V. 279
Goncharsky, Alexander 401
Gorsky, Sergei 289

Goyman, Gordey 379
Grigoriev, Fedor 266
Gubaydullin, Marsel 208
Gumerov, Nail A. 427

Ibrayev, Rashit 159
Ignatov, Andrei 511
Ignatova, Anastasia V. 185
Il'in, V. P. 653
Ilyushin, Yaroslaw 254
Ishanov, Sergey 439
Ivashko, Evgeny 453, 500

Janke, Wolfhard 354

Kalyakin, Alexander 666
Kalyulin, Stanislav L. 641
Kapyrin, Ivan 266
Kataev, Nikita 487
Kaurkin, Maxim 159
Kesarev, Sergey 136
Khaidykov, Valery 3
Khoperskov, Alexander 173
Khrapov, Sergey 173
Kochemazov, Stepan 578
Kondratyuk, Nikolay 218
Konshin, Igor 63, 266
Konyukhov, Sergey 666
Kostin, Victor 331, 343
Kostromin, Roman 289
Kozinov, Evgeny 88
Kramarenko, Vasily 266
Krassilchtchikov, Alexandre 242
Kropotina, Julia 242
Kulikov, Igor 414, 465
Kurochkin, Ilya I. 472
Kutov, Danil 314
Kutuza, Boris 254
Kuzin, Alexey 15

Lebedev, Ilya 50
Lebedev, Sergey 195
Leonenkov, Sergei 631

Levchenko, Vadim   101, 125
Levenfish, Ksenia   242
Levin, Ilya   554
Linev, Alexey   675
Lipavskii, Michael   76
Lisitsa, Vadim   3, 125
Litovchenko, Valentina   500
Lubov, Sergei   379

Massel, Lyudmila   289
Medvedik, Mikhail   114
Melnik, Michael   675
Meyerov, Iosif   195, 523
Mironov, Vladimir   414
Modorskii, Vladimir Ya.   641
Mosin, Sergey   149
Moskaleva, Marina   114
Moskovsky, Alexander   666
Mukhina, Ksenia   675

Nasonov, Denis   675
Neiman-zade, Murad I.   666
Neklyudov, Dmitry   331
Nikitina, Natalia   453, 578
Nikolskiy, Alexander   195

Odintsov, Igor   666
Orlov, Stepan   15
Osinin, Ilya   620
Osipov, Grigory   195

Pavlinov, Alexander M.   641
Perepelkina, Anastasia   101
Petrov, Valentin   195
Petukhov, Evgeniy   367
Petukhova, Margarita   367
Pirova, Anna   195
Pityuk, Yulia A.   427
Pleshchinskii, Ilya   149
Pleshchinskii, Nikolai   149
Pleshkevich, Alexander   125
Pogarskaia, Tatiana   367
Popova, Elena N.   641
Posypkin, Mikhail   511
Prigarin, Vladimir   414, 465

Radchenko, Victor   173
Reshetnikov, Roman V.   279
Reshetova, Galina   3
Romanov, Sergey   388

Sapozhnikov, Sergei B.   185
Seryozhnikov, Sergey   401
Severiukhina, Oksana   136
Shabley, Alexandra A.   185
Shabrov, Nikolay   15
Sharamet, Alexander   439
Shashkin, Vladimir   379
Shchapov, Vladislav A.   641
Shchur, Lev   354
Shirobokov, Dmitrii   76
Shvets, Pavel   675
Sidorov, Ivan   289
Smirnov, Grigory   218
Smirnov, Sergey   687
Smirnov, Yury   114
Sobolev, Sergey   675
Solovyev, Sergey   343
Sovrasov, Vladislav   50
Stefanov, Konstantin   609
Stegailov, Vladimir   218, 543
Sukhanovskii, Andrei N.   641
Sukhoroslov, Oleg   687
Sulimov, Alexey   314
Sulimov, Vladimir   314
Sysoyev, Alexander   523

Terekhov, Kirill   230
Timofeev, Alexey   543
Titov, Alexander   173
Tolstykh, Andrei   76
Tolstykh, Mikhail   379
Toporkov, Victor   531
Tumakov, Dmitrii   149
Turlapov, Vadim   195
Tutukov, Alexander   414
Tyutlyaeva, Ekaterina   666

Vasilyev, Evgeniy   195
Vassilevski, Yuri   230
Vatutin, Eduard   578

Visheratin, Alexandr   675
Vishnevsky, Dmitry   125
Voevodin, Vadim   609
Volkov, Nikita   587
Voloshinov, Vladimir   687
Vshivkov, Vitaly   414, 465
Vshivkova, Lyudmila   465

Weigel, Martin   354
Weins, D. V.   29

Yakimets, Vladimir N.   472
Yemelyanov, Dmitry   531
Yur'ev, Alexander   208

Zaikin, Oleg   578
Zalevsky, Arthur O.   279
Zamarashkin, Nikolai   40
Zheltkov, Dmitry   40
Zhumatiy, Sergey   631
Zinin, Leonid   439